Quality of Life and Climate Change:

Impacts, Sustainable Adaptation, and Social-Ecological Resilience

Kasturi Shukla
Symbiosis Institute of Health Sciences, Symbiosis International University (Deemed), Pune, India

Yogesh B. Patil
Symbiosis Centre for Research and Innovation, Symbiosis International University (Deemed), Pune, India

Ronald C. Estoque
Center for Biodiversity and Climate Change, Forestry and Forest Products Research Institute, Tsukuba, Japan

Pedro Antonio López de Haro
Universidad Autónoma Indígena de México, Mexico

IGI Global
PUBLISHER of TIMELY KNOWLEDGE

A volume in the Practice, Progress, and Proficiency in Sustainability (PPPS) Book Series

Published in the United States of America by
IGI Global
Engineering Science Reference (an imprint of IGI Global)
701 E. Chocolate Avenue
Hershey PA, USA 17033
Tel: 717-533-8845
Fax: 717-533-8661
E-mail: cust@igi-global.com
Web site: http://www.igi-global.com

Library of Congress Cataloging-in-Publication Data

Names: Shukla, Kasturi, 1983- editor. | Patil, Yogesh, 1970- editor. | Estoque, Ronald, 1980- editor. | Lopez de Haro, Pedro, 1990- editor.
Title: Quality of life and climate change : impacts, sustainable adaptation, and social-ecological resilience / edited by: Kasturi Shukla, Yogesh Patil, Ronald Estoque, Pedro Lopez de Haro.
Description: Hershey PA : Engineering Science Reference, 2023. | Includes bibliographical references. | Summary: "The book will add immense value to literature by bringing together resilience enhancing models in maintaining QoL and building sustainable living environments. An effort is made to identify the possible models to identify and develop the solutions to build sustainable and resilient systems in the face of climate change and variability"-- Provided by publisher.
Identifiers: LCCN 2023034647 (print) | LCCN 2023034648 (ebook) | ISBN 9781668498637 (hardcover) | ISBN 9781668498644 (ebook)
Subjects: LCSH: Climatic changes--Social aspects. | Quality of life. | Human beings--Effect of climate on. | Human beings--Effect of environment on. | Sustainable development.
Classification: LCC HN49.C55 Q355 2023 (print) | LCC HN49.C55 (ebook) | DDC 307.1/40286--dc23/eng/20230829
LC record available at https://lccn.loc.gov/2023034647
LC ebook record available at https://lccn.loc.gov/2023034648

This book is published in the IGI Global book series Practice, Progress, and Proficiency in Sustainability (PPPS) (ISSN: 2330-3271; eISSN: 2330-328X)

British Cataloguing in Publication Data
A Cataloguing in Publication record for this book is available from the British Library.

All work contributed to this book is new, previously-unpublished material. The views expressed in this book are those of the authors, but not necessarily of the publisher.

For electronic access to this publication, please contact: eresources@igi-global.com.

Practice, Progress, and Proficiency in Sustainability (PPPS) Book Series

Ayman Batisha
International Sustainability Institute, Egypt

ISSN:2330-3271
EISSN:2330-328X

Mission

In a world where traditional business practices are reconsidered and economic activity is performed in a global context, new areas of economic developments are recognized as the key enablers of wealth and income production. This knowledge of information technologies provides infrastructures, systems, and services towards sustainable development.

The **Practices, Progress, and Proficiency in Sustainability (PPPS) Book Series** focuses on the local and global challenges, business opportunities, and societal needs surrounding international collaboration and sustainable development of technology. This series brings together academics, researchers, entrepreneurs, policy makers and government officers aiming to contribute to the progress and proficiency in sustainability.

Coverage

- Outsourcing
- Green Technology
- Knowledge clusters
- Socio-Economic
- E-Development
- Innovation Networks
- Technological learning
- Intellectual Capital
- ICT and knowledge for development
- Strategic Management of IT

The Practice, Progress, and Proficiency in Sustainability (PPPS) Book Series (ISSN 2330-3271) is published by IGI Global, 701 E. Chocolate Avenue, Hershey, PA 17033-1240, USA, www.igi-global.com. This series is composed of titles available for purchase individually; each title is edited to be contextually exclusive from any other title within the series. For pricing and ordering information please visit http://www.igi-global.com/book-series/practice-progress-proficiency-sustainability/73810. Postmaster: Send all address changes to above address.

Titles in this Series

For a list of additional titles in this series, please visit:
www.igi-global.com/book-series/practice-progress-proficiency-sustainability/73810

Operational Research for Renewable Energy and Sustainable Environments
Joshua Thomas (University of Wollongong Malaysia, Malaysia) Gerhard Wilhelm Weber (Poznan University of Technology, Poland) Roman Rodríguez Aguilar (Universidad Panamericana, Mexico) Elias Munapo (North West University, South Africa) and Pandian Vasant (Ton Duc Thang University, Vietnam)
Engineering Science Reference • copyright 2024 • 350pp • H/C (ISBN: 9781668491300) • US $240.00 (our price)

Sustainable Solutions for E-Waste and Development
Rajesh Kumar K. V. (Woxsen University, India) Hemachandran Kannan (Woxsen University, India) Dmitry Spodarets (Data Phoenix, USA) Parvez Alam Khan (Universiti Teknologi PETRONAS, Malaysia) and Bikash Kumar Pradhan (Woxsen University, India)
Engineering Science Reference • copyright 2024 • 396pp • H/C (ISBN: 9798369310182) • US $265.00 (our price)

Energy Alliance for Realistic and Profitable Energy Transition
Mustapha Hatti (EER-UDES/CDER, Algeria)
Engineering Science Reference • copyright 2024 • 320pp • H/C (ISBN: 9798369308882) • US $275.00 (our price)

Teaching and Learning for a Sustainable Future Innovative Strategies and Best Practices
Chai Lee Goi (Curtin University, Malaysia)
Information Science Reference • copyright 2024 • 341pp • H/C (ISBN: 9781668498590) • US $230.00 (our price)

Fostering Cross-Industry Sustainability With Intelligent Technologies
Brojo Kishore Mishra (NIST Institute of Science and Technology (Autonomous), India & Saveetha College of Liberal Arts and Sciences, SIMATS, Chennai, India)
Engineering Science Reference • copyright 2024 • 603pp • H/C (ISBN: 9798369316382) • US $350.00 (our price)

Global Perspectives on Climate Change, Social Resilience, and Social Inclusion
Aly Abdel Razek Galaby (Alexandria University, Egypt) Mohammed Abo ElEnein (American University, UAE) and Hassan Mohamed (Alexandria University, Egypt)
Engineering Science Reference • copyright 2024 • 279pp • H/C (ISBN: 9781668489635) • US $275.00 (our price)

Natural Resource Management Issues in Human-Influenced Landscapes
Dilini Hemachandra (Faculty of Agriculture, University of Peradeniya, Sri Lanka) and Jeevika Weerahewa (Faculty of Agriculture, University of Peradeniya, Sri Lanka)
Engineering Science Reference • copyright 2024 • 300pp • H/C (ISBN: 9781668470510) • US $225.00 (our price)

IGI Global
PUBLISHER of TIMELY KNOWLEDGE

701 East Chocolate Avenue, Hershey, PA 17033, USA
Tel: 717-533-8845 x100 • Fax: 717-533-8661
E-Mail: cust@igi-global.com • www.igi-global.com

Editorial Advisory Board

Table of Contents

Detailed Table of Contents

Kasturi Shukla, Symbiosis Institute of Health Sciences, Symbiosis International University (Deemed), Pune, India
Yogesh B. Patil, Symbiosis Centre for Research and Innovation, Symbiosis International University (Deemed), Pune, India
Ronald C. Estoque, Center for Biodiversity and Climate Change, Forestry and Forest Products Research Institute, Tsukuba, Japan
Pedro Lopez de Haro, Universidad Autónoma Indígena de México, Mexico

The negative impact of climate change on Quality of Life (QoL) has become an issue of major concern that transcends national boundaries. Unprecedented changes in the Earth's climate have created a variety of effects at different geographical locations, thus impacting the welfare of inhabitants. The impacts are wide ranging—from severe weather that interrupts livelihoods to rising sea levels that endanger coastal towns. Furthermore, food security, water availability, and public health are all impacted by ecological changes brought on by climate change. In order to solve these complex concerns and promote a shared commitment to resilience and sustainable practices, international collaborative efforts are crucial. This will eventually protect the QoL on a global scale from the growing risks posed by climate change. This introductory chapter of the book provides a global perspective on the impact of climate change on the quality of life both in urban and rural settings.

Nibedita Naha, Biochemistry Department, Biological Sciences Division, ICMR-National Institute of Occupational Health (NIOH), Ahmedabad, India
Deepsi Rathore, Biochemistry Department, Biological Sciences Division, ICMR-National Institute of Occupational Health (NIOH), Ahmedabad, India
Somi N. Sharma, Biochemistry Department, Biological Sciences Division, ICMR-National Institute of Occupational Health (NIOH), Ahmedabad, India
Ranjana Hanumant Choudhari, ICMR-National AIDS Research Institute, India

Fishery includes farmed fisheries and wild fisheries; of which, wild fisheries are comprised of marine and freshwater fisheries that provide an important source of food, nutrition, employment and income, mostly

The seas are home to many living things and an excellent resource for the ecosystem. In addition, the seas are a perfect ecosystem component for both the supply of quality food for humans and the continuation of their lives. Due to the ecological and economic importance of the seas, they should be catered to consciously and in a sustainable fashion. Howbeit, the boost in human-induced problems and nature-induced problems may cause some negativities. This causes an increase in nutrients in the seas, sudden changes in water temperature, a decrease in water currents and consequently, an increase in algae and other phytoplankton. Over time, these organisms cause a substance entitled marine saliva or mucilage to cover the seas and damage the marine ecosystem. In this study, a comparative analysis of the mucilage problem, which has become a significant problem in Türkiye in recent years, is carried out. In this context, the prominent reports, the measures taken and the action plans in Türkiye are examined in detail.

Plastic pollution is a global occurrence from villages to urban, farms to deserts, deep oceans to mountaintops, tropical landfills to arctic snow including in food chain. Over the past 60 years, there has been a steady build-up of plastic garbage on land and ocean's surface, mainly because of the unplanned, unrestricted use of plastic in routine use, non-cycling and use and throw habit of a common man. As usage of this strong and cheap material has increased over the past few decades, the waste problem has gotten worse because of non-biodegradation in nature. Since the turn of the century, the amount of plastic produced worldwide has quadrupled, reaching over 400 million metric tonnes annually in 2021. While plastic items typically have a lifespan of ten years, depending on their composition and method of disposal, plastics can take up to 500 years to degrade. The environment is harmed by the removal of all this plastic. Only 16% of plastics are recycled; the remainder is either disposed of in landfills or burned.

This proposed book chapter explores novel legal strategies for holding polluters accountable for climate change contributions and protecting affected communities' quality of life. It highlights the limitations of traditional litigation and advocates for innovative approaches. The first section focuses on technological solutions like satellite imagery, remote sensing, and blockchain to gather evidence and enhance supply chain transparency. The second section examines successful collaborative legal initiatives addressing transboundary pollution and global climate injustices. The third section explores integrating climate science, economics, and law in litigation efforts, quantifying damages, and linking pollution to health impacts. The fourth section discusses future directions and challenges, including accountability for future generations. The chapter concludes by emphasizing interdisciplinary collaboration and legal reforms for climate justice and a better future.

Achieving Climate Action, UN Agenda 2030, is a challenge towards strengthening the resilience and adaptive capacity to climate-related hazards and natural disasters. Climate-proofing of villages can be strategised through climate adaptation and complementary mitigation approaches, leading to climate-resilient rural communities with climate-resilient livelihoods in India. The chapter reviews the scope of rural developmental schemes and strategies that may lead to climate-resilient adaptation practices for rural communities. The first part of the chapter lays out the national perspective of climate action. The second step is to review the problems that rural communities face and investigate how rural development programs like MGNREGS, NRLM, PMAY-G, and other models can be used to converge to create climate resilient communities. Finally, understand how the localization of climate action enhance and strengthens the rural poor's resilience and adaptive capacity.

Globalization and industrialization have exposed the world to the exacerbating issue of sustainability. With the occurrences of climate related extreme events increasing the world over in the last decade, climate change has come to be recognized as the most profound concern staring in the face of mankind. There have been many separate studies addressing climate change and an equally good number of research in the domain of quality of life. But there is a dearth of research linking climate variability and quality of life. Climate uncertainties have led to loss of lives. Settlements and employments, material living conditions, physical and economic security, natural environment, health, basic rights, and social exchanges among communities stand impacted. All of these metrics are important determinants of quality of life (QoL). Also, alarming is the fact that QoL indicators remain poorly connected to climate change related concerns. This is an important research gap and calls for urgent attention.

One of the biggest threats facing mankind in the immediate future is climate change. However, climate change doesn't affect all humans equally; the adverse effects are accentuated for vulnerable groups like women and gender minorities. The SDG goal 13 of the United Nations is about Climate Action and for that goal to be achieved, academic research and societal application must go in concord. While the gender disparities of climate change are starkly manifest in multiple areas like economic opportunity, housing or political action, this chapter will concentrate on the public health perspective only. The authors of this chapter found that there is a dearth of comprehensive academic discourse on the nuanced health effects of climate disasters in different gender groups. While women and LGBTQ+ people are particularly victimized by climate disasters, the men are also not without their own challenges. The chapter discusses its contents within the framework of a conceptual model. Suitable real-life examples from prominent historical events and current events have been depicted. The impacts of climate change on mental health, sexual and reproductive health, gender violence and LGBTQ+ community have also been explored with due diligence. A special section on fragile ecosystems like the coastal areas and the situation in South Asia, the most climate sensitive region of the world, has also been added. Online data resources like PubMed, PLoS, WHO and UNDP were used along with relevant information from several government bodies and NGOs, as available online. In conclusion, the chapter reveals the dual role of males both as perpetrators of climate change and its victims, the situation of females as disproportionate sufferers and LGBTQ+ community facing additional adversities due to the consequential effects of climate change.

Climate change is a long-term shift in temperature and weather patterns. The United Nations states that since the late 18th century, human activities - primarily the combustion of fossil fuels like coal, oil, and gas - have been the primary cause of climate change. Across the different sectors, the greenhouse effect is most prominent which forms a blanket wrapped around the Earth, trapping the sun's heat and raising temperatures. According to the UN, clearing of land and deforestation is responsible for increased carbon dioxide levels in the environment. The main sectors responsible for greenhouse gas emission are energy, transport, construction, agriculture, healthcare, and high land use. The healthcare industry is the largest source of greenhouse gasses, accounting for almost 5% of global greenhouse emissions. This chapter reviews the theories and models proposed by national and international organizations like United Nations and World Health Organization to explore and explain the impact of hospitals on climate change and the short-term and long-term impact on sustainability.

Preface

It is with great pleasure that we present to you this edited reference book, *Quality of Life and Climate Change: Impacts, Sustainable Adaptation, and Social-Ecological Resilience* which is edited by Kasturi Shukla, Yogesh B. Patil, Ronald C. Estoque, and Pedro Antonio López de Haro. This comprehensive volume delves into the intricate interplay between climate change and the quality of life (QoL) experienced by individuals and communities worldwide.

In the 21st century, climate change has emerged as a central concern, capturing the attention of the global scientific community. This book aims to explore the multifaceted ways in which climate change and variability influence the QoL in both rural and urban settings. The chapters within this volume undertake a detailed review of QoL assessments, scrutinizing the extent to which climatic changes and liveability conditions are considered in these evaluations.

Resilience stands out as a critical factor in maintaining urban QoL and fostering sustainable living environments. The book examines various models implemented globally to build resilient and sustainable systems that adapt to the challenges posed by climate change. It not only identifies the impacts but also strives to develop solutions to enhance and maintain QoL in the face of climate change and variability.

The significance of QoL in the context of climatic conditions cannot be overstated. As the synopsis outlines, climatic variations significantly influence the QoL of living beings. This book addresses the identified gaps in QoL assessments, emphasizing the need to incorporate climatic impacts in the evaluation measures. By doing so, it contributes to the creation of aging-friendly and climate-neutral cities that are crucial goals for contemporary societies.

The book's major impact lies in reporting and documenting the association between climatic changes and QoL in rural and urban areas, an aspect that has not been extensively researched. A thorough review of QoL assessments and evaluations, focusing on the coverage of climatic changes, is another key contribution that sets this edited volume apart.

This edited book caters to a diverse audience, including researchers, academicians, policy makers, civil society groups, policy think tanks, government and non-government organizations, international agencies, and other interested parties. It serves as a valuable resource to enhance their capacity and knowledge based on the climate change and its profound impacts on Quality of Life (QoL) and the attainment of Sustainable Development Goals (SDGs).

The interdisciplinary focus of the book centers on climate change and its impact on QoL, with a primary theme of reviewing strategies for achieving SDGs. It further explores the coverage of QoL assessment tools, ensuring the inclusion of consequences related to climate change and variations.

The chapters in this volume cover a wide array of topics, ranging from climatic variations and future projections to technological advances, implications for residential and non-residential areas, industrial solutions, and strategies towards specific SDGs. Models of Urban QoL, resilience, applied research, case studies, and best practices are all explored, providing a comprehensive understanding of the complex dynamics at play.

This edited book, *Quality of Life and Climate Change: Impacts, Sustainable Adaptation, and Social-Ecological Resilience*, provides insights on some of the key challenges of climate change and its impact on QoL along with different strategies of sustainable adaptation and social-ecological resilience across the globalized economy and in regional geographies through a diverse mix of chapters. In general, the chapters of this book can broadly be classified into three important aspects, viz. impacts, sustainable adaptation and social-ecological resilience.

Chapter 1 is the introductory chapter of the book and provides a global perspective on the impact of climate change on the QoL both in urban and rural settings. Further, the first set of chapters (Chapters 2-6) discusses mainly the impact of climate change on QoL in diverse environmental settings. In **Chapter 2**, the authors examine the far-reaching impact of climate change and occupational hazards on the Quality of Life (QoL) of fisherfolk and fish processing communities. Addressing environmental factors, ecological imbalances, and occupational challenges, the chapter explores the diverse challenges faced by communities engaged in the fishery and fish processing industry. In Chapter 3, the authors have explored the economic significance of the *Stenocereus montanus* plant, a wild cactus distributed in the arid and semi-arid territories of northwestern Mexico. The plant provides a fruit that becomes economically profitable for families during the annual season. However, the economic spillover of this activity has not been thoroughly studied. By examining the economic impact, the chapter contributes to a broader understanding of the plant's role in local communities and its potential as a catalyst for positive development. **Chapter 4** delves into the significant threats posed by carbon emissions to the planet's survival. The chapter discusses the multifaceted impacts of climate change on various ecological, environmental, socio-political, and socio-economic aspects. Highlighting the need for strategic management of carbon emissions, the chapter explores potential strategies to address the adverse effects. By offering insights into the global impact of carbon emissions, the chapter advocates for a comprehensive approach to mitigate climate change and protect the planet. In **Chapter 5**, the intricate relationship between marine ecosystems and the challenges posed by climate change and pollution have been discussed. Recognizing the seas as a crucial component of the ecosystem, the chapter emphasizes their ecological and economic significance. However, the surge in human-induced and nature-induced problems has led to negative consequences, including increased nutrients, changes in water temperature, decreased water currents, and the proliferation of algae and phytoplanktons. Over time, these factors contribute to the formation of marine mucilage, a substance that detrimentally affects the marine ecosystem. Focusing on Türkiye, the chapter conducts a comparative analysis of the mucilage problem, a significant issue in recent years. It scrutinizes prominent reports, measures taken, and action plans, providing a comprehensive examination of the situation and contributing to the understanding of marine ecosystem challenges in the face of climate change and pollution. **Chapter 6** tackles the global issue of plastic pollution and its pervasive impact on various environments. Examining the unplanned and unrestricted use of plastic, the chapter traces the steady build-up of plastic garbage on land and in oceans over the past 60 years. With only 16% of plastics being recycled, the chapter emphasizes the urgent need for sustainable management practices to address the increasing production and disposal of plastic. By shedding light on the global impact of plastic pollution, the chapter advocates for comprehensive strategies to achieve sustainable development.

The focus of discussion in the second set of Chapters (Chapters 7-14) is on the diverse adaptation/mitigation strategies and models adopted in rural and urban settings. **Chapter 7** explores groundbreaking legal strategies to hold polluters accountable for their contributions to climate change. It critically examines traditional litigation related limitations and advocates for innovative approaches. The chapter is structured into four sections, each addressing different facets of legal initiatives. These facets include – (i) technological solutions, (ii) collaborative legal initiatives, (iii) integration of climate science, economics, and law in litigation efforts and (iv) future directions, challenges, and the imperative interdisciplinary collaboration and legal reforms. **Chapter 8** focuses on the imperative task of achieving Climate Action in rural India, aligning with the UN Agenda 2030. The chapter reviews various Rural Developmental Schemes and strategies, exploring their potential in fostering climate-resilient practices. It provides a dual perspective, beginning with a national outlook on Climate Action and subsequently delving into the challenges and opportunities at the grassroot level. By highlighting the convergence of initiatives such as MGNREGS, NRLM, PMAY-G, and other models, the chapter advocates for localized approaches that enhance the resilience and adaptive capacity of rural communities. **Chapter 9** addresses the pressing issues arising from globalization and industrialization, particularly in the context of sustainability. Recognizing climate change as a profound global concern, the chapter identifies the dearth of research connecting climate variability and quality of life. It emphasizes the far-reaching impacts of climate uncertainties on various aspects of life, including settlements, employment, health, and basic rights. The chapter serves as a call to action, urging urgent attention to the research gap and advocating for transformative solutions that bridge the gap between quality of life indicators and climate change-related concerns. In **Chapter 10**, the authors address the environmental challenges posed by the rapid expansion of Artificial Intelligence (AI) applications and the consequential demand for data centers. Focusing on the water footprint of AI-embedded data centers, the chapter aims to promote sustainability by exploring techniques and tools to reduce water usage while maintaining efficiency. The research delves into water-efficient cooling approaches, hardware optimization, water recycling, and intelligent water management systems. By providing insights into environment friendly practices for AI-embedded data centers, the chapter contributes to the development and expansion of AI technologies while minimizing their impact on the environment. **Chapter 11** provides valuable insights into the factors critical to the adoption of climate change mitigation strategies in the Indian Iron and Steel Industry (IISI). Employing a systems approach and the TISM framework, the authors analyze key factors and their interrelationships to prioritize mitigation strategies in energy-intensive iron and steel plants. The chapter delves into the institutional and policy processes, exploring how these shape and regulate the adoption of climate change mitigation strategies in the Indian context. By offering a comprehensive understanding of the challenges and solutions in the adoption of climate change mitigation strategies, the chapter serves as a guide for corporations seeking financial stability while contributing to environmental sustainability. **Chapter 12** focuses on the significant issue of uncontrolled veld fires in Zimbabwe, contributing to climate change impacts. To address this challenge, the chapter advocates the utilization of Indigenous Knowledge Systems (IKS) as a practical and effective means of veld-fire reduction. Findings indicate that households employ various IKS strategies, including invoking taboos to protect sacred forests, leading to a reduction in veld fires and the preservation of livelihood sources. The implications of these findings are discussed, highlighting the potential of IKS in governing veld fires and mitigating climate change in Zimbabwe. The focus of **Chapter 13** is on Blockchain Technology which has recently received broad attention. During disaster events, timely and targeted information provision and exchange could provide great help to the stricken population in difficult and complicated environments. This chapter reports a service oriented system, called Blockchain Disaster Rescue Wings, for providing emergency support to sufferers and rescuers in disasters. This system has been tested in several disaster rescue drills, and has been successfully applied. **Chapter 14** addresses the intersection of climate change and caste dynamics in the Indian context, shedding light on the disproportionate impacts faced by marginalized communities. Focusing on Dalits and Adivasi communities, the chapter emphasizes the need to recognize caste as a significant factor in climate action policies. It challenges the oversight in environmental literature that often neglects the ecological perspectives of marginalized castes, considering them within broader categories of poverty and vulnerability. By examining Dalit literature, particularly works by Baby Kamble and Bama, the chapter aims to illuminate the intricate relationship between Dalits and nature. This exploration contributes valuable insights to ecological, literary, and cultural theories, redefining the understanding of environmental paradigms and emphasizing the importance of acknowledging caste dynamics in the climate change discourse.

Chapters from 15 to 20 discuss the social-ecological resilience of communities in response to climate change. **Chapter 15** critically examines the gender-specific impacts of climate change from a public health perspective. Recognizing the non-uniform effects of climate change on different social groups, the authors highlight the vulnerability of women and gender minorities (LGBTQ) to climate events. The chapter emphasizes the pre-existing vulnerabilities and structural inequalities that determine the impacts of natural disasters on these groups. It sheds light on the health issues faced by women and LGBTQ individuals during climate disasters, and the exacerbation of discriminatory factors. The chapter advocates for gender-appropriate affirmative action in climate governance, addressing the unique challenges faced by women and gender minorities in the face of climate change. **Chapter 16** delves into the profound impact of climate change on hospitals and the subsequent repercussions on QoL. With a focus on the healthcare sector, the chapter sheds light on the significant contribution of hospitals to greenhouse gas emissions. Drawing from theories and models proposed by international organizations like the United Nations and the World Health Organization, the chapter explores the short-term and long-term sustainability of hospitals in the face of climate change. This examination is crucial for understanding and developing resilient and sustainable future hospital models that can effectively address the challenges posed by climate change, ensuring the well-being of individuals and communities. **Chapter 17** presents findings from a study assessing the knowledge, effects, and responses to climate change among single female-headed households in Mongu District, Western Zambia. Employing a descriptive research design with 150 participants, the study reveals the participants' varied levels of knowledge on climate change. The chapter explores the causes and effects of climate change as perceived by the participants, shedding light on the challenges faced by single female-headed households. Despite the low levels of knowledge, the study identifies traces of mitigation efforts among these households. The chapter advocates for climate education and affirmative action to enhance the resilience and adaptive capacity of single female-headed families in the face of climate change. **Chapter 18 e**xplores the impact of bamboo production on climate change, carbon capture, and the livelihood of rural populations. Recognizing bamboo's potential for value-added products and its unique botanical classification, the chapter examines its role in carbon capture and trading. By studying the impact of bamboo cultivation on the livelihoods of rural populations, the chapter provides insights into the circular way in which bamboo production can create a social impact. The authors emphasize the need for inclusive definitions in carbon accounting records, recognizing bamboo as a valuable resource for sustainable development. The chapter advocates for the integration of bamboo production into forestry practices, contributing to both climate change mitigation and rural livelihoods. **Chapter 19** takes a unique approach by examining climate change and its possible effects within the framework of a cyborg science theory. The chapter provides a summary of the signs of global climate change and explores the temporal evolution of climate change throughout world history. It delves into whether the evolution process after chaos caused by global warming is chaotic and discusses the conditions for this emergence. The chapter takes a scientific approach to examine the effects of climate change on the Quality of Life (QoL) of individuals and communities. By employing a cyborg science theory, the chapter aims to contribute to the understanding of climate change and its implications for human life. Last but not the least, **Chapter 20** explores the transformative impact of home gardening in Penang, particularly in response to the challenges posed by the COVID-19 pandemic. With the pandemic leading to social distancing, self-isolation, and disruptions in the food supply chain, home gardening emerged as a popular and environment friendly activity. Recognizing the significance of urban gardening initiatives, the state government and organizations like Think City and the Consumers' Association of Penang played a pivotal role in supporting these efforts. The research investigates the multifaceted impact of home gardening on society, highlighting its potential contribution to a socialist society and alignment with sustainable development goals such as zero hunger and climate action. Home gardening is shown to promote self-sufficiency, food security, and sustainable practices, particularly in urban areas.

As the editors of this meticulously crafted edited reference book, we are delighted to present a diverse and insightful collection of chapters that delve into critical issues at the intersection of climate change, sustainability, and human well-being as well as the Quality of Life (QoL). The contributions from the esteemed authors shed light on a range of topics, offering innovative perspectives, evidence-based insights, and actionable solutions.

The chapters in this book collectively contribute to a comprehensive understanding of the intricate relationships between climate change and various facets of human life. From exploring the impacts of climate change on quality of life, to advocating indigenous knowledge systems in veld-fire reduction, and examining the transformative potential of home gardening, each chapter provides a unique lens through which one can view the challenges and opportunities presented by the changing climate.

The interdisciplinary focus of this edited volume is a testament to the interconnected nature of climate change issues. Authors have drawn upon diverse fields, including environmental science, sociology, economics, and law, to offer a holistic perspective on the multifaceted dimensions of climate change mitigation and adaptation. We are confident that this diversity of approaches will enrich the understanding of our readers and contribute to ongoing discussions in academia, policymaking, and community engagement.

One overarching theme that emerges from the chapters is the significance of proactive measures in addressing climate change. From the innovative use of indigenous knowledge systems to the grassroots efforts in home gardening, the chapters exemplify the power of local and community-driven solutions. The relevance of such initiatives becomes even more pronounced in the face of global challenges like the COVID-19 pandemic, which has underscored the importance of resilience and sustainability in our societies.

We hope that this edited reference book serves as a valuable resource for researchers, policymakers, educators, and all those interested in navigating the complex terrain of climate change and its impacts on human life. The insights presented herein not only contribute to the existing body of knowledge but also inspire a call to action for a more sustainable, resilient, and equitable future.

As editors, we extend our gratitude to the dedicated authors who have shared their expertise and passion for creating positive change. We also express our appreciation to readers for engaging with this compilation and joining us in the collective endeavor to address the challenges posed by the climate change. May the knowledge disseminated within these pages inspire meaningful dialogue, informed decision-making, and transformative actions towards a sustainable and climate-resilient world.

We are grateful to our participating writers for their patience and help throughout the compilation of this collection. Simultaneously, we would like to use this occasion to express our profound gratitude to all members of our Editorial Advisory Board (EAB) and the reviewers for sparing their valuable time in thoroughly reviewing the chapters.

Kasturi Shukla
Symbiosis Institute of Health Sciences, Symbiosis International University (Deemed), Pune, India

Yogesh B Patil
Symbiosis Centre for Research and Innovation, Symbiosis International University (Deemed), Pune, India

Ronald C Estoque
Center for Biodiversity and Climate Change, Forestry and Forest Products Research Institute, Tsukuba, Japan

Pedro Lopez de Haro
Universidad Autónoma Indígena de México, Mexico

Chapter 1
Global Perspectives on the Impact of Climate Change on Quality of Life

Kasturi Shukla
https://orcid.org/0000-0001-7006-9001
Symbiosis Institute of Health Sciences, Symbiosis International University (Deemed), Pune, India

Yogesh B. Patil
Symbiosis Centre for Research and Innovation, Symbiosis International University (Deemed), Pune, India

Ronald C. Estoque
https://orcid.org/0000-0001-9681-492X
Center for Biodiversity and Climate Change, Forestry and Forest Products Research Institute, Tsukuba, Japan

Pedro Lopez de Haro
https://orcid.org/0000-0003-3687-8539
Universidad Autónoma Indígena de México, Mexico

ABSTRACT

The negative impact of climate change on Quality of Life (QoL) has become an issue of major concern that transcends national boundaries. Unprecedented changes in the Earth's climate have created a variety of effects at different geographical locations, thus impacting the welfare of inhabitants. The impacts are wide ranging—from severe weather that interrupts livelihoods to rising sea levels that endanger coastal towns. Furthermore, food security, water availability, and public health are all impacted by ecological changes brought on by climate change. In order to solve these complex concerns and promote a shared commitment to resilience and sustainable practices, international collaborative efforts are crucial. This will eventually protect the QoL on a global scale from the growing risks posed by climate change. This introductory chapter of the book provides a global perspective on the impact of climate change on the quality of life both in urban and rural settings.

DOI: 10.4018/978-1-6684-9863-7.ch001

CLIMATE CHANGE AND ITS GLOBAL IMPLICATIONS: A BRIEF OVERVIEW

Climate change has become a significant global challenge caused mainly by human activities, such as deforestation and the combustion of fossil fuels. This has led to unparalleled alterations in its Earth's climate, characterized by a rise in temperatures, increased frequency of extreme weather events, and altered precipitation patterns (Walsh et al. 2020). These changes have far-reaching effects on ecosystems, species diversity, and human societies, compounding pre-existing vulnerabilities. For example, the increase in sea levels poses a threat to coastal regions, while changes in weather patterns can hinder crop production and jeopardize the global food supply (Gomez-Zavaglia et al. 2020; Hauer et al. 2020). Furthermore, climate change can facilitate the transmission of infectious diseases and provoke geopolitical tensions over resource depletion (Dal et al. 2023; Glasser et al. 2022). In order to tackle the climate change effectively, it is essential to gain a thorough understanding of its multifaceted nature and to take action through international cooperation to lessen Green House Gas (GHG) emissions, plan for future changes, and strengthen resilience. This necessitates the adoption of innovative technologies, sustainable practices, and legislative measures to guarantee a sustainable future for future generations and preserve the planet's natural equilibrium (Durán-Romero et al. 2020).

SIGNIFICANCE OF QUALITY OF LIFE (QOL) IN THE FACE OF CLIMATE CHANGE

World Health Organization (WHO) defines QoL as *"an individual's perceptions of their position in life in the context of the culture and value systems where they live and in relation to their goals, expectations, standards, and concerns"* (Veenhoven, 2000; Yusoff, 2020). The term Quality of Life (QoL) encompasses a wide range of factors, including an individual's physical and mental health, degree of independence, social connections, personal beliefs, and their associations with prominent environmental elements (Andrews, 1974; Bowling, 1995; Costanza et al., 2007; Estoque et al., 2019; Theofilou, 2013). The conceptual development of QoL definitions takes into account all multidimensional features of a social-ecological system, including both objective and subjective metrics (Costanza et al., 2007; Estoque et al., 2022; Estoque and Wu, 2024).

The seventeen (17) Sustainable Development Goals (SDGs), which focus on partnerships for the goals, life below water, life on land, good health and well-being, climate action and several others, have underscored the significance of QoL. Comparable recent studies have examined urban transitions while concentrating on Urban Quality of Life, or UQoL. Urban quality of life (UQoL) is the ratio of land usage to land cover (Yadav & Gupta, 2021). Climate variations are not explicitly included in UQoL, and there is disagreement about which components may be broadly used to address all of the SDGs. It has been established that climate change has detrimental effects on natural ecosystems, and maybe more significantly, on people's physical and mental health because of rising temperatures, droughts, floods, and storms thus resulting in an overall diminished QoL (Bhatti, 2017). Therefore, QoL is gaining attention in the context of issues pertaining to climate change and variability.

ELEMENTS OF QUALITY OF LIFE IMPACTED BY CLIMATE CHANGE

There are several aspects of QoL that are greatly impacted by climate change. Elevated temperatures are a contributing factor to severe weather phenomena, endangering both human well-being and assets (Bhatti, 2017). Variations in precipitation patterns influence water availability, which has an impact on daily living and agriculture. Ecosystems are getting affected by the biodiversity loss, which has an impact on food supplies and general well-being (Gomez-Zavaglia et al. 2020). Elevated air pollution exacerbates respiratory conditions, and coastal areas are at risk from increasing sea levels (Hauer et al. 2020). Social and economic inequality is made worse by these environmental changes, which disproportionately affect vulnerable groups (Smith et al. 2022). Preserving the QoL depends on mitigating climate change, which calls for international efforts to cut greenhouse gas emissions, adapt to the changing conditions, and promote sustainable behaviours for a resilient and just future (Abbass et al., 2022; Fekete et al., 2021). Some of the key elements of QoL impacted by climate change include – health and well-being, socio-economic factors and environmental justice (Bolte et al., 2023; Yee et al., 2020). These elements are being discussed in brief in the following sections.

HEALTH AND WELL-BEING

Impact of Changing Climate Patterns on Public Health

It is known that shifting climatic patterns has been seriously threatening to the public health at global level and thus calls for immediate attention and interdisciplinary solutions (Vicedo-Cabrera et al., 2021). Variations in temperature, precipitation patterns, and extreme weather events have increased the likelihood of respiratory disorders, infectious disease onset and dissemination, and heat-related illness prevalence (Xu et al., 2020). The geographic range of diseases like malaria and dengue fever expands due to the favourable conditions created by rising temperatures, which also encourage the growth of mosquitoes, that are transmitters of the diseases (Kulkarni et al., 2022). Furthermore, changes in precipitation patterns and an increase in the frequency of extreme weather events can impair the quantity and quality of water (Puczko & Jekatierynczuk-Rudczyk, 2020), leading to the spread of waterborne illnesses and making sanitation issues worse. Research studies also show that there is an underestimation of climate change and water related parameters creating a huge literature gap in the appraisal tools of QoL that are used to compare cities and living spaces (Ptak-Wojciechowska et al., 2021). These factors impact the QoL to a large extent and are critical goals to be met for creating aging-friendly and climate-neutral cities. Climate Change affects different populations in different ways due to geographical variations and is a huge threat to public health (Cianconi etal., 2020).

In addition to the infectious diseases, shifting climatic patterns increase heatwave frequency and intensity, putting vulnerable populations at risk of heat stress and its related respiratory and cardiovascular problems (Ebi & Hess, 2020). Furthermore, malnutrition may be exacerbated by interruptions in the systems that produce and distribute food, further endangering the health of the community (Gomez-Zavaglia et al. 2020). There are increased hazards for vulnerable groups, such as low-income neighbourhoods and areas lacking proper healthcare infrastructure (Coleman et al., 2020; Hendricks & Van Zandt, 2021). It will take collaborative efforts to mitigate these health effects, combining public health regulations, community involvement, and climate science. In the complex interactions between climate change and public health, preventive measures like early warning systems, enhanced infrastructure resilience, and sustainable public health treatments are critical to ensuring global well-being (Durán-Romero et al. 2020).

Heat-Related Illnesses, Increased Risks of Diseases, and Mental Health Challenges

It is well recognised that the hazard posed by climate change to human health is multidimensional in nature, thereby increasing the likelihood of illnesses, heat-related ailments, and mental health issues (Vicedo-Cabrera et al., 2021; Xu et al., 2020). As the spread of infectious disease increases as vectors like mosquitoes expand their habitats, rising global temperatures play a role in the development of infectious diseases (Kulkarni et al., 2022). Climate change exacerbates heatwaves, which increases the frequency of heat-related illnesses and negatively impacts the vulnerable groups. Furthermore, changed weather patterns can expose people to more extreme occurrences, which would raise the prevalence of respiratory and cardiovascular illnesses. Due to extreme weather events, displacement, and unstable economies brought on by climate change, mental health is also at danger. Vulnerable populations have disproportionate health burdens because they frequently lack the resources to adapt (Ebi & Hess, 2022).

There are many linkages between climate change and mental health issues as well (Cianconi et al., 2020). However, a paradoxical finding is that although most QoL assessments cover the indicators of the liveability of the environment, still these are poorly connected with climate change and variability issues (Barreto Torres et al., 2019; Estoque et al., 2019). This continues to be a critical paucity in literature. Recent research on impact of climate change has shown that heatwaves, extreme precipitation and freezing during cold seasons have a significant impact on urban liveability however, these are not linked to Quality of Life assessment (Liang et al., 2020; Yakoub et al., 2021; Xiao et al., 2022). The intricate relationship between climate change and human health requires immediate mitigation and adaptation measures in order to ensure worldwide well-being in the face of a changing climate.

SOCIO-ECONOMIC FACTORS

Assessment of Climate-Induced Socio-Economic Disruptions

It is well-known that assessment of socio-economic disruptions due to climate change is a complex task and is at the nexus of social policy, economics, and environmental research (Estoque, 2023; Estoque & Wu, 2024; Lefevre et al., 2022; Raihan, 2023). Anthropogenic climate change is gradually changing precipitation patterns, sea levels, and weather patterns, thus leading to a cascading effect on human cultures. Therefore, strict assessment techniques are necessary to identify the complex relationships between changing climates and the socioeconomic consequences they have (Simpson et al., 2021). Elucidating the causative pathways and predicting the magnitude of disruptions need the use of robust climate models, statistical analysis, and integrated assessment frameworks.

It has been well reported that climate fluctuations due to climate change have a significant impact on socio-economic systems, influencing water resource availability, infrastructural resilience, and agricultural production. In such cases, vulnerability assessments are essential for determining which areas and populations are most vulnerable to disruptions brought on by climate change (Birkmann et al., 2022). Furthermore, an investigation into the indirect effects—such as migration trends, the onset of conflicts, and health crises—is necessary for a thorough grasp of the socioeconomic ramifications. Interdisciplinary research projects that cross academic boundaries are essential for developing policy frameworks and adaptive measures to lessen the negative socioeconomic effects of climate change (Estoque & Wu, 2024; Kawamiya et al., 2020). The evaluation of climate-induced socio-economic disruptions emerges as an important frontier in developing resilience and forming informed policies for a more resilient future, as the world community struggles with the pressing need for sustainable development.

Vulnerability Disparities and Adaptive Capacity Among Different Social Groups

Climate change presents a significant issue, revealing differences in susceptibility and adaptive capability across different socioeconomic groups and aggravating already-existing social inequities by disproportionately affecting disadvantaged communities. Scientific investigation of these differences requires a multifaceted approach that takes into account the socio-economic, cultural, and political factors that influence susceptibility in addition to the physical effects of climate change (Chepkoech et al., 2020; Naylor et al., 2020). Risks associated with climate change disproportionately affect marginalized populations, who are frequently defined by poorer socioeconomic position, restricted access to resources, and diminished potential for adaptation (Kosanic et al., 2022). This uneven distribution of rewards and costs within society exacerbates vulnerabilities.

Specific vulnerability profiles are produced by the complex interactions of variables including location, gender, ethnicity, and socioeconomic position (Chakraborty et al., 2020). For example, sea level rise may put coastal communities at greater risk, and changed cultural practices may provide problems for indigenous groups. The uneven distribution of adaptive capacity, or the capability to adapt to and manage the effects of climate change, is a result of differences in access to technology, governance frameworks, and levels of education (Andrijevic et al., 2020; Eriksen et al., 2020). A comprehensive knowledge of the complex web of social, economic, and environmental elements that contribute to vulnerability is necessary in order to address these discrepancies. It is imperative that policy interventions are nuanced, taking into account the many situations in which distinct social groups traverse the intricate terrain of climate change. So the necessary goal required is to minimize environmental consequences and promote equitable adaptation capacities among communities (Owen, 2020).

Some research on sustainability in the context of urbanization highlights the necessity of urban resilience (Zeng et al., 2022). This brings to light many relevant factors that enhance the capacity of urban resilience from adaptive, absorptive and transformative angles and emphasizes the social, economic, and environmental dimensions of urban sustainability (Zeng et al., 2022). All these factors together are likely to progress towards enhanced Quality of Life by focussing on the environment and climate change, in particular. With the advent of smart cities due to increasing urbanization, recent research has published various assessment measures to evaluate the Smart cities and its footprints namely 'project performance index', 'sustainability impact index', and 'sustainability performance index' (Kourtzanidis et al., 2021).

ENVIRONMENTAL JUSTICE

Unequal Impact on Marginalized Populations

In the fields of environmental science and social justice, one of the important task is to assess the unequal effects of climate change on underprivileged groups. Research in this context reveals a concerning trend: vulnerable groups, who frequently live in economically poor areas, are disproportionately affected by the negative consequences of climate change (Ngcamu, 2023). A number of variables come together to cause this occurrence, such as geographic location, socioeconomic inequality, and restricted access to resources and adaptive technology (Kosanic et al., 2022; Ngcamu, 2023).

Based on empirical data, disadvantaged communities—many of whom are located in dry or coastal areas—are more susceptible to extreme weather events, sea level rise, and changes in precipitation patterns (Dube et al., 2022). Their sensitivity to climate-induced difficulties is further intensified by the worsening of pre-existing socio-economic inequities, so jeopardizing their general well-being, food security, and livelihoods (Gomez-Zavaglia et al. 2020; Hauer et al. 2020). The scientific debate around this matter emphasizes how important it is to do multidisciplinary research that integrates sociology, public health, and climatology in order to thoroughly evaluate and solve the complex network of elements that contribute to the varying effects of climate change (Kawamiya et al., 2020; Vicedo-Cabrera et al., 2021). The scientific community's efforts to understand the complexities of this phenomenon are making it more and more clear that in addition to scientific advancements, inclusive policy frameworks that give priority to the resilience and adaptive capacities of people are also necessary to mitigate the disproportionate effects of climate change (Carmona, 2022; Estoque & Wu, 2024).

Intersectionality in Understanding Vulnerability and Resilience

Intersectionality, a notion derived from feminist studies, has emerged as an important paradigm for comprehending the varied character of vulnerability and resilience in the context of climate change (Ha et al., 2023). This paradigm recognizes that individuals and communities do not face environmental difficulties in isolation, but rather as linked entities impacted by several overlapping social characteristics such as gender, race, class, and location. Scientific studies with an intersectional perspective reveal how overlapping oppressive regimes increase the effects of climate change on various demographic groups (Kuran et al., 2020).

For example, women in low-income regions may encounter unique problems that differ from those experienced by their male counterparts or wealthier colleagues. Unequal access to resources, discriminatory legislation, and cultural norms exacerbate vulnerabilities, resulting in distinct routes to resilience (Amorim-Maia et al., 2022). Recognizing intersectionality requires a comprehensive approach to scientific study, policy formation, and community participation because it reveals the complex interplay between social identities and environmental threats (Kuran et al., 2020). By incorporating intersectionality into climate science, scientists and policymakers acquire a more nuanced knowledge of vulnerability, enabling for targeted actions that address the underlying causes of inequality and strengthen various populations' adaptive capabilities in the face of climate change (Singleton et al., 2022; Versey, 2021).

In conclusion, climate change has a significant and varied effects on people's quality of life, which cannot be denied. Significant risks to health, food security, and general well-being are posed by rising temperatures, harsh weather, and environmental degradation. The majority of these changes fall on vulnerable groups, those dealing with resource constraint, dislocation, and elevated health risks. The repercussions of these impacts are widespread, impacting cultures, ecosystems, and economies all around the world. In order to prevent more harm, adapt to the changing climate, and increase resilience, immediate group action is required. The main aim is to protect the standard of living of both the current and future generations by giving priority to sustainable practices, renewable energy, and egalitarian laws.

This book makes an attempt to address the identified gaps in QoL assessments, emphasizing the need to incorporate climatic impacts in the evaluation measures. By doing so, it contributes to the creation of aging-friendly and climate-neutral cities, crucial goals for contemporary societies.

REFERENCES

Abbass, K., Qasim, M. Z., Song, H., Murshed, M., Mahmood, H., & Younis, I. (2022). A review of the global climate change impacts, adaptation, and sustainable mitigation measures. *Environmental Science and Pollution Research International*, *29*(28), 42539–42559. doi:10.1007/s11356-022-19718-6 PMID:35378646

Amorim-Maia, A. T., Anguelovski, I., Chu, E., & Connolly, J. (2022). Intersectional climate justice: A conceptual pathway for bridging adaptation planning, transformative action, and social equity. *Urban Climate*, *41*, 101053. doi:10.1016/j.uclim.2021.101053

Andrews, F. M. (1974). Social indicators of perceived life quality. *Social Indicators Research*, *1*(3), 279–299. doi:10.1007/BF00303860

Andrijevic, M., Crespo Cuaresma, J., Muttarak, R., & Schleussner, C. F. (2020). Governance in socioeconomic pathways and its role for future adaptive capacity. *Nature Sustainability*, *3*(1), 35–41. doi:10.1038/s41893-019-0405-0

Barreto Torres, L. D., Asmus, G. F., & da Cal Seixas, S. R. (2019). Quality of Life and Sustainable Development. Encyclopedia of Sustainability in Higher Education, 1333-1340. Taylor & Francis.

Bhatti, S. S., Tripathi, N. K., Nagai, M., & Nitivattananon, V. (2017). Spatial interrelationships of quality of life with land use/land cover, demography and urbanization. *Social Indicators Research*, *132*(3), 1193–1216. doi:10.1007/s11205-016-1336-z

Birkmann, J., Jamshed, A., McMillan, J. M., Feldmeyer, D., Totin, E., Solecki, W., Ibrahim, Z. Z., Roberts, D., Kerr, R. B., Poertner, H.-O., Pelling, M., Djalante, R., Garschagen, M., Leal Filho, W., Guha-Sapir, D., & Alegría, A. (2022). Understanding human vulnerability to climate change: A global perspective on index validation for adaptation planning. *The Science of the Total Environment*, *803*, 150065. doi:10.1016/j.scitotenv.2021.150065 PMID:34525713

Bolte, G., Dandolo, L., Gepp, S., Hornberg, C., & Lumbi, S. L. (2023). Climate change and health equity: A public health perspective on climate justice. *Journal of Health Monitoring*, *8*(Suppl 6), 3. PMID:38105794

Bowling, A. (1995). What things are important in people's lives? A survey of the public's judgements to inform scales of health related quality of life. *Social Science & Medicine*, *41*(10), 1447–1462. doi:10.1016/0277-9536(95)00113-L PMID:8560313

Carmona, R. (2022). Resilience requires change: Assessing Pehuenche responses to climate change impacts in Southern Chile. *Environmental Justice*, *15*(3), 185–195. doi:10.1089/env.2021.0044 PMID:35686287

Chakraborty, L., Rus, H., Henstra, D., Thistlethwaite, J., & Scott, D. (2020). A place-based socioeconomic status index: Measuring social vulnerability to flood hazards in the context of environmental justice. *International Journal of Disaster Risk Reduction*, *43*, 101394. doi:10.1016/j.ijdrr.2019.101394

Chepkoech, W., Mungai, N. W., Stöber, S., & Lotze-Campen, H. (2020). Understanding adaptive capacity of smallholder African indigenous vegetable farmers to climate change in Kenya. *Climate Risk Management*, *27*, 100204. doi:10.1016/j.crm.2019.100204

Cianconi, P., Betrò, S., & Janiri, L. (2020). The impact of climate change on mental health: A systematic descriptive review. *Frontiers in Psychiatry*, *11*, 74. doi:10.3389/fpsyt.2020.00074 PMID:32210846

Coleman, N., Esmalian, A., & Mostafavi, A. (2020). Equitable resilience in infrastructure systems: Empirical assessment of disparities in hardship experiences of vulnerable populations during service disruptions. *Natural Hazards Review*, *21*(4), 04020034. doi:10.1061/(ASCE)NH.1527-6996.0000401

Costanza, R., Fisher, B., Ali, S., Beer, C., Bond, L., Boumans, R., Danigelis, N. L., Dickinson, J., Elliott, C., Farley, J., Gayer, D. E., Glenn, L. M. D., Hudspeth, T., Mahoney, D., McCahill, L., McIntosh, B., Reed, B., Rizvi, S. A. T., Rizzo, D. M., & Snapp, R. (2007). Quality of life: An approach integrating opportunities, human needs, and subjective well-being. *Ecological Economics*, *61*(2-3), 267–276. doi:10.1016/j.ecolecon.2006.02.023

Dal, T., Ramli, I., & Garaizar, J. (2023). Effect of climate change on nature and human health with a special focus on infectious diseases in the Mediterranean region. *Journal of Infection in Developing Countries*, *17*(11), 1501–1510. doi:10.3855/jidc.17995 PMID:38064385

Dube, K., Nhamo, G., & Chikodzi, D. (2022). Flooding trends and their impacts on coastal communities of Western Cape Province, South Africa. *GeoJournal*, *87*(S4, Suppl 4), 453–468. doi:10.1007/s10708-021-10460-z PMID:34188356

Durán-Romero, G., López, A. M., Beliaeva, T., Ferasso, M., Garonne, C., & Jones, P. (2020). Bridging the gap between circular economy and climate change mitigation policies through eco-innovations and Quintuple Helix Model. *Technological Forecasting and Social Change*, *160*, 120246. doi:10.1016/j.techfore.2020.120246

Ebi, K. L., & Hess, J. J. (2020). Health Risks Due To Climate Change: Inequity In Causes And Consequences: Study examines health risks due to climate change. *Health Affairs*, *39*(12), 2056–2062. doi:10.1377/hlthaff.2020.01125 PMID:33284705

Eriksen, C., Simon, G. L., Roth, F., Lakhina, S. J., Wisner, B., Adler, C., Thomalla, F., Scolobig, A., Brady, K., Bründl, M., Neisser, F., Grenfell, M., Maduz, L., & Prior, T. (2020). Rethinking the interplay between affluence and vulnerability to aid climate change adaptive capacity. *Climatic Change*, *162*(1), 25–39. doi:10.1007/s10584-020-02819-x PMID:33184523

Estoque, R. C. (2023). Complexity and diversity of nexuses: A review of the nexus approach in the sustainability context. *The Science of the Total Environment*, *854*, 158612.doi:10.1016/j.scitotenv.2022.158612

Estoque, R. C., & Wu, J. (2024). The resilience–sustainability–quality of life nexus. *The Science of the Total Environment*, *912*, 169526. doi:10.1016/j.scitotenv.2023.169526

Estoque, R. C., Togawa, T., Ooba, M., Gomi, K., Nakamura, S., Hijioka, Y., & Kameyama, Y. (2019). A review of quality of life (QOL) assessments and indicators: Towards a "QOL-Climate" assessment framework. *Ambio*, *48*(6), 619–638. doi:10.1007/s13280-018-1090-3 PMID:30206898

Fekete, H., Kuramochi, T., Roelfsema, M., den Elzen, M., Forsell, N., Höhne, N., & Gusti, M. (2021). A review of successful climate change mitigation policies in major emitting economies and the potential of global replication. *Renewable & Sustainable Energy Reviews*, *137*, 110602. doi:10.1016/j.rser.2020.110602

Glasser, R., Johnstone, C., & Kapetas, A. (Eds.). (2022). *The geopolitics of climate and security in the Indo-Pacific*. Australian Strategic Policy Institute.

Gomez-Zavaglia, A., Mejuto, J. C., & Simal-Gandara, J. (2020). Mitigation of emerging implications of climate change on food production systems. *Food Research International*, *134*, 109256. doi:10.1016/j.foodres.2020.109256 PMID:32517948

Ha, H. D., Tan, N. Q., Thang, T. N., Que, H. T. H., & Van Tuyen, T. (2023). Towards Inclusive Resilience: Understanding the Intersectionality of Gender, Ethnicity, and Livelihoods in Climate-prone Upland Areas of Central Vietnam. *Environment and Ecology Research*, *11*(4), 630–646. doi:10.13189/eer.2023.110410

Hauer, M. E., Fussell, E., Mueller, V., Burkett, M., Call, M., Abel, K., McLeman, R., & Wrathall, D. (2020). Sea-level rise and human migration. *Nature Reviews. Earth & Environment*, *1*(1), 28–39. doi:10.1038/s43017-019-0002-9

Hendricks, M. D., & Van Zandt, S. (2021). Unequal protection revisited: Planning for environmental justice, hazard vulnerability, and critical infrastructure in communities of color. *Environmental Justice*, *14*(2), 87–97. doi:10.1089/env.2020.0054

Kawamiya, M., Hajima, T., Tachiiri, K., Watanabe, S., & Yokohata, T. (2020). Two decades of Earth system modeling with an emphasis on Model for Interdisciplinary Research on Climate (MIROC). *Progress in Earth and Planetary Science*, *7*(1), 1–13. doi:10.1186/s40645-020-00369-5

Kosanic, A., Petzold, J., Martín-López, B., & Razanajatovo, M. (2022). An inclusive future: Disabled populations in the context of climate and environmental change. *Current Opinion in Environmental Sustainability*, *55*, 101159. doi:10.1016/j.cosust.2022.101159

Kourtzanidis, K., Angelakoglou, K., Apostolopoulos, V., Giourka, P., & Nikolopoulos, N. (2021). Assessing impact, performance and sustainability potential of smart city projects: Towards a case agnostic evaluation framework. *Sustainability (Basel)*, *13*(13), 7395. doi:10.3390/su13137395

Kulkarni, M. A., Duguay, C., & Ost, K. (2022). Charting the evidence for climate change impacts on the global spread of malaria and dengue and adaptive responses: A scoping review of reviews. *Globalization and Health*, *18*(1), 1–18. doi:10.1186/s12992-021-00793-2 PMID:34980187

Kuran, C. H. A., Morsut, C., Kruke, B. I., Krüger, M., Segnestam, L., Orru, K., Nævestad, T. O., Airola, M., Keränen, J., Gabel, F., Hansson, S., & Torpan, S. (2020). Vulnerability and vulnerable groups from an intersectionality perspective. *International Journal of Disaster Risk Reduction*, *50*, 101826. doi:10.1016/j.ijdrr.2020.101826

Lefevre, J., Le Gallic, T., Fragkos, P., Mercure, J. F., Simsek, Y., & Paroussos, L. (2022). Global socio-economic and climate change mitigation scenarios through the lens of structural change. *Global Environmental Change*, *74*, 102510. doi:10.1016/j.gloenvcha.2022.102510

Liang, L., Deng, X., Wang, P., Wang, Z., & Wang, L. (2020). Assessment of the impact of climate change on cities livability in China. *The Science of the Total Environment*, *726*, 138339. doi:10.1016/j.scitotenv.2020.138339 PMID:32325298

Naylor, A., Ford, J., Pearce, T., & Van Alstine, J. (2020). Conceptualizing climate vulnerability in complex adaptive systems. *One Earth*, *2*(5), 444–454. doi:10.1016/j.oneear.2020.04.011

Ngcamu, B. S. (2023). Climate change effects on vulnerable populations in the Global South: A systematic review. *Natural Hazards*, *118*(2), 977–991. doi:10.1007/s11069-023-06070-2

Owen, G. (2020). What makes climate change adaptation effective? A systematic review of the literature. *Global Environmental Change*, *62*, 102071. doi:10.1016/j.gloenvcha.2020.102071

Ptak-Wojciechowska, A., Januchta-Szostak, A., Gawlak, A., & Matuszewska, M. (2021). The Importance of Water and Climate-Related Aspects in the Quality of Urban Life Assessment. *Sustainability (Basel)*, *13*(12), 6573. doi:10.3390/su13126573

Puczko, K., & Jekatierynczuk-Rudczyk, E. (2020). Extreme hydro-meteorological events influence to water quality of small rivers in urban area: A case study in Northeast Poland. *Scientific Reports*, *10*(1), 10255. doi:10.1038/s41598-020-67190-4 PMID:32581301

Raihan, A. (2023). A review of the global climate change impacts, adaptation strategies, and mitigation options in the socio-economic and environmental sectors. *Journal of Environmental Science and Economics*, *2*(3), 36–58. doi:10.56556/jescae.v2i3.587

Simpson, N. P., Mach, K. J., Constable, A., Hess, J., Hogarth, R., Howden, M., Lawrence, J., Lempert, R. J., Muccione, V., Mackey, B., New, M. G., O'Neill, B., Otto, F., Pörtner, H.-O., Reisinger, A., Roberts, D., Schmidt, D. N., Seneviratne, S., Strongin, S., & Trisos, C. H. (2021). A framework for complex climate change risk assessment. *One Earth*, *4*(4), 489–501. doi:10.1016/j.oneear.2021.03.005

Singleton, B. E., Rask, N., Magnusdottir, G. L., & Kronsell, A. (2022). Intersectionality and climate policy-making: The inclusion of social difference by three Swedish government agencies. *Environment and Planning C. Politics and Space*, *40*(1), 180–200.

Smith, G. S., Anjum, E., Francis, C., Deanes, L., & Acey, C. (2022). Climate change, environmental disasters, and health inequities: The underlying role of structural inequalities. *Current Environmental Health Reports*, *9*(1), 80–89. doi:10.1007/s40572-022-00336-w PMID:35338470

Theofilou, P. (2013). Quality of life: Definition and measurement. *Europe's Journal of Psychology*, *9*(1), 150–162. doi:10.5964/ejop.v9i1.337

Veenhoven, R. (2000). The four qualities of life. *Journal of Happiness Studies*, *1*(1), 1–39. doi:10.1023/A:1010072010360

Versey, H. S. (2021). Missing pieces in the discussion on climate change and risk: Intersectionality and compounded vulnerability. *Policy Insights from the Behavioral and Brain Sciences*, *8*(1), 67–75. doi:10.1177/2372732220982628

Vicedo-Cabrera, A. M., Scovronick, N., Sera, F., Royé, D., Schneider, R., Tobias, A., Astrom, C., Guo, Y., Honda, Y., Hondula, D. M., Abrutzky, R., Tong, S., Coelho, M. S. Z. S., Saldiva, P. H. N., Lavigne, E., Correa, P. M., Ortega, N. V., Kan, H., Osorio, S., & Gasparrini, A. (2021). The burden of heat-related mortality attributable to recent human-induced climate change. *Nature Climate Change*, *11*(6), 492–500. doi:10.1038/s41558-021-01058-x PMID:34221128

Walsh, J. E., Ballinger, T. J., Euskirchen, E. S., Hanna, E., Mård, J., Overland, J. E., Tangen, H., & Vihma, T. (2020). Extreme weather and climate events in northern areas: A review. *Earth-Science Reviews*, *209*, 103324. doi:10.1016/j.earscirev.2020.103324

Xiao, Y., Chai, J., Wang, R., & Huang, H. (2022). Assessment and key factors of urban liveability in underdeveloped regions: A case study of the Loess Plateau, China. *Sustainable Cities and Society*, *79*, 103674. doi:10.1016/j.scs.2022.103674

Xu, R., Yu, P., Abramson, M. J., Johnston, F. H., Samet, J. M., Bell, M. L., Haines, A., Ebi, K. L., Li, S., & Guo, Y. (2020). Wildfires, global climate change, and human health. *The New England Journal of Medicine*, *383*(22), 2173–2181. doi:10.1056/NEJMsr2028985 PMID:33034960

Yadav, J., & Gupta, N. (2021, June). Urban quality of life: Domains, dimensions and indicators for Indian cities. [). IOP Publishing.]. *IOP Conference Series. Earth and Environmental Science*, *796*(1), 012032. doi:10.1088/1755-1315/796/1/012032

Yakoub, W. A., Eleinen, O. M. A., Mahmoud, M. F., & Elrayies, G. M. (2021). Developing a holistic green urban meter: An analytical study of global assessment tools for urban sustainability. *International Journal of Sustainable Development and Planning*, *16*(2), 263–275. doi:10.18280/ijsdp.160206

Yee, S. H., Paulukonis, E., & Buck, K. D. (2020). Downscaling a human well-being index for environmental management and environmental justice applications in Puerto Rico. *Applied Geography (Sevenoaks, England)*, *123*, 102231. doi:10.1016/j.apgeog.2020.102231 PMID:34326563

Yusoff, M. M. (2020, August). Improving the quality of life for sustainable development. [). IOP Publishing.]. *IOP Conference Series. Earth and Environmental Science*, *561*(1), 012020. doi:10.1088/1755-1315/561/1/012020

Zeng, X., Yu, Y., Yang, S., Lv, Y., & Sarker, M. N. I. (2022). Urban resilience for urban sustainability: Concepts, dimensions, and perspectives. *Sustainability (Basel)*, *14*(5), 2481. doi:10.3390/su14052481

Chapter 2
Impact of Climate Change and Occupational Hazards on QoL of Fisherfolk/Fish Processing Communities With Probable Mitigation Strategies

Nibedita Naha
ICMR-National Institute of Occupational Health, India

Somi N. Sharma
ICMR-National Institute of Occupational Health, India

Deepsi Rathore
https://orcid.org/0000-0001-7467-5455
ICMR-National Institute of Occupational Health, India

Ranjana Hanumant Choudhari
https://orcid.org/0000-0002-7474-5983
ICMR-National AIDS Research Institute, India

ABSTRACT

Fishery includes farmed fisheries and wild fisheries; of which, wild fisheries are comprised of marine and freshwater fisheries that provide an important source of food, nutrition, employment and income, mostly in the developing countries including India. However, now-a-days, environmental factors, such as climate change, global warming, heat stress, unpredictable weather patterns, ecological imbalance, etc., causes either direct, or indirect effects on the fishery and fish processing industry; thus, encompass numerous challenges to the quality of life (QoL) of the target communities. Also, the release of greenhouse gases into the atmosphere through man-made activities worsen the situation. Besides, occupational exposure is another major factor due to physical, chemical, biological and psychological stressors that affect the overall productivity of this informal sector. Globally, millions of people are employed in the fishery and fish processing industry though women are the majority in the fish processing work in the developing countries; thereby it demands need-based support programs, national policies and action plans for the benefit of the stakeholders. Further, industry's resilience is dependent on the effective governance, sustainable practices and strong mitigation strategies, where fishing industry is not an exception. Hence, the present book chapter scrutinizes a wide range of environmental and occupational issues confronting

DOI: 10.4018/978-1-6684-9863-7.ch002

the QoL of the communities, as well as explores the potential strategies to mitigate the negative impacts of climate change and occupational hazards in the fisherfolk and fish processing communities.

INTRODUCTION

The basic quality of life (QoL) includes physical, mental and spiritual health, relationships, education and social status, belongingness, physical surrounding, workplace environment, wealth, a sense of security and safety, and freedom of speech, movement and decision-making of an individual, which is attached with human needs and well-being (Costanza et al., 2007; Smith & Clay, 2010; Ali et al., 2018). Climate change is a global threat in the 21st century, as several challenges like global warming, heat stress, glacier and Arctic/Antarctic ice melting, rise in sea level and sea temperature, changes in precipitation, etc., directly, or indirectly affect the QoL of the people. The situation become worsen due to anthropogenic activities, workplace malpractices, lack of awareness, safety measures, education, and many more, especially in the developing and underdeveloped countries though data from the underdeveloped countries are scanty. In the developing countries, fishery industry is one of the major sources of living for the common people who are largely dependent on the several climate change issues. Fishery industry includes numerous activities, such as farming, fishing, processing, preserving, storing, transporting and marketing of fishes and related fish products that cover both farm fisheries and wild fisheries; of which, wild fisheries are comprised of marine and freshwater fisheries. In broad sense, United Nations Food and Agricultural Organization (FAO) defined aquaculture, or aquafarming of crustaceans, molluscs, aquatic plants, algae, and other organisms along with fishes (FAO, 2018a); however, the present book chapter deals with fisheries considering the huge production and market demands of fishes and fish products **(Fig. 1)**, as well as the number of skilled and unskilled workers employed in this informal sector.

Globally, fishes and fish products are one of the important sources of animal protein in diet due to low cholesterol and high amino acids, vitamins, minerals, and fatty acids (FAO, 2012), which commercially accounts to 240.99 billion USD in 2017, and is expected to reach 438.59 billion USD by 2026 at a compound annual growth rate (CAGR) of 6.88% (Research & Markets, 2019).The worldwide fish trade is worth of more than 50 billion USD, of which 50% comes from the developing countries (International Labor Organization [ILO], 2004). According to FAO, the rise in global capture of fisheries production from 1990 to 2018 is about 14%, while the total consumption of fishes and fish products increases 122%, suggesting a remarkable growth of the employment and revenue in this profitable sector during the period (FAO, 2022). In 2018, the total global capture of fishes reaches a record of 96.4 million tons, which is equivalent to 5.4% increase from the average since 2015 (ILO, 2004). Further, worldwide around 35 million people is employed in the fishery industry including 27 million full-time, part-time and occasional fishermen, mostly comes from the developing countries of Asia (83%), Africa (9%) and South America (2.5%), while the rest is divided from the fish exporting countries of North America, Europe and Russia (FAO, 2022). India is the 2nd largest country in fish production (6.3%) after China (more than ⅓ of global production), and 6th in the list of fish and seafood export countries (6.6%) with an annual export value of 4.6 billion USD after European Union (36.2%), China (22.7%), Norway (11.9%), Vietnam (9.5%) and USA (6.7%) (Shahbandeh, 2022; Nag, 2017). Though significant changes in the global fish production indicating a major shift in dominance towards the developing countries and China (Delgado et al., 2003); however, the supplies from capture fisheries are slowly being replaced

Figure 1. Worldwide scenario of fishery and aquaculture. A. Types of fisheries. B. Capture fisheries and aquaculture from inland and marine water. C. Diversity of fish capture and aquaculture. D. Utilization of fishes
Source: Author's creation

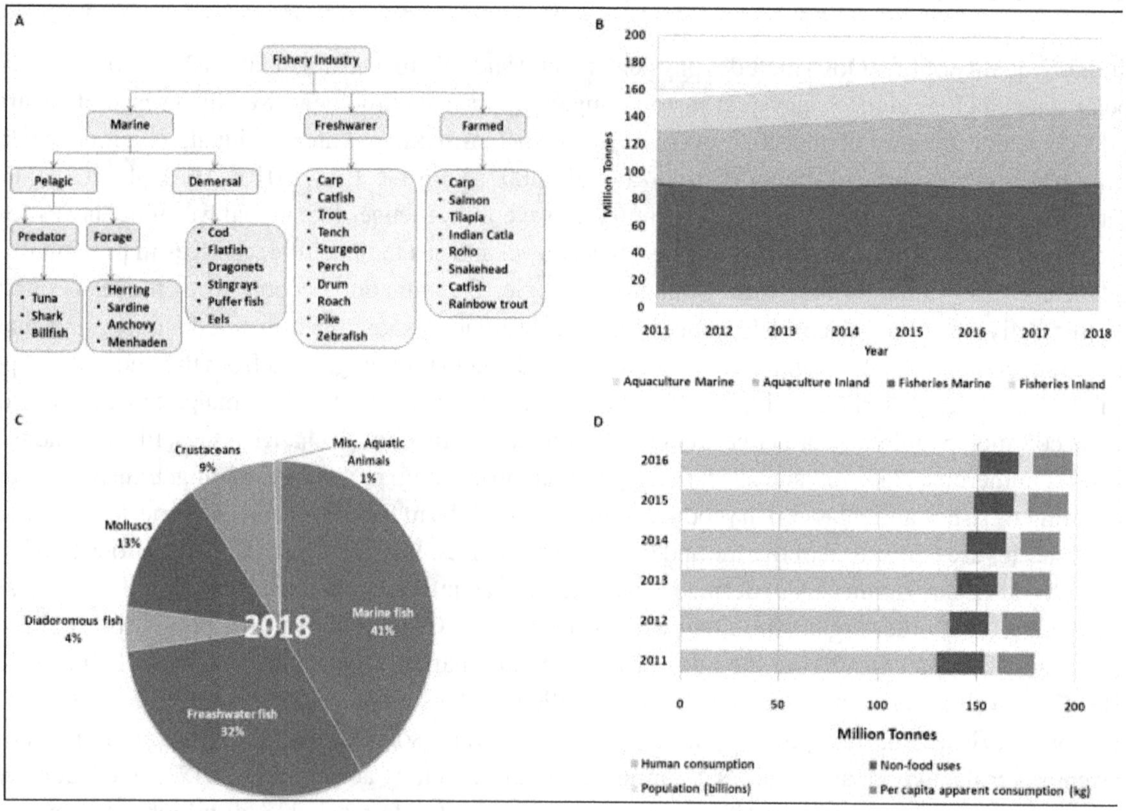

by the farmed and/or cultured supplies i.e., 50% of the present global consumption of fishes and fish products (FAO, 2016).

In India, due to vast marine coastal areas (7500 km), fishing villages (3827) and traditional fish centers (1914), freshwater resources of rivers and canals (195210 km), reservoirs (2.9 million hectors), ponds and lakes (2.4 million hectors), wetlands and water bodies (0.8 million hectors), backwaters (900 km) (FAO, 2016), both the marine and freshwater fisheries contribute to sustainable catch fishing potential of more than 4 million metric tons and farm fishing potential of 3.9 million metric tons (Panda et al., 2016). As per the National Fisheries Development Board, India earned Rs. 334.41 billion ($\approx$4.047 billion USD) in 2020 by export of fish and fish products that involving 14.5 million peoples' livelihoods with the total gross domestic product (GDP) of 1.07% in the country("India's blue economy net getting bigger", 2022). Interestingly, in India and other developing countries, males are primarily contributed to marine fishing and compromised to physical stress (temperature, salinity), injury, etc., whereas women are involved in fish processing and suffered at large mainly due to extreme cold stress and use of different chemicals (solvents, preservatives) (Samantray and Pathak, 2001; Nag et al., 2012; Warrier, 2001). Biological and psychosocial hazards, minimum, or no health benefit, forced migration and social oppression (Warrier, 2001) are also the significant determinants of declining QoL among these population, which affect the overall productivity and gross annual income in this sector. ILO, thus, concerns about the gradual rise of exploitation, trauma, stress and other health hazards among these communities, and highlights the requirement of longitudinal studies, especially, in the Asia-Pacific region (Tomoda, 1997). As compared to the developed countries, till date there are lacunae of adoption of modern technologies, strict regulatory guidelines, policies, and implementation of several acts, such as Environmental Protection Act, Water (Prevention and Control of Pollution) Act, Wildlife Protection Act, etc., in the developing countries including India, which, if taken care, may increase the productivity, revenue and QoL of the communities involved in the fishery and fish processing industry.

Therefore, in the above perspective, the present book chapter is conceptualized to address the climate change issues and challenges, workplace hazards, and their overall impacts on QoL of the fisherfolk and fish processing communities in the developing countries including India, along with the probable mitigation strategies for the benefit of the stakeholders. For this purpose, literature is searched systematically through PubMed, Google, Medline, Research Gate, etc., using numerous combinations of terms pertaining to several aspects of climate change, global warming, heat/cold stress, workplace hazards and risks, and suitable mitigation for improving the QoL of the target population, mostly in the developing countries.

CURRENT SCENARIO OF CLIMATE CHANGE-INDUCED CHALLENGES

Global Warming and Heat Stress

Over the past 100 years, due to a worldwide consequence of high greenhouse gases (GHGs) by anthropogenic activities, the average global temperature has increased at a rate of 0.74±0.18°C that affects both the marine and freshwater fisheries as per the Intergovermental Panel on Climate Change (IPCC, 2007). Moreover, the climate model forecasts the rise in the global surface temperature during the 21st century is likely at a rate of 0.3°C to 1.7°C, even if the lowest emissions scenario is considered, leading to adverse permanent damages of the ecosystem as a whole (IPCC, 2007). Even though to connect definite weather incident to climate change is difficult, the rise in global temperature has been predicted to cause significant shifts in the retreat of glaciers, melting of Arctic/Antarctic Sea ice, rise in sea levels, elevation in sea water temperature, fluctuation of salinity, oxygen and other acidic conditions of oceans, coral bleaching, etc., which is manifested by remarkable negative impact on the marine fish life cycle, and thereby fisheries (Mohanty et al., 2010). It has also been observed that daily as well as seasonal heat stress due to global warming results in lowering of oxygen concentration, leading to shorter incubation period for pathogens; thus, causes a negative impact on the life time of individual fish (Alfonso et al., 2020).

Rise in Sea Temperature

Rise in sea temperature is another most significant environmental variables, which affects the growth and development of the fishery industry. Ocean absorbs the excess heat emitted from the GHGs, resulting in the rise of its temperature; thereby affecting several marine species and overall ecosystems. Increasing mortality, loss of breeding and mass movement in search of favorable environmental conditions, etc., are the prominent threats for marine fishes, sea birds and other marine species (Laffoley & Baxter, 2016). Coral reefs, the resting place for marine species, are also facing high risk of morality due to gradual increasing of sea temperature. However, in the present global scenario, it is also difficult to prevent the massive irreversible damage of ocean warming on the marine fisheries by limiting the global average temperature rise to less than 2°C, above the pre-industrial levels (Allen et al., 2019; Meyer et al., 2022).

Rise in Sea Levels

As a result of the thermal expansion and exchange of water between the other reservoirs like, glaciers, icebergs, etc., as well as anthropogenic changes in land hydrology and atmosphere, sea levels are gradually rising during past few decades (Byrne et al., 2017). The mean yearly rise in global sea level is observed at a rate of 1.8 mm over 1961 to 2003, and 1.06 mm to 1.75 mm rise in India, whereas the total rise in the 20th century is estimated to be 0.17 m (Unnikrishnan & Shankar, 2007). Further, a reduction and disappearance of creek-associated habitats that are common nesting and breeding places for a wide variety of marine fishes, is caused by the flow of the saltwater/freshwater interface (Gillanders & Kingsford, 2002).

Ocean Acidification and Deoxygenation

Seawater absorbs carbon dioxide and generates carbolic acid, thus reducing its pH level, which causes a serious threat to the existence of marine fisheries and sustainable changes in ocean ecosystems. The ocean pH from a pre-industrial age state of 8.179 is dropped to around 8.069; and prediction of a pH change of -0.355 to 7.824 by 2100 (Mallela & Fox, 2018). Ocean acidification reduces the amount of carbonate, a key building block in seawater, resulting in difficulty of survival for calcifying marine organisms and fishes: - exposure during earlier stage of life causes higher risk of disintegration and in some cases, extinction. For example, coral reefs show slower recovery rate and more prone to damage by storms, and planktons and molluscs suffer from reduce growth and development due to their weak, or dissolving shells and skeletons (Long et al., 2016). Loss of habitat and food supply, trouble behavior and larval death are the common challenges due to ocean acidification for Juvenile small and large fishes like herrings, sardines, anchovies, and tuna, bonitos, cod, halibut, respectively (Kroeker et al., 2013; Mathis et al., 2015). These effects become worsen by the combined role of ocean deoxygenation, a condition of decreasing ocean oxygen levels that largely affect the marine life (Long et al., 2016).

Coral Bleaching

Rise in sea water temperature causes coral bleaching i.e., symbiotic zooxanthellae dislodgment in the tissues of corals, which indirectly affects the life cycle of the marine fishes. With global warming and climate change in recent decades, various studies reveal bleaching events (Lix et al., 2016; Glynn, 1991). Individual species react distinctly to variations in thermal environment, with higher degree of mortality, mostly seen in *Acroporaflorida* (Mohanty et al., 2013). In specific global warming hotspots, consecutive bleaching events could cause a decline in the species variations of corals. Coral reef ecosystems reinforce an immense diversity of the aquatic organisms, such as *Zoantharia.* The distribution and community structure of *Zoanthariaon* the Saurashtra coast of Indian state Gujarat, in contrast to their counterparts elsewhere, indicating a great flexibility to changes in the climate and non-living conditions (Kumari et al., 2015).

Changes In Wind Speed and Direction

Changes in wind speed and direction, as well as extreme weather events, in particular, cyclones are also the most important drivers of climatic variability in the aquatic systems, which over time, cause higher damage to the life and availability of fish stock (Zacharia et al., 2016; Vivekanandan et al., 2009). Differences in temperature generate winds, so rise in temperature of the earth's surface slow down the wind speed worldwide due to climatic change. The reduction of wind speed by 1% to 3% is also expected in the coming 50 years, which will likely to affect the marine fishery industry and the life of the fisherfolk (Akinsanola et al., 2021; FAO, 2018b; O'Reilly et al., 2013).

Changes in Precipitation

Changes in normal rainfall, annual rainfall intensity, dry season rainfall, etc., are the climatic variabilities that influences aquatic environment, and thereby fisheries due to increasing temperature and poor adaptation (Dzoga et al., 2018; Begum et al., 2022). This opens an avenue for the policy makers to formulate a better policy framework for sustainable management of marine fisheries by suitable adaptation of the climate change effects. Studies also reveal the loss of commercially important species from their traditional habitats due to El Nino/La Nina effects, which is prominent in Indian Ocean, eastern equatorial Pacific Ocean, and South-east Asian coasts like Indonesia and South American coastal areas, especially Brazil, Chile and Peru (Gasalla et al., 2018; Kemarau & Eboy, 2022). This is occurring because El Nino is an unusually warm ocean current accompanied by heavy rains and flooding that prevents the cold and nutrient-rich deeper water to reach the ocean surface (i.e., place of photosynthesis); thereby breaking of ocean production leading to lowering of food availability and force migration, or collapse the productivity of local fishes like anchovies.

La Nina, on the other hand, are cold ocean temperatures that push warm surface water farther west than usual, creating the opposite effects of El Nino. Seasonal variations of the hydro-climatic factors, such as rainfall, river inflow, snowmelt, and consequent fluctuation in water level also influence the inland fisheries worldwide (Sharma et al., 2023a; Kelkar et al., 2022). Increase in rainfall and water level favors the reproduction, recruitment and immigration of fish in the inland reservoirs, though significantly reduces the fish catch per unit effort due to migration of fishes to newly inundated areas in rainy season escape from the fishing gear (Patrick, 2016). Reduction in the number of fishers, boats, nets and fishing duration during rainy days indirectly promotes fingerling stocking, the common management practices as found in the tropical countries like Sri Lanka (Patrick, 2016).

IMPACT OF CLIMATE CHANGE SCENARIO ON FISHERY HABITATS

Fish Stock Distribution and Life Process

Ocean-atmospheric coupled climate models forecast the variations in the ocean circulation, which activate phytoplankton biomass production in the nutrient-deficient areas of the open ocean that has an impact on the fish stock; and hence, food security (Arteaga & Rousseaux, 2023). Plankton distribution is also concurrent to the changes of sea surface temperature; thus forecasts the productivity of the regional fish stock (Arteaga & Rousseaux, 2023). Physiological factors also affected by the condition of sea water. For example, fish stock distributions and life processes could be affected by a metabolic rise of 10%,which corresponds to a 1°C increase in temperature, implying of sea water temperature as low as 1°C (Mohanty, 2018). This limitation in physiology causes a changes in supplies, recruitments and quantities of fishes, which, if continue for a period of few years to 10 years, become severe in terms of species' ranges and compositions, resulting in the modification of the structures and functions of the ecosystems (Weiskopf et al., 2020). For example, small pelagic fishes having shorter life cycle and expeditious turnover of generations, are most likely to experience such climatic changes (Zacharia et al., 2016). Moreover, variations in the net primary production and its transfer to higher trophic levels are possible in several decades. Numerous tropical fish resources are suffered to intense heat due to continuous exposure of global warming, and regional disappearance/variability as well as migration to furthest distance across the breeding areas/periods, which ultimately affect the overall fish stock, such as threadfin bream, Indian oil sardine and mackerel (Vivekanandan et al., 2009; Zacharia et al., 2016). However, the association between the enhanced primary production of carbon dioxide and the exchange through sea air is not fully identified, so the effect on atmospheric carbon dioxide on fish stock is uncertain.

Fish Stock Availability

Extreme weather events, in particular, cyclones as mentioned above, over time cause higher damage to the availability of fish stock. Studies on the seasonal distribution of skipjack tuna explain the migration in winter towards offshore areas due to low sea surface temperature and inshore areas during warmer months (Zacharia et al., 2016). Global warming-induced variations in sea surface temperature could conclude the changes in the seasonal distribution of the specific species, and eventually the harvesting (Zacharia et al., 2016; Tewabe, 2014). Further, extension of the catchment area between 14°N to 20°N in the Indian Ocean for the past 20 years (mid-80's to 2006), deeply affects the Indian fishery sector by 15% increase of the total Indian oil sardine catch from the southwestern coast of the country (Vivekanandan et al., 2009). Since the catch in the southwestern regions has not reduced over the period, this serves as an extension of the distributional limits of the Indian oil sardine species, which causes a negative impact on the availability of the particular fish stock in their natural habitats.

Fish Harvesting Sectors

Climatic changes influencing the dissemination patterns of the fish stock; thereby affecting the fishing methods, especially in the harvesting sectors of marine, brackish water and inland. For example, the rise in sea surface temperature of the Bay of Bengal facilitates the capture of Indian oil sardine and mackerel after post summer period (monsoon) at which special nets are cast (Zacharia et al., 2016). Variations of wind-velocity have negative impact on the yellow fin tuna of the Indian coastline adjoining Bay of Bengal and Indian Ocean (Lan et al., 2020). The cold southwards wind during autumn and winter seasons are also beneficial for large species like tuna as the respective fish stock moves in water currents with winds from submarine water to surface water towards near shores. This migration as per climatic conditions suggest a positive correlation between the rising sea water temperature and the catch per unit effort in hour, which indicating a greater amount of energy spent to harvest a particular quantity of fish in a specific sea area (Central Marine Fisheries Research Institute-National Innovations on Climate Resilient Agriculture [CMFRI-NICRA], 2016). Though14% of species are sensitive to changes in the sea surface temperature (CMFRI-NICRA, 2016), actions to control the fishing process need to be raised in parallel with conservative management measures to increase the resilience of susceptible species towards such climatic variations. Moreover, cyclones, tsunami, etc., also frequently vandalize fishing equipment and interrupt fishing operations (Ponniah & Muralidhar, 2009; Muralidhar & Vijayan, 2016).

Alteration of oxygen concentration in sea water (i.e., oxygen minimum zones), changes in geophysics and oceanography of water bodies have been related for surge in the severity and prevalence of the disease outburst in the fishery sector (Goggin and Lester, 1995; Harvell et al., 2002; Vilchis et al., 2005). Climate modifications (i.e., strong wind) to the upwelling system and circulation of the northern Indian Ocean and other regions have also intensified the future biogas emissions from the oceans (Sydeman et al., 2014; Zacharia et al., 2016). Life cycle assessments (LCA) shows the majority of the emissions in the fishery industry are produced by the actual harvesting phase, followed by the processing phase (CMFRI-NICRA, 2014). However, up to a certain point, the rise in sea temperatures increase the production of mariculture, but this benefit is short-lived due to adversity of the higher global temperatures in current years (Maulu et al., 2021). Even 1°C deviation of temperature (i.e., 31°C to 32°C) causes significant growth defects, in particular, the musculoskeletal system, bile duct hyperplasia and acute kidney tubular necrosis of the valuable environment-tolerant marine fish species with good market value like silver pompano (Jayakumar et al., 2018; Alfonso et al., 2020).

Salinity is another factor, especially in dry weather during summer as high salinity of the creeks extensively hampers the prawn culture (Jamal et al., 2023). A rapid drop in salinity to levels those are fatal for *Penaeus japonicas* due to heavy showers, causing mass mortality of the farm yield (Preston et al., 2000). Droughts also increase salinity, resulting in intolerable conditions for the cultured species. Alteration of pH (i.e., ocean acidification) has deleterious effects, particularly on the size and exoskeleton of the larvae of the shrimps, prawns and fishes (i.e., dissolution of thin exoskeleton), which, in turn, affect the fish catch and the secondary productivity (CMFRI-NICRA, 2013). Moreover, 50% to 100% loss of productivity due to extreme weather conditions as compared to 20% to 40% seasonal variations are further considered for the risk assessment based on the consequences and livelihood scores in the seaside areas (Muralidhar et al., 2013a; Muralidhar & Jayanthi, 2014; Muralidhar & Vijayan, 2015). Overall, increase in the frequency of extreme weather events and associated damage to the fishing industry including inland fishery at the tropical and subtropical regions of the developing countries in Asia raise global concern with major contributions of China, Indonesia, India and Vietnam (De Silva & Soto, 2009).

Disease Due to Environmental Toxicants

Because of the natural biocidal properties, environmental pollutants viz. heavy metals, pesticides and chemical-discharges, especially, aromatic hydrocarbons in the aquatic environment due to catastrophic climate change events, such as hurricanes, floods, or industrial accidents, cause several diseases in fish stock and mortality of non-target aquatic organisms resulting from the contamination of water bodies with agricultural wastes (Ahmed et al., 2012). Some of the examples are, polycyclic aromatic hydrocarbons (PAHs), carbaryl pesticides, organophosphate parasiticides, antibacterial residues, algicides, coagulants, herbicides, probiotics, secondary metabolites and so on. Low levels of pollutants have bio-accumulation and bio-magnification in the fish bodies, resulting into chronic effects long after the pollutants have passed through the environment, such as epidermal papilloma, fin and tail rot, gill disease, hyperplasia, liver damage, neoplasia, ulceration, immunosuppression and metabolic alteration, etc. (Bleeker & Verbruggen, 2009; Sharma et al., 2012; Takeuchi et al., 2009). Most of these chemicals are xenoestrogens and pharmaceutical estrogens that release into water bodies and mimic the action of estrogen in the fish body, leading to estrogen-responsive and non-responsive breast cancers in human while consumed (Burridge et al., 2010). Surface lesions of *Serratiaplymuthica*, fin and tail rot by *Aeromonashydrophila* and *Pseudomonas fluorescens*, gill disease by the activity of *Flavobacterium sp*., vibriosis by *Vibrio anguillarum*, and enteric red mouth by *Yersinia ruckeri*are the common fish diseases attributed by water pollution (Austin, 1998). Diseases caused by *Aeromonas, Flavobacterium* and *Pseudomonas* also result from the adverse water quality upon contamination with high quantity of organic materials, depletion of oxygen, changes in pH and enhance microbial populations (Dwivedi & Banerjee, 2017). Some infections with *Serratia* and *Yersina* are attributed to the contamination of waterways with domestic sewages e.g., leakage of septic tanks, and vibriosis in fishes is the result of high concentrations of copper, which is reflected by the higher prevalence of the diseased fishes in polluted marine habitats as compared to the non-polluted one (Austin, 1998; Sobhana, 2009; Demeke & Tassew, 2016; Jayasingh et al., 2022).

Unfavorable to Market and Trade

Commercial fisheries are seriously affected due to shifts in distribution of productivity and availability of the fish stocks as a result of climate-induced challenges as mentioned before. Weather change is also likely to modify the forthcoming fisheries' production patterns, either by shifting production as species migrate to new habitats, or by changes in the net oceanic primary production (Brander, 2007; Ficke et al., 2007; Balwan et al., 2021; Raman, 2022; Sridevi et al., 2023). The impacts of climate variation on the output and reliability of farming practices reflect a significant obstacle to the global food security in the developing and underdeveloped countries as they are dependent on the fishes and fish products for 27% of dietary proteins in comparison with 13% of less vulnerable countries (Allison et al., 2009). Thus, lowering of climate change effects, continuous growth of efficiently governed fishery and fish processing industries, increase productivity with no/minimum risk at workplaces/processes, and promotion of resilient social and ecological systems are highly essential for sustainable development, global food security and QoL of the fishing communities (Merino et al., 2012; Dagnachew et al., 2021).

Overall, the impacts of climate change-induced challenges on the productivity of fishery industries are shown in **Fig. 2**.

CLIMATE CHANGE-INDUCED CHALLENGES IN INDIAN FISHERIES

The differences of sea surface temperature along the Indian oceans from 1976 to 2015 reveal a temperature rise by 0.602°C across the northeast India (NEI), while the same for northwest India (NWI), southeast India (SEI) and southwest India (SWI) are 0.597°C, 0.690°C and 0.819°C, respectively (CMFRI-NICRA, 2016). The rate of change of sea surface temperature over the Indian seas causes more impact in the western coast as compared to the eastern coast. Further, the northern Indian Ocean is considered as one of the climate change hotspots among the world oceans that warms faster than 90% of the world oceans, resulting in long standing impact on the marine environment, sustainable fish stock, and stress aggravation on marine fish stocks (Karunarathna et al., 2020). Velmurugan et al. (2015) reported no change of the amount of rainfall in the Andaman and Nicobar Islands in contrast to the patterns of rainfall, and the extreme rainfall events in the country influence Indian fisheries. The possible impact on fish farming and use of water also conflict with other agricultural, industrial and domestic users in India, especially in the areas where aquatic shortage is the result of variations in seasonal length, annual rainfall inten-

Figure 2. Multifaceted impact of climate change on ecosystem of fishery industry. Climate change in catchment area affects behaviors and life cycles of fish population and ecology
Source: Author's creation.

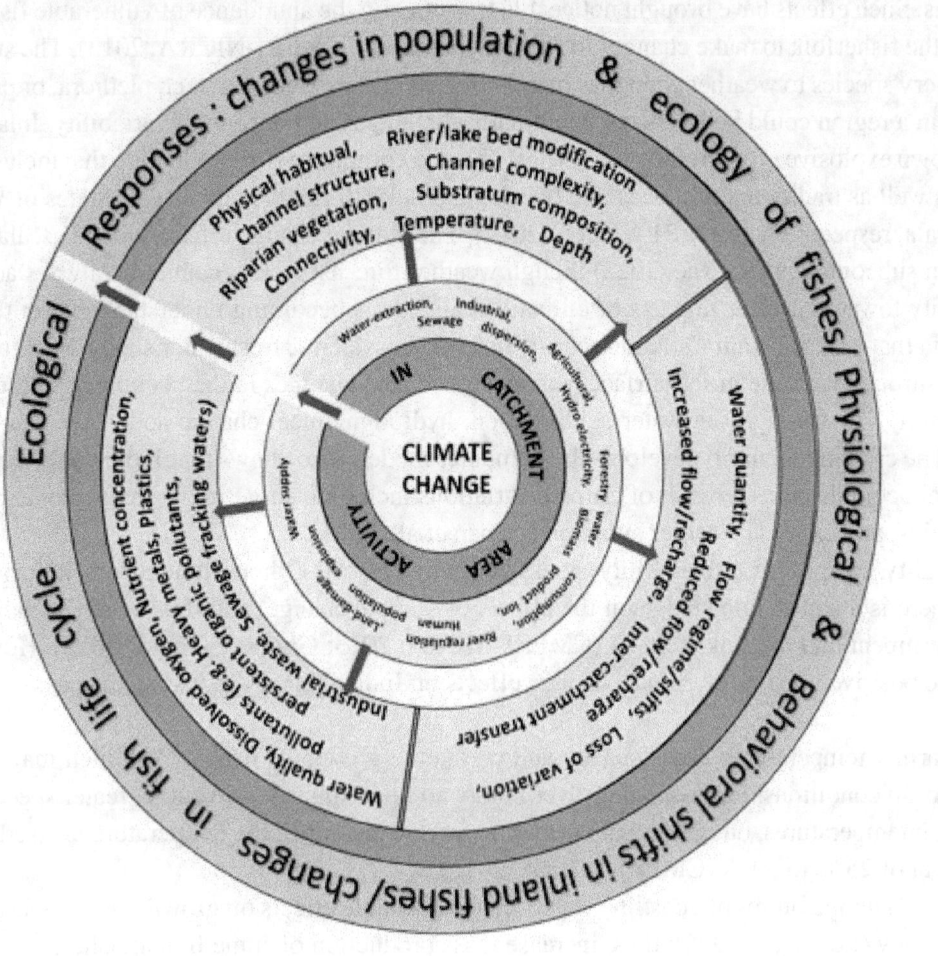

sity and dry season rainfall (Swaminathan, 2012). Moreover, having population explosion, low-lying coastal areas and primarily natural resources-based wealth, India is considered at risk to the adverse effects of the climate change-induced challenges. With an exclusive economic zone (EEZ) of 2.02 million sq.km on the length of entire coastline of 8118 km, India has annually harvestable ability of 4.412 million metric tons of diverse marine fishery wealth, consist of small-scale and traditional fishers; 63% of which is coming from the western coast of the country (Salim & Anuja, 2022), though 13% share of the total fish production in India is contributed by the non-costal fisheries (FAO, 2016). In contrast, shrimp farming is a rapidly growing sector in the brackish water (coastal) areas of alluvial plains and abandoned water bodies (Vijayan, 2018). Overall, fish populations and marine ecosystems respond to the spatial and temporal scale variations in the environment for decade may thus have unpredicted effects, comprising cyclic variations at the production level of marine ecosystems in India, which benefit one species, or group over another.

In addition to the already existing significant challenges, such as unverifiable harvesting, socio-economic strife, low catch per unit effort, poor income, etc., Indian marine fishing sector is mainly facing the issue of climate change-induced sea warming, which is remarkably visible in the Indian EEZ and nearby high seas. Such effects have brought noticeable transition in the abundance of vulnerable fish stocks and forcing the fisherfolk to make changes in fishing operations (CMFRI-NICRA, 2013). The susceptibility of a fishery species to weather change is mainly treated as the extent to which plethora, or production of species in a region could be impacted by climatic challenges and decennial variability. Inland fisheries have shown explosive growth from 46% to 85% share in composite fish production that includes brackish water as well as traditional practices like bheries and pokkali fields in the Indian states of West Bengal and Kerala, respectively (CMFRI-NICRA, 2014). The climate change effects will be escalated by 2050 in Indian subcontinent (McRae, 2022) though weather forecast results cannot be always accurate. The possibility towards greater impacts of climatic challenge is becoming understandable in the countries like India that covers country's food security, natural ecosystems, freshwater supply, health, etc., either directly through increase in sea surface heat and associated changes in the phenology of the organisms, or indirectly i.e., through sea water acidification, hydrodynamical change and rising sea level (FAO, 2015). The climatic scenario developed by computer models also shows that India could experience hot and humid conditions as a result of climate variations including surge in the frequency and severity of heavy rains, and extreme weather situation (Adnan et al., 2021).

In reality, in spite of vulnerability assessment covering 8000 households from Indian coastal villages, there is a gap of knowledge on the effects of climate change, and thus needs further input from the government and the stakeholders (CMFRI-NICRA, 2013; Gupta & Ghosh, 2023). However, there are some positive impacts of climate change effects on Indian fishery industry, such as: -

- Warmer temperatures cause speedy and premature growth of fish stock, which may be useful in certain conditions. For example, silver pompano fingerlings develop at a greater speed by slightly high temperatures, but growth abnormalities are noticed once the temperature crossed the optimal level of 25°C to 30°C (CMFRI-NICRA, 2016).
- Rise in temperatures of coastline waters cause favorable effects on growth rate and feed adaptation efficacy (Lehtonen, 1996); thus, increase mass production of some Indian fishes.
- As per the survey by Indian Council of Agricultural Research-Central Institute of Brackish Water Aquaculture (ICAR-CIBA), more than 800 hector of sea water swamped areas in the Andaman and Nicobar Islands are suitable for brackish water aquafarming in post tsunami as the lands became completely saline and unsuitable for agricultural use (Pillai and Muralidhar, 2006); hence, post 2004 it provides employment opportunities in the fishery industry and nutritional security of the locality.
- Besides, certain species, for example, Indian oil sardine and mackerel have undergone range extensions over the past several decades due to climate change-induced warm waters of the Indian Ocean (Vivekanandan et al., 2009). This is different from a distributional change as it creates disturbances in overall fishing practices and information.
- Conversely, range extensions let the usage of a specific fish stock in a greater number of zones. Because of these positive effects of climate variation, for stocks utilization capability at sustainable levels, is implausible to be harmful.

COMPROMISED QOL OF FISHERFOLK/FISH PROCESSING COMMUNITIES

Rise in sea level due to melting of glaciers/ice barges, and abnormal climatic events like cyclone, tsunami, etc., impose potential risk to fisherfolk and fish processing communities exposed to coastline, leading to the loss of income and livelihood. A 30 cm rise in sea level results in 45 m land erosion in many coastal areas of Asia (IPCC, 2022a), especially in the low islands and coasts with numerous channels that make the area more susceptible, for example, east coast of India, resulting in the vulnerability of whole ecosystems and inhabitants to extreme sea events ("India's blue economy net getting bigger", 2022). Higher risk of inundation is also observed in creeks and backwaters due to easier influx of water ("India's blue economy net getting bigger", 2022). Cyclone devastates fishery industry by averting fisherfolk from carrying out fishing operations besides infrastructural damage to key equipment utilizing in the practice of the craft. Countries in central and western Africa (Malawi, Guinea, Senegal, Uganda), north-western South America (Peru, Columbia) and tropical Asia (Bangladesh, Cambodia, Pakistan, Yemen) are found to be most vulnerable because of the combined effects of global warming, dependency of the national economy and food habits on fisheries, limited social opportunities and poorer economy (Allison et al., 2009).

Further, among the fishing communities, small sector fisheries comprising artisanal and subsistence fishers, are mostly affected by the impacts of climate change issues (FAO, 2022). Studies have shown that socio-economic status, age, workload, chronobiological stress, families, psychological factors, equipment technology, capital, etc., are the determinants of low QoL indicators in artisanal fishermen in Brazil and Mexico (Silva et al., 2019; Bravo-Olivas et al., 2015), whereas it is moderate in deep-sea and coastal fishermen in Malaysia in spite of the competition with foreign fishermen and lack of artificial reefs as fishing spots to increase the fish stock (Zain et al., 2018; Ghani et al., 2017). All these QoL components are critical for the sustainable fisheries' management policies in the developing countries. Uncertain income from fishing business further ends in rigidity for resilience and ability to adapt the economic implication due to climatic variation, and strains on the governmental measures that challenges the time-bound management plans and national policies based on the suitable aids and assistance programs for the fishery and fish processing industry (Akbari et al., 2023; Keese & Marcolin, 2023).

Moreover, a variety of techniques and methods are employed in the fishery industry that broadly classified as industrial/commercial scale, small scale and recreational. Nearly 90% of the world's fisheries are based on oceans and seas as compared to the freshwater fisheries, which is the continuous processes in spite of different climatic conditions (Olaoye et al., 2015). Fish processing, on the other hand, is a seasonal industry, and mainly depends on a variety of sea fishes (Nag & Nag, 2007). Fish processing starts at the time of fishing, or harvesting (i.e., farm fishing) till the final product delivery to the market (Olaoye et al., 2015); hence, maintenance of physical (temperature, salinity) and chemical (use of solvents, preservatives) conditions, minimization of risk and health hazards of the workers (injury, trauma, stress, exploitation), etc., are crucial for assessment of QoL and productivity of this sector. It has also been found that during COVID-19 pandemic, fisherfolk and fish processing communities as informal workers become more vulnerable to low QoL in terms of total and all QoL dimensions (Paskarini et al., 2022).

In India and other developing countries, males are primarily contributed to marine fishing and are prone to injury from loading and unloading tasks, whereas women are involved in fish processing and suffered at large due to exposure of chemicals and extreme cold stress (Dhanya & Mathai, 2017; Nag & Nag, 2007; Nag et al., 2012). Cold-induced peripheral vasoconstriction mediated through autonomic nervous system from the tips of the extremities to the center part is a common physiological response associated with poor workplace organization in the fish processing industry (Yurkevicius et al., 2021). Biological and psychosocial hazards, social oppression, etc., are also the significant determinants of declining QoL among these population, which affect the overall productivity and gross annual income (Asia-Pacific Fishery Commission [APFIC], 2010). Therefore, in the above perspective, occupational hazards related to fishery and fish processing are discussed below, which along with climate change issues need to be considered for successful implementation of the mitigation strategies including proper planning and adaptation along with suitable economic buffers. This is necessary for poverty alleviation and long-term sustainability of QoL of the fisherfolk and fish processing communities, especially in the developing and underdeveloped countries.

OCCUPATIONAL HAZARDS IN FISHERY/FISH PROCESSING INDUSTRIES

Physical Hazards

Workers are exposed to either high heat (45°C to 120°C) due to smoking, frying and drying, or extreme low temperature (below 15°C to cold rooms, even freezers) for processing of fishes and fish products, canning, etc., which cause alteration of body temperature, headache and respiratory disorders by inhalation of smoke, or cold air (Olaoye et al., 2015; Heus et al., 1995; Daanen et al., 1993; Swedish National Board of Occupational Safety and Health, 1991; Amravathi et al., 2016). Smoking kiln and wood burning cause redness of eyes, itching, irritability, blurring of vision and severe burn injuries in the affected population (Ngaruiya et al., 2019). Blanching, or pallor, cyanosis and reactive hyperemia (redness) of the extremities, Raynaud phenomena, tissue necrosis, and other skin disorders, such as erythema, utricarias, chilblains, cold panniculitis and cryoglobulinemias due to vascular insufficiency are observed among the workers of the fish processing industry upon long-term exposure to cold environment without using any protective devices, or gloves (Lehmuskallio et al., 2002). Repetitive lifting, pulling, and transporting heavy fish loads, nets, etc., are resulting in musculoskeletal injuries, bodyache, and discomfort/strain and sprain at the back, knee, ankle, wrist, shoulder and neck, as well as over-use injuries like tenosynovitis from repetitive motion tasks (Nag & Nag, 2007; Chiang et al., 1993; Ohlsson et al., 1994; Nordander et al., 1999; Olafsdottir & Rafnsson, 2000).

In India, Nag & Nag (2007) reports nearly 67% of the female fisherfolk are suffering from musculoskeletal complications. Pain, muscle cramps, or spasms, especially, in hamstrings, and decrease function and movements of joints are manifested by difficulty in walking, forward, or sideways bending, standing, or any awkward sitting posture of these workers along with the feeling of a pop, or tear at the time of injury (Olaoye et al., 2015; Nag & Nag, 2007; Nag et al., 2012). Cuts and scrapes from knife, needle injuries, bruising, fall/twist/blow, fractures, head injuries, etc., due to slippery floors and stairs, tetanus infection, or whitlow characterized by swelling, redness, pain, formation of pus, or abscess adjacent to finger nails, or even loss of the fingers, nails, or hands due to sting from fish spines and teeth, are the common physical hazards in the fishery and fish processing industry (Nag & Nag, 2007; Ngaruiya et al., 2019; Olaoye et al., 2015; Olaoye & Ojebivi, 2021; Sharma et al., 2023b). Moreover, exposure to noise and vibration results in temporary, or permanent hearing loss, and electrocution and high-voltage electrical accidents, in particular, due to water proximities, drowning, etc., are also reported (Nag & Nag, 2007; Olaoye et al., 2015; Ahmed et al., 2012; Ngaruiya et al., 2019). Sunburn due to prolong exposure to open sun is a common climate change-induced challenge to the Indian and other tropical country's fishermen. Collectively, severity of physical hazards depends on the length, intensity and repetition of exposure to extreme physical conditions, such as sunlight, wind, cold, heat, noise, water, and many more.

Chemical Hazards

Several chemical additives, such as sodium disulphite and formaldehyde, disinfectants like chlorine, formalin, gamalin 20, hydrogen sulphide (for anaerobic decomposition of fish), sodium chloride, sodium metabisulphite, and many synthetic chemicals i.e., acriflavin, benzalkonium chloride, hydrogen peroxide, malachite green, tricaine, potassium permanganate, praziquante, etc., are widely used in the fishery and fish processing industry, leading to chemical hazards among the workers. The symptoms are swelling of eyes, redness and watering, occupational asthma, allergy and other acute respiratory diseases, irritation, damage of the mucus membrane, dermatitis and cancer (Dhanya & Mathai, 2017; Idowu & Sogbesan, 2017; Srinivas & Sethy, 2023; Anisha et al., 2023). Respiratory exposure to aerosolized fish and shellfish proteins of high molecular weights and hypertonic saline viz. sea water is responsible for allergy and non-specific bronchial hyper-responsiveness, associated with immunoglobulin E (IgE)-mediated hypersensitivity (Lucas et al., 2010; Jeebhay et al., 2004). Women are more prone to IgE hypersensitivity than males, might be due to more female employment in the fish processing industry (Amravathi et al., 2016).

The chronic recurrent dermatitis, also known as protein contact dermatitis (PCD) due to contact with proteinacious fish material, is further reported in the fishery and fish processing industry. Moreover, common hand cleaners, soaps, detergents and spices like paprika, fluradditives, onion, garlic, mustard, etc., are the causative factors of delayed contact dermatitis (Jeebhay et al., 2004). High prevalence of dry skin, rashes, chapped skin and chronic sores are, especially noticed among the shrimp and salmon processing workers (Aasmoe et al., 2005). Skin contamination of liquid chlorine causes freeze burn depends on the duration and intensity of the exposure. Type II synthetic pyrethroids i.e., deltamethrin, cypermethrin and fenvalerate are critical in the freshwater fishery due to their physico-chemical properties, and metabolic, hematological, histological, behavioral and reproductive toxicities on fishes and human upon consumption (Prusty et al., 2015). Inhalation of smoke containing potential carcinogens like, PAHs during fish processing is another major component associated with chemical hazards (Olaoye et al., 2015).

Biological Hazards

During handling of sea fishes and fish processing, the workers, mainly through skin contacts, are exposed to several bacterial toxins (endotoxin), viruses, marine toxins (histamins), tetanus, fish toxins (pepsin, trypsin, biogenic amines, cadaverine), allergens, etc. (Dhanya & Mathai, 2017; Ngaruiya et al., 2019; Jeebhay et al., 2004; Ahangari et al., 2023). In the freshwater fishery, leeches, nematodes are the common parasites along with fungal and pathogenic infections, such as *Vibrio vulnificus*as infected by the spines of the *Tilapia sp.* (Olaoye et al., 2015; Ahmed et al., 2012). Wound infection during fishing, cleaning, gutting, de-scaling, peeling and processing, pricking and biting by fishes, sharks, etc., also leads to sepsis and even amputation of body parts (Ngaruiya et al., 2019; Kolawole & Bolobilwe, 2019; Olaoye et al., 2015). As compared to marine fishes, the higher body burden of the natural and man-made toxic substances like antibiotics, pesticides, persistent organic pollutants and pathogens (*Salmonella, E. coli, V. cholera, Clostidium, Streptococcus, Mycobacterium*) in farmed fishes impose severe health hazards to the workers and consumers (Ahmed et al., 2012). Antigen and endotoxin concentrations are detected very high in fishmeal loading and bagging activities (Jeebhay et al., 2004) though no threshold limit exists for inhalation of dry dust aerosols-related health hazards in this occupation (Liebers et al., 2020).

The handlings of perishable raw materials that carry microorganisms, exacerbate the risks of exposure (Ahmed, 2020). Food-borne trematodes (*C. sinensis, O. viverrini*) cause enteric diseases in Egypt, Korea, Japan and other Asian communities who ate raw fishes (Ahmed et al., 2012). Skin rash is the highest reported health issues (78% to 81%) among the fisherfolk followed by chronic sores (55.6%), asthma (7% to 10%) and other non-specific allergic symptoms (9% to 15%) (Lopata et al., 2004; Aasmoe et al., 2005). Warts in hands are the consequence of human papilloma virus type 7 (HPV 7) infection in the fishery and fish processing industry where the latent virus become activated and acts as a co-factor (Amravathi et al., 2016). Aflatoxins also have negative impact on human health (El-Sayed & Khalil, 2009). Urinary tract infection, vaginal discharge and other gynecological problems are identified among the women fisherfolk and fish processing workers (Warrier, 2001). Transgenic fishes containing recombinant gene(s), are potentially allergenic and toxic; thus, raising safety concern due to high load of different growth factors (Halleman & Kapuscinski, 1992; Samarasiri et al., 2023). However, epidemiology of the human diseases associated with biological agents and global climatic changing scenario in the fishery and fish processing industry is require to be thoroughly researched on large number of sample population.

Psychosocial Hazards

Mental set-up governs the facial expression and determines the ability, life style and physical condition of an individual (ILO, 2016). Therefore, mental demand is one of the driving forces for QoL indicator. However, mental health, vulnerability and psychosocial hazards are most neglected that manifold increase the occupational injuries among the fisherfolk and fish processing communities (King et al., 2021). The psychosocial hazards are characterized by emotional and mental stress, anxiety and depression, poor concentration, loss of appetite and energy, sleep deprivation, sulking, fatigue, indecisiveness, human errors, and insufficient adaptation, etc., mainly due to long working hours, shift-works and odd-hours job (Olaoye et al., 2015; Jeebhay et al., 2004; Nag et al., 2012). Studies also reported more vulnerability of women fish processers to work related morbidities than their male counterparts despite similar kind of work, which may be one of the reasons for short-lived job period i.e., 5 years (Palsson et al., 1998; Dhanya & Mathai, 2017; Nag & Nag, 2007).

Intoxication, violence, aggression, gender issues, harassment from the customers and owners not only cause the physical hazards, personal discomfort and physiological stress, but also impact the acute and chronic mental health conditions, insecurity and frustration among the fish processing workers, especially sexual overtures of the women and trading of sexual favors at the cost of security in the workplace, forced abortion and poor hygiene from the unhealthy medical practices (Warrier, 2001; Zhao et al., 2013; Michalopoulos et al., 2017). Work stress, poor work perception, ambiguity in job roles and goals, absenteeism, and exploitation at the hands of owner/supervisor like 'zero' incentive, prolonged pace of work and limitation of time, irregular job-rest patterns, verbal and other kind of abuses, involvement of more contractual, or migrant labors with no health benefits, boredom are the potential variables behind the psychosocial hazards and cumulative trauma disorders of the fishing and fish processing communities that affect the productivity in this informal sector (Ngaruiya et al., 2019; Nag & Nag, 2007; Warrier, 2001).

Overall, the younger the fisherfolk, the higher the chances of occupational hazards and risks. The married fishermen risking their own lives for the sake of their family even in the unfavorable weather conditions, found mainly in the developing countries of the Asia-Pacific region, though morbidity and

Table 1. **Overview of occupational hazards in global fishery and fish processing industries.** *The scenario is more aggravated in the developing countries, which need to be taken care;* **Source:** *Author's creation*

Types	Source	Effects
Physical hazards	Cold, heat, sharp equipment (knife, needle), noise, manual peeling, electrical gadgets, wet surface and fish spillage (bile).	Fingers and toes: repeated cutting, piercing; numbness; blanching; pallor, cyanosis, reactive hyperaemia; Raynaud's phenomena, tissue necrosis; falls; fracture; sun burn; shock, electrocution.
Chemical hazards	Chlorine, formalin, gamalin 20, paprika, sodium chloride, sodium metabisulphite, smoke, synthetic and organic chemicals, hypertonic saline.	Asthma; bronchitis; cough; phlegm; local irritation and rashes to skin, nose, throat and eyes; running nose; eye swelling, redness and watering; freeze burn; allergies, dermatitis.
Biological hazards	Bacterial toxins, virus, marine toxins, parasites, fish toxins, spiny fish, pathogens, microorganisms.	Bacterial and viral diseases; asthma; urticaria; dermatitis and allergies; mucous membrane irritation; hypersensitivity; wound infection and sepsis; musculoskeletal disorders, bodyache and discomfort (back, knee, ankle, wrist, shoulder, neck).
Psychosocial hazards	Work stress, prolong working hours, bullying including verbal abuse, sexual overtures,physical exploitation.	Emotional and mental set-back; stress, anxiety and depression; sleep disturbance; fatigue; insecurity; inferiority complex; forced abortion.

mortality are also reported from the developed countries like USA, Australia, Canada, Spain, Korea, Estonia, Italy, Poland and Scandinavia (Ngaruiya et al., 2019). Further, due to involvement of physical, chemical, biological and psychological hazards (**Table 1**), the global fishing and fish processing industry are considered as one of the most dangerous occupations under informal sector (ILO, 1999; El-Saadawy et al., 2014; Rodrigues & Kiran, 2013; Yea & Stringer, 2023).

MITIGATION STRATEGY FOR FISHERY/FISH PROCESSING INDUSTRIES

Climate Change-Oriented Mitigation Strategies

There is a need to develop multifaceted action plans and national policies to enhance the resilience of marine, freshwater, brackish water and inland fisheries. Accordingly, implementation of a suitable need-based arrangement is of prime importance for the stakeholders to take advantage of the system through several vital elements, such as targeted systematic approach, strong coastline ecosystem, community and industrial support system, and weather sensitive technologies with lesser carbon footprints. Continuous mapping of fisheries' habitat coupled with regional models and spatial representation of commercial and vulnerable species along with its linkage to ecosystem are the eventual adaptation alternatives for better conservancy, management and viable utilization of the hydroresources (Georgian et al., 2019; Hu et al., 2022).The development of regional models at species level having representation of the marine and climatic parameter variant could also forecast the assessable changes of the climatic stressors (Gissi et al., 2020). Satellite remote sensing and Geographic Information System (GIS) are successful in space-time distribution of crustacean populace and their habitats in India and other developing countries; the spectral signature of which support coral reef (habitat of several fish species) mapping at an exact scale (Paul et al., 2016; Meaden & Aguilar-Manjarrez, 2013; Pelage et al., 2019).

Also, GIS models can establish obvious knowledge of a water body physics in the particular coastal regions like Gulf of Kachchh, and biological processes like fish larval fetch, egg release, etc. (George et al., 2011). The budding fishing zone could also be point out for sustainable use of oceanic resources associated with cost-effective fishing processes in shorter duration of searching time (Bax et al., 2022). Fish catch forecast models could be established, so as to empower the fisherfolk, self-help groups and stakeholders in the developing countries to manage with the stock shift, as well as new technologies in the perspective of climate change, stock availability and fishing guidelines i.e., flotilla size, net size, spatiotemporal closure, etc. (Scales et al., 2023).

Implementation of minimum legal size (MLS) increases the fishing pressure on vulnerable population. In India, based on the Marine Fishing Regulation Acts (1981) and other scientific studies, CMFRI recommends MLS for fifty-eight commercial species and inspect any violation in this regards (Mohamed et al., 2014). Green fishing protocol for carbon foot prints reduction, a renewable energy usage in fishery, is taken into account that including suitable modification of the fishing boat engines, operational speed, engine design, etc. (Ghosh et al., 2014 & 2015; Munoz et al., 2023). Use of fuel for fishing contributes majority of the total impacts in maximum of the ten environmental impact groups, and engine of marine fleet with a 24% upper value for global warming potential has the higher impact as compared to the mechanized one (ICAR- Central Institute of Fisheries Technology [CIFT], 2016). Identification of stress tolerance species, such as silver pompano, cobia, etc., gain attention because of the presence of genetic factors in their body mass by natural and anthropogenic agents, which attribute to physiological plasticity of bivalves in stressful conditions (Pathak et al., 2022). Purohit et al. (2014) reports the role of specific heat shock proteins (Hsp70, Hsp78, Hsp60) in thermal accustom and stable survival at higher temperature of 36°C. Due to shorter generation time and higher fertility, it would also be possible for selectively breed fishes to tolerate weather change variables like high temperature, drought, salinity and diseases, etc., are likely to impact the fishery and related fish processing industry (Mitra et al., 2023). However, the influence of various abiotic stressors on gene expression, and evolution of physiological and genetic adaptations to osmotic and thermal stress, etc., has been understudied in spite of significant progress in wide range of physiological information available on the link between environmental stress and host response-indicators.

The carbon sequestration potential of mangroves ecosystem mapping, conservation and restoration are important in the perspective of climate variation, as *Rhizophora mangle* provides a substantial habitat, breeding ground and nursery for several species, enhances coastal resilience, abates temperature-induced stress, storms, hurricanes and seismic sea waves, and reduces seaside destruction; therefore, overall beneficial implications for coastal fisheries, income of the fishermen, levitate bio-shields and hedgerow along the coastline areas as well as adjustment to vulnerability and loss of fishing days (Rao, 2009; Suman, 2019). Moreover, stock modeling to boost the resiliency of stocks and recovery from population collapses, prioritization of the vulnerable species for conservation is also important management strategies towards adaptation of fishery stocks (Hilborn et al., 2020). Large scale seaweed farming along the costal water also aiming reduction of carbon footprint, increasing oceanic water pH, abating coastal contamination and farming of fishes along with other aquatic species (Troell et al., 2022). Integrated farming technology i.e., aquafarming, fisheries, agriculture and other productive, or ecosystem management activities in a single platform is considered the future face of the fishery industry. Integration is a fundamental component of the Ecosystem Approach to Aquaculture (EAA) model within a wider ecosystem such an extent it boosts ecological development, rights, and recovery of interconnected socio-environmental system (Soto et al., 2008; Zhou et al., 2022). Also, fish farming with seaweed and mussel, paddy-fish integrated farming are another popular adaptation in the developing countries. Wetland restoration along with scientific fish farming (Panigrahy et al., 2012), evaluation of weather trends and forecasting (Kumar et al., 2013) and inundation of different land resources using geospatial techniques (Rojith & Zacharia, 2015) are prospective climate resilient strategies applicable for both the inland and marine fisheries.

GHGs are the anthropogenic constituents of the fishery and food processing industry, though in any of the GHGs inventory its contribution has neither estimated, nor included. However, GHGs emissions are measured from shrimp and finfish farming systems varying in stocking quantity in different geographical locations in India, which provides significant information on the trends of emission and helps in combating the climate change scenario (Muralidhar et al., 2013b; Vasanth et al., 2016). Development of climate resilient products through need-based management, judicial utilization of energy resources and implementation of microbial interventions can effectively mitigate the emission of GHGs, such as (1) estimation of definite aeration requirement in shrimp culture, (2) use of least nitrous oxide producing anammox bacteria, *Kueneniastuttgartiensis* in culture sediments, (3)increasing carbon capture potential of shrimp culture in pond sediment, (4) biochar synthesis using pyrolysis of mariculture residues containing opulentorganic carbon, and (5) biofuel from micro and macro algae, etc.(Al-Saidi et al., 2021; Godvin et al., 2021; Schneider et al., 2012; Zhang et al., 2022).Unlike freshwater, brackish water aqua culture doesn't face competition with other sectors, and compared to freshwater bears lesser water footprint. Moreover, brackish water is considered as one of the best options for increasing fish production due to gradual decrease of freshwater resources as a result of climate change effects (Maulu et al., 2021).

Further, climate change and consequent irregular rainfall pattern interrupt the production and management of inland fisheries; thus, understanding the influences of climate change challenges in the hydro-climatic factors and predicting the future changes by modeling in a rational manner, are mandatory for sustainable growth and development of the inland fisheries in the developing countries (Patrick, 2016). Last but not the least, based on the global climate change scenario, the mitigation strategies should target the adaptation on vulnerable coastal fishing communities by several Climate Change Preparedness Activities (CPAs) (Daw et al., 2009), and climate change awareness program through suitable scientific interactions and trainings, disaster management planning i.e., early warning system, automatic weather station, providing basic amenities in coastal villages for food security, sanitation, shelters and local infrastructures, etc., which ultimately strengthen sea-walls, or bio-shields (i.e., coastal forestry), and thereby control sea erosion, develop and familiarize e-commerce technologies among these target population and improve the overall QoL indicators (Hurlbert et al., 2019; Lebbe et al., 2021). Moreover, the study of Paskarini et al. (2022) can be used to ascertain the adaptation strategies for improving QoL of the fisherfolk and fish processing communities in the light of post-COVID condition.

Occupational Hazards-Oriented Mitigation Strategies

As occupational health hazards in the work environment significantly contribute to the incidence of work-related injuries among the fish processing workers and fisherfolk; hence development and implementation of preventive measures are the most practical mitigation strategies in this regard. Use of personal protective equipment (PPE), such as protective gloves, masks, gumboots, caps, particular dress code, use of sunglass and sunscreen, especially in outdoor work, etc., are the primary safety measures in the fishery and fish processing industry along with compulsory first-aid kits including user's manual (FAO& World Health Organization [WHO], 2020; U.S. Food and Drug Administration [USFDA], 2022). Jeebhay et al. (2004) opine that gloves and plastic sleeves are fundamental for prevention of infection and allergens among the fisherfolk and workers of the fish processing industry with regard to contact dermatitis, and reduction of hand, or foot temperature, cold related skin syndromes and sharp injuries. Although PPE increases protection against occupational health risks at workplace (Bull et al., 2002; Occupational Safety and Health Administration [OSHA], 2023); however, in reality, very little use of PPE is reported (Ngaruiya et al., 2019; Nag & Nag, 2007; Nag et al., 2012). In the developing country like India, as the fisherfolk and workers of the fish processing industry earn very marginal amount; hence, spending money for a pair of gloves may be a luxury for them (Nag & Nag, 2007; Gopal, 2021). Although some groups, for example, precision job holders like squid processers inform about slowdown of the working process due to PPE, the use of different kinds of well-fitted gloves and foot wares specific to particular task provides a feeling of betterment for work, improve hand and foot temperature, and reduction of cold shock injuries (i.e., Raynaud's phenomena) among these fish processing workers (Nag & Nag, 2007, Amravathi et al., 2016).

Fishery is considered a risky occupation related to injuries, morbidities and mortalities but huge number of people, especially coastal population depends on it for their livelihood. Thus appropriate work design intervention, such as introduction of conveyor belts for transportation, adjustable table, platforms, sitting stools, foot-rests and anti-fatigue mats, ergonomically designed tools, etc., may increases work precision and speed, controls work stress, and decreases physical hazards, health risks and accidents in the fishery and fish processing industry (Jeebhayet al., 2004; Nag et al., 2012).Improved workplace organization like shift work and intermittent resting between work, job rotation, post-injury work management, etc., have great impact in lowering of stress in work environment and trauma disorders among these workers (Chen et al., 2020; Lincoln et al., 2021). Moreover, as it is evidenced that health effects in the working environment are not associated with any single work stressor; therefore, remedial measures for reducing allergen-induced respiratory issues, especially IgE-targeting intervention and avoiding direct respiratory exposure of aerosolized particles, as well as global approach for prevention of occupational hazards play crucial role in reducing the potential risk of occupational diseases like asthma and other respiratory problems, dermatitis, etc. (Amravathi et al., 2016). The sick and dead fishes should be removed from the system, and disposed of as per the rules and regulation of the country to avoid asymptomatic carriers of many diseases (Ahmed et al., 2012). Periodical cleaning and disinfection of the tools, nets, floors, foot-baths, etc., optimization of waste water treatment, effluent discharge and reuse by integrated approach of physical, chemical and biological techniques are also recommended to reduce pollution emitted per ton of fish production (Ahmed et al., 2012; Tom et al., 2021). The direct contact of chlorinated water, salinity, etc., should be advised to avoid (Al-Abri et al., 2019).

The periodical public health awareness program, education and health campaign are essential to reduce the documented health hazards in the fishery and fish processing industry along with the counseling of workers and owners (Black et al., 2019). Also, fisherfolk and fish processors should have the advance knowledge of job-associated risks and accidents, health and safety acts, and thus, reorientation of work profile in terms of safety consciousness is the most essential mitigation strategy in the fishery and fish processing industry. Globally as males are dominated in fishing and females are in fish processing; hence, consideration of gender roles is fundamental to achieve occupational health and safety issues in this informal sector (Ngaruiya et al., 2019), where the active role of non-government organizations (NGOs) and governmental interventions are necessary to prevent exploitation of the vulnerable groups (Warrier, 2001; Samantray & Pathak, 2001; Lentisco & Robert, 2015). Further, the effectiveness of any intervention depends on the strict regulatory guidelines and its implementation where developing countries are comparatively in back foot, for example, regular maintenance of work tools, routine supply and use of PPE, health check-up, counseling, and so on (Cochrane, 2002; Kathleen & Feris, 2022; Williams & Syddall, 2022). Biosecurity through proper management strategies, constant surveillance, and internationally accepted policies and guidelines are another aspect of successful mitigation strategies targeting occupational issues and thereby improving the QoL components in the communities of this particular informal sector viz., fishery and fish processing industry (Mukundan, 2007; Manikfan, 2007; Samantray & Pathak, 2001; Rimmer et al., 2021).

Further, the importance of trained man-power for the shore is last but not the least under occupation-based mitigation strategies that plays a critical role in upliftment of QoL indicators in the fishing communities. In the age of electronic devices, fishermen not completely dependent on the weather for bringing a remunerative catch, rather they are using tools and technologies, which narrow down the time process, though the training of man-power to handle these advancements in the fishery/fish processing industry is a biggest challenge (Temporosa, 1979). Collaborative program of CMFRI and other central organizations, State Fisheries Universities, etc., for imparting man-power training to fish farmers, women

Figure 3. Mitigation strategies targeting climate change and workplace hazards in fishery and fish processing industries. Integrated approaches improve global fish production, ecological balance, livelihood of fisherfolk/fish processing communities and economic growth of the countries
Source: Author's creation.

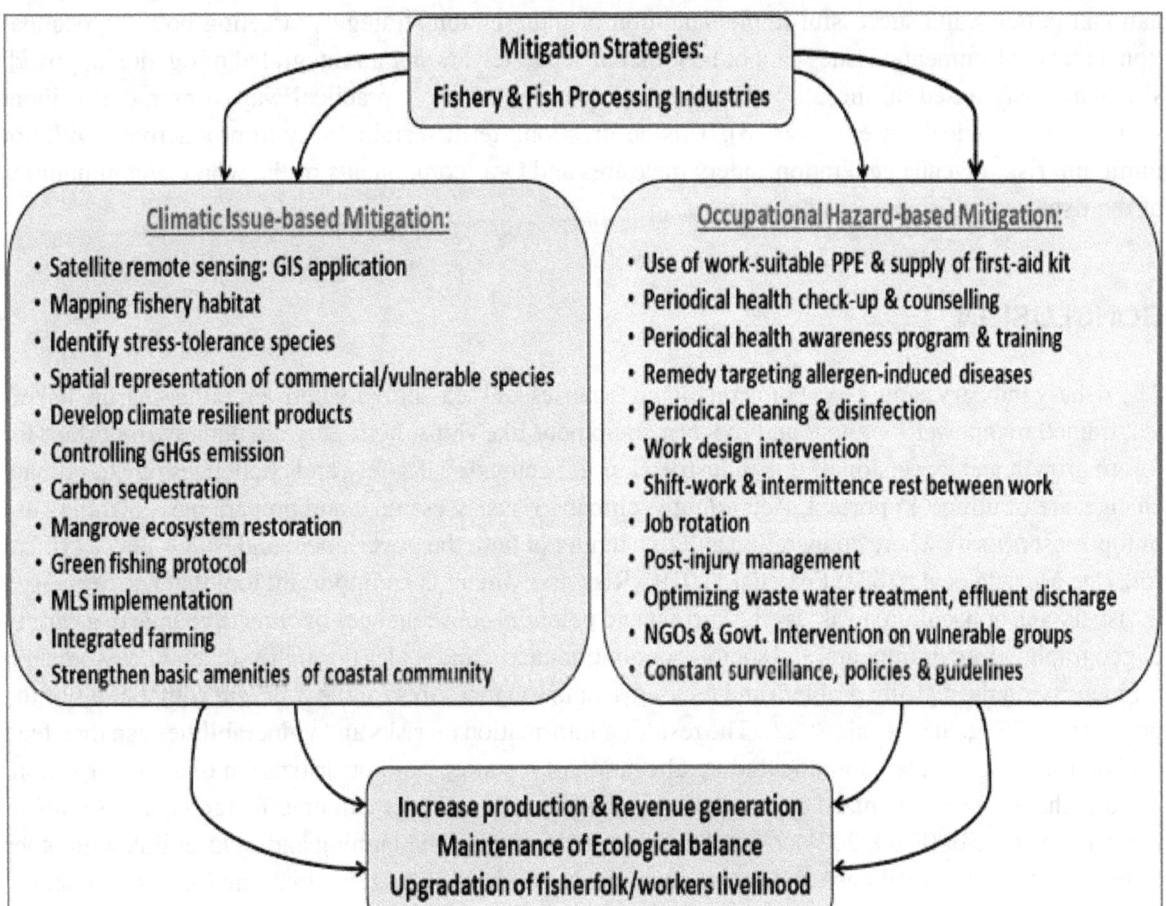

fish processors, State and local level officials, and other related stakeholders can improve deep-sea and sea-shore activities (Handbook on Fisheries Statistics, 2020). Fishery development thus requires not only oceanic vessels and modern canneries but also master fishermen, gear technicians, fishery technicians, marine engineers, oceanographers, and others. Now-a-days, Indian fisheries sector is received attention and policy support owing to the man-power training apart from its nutritional significance. The Government of India has initiated 'Blue Revolution' to augment the fish production and to strengthen the fish-based value chain like fish processing industry near to sea-shore (Lakra and Gopalkrishnan, 2021). However, implementation at grass-route level is itself a major issue due to existence of mostly informal sector, low economic growth, lack of public awareness and mass education, and/or inhibition of new skill adoption by the workers and the owners (Sajesh et al., 2021). The unequipped manpower is also a problem in a vast densely populated country like India that highlights as one of the major deterrents for leveraging the full potential in harvest and post-harvest fishery sector, which is highly skill intensive in nature so far as the marine fishery is concerned (Sajesh et al., 2021). Therefore, demands for the specific type of manpower is highly essential by this fast developing industry where a periodical fishery training programs should be dovetailed to job requirements.

Overall, climate change and occupational hazards are the biggest challenges facing by the fishery and fish processing industry in the developing countries like India, which are closely interlinked with each other, leading to affect the productivity of this particular sector, directly through fish stock and indirectly through health hazards of the target communities. Hence, time-bound management plans, national policies and successful implementation of intervention strategies targeting both the occupational and environmental issues cannot be look out separately, rather an integrated mitigation approach simultaneously based on these two broad key players, is the most practical way to promote resilient social and ecological systems (**Fig. 3**). This ensures long-term sustainability in productivity with no/ minimum risk, revenue generation, safety measures and QoL components of the workers/communities of the fishery and fish processing industry.

CONCLUSION

The fishery industry cannot exist only on inland fisheries and sea-shore manpower, rather marine fisheries, trained manpower for shore and modern equipment like ships, nets, etc., are equally important for future growth and expansion of this industry. Hence, occupational safety and challenges due to climate change, are of utmost important. Accordingly, climatic risks assessment and preparedness planning are of top-most priority where cooperation and fine tuning of both the government and NGOs play a critical role (De Murieta et al., 2021; Fox et al., 2019). Risk assessment is an important tool that has been used in fishery and aquaculture to assess the current and potential consequences of climate change in a variety of geographical, environmental and socio-economic contexts, and scales through a stakeholder workshop for identifying the real-life problem and focused adaptation measures for the different components of the target sector (Engelhard et al., 2022). The resulting information on risks and vulnerabilities can then feed decision-making on adaptation, including allocation of resources and prioritization of areas for action. Though there is no harmonized approach nor methodology to conduct vulnerability and risk assessment in the fishery sector (FAO, 2021), overall trend towards adaptation planning and a closer link with other frameworks, such as resilience and social-ecological systems may largely affect the QoL component.

Preparedness planning is another aspect in this regard, which aims to identify assignments and specific activities covering organizational and technical issues, and ensure successful function of the response systems in the event of a disaster, for example, managing fishing activities with minimum risks even a massive impact of global climatic change on fish stocks, and hence, fishing industry. Thus, fisheries and aquaculture response to emergencies (FARE) e-learning courses by FAO focus on the quality and accountability of preparedness and responses to emergencies for this sector, so that the safety of fishers, fisherfolk livelihoods, infrastructure and associated man-power of the fish-processing industry aligns with sustainable development goal (FAO, n.d.). Also, preparedness to avoid loss of fisheries/fish stock/ shallow fish habitats, as well as early-warning communication and safety-at-sea training, innovative insurance program, disaster risk management cycles, synergistic action with other industries, etc., are some of the highlighted areas of the fisheries governance in this respect (Westland & Poulain, 2007; Thorvaldsen et al., 2023).

Other mitigation strategies like mitigation based on carbon release, GHGs emission, etc., as a part of fishing activities, lies in the effective use of fuel, efficiency of boats and ships structures and gears, improve management of fisheries, reduction of post-harvest losses, increase of waste recycling, and safeguard of stocks with increase resilience to climate change (Bastardie et al., 2022). Artificial reefs made of sand-filled geotextiles also protect the coasts from the impacts of climate change. Now-a-days, GIS, radio, television, etc., are used to inform the fisherfolk about the adverse weather forecasts and warning, and implementation of the effective engineering along the coastal borders reduces the damage of the properties and lives (Indian Meteorological Department [IMD], 2021). In general, climate change policies are integrated with sustainable development strategies in most of the countries. Several schemes are undertaking to set up mariculture farms, parks and hatcheries, capacity building of fisherfolk and entrepreneurs, and development of consumers so as to increase the fish production from the developing and underdeveloped countries including India (Federation of Indian Chambers of Commerce and Industry- PricewaterhouseCoopers [FICCI-PwC], 2022). The fishery research institutes also play a leadership role in this respect, especially, during the time of extreme weather conditions (Mohamed et al., 2017). Moreover, establishment of weather watch groups and decision support systems are functioning on regional basis. The awareness of fisherfolk regarding climate variables and relation to fish catch is necessary, where combination of indigenous knowledge and modern technology reduces the impact of several climate change challenges.

It is a known fact that hazards and risks are potential factors in any occupational sector that overall affect the human activities and QoL, and fishery/fish processing industry are not the exception. However, mostly information of the developed countries is available in the public domain, whereas in the developing countries, stakeholders, such as workers, owners, government, NGOs and consumers largely ignoring the policies and guidelines, where present (Organization for Economic Co-operation and Development [OECD], 2000). Involvement of scientific communities, expert panels, etc., are necessary during such policy designing. Strict guidelines should be there for better environmental protection, occupational hazard analysis and risk assessment, and improve safety standards. The prevention of occupational accidents for fisherfolk and fish processing employees by the respective employers are necessary as work activities share many similarities attribute to poor workplace-associated risks and hazards, as discussed above. Job re-designing, effective use of PPEs and post-injury management, improve working postures, etc., are the effective mitigation strategies and a practical approach to control the common workplace health hazards (OSHA, 2023). Among the listed occupational hazards, the frequency is more with chemical hazards followed by physical and psychological (ILO, n.d.). Hence, in-depth prospective studies need to be explored for predisposition of the associated health risk in the fishery and fish processing industries. For example, the dosage of chlorine and the residual amount in fish processing industry along with the effects on consumers should be studied in detail with respect to the developing and underdeveloped countries. However, the effectiveness of any intervention depends on the regulatory guidelines and the duty provision of the owners to execute appropriate measures, and motivate and help the workers'/target communities for proper implementation of safety standards during work in this vulnerable informal sector.

In our opinion, the stakeholders of this sector should work in union to provide uniform guidelines and policies to promote an environmentally friendly and sustainable industry. Currently, in India, new development schemes for enhancing the skills and capabilities of the traditional fisherfolk for marine fishing are introduced (Ministry of Fisheries, Animal Husbandry & Dairying, 2023). The scope of the Marine Fishing Regulation Acts (MFRAs) include registration of boat building yards, standard design specifications for boats, construction materials, standard procedures for continuous monitoring and control of boat construction, etc., which ensure fuel efficient engines, boats and gears (Mohamed et al., 2017), though Representative Concentration Pathway (RCP), or similar scenarios in 2030, 2050 and 2080 as developed by the IPCC in concert with the predictive modeling techniques to measure the intensity of future changes in the marine fishery sector of different countries (IPCC, 2007; Akhiljith et al., 2019; Gopalakrishnan et al., 2020; Reygondeau et al., 2020; Huang et al., 2020; IPCC, 2022b & 2022c), are yet to be used. Thus, in conclusion, globally the future of the fishery and fish processing industry is bright subject to utilization of the clean and green technology for sustainable development and production in accordance with ecological balance and ensure QoL components in this sector.

ABBREVIATIONS

APFIC, Asia-Pacific Fishery Commission.
CAGR, Compound Annual Growth Rate.
CMFRI-NICRA, Central Marine Fisheries Research Institute-National Innovations on Climate Resilient Agriculture.
CPAs, Climate Change Preparedness Activities.
EAA, Ecosystem Approach to Aquaculture.
EEZ, Exclusive Economic Zone.
FAO, Food and Agricultural Organization.
FICCI-PwC, Federation of Indian Chambers of Commerce and Industry-PricewaterhouseCoopers.
GDP, Gross Domestic Product.
GHGs, Green House Gases.
GIS, Geographic Information System.
HPV7, Human Papilloma Virus type 7.
Hsp, Heat Shock Proteins.
ICAR-CIBA, Indian Council of Agricultural Research- Central Institute of Brackish Water Aquaculture.
ICAR-CIFT, Indian Council of Agricultural Research-Central Institute of Fisheries Technology.
IgE, Immunoglobulin E.
ILO, International Labor Organization.
IMD, Indian Meteorological Department.
IPCC, Intergovermental Panel on Climate Change.

LCA, Life cycle assessments.
MFRAs, Marine Fishing Regulation Acts.
MLS, Minimum Legal Size.
NEI, Northeast India.
NGOs, Non-Governmental Organizations.
NWI, Northwest India.
OECD, Organisation for Economic Co-operation and Development.
OSHA, Occupational Safety and Health Administration.
PAHs, Polycyclic Aromatic Hydrocarbons.
PCD, Protein Contact Dermatitisis.
PPE, Personal Protective Equipment.
QoL, Quality of life.
RCP, Representative Concentration Pathway.
SEI, Southeast India.
SWI, Southwest India.
USFDA, U.S. Food and Drug Administration.
WHO, World Health Organization

AUTHOR CONTRIBUTION

NN, corresponding author; conceptualized, wrote and edited the book chapter. DR, SNS equally contributed to reference arrangement and book chapter editing. RC partially contributed to climate change part of the book chapter during her tenure at ICMR-NIOH along with NN.

CORRESPONDENCE

Dr. Nibedita Naha, Scientist 'E' & Head, Biochemistry Dept.; Biological Sciences Div., ICMR-NIOH, Ahmedabad, India. Email: nibedita.72@yahoo.com; naha.n@gov.in.

DECLARATION

Grant is not received from any funding agency in public, commercial, or not-for-profit sector. Deepsi Rathore and Somi N Sharma have equal contribution. Ranjana Chaudhary is the present employee of ICMR-National Aids Research Institute (NARI).

CONFLICT OF INTEREST

None.

ACKNOWLEDGEMENT

Thanks to ICMR-NIOH for basic infrastructural support, and Vishakha Shrimali, Ph.D. student of Biochemistry Dept., ICMR-NIOH, for image digitalization.

REFERENCES

Aasmoe, L., Bang, B., Andorsen, G. S., Evans, R., Gram, I. T., & Lochen, M. L. (2005). Skin symptoms in the seafood-processing industry in north Norway. *Contact Dermatitis*, *52*(2), 102–107. doi:10.1111/j.0105-1873.2005.00515.x

Adnan, M., Badi, W., Dereczynski, C., Luca, A. D., Ghosh, S., Iskandar, I., Kossin, J., Lewis, S., Otto, F., Pinto, I., Satoh, M., Vicente-Serrano, S. M., Wehner, M., & Zhou, B. (2021). Weather and Climate Extreme Events in a Changing Climate. In V. Masson-Delmotte P, A. Zhai, S. L. Pirani, C. Connors, S. Pean, N. Berger, Y. Caud, L. Chen, M. I. Goldfarb, M. Gomis, K. Huang, E. Leitzell, J. B. R. Lonnoy, T. K. Matthews, T. Maycock, O. Waterfield, R. Y. Yelekci, & B. Zhou (Eds.), *Climate Change 2021: The physical science basis. Contribution of working group I to the sixth assessment report of the Intergovernmental Panel on Climate Change* (pp. 1513–1766). Cambridge University Press. doi:10.1017/9781009157896.013

Ahangari, H., Tavassoli, M., Khezerlou, A., Salmi, N. K., Ehsani, A., & Reza, M. (2023). A review on the determination of biogenic amines in fresh and processed fish products using HPLC, LC-MS/MS and other chromatographic methods. *Combinatorial Chemistry & High Throughput Screening*, *26*(15), 1386–2073. doi:10.2174/1386207326666230316141040

Ahmed, A. A. (2020). Glimpse of fish as perishable staple. *Al-Qadisiyah Journal for Agriculture Sciences, 10*(2), 349–375. doi:10.33794/qjas.2020.167497

Ahmed, Z. A. M., Dosoki, M. I., & Nasr, S. A. A. (2012). Review article; occupational hazards in fish industry. *World Journal of Fish and Marine Sciences*, *4*(2), 201–210. doi:10.5829/idosi.wjfms.2012.04.02.61203

Akbari, N., Failler, P., Pan, H., Drakeford, B. M., & Forse, A. (2023). The impact of fisheries on the economy: A systematic review on the application of general equilibrium and input–output methods. *Sustainability*, *15*(7), 6089–6089. doi:10.3390/su15076089

Akhiljith, P. J., Liya, V. B., Rojith, G., Zacharia, P. U., Grinson, G., Ajith, S., Lakshmi, P. M., Sajna, V. H., & Sathianandan, T. V. (2019). Climatic projections of Indian ocean during 2030, 2050, 2080 with implications on fisheries sector. *Journal of Coastal Research*, *86*(sp1), 198. doi:10.2112/si86-030.1

Akinsanola, A. A., Ogunjobi, K. O., Abolude, A. T., & Salack, S. (2021). Projected changes in wind speed and wind energy potential over West Africa in CMIP6 models. *Environmental Research Letters*, *16*(4), 044033. doi:10.1088/1748-9326/abed7a

Al-Abri, M., Al-Ghafri, B., Bora, T., Dobretsov, S., Dutta, J., Castelletto, S., Rosa, L., & Boretti, A. (2019). Author correction: Chlorination disadvantages and alternative routes for biofouling control in reverse osmosis desalination. *npj Clean Water, 2*(1). doi:10.1038/s41545-019-0033-2

Alfonso, S., Gesto, M., & Sadoul, B. (2020). Temperature increase and its effects on fish stress physiology in the context of global warming. *Journal of Fish Biology, 98*(6), 1496–1508. doi:10.1111/jfb.14599

Ali, N. A. M., Mohamad, M., Juahir, H., Kamarudin, M. K. A., Muhammad, N., & Lee, M. S. M. R. (2018). Source apportionment of social support and quality of life index among drug abuse inmates using multiple linear regression. *Journal of Fundamental and Applied Sciences, 9*(2S), 46. doi:10.4314/jfas.v9i2s.4

Allen, M. R., Dube, O. P., Solecki, W. A., Aragon-Durand, F., Cramer, W., Humphreys, S., Kainuma, M., Kala, J., Mahowald, N., Mulugetta, Y., Perez, R., Wairiu, M., & Zickfeld, K. (2019). Framing and context. In Masson-Delmotte, V., Zhai, P., Portner, H. O., Roberts, D., Skea, J., Shukla, P. R., Pirani, A., Moufouma-Okia, W., Pean, C., Pidcock, R., Connors, S., Matthews, J. B. R., Chen, Y., Zhou, X., Gomis, M. I., Lonnoy, E., Maycock, T., Tignor, M. and Waterfield, T. (Eds.). *Global Warming of 1.5°C. An IPCC Special Report on the impacts of global warming of 1.5°C above pre-industrial levels and related global greenhouse gas emission pathways, in the context of strengthening the global response to the threat of climate change, sustainable development, and efforts to eradicate poverty* (pp. 49-92). Cambridge University Press. doi:10.1017/9781009157940.003

Allison, E. H., Perry, A. L., Badjeck, M. C., Adger, W. N., Brown, K., Conway, D., Halls, A. S., Pilling, G. M., Reynolds, J. D., Andrew, N. L., & Dulvy, N. K. (2009). Vulnerability of national economies to the impacts of climate change on fisheries. *Fish and Fisheries, 10*(2), 173–196. doi:10.1111/j.1467-2979.2008.00310.x

Al-Saidi, M., Das, P., & Saadaoui, I. (2021). Circular economy in basic supply: Framing the approach for the water and food sectors of the Gulf cooperation council countries. *Sustainable Production and Consumption, 27*, 1273–1285. doi:10.1016/j.spc.2021.03.004

Amaravathi, T., Parimalam, P., Premalatha, M., Hemalatha, G., & Ganguli, A. (2016). Health hazard of women employed in small scale seafood processing units. *Indian Journal of Geo-Marine Sciences, 45*(4), 574–582. https://nopr.niscpr.res.in/bitstream/123456789/35094/1/IJMS%2045%284%29%20574-582.pdf

Anisha, P., Athira, P. S., Anagha, B., Charles, P. E., Prabakaran, K., & Rajaram, R. (2023). First report on occurrence of heavy metals in dried fishes from major fishing villages in Kerala coast, Southwest India. *Hygiene and Environmental Health Advances, 6*, 100051. doi:10.1016/j.heha.2023.100051

Arteaga, L. A., & Rousseaux, C. S. (2023). Impact of Pacific Ocean heatwaves on phytoplankton community composition. *Communications Biology, 6*(1), 263. doi:10.1038/s42003-023-04645-0

Asia-Pacific Fishery Commission (APFIC). (2010). *Best practices to support and improve livelihoods of small-scale fisheries and aquaculture households* (pp. 140). RAP Publication 2010/21. ISBN: 978-92-5-106739-0. https://www.fao.org/3/i1966e/i1966e00.pdf

Austin, B. (1998). The effects of pollution on fish health. *Journal of Applied Microbiology, 85*(S1), 234S242S. doi:10.1111/j.1365-2672.1998.tb05303.x

Balwan, W. K., Saba, N., & Rasool, N. (2021). An overview of climate change and food security in India. *Annals of the Romanian Society for Cell Biology, 25*(4), 20124–20137. http://annalsofrscb.ro/index.php/journal/article/view/9066/6625

Bastardie, F., Hornborg, S., Ziegler, F., Gislason, H., & Eigaard, O. R. (2022). Reducing the fuel use intensity of fisheries: Through efficient fishing techniques and recovered fish stocks. *Frontiers in Marine Science, 9*. doi:10.3389/fmars.2022.817335

Bax, N., Novaglio, C., Maxwell, K. H., Meyers, K., Mccann, J., Jennings, S., Frusher, S., Fulton, E. A., Nursey-Bray, M., Fischer, M., Anderson, K., Layton, C., Emad, R., Alexander, K. A., Rousseau, Y., Lunn, Z., & Carter, C. G. (2022). Ocean resource use: Building the coastal blue economy. *Reviews in Fish Biology and Fisheries, 32*, 189–207. doi:10.1007/s11160-021-09636-0

Begum, M., Masud, M. M., Alam, L., Mokhtar, M. B., & Amir, A. A. (2022). The impact of climate variables on marine fish production: An empirical evidence from Bangladesh based on autoregressive distributed lag (ARDL) approach. *Environmental Science and Pollution Research International, 29*(58), 87923–87937. doi:10.1007/s11356-022-21845-z

Black, M., Karki, J., Lee, A. C. K., Makai, P., Baral, Y. R., Kritsotakis, E. I., Bernier, A., & Heckmann, A. F. (2019). The health risks of informal waste workers in the Kathmandu Valley: A cross-sectional survey. *Public Health, 166*, 10–18. doi:10.1016/j.puhe.2018.09.026

Bleeker, E. A. J., & Verbruggen, E. (2009). *Bioaccumulation of polycyclic aromatic hydrocarbons in aquatic organisms*. National Institute for Human Health and the Environment (RIVM). https://www.researchgate.net/publication/281107718_Bioaccumulation_of_Polycyclic_Aromatic_Hydrocarbons_in_Aquatic_Organisms

Brander, K. M. (2007). Global fish production and climate change. *Proceedings of the National Academy of Sciences of the United States of America, 104*(50), 19709–19714. doi:10.1073/pnas.0702059104

Bravo-Olivas, M., Chavez-Dagostino, R., Malcolm, C., & Espinoza-Sanchez, R. (2015). Notes on the quality of life of artisanal small-scale fishermen along the pacific coast of Jalisco, Mexico. *Sustainability, 7*(5), 6046–6068. doi:10.3390/su7056046

Bull, N., Riise, T., & Moen, B. E. (2002). Work-related injuries and occupational health and safety factors in smaller enterprises—a prospective study. *Occupational Medicine, 52*(2), 70–74. doi:10.1093/occmed/52.2.70

Burridge, L., Weis, J. S., Cabello, F., Pizarro, J., & Bostick, K. (2010). Chemical use in salmon aquaculture: A review of current practices and possible environmental effects. *Aquaculture (Amsterdam, Netherlands), 306*(1-4), 7–23. doi:10.1016/j.aquaculture.2010.05.020

Byrne, D., Garcia-Soto, C., Hamilton, G., Leuliette, E., Yu, L., Campos, E., Durack, P. J., Manzella, G. M. R., Tadokoro, K., Schmitt, R. W., Arkin, P., Bryden, H., Nurse, L., Milliman, J., Inniss, L., & Bernal, P. (2017). The Ocean's role in the hydrological cycle. In United Nations (Ed.), *The first global integrated marine assessment world ocean assessment I* (pp. 91–104). Cambridge University Press., doi:10.1017/9781108186148.007

Central Marine Fisheries Research Institute (CMFRI)-National Innovations on Climate Resilient Agriculture (NICRA). (2013). *Marine fisheries*. Report of work done at CMFRI submitted to CRIDA, (pp. 14). http://eprints.cmfri.org.in/9465/1/CMFRI_Annual_Report_2012-13.pdf

Central Marine Fisheries Research Institute (CMFRI)-National Innovations on Climate Resilient Agriculture (NICRA). (2014). Marine fisheries. Report of work done at CMFRI submitted to CRIDA, (pp. 32). http://eprints.cmfri.org.in/10164/1/CMFRI%20Annual%20Report%202013-14.pdf

Central Marine Fisheries Research Institute (CMFRI)-National Innovations on Climate Resilient Agriculture (NICRA). (2016). Marine fisheries. Report of work done at CMFRI submitted to CRIDA (pp. 27. http://eprints.cmfri.org.in/10897/1/CMFRI%20ANNUAL%20REPORT%202015-16.pdf

Chen, Q., Su, H., Yu, X., & Hu, Q. (2020). Livelihood vulnerability of marine fishermen to multi-stresses under the vessel buyback and fishermen transfer programs in China: The case of Zhoushan city, Zhejiang Province. *International Journal of Environmental Research and Public Health, 17*(3), 765. doi:10.3390/ijerph17030765

Chiang, H. C., Ko, Y. C., Chen, S. S., Yu, H. S., Wu, T. N., & Chang, P. Y. (1993). Prevalence of shoulder and upper-limb disorders among workers in the fish-processing industry. *Scandinavian Journal of Work, Environment & Health, 19*(2), 126–131. doi:10.5271/sjweh.1496

Cochrane, K. L. (2002). Fisheries management. In K. L. Cochrane (Eds.), *A Fishery manager's guidebook - Management measures and their application*. FAO Fisheries Technical Paper. No. 424. ISBN: 92-5-10473204. https://www.fao.org/3/y3427e/y3427e00.htm#Contents

Costanza, R., Fisher, B., Ali, S., Beer, C., Bond, L., Boumans, R., Danigelis, N. L., Dickinson, J., Elliott, C., Farley, J., Gayer, D. E., Glenn, L. M., Hudspeth, T., Mahoney, D., McCahill, L., McIntosh, B., Reed, B., Rizvi, S. A. T., Rizzo, D. M., & Simpatico, T. (2007). Quality of life: An approach integrating opportunities, human needs, and subjective well-being. *Ecological Economics, 61*(2), 267–276. doi:10.1016/j.ecolecon.2006.02.023

Daanen, H. A. M., Wammes, L. J. A., & Vrijkotte, T. J. M. (1993). Wind chill and dexterity. Report IZF A-7, TNO Institute for Perception.

Dagnachew, A. G., Hof, A., Soest, H. V., & Vuuren, D. V. (2021). Climate change measures and sustainable development goals. *PBL Netherlands Environmenta l Assessment Agency*. https://sdgs.un.org/sites/default/files/2023-01/pbl-2021-climate-change-measures-and-sustainable-development-goals_4639.pdf

Daw, T., Adger, W. N., Brown, K., & Badjeck, M. C. (2009). Climate change and capture fisheries: potential impacts, adaptation and mitigation. In K. Cochrane, C. De Young, D. Soto and T. Bahri (Eds.). *Climate change implications for fisheries and aquaculture: overview of current scientific knowledge*. FAO Fisheries and Aquaculture Technical Paper. No. 530 (pp. 107-150). https://digitalarchive.worldfishcenter.org/bitstream/handle/20.500.12348/1339/WF_2545.pdf?sequence=1

De Murieta, E. S., Galarraga, I., & Olazabal, M. (2021). How well do climate adaptation policies align with risk-based approaches? An assessment framework for cities. *Cities (London, England), 109*, 103018. doi:10.1016/j.cities.2020.103018

De Silva, S. S., & Soto, D. (2009). Climate change and aquaculture: potential impacts, adaptation and mitigation. In K. Cochrane, C. De Young, D. Soto, & T. Bahri (Eds.), Climate change implications for fisheries and aquaculture: overview of current scientific knowledge (pp. 151–212). Food and Agriculture Organization (FAO) of the United Nations. ISBN: 978-92-5-106347-7

Delgado, C. L., Wada, N., Rosegrant, M. W., Meijer, S., & Ahmed, M. (2003). *Outlook for fish to 2020: Meeting global demand* (pp. 28). International Food Policy Research Institute (IFPRI) and the World Fish Centre. https://ebrary.ifpri.org/utils/getfile/collection/p15738coll2/id/62912/filename/62913.pdf

Demeke, A., & Tassew, A. (2016). A review on water quality and its impact on fish health. *International Journal of Fauna and Biological Studies, 3*(1), 21–31. https://api.semanticscholar.org/CorpusID:54571218

Dhanya, P. R., & Mathai, A. (2017). *Occupational hazards in seafood industry: A case study in Ernakulam district, Kerala.* 6th International conference on recent trends in engineering, science and management, Punjab, India. ISBN: 978-93-86171-21-4

Dwivedi, B., & Banerjee, S. (2017). Fish disease as a monitor of lake water pollution. *International Journal of Research, 5*(3), 209–212. doi:10.5281/zenodo.439573

Dzoga, M., Simatele, D., & Munga, C. (2018). Assessment of ecological vulnerability to climate variability on coastal fishing communities: A study of Ungwana Bay and Lower Tana Estuary, Kenya. *Ocean and Coastal Management, 163*, 437–444. doi:10.1016/j.ocecoaman.2018.07.015

El-Saadawy, M., Soliman, N., El Tayeb, I., & Hammouda, M. (2014). Some occupational health hazards among fishermen in Alexandria city. *Gaziantep Medical Journal, 20*(1), 71. doi:10.5455/gmj-30-44689

El-Sayed, Y. Y., & Khalil, R. H. (2009). Toxicity, biochemical effects and residue of aflatoxin B1 in marine water-reared sea bass (Dicentrarchuslabrax L.). *Food and Chemical Toxicology, 47*(7), 1606–1609. doi:10.1016/j.fct.2009.04.008

Engelhard, G. H., Howes, E. L., Pinnegar, J. K., & Le Quesne, W. J. F. (2022).Assessing the risk of climate change to aquaculture: A national-scale case study for the Sultanate of Oman. *Climate Risk Management 35.* 100416. https://doi.org/10.1016/j.crm.2022.100416

Federation of Indian Chambers of Commerce & Industry (FICCI) - PricewaterhouseCoopers (PwC) (2022). *Championing the blue economy: Promoting sustainable growth of the fisheries sector in India* https://www.pwc.in/assets/pdfs/grid/agriculture/championing-the-blue-economy-promoting-sustainable-growth-of-the-fisheries-sector-in-india.pdf

Ficke, A. D., Myrick, C. A., & Hansen, L. J. (2007). Potential impacts of global climate change on freshwater fisheries. *Reviews in Fish Biology and Fisheries, 17*, 581–613. doi:10.1007/s11160-007-9059-5

Food and Agriculture Organization (FAO) of the United Nations and World Health Organization. (WHO). 2020. *Code of practice for fish and fishery products.* doi:10.4060/cb0658en

Food and Agriculture Organization (FAO) of the United Nations. (2012). *The prevention of losses in cured fish.* FAO.

Food and Agriculture Organization (FAO) of the United Nations. (2018b). *Impacts of climate change on fisheries and aquaculture - Synthesis of current knowledge, adaptation and mitigation options.* ISBN: 978-92-5-130607-9. https://www.fao.org/3/i9705en/i9705en.pdf

Food and Agriculture Organization (FAO) of the United Nations. (2018a). The State of World Fisheries and Aquaculture 2018 - Meeting the sustainable development goals (pp. 200). ISBN: 978-92-5-130562-1

Food and Agriculture Organization (FAO) of the United Nations. (2022). *The State of World Fisheries and Aquaculture 2022. Towards Blue Transformation* (pp. 266). http://www.fao.org/state-of-fisheries-aquaculture

Food and Agriculture Organization (FAO) of the United Nations. (2016). The State of World Fisheries and Aquaculture 2016 - Contributing to food security and nutrition for all (pp. 200). ISBN: 978-92-5-109185-2

Food and Agriculture Organization of the United Nations (FAO). (2015). *Climate change and food security: risks and responses*. ISBN: 978-92-5-108998-9. https://www.fao.org/3/i5188e/I5188E.pdf

Food and Agriculture Organization of the United Nations (FAO). (2021). *Recent advances in climate change vulnerability/risk assessments in the fisheries and aquaculture sector*. ISBN: ISBN 978-92-5-134374-6. . https://doi.org/10.4060/cb4585en.pdf

Food and Agriculture Organization of the United Nations (FAO). (n.d.). *Disaster risk management in fisheries*. CC4 Fias Policy Brief, 1-4. https://www.fao.org/3/cb2729en/cb2729en.pdf

Fox, M., Zuidema, C., Bauman, B., Burke, T., & Sheehan, M. (2019). Integrating public health into climate change policy and planning: State of practice update. *International Journal of Environmental Research and Public Health*, *16*(18), 3232. doi:10.3390/ijerph16183232

Gasalla, M. A., Abdallah, P. R., & Lemos, D. (2018). Potential impacts of climate change in Brazilian marine fisheries and aquaculture. In B. F. Phillips & M. Perez-Ramirez (Eds.), *Climate change impacts on fisheries and aquaculture: a global analysis* (pp. 455–472). Wiley Blackwell., doi:10.1002/9781119154051.ch14

George, G., Vethamony, P., Sudheesh, K., & Babu, M. T. (2011). Fish larval transport in a macro-tidal regime: Gulf of Kachchh, West Coast of India. *Fisheries Research*, *110*(1), 160–169. doi:10.1016/j.fishres.2011.04.002

Georgian, S. E., Anderson, O. F., & Rowden, A. A. (2019). Ensemble habitat suitability modeling of vulnerable marine ecosystem indicator taxa to inform deep-sea fisheries management in the South Pacific Ocean. *Fisheries Research*, *211*, 256–274. doi:10.1016/j.fishres.2018.11.020

Ghani, N. A., Raub, M. A., Adam, F., Abdullah, B., Eusoff, Y. A., & Puteh, D. A. H. M. A. (2017). Quality of Life (QoL) of fishermen in the west coast states of Peninsular Malaysia. *International Journal of Academic Research in Business & Social Sciences*, *7*(4), 313–330. doi:10.6007/ijarbss/v7-i4/2808

Ghosh, S., Rao, M. V. H., Kumar, M. S., Mahesh, V. U., Menon, M., & Zacharia, P. U. (2015). Carbon footprint in life cycle of marine fisheries at major fishing harbours of Andhra Pradesh, India. *Indian Journal of Fisheries*, *62*(4), 37–44. http://eprints.cmfri.org.in/10605/1/IJF_62.4_Ghosh_4.pdf

Ghosh, S., Rao, M. V. H., Kumar, M. S., Mahesh, V. U., Menon, M., & Zacharia, P. U. (2014). Carbon footprint of marine fisheries: Life cycle analysis from Visakhapatnam. *Current Science*, *107*(3), 515–521. http://eprints.cmfri.org.in/10166/1/Ghosh_Current_Science.pdf

Gillanders, B., & Kingsford, M. (2002). Impact of Changes in Flow of Freshwater on Estuarine and Open Coastal Habitats and the Associated Organisms. *Oceanography and Marine Biology - an Annual Review*, *40*, 233–309. doi:10.1201/9780203180594.ch5

Gissi, E., Manea, E., Mazaris, A. D., Fraschetti, S., Almpanidou, V., Bevilacqua, S., Coll, M., Guarnieri, G., Lloret-Lloret, E., Pascual, M., Petza, D., Rilov, G., Schonwald, M., Stelzenmüller, V., & Katsanevakis, S. (2020). A review of the combined effects of climate change and other local human stressors on the marine environment. *The Science of the Total Environment*, *755*(1), 142564. doi:10.1016/j.scitotenv.2020.142564

Glynn, P. W. (1991). Coral reef bleaching in the 1980s and possible connections with global warming. *Trends in Ecology & Evolution*, *6*(6), 175–179. doi:10.1016/0169-5347(91)90208-f

Godvin, V. S., Kumar, D. M., Pugazhendi, A., Bajhaiya, A. K., Gugulothu, P., & Banu, R. J. (2021). Biofuel production from Macroalgae: Present scenario and future scope. *Bioengineered*, *12*(2), 9216–9238. doi:10.1080/21655979.2021.1996019

Goggin, C., & Lester, R. (1995). Perkinsus, a protistan parasite of abalone in Australia: A review. *Marine and Freshwater Research*, *46*(3), 639–646. doi:10.1071/mf9950639

Gopal, N. (2021). Labour challenges in fishing and fish processing industries. In C. N. Ravishankar, K. A. Kumar, L. Edwin, S. Mathew, A. K. Mohanty, A. A. Zynudheen, G. Ninan, T. C. Joseph, V. Chandrasekar, V. K. Sajesh, K. K. Prajith, V. Renuka, S. J. Laly, K. R. Sreelakshmi, K. Ranjit, R. K. Nadella, & S. Murali (Eds.), *Recent advances in harvest and post-harvest technologies in fisheries (Training Manual)* (pp. 407–418). ICAR-Central Institute of Fisheries Technology., http://krishi.icar.gov.in/jspui/handle/123456789/70437

Gopalakrishnan, A., Zacharia, P. U., & Grinson, G. (2020). *Impact, vulnerability and adaptation strategies for marine fisheries of India: IVAS-MFI final report. CMFRI special Publication No.135*. ICAR-Central Marine Fisheries Research Institute., http://eprints.cmfri.org.in/14771/1/CMFRI%20Special%20Publication_135_IVAS.pdf

Gupta, A., & Ghosh, J. (2023). *India's climate research agenda: 2030 and beyond*. https://dst.gov.in/sites/default/files/India%27s%20Climate%20Research%20Agenda%202030%20and%20beyond.pdf

Hallerman, E. M., & Kapuscinski, A. R. (1992). Ecological implications of using transgenic fishes in aquaculture. *International Council for the Exploration of the Seas Marine Science Symposium*, *194*, 56–66.

Handbook on fisheries statistics (2020). Department of Fisheries, Ministry of Fisheries, Animal Husbandry and Dairying, *Government of India*, 99-100. https://dof.gov.in/sites/default/files/2021-02/Final_Book.pdf

Harvell, C. D., Mitchell, C. E., Ward, J. R., Altizer, S., Dobson, A. P., Ostfeld, R. S., & Samuel, M. D. (2002). Climate warming and disease risks for terrestrial and marine biota. *Science*, *296*(5576), 2158–2162. doi:10.1126/science.1063699

Heus, R., Daanen, H. A., & Havenith, G. (1995). Physiological criteria for functioning of hands in the cold: A review. *Applied Ergonomics*, *26*(1), 5–13. doi:10.1016/0003-6870(94)00004-i

Hilborn, R., Amoroso, R. O., Anderson, C. M., Baum, J. K., Branch, T. A., Costello, C., de Moor, C. L., Faraj, A., Hively, D., Jensen, O. P., Kurota, H., Little, L. R., Mace, P., McClanahan, T., Melnychuk, M. C., Minto, C., Osio, G. C., Parma, A. M., Pons, M., & Segurado, S. (2020). Effective fisheries management instrumental in improving fish stock status. *Proceedings of the National Academy of Sciences of the United States of America*, *117*(4), 2218–2224. doi:10.1073/pnas.1909726116

Hu, W., Zheng, X., Li, Y., Du, J., Lv, Y., Su, S., Xiao, B., Ye, X., Jiang, Q., Tan, H., Liao, B., & Chen, B. (2022). High vulnerability and a big conservation gap: Mapping the vulnerability of coastal scleractinian corals in South China. *The Science of the Total Environment*, *847*, 157363. doi:10.1016/j.scitotenv.2022.157363

Huang, Y., Ma, Y., Liu, T., & Luo, M. (2020). Climate change impacts on extreme flows under IPCC RCP scenarios in the mountainous Kaiduwatershed, Tarimriver basin. *Sustainability*, *12*(5), 2090. https://doi.org/10.3390/su12052090

Hurlbert, M., Krishnaswamy, J., Davin, E., Johnson, F. X., Mena, C. F., Morton, J., Myeong, S., Viner, D., Warner, K., Wreford, A., Zakieldeen, S., & Zommers, Z. (2019). Risk Management and Decision making in Relation to Sustainable Development. In P. R. Shukla, J. Skea, E. C. Buendia, V. Masson-Delmotte, H. O. Pörtner, D. C. Roberts, P. Zhai, R. Slade, S. Connors, R. V. Diemen, M. Ferrat, E. Haughey, S. Luz, S. Neogi, M. Pathak, J. Petzold, J. P. Pereira, P. Vyas, E. Huntley, & K. Kissick (Eds.), Climate Change and Land: an IPCC special report on climate change, desertification, land degradation, sustainable land management, food security, and greenhouse gas fluxes in terrestrial ecosystems (pp. 673–800). IPCC. https://doi.org/ doi:10.1017/9781009157988.009

Idowu, T. A., & Sogbesan, O. A. (2017). Synthetic chemical application in aquaculture production. *International Journal of Environment & Agricultural Science, 1*(2), 010.https://www.researchgate.net/publication/322675339_Synthetic_Chemical_Application_in_Aquaculture_Production

India's blue economy net getting bigger! Country ranks third in fisheries and second in aquaculture. (2020, February 14). Financial express.https://www.financialexpress.com/opinion/indias-blue-economy-net-getting-biggercountry-ranks-third-in-fisheries

Indian Council of Agricultural Research (ICAR)-Central Institute of Fisheries Technology (CIFT). (2016). *Annual Report 2015-16* (pp. 160). https://krishi.icar.gov.in/jspui/bitstream/123456789/8266/1/CIFT%20Annual%20Report%202015-16.pdf

Indian Meteorological Department (IMD). 2021. *Standard operation procedure weather forecasting and warning*. https://mausam.imd.gov.in/imd_latest/contents/pdf/forecasting_sop.pdf

Intergovernmental Panel on Climate Change (IPCC). (2007). Climate change 2007: impacts, adaptation and vulnerability: Working Group II contribution to the fourth assessment report of the IPCC Intergovernmental Panel on Climate Change. In M. Parry, O. Canziani, J. Palutikof, P. van der Linden, & C. Hanson (Eds.), (p. 976). Cambridge University Press., https://www.ipcc.ch/site/assets/uploads/2018/03/ar4_wg2_full_report.pdf

Intergovernmental Panel on Climate Change (IPCC). (2022a). Sea Level Rise and Implications for Low-Lying Islands, Coasts and Communities. In *The Ocean and Cryosphere in a Changing Climate: Special Report of the Intergovernmental Panel on Climate Change* (pp. 321–446). Cambridge University Press., doi:10.1017/9781009157964.006

Intergovernmental Panel on Climate Change. (IPCC, 2022b). *Climate change 2022: Impacts, adaptation, and vulnerability. Contribution of working group II to the sixth assessment report of the Intergovernmental Panel on Climate Change.* Cambridge University Press. https://www.ipcc.ch/report/ar6/wg2/

Intergovernmental Panel on Climate Change. (IPCC, 2022c). *Climate change 2022: Mitigation of climate change. Contribution of working group iii to the sixth assessment report of the Intergovernmental Panel on Climate Change.* Cambridge University Press. https://doi.org/10.1017/9781009157926

International Labour Organization. (ILO, 2016). Psychosocial risks, stress and violence in the world of work.*International Journal of Labour Research. 8*(1-2). https://www.ilo.org/wcmsp5/groups/public/---ed_dialogue/---actrav/documents/publication/wcms_551796.pdf

International Labour Organization (ILO). (n.d.). *Psychosocial risks and stress at work.*https://www.ilo.org/global/topics/safety-and-health-at-work/areasofwork/occupational-safety-and-health-management-systems/WCMS_861077/lang--en/index.htm

International Labour Organization (ILO). (1999). *Fishing among the most dangerous of all professions, says ILO. International Labour Office Report.* https://www.ilo.org/global/about-the-ilo/newsroom/news/WCMS_071324/lang--en/index.htm

International Labour Organization (ILO). (2004). *Danger at sea - Working in the fishing sector.* https://www.ilo.org/global/about-the-ilo/mission-and-objectives/features/WCMS_075579/lang-en/index.html

Jamal, Md. R., Kristiansen, P., Kabir, Md. J., & de Bruyn, L. L. (2023). Risks and adaptation dynamics in shrimp and prawn-based farming systems in southwest coastal Bangladesh. *Aquaculture (Amsterdam, Netherlands)*, *562*, 738819. doi:10.1016/j.aquaculture.2022.738819

Jayakumar, R., Sakthivel, M., Abdul Nazar, A. K., Tamilmani, G., Rameshkumar, P., Samal, A. K., Anikuttan, K. K., Anbarasu, M., Balamurugan, V., Thiagu, R., Sirajudeen, S., & Gopakumar, G. (2018). Impact of increase in temperature and light intensity on development and metamorphosis of hatchery reared silver pompano Trachinotusblochii (Lacepede, 1801) larvae. *Indian Journal of Fisheries*, *65*. Advance online publication. doi:10.21077/ijf.2018.65.2.37091-18

Jayasingh, S. I., Janani, M. S., & Sowndharya, R. (2022). Bacterial diseases in fish with relation to pollution and their consequences—a global scenario. In *Bacterial Fish Diseases* (pp. 113–131). Academic Press., https://doi.org/, 10.1016/B978-0-323-85624-9.00022-1

Jeebhay, M. F., Robins, T., & Lopata, A. (2004). World at work: Fish processing workers. *Occupational and Environmental Medicine*, *61*(5), 471–474. doi:10.1136/oem.2002.001099

Karunarathna, H., Maduwantha, P., Kamranzad, B., Rathnasooriya, H., & De Silva, K. (2020). Impacts of global climate change on the future ocean wave power potential: A case study from the Indian Ocean. *Energies*, *13*(11), 3028. doi:10.3390/en13113028

Kathleen, A., & Feris, L. (2022). South Africa's small-scale fishers have been marginalised since apartheid – what needs to change. *The Conversation.*: https://www.researchgate.net/publication/365743989

Keese, M., & Marcolin, L. (2023). *Labour and social policies for the green transition: A conceptual framework. OECD Social, Employment and Migration Working Papers, No. 295.* OECD Publishing., doi:10.1787/028ffbeb-en

Kelkar, N., Arthur, R., Dey, S., & Krishnaswamy, J. (2022). Flood-Pulse variability and climate change effects increase uncertainty in fish yields: Revisiting narratives of declining fish catches in India's Ganga river. *Hydrology*, *9*(4), 53. doi:10.3390/hydrology9040053

Kemarau, R. A., & Eboy, O. V. (2022). How do El Nino Southern Oscillation (ENSO) events impact fish catch in Sarawak water? *Journal of Physics: Conference Series*, *2314*(1), 012013. doi:10.1088/1742-6596/2314/1/012013

King, T. J., Turner, R., Versace, V., Abernethy, K., Kilpatrick, S., & Brumby, S. (2021). Mental health in the commercial fishing industry: Modern uncertainties and traditional risks. *Fish and Fisheries*, *22*(5), 1136–1149. doi:10.1111/faf.12572

Kolawole, O. D., & Bolobilwe, K. (2018). Survival at a cost: How artisanal fishers perceive occupational hazards in the Okavango Delta, Botswana. *The South African Geographical Journal*, *101*(1), 51–71. do i:10.1080/03736245.2018.1541019

Kroeker, K. J., Kordas, R. L., Crim, R., Hendriks, I. E., Ramajo, L., Singh, G. S., Duarte, C. M., & Gattuso, J. P. (2013). Impacts of ocean acidification on marine organisms: Quantifying sensitivities and interaction with warming. *Global Change Biology*, *19*(6), 1884–1896. doi:10.1111/gcb.12179

Kumar, J. A., Mrralidhar, M., Jayanthi, M., & Kumaran, M. (2013). Trend analysis of weather data in shrimp farming areas of Nagapattinam district of Tamil Nadu. *Journal of Agrometeorology*, *15*(2), 129–134. https://doi.org/10.54386/jam.v15i2.1459

Kumari, S., Zacharia, P. U., Kripa, V., Sreenath, K. R., & George, G. (2015). Distribution pattern and community structure of zoanthids (Zoantharia) along the coast of Saurashtra, Gujarat, India. *Journal of the Marine Biological Association of the United Kingdom*, *96*(8), 1577–1584. doi:10.1017/s0025315415001885

Laffoley, D., & Baxter, J. M. (Eds.). (2016). Explaining ocean warming: Causes, scale, effects and consequences. International Union for Conservation of Nature (IUCN) (pp. 456). http://dx.doi.org/doi:10.2305/IUCN.CH.2016.08.en

Lakra, W. S., & Gopalakrishnan, A. (2021). Blue revolution in India: Status and future perspectives. *Indian Journal of Fisheries*, *68*(1), 137–150. doi:10.21077/ijf.2021.68.1.109283-19

Lan, K.-W., Chang, Y.-J., & Wu, Y.-C. (2020). Influence of oceanographic and climatic variability on the catch rate of yellowfin tuna (Thunnusalbacares) cohorts in the Indian Ocean. *Deep-sea Research. Part II, Topical Studies in Oceanography*, *175*, 104681–104681. doi:10.1016/j.dsr2.2019.104681

Lebbe, T. B., Rey-Valette, H., Chaumillon, E., Camus, G., Almar, R., Cazenave, A., Claudet, J., Rocle, N., Meur-Ferec, C., Viard, F., Mercier, D., Dupuy, C., Menard, F., Rossel, B. A., Mullineaux, L., Sicre, M.-A., Zivian, A., Gaill, F., & Euzen, A. (2021). Designing Coastal Adaptation Strategies to Tackle Sea Level Rise. *Frontiers in Marine Science*, *8*, 740602. doi:10.3389/fmars.2021.740602

Lehmuskallio, E., Hassi, J., & Kettunen, P. (2002). The skin in the cold. *International Journal of Circumpolar Health*, *61*(3), 277–286. doi:10.3402/ijch.v61i3.17475

Lehtonen, H. (1996). Potential effects of global warming on northern European freshwater fish and fisheries. *Fisheries Management and Ecology*, *3*(1), 59–71. doi:10.1111/j.1365-2400.1996.tb00130.x

Lentisco, A., & Robert, U. L. (2015). *A review of women's access to fish in small-scale fisheries*. FAO Fisheries and Aquaculture Circular No. 1098. ISBN: 978-92-5-108857-9. https://www.fao.org/3/i4884e/i4884e.pdf

Liebers, V., Brüning, T., & Raulf, M. (2020). Occupational endotoxin exposure and health effects. *Archives of Toxicology*, *94*(11), 3629–3644. https://doi.org/10.1007/s00204-020-02905-0

Lincoln, J. M., Carruth, A., Cherry, D., Kincl, L., & Syron, L. N. (2021). Occupational health research in the commercial fishing industry. *Journal of Agromedicine*, *26*(1), 28–30. doi:10.1080/1059924x.2021.1849494

Lix, J. K., Venkatesan, R., Grinson, G., Rao, R. R., Jineesh, V. K., Arul, M. M., Vengatesan, G., Ramasundaram, S., Sundar, R., & Atmanand, M. A. (2016). Differential bleaching of corals based on El Niño type and intensity in the Andaman Sea, southeast Bay of Bengal. *Environmental Monitoring and Assessment*, *188*(3), 175. doi:10.1007/s10661-016-5176-8

Long, M. C., Deutsch, C., & Ito, T. (2016). Finding forced trends in oceanic oxygen. *Global Biogeochemical Cycles*, *30*(2), 381–397. doi:10.1002/2015gb005310

Lopata, A. L., Baatjies, R., Thrower, S. J., & Jeebhay, M. F. (2004). Occupational allergies in the seafood industry-a comparative study of Australian and South African workplaces. *International Maritime Health*, *55*(1-4), 61–73.https://pubmed.ncbi.nlm.nih.gov/15881544/

Lucas, D., Lucas, R., Boniface, K., Jegaden, D., Lodde, B., & Dewitte, J. A. (2010). Occupational asthma in the commercial fishing industry: A case series and review of the literature. *International Maritime Health*, *61*(1), 13–17. https://pubmed.ncbi.nlm.nih.gov/20496322/

Mallela, J., & Fox, R. J. (2018). The role of parrotfishes in the destruction and construction of coral reefs. In A. S. Hoey& R. M. Bonaldo (Eds.), Biology of Parrotfishes (pp. 161–196). CRC Press. isbn: 978036778140

Manikfan, G. H. (2007). Importance of human resource development for the fishing industry. In S. Girija (Ed.), *Spearheading quality fish processing, Integrated Fisheries Project Souvenir* (pp. 13–16). Integrated Fisheries Project.

Mathis, J. T., Cooley, S. R., Lucey, N., Colt, S., Ekstrom, J., Hurst, T., Hauri, C., Evans, W., Cross, J. N., & Feely, R. A. (2015). Ocean acidification risk assessment for Alaska's fishery sector. *Progress in Oceanography*, *136*, 71–91. doi:10.1016/j.pocean.2014.07.001

Maulu, S., Hasimuna, O. J., Haambiya, L. H., Monde, C., Musuka, C. G., Makorwa, T. H., Munganga, B. P., Phiri, K. J., & Nsekanabo, J. D. (2021). Climate change effects on aquaculture production: Sustainability implications, mitigation, and adaptations. *Frontiers in Sustainable Food Systems*, *5*, 609097. doi:10.3389/fsufs.2021.609097

McRae, H. (2022). *The world in 2050*. Bloomsbury Publishing.

Meaden, G. J., & Aguilar-Manjarrez, J. (Eds.). 2013. *Advances in geographic information systems and remote sensing for fisheries and aquaculture*. Summary version (pp. 425). FAO Fisheries and Aquaculture Technical Paper No. 552. https://doi.org/ doi:10.13140/RG.2.1.4037.7682

Merino, G., Barange, M., Blanchard, J. L., Harle, J., Holmes, R., Allen, I., Allison, E. H., Badjeck, M. C., Dulvy, N. K., Holt, J., Jennings, S., Mullon, C., & Rodwell, L. D. (2012). Can marine fisheries and aquaculture meet fish demand from a growing human population in a changing climate? *Global Environmental Change*, *22*(4), 795–806. doi:10.1016/j.gloenvcha.2012.03.003

Meyer, A. L. S., Bentley, J., Odoulami, R. C., Pigot, A. L., & Trisos, C. H. (2022). Risks to biodiversity from temperature overshoot pathways. *Philosophical Transactions of the Royal Society B: Biological Sciences, 377*(1857). doi:10.1098/rstb.2021.0394

Michalopoulos, L. T. M., Baca-Atlas, S. N., Simona, S. J., Jiwatram-Negrón, T., Ncube, A., & Chery, M. B. (2017). "Life at the River is a Living Hell:" a qualitative study of trauma, mental health, substance use and HIV risk behavior among female fish traders from the Kafue Flatlands in Zambia. *BMC Women's Health*, *17*(1). Advance online publication. doi:10.1186/s12905-017-0369-z

Ministry of Fisheries. Animal Husbandry & Dairying (2023). *Department of Fisheries implements various schemes to develop fisheries sector in the country.* https://pib.gov.in/PressReleaseIframePage.aspx?PRID=1897083#:~:text=The%20Schemes%20being%20implemented%20by,and%20responsible%20development%20of%20fisheries

Mitra, A., Abdel-Gawad, F. K., Bassem, S., Barua, P., Assisi, L., Parisi, C., Temraz, T. A., Vangone, R., Kajbaf, K., Kumar, V., & Guerriero, G. (2023). Climate change and reproductive biocomplexity in fishes: Innovative management approaches towards sustainability of fisheries and aquaculture. *Water (Basel)*, *15*(4), 725. doi:10.3390/w15040725

Mohamed, K. S., Vijayakumaran, K., Zacharia, P. U., Sathianandan, T. V., Maheswarudu, G., Kripa, V., Narayanakumar, R., Rohit, P., Joshi, K. K., Sankar, T. V., Edwin, L., Kumar, A. K., Bindu, J., Gopal, N., & Puthra, P. (2017). Indian marine fisheries code: Guidance on a marine fisheries management odel for India. *Central Marine Fisheries Research InstituteMarine Fisheries Policy Series 4* (pp. 120). http://eprints.cmfri.org.in/11873/1/IMFC%20Code_Revised_2019.pdf

Mohamed, K. S., Zacharia, P. U., Maheswarudu, G., Sathianandan, T. V., Abdussamad, E. M., Ganga, U., Pillai, S. L., Sobhana, K. S., Nair, R. J., Josileen, J., Chakraborty, R. D., Kizhakudan, S. J., & Najmudeen, T. M. (2014). Minimum Legal Size (MLS) of capture to avoid growth overfishing of commercially exploited fish and shellfish species of Kerala. *The Marine Fisheries Information Service, Technical & Extension Series, 220*, 3–7. http://eprints.cmfri.org.in/10282/1/MFIS_220_1.pdf

Mohanty, A. K. (2018). Impact of climate change: A curse to the shrimp farming in India. *International Journal of Advance Research, Ideas and Innovations in Technology, 4*(3), 1151–1158. https://www.ijariit.com/manuscripts/v4i3/V4I3-1559.pdf

Mohanty, B. P., Mohanty, S., Sahoo, J., & Sharma, A. (2010). Climate change: Impacts on fisheries and aquaculture. In S. Simard (Ed.), *Climate Change and Variability* (pp. 199–138). INTECH Open Access Publisher., doi:10.5772/9805

Mohanty, P. C., Mahendra, R. S., Bisoyi, H., Tummula, S. K., Grinson, G., Nayak, S., & Sahu, B. K. (2013). Assessment of the coral bleaching during 2005 to decipher the thermal stress in the coral environs of the Andaman Islands using Remote Sensing. *European Journal of Remote Sensing*, *46*(1), 417–430. doi:10.5721/eujrs20134624

Mukundan, M. K. (2007). Safety for hazards and quality defects in fish processing industry. In S. Girija (Ed.), *Spearheading quality fish processing, Integrated Fisheries Project Souvenir* (pp. 1–12). Integrated Fisheries Project.

Munoz, M., Reul, A., Guijarro, B., & Hidalgo, M. (2023). Carbon footprint, economic benefits and sustainable fishing: Lessons for the future from the Western Mediterranean. *The Science of the Total Environment*, *865*, 160783. doi:10.1016/j.scitotenv.2022.16078

Muralidhar, M., & Jayanthi, M. (2014). Climate change and its implications for brackishwater aquaculture. In *Proceedings of the policy meet on the impact of environmental changes on the livelihoods of coastal women in north Tamil Nadu.* India (pp. 21-31).

Muralidhar, M., & Vijayan, K. K. (2016). Risks associated with extreme weather in aquaculture and their management. In V. U. M. Rao (Ed.), *Agrometeorological techniques for risk assessment and management of extreme events* (pp. 60–67). ICAR-Central Research Institute for Dryland Agriculture.

Muralidhar, M., Kumaran, M., & Ashok Kumar, J. SyamaDayal, J., Jayanthi, M., Jayanthi, R., Lalitha, N., Panigrahi, A., Sreenivasa Rao, A., Muniyandi, B., &Murugan, P. (2013a). Climate change and coastal aquaculture in West Godavari district, Andhra Pradesh: Impacts, vulnerability, adaptations and mitigations for resilience. Journal of Agrometeorology, 15(Special Issue-II), 116–122.https://krishi.icar.gov.in/jspui/bitstream/123456789/1305/1/76-Muralidhar%20M%20-%20J%20Agrometeorology%20paper.pdf

Muralidhar, M., Vasanth, M., Saraswathy, R., Syamadayal, J., Lalitha, N., & Nagavel, A. (2013b). Measurement of greenhouse gas emission from aquaculture. In H. Pathak, R. C. Upadhyay, M. Muralidhar, P. Bhattacharyya, & B. Venkateswarlu (Eds.), Measurement of greenhouse gas emission from crop, livestock and aquaculture (pp. 45–54). Indian Agricultural Research Institute (IARI). ISBN: 978-81-88708-98-7

Muralidhar, M., & Vijayan, K. K. 2015. Climate change and coastal aquaculture: Impacts, mitigation and adaptation measures. In: Souvenir: Aqua Aquaria India. The Marine products Export Development Authority, Kochi, India, 7-13.

Nag, A., Vyas, H., Shah, P., & Nag, P. K. (2012). Risk factors and musculoskeletal disorders among women workers performing fish processing. *American Journal of Industrial Medicine*, *55*(9), 833–843. doi:10.1002/ajim.22075

Nag, O. S. (2017, April 25). Top fish and seafood exporting countries. *Economics*. https://www.worldatlas.com/articles/top-fish-and-seafood-exporting-countries.html

Nag, P. K., & Nag, A. (2007). Hazards and health complaints associated with fish processing activities in India—Evaluation of a low-cost intervention. *International Journal of Industrial Ergonomics*, *37*(2), 125–132. doi:10.1016/j.ergon.2006.10.012

Ngaruiya, F. W., Ogendi, G. M., & Mokua, M. A. (2019). Occupational health risks and hazards among the fisherfolk in KampiSamaki, Lake Baringo, Kenya. *Environmental Health Insights*, *13*(1), 1–11. doi:10.1177/1178630219881463

Nordander, C., Ohlsson, K., Balogh, I., Rylander, L., Palsson, B., & Skerfving, S. (1999). Fish processing work: The impact of two sex dependent exposure profiles on musculoskeletal health. *Occupational and Environmental Medicine*, *56*(4), 256–264. doi:10.1136/oem.56.4.256

Occupational Safety and Health Administration (OSHA). (2023). *Personal Protective Equipment*. https://www.osha.gov/sites/default/files/publications/osha3151.pdf

Ohlsson, K., Hansson, G.-A., Balogh, I., Stromberg, U., Palsson, B., Nordander, C., Rylander, L., & Skerfving, S. (1994). Disorders of the neck and upper limbs in women in the fish processing industry. *Occupational and Environmental Medicine*, *51*(12), 826–832. doi:10.1136/oem.51.12.826

Olafsdottir, H., & Rafnsson, V. (2000). Musculoskeletal symptoms among women currently and formerly working in fish-filleting plants. *International Journal of Occupational and Environmental Health*, *6*(1), 44–49. doi:10.1179/oeh.2000.6.1.44

Olaoye, O. J., & Ojebiyi, W. G. (2021). Occupational health hazards associated with Nigerian fisheries. In Palaniappan, K., & McCauley, P. (Eds.), *Occupational Wellbeing*. https://doi.org/ doi:10.5772/intechopen.94205

Olaoye, O. J., Odebiyi, O. C., & Abimbola, O. T. (2015). Occupational hazards and injuries associated with fish processing in nigeria. *Journal of Aquatic Sciences*, *3*(1), 1–5. doi:10.12691/jas-3-1-1

O'Reilly, C. M., Alin, S. R., Plisnier, P., Cohen, A. S., & McKee, B. A. (2003). Climate change decreases aquatic ecosystem productivity of lake Tanganyika, Africa. *Nature*, *424*(6950), 766–768. doi:10.1038/nature01833

Organisation for Economic Co-operation and Development (OECD). (2000). *Reducing the risk of policy failure: Challenges for regulatory compliance*. https://www.oecd.org/gov/regulatory-policy/46466287.pdf

Palsson, B., Stromberg, U., Ohlsson, K., & Skerfving, S. (1998). Absence attributed to incapacity and occupational disease/accidents among female and male workers in the fish-processing industry. *Occupational Medicine*, *48*(5), 289–295. doi:10.1093/occmed/48.5.289

Panda, S., Panigrahi, G. K., & Padhi, S. N. (2016). *Earning animals*. Anchor Academic Publishing.

Panigrahy, S., Murthy, T. V. R., Singh, J. G., & Patel, T. S. (2012). Wetlands of India: Inventory and assessment at 1:50,000 scale using geospatial techniques. Current Science, 102(6), 852–856. https://currentscience.ac.in/Volumes/102/06/0852.pdf

Paskarini, I., Alwi, M. N. M., & Martiana, T. Mahmudah, Arini, S. Y., &Dwiyanti, E. (2022). The decrease in the quality life of fishermen due to Covid-19 widespread. *Malaysian Journal of Medicine and Health Sciences, 18*, 79-85. https://scholar.unair.ac.id/en/publications/the-decrease-in-the-quality-life-of-fishermen-due-to-covid-19-wid

Pathak, H., Mishra, J. P., & Mohapatra, T. (2022). *Indian agriculture after independence*. Indian Council of Agricultural Research.

Patrick, A. E. S. (2016). Influence of rainfall and water level on inland fisheries production: A review. *Archives of Applied Science Research*, *18*(6), 44–51. https://www.researchgate.net/publication/313529635

Paul, T. T., Dennis, A., & Grinson George, G. (2016). A review of remote sensing techniques for the visualization of mangroves, reefs, fishing grounds, and molluscan settling areas in tropical waters. In C. W. Finkl & C. Makowski (Eds.), *Seafloor mapping along continental shelves* (pp. 105–123). Springer Climate., doi:10.1007/978-3-319-25121-9

Pelage, L., Domalain, G., Lira, A. S., Travassos, P., & Fredou, T. (2019). Coastal land use in north-east Brazil: Mangrove coverage evolution over three decades. *Tropical Conservation Science, 12*, 194008291882241. doi:10.1177/1940082918822411

Pillai, S. M., & Muralidhar, M. (2006). Survey and demarcation of seawater inundated areas for eco-friendly aquaculture in Andaman and Nicobar Islands (pp. 69). Central Institute of Brackishwater Aquaculture (CIBA).

Ponniah, A. G., & Muralidhar, M. (2009). Research requirements to understand impact of climate change on brackish water aquaculture and develop adaptive measures. In *Proceedings of 96th Session of the Indian Science Congress, Part II: Session of Animal Veterinary and Fisheries Sciences* (pp. 46–47). The Indian Science Congress Association (ISCA).

Preston, N. P., Jackson, C., Thompson, P. A., Austin, M., Burford, M. A., & Rothlisberg, P. (2000). *Prawn farm effluent: composition, origin and treatment.* Fisheries research and development corporation report (FRDC). *CRC for Aquaculture*. https://www.frdc.com.au/sites/default/files/products/1995-162-DLD.PDF

Prusty, A. K., Meena, D. K., Mohapatra, S., Panikkar, P., Das, P., Gupta, S. K., & Behera, B. K. (2015). Synthetic pyrethroids (Type II) and freshwater fish culture: Perils and mitigations. *International Aquatic Research.*, *7*(3), 163–191. doi:10.1007/s40071-015-0106-x

Purohit, G. K., Mahanty, A., Suar, M., Sharma, A. P., Mohanty, B. P., & Mohanty, S. (2014). Investigating hsp gene expression in liver of Channastriatus under heat stress for understanding the upper thermal acclimation. *BioMed Research International*, *2014*(381719), 1–10. doi:10.1155/2014/381719

Raman, S. (2022, March 9). Climate change set to upend global fishery agreements, study warns. *Mongabay*. https://news.mongabay.com/2022/03/climate-change-set-to-upend-global-fishery-agreements-study-warns

Rao, R. G. (2009). Climate change mitigation through reforestation in Godavari mangroves in India. *International Journal of Climate Change Strategies and Management*, *1*(4), 340–355. doi:10.1108/17568690911002870

Research & Markets. (2019, February 07). Global Commercial Fishing Industry Market to 2026 - Exhaustion of Marine Resources are Inhibiting Market Growth. *PR Newswire*. https://www.prnewswire.com/news-releases/global-commercial-fishing-industry-market-to-2026---exhaustion-of-marine-resources-are-inhibiting-market-growth-300791788.html

Reygondeau, G., Cheung, W. W. L., Wabnitz, C. C. C., Lam, V. W. Y., Frolicher, T., & Maury, O. (2020). Climate change-induced emergence of novel biogeochemical provinces. *Frontiers in Marine Science, 7*. https://doi.org/10.3389/fmars.2020.00657

Rimmer, M. A., Larson, S., Lapong, I., Purnomo, A. H., Pong-Masak, P. R., Swanepoel, L., & Paul, N. A. (2021). Seaweed aquaculture in Indonesia contributes to social and economic aspects of livelihoods and community wellbeing. *Sustainability*, *13*(19), 10946. doi:10.3390/su131910946

Rodrigues, D. E., & Kiran, U. (2013). A Pilot Study on knowledge & practice regarding prevention of occupational hazards and attitude towards utilisation of safety measures among fishermen working at a selected HA. B.R. *Journal of Health and Allied Sciences NU, 03*(03), 068–071. doi:10.1055/s-0040-1703681

Rojith, G., & Zacharia, P. U. (2015). Enhancing Regional Climate Resilience of Indian Fisheries Through Wetland Restoration and Scientific Fish Farming. *International Journal of Tropical Agriculture*, *33*(4), 3439–3445. https://www.researchgate.net/publication/298788364

Sajesh, V. K., Suresh, A., Mohanty, A. K., Singh, V., & Ravishamkar, C. N. (2021). Skill development in marine fisheries: Some reflections on the issues and way outs. *The Indian Journal of Animal Sciences*, *91*(7), 518–524. https://doi.org/ 10.56093/ijans.v91i7.115894

Salim, S. S., & Anuja, A. R. (2022). *Marine fisheries policies in India: Opportunities and challenges.* In: Souvenir 1st Indian fisheries outlook 2022. ICAR-Central Inland Fisheries Research Institute, Kolkata (pp. 215-224). http://eprints.cmfri.org.in/15861/

Samantray, K., & Pathak, S. C. (2001). An overview of women in the Indian fisheries sectors and suggestions for improvement of their socioeconomic status. In M. J. Williams, M. C. Nandeesha, V. P. Corral, E. Tech, & P. S. Choo (Eds.), *International Symposium on Women in Asian Fisheries* (pp. 121–126). ICLARM- The World Fish Centre. ISBN:983-2346-02-9

Samarasiri, M., Chai, K. F., & Chen, W. N. (2023). Forward-looking risk assessment framework for novel foods. *Food and Humanity, 1*, 500–513. doi:10.1016/j.foohum.2023.06.020

Scales, K. L., Moore, T., Sloyan, B. M., Spillman, C. M., Paige Eveson, J., Patterson, T. A., Williams, A. B., Hobday, A. J., & Hartog, J. R. (2023). Forecast-ready models to support fisheries' adaptation to global variability and change. *Fisheries Oceanography*, *32*(4). Advance online publication. doi:10.1111/fog.12636

Schneider, Y., Beier, M., & Rosenwinkel, K. H. (2012). Effect of substrate availability on nitrous oxide production by deammonification processes under anoxic conditions. *Microbial Biotechnology*, *5*(3), 415–424. doi:10.1111/j.1751-7915.2011.00328.x

Shahbandeh, M. (2022). Leading exporting countries of fish and fishery products worldwide in 2020. *Main exporting countries of fish and fishery products worldwide 2020*. https://www.statista.com/statistics/268269/top-10-exporting-countries-of-fish-and-fishery-products/

Sharma, A., Prusty, S., Rathod, R. K., Arthi, R., Watterson, A., & Cavalli, L. S. (2023b). Occupational hazards of Indian shrimp farm workers. *Aquaculture and Marine Biology, 16*(1), 2225762. doi:10.1080/26895293.2023.2225762

Sharma, A., Upadhayay, V. K., & Ghorai, T. (2023a). Possible impact of climate change on fisheries. In A. Sinha, S. Kumar, & K. Kumari (Eds.), *Outlook of climate change and fish nutrition* (pp. 41–47). Springer., doi:10.1007/978-981-19-5500-6

Sharma, M., Sahni, Y. P., Mandloi, A. K., & Pandey, G. (2012). Toxicity in fish polluted with heavy metals, chemicals or drugs. *Jigyasa, 6*(2), 67–71. https://www.researchgate.net/publication/270213954_Toxicity_in_fish_polluted_with_heavy_metals_chemicals_or_drugs

Silva, B. K. R., Figueiredo, F. W. dos S., Maciel, E. da S., Quaresma, F. R. P., & Adami, F. (2019). Factors associated with perceived quality of life in artisanal fishermen: A cross-sectional study. *BMC Research Notes*, *12*(1), 479. doi:10.1186/s13104-019-4525-4

Smith, C., & Clay, P. (2010). Measuring Subjective and Objective Well-being: Analyses from Five Marine Commercial Fisheries. *Human Organization*, *69*(2), 158–168. doi:10.17730/humo.69.2.b83x6t44878u4782

Sobhana, K. S. (2009). *Diseases of seabass in cage culture and control measures.* National Fisheries Development Board., http://eprints.cmfri.org.in/6087/1/15._Sob.pdf

Soto, D., Aguilar-Manjarrez, J., & Hishamunda, N. (2008). Building an ecosystem approach to aquaculture. FAO/Universitat de les IllesBalears Expert Workshop. In *Proceedings of FAO Fisheries and Aquaculture* (pp. 14). Food and Agriculture Organization of the United Nations (FAO). ISBN: 978-92-5-106075-9

Sridevi, B., Sabira, S., & Sarma, V. (2023). Impact of ocean warming on net primary production in the northern Indian Ocean: Role of aerosols and freshening of surface ocean. *Environmental Science and Pollution Research International*, *30*(18), 53616–53634. doi:10.1007/s11356-023-26001-9

Srinivas, C. R., & Sethy, M. (2023). Occupational dermatoses. *Indian Dermatology Online Journal*, *14*(1), 21. doi:10.4103/idoj.idoj_332_22

Suman, D. O. (2019). Mangrove management: Challenges and guidelines. In Gerardo M. E. Perillo, Eric W., Donald R. C., Charles S. H. (Eds.), Coastal wetlands (Second Edition) (pp. 1055-1079). Elsevier.

Swaminathan, M. S. (2012). Aquaculture and sustainable nutrition security in a warming planet. In *Proceeding of the Global Conference on Aquaculture 2010- Farming the Waters for People and Food*, Food and agriculture organization (pp. 3-895).ISBN: 978-92-5-107233-2

Swedish National Board of Occupational Safety and Health. (1991). *An excerpt from NewsletterBilaga 4*, A quarterly review, provided for ILO's Food and Drink Industries Committee.

Sydeman, W. J., Garcia-Reyes, M., Schoeman, D. S., Rykaczewski, R. R., Thompson, S. A., Black, B. A., & Bograd, S. J. (2014). Climate change and wind intensification in coastal upwelling ecosystems. *Science*, *345*(6192), 77–80. https://doi.org/10.1126/science.1251635

Takeuchi, I., Miyoshi, N., Mizukawa, K., Takada, H., Ikemoto, T., Omori, K., & Tsuchiya, K. (2009). Biomagnification profiles of polycyclic aromatic hydrocarbons, alkylphenols and polychlorinated biphenyls in Tokyo Bay elucidated by δ13C and δ15N isotope ratios as guides to trophic web structure. *Marine Pollution Bulletin*, *58*(5), 663–671. doi:10.1016/j.marpolbul.2008.12.022

Temporosa, R. G. (1979). A study of the manpower skills of fishery workers in the commercial fisheries industry considered important by respondents in Western Visayas. *A thesis presented to the faculty of the school of graduate studies. Central Philippine University*. 6-10. https://repository.cpu.edu.ph/handle/20.500.12852/1701?locale-attribute=ru

Tewabe, D. (2014). Climate change challenges on fisheries and aquaculture. *International Journal of Aquaculture and Fishery Sciences, 1*(1), 006-011. doi:10.17352/2455-8400.000002

Thorvaldsen, T., Salomonsen, C., Ranum, S. A., Trædal, P., Misund, A., & Holmen, I. M. (2023). Prepared for the worst? Emergency preparedness in Norwegian fish farming – status and further improvements. *Aquaculture (Amsterdam, Netherlands)*, *577*, 739921. https://doi.org/10.1016/j.aquaculture.2023.739921

Tom, A. P., Jayakumar, J. S., Biju, M., Somarajan, J., & Ibrahim, M. A. (2021). Aquaculture wastewater treatment technologies and their sustainability: A review. *Energy Nexus*, *4*, 100022. doi:10.1016/j.nexus.2021.100022

Tomoda, S. (1997). *Safety and health of meat, poultry and fish processing workers*. International Labour Office. https://www.ilo.org/wcmsp5/groups/public/---ed_protect/---protrav/---safework/documents/publication/wcms_219576.pdf

Troell, M., Henriksson, P. J. G., Buschmann, A. H., Chopin, T., & Quahe, S. (2022). Farming the ocean – Seaweeds as a quick fix for the climate? *Reviews in Fisheries Science & Aquaculture*, *4*, 1–11. doi:10.1080/23308249.2022.2048792

U.S. Food and Drug Administration (USFDA). 2022. *Fish and fishery products hazards and controls guidance.* https://www.fda.gov/media/80637/download

Unnikrishnan, A. S., & Shankar, D. (2007). Are sea-level-rise trends along the coasts of the north Indian Ocean consistent with global estimates? *Global and Planetary Change*, *57*(3-4), 301–307. doi:10.1016/j.gloplacha.2006.11.029

Vasanth, M., Muralidhar, M., Saraswathy, R., Nagavel, A., Dayal, J. S., Jayanthi, M., Lalitha, N., Kumararaja, P., & Vijayan, K. K. (2016). Methodological approach for the collection and simultaneous estimation of greenhouse gases emission from aquaculture ponds. *Environmental Monitoring and Assessment*, *188*(12), 671. doi:10.1007/s10661-016-5646-z

Velmurugan, A. V., Roy, S. D., Krishnan, P., Swarnam, T. P., Jaisankar, I., Singh, A. K., & Biswas, T. K. (2015). Climate change & Nicobar Islands: Impacts and adaptation strategies. *Andaman Science Association, 20*(1), 7–18. ISSN: 0970-4183

Vijayan, K. K. (2018). Brackishwater aquaculture in India: An overview. https://krishi.icar.gov.in/jspui/bitstream/123456789/20212/1/Training%20Manual%20-%20Final%20-%20Water%20and%20soil%20treatment%20applications%20in%20aquaculture.pdf

Vilchis, L. I., Tegner, M. J., Moore, J. D., Friedman, C. S., Riser, K. L., Robbins, T. T., & Dayton, P. K. (2005). Ocean warming effects on growth, reproduction, and survivorship of southern california abalone. *Ecological Applications*, *15*(2), 469–480. doi:10.1890/03-5326

Vivekanandan, E., Rajagopalan, M., & Pillai, N. G. K. (2009). Recent trends in sea surface temperature and its impact on oil sardine. In A. Surendran (Ed.), *Global Climate Change and Indian Agriculture* (pp. 89–92). Indian Council of Agricultural Research., http://eprints.cmfri.org.in/8435/

Warrier, M. V. S. (2001). Women at work: Migrant women in fish processing industry. *Economic and Political Weekly*, *36*(37), 3554–3562. https://www.epw.in/journal/2001/37/special-articles/women-work.html

Weiskopf, S. R., Rubenstein, M. A., Crozier, L. G., Gaichas, S. G., Griffis, R., Halofsky, J. E., Hyde, K. J. W., Morelli, T. L., Morisette, J. T., Munoz, R. C., Pershing, A. J., Peterson, D. L., Poudel, R., Staudinger, M. D., Sutton-Grier, A. E., Thompson, L., Vose, J., Weltzin, J. F., & Whyte, K. P. (2020). Climate change effects on biodiversity, ecosystems, ecosystem services, and natural resource management in the United States. *The Science of the Total Environment, 733*, 137782. doi:10.1016/j.scitotenv.2020.137782

Westlund, L., & Poulain, F. (2007). *Disaster response and risk management in the fisheries sector. Food and Agriculture Organization of the United Nations*. FAO., https://www.researchgate.net/publication/242239330

Williams, M. J., & Syddall, V. (2022). Women, fisheries technology and development: Toward new research approaches. *Gender, Technology and Development, 26*(1), 1–28. doi:10.1080/09718524.2022.2125456

World Health Organization (WHO). (n.d.). *Occupational hazards in the health sector*. https://www.who.int/tools/occupational-hazards-in-health-sector

Yea, S., & Stringer, C. (2023). The informalisation of precarious work in fishing crew: Experiences of Fijian fishers on distant water vessels. *Marine Policy, 155*, 105709–105709. doi:10.1016/j.marpol.2023.105709

Yurkevicius, B. R., Alba, B. K., Seeley, A. D., & Castellani, J. W. (2021). Human cold habituation: Physiology, timeline, and modifiers. *Temperature, 9*(2), 1–36. doi:10.1080/23328940.2021.1903145

Zacharia, P. U., Dineshbabu, A. P., Thomas, S., Shoba, J. K., Vivekanandan, E., Pillai, L. S., Sivadas, M., Ghosh, S., Ganga, U., Rajesh, K. M., Nair, R. J., Najmudeen, T. M., Koya, M., Chellappan, A., Dash, G., Divipala, I., Akhilesh, K. V., Muktha, M., & Sen, D. S. (2016). *Relative vulnerability assessment of Indian marine fishes to climate change using impact and adaptation attributes* (pp. 192). CMFRI Special Publication (125). http://eprints.cmfri.org.in/id/eprint/11297

Zain, R. M., Kamarudin, M. K. A., & Saad, M. H. M. (2018). Assessment of Quality of Life on fishermen community in Kuala Terengganu, Malaysia: A Review. *International Journal of Academic Research in Business & Social Sciences, 8*(10), 640–650. doi:10.6007/ijarbss/v8-i10/4770

Zhang, Y., Tang, K. W., Yang, P., Yang, H., Tong, C., Song, C., Tan, L., Zhao, G., Zhou, X., & Sun, D. (2022). Assessing carbon greenhouse gas emissions from aquaculture in China based on aquaculture system types, species, environmental conditions and management practices. *Agriculture, Ecosystems & Environment, 338*, 108110. doi:10.1016/j.agee.2022.108110

Zhao, M., Tyzack, M., Anderson, R., & Onoakpovike, E. (2013). Women as visible and invisible workers in fisheries: A case study of Northern England. *Marine Policy, 37*, 69–76. doi:10.1016/j.marpol.2012.04.013

Zhou, X., Fang, J., Shi, S., Yang, L., Xie, B., & Huang, L. (2022). Ecosystem services of ecosystem approach to mariculture: Providing an unprecedented opportunity for the reform of China's sustainable aquaculture. *Frontiers in Marine Science, 9*, 909231. doi:10.3389/fmars.2022.909231

KEY TERMS AND DEFINITIONS

CAGR: Compound Annual Growth Rate is the rate of return required for an investment to grow from its starting balance to ending balance, assuming that the profits are reinvested at the end of each period of the investment's life span. CAGR is used to compare the investments' performance historically, or predict their future expected returns, or gains.

CPAs: Climate Change Preparedness Activities includes a set of standard operative procedures i.e., policies, infrastructures, relationships, behaviors and functions to address the climatic change issues and recovery measures at personal, local, regional and national levels for ensuring protection of life, property and benefit of community.

EAA: Ecosystem Approach to Aquaculture is a holistic approach, or integrated strategy for management of aquaculture that considers all different components within a specific boundary of ecosystem, which characterizes the aquatic animal production and promotes sustainable development, equity, and resilience of interlinked socio-ecological systems.

EEZ: According to UN convention 1982 on the law of the sea, Exclusive Economic Zone is an area that extending 200 nautical miles (230 miles) beyond a nation's territorial sea, within which the said nation has jurisdiction over both living and non-living resources, and a right to explore/use of marine resources/energy production from water and wind.

GDP: Gross Domestic Product represents the total monetary value, or market value, of finished goods and services produced in a country during a particular period, which is an indicator of the income earned from that production, or the total amount spent on that goods and services. GDP is the indicator of the government for measuring the economic health of the country.

GHGs: Carbon dioxide, ozone, methane, nitrous oxide and water vapour are the Green House Gases that absorb infrared radiation and heat emitted from earth's surface and return them back to the surface. The name implies as the glass walls of the green houses works the same way as earth surface upon trapping heat coming from the sun.

GIS: Geographic Information System is designed to capture, evaluate, manipulate, handle, and view all forms of geographical and spatial information, data, etc., for conducting spatio-temporal analysis, managing large data, viewing the data in maps, or graphical formats for presentation, analysis and providing information to an organization for better management.

HPV 7: Human Papilloma Virus type 7is a DNA virus from *Papillomaviridae* family, which causes sexually transmitted infections and cancers of cervix, vulva, vagina, penis, anus, mouth, throat, or tonsils. Prevalence is high among butcher's and fish handlers though 90% infection resolves within two years, but some infection persists, leading to pancreatic lesions, or warts.

Hsp: Heat Shock Proteins are widely distributed in nature and highly conserved molecules of biosphere, which are expressed in cells during stress condition to perform crucial roles in protein folding/ unfolding/translocation as well as assembling/disassembling of the protein complexes. According to molecular weight, Hsp are divided into five categories: - Hsp90, Hsp70, Hsp60 and Hsp100.

IgE: Immunoglobulin E is a type of antibody (immunoglobulins), which is basically a protein in the mammalian immune system. Asthma, allergic rhinitis, atopic dermatitis, cutaneous diseases, neoplastic diseases and in immune deficiencies, IgE levels are elevated in serum, because immune system produces IgE against each of the specific allergen, thus determines the individual susceptibility against different allergic conditions.

LCA: Life Cycle Assessments is a methodology for quantifying environmental impacts arise from material inputs and outputs, such as use of energy, emission of air, extraction of raw material, processing/manufacture of product, distribution/use of product/service, recycling/disposal of product/service, etc., over their entire life cycle. It helps consumers to make a right decision without compromising environmental benefit.

MLS: Minimum Legal Size is a fishery management tool that protects juvenile fish, maintains spawning stocks and controls the sizes of fish that caught by the fisherfolk, or in other words, it sets the smallest size at which a particular species can be legally retained by the fisherfolk. MLS also prevents growth overfishing for major commercially exploited marine fish stocks, maintains healthy stocks, and ensures better income of the fisherfolk on a sustainable basis.

PAHs: Polycyclic Aromatic Hydrocarbons are a class of chemicals that occur naturally in coal, crude oil and gasoline. Burning coal, oil, gas, wood, garbage, tobacco and high heat during cooking of meat and other foods also generate PAHs, which bind to, or form small particles in the air. Exposure to PAHs and its metabolites cause cataracts, kidney and liver damage, jaundice, skin redness, inflammation, RBC lysis and mutagenesis.

PCD: Protein Contact Dermatitis is an allergic skin reaction induced by animal, or plant proteins through foods that causes chronic dermatitis in hands and/or forearms eczema of occupational origin, especially in food handlers, resulting from IgE-mediated allergic reaction. However, it is difficult to differentiate between allergic contact dermatitis and other eczematous dermatoses.

PPE: Personal Protective Equipment, is the specific range of items that worn by the people of different occupations, such as construction workers, miners, healthcare workers, researchers, workers in chemical and pharmaceutical industries, etc., to minimize the workplace exposure and provide protection from workplace hazards, injuries, accidents and illness resulting from the contact with chemical, radiological, physical, mechanical, electrical, and biological agents.

RCP: Representative Concentration Pathway is a greenhouse gas concentration trajectory (not emissions) by IPCC in concert with the predictive modelling techniques to measure the intensity of future changes in the marine fishery sector. For example, RCP predicts the future carbon dioxide emissions and its impact on the fishery sector, as well as possible reduction of its concentration in the environmental throughout a century.

Chapter 3

The Economic Impact of *Stenocereus montanus* Fruit in Rural Areas of the Northeast of Sinaloa, Mexico

Reyna Ocampo
Universidad Autónoma Indígena de México, Mexico

Soledad Ibarra
https://orcid.org/0000-0001-5165-7000
Universidad Autónoma Indígena de México, Mexico

ABSTRACT

The Stenocereus montanus plant is a wild cactus that is widely distributed in the arid and semi-arid territories of northwestern Mexico. It provides the inhabitants of the region with a fruit whose collection and sale is economically profitable for the families that dedicate themselves to it during the almost three months that the annual season lasts; the economic spillover that this entails hasn't been studied. Therefore, its analysis in this chapter serves as a frame of reference for a change of vision regarding its potential use as a trigger for economic, social, and sustainable development.

INTRODUCTION

Concretely, in agrarian aspects, Sinaloa is a state known as "Mexico's barn", due to its estimated agrarian production of 1,029,978.08 hectares, which by 2022, has represented a production value of $70,194,634.88 mexican pesos. (AEPA, 2023). However, despite the region's large amount of farming land and an adequate weather for the growth of products such as corn, beans, wheat, tomato, pepper, mango, among others, the people from mountainous areas can opt for wild flora of temporary growth, this, along with other socio-demographic aspects, limits economic development of rural communities (CONEVAL, 2018).

DOI: 10.4018/978-1-6684-9863-7.ch003

The fact that rural communities use wild life resources as a complement for their economy and subsistence, has been widely studied, as we see in the document from the World Union for the Conservation of Nature, the United Nations Program for the Environment and the World Forum of Wild Life (IUCN/PNUMA/WWF, 1991), that approaches the topic from a standpoint where the communities: look for sustenance and a good life, depend on the local resources and there is egalitarian distribution of the benefits, considering sustainability.

For our region, this is not different, and one of the most appreciated resources for the people are arborescent cacti the *Stenocereus* genus, these plants can be found in arid and semi-arid zones. Many of the populations where we can find this fruit take advantage of its sale and consumption, generating a positive impact for the inhabitants of the community (Lopez Vazquez, 2018).

The inhabitants of the zones where this plant proliferates, work taking advantage and recollecting the pitaya in the months of April, May and June; fulfilling the role of social and economic activity. However, recollection of this fruit is not exempt from difficulties, given that the first challenge to obtain it is to lower it from considerable heights and it has thorns that need to be removed for consumption or sale, moreover, each year, the amount of collected fruit diminishes, given that it gets affected by fires (common in the area) and the change in soil usage for agriculture.

Characteristics of the *S. montanus*

The *Stenocereus* genus is one of the most diverse in the whole world, including 23 species, out of which we find 19 in Mexico, distributed in almost all the country (Bravo and Sanchez, 1991). Eight of them are endemic of Sinaloa, highlighting their usage: *S. martinezii* in Arroyo Grande, Sanalona; *S. montanus* in Choix and Badiraguato, and *S. thurberi* in El Fuerte and parts of Ahome (Hunt, 1999; Salomón, 2018).

The wide distribution and use the fruit has, would make us believe that it's a widely studied species, however, it's not, the amount of available information for the *Stenocereus* genus, and their synonyms in different scientific repositories, can be found only around 315 times, for Latin America, out of which 73 of them correspond to mexican territories and less than 10 of them to Sinaloa, most of them being found in work notes about the distribution of the species.

Table 1. Taxonomic hierarchy, from the Integrated Taxonomic Information System (ITIS) on-line database

Kingdom: Plantae
Subkingdom: Viridiplantae
Division: Tracheophyta
Class: Magnoliopsida
Superorder: Caryophyllanae
Order: Caryophyllales
Family: Cactaceae
Genus: *Stenocereus* (A. Berger) Riccob
Species: *montanus* (Britton & Rose) Buxb.
Synonym(s) *Lemaireocereus montanus* Britton & Rose
Ritterocereus montanus Buxb.

Hierarchal Taxonomy and Common Name

There are some general misconceptions about the name that the pitaya or pitahaya should receive from the different *Stenocereus* species, according to Bravo (1978), the term "pitaya", is antilian in origin and is derived from an original pronunciation that was more similar to "pitahaya", a phonetic version that is widely used in Mexico, in a very general sense for most of the cactus fruits; he also mentions that the term is sometimes used with ortographic variants that result from substituting the "y" for the "ll" or "h" for "j" sounds. Taxonomically, the term "pitaya" or any of its variants is applied, so it's not uncommon for people from rural communities to use this name and its variants to refer to fruits from the *Hylocereus*, *Pachycereus* and *Echinocereus* genus.

As previously mentioned, the terms "pitaya" and "pitahaya" in Mexico encompass a large number of cacti fruits, however, in this regard, the Agro-Food and Fisheries Information Service (SIAP, 2017) explains that "the main difference lies in the genus to which it belongs. The "pitaya" comes from the *Stenocereus* genus, while the "pitahaya" is from the *Hylocereus*".

Morphology

The *Stenocereus* genus comprises candelabriform trees, 6-9 m tall; definite 0.6-1 m tall and 30-40 cm in diameter trunk; more or less numerous branches that branch off into secondaries and tertials, curved at the base, almost all of the same size, 1-5 m long and 13-20 cm in diameter, dark green; 7-9 ribs, acute in cross section, straight in longitudinal section, about 2.7 cm high and 4 cm wide at the base, without horizontal constriction between areoles of the same rib; areoles 1-1.5 cm apart, 6-11 mm long and 6-10 mm wide, circular to elliptical in cross section, with light brown glandular trichomes, eventually forming a dark brown to black crust; thorns not clearly differentiated into radial and central, 9-10, 0.5-1.5 cm long, one of them, the lower one, longer than the others, up to 3 cm long, acicular to subulate, divergent, strong, white, gray with time, the lower one reflects; flowers subapical or lateral, nocturnal, funnel-shaped, 6.6-8.2 cm long and 3.5-5 cm wide; pericarpel ovoid, 1-1.8 cm long and 1-1.5 cm broad, green with purplish tints, with non-prominent podaries and short, triangular foliar primordia, 2-3 mm long and broad at the base, purplish-green with purple edges, provided in the armpits with yellowish-white trichomes, strong, well-developed thorns; 3-3.5 cm long receptacle tube, podaries with scales welded to it, oblong with acute to mucronate apex, of variable length and 3-8 mm wide, with strong thorns; outer segments of perianth narrowly obovate to spatulate, with acute to rounded apex, mucronate, about 2.5 cm long and 1 cm broad, yellowish-green above, green below with purplish tinges; inner segments oblong to spatulate with apiculate apex, edge entire, white to pinkish-white, about 2.5 cm long and 1 cm broad; stamens numerous (Arreola and Terrazas 2003). *Stenocereus* species with brown areoles with the receptacle tube in whorled series; yellowish filaments; basifixed, yellowish-white anthers; style yellowish-white, about 2 mm broad; stigma lobes 8-10, 6.5-8.5 mm long, yellowish-white; nectar chamber semi closed by the curvature of the stamens of the lower whorl, 1.2-1.6 cm long and 4.6-6.8 mm wide, striated; ovary 0.9-1.3 cm long and 0.6-0.8 cm wide; fruit 5-6 cm in diameter, globose to ovoid, green to reddish-green or purple in color, covered by areoles with numerous setose thorns, about 1.5 cm long, pale yellow in color, deciduous when ripe, red pulp, pleasant flavor (Arreola and Terrazas, 2003).

Figure 1. a) Stem and arms. b) Ribs and edges of the tree

Stem

They contain or store enough water to survive in the warm climates in which the space is distributed. The outer layer (epidermis), which is very bulky, with numerous stomata (small sunken holes that allow gas exchange in the plant). The epidermis contains an organic compound with a consistency harder than latex and other substances that regulate the amount of water in the cactus. What differentiates one species from others is the presentation and number of edges or ribs and thorns on the stems. They are 2.7 to 3.4m tall, but occasionally exceed 6.1m. (Vazquez et al., 2009).

Root

The root it is divided into two branches: the main ones are in the ground, at depths between 7 and 30 cm, and 30 cm thick; the secondary ones, also called adventitious, develop during periods of drought, exposed on the soil profile (Vázquez et al., 2009).

Flower

The flowers are subapical or lateral, nocturnal, funnel-shaped, 6.6-8.2 cm long and 3.5-5 cm wide (Arreola and Terrazas, 2003), they are hermaphrodite and have a very striking trumpet-shaped appearance, they

Figure 2. Inflorescence

are white, yellow or rosy. They are born attached to the epidermis between the thorns, open after dark and close shortly after sunrise. These flowers present phototropism (Vázquez, op. cit.).

Thorns

From 9 to 10 thorns not clearly differentiated into radial and central, 0.5-1.5 cm long, one of them, the lower one, longer than the others, up to 3 cm long, acicular to subulate, divergent, strong, white, over time they turn gray. (Arreola and Terrazas, 2003).

Fruit

The fruit is 5-6 cm in diameter, globose to ovoid, green to reddish-green or purple in color, covered by areoles with numerous setose thorns, about 1.5 cm long, pale yellow in color, deciduous when mature (Arreola and Terrazas, 2003).

Pulp

The pulp is juicy red in color, not very fibrous, its flavor is sweet and its texture is wavy, the content of total soluble solids is 10°Bx, when it reaches commercial maturity, it has a bittersweet flavor. The pulp and the seed together are the commercial part and the one with the highest demand in the market for its industrialization (Vázquez et al., 2009).

Figure 3. Thorns (we can see how they detach from the areoles)

Figure 4. Thorn free fruit

Figure 5. Pitaya fruit pulp, we can see some colors, such as orange, yellow and red)

Figure 6. Seeds, abundant and small

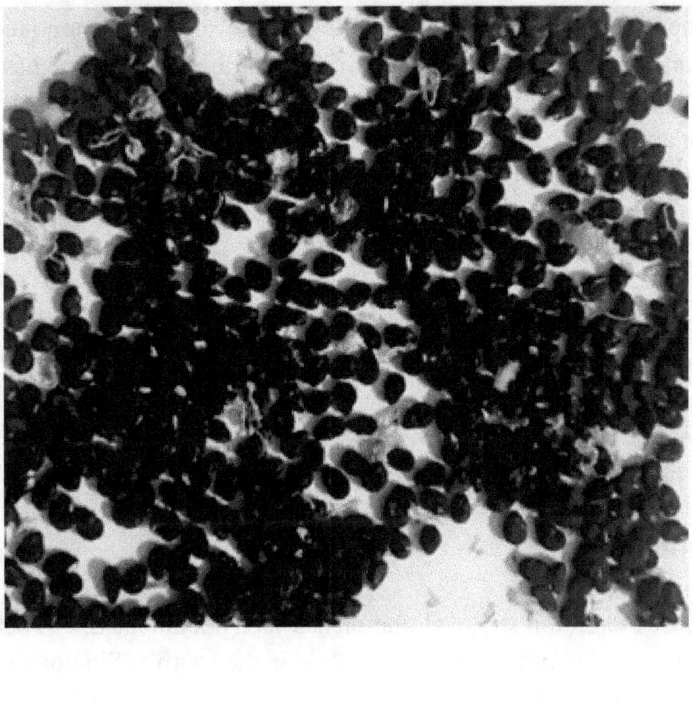

Seeds

Seeds are 2.3-2.5 mm long and 1.62-1.78 mm wide, ovoid, black, opaque, isodiametric test cells, with convex periclinal walls, without interstices, which gives it a finely verrucous appearance, each fruit usually contains about 2000 seeds (Arreola and Terrazas, 2003).

Fenology

The fruiting season for pitaya is highly variable and depends on the species. However, it can be said that there are at least three different seasons:

From early May to June for the *S. montanus, S. pruinosus, S. queretaroensis, Pachycereus weberi, P. pecten-aboriginum* and *P. pringlei* species.

Others; During the rainy season, such as: *S. thurberi, S. fricci, Myrtillocactus geometrizans, Carnegiea* and *Hylocereus undatus.*

And almost at the end of the rainy season (August to the beginning of November) in the late species that have a sourer taste, such as *S. stellatus, S. treleasei, S. gummosus* and *S. eruca* (Vázquez et al., 2009).

Distribution and Habitat

The habitat is mainly in the low deciduous forest ecosystem, in latitudes of 200 to 400 meters, the distribution of *S. montanus* occurs in the south of Sonora and north of Sinaloa, cohabitating with other cactus trees, such as *P. pecten-aboriginum* y *S. thurberi* (Arreola y Terrazas, 2003).

Soils

The best soils for crops are sandy, well-drained soils that prevent the accumulation of moisture in the roots, with a maximum slope of 50%, a pH between 5.5 and 6.5, with high organic matter content, given its high rusticity, the plant It develops well on poor and stony soils (Mora, 2011).

Due to its particular physiology, complex anatomical structure and morphology, this plant offers high potential as a productive alternative for its use in soils in arid and semi-arid regions (Gibson and Horak, 1978).

Weather

The optimal temperature for the development of pitaya ranges between 17 to 20 ºC as an annual average, tolerating extremes of 38 to 40 ºC, in young plants it is common to be affected by high and low temperatures. In general, cacti have greater tolerance to high temperatures and long periods of drought with respect to other plants (Gibson and Nobel, 1986).

Growing the Plant

In Sinaloa, only one case of agroforestry exploitation has been documented, this being the community of San José del Llano, municipality of Badiraguato (located in the south of Sinaloa), which in 2015 decided to transform what was traditionally the collection and sale of the fruit, in what is now the "Sociedad de

Producción Rural de Responsabilidad Limitada" (Society of rural production of limited responsibility), which uses the fruit under the brand "Pitaya Golden", which markets the processed fruit mainly in the form of jam and jelly (CONAFOR, 2018).

Given the characteristics of cacti, from the physiological point of view, the cultivation of this fruit requires little irrigation, fertilizers and pesticides (Pimienta-Barrios and Nobel, 1994). What requires more attention is pruning and healing cuts or damage, because it contributes to keeping the plant healthy by removing sources of infection. It is necessary to highlight that in each and every one of the types of pruning, it is mandatory to use tools such as scissors that provide smooth cuts (INTA, 2014).

It is common that in rural areas, the arms of the tree (called elbows or stakes) are cut when they are still tender (soft) and they are used as fences or land delimiters.

Harvesting Method

The harvesting is performed with rudimentary tools. The most widely used tool by the people of the Guadalupe community, and is made up of a reed stick (plant of the *Phragmites* genus) 2 to 3 meters long, at one of its ends a pointed stick is attached in a "Y" shape "*huichuta*", the tip is made from different types of wood, such as *papachi, guasima* or *guinolo*, among others. The *huichuta* is held with a rope or rubber band. The procedure to get the fruit consists in perforating it transversally, which ensures a good grip on it and allows the fruit to rest on the *huichuta.*

Figure 7. a) Joint of the huichuta with the stick, b) Pitaya fruit being perforated and held by the huichuta during recollection

Plagues

The phytosanitary problems that limit the production of this plant species are: basal rot of the fruit (for which the cause of this agent has not been determined), anthracnose (*Colletotrichum sp.*), the flower bud fly (*Dasiops saltans*), the *Anastrepha sp.* fruit fly., and nematodes (García, 2003).

The flower bud fly can cause losses in flowering that vary between 40 and 80%. The damage is initially caused by the female, who deposits the eggs in the flower bud with the help of her ovipositor; the larvae feed on the flower bud causing rotting of the internal tissue (Delgado et al, 2010).

Diseases

Cacti can be attacked by the fast-spreading *Pythium debaryanum* fungus, which causes basal rot, darkening and ultimately death. These manifestations are always due to excess moisture. To avoid these diseases, it is advisable to give them a treatment with a copper-based fungicide or Triadimenol (25%, 20 cc) as a preventive and curative treatment, we must take into account preventive treatments after or before the rains. (Mora, 2011).

The first symptoms are small yellow to brown spots, which give rise, under favorable conditions, to a soft rot (Hyo Won et al., 2007). In the fruits they appear, initially, in the peduncle; when there is a high degree of severity, the disease can cause their fall. In Colombia, incidences of up to 29.3% have been found (Araujo and Medina, 2008).

RESEARCH METHODS AND TECHNIQUES

Description of the Area of Study

Regarding the socioeconomic sphere, in 2022 the municipality of Choix (where the Guadalupe community is located) appears on the list of the "Decree formulating the Declaration of Priority Attention Areas for the year 2023" of the Diario Oficial de la Federación (Official Gazette of the Federation). and it is classified as the second municipality with the highest percentage of population in poverty (54.2%), second place with the highest percentage of people in extreme poverty (14.2%), it also occupies second place with 30.8% of its population with deficiencies due to educational lag and fifth place as the municipalities whose population does not have access to public services (CONEVAL, 2022).

The town of Guadalupe is located (Choix municipality in the State of Sinaloa, Mexico) at 26° 37' 18.0" Longitude and -108° 19' 16.0" Latitude, its average altitude is 368 meters above sea level (See Image 8). By 2020, it had 950 inhabitants, of which 5.43% self-identified as indigenous (INEGI, 2021; SEDESOL, 2021).

The community has 182 homes complemented by educational institutions (Indigenous education, elementary, tele-secondary, community high school), as well as a communal house and community clinic. Their water supply is through a spring, they are organized through an communal Commissioner and a municipal authority, the main natural phenomenon that causes damage in the community is drought and the main socioeconomic problem reported is unemployment or paid employment insufficient to cover their basic needs.

Figure 8. Location of the Guadalupe community, belonging to the Municipality of Choix, Sinaloa, México (map generated in Google Earth PRO, 2022)

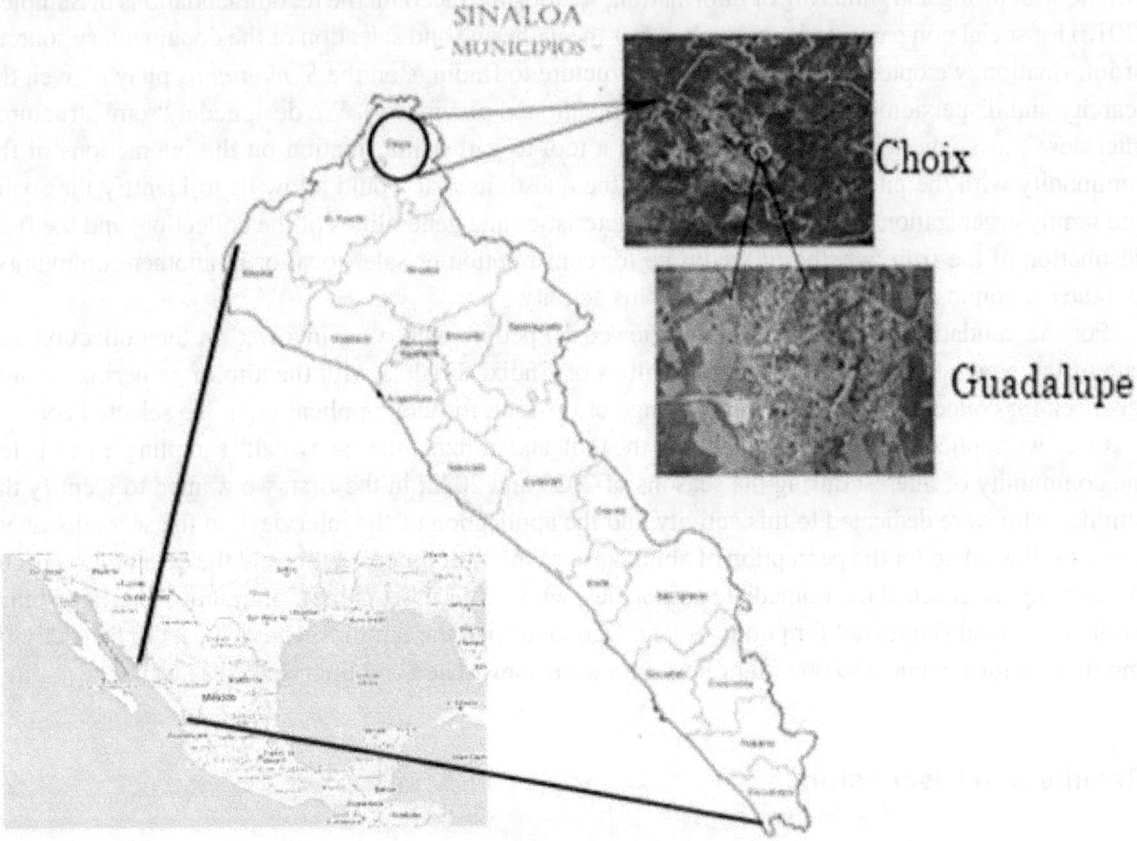

In the region, forest fires and land clearing represent a threat that should not go unnoticed, as Hernández and Godínez (1994) make clear: "Cacti are one of the most unprotected groups in the plant kingdom, the natural populations of many of the species have been affected by the pressures of human development, mainly due to the conversion of land for agricultural uses and the extraction activities of plants from their habitat, for their sale as ornamental plants in national and international markets", in the same way, we must not forget that globally, the conversion of areas of native vegetation to farmland contributes up to 22% of total greenhouse gas emissions (UNDP, 2023).

According to the information available in "field notes" dated 2019, of the Epidometry subject of the Autonomous Indigenous University of Mexico, Campus Choix. In an exploratory sampling of the *Stenocereus* genus in the study area, *S. montanus* was identified as the dominant species. They also mention the presence of a *S. thurberi* tree, but it stands out that the latter was not at the random sampling points, but rather that they reached it thanks to the specific help of a key informant who mentioned knowing a pitaya plant with thorns of a different color than the others, the notes also mention that the identification was made with the document of Arreola-Nava & Terrazas (2003).

Research Methodology

For the structuring and gathering of information, we took into account the recommendations of Sampieri (2018) for social non parametric research. After the gathering and selection of the documentary sources of information, we opted for a monographic structure to findings on the *S. montanus* pitaya, given the scarcity and dispersion of the information (as mentioned previously). We designed a "semi structured interview", as suggested by Geilfus (2002) as a tool to gather information on the interactions of the community with the pitaya tree, emphasizing the questions that would allow us to identify the social and family organization of collectors, the characteristics and generalities of the collection, and the final destination of the fruit, whether it would be for consumption or sale (local or in another community) and the economic and cultural relevance of this activity.

For the validation of the tool, we interviewed 10 people who were involved in the collection and sale of the pitaya fruit, in different communities of Choix, Sinaloa, with the aim of gathering enough feedback for collection, structuring and change of the tool, for their application in the selected zone.

Once we applied the pertinent changes to the tool, and utilizing the "snowball" sampling", we visited the community of interest during the seasons of 2020 and 2022; in the first, we wanted to identify the families who were dedicated to this activity and the application of the interview; in the second season, we just followed up for the perception of abundance of the fruit for collectors and the economic aspects. We initially approached the immediate authorities, who in this case, is the Community Sheriff (administrative political figure, not for public order), who identified the families who dedicate to this activity, and they, in turn, pointed to other families, who were approached and interviewed as well.

Results and Discussion

During the application of the interviews, we identified a total of 10 families dedicated to the activity of collecting and selling pitaya, who presented a good disposition and gave answers to all the questions asked, it should be noted that they were interested in the possible aspects of development that can take the sustainable use of the fruit of *S. montanus*. It stands out that 100% of the families consider the plant as an opportunity to help or improve the family economy during the harvest season, likewise there is no consensus among the families regarding the best collection point, each one goes where consider that there are more fruits, either because in previous seasons they gathered a large number of fruits or someone else told them that the fruit in such and such area is abundant.

All the families agree that since the beginning of May, they are awaiting the ripening of the fruits (coinciding with what the literature indicates), they mention that they make exploratory routes until the ripening is widespread enough to drop several dozen per day. The collectors consider that the season ends with the first rain, because after the rains, the flavor of the fruit changes, losing sweetness (a change that occurs due to hydration of the fruit).

We can also highlight the fact that collectors see the local fauna (bats, birds and rodents) as a direct competitor, some of them even attacking or driving away these animals, however, they do not take into account that this fauna, in addition to serving as pollinators, favor the population increase of the pitaya and its distribution by zoochoric dissemination.

Social Relationships

According to the analysis of the interviews, in the collection and sale of the fruit of the *S. montanus* pitaya in the community of Guadalupe, there are 2 main actors, who are:

The collector. - Who is in charge of going very early to obtain the fruit, adding to this activity, women, young people and older people (in some cases the whole family goes to this activity).

The reseller. - That receives the fruit and goes to the municipal seat of Choix and to other municipalities such as El Fuerte and Ahome to offer the fruit at a higher price.

The fruit is distributed in local markets and thanks to the opening or improvement of communication routes, it is sold in regional and interstate markets. Although the distribution of the fruit outside the area increases its sale price (due to its transfer), this increase is not reflected in the income of the collector since the business model that they currently operate favors the reseller.

We found that approximately 30% of the inhabitants of Guadalupe are dedicated to the activity of collecting pitaya fruits during the season and do not have any type of organization that groups them to establish more competitive prices or access sectoral funds that would allow them to develop a local industry.

The people have a certain degree of appreciation/respect for the plant since the economic benefits are eagerly awaited by the community, the times dedicated to the activity are variable, since everyone works according to their needs. They also mention that there are good times and bad. Likewise, they identified the 2020 season as "bad" with few fruits, which is interpreted as a greater effort to cover the fruit quota and the 2022 season as "good" by obtaining more fruits in less time.

The organization that the local inhabitants have in the process of collection, cleaning and sale includes young people, children, women and the elderly. Regarding the way of collecting and marketing the fruit of the pitaya, the collecting families are organized to obtain the product in a simple and safe way, some lower the fruit and others clean it, this is the way to obtain the product to later take them to the sale or delivery with the reseller, taking 4 to 5 hours of the day, being able to extend it if the family sells at the foot of the road.

ECONOMIC GENERALITIES

It is coherent that all the families agree that the more members of the family participate, the greater the amount of fruits collected and therefore more economic gain, however, they suggest that the different members of the family mostly carry out different activities such as: trapping, cleaning, transportation from the field to home, sales or delivery area, it is noteworthy that they express visiting at least 8 to 10 trees per day. With an average collection of 5 to 10 dozen per day (for both 2020 and 2022 seasons), but they emphasize that in previous years they collected up to 24 dozen per day, making up to 2 rounds per day.

We have that 70% of the families remove and clean the fruit, then deliver it to resellers, at a rate of $35 to $50 mexican pesos per dozen for the 2020 season and 60 to 70 mexican pesos. In the 2022 season, it is these resellers who are in charge of taking the fruits to the municipal capital and its surroundings, where they sold it for $80-$100 mexican pesos per dozen in the 2020 season, and $100-$130 mexican pesos in the 2022 season. The rest of the families sell the fruit at the edge of the road, where buyers generally prefer intense red fruits, since they associate it with a sweeter fruit (data that requires further analysis), this leaves the fruits of paler colors as second in category and difficult to sell, but they obtain a higher profit, investing more time to sell all the fruits and with the risk of losing the non-commercialized fruit.

The incentive to take the risk of greening the fruit for themselves at the roadside is in the prices of $80 to $100 mexican pesos per dozen (2020 season) and from $100 to $120 mexican pesos per dozen (2022 season), as is the case with large quantities of products/fruits of the forest, the sale price is variable, it also lends itself to speculation and/or haggling, most do not bother to establish a price per piece since they only sell them by the dozen reaching previously mentioned prices.

It is evident that the harvesting of the fruit represents the support of the families during the season, because they see in the collection (especially in non-working hours), an opportunity to complement the expense of families whose members have formal work, these being mainly those who decide to deliver their product to resellers. The situation of families whose members do not have a stable job is very different, and the pitaya season represents a period of prosperity or the opportunity to cover family expenses between jobs.

With respect to the quality of the fruits, there are some factors that can affect the quality (good condition of the fruit), the main one is the cutting time, since although most families go out to gathering between 4 and 5 in the morning, some families expressed that they can go at any time and also go in the afternoon when they have a specific or previously established sale or for self-consumption, taking the above as a basis, all the families express that high temperatures are the main factor that causes loss of quality in the fruit (fermentation). In this regard, the people who dedicate themselves to this activity don't associate the damage of the fruit produced by the *huichuta* with the shelf life of the pitaya, nor the cross-inoculation of possible bacteria or pathogens between fruits or trees, since the *huichuta* is rarely cleansed or sterilized during the collection process.

Regarding the processing of the fruit, only one of the families anecdotally mentions having prepared ice cream (sweet and frozen syrup-type preparation) with the fruits in a season prior to 2020.

ECONOMIC ANALYSIS

The fruit of the *S. montanus* pitaya represents a considerable economic income, both for collectors and for resellers in the community of Guadalupe. If we take into account the number of dozens collected per family (from 5 to 10) and their sale price, we have that if they deliver it to the reseller, they can obtain from $175 up to $500 mexican pesos per day, taking this balance to the whole season that can last up to 45 days (a "good" season), we have that the family can obtain from $7,875 up to $22,500 mexican pesos (2020 season) and $13,500 to $32,400 mexican pesos (2022 season).

The previous amount is overshadowed if, on the contrary, the families decide to sell their fruits themselves at the foot of the road where they can obtain between $400 - $1000 mexican pesos (2020 season) and $500 - $1200 mexican pesos (2022 season) daily, in exceptional cases where the fruit that is especially large or colorful may fetch better prices. The above amounts represent an income of 76 to 91 daily minimum wages (mexican pesos) per season for families (CONASAMI, 2023).

However, several factors must be taken into account, for example, the sale at the foot of the road, in addition to the man-hours of investment required, carries a risk of losing the quality or the product itself, generating losses for the families. That is why most prefer to secure the sale and deliver their fruits to resellers and not take the risks of selling them themselves.

It is also very important that not all seasons last the same, as was the case in 2019, where the rains came earlier, shortening the season to 24 days, coupled with the above, the harvesting pressure applied to the pitaya trees decreases the number of fruits that can be collected as the season progresses, together with the dwindling trees that have decreased in density due to human activity.

Despite the fact that it is obvious to the collectors that the amount of fruits they collect is decreasing, there are few efforts to help reforestation or recovery of the population density of *S. montanus* in the communities, highlighting that one of the families, on their own initiative, set up a small backyard garden, without much success as some of the "arms" they planted rotted. In the same vein, another impact on the tree is that, although the plant is classified by CONAFOR as "non-timber", the inhabitants of the community make use of it by using their arms to make fences (without productive intentions and even combining them with the arms of other genera of columnar cacti) around their houses.

CONCLUSION

The trend in the expansion of human populations, the change in land use, low ecological recruitment and overexploitation of the pitaya are evident, highlighting the fragility of the relationship between rural communities and the pitaya plant, which added to the increase in the demand for regional markets (which pressure to increase the volumes of collection), are leading the pitaya to a spiral of unsustainable use for the wild populations of this tree.

In the communities of northwestern Sinaloa with a tradition in the use of the pitaya fruit, the knowledge associated with the characteristics of its harvest has been passed down from generation to generation by local families, fitting perfectly into the category of "biocultural heritage" due to the historical relationship that the community has had with the fruit.

Formally establishing the status of "biocultural heritage" could allow the communities that are dedicated to this activity to access strengthening or empowerment programs of the Secretary of Culture or the Secretary of Tourism, and in turn, to a regulation that favors the conservation of the plant and its associated ecosystem.

Another aspect to take into account, but which can be controversial (especially for anthropologists and sociologists) is the modification of the tool they use to gather the fruit, given that the current method damages the fruit by perforating it, which, added to the high temperatures, results in a considerable loss in quality, shelf life and significantly reduces its fermentation time.

If pitaya cultivation plots are established, it is obvious that the gathering tool will respond to their needs, but a discussion with the indigenous leaders and people more faithful to the tradition could lead to the implementation of some modifications to the design of the "hichuta", which would allow those involved to prolong the shelf life and quality of the product.

Since this fruit is produced naturally in the forest of the region, it is of vital importance to permeate the idea of sustainable use of the plant among the community actors that benefit from it. An alternative that has had little to no implementation yet is to start allocating plots of land for S. *montanus*, which would alleviate the pressure exerted on the plants in their natural environment, allowing for the planning, projection of production and sale of the fruit, since it is evident that the current model for gathering and sale is not sustainable long term.

This opens the doors to production development projects/agroforestry crops as the most viable option to ensure the permanence and profitability of the resource, and joining this effort to the Ten-Year

Framework of Programs on Sustainable Consumption and Production Modalities (SDG 12), and its 12.2 goal would outline this activity as a viable alternative to the traditional use of the fruit, through sustainable management and the efficient use of agroecological production models.

Once again, we could discuss whether an agroforestry use would violate tradition and make the traditional collectors less competitive in the market, but given the highly marginalized situation of the municipalities and mountain communities of northwestern Mexico, an opportunity to maximize the income of the families should be seriously considered as the next step for the sustainable development of these communities.

Regarding the economy derived from the sale of the fruit, even in the most conservative scenarios, the daily income exceeds the national minimum wage for a fraction of the time invested, which is indicative of the potential of the production mentioned in the previous paragraph and it does not stop there, but the conditions for the processing of the fruit must be fostered, increasing its shelf life and its sale value, generating more profits, products such as: jam, sauces, ice cream, among others, are suggested.

There is a lot that can be contributed to 12.8 target of the SDGs, providing those involved with the tools and knowledge; for the responsible and sustainable use of the *S. montanus* fruit, which is why this research is only the starting point for future research projects.

ACKNOWLEDEMENTS

The authors would like to especially thank the inhabitants of the Guadalupe Community for their participation and enthusiasm in the project, to all of the students, professors and administrative people at UAIM, campus Choix, who helped with observations and recommendations. Also, special thanks to Dr. Pedro Antonio López de Haro for his help with the review and translation of this chapter.

REFERENCES

AEPA. (2023). *Anuario Estadístico de la Producción Agrícola.* Servicio de Información Agroalimentaria y Pesquera (SIAP). https://nube.siap.gob.mx/cierreagricola/

Araujo, J., & Medina, O. (2008). *Reconocimiento de patógenos asociados al cultivo de pitahaya amarilla (Selenicereus megalanthus Haw.)* [Thesis, Universidad Popular del Cesar. Facultad de Ciencias de la Salud Microbiología].

Arreola, N. H. (1999). Taxonomía del pitayo Stenocereus B.R. In: Pimienta, B. El pitayo en Jalisco y especies afines en México. Universidad de Guadalajara Fundación Produce Jalisco A. C. P. 234

Arreola-Nava, H. J., & Terrazas, T. (2003). Especies de *Stenocereus* con aréolas morenas: Clave y descripciones. *Acta Botánica Mexicana*, (64), 1–18. doi:10.21829/abm64.2003.923

Barthlott, W., & Hunt, D. R. (1993). Cactaceae. In *Flowering plants· Dicotyledons* (pp. 161–197). Springer. doi:10.1007/978-3-662-02899-5_17

Bravo, H. (1978). *Las cactáceas de México* (Vol. I). Universidad Nacional Autónoma de México.

Bravo, H. & Sánchez Mejorada, H. (1978). *Las cactáceas de México* (No. C/583.470972 B7/v. 1/1978).

Bravo-Hollis, H. (1991). *Las Cactáceas De Mexico* (Vol. II). Biblioteca De Ciencias.

CONAFOR. (2018). *Comisión Nacional Forestal.* Conafor. https://www.gob.mx/conafor/articulos/del-cactus-a-la-mesa?idiom=es

CONASAMI. (2023). *Salarios mínimos 2023: Tabla informativa.* Comisión Nacional de los Salarios Mínimos, Secretaría del Trabajo y Previsión Social (STPS). Conafor. https://www.gob.mx/cms/uploads/attachment/file/781941/Tabla_de_Salarios_M_nimos_2023.pdf

CONEVAL. (2018). *Índice de pobreza del Estado de Sinaloa.* Consejo Nacional de Evaluación de la Política de Desarrollo Social. Conveal. https://www.coneval.org.mx/coordinacion/entidades/Sinaloa/Paginas/principal.aspx

CONEVAL. (2022). Consejo Nacional de Evaluación de la Política de Desarrollo Social. Informe de pobreza y evaluación 2022. Sinaloa. Conveal.

Delgado, A., Kondo, T., Imbachi, K., Quintero, E., Manrique, M., & Medina, J. (2010). Biología y algunos datos morfológicos de la mosca del botón floral de la pitaya amarilla, *Dasiops saltans* (Townsend) (díptera: Lonchaeidae) en el Valle del Cauca. *Boletín del Museo Entomológico de la Universidad del Valle, 11*(2), 1–10.

DOF. (2022). Diario Oficial de la Federación. Secretaria de Bienestar. *DECRETO por el que se formula la Declaratoria de las Zonas de Atención Prioritaria para el año 2023.* DOF.

Drennan, P. M. (2009). *Temperature influences on plant species of arid and semi-arid regions with emphasis on CAM succulents. Perspectives in biophysical plant ecophysiology.*

Garcia Muñoz, M. C. (2003). *Pitaya: Cosecha y postcosecha.* Corporación Colombiana de Investigación Agropecuaria.

Geilfus, F. (2002). *80 herramientas para el desarrollo participativo.* IICA.

Gibson, A. C., & Horak, K. E. (1978). Systematic anatomy and phylogeny of Mexican columnar cacti. *Annals of the Missouri Botanical Garden, 65*(4), 999–1057. doi:10.2307/2398781

Gibson, A. C., & Nobel, P. S. (1986). *The Cactus Primer.* Harvard University Press. doi:10.4159/harvard.9780674281714

Hernandez, H. M., & Godinez, H. (1994). Contribución al conocimiento de las cactáceas mexicanas amenazadas. *Acta Botánica Mexicana,* (26), 33–52. doi:10.21829/abm26.1994.690

Hunt, D. R. (1999). Cactaceae checklist. 2a ed. Royal Botanic Gardens Kew and International Organization for Succulent Plant Study. Lemous Limited. Milborne Port.

Hyo Won Choi, H., & Min Chung, I. (2007). The effect of spent mushroom saw dust compost mixes, calcium cyanamide and solarization on basal stem rot of the captus Hyalocereus trigonus caused by Fusarium oxysporum. Crop Protection, 26, 162-168.

INEGI. (2021). *Instituto Nacional de Estadística, Guadalupe, Choix, Sinaloa (250070086).* INEGI. https://www.inegi.org.mx/app/areasgeograficas/?ag=25

ITIS. (2023). *The Integrated Taxonomic Information System.* [Dataset]. ITIS. www.itis.gov

IUCN/PNUMA/WWF. (1991). Caring for the Earth, A Strategy for Sustainable Living. Gland, Switzerland.

Martinez, G. J. C. (1993). *Caracterización de tipos de pitaya Stenocereus griseus Haworth en la Mixteca.* [Thesis, UACH. Mexico].

Mora Castro, D. P. (2011). *El cultivo de Pitahaya en temporada invernal.* ICA. https://www.ica.gov.co/getattachment/bff8ee09-c032-404b 8fcb-8c5f7d72d532/El-cultivo-de-Pitahaya-en-temporada-invernal.aspx

Ortega Hernández, A., León Andrade, M., & Rosas Vargas, R. (2018). *Producción de pitahaya para promover el desarrollo regional y sustentable*. UNAM.

Pimienta, B. E. (1999). El pitayo en Jalisco y especies afines en México. Universidad de Guadalajara. Fundación produce Jalisco, A.C., Guadalajara, Jalisco, México.

Pimienta-Barrios, E., & Nobel, P. S. (1994). Pitaya (Stenocereus spp., Cactaceae): Anancient and the modern fruit crop of Mexico. *Economic Botany*, *48*(1), 76–83. doi:10.1007/BF02901385

Salomón Montijo, B. (2018). *Estudian la fenología y biología reproductiva de la Pitahaya Roja.* Entreredes. http://entreredes.com.mx/estudian-la-fenologia-y-biologia-reproductiva-de-la-pitahaya-roja

Sampieri, R. H. (2018). *Metodología de la investigación: las rutas cuantitativa, cualitativa y mixta.* McGraw Hill México.

Sánchez, M. H. (1982) algunos usos prehispánicos de las cactáceas entre los indígenas de México.[Dirección de Recursos Naturales, Toluca, México]

Sánchez-Mejorada, R. H. (1984). Breves notas sobre la vegetación y las cactáceas de las Islas Marías. Universidad Nacional Autónoma de México. México.

SEDESOL. (2021). *Secretaria de Desarrollo Social. Sistema de Apoyo para la Planeación del PDZP.* SEDESOL. http://www.microrregiones.gob.mx/catloc/LocdeMun.aspx?buscar=1&tipo=nombre&campo=loc&valor=guadalupe&ent=25&mun=007

SIAP. (2017). *Servicio de Información Agroalimentaria y Pesquera. órgano administrativo desconcentrado de la SAGARPA.* SIAP. https://www.gob.mx/siap/articulos/pitaya-y-pitahaya-no-son-lo-mismo-pero-son-igual?idiom=es

UNDP. (2023). *Programa de las Naciones Unidas para el Desarrollo. Objetivos de Desarrollo Sostenible: Producción y consumo responsable.* UN. https://www.undp.org/es/sustainable-development-goals/produccion-consumo-responsables

Vázquez, V. S., & Manuel, H. E. (2009). *Agroindustrialización de pitaya*. Editorial Universitaria.

Chapter 4
Global Impact of Carbon Emissions and Strategies for Its Management

Amar Prakash Garg
https://orcid.org/0000-0003-0613-9495
Swami Vivekanand Subharti University, India

Monika Chaudhary
Shobhit Institute of Engineering and Technology (Deemed), India

Chitwan Garg
ONGC, India

ABSTRACT

Climate change is a major threat to the survival of life on this planet. It is the major cause of environmental damage besides the accumulation of pollutants, increasing plastic, polythene and e-waste, global warming, and increasing carbon emissions. The increasing use of fossil fuels is resulting in increased amounts of carbon emission and greenhouse gases (GHG), rising temperature, toxic gases, increasing temperature, and air pollutants that are released into the atmosphere. The weather has changed significantly over the past five decades and particularly during the last 5-10 years, from polls to the tropics. With its impact on numerous elements of the ecological, environmental, socio-political, and socio-economic disciplines, climate change (CC) is a multilateral complicated international issue. It is a worldwide threat that has started to strain several industries including agriculture, humans, animals, and biodiversity. In addition to the size of the shift, the likelihood that it will become irreversible determines how harmful human-induced climate change will be.

DOI: 10.4018/978-1-6684-9863-7.ch004

INTRODUCTION

Addressing environmental degradation requires reducing the usage of fossil fuels to lower carbon emissions, efficiently managing non-biodegradable materials like e-waste, plastics, and polythene by promoting their reuse, reduction, and recycling. It involves adopting rainwater harvesting to replenish groundwater levels and minimizing the use of systemic antimicrobial to safeguard microbial diversity and soil fertility. The term "carbon emissions" refers to the overall amount of carbon dioxide (CO_2) emissions generated by an individual, entity (*e.g.*, a country, building, or organization). This encompasses both direct emissions from energy production for powering products and services and indirect emissions from the burning of fossil fuels in manufacturing, heating, and transportation. In some cases, the concept of carbon emissions also considers additional greenhouse gas emissions like methane, nitrous oxide, or chlorofluorocarbons (CFCs).

A rising volume of greenhouse gases has been and continues to be discharged into the Earth's atmosphere due to increased fossil fuel consumption and alterations in land use. These gases, such as carbon dioxide (CO_2), methane (CH_4), and nitrous oxide (N_2O), have seen their concentrations surge, resulting in greater retention of solar heat within the Earth's atmosphere, rather than it being reflected away into space. Methane absorbs more heat than carbon dioxide and the accumulation of methane in the atmosphere cause greater harm and enhance global warming. The strategies to reduce the production of methane or its consumption using methanotrophs may be good strategy for management of carbon emissions.

Climate change is the outcome of the greenhouse effect, which was sparked by this rise in temperature. Increases in average global temperature (global warming), changes in cloud cover and precipitation, especially over land, melting of ice caps and glaciers, reduced snow cover, and increases in ocean temperatures and ocean acidity—caused by seawater absorbing heat and carbon dioxide from the atmosphere—are the main features of climate change (UNFCC, 2022).The extensive use of fossil fuels like coal, natural gas, and oil is a primary driver of the ongoing climate change. When these substances are burned, they release greenhouse gases into the Earth's atmosphere. These gases, in turn, trap heat from the sun, causing a gradual increase in the planet's average temperature, which we refer to as global warming. Global warming, in turn, affects local and regional climates. It's important to understand that the Earth's temperature has naturally fluctuated over geological timescales, a process that has occurred over hundreds of thousands of years. However, what sets the current situation apart is the rapid pace of human-induced climate change. Unlike the gradual temperature variations that have historically taken place, the changes we're witnessing today are occurring at a much faster rate due to human activities.

Human activities, such as the burning of fossil fuels and deforestation, have accelerated the release of greenhouse gases, particularly carbon dioxide (CO_2) and methane (CH_4), into the atmosphere. This heightened concentration of greenhouse gases traps more heat, leading to the phenomenon of global warming. The consequences of this accelerated warming are evident in the form of more frequent and severe weather events, rising sea levels, disruptions to ecosystems, and shifts in weather patterns.

Addressing this accelerated climate change requires collective efforts to reduce greenhouse gas emissions, transition to cleaner energy sources, and implement sustainable land-use practices. This is crucial for mitigating the adverse impacts of climate change and preserving a stable and habitable planet for future generations.

Figure 1. Per capita global carbon emission in CO_2 equivalence in USA, China, European Union (EU), India, and rest countries of the world

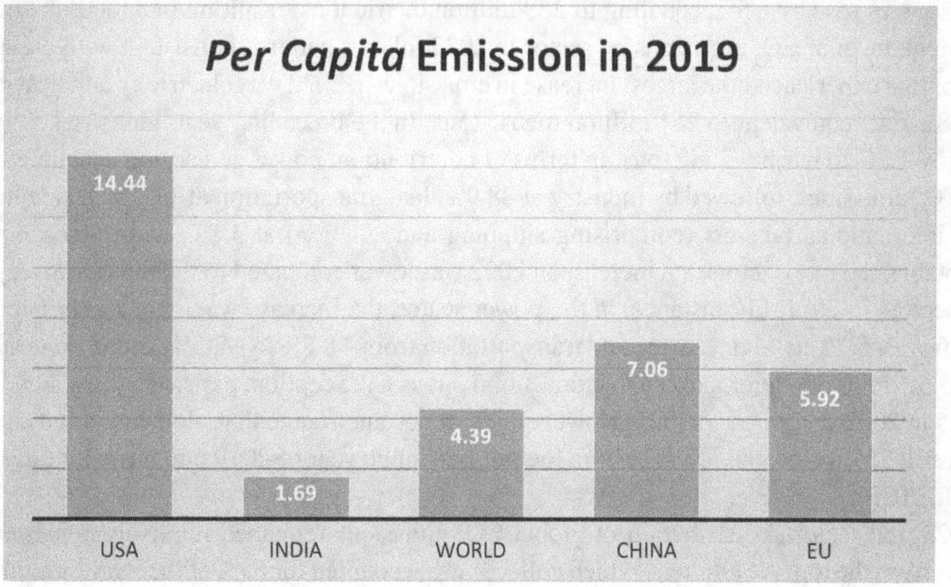

GLOBAL CARBON EMISSION

In 2022, the global emissions of carbon dioxide (CO_2) resulting from activities like cement production and the burning of fossil fuels amounted to 36.1 gig tons of CO_2, with a small margin of error indicated as ±0.3 $GtCO_2$. This represents an increase in emissions compared to previous years, with emissions measured at 35.3 $GtCO_2$ in 2019, 33.4 $GtCO_2$ in 2020, and 35.5 $GtCO_2$ in 2021. These numbers imply an upward trend, with emissions increasing by 2.0% in 2022 when compared to 2019, by 7.9% compared to 2020, and by 1.5% compared to 2021.

The 1.5% increase in emissions from 2021 to 2022, as reported by Carbon Monitor data, falls within a range of 0.9% to 2.6%. This data aligns with other methods of forecasting emissions based on energy consumption. For instance, the Global Carbon Project (GCP) predicted a rise in emissions within the range of 1.0% to 0.9%, while the International Energy Agency (IEA) anticipated a 1.0% increase during the same period. This information indicates that multiple sources and methods of analysis converge on a similar conclusion: global CO_2 emissions increased by approximately 1.5% to 2.0% in 2022, which is consistent with the ongoing concern regarding the rise in carbon emissions and its impact on climate change.

In 2022, global CO_2 emissions from energy sources saw a notable increase of 0.9%, equivalent to 321 million metric tons, reaching a new high of over 36.8 gig tons. This rise was influenced by various factors. Out of the 321 million metric ton increase, approximately 60 million metric tons can be attributed to the greater demand for cooling and heating during extreme weather conditions, while another 55 million metric tons can be linked to the shutdown of nuclear power plants. There was a 1.6% reduction, amounting to 118 million metric tons, in CO_2 emissions from natural gas. However, this decline in natural gas emissions was outweighed by the increased emissions from coal. CO_2 emissions from coal sources increased by 1.6%, totaling 243 million metric tons during the global energy crisis. This exceeded the average growth rate seen in the previous decade and reached a new all-time high of over 15.5 gig tons. This surge was primarily due to a shift from using natural gas to coal. Comparatively, CO_2 emissions from oil sources rose by 2.5%, equating to 268 million metric tons, resulting in a total of 11.2 gig tons.

The breakdown of CO_2 emissions by sector in 2022 follows patterns consistent with previous years. The sector that experienced the largest increase in emissions in 2022 was electricity and heat generation, with a 1.8% rise, equivalent to 261 million metric tons. In the preceding year, industrial emissions had declined by 1.7% to reach 9.2 gig tons. In terms of contribution, power generation accounted for 39.3% of total CO_2 emissions, followed by industry at 28.9%, land transportation at 17.9%, residential sources at 9.9%, international bunkers (comprising shipping and aviation) at 3.1%, and domestic aviation at 0.9%. Notably, the rate of emission increase in 2022 was lower compared to the rapid recovery observed in most sectors in 2021. For instance, in the power sector, the increase went from +6.9% to +0.8%, in industry from +5.7% to +1.1%, in ground transportation from +8.8% to +2.5%, and in domestic aviation from +25.5% to -0.9%. International aviation stood out as an exception, experiencing a 44% increase in emissions in 2022 compared to 2021. However, it's important to note that global aviation carbon emissions are still 25% below the levels seen in the pre-pandemic year of 2019 and have not fully recovered (Liu et al., 2023).

In 2022, the sectorial distribution of global CO_2 emissions remained relatively consistent, but the emissions from the top five emitters, which collectively accounted for 65% of the world's total emissions (23.3 gig tons of CO_2), showed significant changes. China, the largest emitter on the planet, experienced a notable shift in its emissions trajectory. Despite the challenges of the pandemic, its CO_2 emissions had increased by 1.2% in 2020 and a substantial 6.0% in 2021. However, in 2022, there was a surprising decline of 1.5% in emissions. Nevertheless, it's essential to note that even with this reduction, China's 2022 emissions were still 5.6% higher than they were in 2019, before the COVID-19 outbreak.

Figure 2. Total carbon emission country-wise

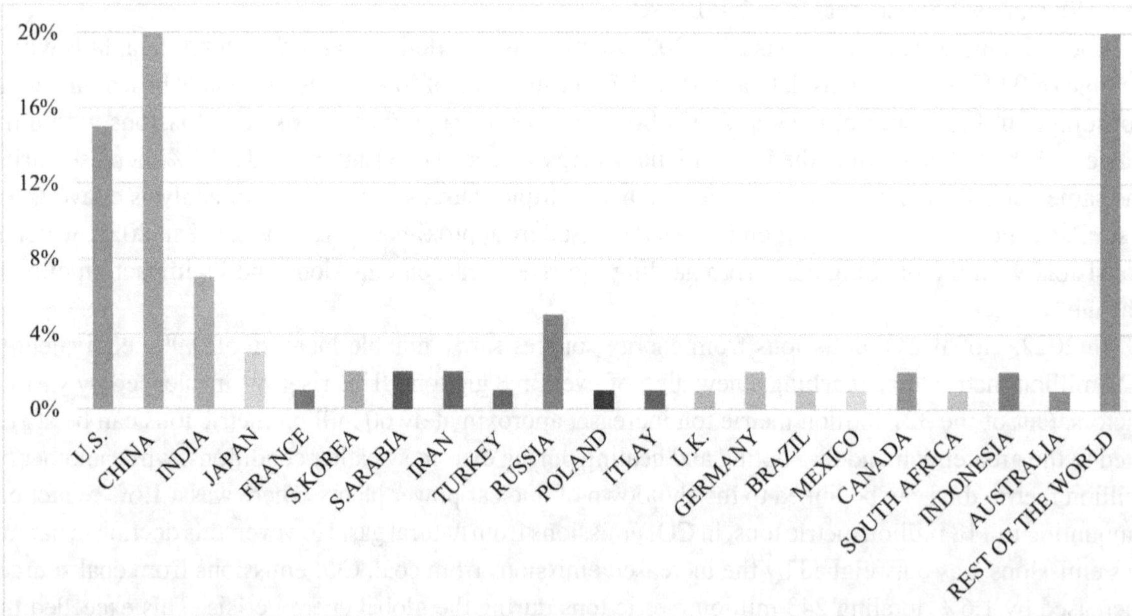

In contrast, the United States and the European Union (EU) saw their emissions increase in 2022. The USA experienced a 3.2% rise in emissions, while the EU had a more modest 0.5% increase. As a result, their emissions in 2022 were very close to the levels seen in 2019, with the USA having just 0.9% less emissions, and the EU having a 0.4% increase compared to 2019.

India's emissions are on a growth trajectory and are expected to surpass the EU as the third-largest emitter in the world in 2023. In 2022, India's emissions increased by 7% compared to 2021, and they were 7.9% higher than they were in 2019. Russia, the fifth-largest emitter globally, experienced yearly increases in emissions from 2019 to 2021. However, in 2022, there was a decline of 1.8% in its emissions. These shifts in emissions among the top emitters reflect the complex interplay of factors, including COVID-19 policies, economic conditions, and energy choices, in shaping global CO_2 emissions patterns (IEA, 2023).

INDIA EMISSIONS

According to a study presented at the United Nations Framework Convention on Climate Change's 27th Conference of Parties (COP27), India's emissions are expected to rise by 6% (within a range of 3.9% to 8%), mostly due to a 5% increase in coal emissions. India already accounts for about 12% of global emissions and is ranked third internationally in terms of total emissions, although having far lower per-person emissions. Oil-related emissions are expected to climb considerably, by 10%, and are likely to reach 2019 levels again. Natural gas emissions are expected to decrease by 4%, but as gas makes up a very tiny portion of India's energy mix, its impact on the overall reduction would be minimal. Cement emissions, which make up 5% of world emissions, are anticipated to decline globally but to rise in India.

India's per capita greenhouse gas emissions in 2020 were 2.4 tCO_2e (tonne carbon dioxide equivalent), which was significantly lower than the global average of 6.3 tCO_2e. At 2.4 tCO_2e, India continues to be considerably below the global average. Least developed nations emit 2.3 tCO_2e yearly per person on average (The wire, 2022).

OTHER GHG EMISSIONS GLOBALLY

Greenhouse gases, including carbon dioxide (CO_2), methane, and nitrous oxide, play a crucial role in maintaining Earth's temperature by preventing it from getting too cold. These gases affect the way energy moves into and out of our planet's atmosphere, essentially acting as a natural insulating layer. When solar energy from the sun reaches Earth, it primarily arrives in the form of visible light. However, the energy that leaves the Earth is in the form of infrared radiation, which we perceive as heat. Greenhouse gases present in our atmosphere interact with this outgoing heat. They have the property of reflecting some of this infrared radiation back towards the Earth's surface, effectively trapping heat.

The term "greenhouse effect" is derived from the way a greenhouse works. In a greenhouse, glass walls and roof allow sunlight to enter but prevent much of the heat from escaping. Similarly, the presence of greenhouse gases in the Earth's atmosphere acts as a metaphorical 'blanket' that traps some of the heat, which would otherwise escape into space. This phenomenon is essential for maintaining a stable and habitable temperature on our planet. However, an excessive increase in greenhouse gases, primarily due to human activities, can intensify this effect, leading to global warming and potentially harmful climate changes (MIT, 2023).

In recent decades, with the growth of the world's population and economy, there has been a significant surge in energy consumption. This increased demand for energy, primarily met by burning fossil fuels, has led to a substantial rise in global greenhouse gas (GHG) emissions. These emissions now exceed 50 billion metric tonnes of carbon dioxide equivalent ($GtCO_2e$) annually. Through their ability to absorb solar energy and stop it from escaping into space, GHGs—which may persist in the atmosphere for centuries—trap heat within the Earth's atmosphere. In order to keep global temperatures from rising more than 1.5 degrees Celsius over pre-industrial levels and to lessen the worst effects of climate change, global emissions must be drastically reduced during the ensuing decades.

While CO_2 is the GHG that receives the most attention, yearly emissions of methane and the frequently disregarded nitrous oxide have also risen to historic levels. Climate experts are extremely concerned about the rise in emissions of these two GHGs since they have a larger potential to cause global warming than CO_2. One of the key objectives of recent climate commitments is the decrease of methane emissions through the end of the decade. As a result, initiatives like the Global Methane Pledge were formed in an effort to reduce emissions of this dangerous gas by 30% from 2020 levels by 2030.

While there are various sources of greenhouse gas (GHG) emissions, coal-fired energy stands out as the most prominent contributor, releasing approximately eight gig tons of CO_2 equivalent ($GtCO_2$) each year. China is a significant player in the global coal-fired energy sector, making it the primary source of these emissions. This has resulted in China surpassing the United States as the world's leading greenhouse gas emitter, largely due to its increased reliance on coal over recent decades. However, it's worth noting that the United States holds a unique position in the historical context of emissions. It has

Table 1: Carbon dioxide equivalent levels in nature (IPCC, 2021)

S.NO	Greenhouse gas	CO_2 equivalent	Description
1	Water vapour	0	Water vapour, which is present in clouds, is really the most prevalent greenhouse gas. However, we don't need to be concerned about our "water emissions" because water vapour swiftly turns into rain. Warmer air, on the other hand, has a greater capacity to store water vapour without producing a downpour. We will thus tend to have more water in the atmosphere at a given period when the globe heats, which warms the earth.
2	Methane	28	Methane, the primary component of natural gas, is the second-leading contributor to climate change. Methane reflects heat at a rate of around 100 times that of CO_2, although it stays in the atmosphere for just about 10 years. Methane emissions are particularly challenging to monitor since they mostly come from sources other than industrial units. Instead, they originate from livestock, alterations to wetlands and woodlands, and gas well and pipe leaks.
3	Nitrous Oxide	273	A potent greenhouse gas that stays in the atmosphere for more than 100 years is nitrous oxide. The commercial application that gives it its most well-known name, laughing gas, accounts for a very small portion of our emissions. By far, using nitrogen-based fertilizers to produce crops is the main source of nitrous oxide that humans release into the atmosphere.
4	Fluorinated gases	Up to 14,600	Fluorinated gases, most frequently hydro fluorocarbons, or HFCs, are employed in a number of industrial operations as well as in refrigerators and air conditioners. There are several of these gases, and the majority of them have a significant warming effect; for example, HFC-23 has a CO_2 equivalent of 14,600.

generated more than 500 $GtCO_2$ from the use of fossil fuels and land since 1850, making it the largest historical emitter in terms of cumulative emissions (Statista, 2023).

The other greenhouse gases are also significant, even though CO_2 is the main factor in climate change that is caused by humans. They originate from various places, persist in the atmosphere for various lengths of time, and may or may not be more or less effective at trapping heat. The measurement of greenhouse gases is often done in "CO_2 equivalents" (CO_2e). An equivalent quantity of CO_2 would be predicted to trap one CO_2e of heat during the following 100 years.

INDIAS GHG EMISSION

India has seen a substantial increase in its greenhouse gas (GHG) emissions in recent decades due to its rapidly growing population and expanding economy. As of 2021, India stands as the third-largest global emitter of GHGs, following China and the United States, with emissions amounting to 3.9 billion metric tonnes of carbon dioxide equivalent ($GtCO_2e$), which represents about 7% of the world's total emissions. It's important to note that India's per capita GHG emissions are notably lower than the global average. With only 2.8 metric tonnes of CO_2 equivalent per person, India's per capita emissions are considerably lower than those of China and the United States, which stand at 9.6 and 17.6 metric tonnes per person, respectively. Furthermore, when considering the historical context, India's cumulative contribution to CO_2 emissions is less than 4%, significantly lower than that of other major economies. This reflects India's relatively lower historical impact on global emissions, despite its current position as a significant emitter.

The most carbon-intensive fossil fuel and the driving force behind India's economic expansion has been coal. India's power mix includes more than 70% coal, and in 2021 the country's coal combustion produced 1.8 $GtCO_2$. The main source of GHGs in India comes from the electric power sector due to the enormous amount of CO_2 that coal-fired power plants release. The agriculture industry, which generates significant volumes of methane (CH_4) from livestock and rice fields, is the second-largest source. CH_4 is a powerful GHG with a global warming potential (GWP) of around 30 times that of CO_2 after 100 years.

In 2021, India's methane emissions peaked at 897 million metric tonnes of carbon dioxide equivalent ($MtCO_2e$). Since 1990, methane emissions in India have grown by more than 30%. The second-largest global emitter of methane is India.

Nitrous oxide (N_2O) emissions in India peaked in 2021 at 272.3 million metric tonnes of carbon dioxide equivalent. Since 1990, nitrous oxide emissions in India have climbed by over 90%. The agriculture industry in India is the biggest contributor of nitrous oxide emissions.

CARBON EMISSION AND CLIMATE CHANGE

The excessive release of carbon dioxide into the atmosphere has a profound and multifaceted impact on our planet, largely contributing to climate change and global warming. These changes have significant consequences, not only for the environment but also for agriculture and wildlife. Rising levels of atmospheric carbon dioxide can have both positive and negative effects on crops. On the positive side, increased carbon dioxide can enhance water-use efficiency in crops and, to some extent, mitigate the adverse effects of climate change on crop yields. However, there are downsides as well. Elevated carbon dioxide levels can disrupt nutrient balances, resulting in reduced essential elements like iron, zinc, and protein in crops. This can lead to potential nutritional deficiencies and health issues in regions that heavily rely on these crops for sustenance.

Climate change, largely driven by carbon emissions, is one of the most significant global challenges. As the average global temperature increases due to the accumulation of greenhouse gases, we witness a warming of the planet. This warming, in turn, leads to a range of extreme weather events, including tropical storms, wildfires, prolonged droughts, and heat waves. These events have devastating effects on ecosystems, human settlements, and agriculture. The impact on agriculture is particularly concerning. While increased atmospheric carbon may offer some benefits to plant growth, climate change can alter landscapes and weather patterns, leading to conditions that are unfavourable for crops. For example, more frequent droughts and erratic weather can reduce agricultural yields, affecting food production and food security. Wildlife is also severely affected by climate change. Changes in temperature and habitat disruption can lead to the decline and even extinction of certain species. Some may not be able to adapt to the changing conditions, while others might thrive and out compete native species, leading to shifts in ecosystems and biodiversity. The excessive carbon emissions leading to climate change have wide-ranging impacts on our planet, affecting everything from agriculture to wildlife. While some aspects of plant growth may benefit from increased carbon dioxide, these potential advantages are outweighed by the negative consequences of climate change, which threaten the stability of ecosystems, the well-being of communities, and the future of our planet.

Climate change refers to the long-term alterations in temperature and weather patterns. While such variations can be triggered by events like major volcanic eruptions or fluctuations in the sun's activity, the primary driver of climate change since the 1800s has been human activities, particularly the burning of fossil fuels such as coal, oil, and gas. This combustion process releases greenhouse gases into the atmosphere, which act like a blanket, trapping heat from the sun and causing global temperatures to rise. The main culprits among these greenhouse gases are carbon dioxide and methane, both of which significantly contribute to climate change. To elaborate, when fossil fuels like coal, oil, and gas are burned for various purposes, including electricity generation, heating buildings, and fueling vehicles, they release carbon dioxide into the atmosphere. Carbon dioxide is a potent greenhouse gas that effectively traps heat, leading to a gradual increase in global temperatures. Additionally, when forests are cleared or land is modified for various human activities, it can release stored carbon into the atmosphere, further contributing to carbon dioxide emissions. Methane, another important greenhouse gas, is released during various human activities, such as agriculture (livestock digestion and rice paddies), the production and transportation of natural gas and oil, and the decay of organic waste in landfills. Methane is even more effective at trapping heat than carbon dioxide, although it is less prevalent in the atmosphere. The consequences of rising global temperatures due to these greenhouse gas emissions are far-reaching and include more frequent and severe weather events, shifts in ecosystems, sea-level rise, and a variety of impacts on human societies, from food security to health concerns. Addressing climate change involves reducing greenhouse gas emissions, transitioning to cleaner energy sources, and adopting sustainable practices to mitigate the ongoing effects of human-induced climate change.

The climate change has been mainly correlated with carbon emissions because the increasing use of fossil fuel. There is a mad race for the reduction of use of fossil fuel with an aim to reduce carbon emissions without thinking on the disadvantages of such technologies like the use of lithium ion batteries in electric cars. The disposal of lithium ion batteries in nature is highly dangerous like the plastic, polythene and e-waste.

Climate change is already affecting every corner of the Earth, and as the planet continues to warm, these impacts will become more pronounced and widespread. These effects go beyond just rising temperatures and encompass a range of changes in various regions, which will intensify with further warming. Here are some examples:

1. Climate change is accelerating the water cycle, leading to more frequent and severe floods due to increased rainfall. At the same time, it is causing drought in many areas.
2. Rainfall patterns are shifting, with a decrease in precipitation expected in most subtropical regions and an increase at higher latitudes. Monsoon patterns are also projected to change, with regional variations.
3. Sea levels are on the rise throughout the 21st century, resulting in coastal erosion and more frequent, severe flooding in low-lying areas. By the end of this century, extreme sea-level events that used to occur once every century may become an annual occurrence.
4. Warming is accelerating the loss of seasonal snow cover, thawing of permafrost, melting of glaciers and ice sheets, and the reduction of summer Arctic sea ice.
5. Human-induced factors, such as ocean warming, increased marine heat waves, ocean acidification, and declining oxygen levels, are having a direct impact on ocean ecosystems and the communities that depend on them, and these changes will persist for at least the remainder of the century.
6. Climate change can exacerbate certain effects in cities, such as urban heat islands, intense precipitation events leading to flooding, and sea-level rise affecting coastal cities.

Climate change is a global phenomenon with far-reaching consequences, affecting everything from the frequency of extreme weather events to the stability of ocean ecosystems. As global temperatures continue to rise, these impacts will become more severe and widespread. (IPCC, 2021).

CAUSES OF CARBON EMISSION

The amount of carbon emissions (CO_2 emissions) in the atmosphere of the Earth is increasing. Global temperatures are rising as a result of these increased carbon emissions, which are also altering the planet's climate. We must first comprehend the origins of carbon emissions, both natural and human-made, in order to fully comprehend their effects (Greentumble, 2023).

HUMAN CAUSES OF CARBON EMISSION

Burning the stored energy in fossil fuels over the past 150 years has allowed humanity to produce many of the amazing things that make up our modern world. The Earth is unable to absorb all of the carbon emissions that humans are producing at the rate that fossil fuel development and use have been taking place. The truth is that carbon dioxide emissions from human activities exceed 35 billion metric tonnes yearly, which is about 135 times greater than the annual emissions from volcanoes, which range from 0.13 to 0.44 billion metric tonnes.

1. Fossil Fuel Burning

When fossil fuels are used for different purposes, such as producing electricity, propulsion for cars, and providing heat for commercial and domestic usage, carbon dioxide is released. Carbon has been stored as fossil fuels for millions of years by ancient plants and animals. The use of fossil fuels for electricity and heat production is mostly responsible for global emissions. The bulk of the world's power is still produced by burning coal, oil, or gas, which releases carbon dioxide and nitrous oxide, two strong greenhouse gases that envelop the earth and trap solar radiation. Renewable energy sources, such as wind, solar, and other natural resources, provide slightly more than 25% of the world's power and produce very little to no greenhouse gases or other air pollutants as compared to fossil fuels (UN, 2020).

Emissions of greenhouse gases occur not only when oil and gas are burned as fuel but also at various stages of their production and consumption. Methane, in particular, is a significant emission associated with oil and gas activities, from drilling and fracking to transportation and refining. This makes fossil fuel development a major contributor to methane emissions. While methane is less prevalent in the atmosphere than carbon dioxide, it is far more potent in trapping heat during the first 20 years after its release. Furthermore, even inactive and abandoned wells, often referred to as "orphaned" wells, continue to release methane. These outdated and abandoned wells, numbering more than 3 million across the country, released over 280,000 metric tonnes of methane in 2018 (NDRC, 2022).

2. Deforestation

Deforestation is a major driver of greenhouse gas emissions and has significant implications for our climate and the environment. Here's a more detailed elaboration on the process and its consequences:

- **Carbon Release:** Trees play a vital role in the carbon cycle. They absorb carbon dioxide (CO_2) from the atmosphere through the process of photosynthesis and store it as carbon in their biomass. When forests are cut down, the carbon stored in trees is released into the atmosphere as carbon emissions, primarily in the form of CO_2. This release of stored carbon contributes to the increase in greenhouse gas concentrations, which in turn leads to global warming and climate change.
- **Burning Forests:** In addition to deforestation, another concerning practice is the intentional burning of forests, often used to clear land for agriculture or other purposes. This practice releases massive amounts of carbon into the atmosphere. It's estimated that around 12 million hectares of forest are intentionally burned each year, further exacerbating carbon emissions.
- **Reduced Carbon Sequestration:** The removal of trees and forests not only results in immediate carbon emissions but also diminishes the Earth's ability to capture and store CO_2. Healthy forests continue to absorb carbon from the atmosphere, helping to offset emissions from various sources. However, when forests are cleared, the planet's natural carbon-absorbing capacity is compromised.
- **Land Use Changes:** Deforestation, combined with changes in land use, such as converting forests into agricultural land or urban development, contributes to a significant portion of global greenhouse gas emissions. These changes can lead to additional emissions from factors like soil degradation and shifts in land management practices.
- **Impact on Climate:** Deforestation is a significant driver of climate change. The increased concentration of CO_2 in the atmosphere acts as a heat-trapping blanket, causing global temperatures

to rise. This warming has a range of cascading effects, including more frequent and severe weather events, disruptions to ecosystems, and threats to biodiversity.

- **Importance of Carbon Sinks:** Forests act as critical carbon sinks, balancing the global carbon cycle. They help absorb and store CO_2, mitigating the impact of human emissions. When forests are intact and well-maintained, they continue to absorb carbon, which is crucial for keeping greenhouse gas concentrations in check.

Deforestation not only leads to immediate carbon emissions but also weakens the Earth's natural mechanisms for capturing and storing carbon. This has far-reaching consequences for climate change and underscores the importance of forest conservation, reforestation, and sustainable land management practices in mitigating the impacts of climate change and reducing greenhouse gas emissions.

3. Agriculture

Agricultural practices emit CO_2 and other greenhouse gases through various farming practices, which is the major method that agricultural activities contribute to carbon emissions.

- **Changing of land:** \When natural ecosystems like forests and grasslands are turned into agricultural fields, the carbon that has been stored in the trees and vegetation is released. Important carbon sinks are lost as a result of this process, which also leads to deforestation.
- **Management of soil:** Tilling and other soil management techniques, such as ploughing, can cause the soil to emit CO_2. By exposing organic material to oxygen, these practices hasten the breakdown process and raise carbon dioxide emissions.
- **Burning of biomass:** Biomass is sometimes burned during agricultural practices to clear land or manage waste. Carbon dioxide and other greenhouse gases are released into the environment when biomass is burned.

The crucial but fragile relationship between soil and climate has been drastically altered by the rise of modern, industrialized agriculture, to the point where, in 2020, agriculture accounted for 11% of greenhouse gas emissions in the United States. This sector is especially well-known for releasing large amounts of methane and nitrous oxide, two strong gases that are good heat trappers. Agriculture is the primary source of nearly three-quarters of the nitrous oxide found in our environment because of the widespread use of artificial fertilizers and certain crop-management practices that prioritise high yields over soil health. Methane is emitted into the atmosphere during the digestion of cattle and other ruminants, and industrialized large-scale livestock production continues to be a significant source of this gas (NDRC, 2022).

However, ranchers and farmers are drawing our attention to the fact that there are other methods to feed the world, especially indigenous farmers who have been managing the land using sustainable practices. By using the principles and practices of regenerative agriculture, we may increase our soil's capacity to retain carbon from the environment, reduce emissions from this business, and create healthier foods.

The burning of agricultural biomass in the fields and the fires in forests also add carbon emissions for which the farmers should be adequately educated to encourage natural composting of lignocarbohydrates in the field using mixed microbial cultures (Sahrawat and Garg, 2023) and there must be international task force to control forest fires using modern technologies.

4. Industry

Fossil fuel combustion for energy production and other manufacturing processes are the main sources of CO_2 emissions from industrial activity. For their energy demands, many businesses rely largely on fossil fuels. In factories and other industrial settings, these fuels are burnt to produce steam, heat, and electricity. As a by-product of combustion, CO_2 is released. A chemical process called as calcination is used to produce cement, an essential building material. Calcination is the heating of limestone (calcium carbonate) to create lime (calcium oxide), which results in the release of carbon dioxide.

Carbon dioxide can be released as a by-product of several industrial processes, including the manufacture of steel, chemicals, and refineries. These emissions may result from a variety of chemical processes and energy-consuming procedures.

Transport and shipping are frequently used by industries to transfer products and raw materials. These activities frequently require using fossil fuels in cars, boats, and aircraft, which produces more CO_2 emissions.

To lessen their carbon footprint, several businesses are attempting to embrace greener technology and procedures. Some of the tactics being pursued to reduce industrial emissions and encourage a greener and more sustainable industrial sector include switching to renewable energy sources, increasing energy efficiency, implementing carbon capture and storage (CCS) technologies, and adopting more sustainable manufacturing processes.

5. Waste Management

The amount of garbage we create is becoming a bigger issue as the world's population quickly approaches 8 billion.

As a result of the waste, carbon dioxide emissions may be released, raising atmospheric levels of greenhouse gases. This gas is released as a result of various waste kinds and disposal techniques. Burning rubbish as part of the waste management process known as incineration has the potential to emit carbon dioxide into the environment. Even though incineration is seen as a type of energy recovery, it still emits carbon dioxide and can worsen air pollution. Waste, including plastic and other items, is sometimes burnt outdoors or in unattended fires. These activities cause the air to become contaminated with dangerous chemicals like carbon dioxide.

On average, each individual on the earth produces 0.74 kilogrammes of garbage per day. The World Bank's estimate, however, takes into account statistics ranging from 0.11 kilogrammes to 4.54 kilogrammes, with high-income nations accounting for more than a third of the 2.01 billion tonnes of garbage produced globally each year. Methane is one of the worst by-products produced by the waste that we squander. It is up to 16% more powerful than CO_2 and contributes to global greenhouse gas emissions. About 30% of the increase in global temperatures since the industrial revolution may be attributed to methane. 20% of the world's methane emissions and 3.3% of its greenhouse gas emissions come from the waste industry (WEF, 2022).

6. Transportation

Vehicles, lorries, ships, and aeroplanes are frequently powered by fossil fuels. As a result, the transportation industry has a big impact on greenhouse gas emissions, especially carbon dioxide emissions. They comprise the majority because internal combustion engines, which are utilized in road vehicles, use fuels derived from petroleum, such as petrol. Ship and aeroplane emissions are still increasing, though. Transportation accounts for the bulk of carbon dioxide emissions related to energy. Furthermore, data indicates that energy use for transportation will increase dramatically over the next few years.

7. Food Production

Apart from deforestation, converting land for farming and livestock grazing, the digestive processes of cows and sheep, the use of fertilizers and manure in crop cultivation, and the energy consumed by farming equipment and fishing vessels, often powered by fossil fuels, all have a collective impact on food production. These activities result in the release of carbon dioxide, methane, and other greenhouse gases. As a consequence, the food production sector plays a substantial role in driving climate change. Moreover, the distribution and packaging of food also add to greenhouse gas emissions. To mitigate this, the food industry must adopt modern technology and practices to reduce carbon emissions.

8. Consuming Too Much

Changes in your lifestyle, energy use, food consumption, waste production, and mobility may all help lower greenhouse gas emissions. The usage of items like clothing, electronics, and plastics may be justified in a similar way. A large amount of the greenhouse gas emissions in the globe are caused by private homes. The environment is significantly impacted by the way we live. The wealthiest 1% of people on Earth bear the brunt of greenhouse gas emissions as their combined emissions are higher than those of the poorest 50%.

NATURAL CAUSES OF CARBON EMISSION

1. Decomposition

Different microorganisms, including bacteria and fungus, break down dead plants, animals, and other organic substances during decomposition. The carbon that is included in these organic compounds is released as carbon dioxide when they decompose. The reason for this is that the microbes create CO_2 as a by-product of their living activities, much like people and other animals do when they breathe.

Nutrient cycling in ecosystems depends on the decomposition process. It facilitates the recycling of vital components like carbon, nitrogen, and phosphorus so that new plants and creatures can use these resources for survival. The carbon cycle is completed when dead organic matter is broken down and carbon is released back into the atmosphere.

2. Ocean Outgassing

While the seas are essential for removing carbon dioxide from the atmosphere, under some circumstances they may also emit CO_2.

As a large carbon sink, the seas take in a lot of carbon dioxide from the atmosphere. Seawater contains CO_2 that dissolves to create carbonic acid, which is subsequently broken down into bicarbonate and carbonate ions. This procedure, sometimes referred to as oceanic carbon absorption or ocean carbon sequestration, aids in controlling the atmospheric concentration of carbon dioxide. The seas, however, do not solely act as a one-way carbon sink. When certain conditions are met, they return carbon dioxide to the atmosphere in select areas. This happens when the ocean's circulation patterns alter or the water gets warmer.

CO_2 is released into the atmosphere as a result of the water's decreased ability to contain dissolved gases when it heats. Furthermore, regions with up welling currents can transport dissolved carbon-rich deep ocean waters to the surface, which aids in out gassing.

For the general maintenance of the carbon cycle and control of atmospheric carbon dioxide levels, the equilibrium between carbon dioxide absorption and out gassing in the seas is essential. The oceans have the ability to release CO_2 in reaction to changes in temperature and circulation patterns, even though they are currently absorbing a sizable percentage of CO_2 emissions produced by humans.

3. Volcanic Activity

Molten rock, known as magma, rises to the surface as volcanoes erupt. Gases have been dissolved in this magma, including carbon dioxide, sulphur dioxide, and water vapour. The pressure drops as the magma rises to the surface, resulting in the release of the gases as volcanic emissions.

In comparison to emissions caused by human activity, the quantity of CO_2 released by volcanic activity is rather tiny. Around 200 million to 300 million metric tonnes of carbon dioxide are thought to be released annually from volcanic eruptions, which is considerably less than the billions of metric tonnes released annually by human activity.

4. Wildfires

Carbon emissions come naturally from wildfires. The carbon held in plants and organic matter is released as CO_2 when wildfires consume that material.

Similar to burning fossil fuels or other organic materials, the burning process that occurs during wildfires is fundamentally combustion. The carbon in the trees, plants, and other flora that the fire burns reacts with the oxygen in the air to form carbon dioxide. This procedure returns the carbon that has been stored to the environment. Wildfires have happened throughout the Earth's history and are a natural component of many ecosystems. However, factors like as climate change, land management techniques, and human activities can affect their frequency and intensity. Due to a combination of these reasons, wildfires have recently grown more intense and widespread in some locations.

The natural carbon cycle includes the carbon dioxide that is emitted as a result of wildfires. The carbon released during wildfires is eventually reabsorbed by new vegetation when it regrows in healthy ecosystems. Wildfires, however, can overrun the natural carbon balance and result in a net increase in atmospheric carbon dioxide when they occur too frequently or intensely.

EFFECTS OF CARBON EMISSION

Figure 3. Effects of carbon emission

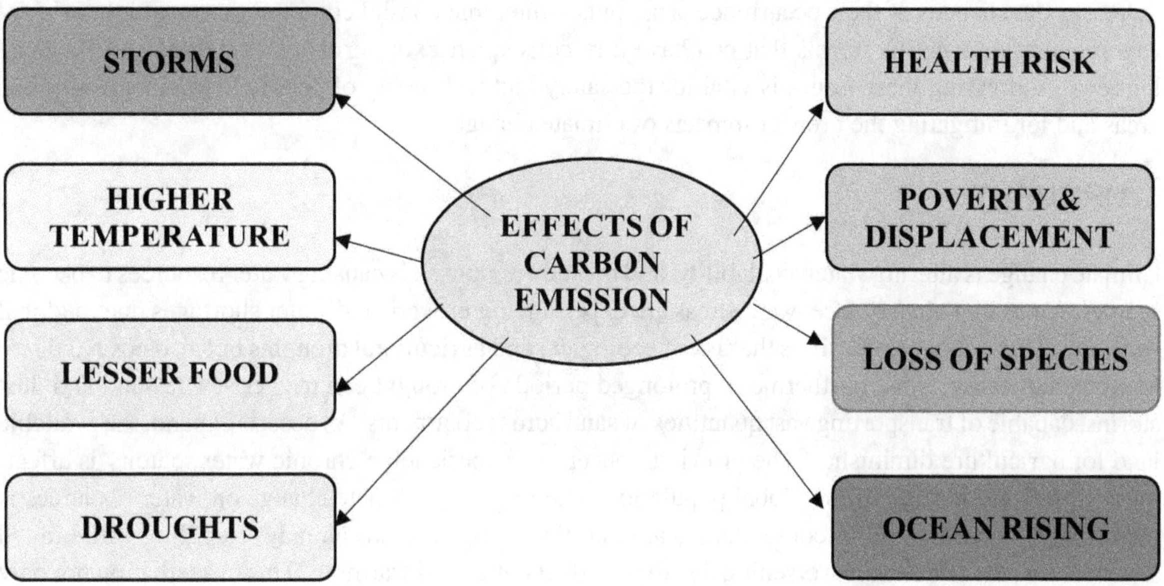

1. HIGHER TEMPERATURE

Global temperatures have been increasing alongside the build-up of greenhouse gases in the atmosphere. The period from 2011 to 2020 stands out as the warmest decade ever recorded. In fact, each successive decade since the 1980s has been hotter than the preceding one. Most regions are experiencing an increase in the number of hot days and heat waves. Elevated temperatures are worsening heat-related health issues and making outdoor work more difficult. Warmer conditions also contribute to the easier ignition and rapid spread of wildfires. Notably, the Arctic has warmed at a rate at least twice as fast as the average global warming rate.

2. STORMS

Destructive storms have become more frequent and intense in various regions. This escalation is often linked to rising temperatures, which cause increased evaporation of moisture, leading to more severe rainfalls and subsequent flooding. The warming of ocean waters also plays a crucial role in influencing the characteristics of tropical storms, affecting both their frequency and intensity. Warm surface waters of the oceans serve as the primary source of energy for cyclones, hurricanes, and typhoons, making them more powerful and frequent. Consequently, these storms often wreak havoc by causing extensive damage to homes and communities, resulting in loss of life and substantial economic consequences.

The connection between climate change and the increased severity of storms is a concerning trend. As temperatures rise, the atmosphere can hold more moisture, which, when released, leads to heavy and prolonged rainfall events. This, in turn, can trigger more frequent and devastating floods. The warming of the oceans is particularly worrisome, as it not only fuels the intensity of tropical storms but also increases the chances of their occurrence. This puts vulnerable coastal communities at a higher risk of experiencing destructive storms that can have dire consequences in terms of casualties and financial burdens. Addressing these issues is vital for the safety and well-being of people living in storm-prone areas and for mitigating the broader impacts of climate change.

3. DROUGHTS

Climate change is altering water availability, and in many regions, it is causing water resources to become scarcer. Areas that already face water stress are experiencing exacerbated water shortages due to global warming. This not only intensifies the risk of ecological and agricultural droughts but also poses a threat to crops and ecosystems. Furthermore, prolonged periods of drought can trigger severe sand and dust storms, capable of transporting vast quantities of sand across continents. As deserts expand, the available land for agriculture diminishes. The growing concern of experiencing chronic water scarcity is affecting a significant portion of the global population. The impact of climate change on water resources is increasingly apparent, and its consequences are multifaceted. In regions already struggling with limited water access, the situation is worsening due to the effects of global warming. This exacerbation not only leads to more frequent water shortages but also raises the risk of ecological and agricultural droughts. These prolonged drought periods can have devastating effects on both crops and vulnerable ecosystems.

One concerning consequence of this changing water landscape is the potential for severe sand and dust storms. Droughts create the conditions for these storms, which can carry enormous amounts of sand and dust across entire continents. Not only do these events have immediate health and environmental impacts, but they also contribute to the expansion of deserts, further limiting the available land for agriculture. The fear of regularly facing water scarcity is a growing concern that affects a substantial portion of the global population. As such, it underscores the urgency of addressing climate change and developing sustainable water management strategies to ensure a more secure and resilient future.

4. OCEAN RISING

The majority of the heat generated by global warming is taken up by the Earth's oceans. Over the last two decades, there has been a substantial increase in ocean warming across all depths. As the temperature of water rises, it expands, leading to an overall growth in the volume of the ocean. This expansion, coupled with the melting of ice sheets, contributes to rising sea levels, posing a significant threat to coastal and island communities. Furthermore, the oceans absorb carbon dioxide from the atmosphere, which helps mitigate its effects. However, the excessive absorption of carbon dioxide makes the water more acidic, putting coral reefs and marine ecosystems in jeopardy.

5. LOSS OF SPECIES

Climate change poses a significant risk to both terrestrial and marine animal species. As temperatures continue to increase, these threats are becoming more pronounced. The current rate of extinction on our planet is unprecedented, standing at a staggering 1,000 times higher than any recorded in human history. Climate change exacerbates this crisis. In the coming decades, it is estimated that around one million species may be pushed to the brink of extinction. The dangers stemming from climate change are diverse and include the introduction of new invasive pests and diseases, the increased incidence of forest fires, and the occurrence of severe and unpredictable weather events. While some species may have the capacity to adapt by shifting their habitats or behaviour, others may not, leading to a dire situation for many vulnerable populations.

The unprecedented rate of species extinction is alarming, and it emphasizes the urgent need for collective action to mitigate the impacts of climate change. It is not just a matter of protecting individual species; it's about safeguarding the delicate balance of ecosystems and biodiversity upon which human well-being also depends. The repercussions of inaction are far-reaching, affecting not only the natural world but also human society in complex and interconnected ways.

6. LESSER FOOD

The world is witnessing a concerning escalation in global hunger and inadequate nutrition, and this alarming trend can be attributed to various factors, including climate change and the increasing frequency of extreme weather events. These challenges are impacting the production and effectiveness of crops, livestock, and fisheries. Moreover, the rising acidity in the oceans is posing a significant threat to marine resources, which serve as a crucial food source for billions of people. In particular, the Arctic region has experienced disruptions in traditional food sources stemming from herding, hunting, and fishing due to shifts in snow and ice cover patterns. Climate change-induced heat stress is also a critical factor, as it not only leads to reduced water availability but also diminishes grazing areas. These combined factors have a detrimental effect on crop yields and livestock, ultimately affecting the livelihoods of communities that rely on these resources.

The growing challenges related to global hunger and malnutrition are deeply interconnected with the impacts of climate change. Climate-induced shifts in weather patterns, increased acidity in the oceans, and changes in Arctic ecosystems are all contributing to a precarious situation where food security is at risk. Addressing these issues requires a multifaceted approach that includes sustainable agricultural practices, conservation of marine resources, and strategies to adapt to changing environmental conditions, all aimed at ensuring that people have reliable access to nutritious food. Failure to address these challenges could have far-reaching consequences for both human well-being and the ecosystems on which we depend.

7. SEVERE HEALTH RISKS

Climate change is the biggest danger to human health, and scientists are already working to lessen its detrimental impacts on health globally. Climate change is the biggest risk to human health. A few of the negative health implications of climate change include air pollution, illness, severe weather, forced relocation, mental health stress, rising rates of hunger, and poor nutrition in places where people cannot grow or receive enough food. An estimated 13 million people are murdered annually by environmental factors. Severe weather occurrences lead to a higher death toll and complicate health care systems' efforts to treat the increasing number of illnesses brought on by altered weather patterns.

The implications of climate change on health already include disruption of food systems, increase in zoo noses and vector-, food-, and water-borne illnesses, as well as mental health issues. Storms, floods, and heat waves are examples of extreme weather occurrences that are occurring increasingly frequently. Furthermore, climate change is weakening a number of socioeconomic determinants of health, including livelihoods, equality, and access to healthcare. Climate-related health risks disproportionately impact the most vulnerable and disadvantaged groups of people: women, children, migrant or displaced persons, poor communities, older populations, those with underlying health difficulties, and minorities (WHO, 2021).

8. POVERTY AND DISPLACEMENT

Climate change contributes to an increased risk of individuals becoming impoverished and remaining in poverty. Urban slums are particularly vulnerable to the destructive impacts of floods, which can lead to the loss of homes and livelihoods. High temperatures make outdoor tasks more difficult to carry out, affecting income-generating activities. Additionally, water scarcity can harm crop yields, impacting the livelihoods of agricultural communities. Over the past decade (2010-2019), an estimated 23.1 million people, on average, were displaced annually due to weather-related disasters, leaving millions more susceptible to the cycle of poverty. Most of these displaced individuals come from countries that have limited resources and readiness to cope with the consequences of climate change.

NET ZERO AND SUSTAINABLE DEVELOPMENT GOALS

In climate policy, net-zero emissions are the new focus. Historically, the objective of lowering emissions has been stated as a fixed level of atmospheric concentrations or as a percentage. These days, the most common definition of climate ambition is a specific date for reaching net-zero emissions, typically associated with the peak temperature goals of the Paris Agreement. For around two thirds of worldwide emissions and a somewhat higher proportion of the global gross domestic product, net-zero targets have already been met (Black et al., 2021).

Fundamentally, net zero is a scientific concept. In order to keep the rise in global average temperatures within certain limitations, physics dictates that there is a finite budget of carbon dioxide and other greenhouse gases that are allowed into the atmosphere. Once this budget is met, all further releases must be countered by removal into sinks. But net zero is much more than just a theoretical concept or a technological objective. It provides a framework for organizing and comprehending global initiatives to address climate change, which are occurring more and more often (Fankhauser et al., 2022).

Climate research supports the socially determined acceptable temperature rise. In the Paris Agreement, 197 countries agreed to keep global warming well below 2 °C and strive towards 1.5 °C. With a 400–800 $GtCO_2$ carbon budget, there is a 50% likelihood of achieving the 1.5 °C target. To keep within this carbon budget, CO_2 emissions must peak before 2030 and then begin to fall, reaching zero by around 2050 (IPCC, 2018).

Emissions, including those absorbed by oceans and forests, play a crucial role in the fight against climate change. Scientific consensus underscores the necessity of limiting global temperature rise to 1.5°C to avert the most severe climate impacts and ensure a habitable planet. Currently, the Earth is 1.1°C warmer than it was in the late 19th century, and emissions continue to increase. To meet the Paris Agreement's goal, emissions must be reduced by 45% by 2030 and achieve net-zero status by 2050. Transitioning to a

Figure 4. Global climate agreements through the years

MONTREAL PROTOCOL (1987)-The United Nations' member nation's band together to control roughly 100 compounds classified as ozone depleting substances in response to the depletion of the ozone layer. The Montreal Protocol is the only UN pact that has ever received ratification from all 197 member nations. Although the accord was primarily aimed at restoring the ozone layer, it showed that the whole community is capable of coming together to take urgent environmental action and produce results.

EARTH SUMMIT (1992) -The UNFCCC was made available for signing during the Rio de Janeiro Earth Summit, an event hosted by the United Nations Conference on Environment and Development. 158 countries had ratified UNFCCC by year's end, clearing the path for yearly Conferences of the Parties (COP).

KYOTO PROTOCOL (1997) - With industrialized nations agreeing to collectively reduce their greenhouse gas emissions to 5% below 1990 levels within 10-15 years, COP 3 resulted in the first international treaty demanding the reduction of global warming pollution. Notably, India and China were not deemed industrialized countries for the purposes of the agreement and thus exempt from the standards since they had not yet experienced their respective economic booms. The Kyoto Protocol would eventually be ratified by 192 countries, but the United States has not, undercutting the accord.

PARIS AGREEMENT (2015) - The Paris accord, the most ambitious global climate accord ever, was reached at COP 21 after years of stumbling to reach a broad agreement that could expand upon the Kyoto Protocol. The Paris Agreement featured each country providing precise, demanding objectives that they believed were possible, even though they have so far been insufficient, and established an overall aim to keep warming "well below" 2 degrees Celsius while "pursuing efforts" to keep it under 1.5 degrees. This historic agreement offered the world the structure to stop global temperatures from increasing as well as the diplomatic clout to achieve it with China, India, and the US on board at its signing.

COP-26 (2021) - The Paris Agreement's 1.5°C warming limit is "alive" because to conference commitments, but its "pulse is weak." The final deal, the Glasgow Climate Pact, asks states to submit more aggressive emissions-reduction targets by the end of 2022 and calls for countries to cut coal usage and fossil fuel subsidies—both firsts for a UN climate pact. Delegates also finalize the regulations for a world carbon market. Significant side agreements on issues like coal, methane emissions, and deforestation are made by smaller groupings of nations. However, scientists point out that even if nations adhere to their commitments for 2030 and beyond, global temperatures would still rise by an

COP-27 (2022)-Nations agree to create a fund to compensate poor and vulnerable countries for losses and disasters brought on by climate change at COP27 in Sharm el-Sheikh for the first time, while the specifics are still up in the air. The conference's final communiqué also for the first time urges international financial institutions to change their procedures in order to solve the climate catastrophe. The communiqué removes the target of reaching peak emissions by 2025 and does not require nations to phase out the use of all fossil fuels. Continued reliance on fossil fuels,

net-zero world is one of the most pressing challenges humanity faces, demanding a radical transformation in how we produce, consume, and travel. The energy sector, responsible for roughly three-quarters of greenhouse gas emissions, holds the key to mitigating climate change's worst consequences. Replacing fossil fuel-based power with clean energy sources like wind and solar could significantly reduce carbon emissions. This transition is vital to addressing the climate crisis effectively. More and more countries, cities, businesses, and other organizations have committed to achieving net-zero emissions. With the three biggest polluters being the United States, China, and the European Union, together accounting for roughly 76% of global emissions, more than 70 countries have set a net-zero goal. More than 3,000 businesses and financial institutions are cutting their emissions in line with climate research according to the research-Based Targets Initiative. Furthermore, the Race to Zero initiative has been endorsed by over 400 financial organizations, over 1000 educational institutions, and over 1000 cities, all of which have committed to taking stringent, immediate action to cut global emissions in half by 2030.

THE PARIS AGREEMENT

The third generation of international treaties created to address the threat of climate change is the Paris Agreement. The UN Framework Convention on Climate Change (UNFCCC) was the first, having been approved in 1992, and right before the Rio de Janeiro United Nations Conference on Environment and Development.2 Stabilizing greenhouse gas concentrations in the atmosphere at safe levels is a key aim of the Convention. It created the key organizations required for the UN's climate policy to operate, notably the UNFCCC Secretariat and its annual Conference of the Parties (COP) to elect its executive branch (Delbeke et al., 2019).

It marks a step forward in bridging the various gaps that had divided the Kyoto region, including those between rich and developing nations, industrialized countries inside and outside the Protocol, and those who supported and violently opposed market processes (Streck et al., 2016).

The Paris Climate Agreement, officially known as the Paris Agreement or COP21, was adopted in December 2015 in Paris, France, under the United Nations Framework Convention on Climate Change. This international treaty aimed to restrict the emissions of greenhouse gases that contribute to global warming. It was designed to build upon and succeed the Kyoto Protocol, an earlier international agreement with similar goals. The Paris Agreement became operational on November 4, 2016, with 195 nations signing it and 190 of them ratifying it (UNFCC, 2021).

How Does the Paris Agreement Help in Climate Change?

The Paris Agreement outlines a critical requirement for governments to continuously escalate their efforts in addressing climate change over five-year intervals. To meet this commitment, each country is obligated to provide updated Nationally Determined Contributions (NDCs), which are essentially national climate action plans, every five years. In these NDCs, nations specify the measures they intend to undertake to reduce their greenhouse gas emissions, aligning with the overarching goals of the Paris Agreement. Furthermore, these updated NDCs go beyond mitigation efforts and also include strategies to enhance a country's ability to adapt to the impacts of global warming. This adaptation aspect of the NDCs is crucial, as it addresses the real-world consequences of climate change and underscores the importance of not only reducing emissions but also building resilience to the changing climate. In essence, the Paris Agreement establishes a dynamic and evolving framework where countries commit to regular, more ambitious climate actions, both in reducing emissions and in adapting to the effects of climate change. This process ensures that nations remain accountable and responsive to the ever-evolving challenges posed by global warming while working collectively to achieve the Agreement's objectives.

The first "*global stocktake*" will evaluate the Paris Agreement targets in 2023. This procedure will provide nations more motivation to adopt aggressive climate change measures to keep global warming below 1.5 degrees Celsius. The Paris Agreement urges nations to develop and submit long-term strategies in order to properly define their efforts towards the long-term objective. They are not required, in contrast to NDCs.

The Paris Rule book, commonly referred to as the operational guidelines for the practical implementation of the Paris Agreement, was agreed upon during the UN Climate Change Conference (COP24) in Katowice, Poland, in December 2018, and was finalized at COP26 in Glasgow, Scotland, in November 2021 (UN, 2022).

While there is ongoing effort required to address climate change, the period following the implementation of the Paris Agreement has seen the emergence of low-carbon technologies and new economic opportunities. An increasing number of countries, regions, cities, and businesses are committing to carbon neutrality objectives. In all sectors of the economy, responsible for a quarter of emissions, zero-carbon solutions are becoming increasingly cost-competitive. The most notable advancements in this direction are observed in the electricity and transportation industries, which have opened up new business opportunities for pioneers in adopting clean energy solutions.

In industries responsible for more than 70% of global emissions by 2030, zero-carbon solutions could be competitive.

THE EGYPT AGREEMENT (COP 27)

The establishment and operationalization of a loss and damage fund was the historic conclusion of the 27th Conference of the Parties to the United Nations Framework Convention on Climate Change (COP27), which was held in the Egyptian beach city of Sharm el-Sheikh, on November 20. In COP-27 in Egypt, the policy makers agreed to establish loss and damage fund by making a draft decision for the first time as "*Matters related to funding arrangements responding to loss and damage associated with the adverse effects of climate change including a focus on addressing loss and damage*" (COP27 overarching decision Version 18/11/2022). It shows the willingness of world polity to address the issues related to climate change and environmental degradation. We strongly feel and propose that the environmental damage by different countries be assessed by modern scientific technologies drawing a mix of the panel of scientific experts from developing and developed nations both and the "contributions of the loss and damage causing countries" should be imposed in proportion to the damage and losses caused by them to the environment. The fund so generated be distributed to and used by the adversely affected countries only and only for the protection and preservation of environment and for sustainable development of earth. It will greatly help the developing countries to protect themselves against the damages caused by developed nation, particularly during the industrialization era after 1970.

FINDINGS OF THE REPORT

As per the report's forecasts, China and the European Union (EU) are anticipated to reduce their emissions by approximately 0.9% and 0.8% in 2022, respectively. In contrast, India's emissions are projected to increase by 6%, and the United States' emissions are expected to rise by 1.5%. The combined estimate for emissions growth in the rest of the world is approximately 1.7%.

The survey noted that India and China have the highest rate of emission growth during the previous ten years. According to the report, "China's emissions climbed by 1.5% annually on average over the previous 10 years, leading the global trend, and India's emissions increased by 3.8% annually, while emissions declined in the EU by 1.8% annually and the US by 1.1% annually. The report was created by the worldwide Carbon Project, a study aiming to calculate worldwide greenhouse gas emissions and determine their sources. It involved a team of over 100 experts from throughout the world.

India currently stands as the world's fourth-largest emitter of greenhouse gases, trailing behind only China (31%) and the United States (14%). However, recent data reveals that India is rapidly approaching the European Union's (EU) share, which is composed of 27 member nations. In 2021, India was expected to emit a total of 2.7 billion tonnes of carbon dioxide, slightly below the EU's projected total of 2.8 billion tonnes. India's contribution to global emissions in 2021 accounted for 7.5%, just slightly below the EU's 7.7%, but it is projected to rise to 8% in 2022.

It's important to note that India's per capita emissions remain significantly lower than those of the top three emitters. Despite India's relatively higher rate of emissions increase, per capita emissions in India were reported at 1.9 tonnes in 2021, placing the country at the bottom of the list among the top 20 global polluters. In comparison, countries like Indonesia (2.3), Brazil (2.3), Mexico (3.2), and Vietnam (3.3) had higher per capita emissions. In contrast, nations with the highest per capita emissions included Saudi Arabia (18.7), the United States (14.9), Australia (15.1), Canada (14.3), Russia (12.1), and South Korea (11.9). This data highlights a significant disparity in per capita emissions across nations. It indicates that, relative to other major developed or emerging regions, individuals in India typically have less access to power and contribute less to global emissions on an individual basis.

Furthermore, when considering fossil fuel emissions per person, India lags far behind. The per capita fossil CO_2 emissions in 2021 were considerably lower in India compared to the United States, China, and the EU. This emphasizes the importance of considering both total emissions and per capita emissions when evaluating a country's contribution to climate change and its access to energy resources.

The Five Important Takeaways From COP-27 (UNFCC, 2022)

1. Establishing a Dedicated Fund for Loss and Damage

The conclusion of COP27 saw a landmark agreement aimed at delivering assistance for nations that are most vulnerable to the severe consequences of climate-related disasters such as floods and droughts. This agreement signifies a significant step forward, as it marks the first time that countries have collectively committed to establishing a dedicated fund and the necessary financial mechanisms to address the issue of loss and damage arising from the destructive impacts of climate change. The importance of this agreement lies in its recognition of the pressing need for financial support to aid those nations that are disproportionately affected by climate-related disasters. Such events can have devastating consequences on vulnerable communities, and this commitment to creating a fund and financing mechanisms demonstrates a shared responsibility among nations to address the aftermath of climate disasters. It also emphasizes the need for global cooperation and financial assistance to help affected nations cope with the loss and damage caused by climate change, further underscoring the significance of collective action in tackling the challenges of a changing climate.

2. Maintaining a Clear Intention to Keep 1.5 C Within Reach

At COP27, countries reaffirmed their commitment to the crucial target of limiting the global temperature increase to 1.5°C above pre-industrial levels. This objective aligns with the scientific consensus that outlines the necessary actions to be taken by 2030. To achieve this goal, the global economy must actively "mitigate" climate change, which entails efforts to reduce or prevent the release of greenhouse gases into the atmosphere.

In response to this imperative, a mitigation work program was initiated in Sharm el-Sheikh, with an urgent focus on enhancing both the ambition and implementation of climate mitigation measures. This program will kick-start without delay and extend until 2026, at which point it will undergo a review to assess its progress and determine if it should be extended further.

This initiative reflects the global acknowledgment of the need to accelerate mitigation efforts to combat climate change effectively. It underlines the sense of urgency in taking concrete actions to curb greenhouse gas emissions and transition towards a more sustainable and climate-resilient future. The periodic reviews will ensure that the program remains adaptive and aligned with the evolving requirements of mitigating climate change and limiting global temperature rise.

Governments were also urged at COP27 to review and enhance the 2030 objectives in their national climate plans by the end of 2023, as well as to step up efforts to phase out wasteful fossil fuel subsidies and phase down unabated coal power. In order to diversify energy sources and systems, governments emphasized the significance of a clean energy mix, including low-emission and renewable energy.

3. Holding Businesses and Institutions to Account

In this next stage of implementation, there is a heightened focus on holding industries, corporations, and organizations accountable for the commitments they have made. It's essential to ensure that the progress achieved is not forgotten once the public attention wanes. In 2023, the UN Climate Change will place a strong emphasis on transparency in the commitments made by these entities.

4. Mobilizing More Financial Support for Developing Countries

Finance plays a pivotal role in all global efforts to combat climate change. Whether it's actions related to mitigation, adaptation, addressing loss and damage, or developing climate-friendly technology, adequate funding is essential for these initiatives to be effective. COP27 has established a road map to guide a more substantial flow of financial resources towards low-emission development and climate resilience. The Sharm el-Sheikh Implementation Plan, a key decision of COP27, underscores the need for a substantial annual investment of approximately USD 4-6 trillion to facilitate a global transition to a low-carbon economy. Realizing this level of financial support requires a rapid and comprehensive overhaul of the financial system, involving institutional investors, commercial banks, governments, central banks, and other financial stakeholders. Among the significant outcomes related to climate financing at COP27 is the call for developed nations to contribute funds for the second replenishment of the Green Climate Fund. Furthermore, countries welcomed recent contributions made to the Special Climate Change Fund, the Least Developed Countries Fund, and the Adaptation Fund. These contributions totaled millions of USD and are crucial for supporting climate-related initiatives in vulnerable regions. Discussions during COP27 also focused on the development of a "new collective quantified goal on climate finance" for 2024, which takes into account the specific needs and interests of less affluent nations. This reflects the ongoing commitment to ensure that the financial resources necessary for tackling climate change are accessible and sufficient, especially for the countries that are most vulnerable to its impacts.

However, there is grave concern that the developed countries have not yet reached their joint goal of mobilizing $100 billion annually by 2020. The developed countries have been urged to do so, and multilateral development banks and other international financial institutions have been urged to mobilize climate finance. Both the biannual progress report on the USD 100 billion aim starting in 2024 and a report on doubling adaptation funds have been requested by the UN Climate Change Standing Committee on funds for presentation at COP28 the following year.

5. Making the Pivot Toward Implementation

The decisions adopted during COP27 are geared towards enhancing efforts at the national level to cut down on greenhouse gas emissions and prepare for the inescapable impacts of climate change. Additionally, these decisions are designed to ramp up the financial, technological, and capacity-building assistance required by developing nations to address climate change. What's particularly noteworthy is that governments have committed to pursuing a transition toward development that is not only low in emissions and climate-resilient but also characterized by ambition, fairness, and equity. At COP27, there was a further step taken by introducing a work program focused on the concept of a "just transition." This initiative is intended to provide support and complement the imperative scaling up of mitigation efforts and their practical implementation. A just transition underscores the importance of ensuring that the shifts toward a sustainable, low-carbon future are not only environmentally responsible but also considerate of the social and economic implications, particularly for workers and communities impacted by the transition. This signifies a holistic approach to addressing climate change that prioritizes both environmental goals and social equity.

Signatories of the Paris Agreement are obligated to consider "the imperatives of a just transition of the workforce and the creation of decent work and quality jobs in accordance with nationally defined development priorities." According to the International Institute for Sustainable Development, this means that a just transition is not a rigid set of guidelines but rather "a vision and a process based on dialogue and an agenda shared by workers, industry, and governments that need to be negotiated and implemented in their geographical, political, cultural, and social contexts."

"The heart of implementation is: Everybody, everywhere in the world, every single day, doing everything they possibly can to address the climate crisis".

INDIAN INITIATIVE TO REDUCE CARBON EMISSION

1. NDRC's India's Initiative on Climate Change and Clean Energy

The Natural Resources Defence Council launched an effort in 2009 to create a sustainable economy based on minimal carbon emissions. The National Resources Defence Council collaborates with its Indian counterparts on four significant projects:

- Improving structure effectiveness.
- Improving climate change cooperation between the US and India.
- Getting ready for how climate change will affect public health.
- Improving environmental management.

2. PepsiCo India's Waste to Wealth Initiative

This income-generating project is being worked on by PepsiCo India in collaboration with Exnora, an environmental NGO, and benefits more than 5,000 individuals throughout India. In accordance with this programme, biodegradable trash is transformed into organic manure and 80% of people's domestic rubbish is recycled. A community awareness programme is also undertaken by it. Participation from the local community is the foundation of this initiative. An estimated 35,000 tonnes of waste that would have otherwise been sent to landfills have already been recycled as a result of the project.

3. Save Himalaya Campaign

It is a project of the Indian Environmental Society that strives to protect the Himalayas and enlists the aid of local populations in the area to accomplish so. There is little question that India's Protector Mountains are the Himalayas. The programme focuses on comprehending the ecosystem of the Himalayas and finding legislative solutions to maintain this system. The mission makes an effort to solve several crucial concerns regarding:

- The hydrological effects of the Himalayan glaciers.
- The preservation and protection of biodiversity.
- The preservation and protection of wildlife.
- Planning for the sustainability of the Himalayan Ecosystem.
- Traditional knowledge societies and their means of subsistence.

4. National Action Plan on Climate Change

The plan of action, which was initially implemented in 2008, was centred on determining the primary environmental issues and solutions to those issues. This strategy promises to minimize India's carbon emissions while pursuing development goals, acknowledging that the country's economic growth is equally essential. There are 8 main "national missions" included in this strategy.

- National Solar Mission.
- National Mission for Enhanced Energy Efficiency.
- National Mission on Sustainable Habitat.
- National Water Mission
- National Mission for sustaining the Himalayan Eco system
- National Mission for a Green India.
- National Mission for Sustainable agriculture.
- National Mission on Strategic Knowledge for Climate Change.

5. National Wetland Conservation Programme

"Areas of marsh, fen, peat-land, or water, whether natural or artificial, permanent or temporary, with water that is static or flowing, fresh, brackish, or salt, including areas of marine water the depth of which at low tide does not exceed six meters," according to the Ramsar Convention on Wetlands.

In 1985, the Government of India announced the National Wetland Conservation Programme after identifying 115 wetlands that need immediate attention and developing a management action plan for the following four to five years. The State/Union Territories are required to submit long-term comprehensive Management Action Plans (MAPs) for a period of 3-5 years, preferably 5 years, corresponding with the Plan period after wetlands under the Scheme have been identified (YKA, 2019).

CARBON CAPTURE AND STORAGE TECHNOLOGY (CCS)

CCS is a method of limiting carbon emissions, which may be essential in the fight against global warming. It involves collecting the carbon dioxide produced by power plants or other industrial processes, such the production of steel or cement, transferring it, and then storing it underground. Carbon dioxide (CO_2) emissions from industrial operations, such as the manufacture of steel and cement, or from the burning of fossil fuels in the creation of electricity are captured using CCS. Then, after being shipped or piped from the site of production, this carbon is taken to be buried deep below in geological formations (National grid, 2023).

The Intergovernmental Panel on Climate Change (IPCC) stressed that, in addition to stepping up our efforts to decrease emissions, we also need to deploy technology to remove carbon from the atmosphere if we are to reach the goals of the Paris Agreement and limit future temperature increases to 1.5°C (2.7°F). CCS is one of these technologies that might be very important in the battle against global warming.

There are three steps to the CCS process:

1. Storing carbon dioxide by capturing it
   The CO_2 is separated from other gases generated in industrial operations, such as those at steel or cement factories, or coal- and gas-fired power plants.
2. Transport
   After compression, the CO_2 is sent to a location for storage using pipelines, ships, or land vehicles.
3. Storage
   Finally, the CO_2 is pumped into subterranean rock formations for long-term storage.

CARBON CAPTURE UTILIZATION AND STORAGE (CCUS)

In addition to CCS, there is a similar idea called CCUS, which stands for Carbon Capture Utilization and Storage (or, alternately, "usage"). The concept is that carbon might be recycled in industrial processes by being transformed into things like plastics, concrete, or bio fuel instead of being stored. In CCUS, CO_2 is usually captured from large point sources, such as industrial sites or power plants that burn either fossil fuels or biomass as fuel. If not utilized immediately, the CO_2 is compressed and either injected into deep geological formations like depleted oil and gas reservoirs or salty aquifers, or it is transported by pipeline, ship, rail, or truck for use in a variety of applications (IEA, 2021).

Currently, there are approximately 40 commercial carbon capture systems in operation worldwide, collectively capable of capturing over 45 million metric tons of CO_2 annually. Among the recently launched large-scale projects, seven stand out with capture capacities exceeding 100,000 metric tons of CO_2 per year, which includes the Red Trail Energy Project in the United States, the Arcelor LanzaTech Carbalyst

(Steelanol) project in Belgium, the Global Thermostat DAC pilot in the US, and four projects in China, located at the Sinopec Qilu Petrochemical Shengli facility, the Jiling Petrochemical CCUS facility, and the CNOOC Enping oil field. However, despite over 50 additional carbon capture facilities having been announced with a starting date set for 2030 or later since January 2022, it's important to note that the existing pipeline of ongoing projects only marginally fulfills the 2030 requirements outlined in the Net-Zero Emissions (NZE) Scenario. This indicates that while progress has been made in the development of carbon capture initiatives, there is still a considerable gap in meeting the targets necessary to achieve a net-zero emissions future. Addressing this gap will require significant investment and expansion in carbon capture technologies and infrastructure to align with the emission reduction goals.

Since about the beginning of 2018, CCUS has gained momentum. Developers of the projects have stated goals for approximately 50 additional collection units to be operational by 2030, catching over 125 Mt CO_2 annually, since January 2022. Only around 20 of the commercial capture projects in development as of June 2023 have, however, made a final investment decision (FID). Activities and developments specific to CO_2 transport and storage are detailed in CO_2 Transport and Storage (IEA, 2022).

ROLE OF CARBON CAPTURE AND STORAGE IN ACHIEVING INDIA 2070 NET ZERO TARGET

As per IPCC and IEA, CCS will play a critical role in decarbonisation of industrial sectors toward achieving Paris Agreement ambition of keeping the global temperature within 1.5/2.0 °C as compared to pre-industrial levels. CCS has potential to address overall 10-13% of global CO_2 emissions by 2050. CCS brings strategic value in 4 key areas:

1. Reduce emissions from existing Energy infrastructure such as coal and gas fired plants by retrofitting CCS solution.
2. Decarbonise hard to abate industrial sectors such as cement, steel and other industrial sectors.
3. Provide a cost-effective pathway for Blue Hydrogen.
4. Negative emissions by Direct Air Capture and BECCS (Bioenergy CCS).

As per IEA report on Net Zero, there is a need to enhance CCS capacity from 40 million Tons of CO_2 Injection per year in 2021 to over 7.6 billion Tons of CO_2 injection per year by 2050, which is creating 190 times more capacity over next 30 years. Today, more than 150 new projects have been announced totaling over 500 million Tons of capacity generation by 2030. These projects have been driven by global commitments towards decarbonisation, tightening emission norms as well as supported by government policies and incentives around the globe. Permanent storage of CO_2 in geological formations does not generate a direct revenue. Therefore, business case for CCS project relies on some form of carbon pricing, which is a prerequisite for the CCS industry to take off successfully. These carbon pricing can be in the form of incentives such as 85$/ Ton of tax credit under 45Q scheme in USA or carbon tax such as in Norway or license to operate (such as for many LNG/ new gas field development projects in Australia). Cap and trade system such as European Trading System (ETS) also enhance the attractiveness of CCS projects as an alternative mechanism to abate emissions.

India is the third largest GHG emitter today and with significant growth in economic and industrial activity, its emissions will go up unless various pathways such as renewable energy, electrification, green hydrogen and other measures such as CCS are put in place. India announced its commitment to be net zero by 2070. CCS, in addition to other measures, can support India towards its decarbonisation goals while taking it on a low carbon growth. However today India does not have any CCS project (whether commercial or pilot) due to lack of regulatory framework and financial support/ incentive/ carbon pricing not existing.

On 29 July 2022, the Indian Lower House of Parliament published an amendment bill to the 2021 Energy Conservation Act setting out the framework of a voluntary carbon credit trading scheme. India's Bureau of Energy Efficiency (BEE) presented a draft blueprint for the phased introduction of a national Cap-and-Trade system in India, providing for the introduction of a voluntary market in the first phase and moving to a mandatory cap and trade emissions scheme in third phase. This will provide a significant impetus to carbon financing in India and propel industries towards reducing their carbon emissions.

Given very nascent stage of CCS in India, Oil& Gas leaders are in the best position to champion this technology and provide a decarbonization solution for industrial sectors. Most of the early projects in CCS in several countries have been funded/ supported by government and/or executed by state owned entities. CCS projects have a long cycle and from feasibility studies to pilot to commercial projects, it can take anywhere from 5-10 years. By investing in CCS at an early stage, the leading players in India can play a key role in helping India achieve the ambition of being a Net Zero country by 2070 and provide additional economic value to the industries and country.

CARBON CREDIT: THE FUTURE OF SUSTAINABLE DEVELOPMENT

Environmental advocates who prioritize green initiatives are actively seeking to promote businesses and legislative measures that work in favor of the environment. This is particularly important because the consequences of global warming are becoming increasingly costly. One of the key drivers of climate change is the alarming rise in the levels of carbon dioxide in the Earth's atmosphere, primarily due to the combustion of fossil fuels. In response to this challenge, a global "carbon market" has emerged, allowing for the exchange of carbon credits both within and beyond regulated regions. The functioning of this carbon trading system is based on the principles outlined in the Kyoto Protocol, which places restrictions on the emissions of greenhouse gases (GHGs). Under this international agreement, predetermined emission limits are allocated to individual countries, which are then responsible for regulating and reducing greenhouse gas emissions originating from various industries and commercial enterprises operating within their borders. In essence, the goal is to encourage a shift towards more sustainable practices by incentivizing emission reductions and creating a mechanism for countries to work collectively to combat climate change. By reducing greenhouse gas emissions through various means, nations can earn carbon credits, which can be traded on the global carbon market. This system not only helps mitigate the environmental impact but also promotes economic activities that are less carbon-intensive, contributing to a greener and more sustainable future.

Although they contain a "tradable" element, carbon credits, also known as carbon allowances, may be considered of as a measurement. Permits that allow the owner to emit a specific quantity of carbon dioxide or other greenhouse gases are known as carbon credits, commonly referred to as carbon offsets. One credit allows for the emission of one tonne of carbon dioxide or another greenhouse gas' equivalent.

The carbon credit is one component of a so-called cap-and-trade programme. Companies who pollute are given credits that enable them to do so up to a limit that is regularly decreased. In the meanwhile, the business may sell any credits it doesn't require to another business that does. Thus, private businesses have two incentives to lower greenhouse gas emissions. If their emissions exceed the cap, they must first pay extra for credits. Second, by cutting their emissions and reselling their extra permits, they may generate income. The carbon credit system's proponents assert that it causes certified climate action initiatives to decrease, eliminate, or prevent greenhouse gas (GHG) emissions in a way that is measured and verifiable. Companies have a financial incentive to cut their carbon emissions thanks to carbon credits. Those that find it difficult to cut emissions can continue function, albeit at a greater cost. The creation of a worldwide carbon credit offset trading market was decided upon by negotiators in the Glasgow COP26 climate change summit in November 2021 (Gupta, 2011).

CARBON OFFSET

A different financial option that employs a similar methodology to minimize greenhouse gas emissions is carbon offset. One metric tonne of CO_2 or an equivalent greenhouse gas is reduced by the value of a carbon offset credit. It makes a significant contribution to the promotion of green energy sources including wind and solar energy as well as the support of initiatives for reforestation and wildlife preservation. This significant reduction is made possible by using more sustainable energy sources, such as wind and tidal energy.

Even individuals are employing this strategy and purchasing carbon offsets to improve the environment and raise public awareness of environmental preservation. For example-In order to absorb all the CO_2 produced by a driver who travels 400 km each week on a 2L petrol vehicle, 19 trees should be planted. The trees expand as the automobile travels. As long as the trees are not cut down, planting 19 trees will eliminate all of your vehicle's emissions for the duration of your life.

Global Initiatives for Carbon Credit

In 1997, the Intergovernmental Panel on Climate Change (IPCC) established a carbon credit program as part of the Kyoto Protocol, aimed at reducing global carbon emissions. This landmark international agreement set legally binding targets for reducing emissions among its signatory nations. To clarify how this system would operate, the Marrakesh Accords provided detailed guidelines. Under the Kyoto Protocol, countries were categorized into industrialized and developing economies. The industrialized nations, referred to as Annex 1, had their own emissions trading market. Through a mechanism called an Emission Reduction Purchase Agreement (ERPA), a country that produced fewer hydrocarbon emissions than it had planned for could sell the surplus credits to nations that had fallen short of their Kyoto targets.

For developing countries, a carbon credit known as a Certified Emission Reduction (CER) was awarded through the Clean Development Mechanism. These credits could be earned by developing nations that supported sustainable development programs. CER trading took place on a separate market. The initial commitment period of the Kyoto Protocol concluded in 2012, with some nations having made substantial progress in reducing emissions. Notably, the United States had withdrawn from the agreement in 2001. This international framework marked a significant step in addressing climate change by introducing mechanisms to incentivize emission reductions and encourage global cooperation in the fight against global warming (UNCC, 2021).

The Glasgow Agreement COP 26

In a significant development, nearly 200 nations came together to implement Article 6 of the 2015 Paris Agreement, following a pact reached during the conference in November 2021. This agreement creates a framework that enables countries to make progress towards their climate commitments by purchasing offset credits, which represent emission reductions achieved by other nations. The primary aim of this agreement is to incentivize countries to invest in projects and technologies that help conserve forests and build infrastructure for renewable energy sources, thereby contributing to global efforts to combat climate change. Several key measures within this agreement work to reduce global emissions. One of them involves the cancellation of 2% of all credits, essentially taking them out of circulation, which helps ensure a net reduction in greenhouse gas emissions. Another important provision is the absence of taxes on bilateral offset swaps between nations, which encourages international cooperation and trading of offset credits to meet climate goals. Furthermore, a noteworthy aspect of the agreement is the commitment to allocate 5% of the revenue generated from the sale of offset credits to a fund dedicated to supporting underdeveloped nations in adapting to the impacts of climate change. This allocation recognizes the importance of global solidarity and assistance for vulnerable regions as they address the challenges posed by a changing climate.

Additionally, to expand the new market for offset credits, negotiators agreed to allow the inclusion of up to 320 million credits that have been generated and filed since 2013. This step serves to broaden the opportunities for countries to engage in offset trading and promote emission reductions on a global scale (UN, 2021, Carbon credit, 2021).

Estimating the size of the carbon credit market can be challenging due to variations in regional regulations and other factors. The market can be broadly categorized into two segments: the voluntary carbon market and the compliance credit market.

Voluntary Carbon Market: This segment primarily consists of businesses and organizations that purchase carbon offsets for reasons related to corporate social responsibility (CSR) and sustainability. In 2021, this market was estimated to be worth approximately $1 billion. These businesses voluntarily choose to offset their emissions by investing in projects that reduce or capture greenhouse gases, such as reforestation, renewable energy, or methane capture projects. They do this as part of their commitment to environmental responsibility and to demonstrate their dedication to mitigating climate change.

Compliance Credit Market: This segment is linked to regulated carbon restrictions, often driven by government mandates and international agreements such as the Kyoto Protocol and the Paris Agreement. Companies and industries subject to emissions limits may purchase compliance credits to meet their regulatory obligations. The value of this market can vary significantly, but estimates for the year 2020 suggest it could be as substantial as $272 billion. The compliance credit market plays a crucial role in helping countries and regions achieve their emissions reduction targets by allowing for flexibility in meeting their obligations.

It's important to note that these estimates are subject to change over time and can be influenced by evolving climate policies, market dynamics, and international agreements. Additionally, as the global community places greater emphasis on addressing climate change, it is expected that both segments of the carbon credit market will continue to evolve and expand to facilitate greater emissions reductions and sustainability efforts.

CONCLUSIONS

The issue of carbon dioxide emissions and their impact on our planet's climate is of paramount concern, and the consequences of these emissions are indeed already irreversible to some extent. Climate change has, and will continue to have, far-reaching and long-term effects on our world. However, the important message here is that while we cannot undo the past, there is still an opportunity to mitigate and minimize the worst outcomes if we act swiftly and decisively. Scientific consensus overwhelmingly underscores the urgency of addressing climate change. It is a global problem that demands immediate international cooperation. The choices made by modern society will have a profound influence on the planet's future. While it can be daunting to consider the scale of the issue, individual and collective actions can bring about meaningful change.

Simple steps, such as conserving energy by turning off lights, opting for public transportation to reduce individual carbon footprints, and adopting more sustainable dietary habits like reducing meat consumption, are impactful measures that, when practiced on a societal level, can bring about substantial change. While no single individual or action can change the entire world on their own, every positive effort contributes to a collective momentum for change. Being a part of this positive change, even on a smaller scale, can help pave the way for more comprehensive global efforts. Together, we can make a difference and work toward a more sustainable and resilient future.

CONFLIC OF INTEREST: The authors declare no conflict of interest.

Note: A part of the present work was presented as keynote address in 108th Session of Indian Science Congress at RTM University, Nagpur from 3-7th January, 2023 by Chitwan Garg

REFERENCES

Anonymous. (2019) *The ins and outs of carbon dioxide*. Sequestration. http://www.sequestration.org/science/carbondioxide.html

Black, R. (2021). *Taking Stock: A Global Assessment of Net Zero Targets* (ECIU and Oxford Net Zero, 2021).

Carbon credit (2021), Cop26-Key-Outcomes-Agreed-At-The-Un-Climate-Talks-In-Glasgow https://www.carbonbrief.org/cop26-key-outcomes-agreed-at-the-un-climate-talks-in-glasgow/

Delbeke, J., Runge-Metzger, A., Slingenberg, Y., & Werksman, J. (2019). The paris agreement. In *Towards a climate-neutral Europe* (pp. 24–45). Routledge. doi:10.4324/9789276082569-2

Dendooven, L., Patiño-Zúñiga, L., Verhulst, N., Luna-Guido, M., Marsch, R., & Govaerts, B. (2012). Global warming potential of agricultural systems with contrasting tillage and residue management in the central highlands of Mexico. *Agriculture, Ecosystems & Environment*, *152*, 50–58. doi:10.1016/j.agee.2012.02.010

Fankhauser, S., Smith, S. M., Allen, M., Axelsson, K., Hale, T., Hepburn, C., Kendall, J. M., Khosla, R., Lezaun, J., Mitchell-Larson, E., Obersteiner, M., Rajamani, L., Rickaby, R., Seddon, N., & Wetzer, T. (2022). The meaning of net zero and how to get it right. *Nature Climate Change*, *12*(1), 15–21. doi:10.1038/s41558-021-01245-w

Greentumble. (2023). *What Causes Carbon Dioxide Emissions in the Air?* Greentumble. https://greentumble.com/what-causes-carbon-emissions

Gupta, Y. (2011). Carbon credit: A step towards green environment. *Global Journal of Management and Business Research, 11*(5), 16–19. https://www.statista.com/statistics/1284586/india-methane-emission/

IEA. (2023). *Carbon capture, utilisation and storage*. IEA. https://www.iea.org/energy-system/carbon-capture-utilisation-and-storage

IEA. (2023). *CO_2 Emissions in 2022*. IEA, Paris https://www.iea.org/reports/co2-emissions-in-2022, License: CC BY 4.0

Liu, Z., Deng, Z., Davis, S., & Ciais, P. (2023). Monitoring global carbon emissions in 2022. *Nature Reviews. Earth & Environment*, *4*(4), 205–206. doi:10.1038/s43017-023-00406-z PMID:37065615

MIT. (2023). *Greenhouse Gases* MIT. https://climate.mit.edu/explainers/greenhouse-gases

National grid. (2023). *What is carbon capture and storage?* National Grid. www.nationalgrid.com/stories/energy-explained/what-is-ccs-how-does-it-work

NDRC. (2022). *What Are the Causes of Climate Change?* NDRC. https://www.nrdc.org/stories/what-are-causes-climate-change#natural

Outlook. (2022). *Report At COP27: India Records Highest Emission Increase Among Top Global Contributors*. Outlook. Www.Outlookindia.Com/International/Report-At-Cop27-India-Records-Highest-Emission-Increase-Among-Top-Global-Contributors-News-236452

Sahrawat, R., & Garg, A. P. (2023). Rice straw degradation with mixed culture of microfungi and their enzymes. *International Journal of Environment and Climate Change.*, *13*(8), 2097–2104. doi:10.9734/ijecc/2023/v13i82167

Streck, C., Keenlyside, P., & Von Unger, M. (2016). The Paris Agreement: A new beginning. *Journal for European Environmental & Planning Law*, *13*(1), 3–29. doi:10.1163/18760104-01301002

The Wire. (2022). India's Per Capita Greenhouse Gas Emissions Far Below World Average: UNEP Report. UN. https://thewire.in/environment/india-per-capita-greenhouse-gas-emissions-unep

UN. (2015). *The Paris Agreement*. UN. https://www.un.org/en/climatechange/paris-agreement

UN. (2020). *Causes and Effects of Climate Change*. UN. https://www.un.org/en/climatechange/science/causes-effects-climate-change

UNFCC. (2015). *The Paris Agreement.* UNFCC. https://unfccc.int/process-and-meetings/the-paris-agreement

UNFCC. (2022). *Making the pivot toward implementation*. UNFCC. https://unfccc.int/making-the-pivot-toward-implementation

United Nations. (2021). *COP26: Together for our Planet.* UN. https://www.un.org/en/climatechange/cop26

United Nations. (2021). *The Guidelines to Implement the Kyoto Protocol: the Marrakesh Accords and the 5,7&8 Implications*. UN. https://unfccc.int/process-and-meetings/transparency-and-reporting/reporting-and-review-under-the-kyoto-protocol/overview/background-and-resources/the-guidelines-to-implement-the-kyoto-protocol-the-marrakesh-accords-and-the-578-implications

United Nations. (2021). "*What Is the Kyoto Protocol?*" UN.https://unfccc.int/kyoto_protocol

WEF. (2022). *This is how cities can reduce emissions with waste reduction solution*. WEF. https://www.weforum.org/agenda/2022/11/waste-emissions-methane-cities/

WHO. (2021). *Climate change and health*. WHO. https://www.who.int/news-room/fact-sheets/detail/climate-change-and-health

WMO. (2018). *IPCC Special Report on Global Warming of 1.5 °C* (V. Masson-Delmotte, Eds.). WMO..

YKA. (2019). *How 5 Initiatives are helping india reduce its carbon emission*. YKA. https://www.youthkiawaaz.com/2019/11/5-initiatives-helping-india-reduce-carbon-emissions/

Chapter 5
Marine Mucilage as an Outcome of Climate Change and Pollution:
Understanding the Case of Türkiye Through Salient Documents

Adem Yilmaz
https://orcid.org/0000-0002-1424-8934
Kastamonu University, Turkey

ABSTRACT

The seas are home to many living things and an excellent resource for the ecosystem. In addition, the seas are a perfect ecosystem component for both the supply of quality food for humans and the continuation of their lives. Due to the ecological and economic importance of the seas, they should be catered to consciously and in a sustainable fashion. Howbeit, the boost in human-induced problems and nature-induced problems may cause some negativities. This causes an increase in nutrients in the seas, sudden changes in water temperature, a decrease in water currents and consequently, an increase in algae and other phytoplankton. Over time, these organisms cause a substance entitled marine saliva or mucilage to cover the seas and damage the marine ecosystem. In this study, a comparative analysis of the mucilage problem, which has become a significant problem in Türkiye in recent years, is carried out. In this context, the prominent reports, the measures taken and the action plans in Türkiye are examined in detail.

INTRODUCTION

Arguably, amongst the issues threatening all the living things today either directly and/or indirectly in the world is climate change which entails global cooperation toward assuring social-ecological resilience in diverse regions. In fact, climate change has brought about a fair number of undesired outcomes for all the living beings, and what is more, some of these results have not hitherto been understood to the full

DOI: 10.4018/978-1-6684-9863-7.ch005

extent. With that being said, there is a lot to do for all the involved parties and stakeholders to combat with what is awaiting in a fair, equitable and inclusive manner. One of these attempts can be coming up with action plans at national and international levels so that a step-by-step approach can be adopted when fighting against the undesired consequences of climate change, which carry the potential to impact humans by harming oceans and lands even causing drought and famine (Yanarateş & Yılmaz, 2020).

As a matter of fact, even though addressing the causes of climate change may necessitate the working of countries together at larger scales more, the effects of climate change can be dealt with at relatively more local levels where 'the success stories' can still act as models for the other countries across the world. In this direction, scrutinizing the issue of marine mucilage, which has been on the agenda in Türkiye in particular for two years with the recent problems occurring in several parts of the country, but especially accumulating in the Marmara Sea, will both be a meaningful and purposeful attempt toward paving the way for collaboration and collective thinking in the longer run. "The presence of large mucilaginous masses, referred to as mucilage" (Penna et. al, 2000: 299) is indeed a crucial matter to be handled in seas. The line of literature points to the vitality of the issue e.g., by stating that *"although the mucilage did not occur in the Sea of Marmara for the first time, nor is the occurrence unique to the area, it is thought to be one of the worst mucilage outbreaks ever"* (Savun-Hekimoglu & Gazioglu, 2021: 402). It has also been underpinned vastly in the bulk of the related literature that the elevation of the levels of pollution along with the water temperature rise in the Marmara Sea have yielded the sea saliva i.e., mucilage to become more and more visible (e.g., Koncagul, Dulger & Yinanc, 2022) after the first mucilage occurrence in Türkiye was witnessed in the mid-autumn of the year 2007 along the northeastern coasts of the Marmara Sea (Karadurmus & Sari, 2022). It is reported too that *"Marmara Sea has a rather complex hydrologicalsystem, in a zone of transition betweendense (salinity 37- 38.5 ‰) and warmer waters originating in the Mediterranean Sea, and cold, lower-salinity water (20-22 ‰) coming from the Black Sea"* (Aktan, Dede & Ciftci, 2008: 1). To this end, it might be reiterated herein that focusing on the mucilage problem in the Marmara Sea will be a valuable endeavor to cast more light on the issue observed in some other corners of the world as well.

In light of all these, the present study aims to portray and evaluate the current status in Türkiye respecting the mucilage problem by checking the recent efforts to establish and maintain plans including action plans and strategic plans via resorting to the available reports produced as part of the initiatives in this respect. The documents in question can be described as follows: The first one is the report by Turkish Academy of Sciences belonging to year 2021 entitled 'Ecology of the Marmara Sea: Formation and Interactions of Marine Mucilage, and Recommendations for Solutions'. The second one, on the other hand, is a follow-up report which discusses the mucilage issue of Türkiye in the form of meeting notes of a convention took place in the premises of Council of Higher Education of Türkiye (CoHE) offering a number of scientific/academic solutions to the problem. The third one is a strategic plan of integrated sort by Republic of Türkiye Ministry of Environment, Urbanization, and Climate Change covering the years 2021- 2024 concentrating on the causes and effects of climate change and mucilage in the Marmara Sea. The fourth of these is a conservation action plan for the Marmara Sea by the said Ministry. The last one, in other words, the fifth document is the report called: 'Mucilage Treatment and Its Conversion to Plant Growth' by Türkiye Energy, Nuclear and Mineral Research Agency.

So as to be able to carry out the document analysis adopting a qualitative paradigm, namely, with a view to reviewing the aforementioned institutional documents, it is intended that eliciting meanings, gaining insights, and thusly obtaining empirical knowledge will be made possible (Corbin & Strauss, 2008). It is well known that, *"documents of all types can help the researcher uncover meaning, develop understanding, and discover insights relevant to the research problem"* (Merriam, 1988: 118) and making use of documents that pre-exist as a form of data would mean lesser number of ethical problems when compared to the other qualitative methods (Merriam & Tisdell, 2016; Yılmaz & Yanarateş, 2020) and public records' being accessible for everyone to check (Morgan, 2022) would be convenient for scholars.

It is hoped that the present study will not only shed further light to the snapshot of the problem from the Turkish perspective, but it will also reveal opportunities for further development and innovation toward the future with the help of the recommendations to be made for international authorities, decision-making bodies, Non-Governmental Organizations (NGOs), and field members.

LITERATURE REVIEW ON MUSILAGE RESEARCH

Mucilage (otherwise known as sea snot) is a sticky substance and a thick mucus-like waste material that almost all plants and microorganisms can produce (Mecozzi et al., 2005). For mucilage to occur, some conditions, both biological and chemical, must be present (Danovaro et al., 2009). In recent years, especially in the Marmara Sea in Türkiye, mucilage has started to be seen frequently, and fighting against it has become a mandatory issue to be taken on the agenda (Şeker & Yalçın, 2021). In general, three types of reasons can be mentioned for mucilage formation. These reasons are the increase in the average temperature in the seas, the growth of pollution levels in the seas and the stagnation of the seas (Keleş et al., 2020). These reasons cause many different issues to take place. One of these is the escalation of some plankton species and the increase in sea saliva rates accordingly. When the studies on mucilage are analyzed, it is seen that interdisciplinary studies such as "Zoology, Plant Sciences, Oceanography, Marine & Freshwater Biology, Limnology, Fisheries, Environmental Sciences and Ecology" are usually carried out together. In addition, the number of studies on mucilage in Italy, France, China, Croatia, and the USA countries is also noteworthy (Şeker & Yalçın, 2021).

There exist several eminent studies on mucilage events in the literature. According to these, firstly, plant plankton and underwater algal blooms cause sea slobber formation to appear. It has been reported that sea slime formation first appeared on the northern shores of the Adriatic Sea. Although this has decreased over time, it came to the agenda again in 1988, and after 1991, it was observed again off the coast of the Sea of Tiran in the west of Italy. Another major study is the observation of the occurrence of mucilage in the Liguria Sea, also in Italy, in 2003. It was determined that it was different from the microorganisms that had previously caused mucilage in the Sea of Tiran (Schiaparelli et al., 2007). With the formation of mucilage, the ecosystem of underwater organisms is disrupted, and this situation greatly affects them. In 2003, when the causes of the increase in mucilage were investigated, it was stated that it could be associated with the heat wave experienced in Europe, the rise in seawater temperature and the decrease in the flow of clean water flowing from freshwater sources to the seas with drying rivers (Schiaparelli et al., 2007).

Another study in the literature is the algal bloom and the resulting mucilage problem in the North Sea between England and Scandinavian countries. In this study, attention was drawn to eutrophication, i.e., the increase in the amount of nitrogen and phosphate. At the same time, it is known that algal blooms intensify with eutrophication and cause the death of underwater organisms by reducing the oxygen level. As a result of the related research, it was observed that unfair agricultural practices increase the nitrogen content in water. The increase in nitrogen and, consequently, in the levels of carbon cause more intense algal blooms and mucilage formation (Johns & Reid, 2001). Another study on mucilage was conducted

Figure 1. Interrelation of the studies on mucilage

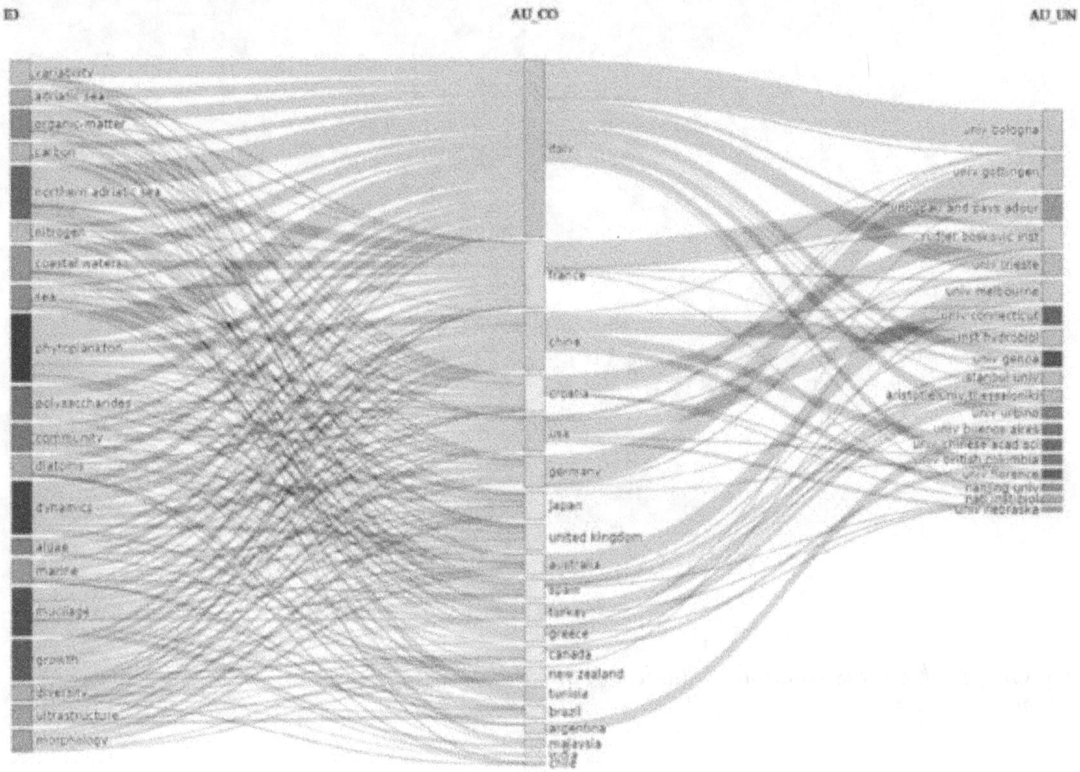

in the Gulf of Mexico. The oil spill from the Deep Horizons oil exploration/discovery platform operated by British Petroleum (BP) caused a considerable number of adverse effects. As a result of the oil spill, it was understood that plant plankton were under stress and secreted more mucilage (Abbriano et al., 2011). In parallel with these developments witnessed worldwide, marine saliva has begun to be spotted in the Marmara Sea of Türkiye. Scholars have stated that the causes of sea saliva in the Marmara Sea are due to global climate change, the increase in the temperature of the sea waters in the Mediterranean basin and the stagnation of the seas (Abbasnia & Toros, 2019; Orhon et al., 2021). In addition to these studies, it appears that many studies have been conducted on mucilage in the literature. Figure 1 shows the countries, universities and research topics that have conducted studies on mucilage (Şeker & Yalçın, 2021: 76).

When Figure 1 is analysed, it is clear that Italy, France, China, Croatia, Croatia, the USA, Germany, Japan and the United Kingdom are the countries with the highest number of studies on mucilage. This also indicates that the mucilage phenomenon is encountered more intensively in the countries in question. Figure 2 below shows the rates of studies carried out from 1950 to the present day (Şeker & Yalçın, 2021: 75).

Through Figure 2 it may be understood that the first studies on mucilage were designed at the end of the 1950s and progressed at a static level until 1975. After the year 1990, the studies gained momentum and started to increase regularly. It is seen that the highes number of studies were between the years

Figure 2. Distribution of studies on mucilage according to years

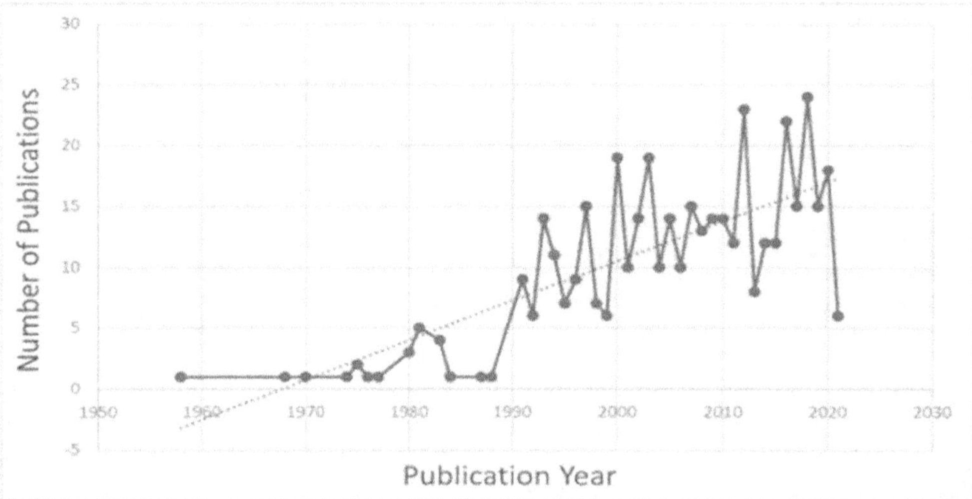

Figure 3. Geographical distribution of studies on mucilage
[Source: Author's creation]

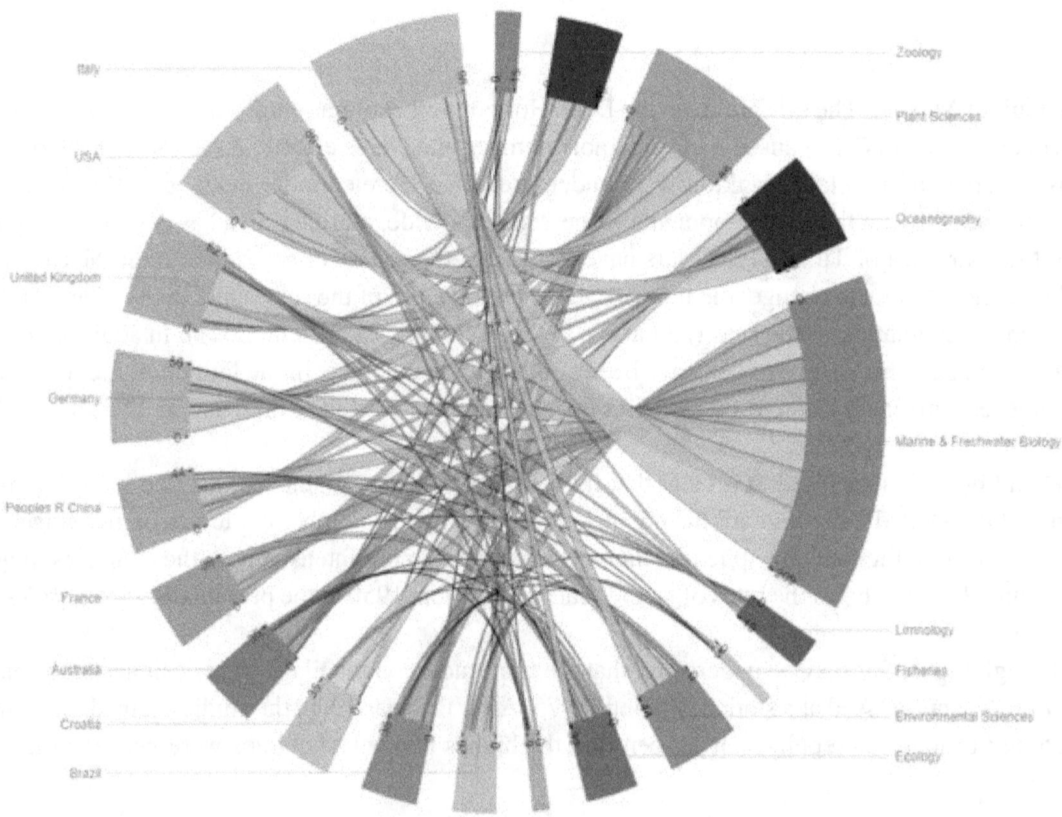

2010 and 2020. Figure 3 displays the geographical distribution of the countries working on mucilage (Şeker & Yalçın, 2021: 77).

Figure 3 exhibits the countries working on mucilage and the areas in which their work concentrates upon. Accordingly, it is seen that the key countries are Italy, the USA, the UK, Germany, and China in this regard. When the fields of study are explored, it is detected that "Marine & Freshwater Biology, Plant Sciences, Zoology, Oceanography, Ecology" are the areas.

FORMATION OF MUSILAGE, CAUSES AND RECOMMENDATIONS: THE EXAMPLE OF THE SEA OF MARMARA

It has already been mentioned that climate change, increases in seawater temperature and anthropogenic factors are highly effective on mucilage. Apart from these factors, some other factors cause mucilage. In particular, chemicals that are mixed into groundwater as a result of agricultural activities are influential in this sense. It has been pinpointed that the intensive use of nutrients such as N and P in chemicals, pesticides, polychlorinated biphenyls (PCB), polyaromatic hydrocarbons (PAH), polybrominated diphenyl ethers (PBDE), heavy metals and microplastics have strong effects in this manner (Aydın, 2021). These substances are generally called microcontaminants. These accumulate in the food chain over time and may lead to an irregular increase and decrease in some species. This situation causes the number of marine organisms that feed on phytoplankton to diminish and the number of phytoplankton to augment.

The increasing number of phytoplankton results in an uncontrolled proliferation over time, and more marine saliva is secreted. At the same time, unregulated fishing by humans reduces the number of fish feeding on plankton, and this accelerates the formation of mucilage. As a matter of fact, the issue with mucilage is an old problem. The mucilage problem dates back to nearly ancient times, and it was first marked in the Adriatic Sea in the 1870s. In Türkiye, it started to be observed in the Aegean Sea between the years 1990 and 2010 and in the Marmara Sea between the years 2007 and 2010 and then more intensively in the year 2021 (Danovaro, Umani & Pusceddu, 2009). Figure 4 pinpoints the seas where mucilage is found and the years when it began (Aydın, 2021: 53).

When Figure 4 is inquired into, it is discovered that mucilage is a relatively new phenomenon in the Marmara Sea and thus does not have a long history. On the other hand, it is unearthed that it actually is a marine phenomenon that has been present in the Adriatic Sea for a lot of years and has become chronic. The mucilage problem, observed in the Marmara Sea since 2007, has been evaluated as a result of environmental pollutants and climate change. Thereby, it was unveiled that polysaccharides and carbohydrates feed phytoplankton in mucilage formation. When the mobility of sea waters is further scrutinized, it turns out that not much phytoplankton is formed in the harsh winter months. This is owing to the fact that during the harsh winter months, sea mobility increases, and sea temperature can remain constant. That said, calmer winter months followed by hot summer months end up with sea currents' slowing down and an increase in mucilage formation (Mecozzi et al., 2001). Numerous studies have put forth that phytoplankton, diatoms, benthic macroalgae and bacteria effectively form mucilage (e.g., Aydın, 2021). Mucilage contains bacteria and viruses due to its structure. The organic substances in the tangles are broken down on account of the bacteria. In the meantime, dissolved organic substances are formed, and nutrient increase is accelerated. The increase in nutrients causes plankton to reproduce, multiply irregularly, and form sea slobber. Table 1 shows the rates of bacteria and viruses in mucilage and water environments (Aydın, 2021: 54).

Figure 4. Seas where mucilage occurs and years of the occurrence
[Source: Author's creation]

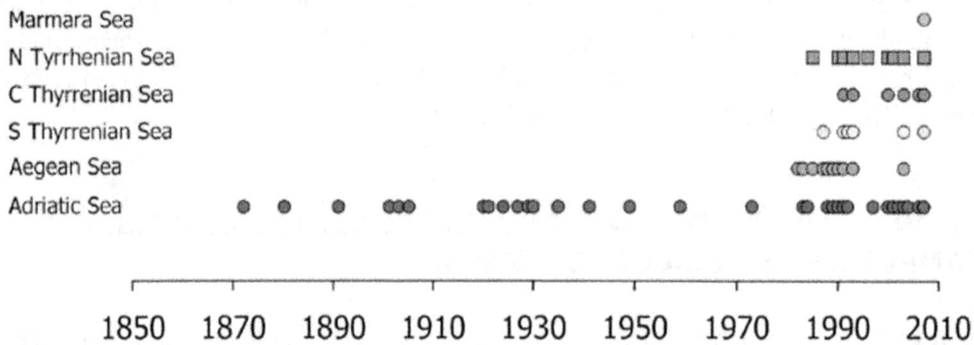

Table 1. Bacteria and virus rates in mucilage and water environment

	n	Virus (10^{10} VLP/L)	Bacteria (10^{9} Cell/L)	Virus/Bacteria Ratio
Ambient Water	7	0.3 ± 0.1	0.7 ± 0.1	4.7 ± 0.1
Water in the mucilage cavities	16	25.7 ± 4.2	88.9 ± 9.8	2.9 ± 0.1
Mucilage	17	92.2 ± 69.0	87.9 ± 1.0	72.0 ± 0.2

VLP/L: Virus-like Particle / Litre

When Table 1 is studies, it is accesible that the virus/bacteria ratio in a normal environment is 4.7. That being saisd, in an environment with mucilage, this ratio is 72.0. In other words, the rate of viruses and bacteria increases tremendously.

Mucilage has many direct and indirect harmful effects. Tourism and fisheries sectors are affected at the beginning of all these, and severe economic losses become inevitable. The width of the mucilage can start from the surface and progress to a depth of approximately 30 m of water. This affects the marine ecosystem and triggers the death of plentiful marine organisms. When these issues are inspected in terms of the Marmara Sea, reducing the pollutant load to the sea is essential. The C, N, and P load in the structure of these pollutants should be reduced. This reduction is significant toward ensuring the

Table 2. Total pollution load in the Marmara Sea

Discharge Source	SSM (kg/day)	BOD (kg/day)	COD (kg/day)	TN (kg/day)	TP (kg/day)
Domestic	1.341.334	1.184.041	2.755.797	213.493	60.226
Industrial	33.277	34.055	68.366	3.244	365
Streams	858.971	140.033	599.248	116.902	41.382
Total	2.233.582	1.358.129	3.423.411	333.639	101.973

SSM: Suspended solid matter, BOD: Biochemical oxygen demand, COD: Chemical oxygen demand, TN: Total nitrogen, TP: Total phosphorus ecological balance and toward not disrupting the food chain. Table 2 presents the total pollution load in the Marmara Sea (Aydın, 2021: 56; OSB, 2013).

As a result of agricultural and animal husbandry practices, very high levels of N, P and pesticides are mixed into the Marmara Sea basin. This ignites pollutants such as PAH, PCB and PBDE to be mixed into the air with domestic and industrial activities. Aside from C, N, and P pollutants, micropollutants such as metals, pharmaceuticals and microplastics are also mixed into the waters. These pollutants reach the Marmara Sea through rainwater and wet sedimentation (OSB, 2013). In this direction, some suggestions for the prevention of mucilage in the Marmara Sea are presented below (Aydın, 2021):

- Treatment of water up to the third-degree advanced biological treatment level in all settlements with a population exceeding 5000 in the Marmara Sea Basin
- Mobilisation and dissemination of good agricultural practices
- Proper disposal of solid and hazardous wastes with the latest technology
- Controlled use of pesticides, fertilizers, and chemicals
- Prevention of unregulated dumping of petrochemical industry wastes into the environment
- Identification, monitoring and prevention of sources that release micro-pollutants to nature and the environment
- Preventing all kinds of fishing that disrupt the marine ecosystem and adversely affect the balance of the food chain
- Prevention of entry of pollutants in maritime transport and port activities
- Establishment of quality monitoring stations in the sea, at river mouths and specific critical points of rivers
- Strict monitoring of micro-pollutants and microplastics in the sea basin
- International co-operation, projects, and studies should be carried out to reduce the pollution entering the Danube River and, thusly, into the Marmara Sea.

MARMARA SEA INTEGRATED STRATEGIC PLAN (2021-2024) ANALYSIS

As a result of the mucilage problem in the Marmara Sea reaching profound dimensions, an integrated strategic plan (MSISP) covering the years 2021-2024 was prepared by the Republic of Türkiye's Ministry of Environment, Urbanisation, and Climate Change. This plan discusses the general characteristics, physical parameters, ecological quality indicators, temperature and chlorophyll changes, mucilage, and pollution status of the Marmara Sea. Figure 5 shows the focal points of the strategic plan (MSISP, 2021: 17).

Figure 5 italicizes the comprehensive plan prepared to protect and improve the Marmara Sea, in particular towards the mucilage problem. Within the scope of this plan, a good number of implementations have so far been completed at the level of academic and official institutions. Figure 6 highlights some visuals of the Marmara Sea covered with mucilage (MSISP, 2021: 39).

When Figure 6 is checked, it turns out that the mucilage problem has reached severe dimensions in the Marmara Sea. To this end, various studies are planned by the Republic of Türkiye's Ministry of Environment, Urbanisation, and Climate Change within the framework of the strategic plan. These are as follows:

Figure 5. Focus points of the strategic plan

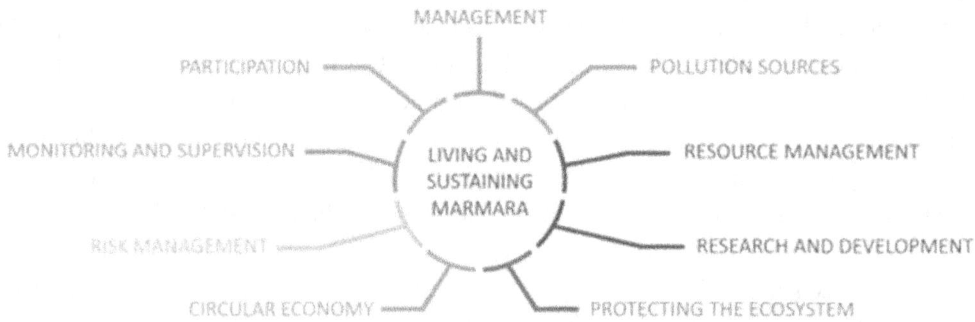

Figure 6. Images for covering the Marmara Sea with mucilage

- In June 2021, the studies were initiated, and the sediment cores were taken in the deep layer (30-200m) and at depths of 50 and 1200 metres.
- Since the oxygen levels in the water decreased because of the effects of mucilage, oxygen was pumped temporarily, and normalisation started with the cessation of mucilage production.
- Phosphorus and nitrogen values were monitored frequently in June.

- In 2007, it was figured out that the organisms in the mucilage problem that occurred in the Marmara Sea and the mucilage problem that occurred in 2021 were caused by similar organisms.

Within the scope of MSISP (2021-2024), an external environmental analysis was carried out in the Marmara Sea. This analysis has been enquired into in terms of economic, political, socio-cultural, technological, legal, and environmental aspects. Together with these, a SWOT analysis was also designated as a follow up of the both studies; weaknesses, strengths, opportunities and threats were identified in this frame.

Strengths:

- Having competent researchers and well-trained human resources
- The existence of an effective cooperation mechanism between stakeholders
- Widespread use of environmentally orientated production technologies and environmentally friendly products
- The existence of environmental regulations and the widespread use of international standards
- Increased individual and social awareness and high public demand for environmental protectio,
- Increased adoption of sustainability policies by businesses in the private sector

Weaknesses:

- Difficulties in managing environmental infrastructure costs
- Limited quantitative and qualitative capacities of wastewater treatment plants
- Limitations in environmental monitoring and inspection capacity
- Lack of an integrated environmental risk management system
- Difficulties encountered during the implementation of the plans

Opportunities:

- Increase in international environmental funding
- Environmental production technologies and popularisation of environmentally friendly products
- Green development trend is becoming widespread in the world
- Global consensus has been achieved in terms of adaptation to climate change

Threats:

- Acceleration and widespread global warming and climate change
- Population growth and the challenges of urbanisation
- Problems with financing sources in meeting environmental costs
- The existence of conflicts of duties and authorisations between institutions
- Difficulties in the competitiveness of environment-oriented production in enterprises

In addition to these studies, the Republic of Türkiye's Ministry of Environment, Urbanisation, and Climate Change has prepared the Marmara Sea Protection Action Plan comprising 22 articles. Within the frame of reference of this action plan, it was decided to realise the following actions (URL-1, 2023).

Actions

Action-1: A coordination committee consisting of the Ministry of Environment and Urbanisation, relevant institutions and organisations, universities, chambers of industry and NGOs will be established.		**Action-12:** Wastewater treatment plants will be monitored 24/7 online. 91 monitoring points in the Marmara Sea will be increased to 150.	
Action-2: Marmara Sea Integrated Strategic Plan will be prepared within three months, and the works will be carried out within the framework of this plan.		**Action-13:** A regional waste management action plan and marine litter action plan will be prepared and implemented within three months.	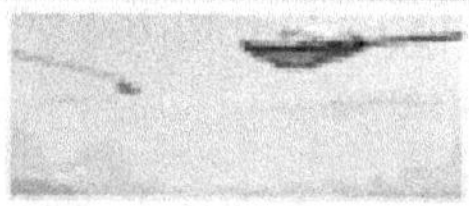
Action-3: Efforts to designate the entire Marmara Sea as a protected area will be initiated.		**Action-14:** Good agriculture and organic agriculture practices and pressurised and drip irrigation systems will be popularised.	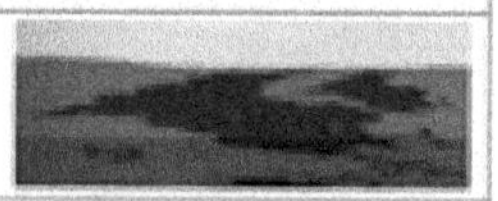
Action-4: Studies will be initiated to clean the mucilage in the Marmara Sea with scientifically based methods.		**Action-15:** In the basins associated with the Marmara Sea, artificial wetlands and buffer zones will be created in the river beds to prevent pollution from reaching the sea.	
Action-5: All existing regional wastewater treatment plants will be converted into advanced biological treatment plants.		**Action-16:** To prevent the pollution caused by olive, black water and cheese wastewater, technological transformations to reduce wastewater will be provided.	
Action-6: Discharge standards of wastewater treatment plants will be updated and re-implemented within three months.		**Action-17:** The use of cleaning agents containing phosphorus and surfactants will be gradually reduced. Organic cleaning products will be encouraged.	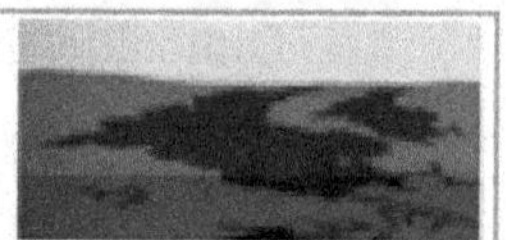
Action-7: Reuse of treated wastewater will be increased and supported wherever possible.		**Action-18:** All ghost nets in the Marmara Sea will be cleaned within one year.	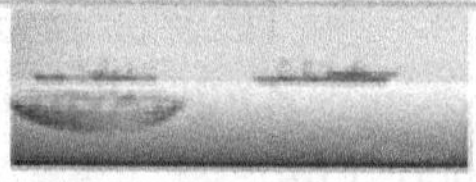
Action-8: The transition of OIZs to advanced treatment technologies will be accelerated through rehabilitation and improvement works.		**Action-19:** Ecosystem-based fishing activities will be ensured, and protected areas will be developed.	
Action-9: Public-private sector cooperation models will be implemented to make the construction and operation of wastewater treatment plants much easier.		**Action-20:** Economic support will be provided to fishermen who suffered losses due to mucilage.	
Action-10: Arrangements will be made within three months to prevent wastewater discharge from ships into the Marmara Sea.		**Action-21:** Activities will be carried out to prevent marine pollution and to raise awareness among citizens.	
Action-11: Clean production techniques will be popularised in shipyards.		**Action-22:** Measures will be taken to minimise the effects of hot waters from cooling waters and thermal facilities on the Marmara Sea.	

*All images were accessed from (URL-1, 2023).

CONCLUSION AND RECOMMENDATIONS

Within the framework of the studies, the existing reports and action plans prepared for the mucilage problem occurring in the Marmara Sea, some recommendations have been made regarding the causes and dimensions of mucilage. The primary points pertaining to the mucilage problem are as follows (TAoS, 2021):

- Mucilage is a condition that causes problems for all the living things. It also negatively affects seafood fishing and species diversity.
- The main reason for mucilage is the overload of industrial and agricultural nutrients and excessive biological production far above the carrying capacity of the Marmara Sea ecosystem.
- Mucilage causes adverse conditions such as the lack of oxygen in the seas and ecosystem destruction.
- Marmara region is the first region of Türkiye respecting population density and economic activities.
- There is an increase in the wastewater load of cities and the number of pollutants depending on the rapid population growth.
- Inadequacy of wastewater treatment plants is another issue.
- Solving contaminations such as mucilage entails a not short term but a long term approach.
- Research should not be seasonal, and studies should be carried out in every period of the year.
- Preventing overfishing and protecting the marine ecosystem are amongst the significant precautions.
- Systematic examination of terrestrial inputs, continuous control of wastewater and nutrient values are all of utmost importance.

It goes without saying that the practicum and other points to consider are also worth mentioning here. These can be accessed below (URL-2, 2023):

- Multidisciplinary approaches should be employed whilst fighting against difficult situations faced during the elimination the mucilage problem.
- The Marmara Sea environment should be declared a "protection zone" within a specific plan.
- Measures should be taken urgently to prevent problems in the fisheries sector.
- Emergency action plans should be prepared for the accidents in the straits.
- The necessary arrangements should be made to assure that wastewater is not returned to the sea even after being cleaned but be used in irrigation.
- The inputs of terrestrial origin should thoroughly be grasped, advanced biological treatment methods should be utilized, and deep discharge operations should be in action.
- Temporal and spatial distribution of mucilage should be found out, and the mobile platforms on the Marmara should make sure reactive oxygen dosing is maintained and supplied.
- The integration of the present treatment plants with more advanced treatment technologies should be realised.
- Higher education policies for the seas should be established to meet the need for infrastructure and trained people taking an academic outlook on the problem.
- The areas of education and research should be prioritised in project calls, and researchers should be funded with larger budgets to be sure of a scientific understanding.

REFERENCES

Abbasnia, M., & Toros, H. (2019). Analysis of long-term changes in extreme climatic indices: A case study of the Mediterranean climate, Marmara Region, Turkey. İçinde Meteorology and climatology of the Mediterranean and Black Seas (ss. 141-153). Springer.

Abbriano, R. M., Carranza, M., Hogle, S., Levin, R., Netburn, A., Seto, K., Snyder, S., & Franks, P. (2011). Deepwater Horizon oil spill: A review of the planktonic response. *Oceanography (Washington, D.C.)*, *24*(3), 294–301. doi:10.5670/oceanog.2011.80

Aktan, Y., Dede, A., & Ciftci, P.S. (2008). *Mucilage event associated with diatoms and dinoflagellates in Sea of Marmara, Turkey*. Harmful Algae News.

Aydın, M. E. (2021). Mucilage formation in Marmara Sea possible causes and recommendations. I. Ozturk & M. Şeker (Eds.). In Ecology of the Marmara Sea: Formation and Interactions of Marine Mucilage, and Recommendations for Solutions (pp.49-68). Tek Ses Ofset Matbaacılık. Ankara.

Corbin, J., & Strauss, A. (2008). *Basics of qualitative research: Techniques and procedures for developing grounded theory* (3rd ed.). Sage. doi:10.4135/9781452230153

Danovaro, R., Fonda Umani, S., & Pusceddu, A. (2009). Climate change and the potential spreading of marine mucilage and microbial pathogens in the Mediterranean Sea. *PLoS One*, *4*(9), e7006. doi:10.1371/journal.pone.0007006 PMID:19759910

Johns, D.G., & Reid, P.C. (2001). *An overview of plankton ecology in the North Sea* (TR_005; Strategic Environmental Assessment – SEA2). SAHFOS.

Karadurmus, U., & Sari, M. (2022). Marine mucilage in the Sea of Marmara and its effects on the marine ecosystem: Mass deaths. *Turkish Journal of Zoology*, *46*, 93–102.

Keleş, G., Yılmaz, S., & Zengin, M. (2020). Possible economic effects of mucilage on Sea of Marmara fisheries. *International Journal of Agriculture Forestry and Life Sciences*, *4*(2), 173–177.

Koncagul, M., Dulger, N. E., & Yinanc, A. (2022). Formation and the effect of musilage in Marmara Sea and in the world. *European Journal of Engineering and Applied Sciences*, *5*(2), 73–79.

Marmara Sea Integrated Strategic Plan. (2021). Republic of Türkiye Ministry of Environment, Urbanization, and Climate Change. Ankara.

Mecozzi, M., Acquistucci, R., Noto, V. D., Pietrantonio, E., Amici, M., & Cardarilli, D. (2001). Characterization of mucilage aggreagtes in Adriatic and Tyrrhenian Sea: Structure similarities between mucilage samples and the insoluble fractions of marine humic substances. *Chemosphere*, *44*(4), 709–720. doi:10.1016/S0045-6535(00)00375-1 PMID:11482660

Mecozzi, M., Pietrantonio, E., Di Noto, V., & Pápai, Z. (2005). The humin structure of mucilage aggregates in the Adriatic and Tyrrhenian seas: Hypothesis about the reasonable causes of mucilage formation. *Marine Chemistry*, *95*(3), 255–269. doi:10.1016/j.marchem.2004.09.005

Merriam, S. B. (1988). *Case study research in education: A qualitative approach*. Jossey-Bass.

Merriam, S. B., & Tisdell, E. J. (2016). *Qualitative research: A guide to design and implementation* (4th ed.). Jossey Bass.

Morgan, H. (2022). Conducting a qualitative document analysis. *The Qualitative Report*, *27*(1), 64–77. doi:10.46743/2160-3715/2022.5044

Orhon, D., Sözen, S., Kirca, V. S. O., Duba, S., Mermutlu, R., & Sumer, B. M. (2021). Pollutant Dynamics between The Black Sea and The Marmara Sea: Basis for wastewater management strategy. *Marine Pollution Bulletin*, *168*, 112388. doi:10.1016/j.marpolbul.2021.112388 PMID:33940368

Orman ve Su İşleri Bakanlığı (OSB), (2013). Su Yönetimi Daire Başkanlığı. *Marmara Denizi kirlilik raporu.*

Penna, N., Berluti, S., Penna, A., & Ridolfi, F. (2000). Study and monitoring of mucilage in the Adriatic Sea. *Water Science and Technology*, *42*(1-2), 299–304. doi:10.2166/wst.2000.0329

Savun-Hekimoğlu, B., & Gazioğlu, C. (2021). Mucilage problem in the semi-enclosed seas: Recent outburst in the Sea of Marmara. [IJEGEO]. *International Journal of Environment and Geoinformatics*, *8*(4), 402–413. doi:10.30897/ijegeo.955739

Schiaparelli, S., Castellano, M., Povero, P., Sartoni, G., & Cattaneo-Vietti, R. (2007). A benthic mucilage event in North-Western Mediterranean Sea and its possible relationships with the summer 2003 European heatwave: Short term effects on littoral rocky assemblages. *Marine Ecology (Berlin)*, *28*(3), 341–353. doi:10.1111/j.1439-0485.2007.00155.x

Şeker, M., & Yalçın, H. (2021). An analysis on mucilage (sea snot) research. I. Ozturk & M. Şeker (Eds.). In Ecology of the Marmara Sea: Formation and Interactions of Marine Mucilage, and Recommendations for Solutions (pp.69-84). Tek Ses Ofset Matbaacılık. Ankara.

Turkish Academy of Sciences. (2021). *TÜBA Müsilaj-Deniz Salyası Değerlendirme Raporu (3 Haziran 2021).* Turkish Academy of Sciences. https://www.tuba.gov.tr/tr/yayinlar/suresiz-yayinlar/bilim-ve-dusunce/marmara-denizinin-ekolojisi-deniz-salyasi-olusumu-etkilesimleri-ve-cozum-onerileri-1.

URL-1. (2023). *Marmara denizi koruma eylem planı.* URL-1. https://webdosya.csb.gov.tr/db/marmarahepimizin/eylemplani/ Retrieved Date: 29.05.2023

URL-2. (2023). *Marmara belediyeler birliği sonuç bildirgesi.* URL2. https://www.marmara.gov.tr/ Retrieved Date: 29.05.2023

Yanarateş, E., & Yılmaz, A. (2020). Metaphorical perceptions of prospective teachers towards the concept of environmental sensitivity. *Gazi University Journal of Gazi Educational Faculty*, *40*(3), 1019–1050. doi:10.17152/gefad.699406

Yılmaz, A., & Yanarateş, E. (2020). Determination of metaphorical perceptions of prospective teachers on the concept of "water pollution" through triangulation. *Kastamonu Education Journal*, *28*(3), 1500–1528. doi:10.24106/kefdergi.722554

Chapter 6
Global Impact of Plastic Pollution and Its Management for Sustainable Development

Monika Chaudhary
Shobhit Institute of Engineering and Technology (Deemed), India

Amar Prakash Garg
https://orcid.org/0000-0003-0613-9495
Swami Vivekanand Subharti University, India

ABSTRACT

Plastic pollution is a global occurrence from villages to urban, farms to deserts, deep oceans to mountaintops, tropical landfills to arctic snow including in food chain. Over the past 60 years, there has been a steady build-up of plastic garbage on land and ocean's surface, mainly because of the unplanned, unrestricted use of plastic in routine use, non-cycling and use and throw habit of a common man. As usage of this strong and cheap material has increased over the past few decades, the waste problem has gotten worse because of non-biodegradation in nature. Since the turn of the century, the amount of plastic produced worldwide has quadrupled, reaching over 400 million metric tonnes annually in 2021. While plastic items typically have a lifespan of ten years, depending on their composition and method of disposal, plastics can take up to 500 years to degrade. The environment is harmed by the removal of all this plastic. Only 16% of plastics are recycled; the remainder is either disposed of in landfills or burned.

INTRODUCTION

Plastics are all around us and have gained popularity in scholarly circles. Numerous research have examined the analytical techniques, sources, abundance, travel, destiny, and environmental degradation, hazards to the natural world, animals, and even human health. The properties of plastic pollution, which are essential to comprehending this new issue, are yet unclear (Li et al, 2021).Plastic pollution of the environment has attracted a lot of public attention, but the issue has also been the subject of debate, with

DOI: 10.4018/978-1-6684-9863-7.ch006

arguments suggesting money would be better used on other issues, such global warming (Lavers et al., 2022). Global attention is presently being paid to environmental plastic pollution. The ecosystem gets contaminated when used or abandoned plastic garbage is dumped improperly. A critical problem and a key source of worry for soil contamination is the disposal of municipal wastewater effluent, sewage sludge landfills, and plastic mulch from agricultural operations (Chae et al., 2018).

Global plastic production increased from 1.5 million tonnes in the 1950s to 335 million tonnes in 2016 (Li et al. 2016; Lebreton et al. 2017; Statista 2018). Reports indicate that a staggering 79% of plastic items have not undergone appropriate recycling procedures and, instead, have been disposed of in landfills or released into the environment. The presence of plastic waste in coastal waterways was initially observed by researchers in the 1970s (Carpenter and Smith 1972). However, the problems weren't given much attention until the term "micro plastic" (with an effective size smaller than 5 mm) was proposed in 2004 by Thompson et al. (2004). Researchers have since written an increasing number of pertinent publications on topics including the origin, prevalence, destiny, incidence, and sink of micro plastics in natural settings. The so-called "micro plastic pollution" is most likely this (Karbalaei et al. 2018).

The idea of plastic pollution is when plastics—regardless of their sizes, forms, or types—are introduced into the environment and pose a risk to the environment, species, or even human health. Plastics, when introduced into natural habitats, have the potential to travel from the land to rivers and ultimately the ocean (Jambeck et al. 2015). Plastics can have a negative effect on the environment, animals, and even human health throughout the trip (Galloway and Lewis 2016; Sussarellu et al. 2016). In the meanwhile, persistent organic pollutants (POPs), heavy metals, and microbes in the environment might travel through plastics (Rodrigues et al. 2018; Massos and Turner 2017; Neto et al. 2019). Additionally, it has been shown that compounds such as monomer, oligomer, and additives are released into the environment by plastics, suggesting that these materials may potentially be sources of some harmful substances (Chen et al 2019).

Diversities (such as sizes, kinds, colours, and origins), resistance to degradation, combined pollution, and possible dangers are all characteristics of plastic pollution. Recent research, however, has only emphasized a few facets of environmental plastic contamination. A wide array of plastic items plays a significant role in shaping the extensive scope of plastic pollution. In our daily routines, we depend on a multitude of plastic products from the consumer market. To illustrate, some of the most widely used plastics in this market encompass polyethylene (PE), polypropylene (PP), polyvinyl chloride (PVC), polyethylene terephthalate (PET), and polystyrene (PS). These plastics, in particular, are found in various everyday items, such as packaging materials, containers, and consumer goods. As a result, the diversity of plastic pollution is greatly influenced by the prevalence and widespread use of these plastic materials, which can persist in the environment for extended periods, causing environmental and ecological concerns (Bond et al., 2018). Despite the variety of plastics, they were always used in combinations that included monomers, oligomers, and additives in daily life (Hermabessiere et al., 2017).

Moreover, the sources of plastic pollution are intricate and primarily originate from various human actions. It has been determined that plastics can enter aquatic, terrestrial, and atmospheric ecosystems either directly or indirectly via a range of pathways. These pathways encompass inappropriate disposal of plastic waste, sewage treatment facilities, household sewage, landfill sites, agricultural practices, and even urban dust (Windsor et al., 2019, Bai and Li 2020).

This indicates that plastics have diverse sources, and the environment exhibits a wide range of plastic materials in terms of their colors, sizes, and shapes. Nevertheless, due to the absence of standardized analytical methods and effective tools, we remain uncertain about both the rate at which plastics break down and the quantity of plastics within the environment. These discrepancies likely contribute to the variability in plastic pollution found in the environment, making it difficult to establish uniform and consistent analytical methods. This, in turn, hinders the development of relevant studies. To gain a deeper understanding of plastic contamination in the ecosystem, further research is imperative.

Because of their long-lasting nature, plastics have likely had the opportunity to interact with various aspects of the environment. There is a growing consensus among experts that plastics can act as carriers for other environmental pollutants (Wang et al. 2018; Engler 2012). For example, the surface of plastic debris contained heavy metals (Cd, Pb, Cu, Zn, etc.) and polycyclic aromatic hydrocarbons (PAHs), polychlorinated biphenyls (PCBs), tetracycline, and dichlorodiphenyltricgloroethane (DDT). As a result, mixtures of plastic and pollutants were released into the environment (Tang et al. 2018; Velzeboer et al. 2014; Brennecke et al. 2016).

Plastics are ubiquitous in aquatic, atmospheric, and terrestrial systems, according to research, and given their stability and endurance in the environment, these particles may stay there for a long time, which probably adds to the worldwide spread of plastics (Do Sul and Costa 2014). Because different systems have different characteristics, different polymers enter different systems by quite diverse paths. Examples of the main contributors of plastic debris in the aquatic environment are wastewater treatment plants, residential trash, urban pollution, industrial operations, and storms (Koelmans et al. 2019).

Due to long-distance transportation through many systems, plastic pollution is a problem that affects everyone in the globe. Fortunately, this environmental problem on a worldwide scale is receiving more attention.

GLOBAL PLASTIC PRODUCTION

Plastics are a broad category of synthetic organic polymers that have a crucial and pervasive role in how modern civilization functions. The amount of plastic created has significantly expanded since the 1940s, when mass production of plastic items first started. The worldwide plastic market was estimated to be worth 593 billion dollars in 2021, and it is anticipated to grow significantly during the next ten years. ExxonMobil, Sinopec, and Lyondell Basell are a few of the major companies in the plastics sector.

With a compound annual growth rate (CAGR) of 5.8% from 2009 to 2021, annual worldwide plastic output peaked at 390.7 million metric tonnes. China has taken over as the world's top manufacturer of plastics in recent years, making up around 32% of worldwide output. Despite the pandemic's negative economic effects, China has regularly produced seven million metric tonnes or more of plastic items each month. Comparatively, the output of plastic in Europe fell in 2020 before rising once again to 57.2 million metric tonnes in 2021. The United States will generate around 56.89 million metric tonnes of plastic in 2022, making North America (NAFTA) the second-largest producer.

Plastics are flexible materials that may be used for many different products and applications. However, not all plastic is made equally. The roughly seven primary varieties of plastic that are made and used in substantial amounts include polyethylene terephthalate (PET), high-density polyethylene (HDPE), polyvinyl chloride (PVC), low-density polyethylene (LDPE), polypropylene (PP), polystyrene/styrofoam (PS), and other plastics. With a production volume of 75.40 million metric tonnes in 2020 and a predicted increase to 107.20 million metric tonnes by 2050, polypropylene (PP) stands out among them. The automotive, packaging, textile, and home goods sectors are just a few of the many areas where polypropylene (PP) is utilized. Due to its outstanding mix of qualities, including toughness, chemical resistance, and thermal stability, it may be used to create consumer goods, food packaging, and automobile parts.

Although one of the most talked-about anthropogenic environmental concerns of our day is the potential ecological harm that might follow from negligent plastic disposal, plastic is still a very significant commodity that is utilized in almost all facets of our daily life. As a result, the need for plastic is expected to increase for the foreseeable future, with 589 million metric tonnes of plastic expected to be produced in 2050. In 2020, there were 27 million metric tonnes of polyethylene terephthalate (PET), a thermoplastic polymer most commonly found in plastic bottles. By 2030, 42 million metric tonnes are expected to be demanded globally. Additionally anticipated rise is the desire for environmentally friendly products like long-lasting plastic packaging. However, COVID-19 has had an effect on industry-wide sustainability objectives. Due to increased demand for single-use plastic items like face masks and gloves as a consequence of hygiene concerns, some localities had to temporarily lift their restrictions on these products (Statista, 2023).

China was by far the world's greatest manufacturer of plastic in 2021, contributing 32% of the world's total production of plastic materials. With over 20% of worldwide output, North America was the second-largest producer of plastics in that year. 391 million metric tonnes of plastic were produced globally in total in 2021. Almost every aspect of life uses plastics, from daily activities to industrial uses. Many conventional materials, including wood, stone, horn and bone, leather, paper, metal, glass, and ceramic, have been replaced by plastics on a worldwide scale. The usage of plastics worldwide is anticipated to increase over the ensuing decades, reaching more than 1.2 billion metric tonnes by 2060. It is not unexpected that China is also the world's top producer of plastic given its position as the world's largest manufacturer and exporter of commodities. China produces between six and eight million metric tonnes of plastics per month on average.

INDIAN PLASTIC PRODUCTION

When India started producing polystyrene, the country created the plastic industry. With more than 20,000 processing units, India's plastics sector has expanded quickly since its start and is now a major producer of plastics worldwide. It is a multi billion dollar business that employs around four million people and is a significant contribution to India's economy. With about 1.5 million metric tonnes of polymer exports from India in 2021, it is also a major provider to other countries.

India's usage of plastic surged 23 times between 1990 and 2021, reaching almost 21 million tonnes. With this quick growth, the amount of plastic used per person increased from one kilogramme to 15 kilogrammes. India stands as the third-largest consumer of plastics globally, following China and the United States, utilizing approximately 6% of the world's total plastic supply. Predictions for the coming decades suggest that factors such as population growth and economic expansion will further boost the demand for plastics. Estimates indicate that by 2060, India's plastic consumption could exceed 160 million metric tonnes. This would signify a substantial increase, more than tripling the country's share of global plastic consumption compared to its current levels.

In the fiscal year 2022, India's total thermoplastic demand was 19.3 million tonnes. 32 percent or 6.1 million tonnes of the demand that year went to polypropylene (PP). Following closely behind with a combined share of 29% were high-density polyethylene (HDPE) and linear low-density polyethylene (LLDPE). Numerous other sectors, including those that make home goods, cars, packaging, electrical and electronics, among many others, employ polymers to create products that touch practically every aspect of life. For rigid packaging, polyethylene terephthalate (PET) is a common material choice. Water and soft drink bottles are typical PET packaging materials. Because it is strong, colourless, light, and inexpensive to make compared to glass packaging, PET is the perfect material for numerous packaging applications. Additionally, it is entirely recyclable (Statista, 2023).

GLOBAL PLASTIC POLLUTION

One of the major problems now affecting the globe is plastic garbage, which has far-reaching effects. It may impair human health, degrade ecosystems, and harm animals, especially marine species. It can be found fouling everything from the highest mountains to the deepest ocean trenches. Over the past several decades, as more of this flexible material has been used, the waste issue has become worse. Over 400 million metric tonnes of plastic will be manufactured yearly in 2021, having tripled since the year 2000. Polymers can take up to 500 years to decompose, although plastic objects normally have a 10-year lifetime, contingent upon their composition and disposal technique.

More than 350 million metric tonnes of plastic waste are presently produced annually by humans. Global plastic trash production is anticipated to quadruple by 2060, reaching a startling one billion metric tonnes, if existing policies remain unchanged. The amount might fall to fewer than 700 million metric tonnes if new laws and regulations are passed, such as higher charges on the usage of plastics and a greatly enhanced worldwide recycling system. Though less than 10% of plastic garbage is presently recycled annually, this objective is still lofty. The great majority of produced plastic garbage is either land filled or burned, which releases dangerous chemicals. Another quarter is mishandled or left in disarray.

Mishandled plastic garbage is primarily burnt in open pits or dumped illegally, but a sizeable quantity also escapes into waterways including rivers and seas. An estimated 30 million metric tonnes of plastics are predicted to have collected in the ocean between 1970 and 2019, while more than 100 million tonnes are anticipated to have accumulated in rivers and lakes. Such significant plastic trash contamination in rivers can have catastrophic effects on ecosystems and marine life. More than 80 percent of the plastic debris dumped into the ocean worldwide comes from Asia. While wealthy places contribute around 5% of the ocean's plastic garbage, they frequently export massive volumes of plastic debris to underdeveloped countries for processing. Several of these places lack the capacity (Statista, 2023).

INDIAN PLASTIC POLLUTION

According to information supplied by 35 States/UTs, the predicted amount of plastic garbage produced in 2020–21 is 4126997TPA.the trend in garbage production per person during the previous six years (2016–21). Over the past five years, it has been noted that the amount of plastic garbage produced per person has virtually doubled. The states with the greatest per capita plastic trash creation were Delhi, Goa, and Telangana, whereas the states with the lowest per capita plastic waste generation were Sikkim, Meghalaya, and Tripura (CPCB, 2021).

Figure 1. State/UT wise plastic waste generation
[Source: Author's creation]

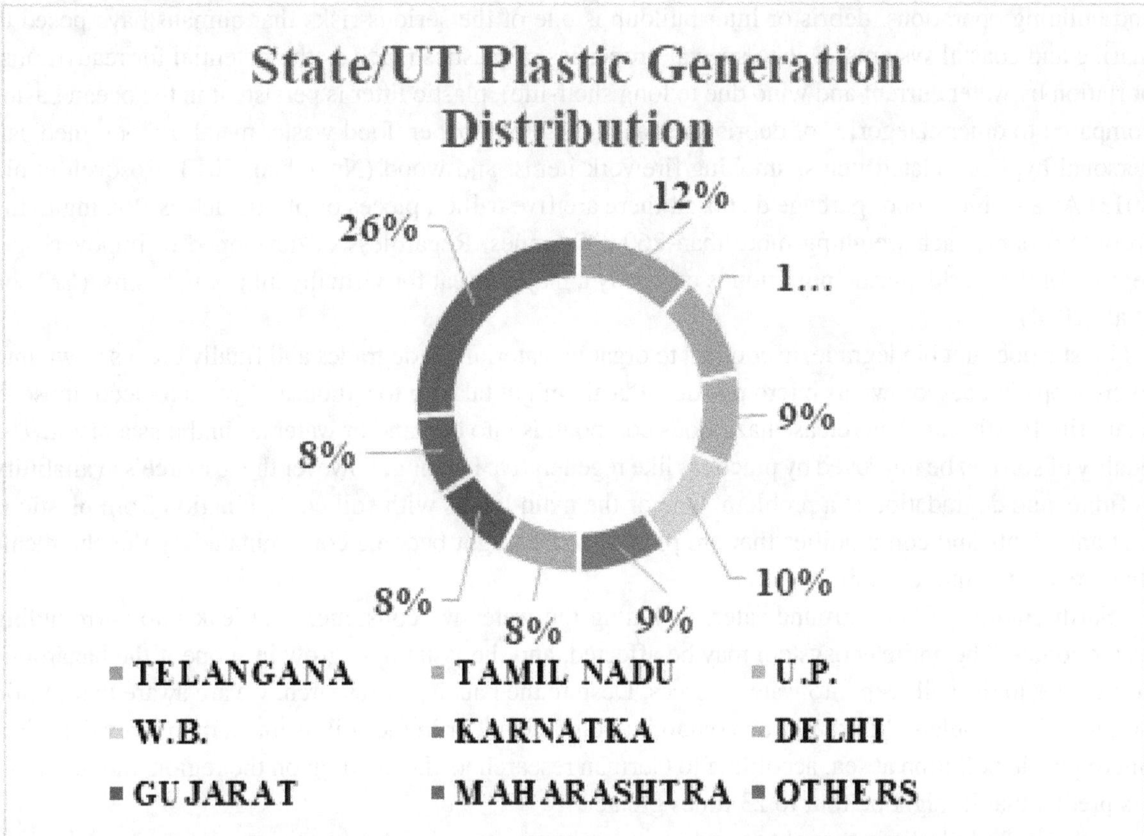

PLASTIC POLLUTION IN SOIL AND MARINE ENVIRONMENT

For many years, there has been an increasing global concern over marine plastic pollution (Barnes et al., 2009). Plastic pollution is a chronic worldwide pollution issue that requires several solutions to limit its origins (i.e., leakage) at the local, national, and global levels since it knows no borders (Adam et al., 2020; Clayton et al., 2020; Bezerra et al., 2021).

On the basis of differences in size, historical, cultural, geographical, and scientific aspects, the world's oceans are divided into five major oceans and numerous seas. The most well-known marine systems that humans have invaded include those in the Atlantic, Pacific, Indian, Arctic, and Antarctic ocean basins. The International Hydrographic Organization designated the Southern Pole (Antarctic) ocean basin as the fifth ocean basin. For the benefit of mankind, all ocean basins function as significant biological and economic systems. Lagoons and estuaries are examples of distinctive, transitional ecosystems formed when freshwater lotic systems interact with oceans and seas. Since seawater and freshwater combine on the continental shelf of the marine environment, this region develops a distinctive coastal ecology. (Reddy et al., 2018).

Because aquatic ecosystems and the terrestrial environment are intertwined, changes in one system have an effect on the other. The anthropogenic activities and other causes have strained the coastal and marine ecosystems for many years. The environment is being physically destroyed and pollution are two examples of these stressors (Adams, 2005; Richmond, 2015). As a result of unsustainable development and building operations, debris or litter buildup is one of the serious risks that humans have posed to marine and coastal systems. Due to special properties of plastics (such as the potential for ready transportation by water current and wind due to long shelf-life), plastic litter is persistent in the ocean basins compared to other categories of debris such as glass, cloth, paper, food waste, metal, rubber, medical/personal hygiene-related items, smoking/firework items, and wood (Nualphan, 2013; Rosevelt et al., 2013).As a result of poor garbage disposal, there are five trillion pieces of plastic debris floating in the world's oceans, each weighing more than 260,000 tonnes. Regardless of developed or impoverished regions of the world, plastic pollution is currently a major threat for virtually all ocean basins. (Eriksen et al., 2014).

Plastic does not biodegrade, in contrast to organic materials. It degrades and finally breaks down into microscopic pieces known as micro plastics. Plastic might take up to a thousand years to decompose in a landfill. Plastic can then release hazardous compounds into the land or water around it as a result. The quality of soil can be improved by practices like regenerative farming, however the resource's availability is finite, and degradation is a problem. One of the main issues with soil contamination from plastic is that any plants and commodities that are planted there might become contaminated by the chemicals that are leaking into the soil.

Furthermore, polluted groundwater, including the water we consume, may leak into surrounding water bodies. The entire ecosystem may be affected, and the issue may grow in scope if the hazardous substances in the soil seep into water sources. Despite the paucity of research, we are aware that plastic pollution is a problem. Micro plastic contamination on land, or in the soil, is much more prevalent than micro plastic pollution at sea, according to German researchers. Depending on the region, the researchers predict that it might be four to 23 times greater.

A third of all plastic is thought to end up in freshwater or soil. This plastic eventually degrades into micro plastics, which further degrade into nanoparticles smaller than 0.1 micrometers, which then enter the food chain through crops. Plastic particles acquire new chemical characteristics as they degrade, increasing the possibility that they will turn harmful to the creatures around. We are aware that chemicals added to plastic, such as biphenyl A and phthalates, can leak and affect the hormones of both invertebrates and vertebrates, such as humans. These nanoparticles can penetrate the blood-brain barrier and the placenta, triggering inflammation in humans and animals, as well as altering gene expression.

SOURCES OF PLASTIC WASTE ON SOIL AND MARINE ENVIRONMENT

1. Packaging

There are plastic packages everywhere. Many of the items we purchase from stores and practically all items that are delivered to our front door are covered by it. Plastic packaging has taken the place of paper, glass, metal, and other recyclable materials since the 1950s. Between 1950 and 2017, 3.4 billion tonnes of plastic packaging are thought to have entered our lives (Geyer, 2020).

To transition from plastic packaging to environmentally sustainable alternatives, it's crucial to consider the advantages of plastics as well as any potential drawbacks throughout the entire life cycle of the product. Presently, plastic resins are primarily utilized in packaging, constituting 36% (equivalent to 158 million tonnes) of the total global plastic production (UNEP, 2021). Commercial, retail, home, travel, and agricultural industries all use plastic packaging. Consumption rates differ between and within nations, with emerging nations often relying less on packaging. China consumes about 14 kilogrammes of plastic packaging year per person, compared to an average of 174 kg in Europe (Eurostat 2020).

The majority of plastic packaging is thrown away in a quite short amount of time. 46% (158 million tonnes) of the entire yearly plastic waste production comes from discarded packaging. It is believed that residential garbage accounts for the majority of plastic packaging waste. A 2010 evaluation conducted by the Waste and Resources Action Programme (WRAP) revealed that households in the United Kingdom were accountable for 73% of the total plastic packaging waste. Moreover, aquatic debris consists predominantly of discarded plastic packaging, with 15.9% found in oceans and 74.5% in rivers (Schwarz et al. 2019).

The primary factors behind plastics' steadily rising use are their affordability and ease. Plastic packaging may help keep food fresher for longer, reduce food waste, and provide customers access to a wider range of foods. Plastic packaging also makes it simple to ship and distribute products (UNEP, 2021).

Plastic packaging contributes significantly to global food security in this way. It might not always be feasible or sustainable to replace plastic packaging. To lessen reliance on plastic packaging without jeopardising food security, efforts are still being made (Guillard et al. 2018).

2. Agriculture

In agriculture, plastic items are widely employed, for instance to boost crop output and boost the effectiveness of water and agrochemical usage. The term "agriplastic" encompasses a range of agricultural products, such as films used for greenhouse and tunnel coverings, soil mulch (which serves purposes like weed control, water conservation, soil temperature management, and fertilizer application), shading materials, trays for seedlings, containers for pesticides, protective mesh, and tubing for irrigation (Scarascia-Mugnozza et al. 2012). Low-density polyethylene (LPDE), linear low-density polyethylene (LLDPE), polypropylene (PP), and polyvinyl chloride (PVC) are the polymers most frequently employed in these items (Briassoulis 2013; Plastics Europe 2019). Estimating the total quantity of plastics employed in agriculture poses a difficulty. In 2012, worldwide usage was reported to be around 6.5 million tonnes annually, and projections indicated that demand could rise to a range between 7.3 million and 9 million tonnes by 2015 (Cassou 2018). China increased its use of plastic mulch from 0.11 million tonnes to 1.4 million tonnes between 1981 and 2014, covering an area of around 18.14 million hectares. In the EU, 708,000 tonnes of plastics are thought to have been used in crop and livestock production in 2019 (Liu et al. 2014; World Agriculture 2017).75% of these were plastic films, which were mostly used for bale wrapping for silage preparation, low tunnels, and plastic mulch (EIP-AGRIE Focus Group 2021).

Mulch residue has been produced in significant quantities as a result of the widespread usage of plastic mulch and a lack of systematic collection and treatment. The mulch gradually becomes fragmented as a result of weathering and deterioration. Studies show that bigger plastic particles and these fragments build up in soil. In locations where the mulch has been utilized for more than 10 years, topsoil mulch residue has been detected at levels of 50 to 260 kg per hectare, confirming that mulching is a substantial source of soil pollution with both micro plastics and macro plastics (Huang et al. 2020).

Due to high contamination levels (up to 40–50% by weight contamination by pesticides, fertilizers, soil and debris, wet vegetation, silage juice water, and UV stabilizers) and collecting challenges, agricultural plastics, particularly plastic films, are difficult to recycle. As a result, they are frequently buried or left in rivers and fields (Kasirajan and Ngouajio 2012). These disposal techniques cause soil deterioration, and precipitation runoff and tidal washing can contaminate soils and cause micro plastic leakage into the marine ecosystem (Li et al. 2018; SAPEA 2019; Hurley et al. 2020).

Additionally, chemicals in leftover plastic film might harm crop development, soil structure, nutrient delivery, and salt content. Plastic mulch has the risk of depleting soil organic matter reserves, increasing soil water resistance, emitting greenhouse gases, and deteriorating soil quality (Steinmetz et al. 2016). Additionally, pollutants that can be transferred up the food chain can be absorbed and concentrated in the micro plastics that are generated when agricultural plastics are fragmented (Galloway and Lewis 2016).

3. Fishing and Aquaculture

The output of fish worldwide in 2018 was above 179 million tonnes. Aquaculture output has increased in recent years, but capture fisheries production has stayed mostly stable. Presently, it accounts for about half of all fish output. The global fishing industry employs roughly 60 million fishermen, operating a fleet of over 4.5 million boats. Of these boats, around 65.5% (equivalent to 39 million) are engaged in capture fisheries, while the rest are involved in aquaculture (FAO 2020). Even though there is more study being done to define the origins of marine litter, it is still difficult to estimate how much comes from fisheries and aquaculture. (GESAMP, 2020).

Fish aggregating devices (FADs), constructed from non-biodegradable materials, fall under the category of abandoned, lost, or otherwise discarded fishing gear (ALDFG). ALDFG encompasses various items such as nets, mono filament and multi filament lines, hooks, ropes, floats, buoys, sinkers, and anchors, made from materials like metal, plastic, and concrete. It is estimated that approximately 5.7% of all fishing nets, 8.6% of all traps, and 29% of all fishing lines are lost annually on a global scale (MEPC 2020; FAO ISSCFG, 2013). ALDFG may entangle and ingest marine species, which can have detrimental effects (Richardson et al. 2019).

Based on the product and location, aquaculture generates a variety of marine waste. Extensively utilized in floats and marine cage collars, expanded polystyrene (EPS) is the plastic kind that has been reported the most frequently. Cage nets and plastic harvest bins are some other typical waste materials (Huntington, 2019). A survey of 64 distinct objects, 19 of which were specifically related to aquaculture, was found in the North, Baltic, and Mediterranean Seas. Estimates of the quantity of aquaculture waste flowing into the oceans exhibit significant variations depending on the approach and methodology employed (Sandra et al., 2020). For example, losses in the European Economic Area have been estimated to have ranged from 3,000 tonnes to 41,000 tonnes each year (Sherrington et al., 2016).

4. Unsustainable Production

The production of plastic is an unsustainable practice, and it constitutes a significant contributor to the problem of plastic pollution. In recent years, there has been a substantial surge in the annual production of millions of tons of plastic. Unfortunately, a substantial portion of this plastic is not either recycled or disposed of properly, leading to a significant amount of plastic waste finding its way into the environment. This situation highlights the environmental challenges associated with plastic production and management. The rapid increase in plastic production exacerbates the problem, as much of the plastic waste is not effectively managed, resulting in environmental pollution. To combat this issue, it is imperative to promote sustainable plastic production, recycling, and proper waste disposal practices.

5. Single Use Plastic

The production and usage of single-use plastics stand out as one of the most prevalent origins of plastic pollution. This category encompasses items like disposable plastic water bottles, bags, and straws, which are designed for a single use and then discarded. While these single-use plastics may seem convenient, they pose a substantial threat to the environment. This is because they make a significant contribution to the immense volumes of plastic waste that is challenging to recycle and can persist in the environment for many years.

Single-use plastics, despite their short-lived utility, carry long-lasting consequences for the environment. Their widespread consumption results in an accumulation of plastic waste that can have adverse effects on ecosystems, wildlife, and even human health. This underscores the importance of reducing the use of single-use plastics and adopting more sustainable alternatives to mitigate the environmental impact of plastic pollution.

6. Poor Waste Management

Ineffective waste management is a significant contributor to plastic pollution. If garbage is not disposed of correctly, it may enter the ecosystem and contaminate the land, the air, and the water. In poor nations with little resources and weak waste management systems, this is a significant problem. As a result, there is a substantial amount of plastic pollution since plastic debris is frequently burnt or deposited in landfills, rivers, and seas.

7. Improper Disposal

Inadequate disposal practices play a substantial role in exacerbating plastic pollution. Many individuals, often unaware of the critical importance of responsible disposal, tend to discard plastic waste haphazardly, leading to its accumulation in the environment. This negligent disposal encompasses instances where plastic is carelessly left on the ground or deposited into water bodies. The consequences are profound: wildlife bears the brunt of this pollution, and it permits plastic to infiltrate the food chain, posing a significant risk to both ecosystems and human health.

The failure to dispose of plastic waste properly not only mars the natural environment but also endangers various species. Wildlife can ingest or become entangled in discarded plastic items, leading to injury or death. Furthermore, when plastic waste breaks down into micro plastics, it can enter the food web, potentially harming aquatic life and subsequently affecting human consumption. Recognizing the importance of responsible plastic disposal and adopting eco-friendly practices is crucial to mitigating these adverse impacts and addressing the plastic pollution crisis.

8. Incorrect Recycling Practices

Plastic pollution can also arise due to inadequate recycling methods, as a result of individuals placing inappropriate items in their recycling bins, often due to a lack of awareness regarding which types of plastic are suitable for recycling and which are not. This may result in garbage that is not recycled and instead polluting the environment or ending up in landfills. Consumers should be in charge of properly disposing of plastic waste, but we fail to do so, which exposes a large amount of plastic to the environment and poisons the planet. As a result, the planet's conditions continue to deteriorate over time.

In India, different cities generate between 0.0352 and 0.21 kg of plastic garbage per person every day, and plastic accounts for 8.9% of the waste stream's constituent components. Nowadays, the majority of goods are packaged in polyethylene film, which accounts for 70% of municipal garbage in streams and contributes significantly to global pollution (Soloman, 2014).

9. Microplastics

Another significant source of micro plastic contamination is plastic waste. Micro plastics are minute fragments of plastic, less than 5 millimeters in size, produced when larger plastic pieces fragment into smaller ones.

Micro plastics are mainly derived from two sources: the shedding of tiny fibers from textiles like clothing and fishing nets and the production of minuscule particles for commercial applications, such as those used in cosmetics. Another category of micro plastics, called secondary micro plastics, forms when larger plastic items, like water bottles, deteriorate. Environmental factors like sunlight and ocean waves play a pivotal role in this breakdown process. Much like their larger plastic counterparts, micro plastics share the concerning characteristic of not easily breaking down into harmless components. The decomposition of plastics can span hundreds or even thousands of years, during which they inflict significant environmental harm. On beaches, micro plastics can manifest as tiny, colored specks embedded in the sand. In aquatic environments, marine organisms are prone to ingesting micro plastic pollution, which can have detrimental effects on their health. As micro plastics infiltrate ecosystems, they underscore the urgency of addressing the persistent issue of plastic pollution and its far-reaching consequences for both terrestrial and aquatic life.

The presence of micro- and Nanoplastic in marine ecosystems has been widely documented. They have been found in large "garbage patches" in the Pacific, fish taken for commercial purposes in the North Atlantic, and even corals on the Great Barrier Reef. But the plastic pollution problem may go far deeper than just the oceans. Approximately 32% of all plastic garbage, according to some estimates, ends up in soil or inland waterways. One significant source of plastics is wastewater sludge spread on agricultural fields. Another is plastic that has been aerosolized by factories. Micro- and Nanoplastic in soil have mainly gone un-searched despite various worrying findings regarding their influence on terrestrial habitats.

10. Industrial Activities

Industries and factories are also significant contributors to plastic pollution. These operations frequently generate enormous volumes of plastic garbage, which, if improperly disposed of, may wind up in the environment. Furthermore, a lot of firms and enterprises employ plastic items in their manufacturing processes, which increases the amount of plastic trash that enters the environment. Another significant cause of plastic pollution is construction. Plastic wrap can be thrown away in landfills during construction, where it may take hundreds of years to degrade. Additionally, the usage of plastic pipes and sheets during construction might release hazardous chemicals into the environment (Climate café, 2022).

ECOLOGICAL IMPACT OF PLASTIC POLLUTION ON SOIL AND MARINE LIFE

Numerous mechanisms may have an impact on organisms and ecological systems when exposed to plastic. Although the majority of the work to date focuses on physical effects on biota or ecosystem function, chemically linked effects are also feasible and may be assisted by the adsorption characteristics of plastic surfaces, the build-up of hydrophobic chemicals, the leaching of additives in particles, and the accumulation of hydrophobic chemicals (Windsor et al., 2019).

Plastic pollution in marine environments results in a range of hazardous and ecologically damaging consequences. It directly threatens wildlife, leading to adverse impacts on a diverse array of animals. Among these species, the most significant risks are associated with entanglement in plastic debris and the ingestion of plastic materials (Derraik, 1992, Pembertton et al., 1992). Young animals, in particular, are often ensnared by plastic waste, leading to significant harm as they grow and develop. These entanglements can hinder their movement, making it challenging for them to feed properly and, in the case of marine mammals, affecting their ability to breathe (Sazima et al., 2002, Gregory, 2009).

Several well-documented species, including marine birds, sea turtles, cetaceans, fur seals, sharks, and filter-feeding creatures, have been notably affected by plastic debris. Marine birds are particularly vulnerable to ingesting plastic objects, often confusing them for food. When these animals consume plastic, it remains in their digestive systems, leading to adverse effects such as reduced appetite, gastrointestinal blockages, diminished stomach enzyme production, and altered steroid hormone levels, all of which can impair their reproductive processes (Baird & Hooker, 2000, Barreiros & Barcelos, 2001).

Plastic debris in the ocean has been discovered to contain elevated levels of organic pollutants. This includes the presence of toxic compounds such as polychlorinated biphenyls (PCBs), nonylphenol (NP), organic pesticides like dichlorodiphenyltrichloroethane (DDT), polycyclic aromatic hydrocarbons (PAHs), polybrominated diphenyl ethers (PBDEs), and bisphenol A (BPA) (Mato et al., 2001, Rios et al., 2007). Furthermore, a significant number of these chemicals have the potential for substantial bio magnification, which could pose a direct threat to human health. This compounds the risks associated with wildlife consuming plastic waste (Schecter et al., 2010,Harai et al., 2011). A wide range of health concerns, encompassing issues like developmental disorders (including neurological problems, growth irregularities, and hormonal imbalances), cancer, disruptions to the endocrine system, alterations in neuro-behavior, arthritis, breast cancer, diabetes, and DNA hypomethylation, have been correlated with and are linked to these harmful substances (Chung et al., 2011, Trudel et al., 2011).

Additionally, the chance for potentially invasive and dangerous wildlife to move to new settings is boosted by plastic litter. Marine organisms often attach to and then transport plastic debris, with examples of barnacles, bryozoans, polychaetes, dinoflagellates, algae, and mollusks adhering to plastic waste in the environment. Additionally, terrestrial animals have been observed using maritime debris as a means of travel. Ants have been observed travelling on trash from the Brazilian mainland to San Sebastian Island, which is many miles distant. Iguanas and other big animals have also been observed travelling on garbage to new Caribbean islands (Webb et al., 2013).

Plastic may contaminate groundwater supplies in addition to harming the ocean. While plastic pollution is already a concern in many regions, it poses a global threat to the world's water sources. Plastic waste and litter can contaminate groundwater, which is a vital source of drinking water for people. On land, wind can disperse plastic debris throughout the environment, leading to its entanglement in trees, fences, traffic signals, and various objects. When animals encounter this plastic debris, they face the risks of ingesting toxins or becoming entangled, which can lead to suffocation. If an animal consumes a piece of plastic, it may block their stomach and introduce harmful chemicals. Notably, approximately 200 different species of animals have been observed consuming plastic waste.

Another problem that both people and animals face is air pollution. Large amounts of pollutants are released when plastic is burned outside, which contaminates the atmosphere. Inhaling the poisons over an extended length of time might cause respiratory issues. Understanding the possible effects of plastic particles in the environment is of significant study interest due to the ubiquity of plastic pollution. However, compared to research on the aquatic ecosystem, relatively little is known about the possible effects and repercussions of plastic particles on the soil ecosystem or inhabitants.

The build-up of plastic particles in soils may be harmful to the soil biota. Plastic particles that reach the soil environment have the potential to act as harmful chemical transport vectors. They could include chemical additives that are applied when making parent plastic particles and that can/may leak from plastic particles, such as plasticizers (such phthalates), ame-retardants, UV stabilizers, and antioxidants. Additionally, harmful organic contaminants like metals and hydrophobic organic pollutants can adsorb and enrich on the surfaces of plastic particles in the soil environment before being desorbed and discharged into the surrounding environment. Plastic particles have the potential to be easily swallowed by soil organisms, potentially acting as a significant pathway for the movement of these hazardous compounds through the ecosystem and food chain, with potentially harmful consequences for human health (Okoffo et al., 2021).

Due to their Eco toxicity and enduring presence in the environment, micro and Nanoplastic pose a hazard to the ecosystem. According to reports, micro and Nanoplastic are present in microorganisms' digestive systems because of their tiny size and shape, which makes them easy to swallow and collect in the digestive systems of non-discriminating feeders who mistake them for food. In contrast to bigger particles, smaller Micro & Nanoplastic may enter cells and tissues more readily and are more difficult to eliminate. Because the body's free radicals and antioxidants are out of balance, micro and Nanoplastic will also produce oxidative stress (Wang et al., 2020a). Additionally, it was discovered that Micro & Nanoplastic can result in inappropriate immunological responses that threaten the health of living things as well as abnormal chemical processes in the body that are involved in catabolism and anabolism(Blair et al., 2017).

The high surface to volume ratio of Nanoplastic is believed to make it more dangerous to soil bacteria, fungus, and microorganisms. Micro plastics appear to cause the Proteobacteria to become the dominant microbial phylum in soil, disrupting the food chain and the microbial community. The quantity of greenhouse gas (GHG) emission, nitrogen cycling, and carbon levels can all be impacted by the interaction between microorganisms and micro and Nanoplastic. One of the GHGs is nitrous oxide, or N2O, and micro plastics will have an impact on N2O emission (Ren et al., 2020).

It is well known that plastics contain a significant amount of inert carbon, which is difficult to decompose. Carbon concentration steadily rises over time, altering the C:N ratio, which ultimately causes N to become immobilized by microorganisms. These changes won't have any biological consequences during the lifetime of microbes since plastics take hundreds to thousands of years to disintegrate. MNPs particles have an impact on the structure of the soil and the cycling of nutrients. Due to these modifications, the soil's fertility is also impacted and decreased, which inhibits the development of microbial communities in the rhizosphere (Rillig et al., 2019).

Due to their important function in plant growth, these microorganisms' decreased activity may have detrimental effects on crop growth and plant health. Additionally, the oxygen movement in soil as well as aerobic and anaerobic microbes can be impacted by changes in soil moisture and porosity brought on by micro & Nano plastics. The habitat of microorganisms is threatened by any alterations to the gaps between soil particles. HDPE (high-density polyethylene) lowers the pH of the soil. On the other side, polylactic acid (PLA), low-density polyethylene (LDPE), and PE raise pH (Allouzi et al., 2021).

The size of the MNPs and the texture of the soil both have an impact on their movement in the soil. Transport often took place through the soil's crevices, as well as through bioturbation, tilling, and ditching. To improve the plastic's characteristics, various hazardous and cancer-causing substances were added during manufacture. These micro and Nanoplastic chemicals can alter their physical and chemical characteristics as they decompose in the soil, as demonstrated in. Depending on the molecular surface charge, these additives will either be discharged into the soil or accumulate in the plants. The roots of the plant were most impacted, then the leaves, shoots, and finally the stem (Khalid et al., 2020).

Roots were discovered to have suffered physical harm when exposed to micro and Nanoplastic with sharp edges. Some microfibers twisted young roots and hampered their development. It was still unclear if the decreased root development was brought on by a modification in the soil's composition or by the harmful substances included in micro and Nanoplastic (Iqbal et al., 2020).

CLIMATE CHANGE AND PLASTIC POLLUTION

Over their entire life, plastics generate greenhouse gas emissions, starting with their production, which primarily depends on energy-intensive procedures including the processing of fossil fuels, and ending with their disposal, which requires energy and generates emissions immediately. Given that the majority of polymers originate from fossil fuels and need significant energy to refine, 90% of the greenhouse gas emissions associated with plastics may be attributed to the production and conversion process.

Fossil-based plastics generated 1.8 giga tonnes of carbon dioxide equivalent ($GtCO_2e$) across their whole life cycle in 2019, accounting for 3.7% of global emissions. These emissions are expected to more than double and reach 4.3 $GtCO_2e$, or 4.5% of global GHG emissions, by 2060 if new laws are not implemented.

These calculations do not take into account how plastic leakage affects GHG emissions. But according to current study based on experimental data, environmental deterioration and unhygienic land filling result in methane emissions of around 2 million metric tonnes (Mt) CO_2e annually. In addition, plastics may help lower certain GHG emissions from the use phase of items due to their light weight, but they may also lead to additional environmental problems with reference to trash management. However, these factors can only be taken into account for each individual product.

Primary polymers derived from fossil fuels account for 93% of all plastics produced in 2019—by far the greatest percentage. Secondary (i.e. recycled) plastics still make up a small portion of plastics despite having significantly expanded over the past 20 years, making up around 6% of plastic output in 2019. Bio based plastics, or polymers made from biomass like maize, sugarcane, wheat, or leftovers from other processes, still make up just 0.6% of all plastics produced. Even though secondary plastics are anticipated to increase more quickly than primary plastics, current regulations predict that they will only account for 12% of all plastics usage in 2060. Plastic emissions also differ by polymer. The largest emitters are fibers from textiles and apparel, followed by polypropylene (PP), which is used in many different items, including as food packaging and moulded auto parts. Producing low-density polyethylene (LDPE), which is utilized in dispensing bottles and plastic bags, is the third-highest emitter.

Climate Change Mitigation is Enhanced by Initiatives to Limit Plastic Pollution

Policies that target the whole plastics life cycle, including limiting demand, promoting recycling, and blocking leakage pathways, are necessary to reduce plastic leakage into the environment. Avoiding a significant rise in the usage of primary plastics and lowering plastic waste are also included in a scenario with the Global Ambition to reduce plastic leakage to aquatic habitats by 2060. Without specifically attempting to minimize GHG emissions, this scenario might cut the GHG emissions from the lifespan of plastics by 50% (2.1 Gt CO_2e) in 2060 compared to the Baseline scenario. Because of measures that limit the use of plastics (such as plastic levies), the life cycle emissions of plastics are reduced, but secondary plastics are also becoming more popular.

Mismanaged waste is anticipated to decrease as a result of legislation aimed at enhancing waste management, but more garbage is forecast to be burned, dumped, and recycled. To achieve highly ambitious climate scenarios, such as the Net-Zero emissions scenarios, efforts to address plastic pollution must reduce the life cycle emissions of plastic. In contrast to emissions relating to plastics, global emissions consider all industries and are influenced by several factors. Global emissions would rise if the adoption of new GHG-intensive materials severely offset the decline in plastics use. However, the Global Ambition scenario is anticipated to result in a 0.8% (0.8 Gt CO_2 e) reduction in global GHG emissions in 2060 as compared to the Baseline scenario.

Together, Climate Mitigation and Plastics Policy Can Achieve Significant Synergy

Different channels—the GHG intensity of plastics manufacturing for climate change mitigation policies, and the decline in plastics usage for plastics policies—influence the life cycle GHG emissions of plastics. This indicates that the two sets of laws work in concert to reduce greenhouse gas emissions from the life cycle of plastics as much as possible.

Combining the two policy packages decreases GHG emissions from the plastics life cycle by 2.8 Gt CO_2e, or by two-thirds when compared to Baseline. Primary and secondary plastics usage both decline, although primary plastic use declines more than secondary due to primary plastics being more energy intensive. The combined policy package raises the percentage of secondary plastics in overall plastics use even higher. This demonstrates how climate mitigation measures may reduce the production of primary plastics, which might be further improved with stricter climate regulations.Beyond the GHG emissions from the life cycle of plastics, the Climate Mitigation scenario decreases world GHG emissions by 31.6 Gt CO_2e in 2060, or 33%.

What Are the Options for Stakeholders and Policymakers?

1. Strengthen the Ambition of Domestic Policies to Address Plastics Use

These results suggest that the simplest way to lower greenhouse gas emissions from the plastics life cycle is to cut down the growth in plastics use and waste worldwide. Countries may go closer to achieving their climate targets in addition to lessening the environmental consequences of plastic leakage.

2. Support Decarbonisation of the Plastics Life Cycle

Reducing the energy intensity of plastics manufacturing and increasing the share of secondary plastics can result in significant reductions in greenhouse gas emissions. Increasing the availability and use of recycled plastics, decarbonizing production and conversion, and enhancing waste-treatment practices—such as by substituting more electricity for fossil fuels and lowering the carbon footprint of electricity generation—are further strategies to cut emissions.

3. Foster Innovation in Plastics Production and Waste Management

More aggressive steps are needed to direct technical innovation in the direction of closing the plastics loop and reducing environmental leakage in a way that is climate-friendly. To guarantee a strong market for circular plastics, investments in innovation should be paired with financial incentives for changing behaviour, environmental awareness campaigns, and education. Additionally, legally binding laws that are customized for the region should be in place.

4. Strengthen the Secondary Plastics Market

The secondary plastics markets remain vulnerable. Policies that incentivize the production and demand for secondary plastics include taxes on non-recycled plastic waste and goals for recycled content.

5. Tap Into the Complementarity of Plastics and Climate Policies

When implemented, laws restricting plastic leakage and climate policies combine to lower greenhouse gas emissions from plastics. Plastics policies, however, cannot be used *in lieu* of particular climate mitigation methods to prevent plastic leakage, just as plastics policies cannot be used in place of specific climate mitigation initiatives (OECD, 2023).

GLOBAL INITIATIVE ON PLASTIC POLLUTION

Governmental organizations at all levels are enforcing prohibitions and putting recycling technologies into practise, from India to Maine. To safeguard the conservation of our environment, customers are making wiser purchasing decisions and businesses are improving construction. As other nations follow the example, the globe is gradually preparing for a better future. However, there is a lot of work that has to be completed or reversed. Since the 1950s, more than 8.3 billion metric tonnes of plastic have been produced, according to research. Furthermore, sixty percent of the trash was disposed of in landfills or the open air. That number has increased dramatically over time. The amount of plastic produced worldwide quadrupled between 1976 (50 million metric tonnes) and 1989 (100 million metric tonnes). Since then, the annual number has grown significantly, reaching a record of 368 million in 2019 and a subsequent decline to 367 million in 2020. One million metric tonnes less plastic produced year will not address the growing environmental problems brought on by plastic. It is currently unknown how long plastic takes to break down. That is to say, the 12 million metric tonnes of plastic waste that entered the ocean in 2010 are still breaking down into small pieces of plastic and building up within the bellies of birds and other water creatures (Global Citizen, 2021).

We have released plastic into the environment, and it will be there. The greatest method to address our environmental concerns is *via* reduction, even though recent research of plastic-eating enzymes and bacteria provide some promise for the management of plastic trash in the future.

Here are seven instances where towns and nations achieved historic firsts in the drive to reduce plastic trash globally:-

1. Ban on Plastic Bags

When New York City decided to ban plastic bags in 2020, the US saw a nationwide uproar. However, were you aware that this was not the first US state to try to ban plastic bags? Actually, this idea wasn't even first presented in the United States. In 2002, Bangladesh became the first country to outlaw plastic bags. Since then, a number of countries have enacted such laws in an attempt to reduce the widespread use of plastic bags. Plastic bag bans have not only brought up new concerns, such as the environmental impact of biodegradable or reusable alternatives, but they have also improved plastic usage overall. In a decision by the High Court of Bangladesh, it was mandated that all seaside hotels and eateries must completely phase out the use of single-use plastics by the year 2020.

2. No Micro Beads

The tiny pieces of plastic, called micro beads, makes toothpaste, home cleaners, and makeup cleansers more abrasive. Plastic was also brought by them into our natural streams. 2014 saw the introduction of the first country to prohibit the use of micro beads in personal hygiene products: the Netherlands. Since then, countries have rallied around the cause, with the UK having some of the strictest legislation, followed by South Korea and Australia. The 2018 UK prohibition stands out from earlier bans since it forbids the manufacture of micro bead-containing goods as well as the usage of "biodegradable" micro beads. There is still a long way to go, however, since only a small number of nations have passed laws addressing the plastic bead issue, and nations like the US have exceptions to the laws established by their 2015 micro beads ban.

3. Roads by Recycled Plastics

India generates approximately 26,000 metric tonnes of plastic waste every day. However, rather than discarding it, the country is actively exploring innovative methods for its reuse. Since the early 2000s, India has been pioneering the use of plastic in road construction, transforming its plastic waste problem into a sustainable and robust path to the future.

In 2002, scientist and researcher Rajagopalan Vasudevan introduced the ground-breaking idea of using plastic to bind gravel for road construction. Vasudevan claims that for each kilometre of road built using recycled plastic, about three metric tonnes of carbon dioxide emissions are saved compared to the traditional method of burning plastic waste. Since that initial experiment, this approach to plastic recycling has been implemented on more than 6,000 miles of roads across India. Since then, other nations have started to imitate India. This revolutionary, environmentally friendly road technology is paving the path to a brighter future everywhere from Ghana to the UK to the US.

4. Making Companies Pay

In 2021, Maine took a pioneering step by becoming the first U.S. state to introduce a ground-breaking law that holds businesses responsible for their packaging waste. The Extended Producer Responsibility (EPR) program was designed to transfer the financial burden of recycling and disposing of materials such as cardboard, plastic containers, and non-recyclable packaging from consumers to the producers themselves. Just a few weeks later, Oregon became the second state in the United States to enact similar packaging legislation. It's worth noting that the concept of Extended Producer Responsibility is not new. France has had its own set of producer regulations in place for the past two decades. According to The New York Times, packaging regulations are quite consistent across numerous European Union member states, Japan, South Korea, and five Canadian provinces. While Maine has set a precedent in the United States, it's anticipated that ten other U.S. states may soon follow suit in implementing similar measures.

5. Battle Against Styrofoam

Polystyrene, more frequently referred to as Styrofoam, is not biodegradable and is challenging to re-cycle, in contrast to other types of plastic. The cup itself isn't empty when a cup of coffee is. It will persist for many years until it disintegrates into little particles and contaminates the nearby ecosystems and waterways. Not only does polystyrene endure in our environment, but the controversy around it has endured as well. The first US state to outlaw Styrofoam was New York in 1988, but it was overturned two years later. But perhaps this is the time for the world to confront Styrofoam. Polystyrene distribution and importation were outlawed in Costa Rica in 2021, and sanctions were put in place for anyone who broke the law. Despite other towns and counties around the US already having such restrictions in place, Maine became the first state in the US to outlaw the use of polystyrene food containers in commercial settings. And New York State reintroduced plans to outlaw polystyrene in 2019. The world is prepared to witness significant advancements in both trash management and reduction as more US states and other nations enforce the banning of Styrofoam.

6. Circular Plastic Economy

In 2018, the EU unveiled its first-ever pan-European strategy to reduce plastics use and pollution. The EU aspires to construct the first "circular" plastics economy by outlawing single-use plastics, developing a comprehensive reuse system, and setting up a market for recovered plastics. By tackling the disposable objects most frequently seen on local beaches, including as cigarette butts, wrappers, cotton bud sticks, food and beverage containers, and plastic bags, the initiative seeks to prevent marine litter. In an effort to transition to a low-carbon economy, the EU is working to shorten the life cycle of plastic while also lowering its consumption. The total programme is being adopted by the 27 EU members, whose plastic recycling rates are already three times higher than those of the US.

7. Import/Export of Plastic Waste

The practice of importing and exporting plastic waste is not a new concept. Nevertheless, there is a grow-ing trend among various countries to reject such practices. Wealthier nations have traditionally relied on less developed countries to manage their waste, shipping garbage to destinations like Vietnam or

Ethiopia, where it is either recycled or incinerated. In 2018, it is estimated that the United States exported plastic waste equivalent to 68,000 shipping containers. This substantial amount of waste is disposed of in foreign landfills, leading to a range of problems related to strained infrastructure and waste management systems in those host countries. China, the largest consumer of plastic in the world, imposed a ban on the importation of plastic garbage in 2017. And other nations did the same. 180 countries enacted new international regulations that restricted the trade of plastic garbage in 2020. However, the practise of wealthy nations shifting their issues to less developed nations persisted. The US has not signed the accord, but it still exports plastics to other nations.

PLASTIC POLLUTION AND SUSTAINABLE DEVELOPMENT GOALS

In light of the serious threat that plastic waste poses to our world, a new UN Environment Programme report provides a road map for a sustainable future.

The threat posed by plastic contamination to our environment is serious. However, a recent UN Environment Programme (UNEP) report gives optimism for a sustainable future. The paper provides a plan for changing market conditions and public policies to reduce plastic pollution by 80% by 2040. Reuse, recycling, reorienting, and diversifying approaches can help with this. We can significantly reduce plastic pollution by increasing reuse choices, improving recycling economics, and switching out single-use plastics for alternatives. Moving towards a circular economy has advantages for the economy as well as the environment. According to UNEP experts, there might be $1.27 trillion in potential savings when expenses and recycling income are taken into account, in addition to $3.25 trillion in savings from preventing environmental and health problems.

By 2040, implementing these improvements might generate 700,000 employment, primarily in low-income nations, enhancing living conditions and lowering inequality. Even while investment expenditures are high, they are still less than the anticipated costs of inaction. It is imperative to act quickly since, by 2040, there might be an additional 80 million metric tonnes of plastic pollution as a result. Policymakers are advised to adopt a lifetime approach to plastic management and include regulatory measures. Nations may obtain significant economic, social, and environmental advantages by implementing this path. Governments and corporations must act and collaborate to create a sustainable future free from plastic waste (UNEP, 2023).

1. SDG 3: Good Health and Well-Being

Polymers and additives come together to form plastics. Chemicals like plasticizers and flame retardants are examples of additives that are added to plastic to give it the necessary qualities for industrial or commercial application. Some of these drugs mess with both humans and animals' hormone balances. The impact of Nanoplastic represent a growing concern, primarily due to their minuscule size, which allows them to easily infiltrate the human body. These particles, given their dimensions, can potentially penetrate cells and various organ tissues, whether it be in the stomach or even the brain. The toxic nature of these particles can trigger a range of physiological effects, some of which remain insufficiently researched, including the possibility of localized inflammation.

2. SDG 6: Clean Water and Sanitation

The sixth sustainability goal requires a focus on improving the quality of freshwater resources. This entails reducing pollution in inland water bodies and significantly increasing water purification efforts to make more safe drinking water accessible.

3. SDG 11: Sustainable Cities and Communities

Even though it is considered an essential and vital service, many urban and residential areas often face challenges in achieving sufficient garbage collection and processing. This results in various problems, particularly in densely populated areas, including:

1. Mitigating the environmental impact of plastic pollution.
2. Reducing health risks.
3. Achieving a significant reduction in overall plastic usage.

4. SDG 12: Responsible Consumption and Production

The primary source of plastic pollution, both in our oceans and on land, can be attributed to the extensive production and consumption of plastic, especially single-use packaging materials. This pollution poses a significant threat to wildlife and impacts the food supply for large human populations, while also disrupting ecosystem functioning. Incinerating all plastic as a waste management strategy leads to harmful air pollution. The most effective approach to advance Sustainable Development Goal 12 (SDG 12) is a comprehensive reduction in plastic usage. Currently, efforts to promote and improve recycling are inadequate and do not provide a practical solution to combat the issue of plastic pollution.

5. SDG 13: Climate Action

The production of plastic heavily relies on fossil fuels, especially oil and shale gas. The energy required for plastic manufacturing is significant, accounting for approximately 10% of the world's annual fossil fuel consumption. Half of this energy is directed towards plastic production, while the other half is used for fuel. If our current patterns of plastic usage persist, this percentage is projected to increase to 20% by 2050. In light of the urgent need to reduce CO_2 emissions and limit the global temperature rise to within two degrees, it is crucial to prioritize environmental efforts. Comparatively, when considering carbon footprints, plastic manufacturing and usage are more favorable than materials like paper and metal. However, it's essential to acknowledge that this decision should not ignore the environmental impact of plastic once it reaches the end of its useful life.

6. SDG 14: PROTECTION of Seas and Oceans

The fourteenth sustainability goal is dedicated to preventing and substantially reducing marine pollution, particularly that stemming from land-based sources. It is undeniably crucial to achieve a significant reduction in the amount of plastic waste that enters our oceans, especially considering the rough estimate that approximately 80% of marine debris originates from land-based sources, resulting in billions of kilograms of waste annually. To realize substantial benefits, it is imperative to establish effective waste collection systems universally, alongside reducing plastic usage. While it's essential to clean up plastic waste from beaches and collect it at sea, these efforts will ultimately prove ineffective if the inflow of plastic into the oceans is not simultaneously addressed.

7. SDG 15: Repair Ecosystems and Retain Biodiversity

(Micro) plastics and chemical additives pose a threat to both marine and terrestrial ecosystems. Plastics can lead to various adverse consequences, including the suffocation of animals, reduced food intake, increased vulnerability to predation, and harm to coral reefs, among other issues. The recovery of ecosystems and biodiversity is achievable with a substantial reduction in plastics and micro plastics (Plastic Soup, 2023).

PLASTIC TO ENERGY AND WEALTH

As public awareness of the significant environmental damage caused by single-use plastics and people's subpar recycling practices continues to grow, researchers are exploring alternative approaches to manage the increasing plastic waste. In the energy sector, plastic-to-fuel projects are beginning to gain prominence. These endeavour harness the chemical energy stored in the hydrocarbon structure of plastics to generate fuel. While this technique has received recognition for its positive economic and environmental impacts, it has largely remained in the research and development phase.

Estimates suggest that less than 5% of the plastic produced annually is recycled. Plastic production is expected to increase by 3.8% per year until 2030, adding to the 6.3 billion tonnes produced in the past six decades. Unfortunately, a significant portion of this plastic ends up in our oceans, causing disruptions to marine ecosystems and taking at least 450 years, if not longer, to biodegrade, as indicated by studies.

Plastics-to-fuel technology has the potential to substantially reduce pollution and offer significant economic benefits to local communities. The American Chemistry Council anticipates that plastic-to-fuel plants in the United States alone could generate approximately $9 billion in economic output and provide nearly 39,000 jobs, highlighting the vast potential of this industry on a global scale (Power Technology, 2018).

Plastic to Hydrogen

Recently, researchers at Swansea University have made a significant breakthrough by converting plastic waste into hydrogen fuel, potentially for future use in powering vehicles. Their approach involved the modification of plastic objects using a photo catalyst that harnesses light and transforms it into chemical energy through a process called "photo reforming." This plastic-catalyst blend was immersed in an alkaline solution and exposed to sunlight, resulting in the breakdown of the plastic and the generation of hydrogen gas bubbles. Given that any plastic may be utilized without having to be cleaned, the new process would be less expensive than the recycling solutions now available. Even if it would take some time until this plastic-to-fuel process is implemented on a large-scale, the introduction of hydrogen-powered cars would be a good time for its development.

Plastic to Diesel

Untreated polyethylene can be broken down, but doing so necessitates using a lot of heat or toxic, reactive chemicals, which causes the atomic bonds to shatter in an unusable fashion. The researchers' method, however, consumes far less heat and permits the conversion of the finished product into a new fuel source.

They utilized an alkane, a type of hydrocarbon molecule that is often used to create polymers but was used in this instance to dissolve polymers. The researchers succeeded in reassembling the polyethylene into a liquid fuel suitable for use in automobiles and other industrial applications by progressively removing and introducing hydrogen and carbon bonds inside the polymer.

Plastic to Crude Oil

To make plastic crude oil (PCO), high-density polyethylene bags obtained from local stores were processed in a pyrolysis machine. The PCO was then subjected to a refining process, resulting in the production of two distinct types of diesel and petrol. After incorporating antioxidants, the resulting materials exhibited enhanced lubrication properties and achieved a higher octane number, indicating improved ignition quality compared to conventional diesel fuels.

Plastic to Sulphur

An American company known as Plastic2Oil has developed a method to convert discarded plastic into low-sulphur fuel. They use this waste as a raw material to produce "ultra-low sulphur diesel" with a sulphur content of 15 ppm or less. While most ultra-low sulphur diesel is currently derived from petroleum, Plastic2Oil provides an eco-friendly alternative by creating fuel from plastic. The company's processing system accepts unwashed and unsorted plastic, converting approximately 8.3 pounds of plastic into approximately one gallon of gasoline. What's noteworthy is that the processor operates on its own off-gases, constituting 10%–12% of the process output, thus consuming minimal energy. Additionally, there's no need for a distillation tower to refine and separate the produced fuel.

Thriving Indian Businesses that Convert Waste Plastics to Energy

Waste plastics has emerged as the solution to WTE plants' drawbacks, allowing us to go forward and convert end-of-life plastics into renewable fuels by implementing straightforward and affordable methods. Numerous techniques have been developed to convert plastic trash into fuels in response to the increased interest in producing something new from plastic garbage. Although plastics are thought to be harmful, they may be vaporized rather than burned and oxidized by using suitable waste management practices.

Table 1: Indian businesses that convert waste plastics to energy (Electronics Foru, 2021)

S.NO	NAME	WORK DESCRIPTION
1	Indian Institute of Petroleum (IIP), Dehradun & Council for Scientific & Industrial Research (CSIR)	The Council for Scientific & Industrial Research (CSIR) oversees the Indian Institute of Petroleum (IIP), Dehradun, which has established India's first government research institute plant. Using pyrolysis, it transforms waste plastics into beneficial hydrocarbons like fuel or diesel. By heating the separated plastic wastes without the presence of air, rather than burning them, the complex hydrocarbons are broken down into smaller molecules. These tiny molecules are given the proper amount of catalysts before being transformed into valuable hydrocarbons in the form of vapour. The final products, such as liquid petrol, diesel, or aromatics, are subsequently produced by condensing this vapour. One tonne of plastic trash may be converted at the IIP facility into 850 liters of diesel, 700 liters of petrol or 500 liters of aromatic chemicals. The fuels that are produced are entirely environmentally benign because they contain no hazardous ingredients. Indian Railways is happy with the final IIP goods since they have no sulphur and hence satisfy Euro-III criteria. The aromatics include a lot of BTX.
2	AgileProcess Chemicals(APChemi)	Leading solution supplier in the field of pyrolysis-based chemical recycling is APChemi. The business is one of the eleven winners of the Asia-Pacific End Plastic Waste Innovation Platform, which encourages entrepreneurs' solutions in eliminating waste plastics and is run by Plug and Play and the Alliance to End Plastic Waste. The plastic from the municipal solid wastes is properly cleaned, dried, and shredded at APChemi before being fed into a gas-powered reactor where the plastic's complex hydrocarbons are liquefied by heat. It produces petrol and oil that are efficient and clean hydrocarbons that are used to make petrochemicals and fuels, decreasing reliance on crude oil and opening the door to an ecological circular economy.
3	Rudra Environmental Solutions	The Dynamic Asia Global Award went to Rudra Environmental Solutions for their excellent contribution to the recycling sector. Rudra use the Thermo Catalytic Depolymerization (TCD) method, which involves heating shredded plastics at 380°C to 430°C without any oxygen present. The polymer chain is broken by a catalyst, releasing condensed hydrocarbon vapour. The condensed vapour result in the production of syngas and polyfuel, two synthetic gases. Before being collected, the polyfuel is filtered, and the syngas is cleaned before being used in a reactor.
4	PolyCycl Private Limited	The Confederation of Indian Industry (CII) has given PolyCycl Private Limited the Indian Innovation Initiative award for its contribution to a sustainable circular economy through the conversion of waste plastic into high-value petroleum fuels. The Alliance to End Plastic Waste & Plug and Play has also selected PolyCycl as one of the eleven winners of the Asia-Pacific End Plastic Waste Innovation Platform, which recognize entrepreneurs' creative uses of plastic waste. The fuels produced by PolyCycl's process are lighter, less viscous, and have a high GCV (global value chain), making them highly effective and valuable, in contrast to the batch-based approach, which is tedious and results in low-value fuels, according to PolyCycl. These fuels are prepared for sale as an alternative to industrial diesel and don't need any additional distillation.
5	Paterson Energy	Through a sustainable and environmentally friendly process, Paterson Energy offers a comprehensive solution to plastic trash. In its recycling facility, Paterson breaks the big plastic molecules using continuous thermochemical depolymerization (TCD), producing a fuel they call "verte-x" that is devoid of harmful substances similar to sulphur. Verte, which translates to "green" in French, has a "x" added by the corporation since it has multiple industrial uses.
6	MK Aromatics	The producer of different aromatic hydrocarbon solvents, MK Aromatics, also offers a technological means of synthesizing waste plastics or multi-layered plastics, such as sachets. The business has a fully automated process that eliminates the need to separate, wash, or dry the plastics. Seven to ten tonnes of waste plastics are converted daily at the factory.

CONCLUSION AND FUTURE PRESPECTIVES

No country or region is immune to the damaging effects of pollution since it is a worldwide problem. Particularly dangerous to the environment's living and non-living components is plastic pollution. The harmful impact of plastics, including macro-, micro-, and Nanoplastic, on the environment and living organisms arises from a combination of their toxic characteristics, the release of additives or chemical components, and the persistent release of absorbed pollutants. Current literature underscores the need to re-evaluate the entire plastic economy value chain to ensure environmental sustainability. It's important to note that research on the impact of plastic materials on various ecosystems, including soil and air, is relatively limited. To achieve environmental sustainability, plastic pollution, a global phenomenon that exacerbates climate change and floods, must be reduced. Although plastic pollution poses a major danger to the environment, there are several options that may be used to manage and reduce this widespread issue. To completely comprehend this issue, more research is necessary as our understanding of plastic contamination is still limited.

The research and study potential for recycling techniques is enormous, and they might be essential in reducing post-consumer plastic waste. We want to talk about the negative consequences that post-consumer plastic waste has on the environment as well as new methods that are being developed to handle this garbage in a way that promotes sustainability and the circular economy. Linking the SDGs, UNFCCC, and future international instrument with an overview to emphasize the benefits of collaboration and coordination. We should not wait for a treaty before taking action even though the global negotiating process is moving along steadily. We are unable to wait. Every minute, one garbage truck full of plastic debris is thrown into our ocean. It is necessary to set unified standards. In the emerging field of plastic pollution research, there is a growing need for standardized criteria encompassing plastic classification, analytical methodologies (including sample collection, per-treatment, quantification, and identification procedures), units of measurement, and even a universally accepted method for toxicity testing. Since plastic pollution research is still in its infancy, various existing classifications, analytical approaches, and environmental plastic measurement units can serve as a foundation for exploring and identifying new contaminants. Unified standards help advance our knowledge of plastic contamination in the environment as quickly as feasible.

CONFLIC OF INTEREST: The authors declare no conflict of interest.

REFERENCES

Adam, I., Walker, T. R., Bezerra, J. C., & Clayton, A. (2020). Policies to reduce single-use plastic marine pollution in West Africa. *Marine Policy*, *116*, 103928. doi:10.1016/j.marpol.2020.103928

Adams, S. M. (2005). Assessing cause and effect of multiple stressors on marine systems. *Mar. Pollut. Bull.*, *51*, 8–12. https://www.sciencedirect.com/science/article/pii/ S0025326X04004667.

Allouzi, M. M. A., Tang, D. Y. Y., Chew, K. W., Rinklebe, J., Bolan, N., Allouzi, S. M. A., & Show, P. L. (2021). Micro (nano) plastic pollution: The ecological influence on soil-plant system and human health. *The Science of the Total Environment*, *788*, 147815. doi:10.1016/j.scitotenv.2021.147815 PMID:34034191

Bai, M., & Li, D. (2020). Quantity of plastic waste input into the ocean from China based on a material fow analysis model. *Anthropoc Coasts*, *3*(1), 1–5. doi:10.1139/anc-2018-0028

Baird, R. W., & Hooker, S. K. (2000). Ingestion of plastic and unusual prey by a juvenile harbour porpoise. *Marine Pollution Bulletin*, *40*(8), 719–720. doi:10.1016/S0025-326X(00)00051-5

Barnes, D. K., Galgani, F., Thompson, R. C., & Barlaz, M. (2009). Accumulation and fragmentation of plastic debris in global environments. *Philosophical Transactions of the Royal Society of London. Series B, Biological Sciences*, *364*(1526), 1985–1998. doi:10.1098/rstb.2008.0205 PMID:19528051

Barreiros, J. P., & Barcelos, J. (2001). Plastic ingestion by a leatherback turtle Dermochelys coriacea from the Azores (NE Atlantic). *Marine Pollution Bulletin*, *42*(11), 1196–1197. doi:10.1016/S0025-326X(01)00215-6 PMID:11763235

Bezerra, J. C., Walker, T. R., Clayton, C. A., & Adam, I. (2021). Single-Use Plastic Bag Policies in the Southern African Development Community. *Environmental Challenges*, *3*, 100029. doi:10.1016/j.envc.2021.100029

Blair, R. M., Waldron, S., Phoenix, V., & Gauchotte-Lindsay, C. (2017). Micro- and nanoplastic pollution of freshwater and wastewater treatment systems. *Springer Science Reviews*, *5*(1-2), 19–30. doi:10.1007/s40362-017-0044-7

Bond, T., Ferrandiz-Mas, V., Felipe-Sotelo, M., & van Sebille, E. (2018). The occurrence and degradation of aquatic plastic litter based on polymer physicochemical properties: A review. *Critical Reviews in Environmental Science and Technology*, *48*(7-9), 685–722. doi:10.1080/10643389.2018.1483155

Brennecke, D., Duarte, B., Paiva, F., Caçador, I., & Canning-Clode, J. (2016). Microplastics as vector for heavy metal contamination from the marine environment. *Estuarine, Coastal and Shelf Science*, *178*, 189–195. doi:10.1016/j.ecss.2015.12.003

Briassoulis, D., Babou, E., Hiskakis, M., Scarascia, G., Picuno, P., Guarde, D., & Dejean, C. (2013). Review, mapping and analysis of the agricultural plastic waste generation and consolidation in Europe. *Waste Management & Research*, *31*(12), 1262–1278. doi:10.1177/0734242X13507968 PMID:24293230

Carpenter, E. J., & Smith, K. L. Jr. (1972). Plastics on the Sargasso Sea surface. *Science*, *175*(4027), 1240–1241. doi:10.1126/science.175.4027.1240 PMID:5061243

Cassou, E. M. (2018). *Plastics. Agricultural Pollution: World Bank*. World Bank. https://openknowledge.worldbank. org/handle/10986/29505.

Chae, Y., & An, Y. J. (2018). Current research trends on plastic pollution and ecological impacts on the soil ecosystem: A review. *Environmental Pollution*, *240*, 387–395. doi:10.1016/j.envpol.2018.05.008 PMID:29753246

Chen, Q., Allgeier, A., Yin, D., & Hollert, H. (2019). Leaching of endocrine disrupting chemicals from marine microplastics and mesoplastics under common life stress conditions. *Environment International*, *130*, 104938. doi:10.1016/j.envint.2019.104938 PMID:31234003

Chung, S. Y., Yettella, R. R., Kim, J. S., Kwon, K., Kim, M. C., & Min, D. B. (2011). Effects of grilling and roasting on the levels of polycyclic aromatic hydrocarbons in beef and pork. *Food Chemistry, 129*(4), 1420–1426. doi:10.1016/j.foodchem.2011.05.092

Clayton, C. A., Walker, T. R., Bezerra, J. C., & Adam, I. (2020). Policy responses to reduce single-use plastic marine pollution in the Caribbean. *Marine Pollution Bulletin, 162*, 111833. doi:10.1016/j.marpolbul.2020.111833 PMID:33213855

CPCB. (2021). *Annual Report 2020-21 on Implementation of PWM Rules 2016*. CPCB. https://www.cpcb.nic.in/

Derraik, J. G. B. (2002). The pollution of the marine environment by plastic debris: A review. *Marine Pollution Bulletin, 44*(9), 842–852. doi:10.1016/S0025-326X(02)00220-5 PMID:12405208

Do Sul, J. A. I., & Costa, M. F. (2014). The present and future of microplastic pollution in the marine environment. *Environmental Pollution, 185*, 352–364. doi:10.1016/j.envpol.2013.10.036 PMID:24275078

EIP-AGRIE Focus Group. (2021). *Reducing the plastic footprint of agriculture, Final Report.* EIP-AGRIE Focus Group.

Electronics forum. (2021). P*lastic Waste To Renewable Energy–Implementing A Circular Economy*. Electronics Forum. https://www.electronicsforu.com/market-verticals/power-electronics/plastic-waste-renewable-energy-implementing-circular-economy

Engler, R. E. (2012). The complex interaction between marine debris and toxic chemicals in the ocean. *Environmental Science & Technology, 46*(22), 12302–12315. doi:10.1021/es3027105 PMID:23088563

Eriksen, M., Lebreton, L. C. M., Carson, H. S., Thiel, M., Moore, C. J., Borerro, J. C., Galgani, F., Ryan, P. G., & Reisser, J. (2014). Plastic pollution in the world's oceans: More than 5 trillion plastic pieces weighing over 250,000 tons afloat at sea. *PLoS One, 9*(12), e111913. doi:10.1371/journal.pone.0111913 PMID:25494041

Eurostat. (2020). *Packaging waste statistics*. Europea. https://ec.europa.eu/ eurostat/statistics-explained/index.php?title=Packaging_waste_ statistics.

FAO, ISSCFG. (2013). *International standard statistical classification of fishing gear*. FAO.

FAO (2020). *The State of World Fisheries and Aquaculture 2020. Sustainability in action*. Rome. . doi:10.4060/ca9229en

Galloway, T., & Lewis, C. (2016). Marine microplastics spell big problems for future generations. *Proceedings of the National Academy of Sciences of the United States of America, 113*(9), 2331–2333. doi:10.1073/pnas.1600715113 PMID:26903632

Galloway, T. S., & Lewis, C. N. (2016). Marine microplastics spell big problems for future generations. *Proceedings of the National Academy of Sciences of the United States of America, 113*(9), 2331–2333. doi:10.1073/pnas.1600715113 PMID:26903632

GESAMP. (2020). *Sea-Based Sources of Marine Litter – A Review of Current Knowledge and Assessment of Data Gaps.* Second Interim Report of GESAMP Working Group 43.

Geyer, R. (2020). Production, Chapter 2- Production, Use and Fate of Synthetic Polymers. In Letcher, T.M. (ed.). Plastic Waste and Recycling. Cambridge, MA: Academic Press.

Gregory, M. R. (2009). Environmental implications of plastic debris in marine settings—Entanglement, ingestion, smothering, hangers-on, hitch-hiking and alien invasions. *Philosophical Transactions of the Royal Society of London. Series B, Biological Sciences, 364*(1526), 2013–2025. doi:10.1098/rstb.2008.0265 PMID:19528053

Guillard, V., Gaucel, S., Fornaciari, C., Angellier-Coussy, H., Buche, P., & Gontard, N. (2018). The next generation of sustainable food packaging to preserve our environment in a circular economy context. *Frontiers in Nutrition, 5*, 121. doi:10.3389/fnut.2018.00121 PMID:30564581

Hermabessiere, L., Dehaut, A., Paul-Pont, I., Lacroix, C., Jezequel, R., Soudant, P., & Duflos, G. (2017). Occurrence and efects of plastic additives on marine environments and organisms: A review. *Chemosphere, 182*, 781–793. doi:10.1016/j.chemosphere.2017.05.096 PMID:28545000

Hirai, H., Takada, H., Ogata, Y., Yamashita, R., Mizukawa, K., Saha, M., Kwan, C., Moore, C., Gray, H., Laursen, D., Zettler, E. R., Farrington, J. W., Reddy, C. M., Peacock, E. E., & Ward, M. W. (2011). Organic micropollutants in marine plastic debris from the open ocean and remote and urban beaches. *Marine Pollution Bulletin, 62*(8), 1683–1692. doi:10.1016/j.marpolbul.2011.06.004 PMID:21719036

Huang, Y., Liu, Q., Jia, W., Yan, C., & Wang, J. (2020). Agricultural plastic mulching as a source of microplastics in the terrestrial environment. *Environmental Pollution, 260*, 114096. doi:10.1016/j.envpol.2020.114096 PMID:32041035

Huntington, T. (2019). *Marine Litter and Aquaculture Gear – White Paper*. Report produced by Poseidon Aquatic Resources Management Ltd for the Aquaculture Stewardship Council.

Hurley, R., Horton, A., Lusher, A., & Nizzetto, L. (2020). Plastic waste in the terrestrial environment. Chaper 7. *Plastic Waste and Recycling Environmental Impact, Societal Issues, Prevention, and Solutions., 2020*, 163–193. doi:10.1016/B978-0-12-817880-5.00007-4

Iqbal, S., Xu, J., Allen, S. D., Khan, S., Nadir, S., Arif, M. S., & Yasmeen, T. (2020). Unraveling consequences of soil micro- and nano-plastic pollution on soil-plant system: Implications for nitrogen (N) cycling and soil microbial activity. *Chemosphere, 260*, 127578. doi:10.1016/j.chemosphere.2020.127578 PMID:32683024

Jambeck, J. R., Geyer, R., Wilcox, C., Siegler, T. R., Perryman, M., Andrady, A., Narayan, R., & Law, K. L. (2015). Plastic waste inputs from land into the ocean. *Science, 347*(6223), 768–771. doi:10.1126/science.1260352 PMID:25678662

Karbalaei, S., Hanachi, P., Walker, T. R., & Cole, M. (2018). Occurrence, sources, human health impacts and mitigation of microplastic pollution. *Environmental Science and Pollution Research International, 25*(36), 36046–36063. doi:10.1007/s11356-018-3508-7 PMID:30382517

Kasirajan, S., & Ngouajio, M. (2012). Polyethylene and biodegradable mulches for agricultural applications: A review. *Agronomy for Sustainable Development, 32*(2), 501–529. doi:10.1007/s13593-011-0068-3

Khalid, N., Aqeel, M., & Noman, A. (2020). Microplastics could be a threat to plants in terrestrial systems directly or indirectly. *Environmental Pollution*, *267*, 115653. doi:10.1016/j.envpol.2020.115653 PMID:33254725

Lavers, J. L., Bond, A. L., & Rolsky, C. (2022). Far from a distraction: Plastic pollution and the planetary emergency. *Biological Conservation*, *272*, 109655. doi:10.1016/j.biocon.2022.109655

Lebreton, L. C. M., Van der Zwet, J., Damsteeg, J. W., Slat, B., Andrady, A., & Reisser, J. (2017). River plastic emissions to the world's oceans. *Nature Communications*, *8*(1), 15611. doi:10.1038/ncomms15611 PMID:28589961

Li, J., Liu, H., & Paul Chen, J. (2018). Microplastics in freshwater systems: A review on occurrence, environmental effects, and methods for microplastics detection. *Water Research, 137*, 362-374. https://www. sciencedirect.com/science/article.

Li, P., Wang, X., Su, M., Zou, X., Duan, L., & Zhang, H. (2021). Characteristics of Plastic Pollution in the Environment: A Review. *Bulletin of Environmental Contamination and Toxicology*, *107*(4), 577–584. doi:10.1007/s00128-020-02820-1 PMID:32166334

Li, W. C., Tse, H. F., & Fok, L. (2016). Plastic waste in the marine environment: A review of sources, occurrence and efects. *The Science of the Total Environment*, *566*, 333–349. doi:10.1016/j.scitotenv.2016.05.084 PMID:27232963

Liu, E. K., He, W. Q., & Yan, C. R. (2014). White revolution to white pollution—Agricultural plastic film mulch in China. *Environmental Research Letters*, *9*(9), 091001. doi:10.1088/1748-9326/9/9/091001

Massos, A., & Turner, A. (2017). Cadmium, lead and bromine in beached microplastics. *Environmental Pollution*, *227*, 139–145. doi:10.1016/j.envpol.2017.04.034 PMID:28458244

Mato, Y., Isobe, T., Takada, H., Kanehiro, H., Ohtake, C., & Kaminuma, T. (2001). Plastic resin pellets as a transport medium for toxic chemicals in the marine environment. *Environmental Science & Technology*, *35*(2), 318–324. doi:10.1021/es0010498 PMID:11347604

MEPC 75/INF.23 (2020). Progress report of the GESAMP Working Group on Sea-based Sources of Marine Litter.

Neto, J. A. B., Gaylarde, C., & Beech, I. (2019). Microplastics and attached microorganisms in sediments of the Vitória bay estuarine system in SE Brazil. *Ocean and Coastal Management*, *169*, 247–253. doi:10.1016/j.ocecoaman.2018.12.030

Nualphan. (2013). *Types and Sources of Marine Debris in BangSaen Beach. Chonburi Province*, [Master Thesis, Chulalongkorn University, Thailand].

OECD. (2023). *Climate Change and Plastics Pollution.* OECD. https://www.oecd.org/environment/plastics/Policy-Highlights-Climate-change-and-plastics-pollution-Synergies-between-two-crucial-environmental-challenges.pdf

Okoffo, E. D., O'Brien, S., Ribeiro, F., Burrows, S. D., Toapanta, T., Rauert, C., O'Brien, J. W., Tscharke, B. J., Wang, X., & Thomas, K. V. (2021). Plastic particles in soil: State of the knowledge on sources, occurrence and distribution, analytical methods and ecological impacts. *Environmental Science. Processes & Impacts*, *23*(2), 240–274. doi:10.1039/D0EM00312C PMID:33514987

Pemberton, D., Brothers, N. P., & Kirkwood, R. (1992). Entanglement of Australian fur seals in man-made debris in Tasmanian waters. *Wildlife Research*, *19*(2), 151–159. doi:10.1071/WR9920151

PlasticsEurope. (2019). *Plastics – the Facts 2019: An analysis of European plastics production, demand and waste data*. Plastics Europe. https:// www.plasticseurope.org/application/files/9715/7129/9584/FINAL_ web_version_Plastics_the_facts2019_14102019.pdf. Accessed 21 June 2021.

Power technology. (2018). *Turning waste into power: the plastic to fuel projects*. Power Technology. https://www.power-technology.com/comment/plastic-to-fuel/

Ren, X., Tang, J., Liu, X., & Liu, Q. (2020). Effects of microplastics on greenhouse gas emissions and the microbial community in fertilized soil. *Environmental Pollution*, *256*, 113347. doi:10.1016/j.envpol.2019.113347 PMID:31672352

Richardson, K., Hardesty, B. D., & Wilcox, C. (2019a). Estimates of fishing gear loss rates at a global scale: A literature review and metaanalysis. *Fish and Fisheries*, *20*(6), 1218–1231. doi:10.1111/faf.12407

Richmond, R. (2015). Coral reefs: Present problems and future concerns resulting from anthropogenic disturbance. *Integrative and Comparative Biology*, *33*(6), 524–536. https://academic .oup.com/icb/article/33/6/524/2107143

Rillig, M. C., Lehmann, A., de Souza Machado, A. A., & Yang, G. (2019). Microplastic effects on plants. *The New Phytologist*, *223*(3), 1066–1070. doi:10.1111/nph.15794 PMID:30883812

Rios, L. M., Moore, C., & Jones, P. R. (2007). Persistent organic pollutants carried by synthetic polymers in the ocean environment. *Marine Pollution Bulletin*, *54*(8), 1230–1237. doi:10.1016/j.marpolbul.2007.03.022 PMID:17532349

Rodrigues, J. P., Duarte, A. C., Santos-Echeandía, J., & Rocha-Santos, T. (2018). Signifcance of interactions between microplastics and POPs in the marine environment: A critical overview. *Trends in Analytical Chemistry*, *111*, 252–260. doi:10.1016/j.trac.2018.11.038

Rosevelt, C., Los Huertos, M., Garza, C., & Nevins, H. M. (2013). Marine debris in central California: Quantifying type and abundance of beach litter in Monterey Bay, CA. *Marine Pollution Bulletin*, *71*(1-2), 299–306. doi:10.1016/j.marpolbul.2013.01.015 PMID:23499538

Sandra, M., Devriese, L., De Raedemaecker, F., Lonneville, B., Lukic, I., & Altvater, S. (2020). *Knowledge wave on marine litter from aquaculture sources. D2.2 Aqua-Lit project*. Oostende.

SAPEA (Science Advice for Policy by European Academies). (2019). *A Scientific Perspective on Microplastics in Nature and Society*. SAPEA. doi:10.26356/microplastics

Sazima, I., Gadig, O. B. F., Namora, R. C., & Motta, F. S. (2002). Plastic debris collars on juvenile carcharhinid sharks (Rhizoprionodon lalandii) in southwest Atlantic. *Marine Pollution Bulletin*, *44*(10), 1147–1149. doi:10.1016/S0025-326X(02)00141-8 PMID:12474977

Scarascia-Mugnozza, G., Russo, G., & Sica, C. (2012). Plastic materials in European agriculture: Actual use and perspectives. *Journal of Agricultural Engineering*, *3*(3), 15–28. doi:10.4081/jae.2011.28

Schecter, A., Colacino, J., Haffner, D., Patel, K., Opel, M., Päpke, O., & Birnbaum, L. (2010). Perfluorinated compounds, polychlorinated biphenyls, and organochlorine pesticide contamination in composite food samples from Dallas, Texas, USA. *Environmental Health Perspectives*, *118*(6), 796–802. doi:10.1289/ehp.0901347 PMID:20146964

Schwarz, A. E., Ligthart, T. N., Boukris, E., & Van Harmelen, T. (2019). Sources, transport, and accumulation of different types of plastic litter in aquatic environments: A review study. *Marine Pollution Bulletin*, *143*, 92–100. doi:10.1016/j.marpolbul.2019.04.029 PMID:31789171

Sherrington, C.S., Hann, G., Cole, G., & Corbin, M. (2016). Study to support the development of measures to combat a range of marine litter sources. *Report for European Commission DG Environment.*

Solomon, T. (2014). ENVIRONMENTAL ISSUES DUE TO IMPROPER PLASTIC DISPOSAL IN INDIAN CITIES. *International Journal of Research in Engineering and Technology*, *03*(23), 286–289. doi:10.15623/ijret.2014.0323061

Statista. (2023). *Global plastics industry*. Statista, https://www.statista.com/topics/5266/plastics-industry/#topicOverview

Statista. (2023). *Plastic industry in India*. Statista. https://www.statista.com/topics/6902/plastic-industry-in-india/#topicOverview

Statista. (2023). *Plastic waste worldwide*. Statista. https://www.statista.com/topics/5401/global-plastic-waste/#topicOverview

Steinmetz, Z., Wollmann, C., Schaefer, M., Buchmann, C., David, J., Troger, J., Muñoz, K., Frör, O., & Schaumann, G. E. (2016). Plastic mulching in agriculture. Trading shortterm agronomic benefits for long-term soil degradation? *The Science of the Total Environment*, *550*, 690–705. doi:10.1016/j.scitotenv.2016.01.153 PMID:26849333

Sussarellu, R., Suquet, M., Thomas, Y., Lambert, C., Fabioux, C., Pernet, M. E. J., Le Goïc, N., Quillien, V., Mingant, C., Epelboin, Y., Corporeau, C., Guyomarch, J., Robbens, J., Paul-Pont, I., Soudant, P., & Huvet, A. (2016). Oyster reproduction is afected by exposure to polystyrene microplastics. *Proceedings of the National Academy of Sciences of the United States of America*, *113*(9), 2430–2435. doi:10.1073/pnas.1519019113 PMID:26831072

Tang, G., Liu, M., Zhou, Q., He, H., Chen, K., Zhang, H., Hu, J., Huang, Q., Luo, Y., Ke, H., Chen, B., Xu, X., & Cai, M. (2018). Microplastics and polycyclic aromatic hydrocarbons (PAHs) in Xiamen coastal areas: Implications for anthropogenic impacts. *The Science of the Total Environment*, *634*, 811–820. doi:10.1016/j.scitotenv.2018.03.336 PMID:29653425

Thompson, R. C., Olsen, Y., Mitchell, R. P., Davis, A., Rowland, S. J., John, A. W. G., McGonigle, D., & Russell, A. E. (2004). Lost at sea: Where is all the plastic? *Science*, *304*(5672), 838–838. doi:10.1126/science.1094559 PMID:15131299

Trudel, D., Scheringer, M., von Goetz, N., & Hungerbühler, K. (2011). Total consumer exposure to polybrominated diphenyl ethers in North America and Europe. *Environmental Science & Technology*, *45*(6), 2391–2397. doi:10.1021/es1035046 PMID:21348481

UNEP. (2021). *Addressing Single-Use Plastic Products Pollution using a Life Cycle Approach.* UNEP. https://www.unep.org/resources/publication/ addressing-single-use-plastic-products-pollution-using-life-cycleapproach. Accessed 6 July 2021.

UNEP. (2023). *Sustainable Roadmap.* UNEP. https://unpartnerships.un.org/news/2023/roadmap-sustainable-future-ending-plastic

United Nations Environment Programme. (2021). *Drowning in Plastics – Marine Litter and Plastic Waste Vital Graphics*. UN.

Velzeboer, I., Kwadijk, C. J. A. F., & Koelmans, A. A. (2014). Strong sorption of PCBs to nanoplastics, microplastics, carbon nanotubes, and fullerenes. *Environmental Science & Technology*, *48*(9), 4869–4876. doi:10.1021/es405721v PMID:24689832

Wang, F., Wong, C. S., Chen, D., Lu, X., Wang, F., & Zeng, E. Y. (2018). Interaction of toxic chemicals with microplastics: A critical review. *Water Research*, *139*, 208–219. doi:10.1016/j.watres.2018.04.003 PMID:29653356

Wang, W., Ge, J., Yu, X., & Li, H. (2020a). Environmental fate and impacts of microplastics in soil ecosystems: Progress and perspective. *The Science of the Total Environment*, *708*, 134841. doi:10.1016/j.scitotenv.2019.134841 PMID:31791759

Webb, H. K., Arnott, J., Crawford, R. J., & Ivanova, E. P. (2013). Plastic Degradation and Its Environmental Implications with Special Reference to Poly(ethylene terephthalate). *Polymers*, *5*(1), 1–18. doi:10.3390/polym5010001

Windsor, F. M., Durance, I., Horton, A. A., Thompson, R. C., Tyler, C. R., & Ormerod, S. J. (2019). A catchment-scale perspective of plastic pollution. *Global Change Biology*, *25*(4), 1207–1221. doi:10.1111/gcb.14572 PMID:30663840

Windsor, F. M., Durance, I., Horton, A. A., Thompson, R. C., Tyler, C. R., & Ormerod, S. J. (2019). A catchment-scale perspective of plastic pollution. *Global Change Biology*, *25*(4), 1207–1221. doi:10.1111/gcb.14572 PMID:30663840

Chapter 7
Climate Litigation for Quality of Life:
Exploring Innovative Legal Avenues to Hold Polluters Accountable

A. Karthiayani
Symbiosis Law School, Symbiosis International University (Deemed), Pune, India

Shashikala Gurpur
Symbiosis Law School, Symbiosis International University (Deemed), Pune, India

Rupal Rautdesai
Symbiosis Law School, Symbiosis International University (Deemed), Pune, India

ABSTRACT

This proposed book chapter explores novel legal strategies for holding polluters accountable for climate change contributions and protecting affected communities' quality of life. It highlights the limitations of traditional litigation and advocates for innovative approaches. The first section focuses on technological solutions like satellite imagery, remote sensing, and blockchain to gather evidence and enhance supply chain transparency. The second section examines successful collaborative legal initiatives addressing transboundary pollution and global climate injustices. The third section explores integrating climate science, economics, and law in litigation efforts, quantifying damages, and linking pollution to health impacts. The fourth section discusses future directions and challenges, including accountability for future generations. The chapter concludes by emphasizing interdisciplinary collaboration and legal reforms for climate justice and a better future.

INTRODUCTION

In the twilight of the 21st century, humanity stands at a crossroads, confronting an existential threat that has materialized with alarming speed and tenacity - the global climate crisis. With unwavering

DOI: 10.4018/978-1-6684-9863-7.ch007

persistence, climate change has infiltrated the very fabric of our planet, leaving no corner of the Earth untouched. Rising sea levels imperil coastal communities, erratic weather patterns devastate agriculture, and biodiversity loss threatens the delicate ecosystems that sustain life. The consequences of this crisis are far from hypothetical, as they reverberate through every aspect of human existence, impinging on the very quality of life we hold dear.

In the face of this unprecedented challenge, the legal realm has endeavoured to hold those responsible accountable. Traditional litigation approaches have laid the groundwork for acknowledging the culpability of polluters, providing some semblance of justice for the communities ravaged by environmental degradation. These efforts are commendable, having brought corporate malfeasance to the forefront of public consciousness and courtroom battles. However, the tools and methods of traditional litigation have their limitations, and the scale and urgency of the climate crisis necessitate innovative, far-reaching legal strategies.

As the climate crisis deepens, so must our resolve to redefine the boundaries of justice. The traditional contours of legal practice are no longer sufficient to address the complexities and scale of environmental harm. It is time to push the boundaries and embark on a journey that explores new, bold legal avenues, where imagination and creativity interlace with the rigors of law to seek unprecedented solutions.

This chapter serves as a compass to navigate the uncharted waters of climate justice, daring to explore unorthodox avenues while anchoring itself in the values of equity and accountability. By embracing innovation, technology, collaboration, and interdisciplinary approaches, we endeavour to revolutionize the landscape of legal responses to climate change. The central mission of this chapter is twofold: to hold polluters accountable for their actions and to safeguard the quality of life for communities adversely affected by climate change.

At the heart of the climate crisis lies an intricate symphony of ecological relationships that have evolved over eons. As human activities disrupt this delicate balance, entire ecosystems unravel, and the symphony turns discordant. To fully comprehend the implications of the crisis and devise effective legal strategies, we must recognize the inextricable links between the environment, society, and the economy.

The climate crisis is not just a matter of rising temperatures and melting ice; it is a human crisis. Vulnerable communities bear the brunt of environmental harm, exacerbating inequalities and infringing on human rights. Indigenous peoples witness the erosion of their cultures and lands, while marginalized populations are disproportionately affected by extreme weather events and ecological disruptions. Any legal approach must prioritize the protection of human dignity and inclusivity, seeking justice for those most impacted by climate change.

Traditional litigation, while a vital tool in the fight for justice, often operates reactively, addressing harm after it has occurred. To break free from this conundrum, we must foster proactive legal measures that anticipate and prevent environmental degradation. This forward-looking approach entails establishing regulations, standards, and responsibilities that act as deterrents and incentives for environmentally responsible practices.

Technology presents a potent force in our quest for innovative climate justice solutions. From satellite imaging to blockchain-based contracts, technological advancements offer unprecedented ways to gather evidence, monitor ecological changes, and enforce accountability. By integrating technology into the legal landscape, we empower communities, facilitate access to justice, and amplify the voices of those most affected by environmental harm.

One example of how technology is being used to hold polluters accountable is the use of satellite imagery. Satellite imagery can be used to track emissions from polluters, and this information can be used to build legal cases against them. For example, in 2019, a group of environmental organizations used satellite imagery to track emissions from the Dakota Access Pipeline. This information was used to file a lawsuit against the pipeline company, and the lawsuit was eventually settled for $1.9 billion (Volcovici, 2017).

Another example of how technology is being used to hold polluters accountable is the use of data analytics. Data analytics can be used to identify patterns of harm caused by climate change. For example, a recent study used data analytics to show that climate change is causing more extreme weather events in the United States (Hersher, 2023). This information could be used to build legal cases against polluters who are contributing to climate change.

Confronting the climate crisis demands collective action. Governments, NGOs, businesses, scientists, communities, and legal practitioners must form a symphony of stakeholders, each playing a unique role in orchestrating a harmonious response. Collaborative partnerships are essential to break down silos, share expertise, and design multifaceted solutions that transcend boundaries and jurisdictions (Erkic, 2022)

Communities on the frontlines of climate change possess invaluable insights into local environmental challenges and possess a vested interest in safeguarding their homes. By empowering these communities with legal knowledge and resources, we strengthen their capacity to advocate for their rights, engage with legal processes, and influence decisions that impact their lives directly (WIPO, 2020)

True climate justice transcends the traditional notions of legal victories or compensatory damages. It calls for a comprehensive, holistic approach that integrates environmental, social, and economic considerations. Environmental justice must coalesce with social justice, recognizing the intersectionality of climate impacts and addressing systemic inequalities that amplify vulnerability to environmental harm (Amorim-Maia et.al, 2022)

The link to climate litigation for quality of life lies in the transformative potential of innovative legal strategies to address the global climate crisis and its far-reaching impacts on human existence (Amorim-Maia et.al,2022). Climate litigation, while already a commendable tool for holding polluters accountable and seeking justice for affected communities, needs to evolve to meet the scale and urgency of the challenges posed by climate change (Preston,2018).

The consequences of the climate crisis have significant implications for the quality of life of individuals and communities worldwide. Rising sea levels, erratic weather patterns, biodiversity loss, and extreme weather events all directly and indirectly affect people's well-being, health, livelihoods, and access to essential resources (European Commission,2023) Vulnerable populations, such as marginalized communities and indigenous peoples, are particularly impacted, leading to increased inequalities and violations of human rights (WIPO,2020).

Traditional litigation approaches have limitations, as they often operate reactively and may not fully consider the complex and interconnected nature of ecological relationships (Preston,2018). To address these limitations, new legal avenues that are proactive, collaborative, and technologically advanced must be explored. By embracing innovation and interdisciplinary approaches, the legal realm can better anticipate and prevent environmental harm, safeguarding the quality of life for those most affected by climate change.

Technology plays a crucial role in the pursuit of climate justice, offering powerful tools to gather evidence, monitor ecological changes, and hold polluters accountable. For instance, satellite imagery can track emissions from polluters, providing valuable evidence for legal cases. Data analytics can identify patterns of harm caused by climate change, assisting in building legal arguments against those contributing to the crisis.

To achieve climate justice, collective action is essential. Governments, NGOs, businesses, scientists, communities, and legal practitioners must collaborate and share expertise to design multifaceted solutions that transcend boundaries and jurisdictions. Empowering communities on the frontlines of climate change with legal knowledge and resources strengthens their ability to advocate for their rights and influence decisions that impact their well-being directly.

True climate justice necessitates integrating environmental, social, and economic considerations, recognizing the intersectionality of climate impacts and addressing systemic inequalities (Sultana, 2022). By reimagining and redesigning legal systems to weave justice into the fabric of society, the law can become a transformative force for environmental healing and collective well-being. This comprehensive and holistic approach to climate litigation is crucial in protecting and enhancing the quality of life for current and future generations amidst the existential threat of the global climate crisis

In the pursuit of climate justice, we are not beholden to the status quo. Instead, we must seize this pivotal moment to reimagine and redesign our legal systems, weaving justice into the very fabric of society. As we traverse the uncharted territory of innovative legal strategies, we must hold steadfast to our dreams of a future where the law becomes a transformative force for environmental healing and collective well-being.

TECHNOLOGICAL SOLUTIONS

In the grand tapestry of climate justice, technology emerges as a potentially, transforming the conventional legal landscape into an arena of unprecedented possibilities. As we delve into the realm of innovation, this chapter embarks on a journey through the interwoven threads of emerging technologies - satellite imagery, remote sensing, data analytics, blockchain, geospatial mapping, and predictive modelling - unveiling their transformative potential in the pursuit of climate accountability.

The Eyes in the Sky: Satellite Imagery and Remote Sensing

Like watchful sentinels above, satellites offer a celestial vantage point that transcends borders and allows us to observe the planet as a unified whole. Satellite imagery and remote sensing technologies become our lenses to scrutinize climate-related harm, providing irrefutable evidence of environmental degradation (Louw et.al, 2022). High-resolution imagery uncovers the scarred landscapes, deforested terrains, and receding glaciers, capturing the footprints of those responsible for ecological plunder (Weigand et.al, 2019).

In legal battles, such technological prowess becomes a sword of truth, slicing through the veils of denial and deception. Satellite data stands as a silent witness, a testament to the destruction wrought upon our Earth. By integrating this treasure trove of information into climate litigation, the burden of proof shifts from affected communities to the polluters themselves, laying bare their responsibility and ensuring justice is met.

The use of satellite imagery and remote sensing technologies in climate litigation is a powerful tool that strengthens the quest for justice and protection of quality of life in the face of the global climate crisis. These technologies offer an unparalleled view of the Earth, transcending borders and providing crucial evidence of environmental harm (Dwivedi et.al, 2022).

Satellite imagery and remote sensing act as watchful sentinels, capturing the consequences of climate change with high-resolution precision. From scarred landscapes to deforested terrains and receding glaciers, these images become undeniable documentation of the ecological plunder caused by human activities. The evidence they provide stands as a silent witness to the destruction wrought upon the planet (Yager, 2015).

In legal battles, this technological prowess becomes a sword of truth that can cut through denial and deception. By integrating satellite data into climate litigation, the burden of proof shifts from affected communities to the polluters themselves. The irrefutable evidence from satellite imagery lays bare the responsibility of those who contribute to environmental degradation.

In the case of ***The People of the State of California v. Plains All American Pipeline (2020)***, which involved an oil spill off the coast of California in 2015, satellite imagery was used to track and assess the extent of the environmental damage caused by the spill. The satellite images provided crucial evidence in the legal proceedings and helped determine the responsibility of the pipeline company in the incident.

The case ***Lliuya v. RWE AG (2015)*** involved a Peruvian farmer, Saul Luciano Lliuya, suing the German energy company RWE AG for contributing to climate change and its impact on his community due to melting glaciers and the risk of glacial lake outburst floods. Remote sensing data, including satellite imagery, was used to demonstrate the connection between RWE's greenhouse gas emissions and the threat faced by Lliuya's community in the Andes.

Integrating satellite imagery into climate litigation can lead to a more effective pursuit of justice (The Geneva Association,2021). Affected communities no longer bear the sole responsibility of proving harm caused by climate change; instead, the comprehensive and objective evidence from satellites strengthens their cases (Diezmartínez & Gianotti,2022). Holding polluters accountable becomes more feasible and just, ensuring that those responsible for environmental harm are held to task for their actions (Economist Impact, 2022).

By using satellite imagery and remote sensing as a critical component in climate litigation, the legal realm can play a vital role in safeguarding the quality of life for communities adversely affected by climate change. This technology becomes a powerful ally in the fight for climate justice, helping to protect vulnerable populations, preserve ecosystems, and hold accountable those who perpetrate ecological harm. Ultimately, the use of satellite imagery and remote sensing brings us closer to a future where justice is served and the quality of life is upheld for current and future generations in the twilight of the 21st century.

The Blockchain Symphony: Transparency and Accountability

In a world rife with intricate supply chains and labyrinthine networks, tracing the origins of environmental harm can be a Herculean task. Enter blockchain technology - the ledger of immutable truth, intertwining transparency and traceability. As goods traverse vast distances, blockchain creates an indelible record, leaving no room for opacity or subterfuge.

By anchoring supply chains on blockchain platforms, we unlock the ability to trace the provenance of goods and services, illuminating the path from manufacturer to consumer. In the realm of climate justice, this presents a transformative opportunity to hold polluters accountable for their actions throughout the supply chain (Bhatia & Albarrak, 2023). Corporations must own up to their environmental footprint, from sourcing raw materials to the end-user, fostering a new era of corporate responsibility and environmental stewardship.

The application of blockchain technology in climate litigation holds immense potential for enhancing transparency and accountability in complex supply chains, thereby safeguarding the quality of life for communities affected by environmental harm (Friedman & Ormiston, 2022). By creating an immutable ledger that traces the origins and movements of goods and services, blockchain can become a powerful instrument in holding polluters accountable throughout the supply chain (Friedman & Ormiston, 2022).

In climate litigation, establishing the link between environmental harm and specific polluters can often be challenging due to the complexity and opacity of supply chains (Wang, 2020). However, blockchain's ability to provide a transparent and tamper-proof record of every step in the supply chain offers a transformative opportunity (Park, 2021). This technology allows for the tracing of environmental impacts from the source of raw materials, through manufacturing processes, to the end consumers (Dutta, 2020).

Through blockchain's transparency and traceability, corporations can be compelled to take responsibility for their actions and environmental footprint. This comprehensive understanding of their impact can drive businesses to adopt environmentally responsible practices and embrace a new era of corporate accountability and environmental stewardship.

For instance, some industries have already begun to use blockchain to track the sustainability and authenticity of products, such as the fashion and food industries.

Although not directly linked to climate litigation, examples of blockchain's role in enhancing transparency and accountability in supply chains include:

1. Walmart's Use of Blockchain for Food Traceability: Walmart implemented blockchain technology to track the journey of certain food products from the farm to the store shelves (Vitasek et.al, 2022). By doing so, they aimed to improve food safety and transparency in their supply chain, which could have implications for environmental sustainability and quality of life through safer food practices.
2. IBM and Maersk's TradeLens Platform: IBM and Maersk collaborated on TradeLens, a blockchain-based platform that enhances supply chain visibility and efficiency for the global shipping industry (IBM Newsroom, 2018). While not strictly climate-related, this platform's adoption has implications for reducing carbon emissions associated with shipping and improving sustainability in the logistics sector.

As the use of blockchain in supply chain management gains momentum, it can play a crucial role in climate litigation, ultimately contributing to the protection of quality of life for communities affected by environmental degradation.

The Cartographer's Dream: Geospatial Mapping and Predictive Modelling

In the intricacies of climate change, patterns emerge, hidden to the naked eye but discernible through geospatial mapping and predictive modelling. The dynamic dance of ecosystems and climate data unfolds before us, revealing the vulnerabilities and risks faced by communities and environments alike.

Geospatial mapping brings the power of visualization, portraying the spatial distribution of climate risks with astonishing clarity (Boulos & Wilson, 2023). By overlaying environmental and socioeconomic data, we gain profound insights into the communities most at risk, empowering legal teams to tailor their strategies to specific contexts.

Predictive modelling takes us into the realm of future possibilities, forecasting the trajectory of climate events with increasing accuracy. Armed with such knowledge, climate litigation becomes a proactive endeavour, anticipating harm and securing injunctions to prevent further damage (Collins et.al, 2021).

In a symphony of data and algorithms, geospatial mapping and predictive modelling breathe life into legal strategies, transforming advocacy into a symphony of foresight and precision.

Geospatial mapping and predictive modelling play a crucial role in climate litigation, enhancing the pursuit of justice and protection of quality of life for communities affected by the impacts of climate change (Bene & Ávila, 2023). These powerful tools offer insights into the spatial distribution of climate risks and provide forecasts of future possibilities, enabling legal teams to proactively tailor strategies to address environmental harm.

Climate litigation often involves complex and interconnected environmental and socioeconomic factors. Geospatial mapping allows legal teams to visualize and understand the spatial patterns of climate risks and vulnerabilities. By overlaying various data layers, such as climate data, population density, infrastructure, and ecological information, lawyers gain a comprehensive view of the communities most at risk. This knowledge empowers legal practitioners to focus their efforts on specific contexts and communities that require immediate attention and support.

Predictive modelling takes climate litigation into the realm of foresight and precision. By using algorithms and historical data, legal teams can forecast the trajectory of climate events with increasing accuracy (Bochenek & Ustrnul, 2022). Armed with this knowledge, they can take proactive measures to prevent further damage and secure injunctions to protect vulnerable communities and ecosystems. This forward-looking approach is crucial in mitigating the impacts of climate change and preserving the quality of life for current and future generations.

Case laws related to the use of geospatial mapping and predictive modelling in climate litigation may not be as common or widely reported as other types of climate litigation cases. However, the application of these tools in climate-related legal actions is becoming increasingly prevalent.

In the case ***Conservation Law Foundation v. Exxon Mobil Corp. (2016)***, the Conservation Law Foundation (CLF) filed a lawsuit against Exxon Mobil, alleging that the company failed to protect its oil storage facility from climate change impacts. The CLF used geospatial mapping to illustrate how rising sea levels and increased storm surges could potentially flood the facility, causing environmental harm and endangering nearby communities.

Center for Biological Diversity v. Bureau of Land Management (2014) involved a lawsuit against the Bureau of Land Management (BLM) for failing to consider the potential impacts of climate change in the leasing of public lands for oil and gas extraction. The plaintiffs used predictive modelling to demonstrate how such leasing decisions could contribute to greenhouse gas emissions and exacerbate climate change.

While these examples do not solely focus on geospatial mapping and predictive modelling, they highlight the importance of data-driven approaches in climate litigation. As technology continues to advance, the integration of geospatial mapping and predictive modelling in climate litigation will likely become more prevalent, enabling legal teams to advocate for climate justice with greater foresight and precision.

The Conductor's Baton: Ethical Integration of Technology

As technology surges forward, the conductor's baton must guide the symphony of innovation with ethical purpose. We must address concerns of privacy, data security, and consent, ensuring that the transformative power of technology does not infringe upon the rights of individuals and communities.

Ethical considerations encompass the responsible use of data, ensuring its anonymization, and safeguarding against unintended consequences. We must be vigilant, striking a harmonious balance between technological prowess and ethical integrity, for the path to climate justice is not merely measured by the means but also by the principles we uphold.

The Ensemble: Collaboration and Capacity Building

The symphony of technological solutions requires a harmonious ensemble of experts, stakeholders, and communities. Collaborative efforts foster a shared understanding, driving innovation while promoting capacity building in affected communities.

By nurturing a diverse array of voices, we expand the repertoire of technological solutions, ensuring that the benefits of innovation are accessible to all. Partnerships between scientists, legal experts, and communities create a resounding echo of change, amplifying the impact of technology in the pursuit of climate justice.

The Overture of Hope

At the end of this section, the overture of hope resounds, transcending the limitations of the present and embracing the promise of the future. Emerging technologies have unlocked a new symphony of possibilities, where accountability and justice entwine, shaping a world where ecological guardianship and legal acumen unite.

As we venture into the uncharted realms of technology and climate justice, let us tread with reverence and audacity, for the symphony of innovation echoes not just in the courtrooms but in the hearts of every individual determined to heal our planet and secure a flourishing future for generations to come. The transformative alchemy of technology and law ignites the beacon of hope, illuminating the path to a world where humanity and nature dance in harmony once more.

COLLABORATIVE LEGAL INITIATIVES

In the crescendo of the climate crisis, a new symphony of justice emerges - one orchestrated by collaborative legal initiatives that unite individuals, communities, NGOs, and legal professionals in a harmonious pursuit of climate litigation. Breaking free from the constraints of traditional legal boundaries, these initiatives resonate with the power of unity, pooling resources, and expertise to confront the complex challenges posed by climate change.

The Rise of Collaborative Legal Initiatives

In the face of a global crisis, collective action becomes an imperative. Collaborative legal initiatives embody the spirit of togetherness, drawing strength from the synergy of diverse stakeholders. These initiatives go beyond individual cases, forming networks that amplify the impact of climate litigation. In a world where climate change transcends borders, these initiatives create a united front against powerful polluting industries, harnessing the strength of unity to drive change.

At the heart of these endeavours lie the affected communities, NGOs, and legal professionals, each contributing their unique skills and experiences. Empowered by shared goals and mutual support, these initiatives infuse new energy into the fight for climate justice, transforming adversarial litigation into a collective quest for a sustainable future.

Collaborative legal initiatives are vital in climate litigation as they foster collective action, amplify the pursuit of climate justice, and ultimately protect and enhance the quality of life for communities impacted by the global climate crisis (Peel & Lin,2019). These initiatives bring together diverse stakeholders, including affected communities, NGOs, and legal professionals, to work together towards common goals and address the complex and interconnected challenges of climate change.

Collaborative legal initiatives prioritize the well-being and quality of life for communities adversely affected by climate change. By uniting different stakeholders, these initiatives can pool resources, expertise, and efforts to achieve more comprehensive and impactful outcomes. By working together, they strengthen the pursuit of justice for vulnerable populations, advocate for equitable solutions, and protect the rights of those most impacted by environmental harm.

In this ground-breaking case, the ***Urgenda Foundation v Netherlands (2019)***, a Dutch NGO representing citizens and environmental groups, collaborated with legal professionals to sue the Dutch government for inadequate climate action. The initiative argued that the government's failure to take sufficient measures to reduce greenhouse gas emissions violated its duty of care to protect citizens' lives and their right to a safe and healthy environment. The Dutch court ruled in favour of Urgenda, ordering the government to increase its emissions reduction targets significantly. This case demonstrated the power of collaborative legal efforts in holding governments accountable for their climate commitments and safeguarding the well-being of citizens.

Our Children's Trust, a non-profit organization, has been involved in numerous climate lawsuits in the United States. These lawsuits, filed on behalf of young plaintiffs, argue that the government's failure to act on climate change infringes on the constitutional rights of the younger generation, including the right to life, liberty, and property. While not a single initiative, the collaboration between Our Children's Trust, legal professionals, and youth activists exemplifies the impact of collective legal action in advocating for climate justice and the protection of future generations' quality of life (Our Children's Trust,2019).

Collaborative legal initiatives represent a shift from adversarial litigation to a united and coordinated effort, recognizing the interconnected nature of climate change and the need for collective action. By joining forces, these initiatives enhance their capacity to drive change, challenge powerful polluting industries, and work towards a more sustainable and equitable future for all. In the face of a global crisis, collaborative legal initiatives are essential in driving meaningful climate litigation and securing a better quality of life for current and future generations.

Bridging Borders: Cross-Border and International Collaborations

The battle against climate change is borderless, transcending the confines of nations and legal jurisdictions. Cross-border and international collaborations forge a tapestry of global solidarity, addressing transboundary pollution and climate injustices that demand unified efforts.

Successful examples of cross-border collaborations paint a vibrant picture of international solidarity. Legal teams from different nations join forces, leveraging their diverse expertise to navigate complex legal systems and jurisdictional challenges (OECD-OPSI, 2021). Shared knowledge, resources, and legal precedents become bridges over which the pursuit of climate justice flows seamlessly.

Beyond national borders, international collaborations craft an anthem of unity that resonates in the halls of global governance. In the United Nations Climate Change Conferences and other international forums, collaborative legal initiatives amplify the voices of the most vulnerable, demanding accountability and action from the international community (UNDP, 2022).

The Strains and Triumphs of Collective Legal Action

Collective legal action brings forth both challenges and triumphs in the struggle for climate justice. Assembling diverse stakeholders with disparate interests can be a delicate symphony, requiring open communication and mutual understanding. Navigating differing legal systems, languages, and cultural contexts demands deft conductors to harmonize the orchestra of collaborative initiatives.

Yet, in these challenges lie the seeds of triumph. The convergence of diverse perspectives fuels innovation, enriching legal strategies with nuanced approaches that resonate with local and global audiences alike. By leveraging collective resources, these initiatives wield a power that surpasses the sum of individual efforts, standing as a formidable force against corporate giants and vested interests.

Through the crucible of collaboration, communities find a collective voice, raising it to crescendo as a symphony of resilience, determination, and hope. In the harmony of their united efforts, the world witnesses the emergence of climate justice as an unyielding force that knows no boundaries.

The Anthem of Change

In the climactic finale of this section, the anthem of change resounds, echoing the triumphs and trials of collaborative legal initiatives in the fight for climate justice. As these initiatives sweep across nations and traverse continents, they weave a new narrative of resilience and unity.

In the unison of individuals, communities, NGOs, and legal professionals, the symphony of collaborative legal initiatives calls forth a future where accountability knows no borders, where the powerful are held responsible, and where justice thrives. Together, they orchestrate a melody of transformation that reverberates through the halls of power, echoing the collective will of humanity to safeguard our planet and secure a sustainable future for generations to come.

In this anthem of change, the spirit of collaboration becomes a rallying cry, summoning the world to stand as one in the pursuit of climate justice. As the symphony of unity plays on, may its chords resonate in the hearts of all, inspiring a global movement that marches forward, unwavering in its determination to heal our Earth and create a harmonious world for all living beings.

INTERDISCIPLINARY APPROACHES IN CLIMATE LITIGATION

In recent times, the urgency to address the adverse impacts of climate change has prompted the integration of various disciplines to strengthen legal efforts in seeking justice and compensation for those affected. This section delves into the innovative and unconventional methods of combining climate science, economics, and law in litigation endeavours. It further explores the application of cost-benefit analysis and economic valuation techniques to quantify the damages arising from climate change. Additionally, it highlights the crucial role of public health research and epidemiological evidence in climate litigation, emphasizing the need for holding polluters accountable for their actions.

Intersection of Climate Science, Economics, and Law

Climate change is a complex, multi-faceted challenge that requires a comprehensive understanding of its scientific basis, economic implications, and legal ramifications (Hoffman & Lott, 2022). By merging these disciplines, legal experts, economists, and climate scientists can collectively address the intricate issues at hand.

In the ground-breaking case of ***Juliana v. United States (2015)***, young plaintiffs filed a lawsuit against the federal government, arguing that its actions have violated their constitutional rights to a stable climate system. This lawsuit exemplifies the powerful synergy between climate science and law, as expert testimonies from renowned climate scientists provided robust evidence of the government's contributions to climate change through inadequate policies and regulations.

Utilizing Cost-Benefit Analysis and Economic Valuation Methods

Quantifying the damages caused by climate change is often a daunting task. However, the application of cost-benefit analysis and economic valuation methods provides a systematic approach to assess and present the economic impact of climate-related disasters and ecological losses (Rising et.al, 2022).

In the case of ***Comer v. Murphy Oil USA (2005)***, plaintiffs in Mississippi sued several energy companies for their contribution to global warming and its alleged role in intensifying Hurricane Katrina. The court dismissed the case, but it demonstrated the potential of incorporating cost-benefit analysis to establish the economic costs borne by communities affected by extreme weather events linked to climate change.

Integration of Public Health Research and Epidemiological Evidence

Health impacts resulting from climate change and pollution have garnered significant attention in recent years. By integrating public health research and epidemiological evidence into climate litigation, plaintiffs can establish a direct link between polluters' actions and adverse health consequences (Swaen et.al, 2018).

While not directly related to climate change, the case of ***Lujan v. Defenders of Wildlife (1992)*** highlights the significance of public health evidence in litigation. The Supreme Court ruled that plaintiffs must show concrete and particularized injuries caused by a defendant's actions. In climate litigation, presenting epidemiological evidence of health impacts linked to pollution strengthens the argument for polluter accountability and the duty to compensate affected communities.

Exploring Climate Refugees and Human Rights

As climate change intensifies, its impact on vulnerable communities leads to the concept of climate refugees. The unprecedented number of people forced to leave their homes due to climate-related factors raises pertinent questions about human rights and international law. Incorporating human rights perspectives in climate litigation adds a new dimension to the fight for justice (McAdam, 2020).

In the case of ***Ioane Teitiota v. New Zealand (2020)***, a man from Kiribati sought asylum in New Zealand, claiming that the effects of sea-level rise and other climate-related issues had made his homeland uninhabitable. The case highlighted the complexities of determining climate refugee status and the potential implications for international law. Although the court ultimately rejected Teitiota's claim, it brought the issue of climate refugees to the forefront of global discussions.

Novel Application of Geoengineering and Liability

As the consequences of climate change become more severe, discussions around geoengineering, the deliberate manipulation of Earth's climate etc., have gained traction. Some proposals aim to mitigate climate change, while others pose risks and uncertainties. Analysing the legal implications and potential liability arising from geoengineering projects opens up innovative avenues for climate litigation.

While geoengineering is still a largely theoretical concept, its potential implementation raises questions about liability for unforeseen consequences. Anticipating future legal challenges and developing frameworks for holding geoengineering projects accountable can prepare the legal system for emerging environmental issues (Reynolds, 2019).

Digital Climate Activism and Citizen Participation

The digital age has revolutionized how people engage with climate issues. Social media and online platforms have provided avenues for climate activism and citizen participation in ways previously unimaginable. Examining the role of digital activism in climate litigation allows for a better understanding of its influence and potential to shape legal strategies.

Digital climate activism is relatively new, and its full impact on climate litigation is yet to be seen. However, analysing its role in mobilizing public support, influencing public opinion, and fostering global solidarity can be invaluable in shaping future legal approaches.

Conventional legal systems treat nature as property, leading to the exploitation of natural resources without adequate consideration for ecological well-being. The emerging concept of "rights of nature" challenges this anthropocentric approach and seeks legal recognition of the inherent rights of ecosystems and the environment.

In Ecuador, the Vilcabamba River was granted legal rights in the landmark case of ***Vilcabamba River v. Provincial Government of Loja (2011).*** The court ruled that the river possessed the right to exist, flourish, regenerate, and perform its natural cycles. This case exemplifies the growing recognition of ecocentrism in law and provides a precedent for including nature's rights in climate litigation.

Alternative Dispute Resolution for Climate-Related Conflicts

Traditional legal avenues often involve lengthy court battles and adversarial approaches, which may not always be the most effective means of resolving climate-related conflicts. Considering the urgency of addressing climate change, exploring alternative dispute resolution (ADR) methods can offer faster and more collaborative solutions (McDonald, 2021).

As ADR in climate litigation is a relatively new area, there are no specific case laws to cite. However, the potential for utilizing mediation, arbitration, or negotiation in resolving climate disputes warrants further research and exploration.

ADR processes can involve all stakeholders, including governments, corporations, and affected communities, to jointly seek solutions that balance environmental protection with socio-economic interests.

Precautionary Principle and Climate Litigation

The precautionary principle suggests that if an action or policy has the potential to cause harm to the public or the environment, in the absence of scientific consensus, the burden of proof falls on those advocating the action. Incorporating this principle into climate litigation can shift the burden of proof from the plaintiffs to the defendants, encouraging proactive measures to prevent climate change harm (Pinto-Bazurco, 2021)

In ***Friends of the Earth v. Laidlaw Environmental Services (2000)***, the court ruled that the defendant's failure to follow the Clean Water Act's reporting requirements justified applying the precautionary principle. This case illustrates how the precautionary principle can be invoked to protect the environment and hold polluters accountable, setting a precedent for its application in climate litigation.

Indigenous Rights and Climate Justice

Indigenous communities often bear the brunt of climate change impacts due to their close relationship with the environment and dependence on natural resources. Recognizing the rights of Indigenous peoples in climate litigation acknowledges their traditional knowledge and fosters climate justice (ILO, 2017).

The case of ***Arctic Athabaskan Council v. United States (2013)*** highlighted the adverse impact of climate change on Indigenous communities in the Arctic. The court's recognition of the rights and interests of the Athabaskan Council in protecting their environment demonstrates the importance of incorporating Indigenous perspectives in climate litigation.

Interdisciplinary approaches in climate litigation continue to evolve, shaping the landscape of environmental law and its potential to address the global challenge of climate change (Peel & Osofsky, 2020). By exploring unconventional avenues such as ADR, the precautionary principle, and indigenous rights, the legal system can adapt to the complexities of climate-related disputes (ICC, 2019). As climate change intensifies, innovative and unique legal strategies will be crucial in seeking justice, promoting sustainable practices, and safeguarding the planet for future generations. Through the integration of diverse disciplines and perspectives, climate litigation can become a powerful force in mitigating climate change's impacts and securing a more sustainable and just future.

Interdisciplinary approaches in climate litigation mark a paradigm shift in the pursuit of environmental justice. By intertwining climate science, economics, and law, legal claims gain robustness and credibility (Setzer, 2022). Cost-benefit analysis and economic valuation methods provide a structured way to quantify damages, while the integration of public health research strengthens the case for polluter accountability (O'Mahony, 2021). As climate change continues to affect communities worldwide, adopting unconventional approaches becomes imperative to address the challenges ahead and foster a sustainable future.

The ever-evolving landscape of climate change demands continuous innovation and adaptability in legal approaches. Exploring unconventional avenues, such as climate refugees, geoengineering liability, digital activism, and rights of nature, enriches climate litigation by encompassing diverse perspectives and novel concepts. As the world grapples with the challenges of climate change, interdisciplinary approaches will be instrumental in securing justice for impacted communities and safeguarding the planet for future generations.

FUTURE DIRECTIONS AND CHALLENGES IN CLIMATE LITIGATION

As the consequences of climate change continue to unfold, the question of responsibility and accountability for future generations arises. This section delves into the innovative concept of legal frameworks that seek to hold future generations accountable for the impacts of climate change. It explores the ethical dimensions of intergenerational climate justice and the potential for legal mechanisms to address the long-term consequences of present-day actions.

The idea of holding future generations accountable through legal frameworks is still in its infancy and lacks specific case laws. However, various international agreements and declarations, such as the Paris Agreement, emphasize the importance of considering the needs and rights of future generations in climate decision-making.

Environmental Justice and Equitable Outcomes

Climate change disproportionately affects marginalized and vulnerable communities worldwide. Addressing the ethical implications of climate litigation and the role of environmental justice becomes vital to ensure equitable outcomes for those most impacted. This section examines how legal approaches can incorporate principles of environmental justice and strive for fairness in addressing climate-related harms.

In the case of ***Standing Rock Sioux Tribe v. U.S. Army Corps of Engineers (2021),*** the Standing Rock Sioux Tribe challenged the Dakota Access Pipeline, citing violations of environmental justice principles. The court acknowledged the need to consider the impacts on Indigenous communities and their sacred sites, highlighting the significance of environmental justice in climate litigation.

Challenges of Innovative Legal Approaches

As legal systems adapt to address climate change, innovative approaches may encounter various challenges and criticisms. This section identifies potential hurdles related to privacy concerns, admissibility of new evidence, and the need for judicial education and expertise to effectively implement unconventional legal strategies.

While not directly related to climate litigation, the case of ***Carpenter v. United States (2018)*** raised significant privacy concerns. The Supreme Court ruled that obtaining historical cell phone location data without a warrant violated the Fourth Amendment. In climate litigation, the use of novel data collection methods to establish climate impacts may raise similar privacy concerns that require careful consideration.

Climate Migration and Legal Protections

As the impacts of climate change intensify, millions of people are projected to become climate migrants, seeking refuge from the effects of rising sea levels, extreme weather events, and ecological disruptions. This section delves into the legal challenges and protections required to address the complex issue of climate migration.

Climate migration is an emerging and evolving issue, and there are no landmark case laws yet. However, several international conventions, such as the Global Compact for Safe, Orderly, and Regular Migration, provide a foundation for addressing the rights and protections of climate migrants (Dhingra et.al, 2019).

The section explores the need for legal frameworks that recognize climate migrants' unique circumstances, ensure their rights, and establish mechanisms for international cooperation to address this humanitarian challenge.

Corporate Climate Accountability and Liability

Large corporations and industries play a significant role in contributing to climate change through greenhouse gas emissions and unsustainable practices (Filho et.al, 2022); (Ali, 2020). This section examines the potential for innovative legal strategies to hold corporations accountable for their climate impacts and explore the concept of climate liability.

In the case of ***Kivalina v. ExxonMobil Corp (2009)***, a Native Alaskan village sued several major energy companies, alleging that their greenhouse gas emissions contributed to climate change, which caused erosion and threatened the village's existence. While the case was dismissed on jurisdictional grounds, it brought attention to the concept of climate liability and corporate responsibility for climate impacts.

Climate Disinformation and Legal Remedies

The spread of climate disinformation and denial of climate science poses a significant challenge to climate action. This section discusses the potential for legal remedies to address climate disinformation and hold those spreading false or misleading information accountable.

Climate disinformation cases are relatively new, and there are no landmark cases yet. However, some jurisdictions have explored the potential for legal actions against individuals or entities that deliberately disseminate false climate information (Wentz & Franta, 2022)

The Role of Artificial Intelligence in Climate Litigation

As technology continues to advance, the potential for leveraging artificial intelligence (AI) in climate litigation emerges as a transformative force (Cowls et.al, 2021). This section explores how AI can be harnessed to analyse vast amounts of climate data, model complex climate scenarios, and aid in assessing the causal link between climate change and specific events or actions.

Swarm Intelligence AI for Climate Pattern Analysis:

Implementing swarm intelligence AI algorithms can simulate the collective behaviour of natural systems, such as bird flocks or fish schools (Rath et.al, 2020). These algorithms can be applied to analyse vast climate datasets, identifying patterns and anomalies that traditional methods might overlook (Yang & Karamanoglu, 2013). By mimicking nature's adaptive and decentralized approach, swarm intelligence AI can unveil hidden climate patterns and offer novel insights into climate change dynamics.

Quantum AI for Climate Modelling:

Leveraging quantum computing's immense processing power, researchers can develop quantum AI models to simulate complex climate scenarios more accurately (Cooper et.al, 2022). Quantum AI's ability to handle exponentially larger datasets and perform complex calculations could lead to more precise predictions and better understanding of how climate change influences different regions and ecosystems (Frąckiewicz, 2023)

Emotional AI to Understand Climate Sentiments:

Emotional AI, which analyses emotions from textual data, can be applied to assess public sentiment on climate change (Nandwani, 2021). By analysing social media posts, news articles, and public opinion surveys, emotional AI can help gauge people's emotional responses to climate events, policies, or actions. Understanding the emotional aspect of climate discourse can aid in designing more effective climate communication strategies (Kirelli & Arslankaya, 2020)

Generative Adversarial Networks (GANs) for Climate Data Augmentation:

GANs are a type of AI that can generate realistic synthetic data (Aggarwal et.al, 2021). In the context of climate analysis, GANs can augment limited or sparse climate datasets, creating additional synthetic data points. These augmented datasets can then be used to train climate models more effectively, enhancing the accuracy and generalization capabilities of AI-driven climate simulations.

Explainable AI for Climate Event Attribution:

Explainable AI techniques can help establish causal links between specific events or actions and climate change outcomes. By providing human-readable explanations for AI-driven predictions, researchers can better understand how certain events or human activities contribute to climate change or extreme weather events, aiding policymakers in making informed decisions to mitigate and adapt to climate change (Miller, 2019).

Bio-Inspired AI for Climate Resilience:

Bio-inspired AI, inspired by biological systems, can be utilized to develop climate-resilient solutions (Nguyen et al., 2021). For instance, algorithms based on plant or animal behaviour can optimize energy distribution systems, design more efficient cooling strategies, or plan adaptive infrastructure based on nature's resilience strategies in the face of changing environmental conditions.

AI Swarm Robotics for Climate Monitoring:

Swarm robotics, where multiple robots work together in a coordinated manner, can be employed to monitor and collect climate data over vast and remote areas. These intelligent robotic swarms can efficiently gather real-time information on various climate parameters, enabling continuous monitoring of critical regions and ecosystems, aiding in more proactive climate change mitigation strategies (Schranz et.al, 2020).

Meta-learning and Transfer Learning for Climate Model Tuning:

Meta-learning and transfer learning techniques can be applied to accelerate climate model tuning and adaptation across different geographical regions (Pinto et.al, 2022). Instead of training individual models for each location, AI models can leverage knowledge learned from other regions with similar climatic conditions. This approach can save computational resources and provide more accurate predictions for lesser-studied areas (Boukabara et.al, 2019).

Reinforcement Learning for Climate Policy Optimization:

Reinforcement learning algorithms can optimize climate policies by simulating various policy interventions and evaluating their long-term impacts on the climate. By using reinforcement learning to find optimal climate policies, governments and organizations can make data-driven decisions to mitigate climate change and achieve sustainable development goals (Nishant et.al, 2020).

Sentiment Analysis for Climate Change Adaptation:

Applying sentiment analysis to data from climate-affected communities can help understand the emotional and psychological impact of climate change. By assessing the sentiments of vulnerable populations, policymakers and aid organizations can design more targeted and empathetic climate change adaptation strategies, ensuring that solutions are contextually appropriate and inclusive (Wamsler & Bristow, 2022).

Incorporating these unique and unconventional AI approaches can expand our understanding of climate change, improve climate modelling accuracy, and assist in identifying the causal connections between human actions, climate change, and its consequences. This integration of AI and climate science holds the potential to accelerate climate action and build a more sustainable future.

The integration of AI in climate litigation is a relatively nascent field, and there are no landmark cases yet. However, there is growing interest in utilizing AI-driven tools to strengthen the scientific evidence presented in climate-related disputes.

The section further delves into the challenges and benefits of incorporating AI in climate litigation, including issues of data privacy, bias in AI algorithms, and the need for AI regulation to ensure its responsible use in legal proceedings.

Benefits of Incorporating AI in Climate Litigation

a) **Improved Data Analysis and Evidential Support:** AI can process and analyse vast and complex climate datasets efficiently, helping legal teams uncover relevant patterns, correlations, and causations. This can strengthen the evidential support for climate-related claims and facilitate more informed legal arguments (Cho, 2018).

b) **Enhanced Decision-Making and Risk Assessment**: By leveraging AI, litigators can assess the potential impact of climate change on specific regions or industries more accurately. This data-driven risk assessment can help stakeholders make better-informed decisions and allocate resources effectively to address climate-related challenges (GPAI, 2021).

c) **Cost and Time Efficiency**: AI-powered document review and data analysis can significantly reduce the time and cost involved in climate litigation. Automation of repetitive tasks allows legal teams to focus on higher-value strategic aspects, expediting the legal process (LSE, 2023).

d) **Predictive Insights and Early Warning Systems:** AI can develop predictive models for identifying potential climate-related risks, such as extreme weather events or environmental degradation.

Early warning systems based on AI predictions can enable proactive measures, minimizing the likelihood of future legal disputes (Chen et.al, 2023).

Challenges of Incorporating AI in Climate Litigation:

a) **Complexity of Climate Data Interpretation:** AI can analyse vast amounts of climate data, but interpreting and presenting the results in a legal context can be challenging. Lawyers and judges may struggle to comprehend the technical intricacies of AI-generated evidence, leading to potential misinterpretations or disputes over the validity of AI-driven findings (LSE,2023).
b) **Attribution of Climate Impact:** Establishing causation between specific actions or events and climate change outcomes is complex. AI models might identify correlations but may not fully determine causation, raising concerns about how such correlations are presented as evidence in court. Ensuring the reliability and transparency of AI algorithms in determining causality becomes crucial (Chen et.al, 2023).
c) **Adversarial Attacks on AI Evidence:** In climate litigation, adversaries might attempt to manipulate AI algorithms or introduce biased data to undermine the opposing party's arguments. Protecting AI systems from adversarial attacks and ensuring their robustness and integrity are essential to maintain the credibility of AI-generated evidence in court (Vinuesa et.al, 2020).
d) **Data Privacy and Security Concerns:** Climate data often involves sensitive information about individuals, businesses, or governments. AI systems must adhere to strict data privacy regulations and ensure that confidential data is appropriately anonymized or protected to prevent misuse or breaches during litigation (Cho, 2018).

Regulations should mandate that AI models used in climate litigation provide transparent and interpretable explanations for their decisions. This ensures that lawyers, judges, and stakeholders can understand and challenge the basis of AI-generated evidence effectively. AI algorithms must undergo rigorous testing to identify and mitigate biases that could influence their outcomes. Legal regulations should require AI developers to implement bias detection and correction mechanisms to prevent unfair impact in climate litigation.

Third-party organizations specializing in AI validation should assess the performance and accuracy of AI systems used in climate litigation. Regulatory bodies can then ensure that AI algorithms meet the required standards of reliability and precision. Legal frameworks should outline strict guidelines for the collection, storage, and usage of climate data in AI systems to safeguard individuals' and entities' privacy. Compliance with data protection laws is crucial to prevent unauthorized access or misuse of sensitive information. Regulations should include ethical guidelines for AI deployment in climate litigation to ensure that the technology is used responsibly and with consideration of potential societal impacts. Regular audits of AI systems used in climate litigation can help identify any potential issues or deviations from regulatory requirements, ensuring that AI remains a trustworthy and reliable tool in legal proceedings.

By addressing these challenges and implementing responsible AI regulations, the incorporation of AI in climate litigation can be more beneficial, providing a powerful tool to support climate justice, accountability, and sustainable environmental practices.

Climate Remedies and Global Enforcement

Climate litigation often seeks remedies such as injunctive relief, monetary damages, or the implementation of policies to address climate change. However, climate impacts transcend national borders, requiring international cooperation and global enforcement of legal decisions.

In the case of ***Urgenda Foundation v. Netherlands (2019)***, a Dutch court ordered the government to increase its climate action targets to protect its citizens from the impacts of climate change. The case is a landmark example of successful climate litigation that highlights the need for global enforcement of climate remedies.

The future of climate litigation is a dynamic landscape where innovation, ethical considerations, and equitable outcomes will play pivotal roles. Legal frameworks that address intergenerational climate justice can instil a sense of responsibility in climate decision-making for the well-being of future generations. Emphasizing environmental justice principles ensures that vulnerable communities receive fair treatment in climate-related disputes.

However, unconventional legal approaches come with challenges. Striking the right balance between privacy rights and data-driven evidence, as well as ensuring the competency of judges in dealing with complex scientific and technological issues, will require concerted efforts from legal experts, policymakers, and stakeholders.

By embracing innovation and inclusivity in climate litigation, legal systems can adapt to the complexities of climate change impacts and foster climate justice. Collaborative approaches and interdisciplinary engagements will be crucial in shaping effective legal responses to climate change and securing a sustainable and equitable future for all.

RECOMMENDATIONS

1. ***Foster Collaborative Partnerships***: Encourage collaboration between governments, NGOs, businesses, scientists, communities, and legal practitioners to pool resources, share expertise, and design comprehensive climate justice solutions that transcend borders and jurisdictions.
2. ***Promote Interdisciplinary Training***: Enhance interdisciplinary training for legal professionals, equipping them with the knowledge and skills to navigate complex climate-related issues, integrate scientific evidence, and collaborate with experts from various fields.
3. ***Invest in Climate Technology Research***: Allocate resources to research and development of climate technology, including AI-driven tools, geospatial mapping, and remote sensing, to strengthen the scientific evidence presented in climate litigation.
4. ***Integrate Human Rights into Climate Litigation:*** Recognize the intersectionality of climate impacts and human rights, ensuring that climate litigation prioritizes the protection of human dignity and inclusivity, seeking justice for marginalized and vulnerable communities.
5. ***Explore Climate Disinformation Legislation:*** Consider the implementation of legal frameworks to address climate disinformation, holding those spreading false or misleading information accountable while carefully balancing free speech rights.
6. ***Expand Rights of Nature Laws:*** Encourage the recognition of nature's inherent rights and expand rights of nature laws to provide legal protection for ecosystems and the environment.

7. ***Encourage Proactive Legal Measures:*** Advocate for the incorporation of the precautionary principle in climate litigation, shifting the burden of proof to polluters and encouraging proactive measures to prevent environmental harm.
8. ***Enhance Judicial Education:*** Provide training and education for judges to enhance their understanding of complex scientific and technological issues related to climate change, ensuring fair and informed decisions.
9. ***Strengthen Global Enforcement Mechanisms:*** Promote international cooperation to enforce climate remedies, ensuring that legal decisions transcend borders and hold responsible parties accountable on a global scale.
10. ***Empower Vulnerable Communities:*** Support capacity-building efforts to empower vulnerable communities with legal knowledge and resources, enabling them to advocate for their rights and engage in legal processes that affect their lives directly.
11. ***Pursue Legal Protections for Climate Migrants:*** Advocate for legal protections and frameworks that address the unique circumstances of climate migrants, ensuring their rights and fostering international cooperation to address climate-induced displacement.
12. ***Foster Ethical Integration of Technology:*** Emphasize ethical considerations in the integration of technology in climate litigation, safeguarding privacy, data security, and consent in data-driven evidence.

By implementing these recommendations, the legal community can pave the way for innovative, equitable, and effective climate litigation strategies that address the urgent challenges of the global climate crisis and secure a sustainable and just future for all.

CONCLUSION

As the curtain falls on this chapter, the symphony of climate justice resounds with innovative and unconventional legal strategies, woven together by the threads of collaboration, technology, and interdisciplinary approaches. In the twilight of the 21st century, humanity faces the profound challenge of the global climate crisis. The traditional boundaries of legal practice have proved insufficient to address the scale and urgency of environmental harm. It is now time to embark on a transformative journey that dares to explore unorthodox avenues, challenging the status quo and embracing imagination and creativity in the pursuit of justice.

The orchestra of climate justice harmonizes the diverse instruments of satellite imagery, remote sensing, blockchain, geospatial mapping, predictive modelling, and more to paint a vivid picture of environmental degradation and hold polluters accountable. Technological solutions stand as watchful sentinels above, providing the truth of ecological harm, while the blockchain symphony ensures transparency and traceability in complex supply chains. The cartographer's dream reveals vulnerable communities most at risk, empowering legal strategies with precision and foresight.

As the symphony plays on, the conductors wield ethical integration of technology, safeguarding privacy, and ensuring data security. Collaborative partnerships and capacity building unite diverse stakeholders, amplifying the impact of climate litigation and fostering a shared understanding of the challenges we face. Empowered communities find their voice, raising it as a symphony of resilience and determination, demanding justice for the environmental injustices they endure.

The overture of hope resonates, echoing in the hearts of every individual determined to heal our planet and secure a flourishing future for generations to come. Collaborative legal initiatives form a harmonious ensemble, breaking free from the constraints of traditional litigation and creating networks that amplify the impact of climate litigation globally.

Interdisciplinary approaches enrich the symphony of climate justice, infusing legal claims with robustness by intertwining climate science, economics, and law. Cost-benefit analysis and economic valuation methods quantify damages, while public health research establishes direct links between polluters' actions and adverse health consequences. Rights of nature challenge the anthropocentric approach to law, recognizing the inherent rights of ecosystems and the environment. Alternative dispute resolution methods provide faster and more collaborative solutions, while the precautionary principle shifts the burden of proof to polluters, fostering proactive measures to prevent harm.

As the symphony reaches its climactic finale, the anthem of change resounds, calling the world to stand united in the pursuit of climate justice. Innovative legal frameworks consider the needs and rights of future generations, while environmental justice principles ensure equitable outcomes for vulnerable communities. Climate migration demands legal protections and international cooperation, recognizing the humanitarian challenges ahead.

Corporate climate accountability and liability hold powerful industries responsible for their contributions to climate change, and legal remedies seek redress for climate disinformation and its detrimental impact. Artificial intelligence adds a transformative force to climate litigation, aiding in data analysis, climate modelling, and causal link assessments. Global enforcement ensures that climate remedies transcend borders, requiring international cooperation in the face of a global challenge.

In this harmonious symphony, the law becomes a transformative force, weaving justice into the very fabric of society. Collaborative, interdisciplinary, and innovative approaches unite in a resounding melody of resilience, determination, and hope. As the world adapts to the consequences of climate change, legal systems must rise to the occasion, embracing unconventional strategies to secure a sustainable and just future for all living beings.

The baton of change is in our hands, and as we move forward, let us heed the call of the symphony of climate justice. With audacity and reverence, let us compose a future where humanity and nature dance in harmony once more, and where the law becomes a guiding force in healing our planet for generations yet to come. Together, we shall play this symphony of hope, echoing in the hearts of all who believe in a brighter and greener tomorrow.

REFERENCES

Achieving Cross-Border Government Innovation. (2021, October 18). OECD-OPSI. https://cross-border.oecd-opsi.org/

Aggarwal, A., Mittal, M., & Battineni, G. (2021). Generative Adversarial Network: An overview of theory and applications. *International Journal of Information Management Data Insights*, *1*(1), 100004. doi:10.1016/j.jjimei.2020.100004

Ahmed Ali, K., Ahmad, M. I., & Yusup, Y. (2020). Issues, impacts, and mitigations of carbon dioxide emissions in the building sector. *Sustainability (Basel)*, *12*(18), 7427. doi:10.3390/su12187427

Aiming higher: Elevating meaningful youth engagement for climate action: United nations development programme. (2022). UNDP. https://www.undp.org/publications/aiming-higher-elevating-meaningful-youth-engagement-climate-action

Amorim-Maia, A. T., Anguelovski, I., Chu, E., & Connolly, J. (2022). Intersectional climate justice: A conceptual pathway for bridging adaptation planning, Transformative Action, and social equity. *Urban Climate*, *41*, 101053. doi:10.1016/j.uclim.2021.101053

Bhatia, S., & Albarrak, A. S. (2023). A blockchain-driven food supply chain management using QR code and XAI-faster RCNN architecture. *Sustainability (Basel)*, *15*(3), 2579. doi:10.3390/su15032579

Bochenek, B., & Ustrnul, Z. (2022). Machine learning in weather prediction and climate analyses—Applications and perspectives. *Atmosphere (Basel)*, *13*(2), 180. doi:10.3390/atmos13020180

Chen, L., Chen, Z., Zhang, Y., Liu, Y., Osman, A. I., Farghali, M., Hua, J., Al-Fatesh, A., Ihara, I., Rooney, D. W., & Yap, P.-S. (2023, October). Yap, P.-S. (2023). *Artificial Intelligence-based solutions for climate change: A Review. Environmental Chemistry Letters*, *21*(5), 2525–2557. doi:10.1007/s10311-023-01617-y

Cho, R., Wiseman, J., Amblee, R., Amaan, A., Green, J., & Sharma, Y. Hakim, Basak, S., Singh, R., (2022, September 13). Artificial Intelligence-a game changer for climate change and the environment. *State of the Planet*. https://news.climate.columbia.edu/2018/06/05/artificial-intelligence-climate-environment/

Climate Change Litigation – insights into the evolving Global Landscape. (2021, July 5). Grantham Research Institute on climate change and the environment. https://www.lse.ac.uk/granthaminstitute/publication/climate-change-litigation-insights-into-the-evolving-global-landscape/

Collins, M., Ashok, K., Barreiro, M., Roxy, M. K., Kang, S. M., Frölicher, T. L., Wang, G., & Tedeschi, R. G. (2021). Editorial: New techniques for improving climate models, predictions and projections. *Frontiers in Climate*, *3*, 811205. doi:10.3389/fclim.2021.811205

Cooper, P., Ernst, P., Kiewell, D., & Pinner, D. (2022, May 19). *Quantum computing just might save the planet*. McKinsey & Company. https://www.mckinsey.com/capabilities/mckinsey-digital/our-insights/quantum-computing-just-might-save-the-planet

Cowls, J., Tsamados, A., Taddeo, M., & Floridi, L. (2021). The AI Gambit: Leveraging artificial intelligence to combat climate change—opportunities, challenges, and recommendations. *AI & Society*, *38*(1), 283–307. doi:10.1007/s00146-021-01294-x PMID:34690449

Del Bene, D., &; Ávila, S. (2023). Critical mapping for researching and acting upon environmental conflicts – the case of the EJAtlas. *Studies in Ecological Economics*, 195–210. doi:10.1007/978-3-031-22566-6_17

Diezmartínez, C. V., & Short Gianotti, A. G. (2022). US cities increasingly integrate justice into climate planning and create policy tools for climate justice. *Nature Communications*, *13*(1), 5763. Advance online publication. doi:10.1038/s41467-022-33392-9 PMID:36180440

Dwivedi, Y. K., Hughes, L., Kar, A. K., Baabdullah, A. M., Grover, P., Abbas, R., Andreini, D., Abumoghli, I., Barlette, Y., Bunker, D., Chandra Kruse, L., Constantiou, I., Davison, R. M., De', R., Dubey, R., Fenby-Taylor, H., Gupta, B., He, W., Kodama, M., & Wade, M. (2022). Climate change and COP26: Are Digital Technologies and information management part of the problem or the solution? an editorial reflection and call to action. *International Journal of Information Management*, *63*, 102456. doi:10.1016/j.ijinfomgt.2021.102456

Erkic, A. (2022, July 7). *How to break down team silos and improve collaboration*, Pumble. https://pumble.com/blog/break-down-team-silos/

European Commission. (2023). *Consequences of climate change*. Climate Action. https://climate.ec.europa.eu/climate-change/consequences-climate-change_en

Friedman, N., & Ormiston, J. (2022). Blockchain as a sustainability-oriented innovation?: Opportunities for and resistance to blockchain technology as a driver of sustainability in global food supply chains. *Technological Forecasting and Social Change*, *175*, 121403. doi:10.1016/j.techfore.2021.121403

Global trends in climate change litigation: 2022 snapshot. (2023, June 7). Grantham Research Institute on climate change and the environment. https://www.lse.ac.uk/granthaminstitute/publication/global-trends-in-climate-change-litigation-2022/

Hersher, R. (2023, January 9), *Climate change makes heat waves, storms and droughts worse, climate report confirms*. NPR. https://www.npr.org/2023/01/09/1147805696/climate-change-makes-heat-waves-storms-and-droughts-worse-climate-report-confirm

Hoffman, J., & Lott, C. (April 20, 2022). *The intersection of economics, environment and equity*. The Crimson White. https://thecrimsonwhite.com/98316/culture/the-intersection-of-economics-environment-and-equity/

How Walmart Canada uses blockchain to solve supply-chain challenges. (2022, January 5). Harvard Business Review. https://hbr.org/2022/01/how-walmart-canada-uses-blockchain-to-solve-supply-chain-challenges

Kamel Boulos, M. N., & Wilson, J. P. (2023). Geospatial techniques for monitoring and mitigating climate change and its effects on human health. *International Journal of Health Geographics*, *22*(1), 2. doi:10.1186/s12942-023-00324-9 PMID:36707823

Kirelli, Y., & Arslankaya, S. (2020). Sentiment analysis of shared tweets on global warming on Twitter with data mining methods: A case study on Turkish language. *Computational Intelligence and Neuroscience*, *2020*, 1–9. doi:10.1155/2020/1904172 PMID:32963511

Leal Filho, W., Perry, P., Heim, H., Dinis, M. A., Moda, H., Ebhuoma, E., & Paço, A. (2022). An overview of the contribution of the textiles sector to climate change. *Frontiers in Environmental Science*, *10*, 973102. doi:10.3389/fenvs.2022.973102

Louw, A. S., Fu, J., Raut, A., Zulhilmi, A., Yao, S., McAlinn, M., Fujikawa, A., Siddique, M. T., Wang, X., Yu, X., Mandvikar, K., & Avtar, R. (2022). The role of remote sensing during a global disaster: Covid-19 pandemic as case study. *Remote Sensing Applications: Society and Environment*, *27*, 100789. doi:10.1016/j.rsase.2022.100789 PMID:35774725

Maersk and IBM introduce TradeLens blockchain shipping solution. (n.d.). IBM Newsroom. https://newsroom.ibm.com/2018-08-09-Maersk-and-IBM-Introduce-TradeLens-Blockchain-Shipping-Solution

McAdam, J. (2020). Protecting people displaced by the impacts of climate change: The UN Human Rights Committee and the principle of non-refoulement. *The American Journal of International Law*, *114*(4), 708–725. doi:10.1017/ajil.2020.31

McDonald, P. (2021, March 31). *Resolving climate change disputes through arbitration.* Pinsent Masons. https://www.pinsentmasons.com/out-law/analysis/resolving-climate-change-disputes-through-arbitration

Miller, T. (2019). Explanation in artificial intelligence: Insights from the Social Sciences. *Artificial Intelligence*, *267*, 1–38. doi:10.1016/j.artint.2018.07.007

Nandwani, P., & Verma, R. (2021). A review on sentiment analysis and emotion detection from text. *Social Network Analysis and Mining*, *11*(1), 81. doi:10.1007/s13278-021-00776-6 PMID:34484462

Nguyen, T.-H., Nguyen, L. V., Jung, J. J., Agbehadji, I. E., Frimpong, S. O., & Millham, R. C. (2020). Bio-inspired approaches for Smart Energy Management: State of the art and Challenges. *Sustainability (Basel)*, *12*(20), 8495. doi:10.3390/su12208495

Nishant, R., Kennedy, M., & Corbett, J. (2020). Artificial Intelligence for Sustainability: Challenges, opportunities, and a research agenda. *International Journal of Information Management*, *53*, 102104. doi:10.1016/j.ijinfomgt.2020.102104

O'Mahony, T. (2021). Cost-benefit analysis and the environment: The time horizon is of the essence. *Environmental Impact Assessment Review*, *89*, 106587. doi:10.1016/j.eiar.2021.106587

Peel, J., & Lin, J. (2019). Transnational Climate Litigation: The contribution of the Global South. *The American Journal of International Law*, *113*(4), 679–726. doi:10.1017/ajil.2019.48

Peel, J., & Osofsky, H. M. (2020). Climate Change Litigation. *Annual Review of Law and Social Science*, *16*(1), 21–38. doi:10.1146/annurev-lawsocsci-022420-122936

Pinto, G., Wang, Z., Roy, A., Hong, T., & Capozzoli, A. (2022). Transfer learning for smart buildings: A critical review of algorithms, applications, and future perspectives. *Advances in Applied Energy*, *5*, 100084. doi:10.1016/j.adapen.2022.100084

Pinto-Bazurco, J. F. (2022, June). *Still Only One earth: Lessons from 50 years of UN Sustainable Development Policy.* International Institute for Sustainable Development. https://www.iisd.org/projects/still-only-one-earth-lessons-50-years-un-sustainable-development-policy

Preston, B. J. (2018). The evolving role of environmental rights in Climate Change Litigation. *Chinese Journal of Environmental Law*, *2*(2), 131–164. doi:10.1163/24686042-12340030

Rath, M., Darwish, A., Pati, B., Pattanayak, B. K., &; Panigrahi, C. R. (2020). Swarm intelligence as a solution for technological problems associated with internet of things. *Swarm Intelligence for Resource Management in Internet of Things*, 21–45. doi:10.1016/B978-0-12-818287-1.00005-X

Reva Dhingra, E. F., & Simeon, K. (2022, March 9). *The climate crisis, migration, and refugees.* Brookings. https://www.brookings.edu/articles/the-climate-crisis-migration-and-refugees/

Reynolds, J. L. (2019). Solar geoengineering to reduce climate change: A review of governance proposals. *Proceedings - Royal Society. Mathematical, Physical and Engineering Sciences*, *475*(2229), 20190255. doi:10.1098/rspa.2019.0255 PMID:31611719

Rising, J. A., Taylor, C., Ives, M. C., & Ward, R. E. T. (2022). Challenges and innovations in the economic evaluation of the risks of climate change. *Ecological Economics*, *197*, 107437. doi:10.1016/j.ecolecon.2022.107437

Schranz, M., Umlauft, M., Sende, M., & Elmenreich, W. (2020). Swarm robotic behaviors and current applications. *Frontiers in Robotics and AI*, *7*, 36. doi:10.3389/frobt.2020.00036 PMID:33501204

Sultana, F. (2021). Critical climate justice. *The Geographical Journal*, *188*(1), 118–124. doi:10.1111/geoj.12417

Swaen, G. M. H., Langendam, M., Weyler, J., Burger, H., Siesling, S., Atsma, W. J., & Bouter, L. (2018). Responsible epidemiologic research practice: A guideline developed by a working group of the Netherlands Epidemiological Society. *Journal of Clinical Epidemiology*, *100*, 111–119. doi:10.1016/j.jclinepi.2018.02.010 PMID:29432862

The Economist Newspaper. (n.d.). Holding polluting sectors accountable for the Climate Crisis. *The Economist*. https://impact.economist.com/sustainability/circular-economies/how-to-hold-polluting-sectors-accountable-for-the-climate-crisis

Volcovici, V. (2017, August 22), *Energy Transfer sues Greenpeace over Dakota pipeline.* Reuters. https://www.reuters.com/article/us-energy-transf-lawsuit-idUSKCN1B2245

Weigand, M., Wurm, M., Dech, S., & Taubenböck, H. (2019). Remote Sensing in environmental justice research—A Review. *ISPRS International Journal of Geo-Information*, *8*(1), 20. doi:10.3390/ijgi8010020

Wentz, J. A., & Franta, B. (2022, December). Liability for Public Deception: Linking Fossil Fuel Disinformation Liability for Public Deception: Linking Fossil Fuel Disinformation to Climate Damages to Climate Damages. *Environmental Law Reporter*, *52*, 10995. https://scholarship.law.columbia.edu/cgi/viewcontent.cgi?article=1196&context=sabin_climate_change

WIPO. (2020, January), *Climate Action and Sustainability: Indigenous Peoples Are Part of the Solution.* WPO. https://www.wipo.int/wipo_magazine/en/2020/01/article_0007.html

Yager, K. (2015). Satellite imagery and community perceptions of climate change impacts and landscape change. *Climate Cultures*, 146–168. doi:10.12987/yale/9780300198812.003.0007

Yang, X.-S., &; Karamanoglu, M. (2013). Swarm intelligence and bio-inspired computation. *Swarm Intelligence and Bio-Inspired Computation*, 3–23. doi:10.1016/B978-0-12-405163-8.00001-6

Chapter 8
Prospects of Accomplishing Climate Action Through Rural Development Initiatives in the Indian Context

Ravindra Sopan Gavali
National Institute for Rural Development and Panchayati Raj, India

Nissy John
National Institute for Rural Development and Panchayati Raj, India

Goundla Pradeep Goud
https://orcid.org/0009-0002-0267-1652
National Institute for Rural Development and Panchayati Raj, India

ABSTRACT

Achieving Climate Action, UN Agenda 2030, is a challenge towards strengthening the resilience and adaptive capacity to climate-related hazards and natural disasters. Climate-proofing of villages can be strategised through climate adaptation and complementary mitigation approaches, leading to climate-resilient rural communities with climate-resilient livelihoods in India. The chapter reviews the scope of rural developmental schemes and strategies that may lead to climate-resilient adaptation practices for rural communities. The first part of the chapter lays out the national perspective of climate action. The second step is to review the problems that rural communities face and investigate how rural development programs like MGNREGS, NRLM, PMAY-G, and other models can be used to converge to create climate resilient communities. Finally, understand how the localization of climate action enhance and strengthens the rural poor's resilience and adaptive capacity.

DOI: 10.4018/978-1-6684-9863-7.ch008

INTRODUCTION

Climate Change is the most crucial issue seeking global attention at all levels, irrespective of geographical barriers ranging from rising sea levels to the extreme events of floods and drought, decline in agricultural productivity and damaging infrastructure to the loss of livelihoods. The climate variabilities and extremities have posed negative impacts in an uncontrollable way that present limitations to expanding the adaptation and mitigation strategies, sometimes make them ambiguous and incorrigible. Although the problem is widespread, it is more pronounced in ecologically weak regions, economically unstable and socially deprived environments. The coastal regions often suffer the most due to the damage caused by floods, cyclones, storms and strong winds. Hence, anticipating future climate events is a crucial challenge to act mindfully and tackle the issue at the grassroot level, as warned by the scientific communities. If the problem is not addressed, it is likely to push humanity to suffer the worst of its kind with significant economic loss, damaging the whole tangible and intangible developmental progress that had been gained for many years. Then the world would have to face issues satisfying the basic requirements such as good food, clean water, fresh air and proper home for a common man's living. So, Climate change can be seen as a long-term change that can be identified through temperature variability and reflected with severe droughts and floods in various regions across the globe.

The 'IPCC Assessment Report 6' states that there has been a rise of 1.06 degrees Celsius in temperature from 1850-1900 to 2010-2019, attributed to human-induced causes (Delmotte, 2022. Roome (2010), projects that by the end of the century, the majority of the world's incidents like flood and drought events will have increased not only in terms of magnitude and intensity, but also the arrivals of catastrophic events would significantly increase in frequency and strength with unpredictable weather changes. The tropical Cyclone Fani killed 20 people and affected 28 million in 2019 (Chauhan et al., 2021). With the persisting temperature at 1.5°C of warming, river flooding is predicted to rise by 49% (The India, 2022) in the coming years. The Ministry of Agriculture states that about 33.9 million hectares of cropland have been destroyed due to hydro-meteorological catastrophes like floods in just six years (2015-2022), and the most witnessed states are Assam, Bihar, Uttar Pradesh (IMP Report, 2021). These vulnerability effects of climate change is expected to have significant repercussions for the national economy and food security and a steep decline is projected for rice and wheat productions by the midcentury (1.8-3.4% of GDP) (Leone, 2022). The extreme climate events would put people in trouble bringing huge losses to agricultural and livestock productivity, soil erosion, loss of livelihoods, severe health risks, water and sanitation problems, water management problems, destruction of infrastructure, interruption of essential services including education, energy and transportation, critically leading to unbalanced sustainability concerns.

It is the massive use of fossil fuels like coal, petroleum and natural gas since 1800 (pre-industrial period), are known to have recognized to be the primary contributors to Greenhouse Gas Emissions (anthropogenic) rather than the natural causative factors. But erasing drop of GHG emissions during the pandemic (2019), when the issues were seemingly less intense, has now turned out to have global emissions escalate to 6 percentage, showing a record rise than ever before (Bhanumati et al., 2022).

Added to this, the current patterns of anthropogenic use of natural resources and the adverse impacts of climate change have amplified the problem within the rural sector, leading to the emergence of a rural-urban divide or migration. The rate of urbanization is likely to grow from 35% in 2020 to more than 50% in 2050, the reason which is ascribed to be the climate impacts. The expected trend in the urban areas is 340 million (2008) to 590 million (2030) (Leone, 2022). The projected climate impacts

Figure 1. Global climate risk index, 2020
Source: Germanwatch, 2021

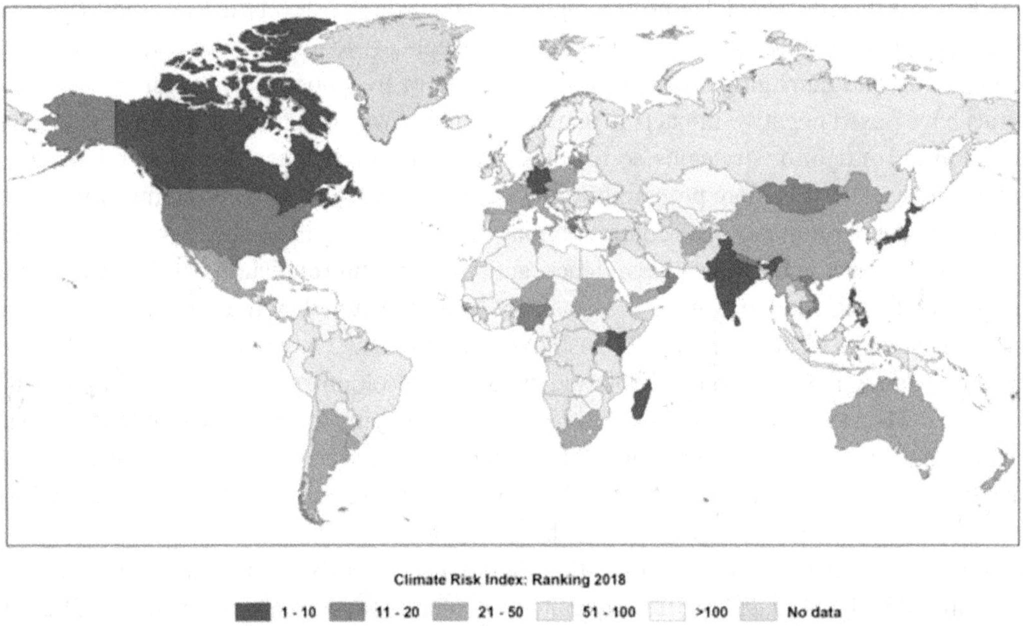

can be linked to the most noticeable disparity in social groups in the agro-climatic zones where small-crop farmers and unemployed youth are dismantled by socio-economic instability as well as destruction of public infrastructure and services. Areas like the high and low land, are particularly most driven by climate events proliferating site-specific local problems. The cascading crisis takes along with it a spin of issues that limit the choice of rural livelihoods, pushing them to enormous obstacles to respond to climate vulnerabilities. Seemingly, about 100 million people may fall into poverty by 2030 due to climate tremors (Roome, 2015) and a long-term shift in temperature and weather patterns would destroy the quality of life.

In this context, India ought to take steps seriously. As of right now, India is the seventh most vulnerable country in the world (Eckstein et al, 2021) affected by this phenomenon. Escalating impacts of climate change and its frequent signs cannot be ignored in any region in the world, particularly the events which affect the poorest countries. The Climate Risk Index (CRI) from **Figure 1** indicates the level of exposure and vulnerability to extreme events and the countries are warned for preparation against severe events in future. The index centers around the climate extremities but does not take into account of the slow processes such as the melting of polar ice caps, sea level rise or ocean acidification.

Climate change cannot be avoided through adaptation strategies, and no mitigation strategies can prevent it completely. Developing nations like India face numerous challenges as their impoverished populations depend on locally available natural sources (Mertz, et al., 2009). Approximately 70% of the population resides in more than 6 lakh villages, out of which 26% are disadvantaged (Census, 2011) and 57% of the population is directly dependent on agriculture for their livelihood. Thus, future climate change is expected to pose intense negative impacts on the biophysical environment affecting the total well-being of human beings. In the current scenario, tackling climate change at the local level in a combined bottom-top and top-to-bottom approach is the easiest and most effective strategy. This is possible by developing programme of action through adopting climate-resilient, sector-specific innovative solutions by bringing together the scientific communities and inventing new tactics to reduce and weaken the effects of climate change by increasing adaptive capacity and enabling vulnerability management, to lessen the adversities.

Climate action signifies the measures to 'Take urgent action to combat climate change and its impacts' through building resilience and adaptive capacity by collective efforts from individuals, governments and nations to reduce the impacts anticipating the climate events. "Transforming our world: The 2030 Agenda for Sustainable Development" is the universal call for resolution with 169 targets and 304 indicators to end poverty, protect the planet, and ensure that by 2030, all people enjoy peace and prosperity, leaving no one behind.

Although many climate interventions involve policy on national and international level, a smaller involvement that is accessible such as tree planting by the community can be functional (Lin, 2020; Riondato et al., 2020). However, it is indispensable to expand the limits of adaptation through a clear adaptation policy and overcome the limitations of devising a proper policy framework. The probability of limits to adaptation increases with greater frequency and magnitude of the climate impacts and may lessen the range of positive synergies between adaptation and mitigation fading the effectiveness of reaching the desired goal.

Looking ahead to the urgency of attaining climate action and understanding the present status and the targets to be achieved, it is possible that through a participatory and integrated approach of various RD schemes, the two-fold approach – Adaptation and Mitigation can build climate-proofing of the villages by adopting sustainable usage mechanisms and management of natural resources (NRM). Meanwhile, awareness generation to the rural inhabitants at all levels about the scientific concepts of climate change saves them from climate hardships. Certainly, efforts pooled under all the concerned departments and their schemes can support the achievement of climate action. The Department of Rural Development has been ensuring the social security of the rural population through providing diverse rural livelihood options, employment generation, and skill development to the youths and women establishing better infrastructure facilities of governmental offices, other public assets and transportation. Though the social and economic indicators also show a slight positive growth rate in rural areas in existing scenarios in every dimension, the strategies to fulfil the climate targets under most rural development initiatives are not precisely specified with the climate objectives but mostly come as an added environmental benefit.

So, planning climate-resilient development pathways in the face of climate change and related disasters should embed flexibility and open ways to bridge smooth connectivity between development and climate, causing the most negligible impact on vulnerable communities.

METHODOLOGY

The chapter followed a non- systematic approach to locate and examine the peer-reviewed journal articles based on approaches to climate adaptation and mitigation and the reports on Mahatma Gandhi National Rural Employment Guarantee Program/ Act (MGNREGS), flagship programs (PMKSY, NRLM, and PMAY-G) associated with the rural development activities towards natural resource management and climate resilient actions implemented across different parts of the country. The electronic retrieval of the considered reports was included from multidisciplinary databases: Web of Science, PubMed, Google Scholar, Science Direct, Taylor and Francis, Scopus and Springer Link. After screening the references, a general google search was conducted to obtain documents and the reports from national, regional and subnational agencies.

India's Perspective on Climate Action

India abides with the Paris Agreement protocol, the international effort to limit the emission of gases by 5% lower than in 1990 by 2012 and mitigate climate change, towards tackling the problem. In Glasgow COP26, India has agreed to cut its emissions (mitigation strategy) to net zero by 2070. In contrast, in the COP27 forum, India has declared to increase the carbon sink (adaptation strategy) to 2.5 to 3 billion tonnes of carbon sequestration equivalent through additional forest and tree cover. Given the Nationally Determined Contribution (NDC), the achievable goal by 2030 is to reduce the emission intensity by 45 per cent (Roelefsema et al., 2020). This update to India's existing NDC is a step forward towards our long-term goal of reaching net zero by 2070 (Cohen et al., 2022). India also reaffirms its commitment to the UNFCCC and 'LiFE' Lifestyle for Environment approach (NDC, 2022) to mobilise a minimum of 1 billion Indian or global citizens towards environmental protection and conservation and at least 80% of Indian villages and urban local bodies to become environment friendly by 2028.

Foreseeing the magnitude of the impacts, steps for preparedness and readiness to face the challenges, India needs to formulate adaptation strategies and measures that can be included in the system. A supportable way of living by preserving indigenous conservation methods and controlling the environment by incorporating a cleaner and climate-friendly pathway like native tree plantation, community-managed natural forests, are advisable to inculcate. Many communities have learned to adapt to the situations and have built resilience against climate shocks. The Government of India, apart from National Missions and National Action Plans on Climate Change, in collaboration with many international organizations, has devised ways to deal with the unavoidable impact of climate change. India plays a crucial role in determining this agenda's relative success or failure as it is home to 17% of the world's population.

Climate Initiatives in India

India has pledged the National Action Plan on Climate Change (NAPCC), which encompasses climate protection and adaptation policies. This Action Plan recognizes eight core national missions: (i) solar energy, (ii) Enhanced Energy Efficiency, (iii) Water, (iv) Sustaining the Himalayan ecosystem, (v) Green India, (vi) Sustainable Agriculture and (vii) strategic knowledge for climate change.

The other significant initiations include the Low Carbon Strategy for 12th Five Year Plan, National Action Plan on Climate Change (NAPCC), 2008, State Action Plan on Climate Change (SAPCC), Network-based scientific programme "Indian Network of Climate Change Assessment" (INCCA), 2009, strengthening collaborations (national and regional; multilateral and bilateral). For green initiation under the rural areas, as per Section 52(1) (b) of the New Panchayat Raj Act-2018, it is also recommended to raise nurseries in all Gram Panchayats (GPs). Accordingly, in every Gram Panchayat, a Green Action Plan is to be prepared where 10% of the total Budget released to the Gram Panchayat is spent as a Green Budget per the Gram Panchayat (GP) Demand.

India has signed and committed to many international organisations to address the issue comprehensively through undertaking green technology initiatives. In the spirit of 'leaving no one behind', India has entered the 'Decade of Action' to address the challenges for sustainable development. Even though SDG has been off track since the COVID-19 pandemic, India advocates behavioural changes in the utilisation of natural resources, promoting renewable energy sources, and increasing forest cover (UNHLPF, 2020).

India's Status on SDG Climate Action

The UN SDG Agenda 2030 serves as a blueprint to achieve all 17 SDG goals and has redefined the governmental policy frameworks, priorities and metrics to measure its achievement. NITI Aayog has extensively measured India and its States' progress towards the SDGs for 2030 through the SDG India Index. MoSPI has drafted the National Indicator Framework (NIF) in consultation with the Ministries and States/UTs. Several States/UTs have mapped the Central and State sponsored schemes vis-à-vis the SDGs and undertaken long-term (visioning), medium-term (strategy development) and short-term (action plan) exercises towards achieving SDG targets. Based on the data of the indicators achievable under each target, the status of each goal is evaluated, and thus, the overall performance of all the SDGs are measured.

A detailed review of the achievable targets would help identify and evaluate the existing schemes and programmes that fall under different Ministries of the Government of India to collaborate to achieve climate action. The output thus obtained develops a transformation of the existing strategies and policy frameworks. It occasionally fixes the existing gap areas and the challenges for effective implementation of the schemes. A broad strategy employed by the Ministry of Rural Development is to address poverty eradication SDG Goal 1 (End Poverty in all forms everywhere) and SDG Goal 2 (End hunger, achieve food security, improved nutrition and promote agriculture) but not specifically SDG 13 (Climate Action). However, the existing and planned works under the Ministry of Rural Development indirectly aim at integrating the various community-grounded climate change resilient practices.

According to the national progress report SDG Index, 2020-2021, it takes 16 goals with 70 targets and 115 indicators for assessment (MoSPI, 2022). SDG 13 covers five targets to measure and monitor the changing impacts of climate-related hazards (SDG NIF, 2021). As of now, according to the Ministry of Environment, Forest and Climate Change, there are 33 States/UTs with strategies for enhancing adaptive capacity and dealing with climate extreme weather events in India in 2017-2018 (MoEFCC) (SDG Progress Report 2019), and the percentage of reduction in emission intensity of GDP over 2005 level by 2020 is 21 percentage (Pre 2020 action achievements of pre-2020 Goals as per country priority) (SDG-NIF, 2020). India has improved the score from 60 (2019) to 66 (2020).

From the 2021 report, most of the states have shown an improvement in their scores and have upgraded from performer to frontrunner.

The Mahatma Gandhi National Rural Employment Guarantee Act (MGNREGA) implemented under the Rural Development Department is recognized to play a significant role (SDG India Index, 2021) towards achieving climate action. **Figure 2** depicts the state-wise status of the SDG score classified under the category (i) Achiever (blue) -100 (ii) Front Runner (green) - 65-99 (iii) Performer (Yellow) 50-64 (iv) Aspirant (red) 0-49. The overall SDG score of India is 66. From among the states and the UTs, the top five states in achieving the SDG goals include Kerala (75), Himachal Pradesh (74), Tamil Nadu (74), Andhra Pradesh (72) and Goa (72). Considering SDG 13, the performance level is 'no change' from 2021 to 2022, with an overall score of 54. The top 5 states in the performance category towards achieving SDG 13 include Rajasthan (49), Goa (44), Telangana (43), Tripura (41) and Uttar Pradesh (39) (SDG India Index, 2022) (**Figure 2**). Nevertheless, the UN announces that the people's action inclusive of all age groups, particularly the youth and women in a civic society and the collective effort of public and

Figure 2. Yearly SDG index, state-wise performance
Source: Niti Ayog 2021-2022

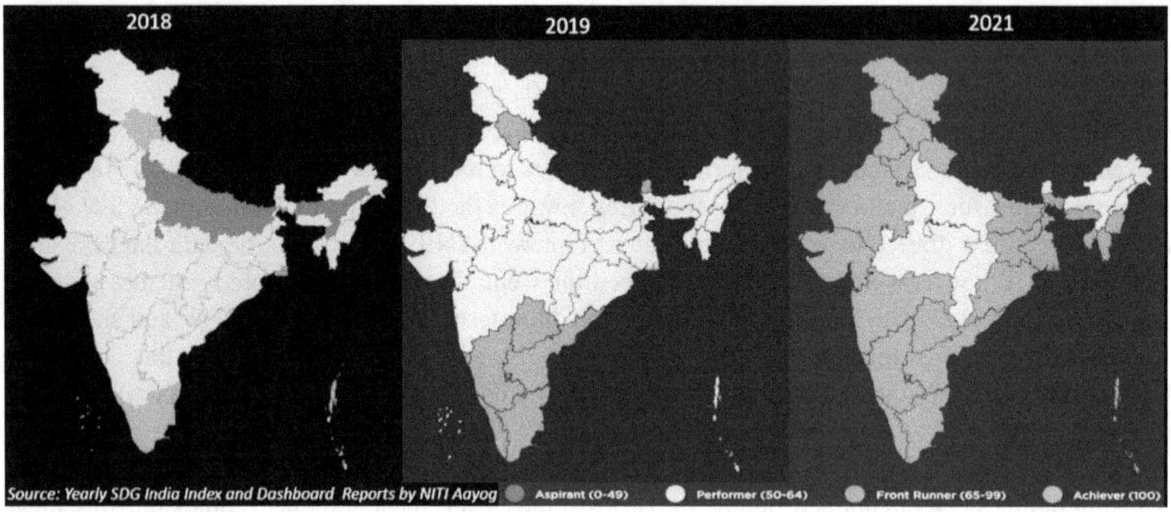

private sector, media and communication, unions, academia and the stakeholders can make the world far better to combat the climate issues.

Approach to Deal With Vulnerability to Climate Change

Why does India have a high risk of climate change? India's vulnerability is primarily ascribed to the rise in Green House Gases (GHGs), demographic, socio-economic and physiographic situations and in fact, the presence of the Arabian Sea, the Western Ghats and the geographically slanting terrain making the country a high-risk area for climate change disasters. The southernmost part of India is more vulnerable to climate due to the exposure of people, livelihood, and infrastructures to climate events **(Figure 3).**

The degree of Climate vulnerability varies from place to place; some are manageable, while some may exceed the threshold limit. Therefore vulnerability can be defined as the function of exposure and sensitivity to the adaptive capacity of a region, which may arise due to the climate extremities (flood, cyclone) or (climate variabilities (drought, winds, coastal erosion).

Vulnerability (f) = (Exposure x sensitivity)/ (Adaptive Capacity) (Mohanty et al., 2021).

The hazard's significant drivers include the exposure of the population and the regions, ecological fragility or exposure of the sensitive ecosystems, their vulnerability and the weak interaction between the components in the environment (IPCC). In India, three out of four districts are exposed to the severity of climate events, whereas 5 out of 20 people are exposed to the extremities of hydro-met disaster (Wadhawan, 2021). The top five states include Assam, Andhra Pradesh, Maharashtra, Karnataka and Bihar. The current land use and land change pattern has also abruptly led to climate incidences in different parts of the country.

Depending on the sensitivity and adaptive capacity, the degree of vulnerability of the population to varying climate events can be minimised. A broad approach to understanding the existing situation of the land area, the land utilization pattern and the availability of water (NRM) enables effective uptaking of the climate-proofing of villages.

Figure 3. District-wise vulnerability to climate change in India
Source: *National Climate Vulnerability Assessment Report*

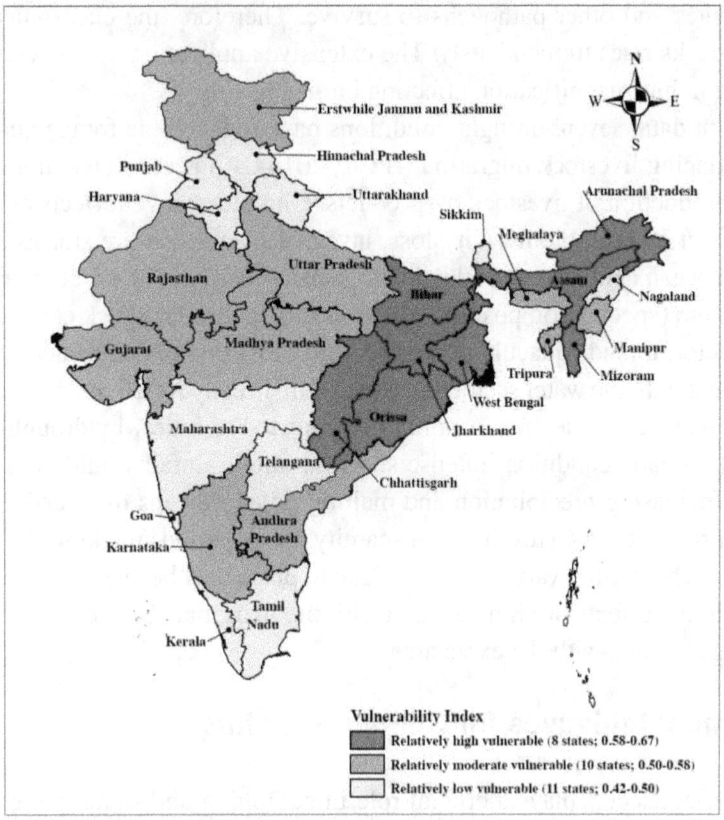

'Climate proofing' refers to the identification of risk to a natural or manmade asset as a consequence of current or future climate change extremes or climate variability and ensuring that those risks are reduced to an acceptable level through environmentally sound, economically viable and socially acceptable changes implemented' (Waterman et al., 2009).

Climate Impacts in Rural Settlement Areas

Food safety is one of the major significant challenges the farmers may have to face due to the frequent visit of droughts and floods (Ziska et al., 2016). More conspicuously, the delayed rainfall and early withdrawal of rain which affects the agrarian community with a decline in crop productivity, particularly the staple food crops like rice, wheat, sorghum, etc. by ten to forty percentage (IPCC, 2014). The situation leads to inadequate food/hunger, malnutrition, health issues among the poorer sections of the society, surging with the emergence of pests and diseases (Reddy et al., 2019). Consequently, the demand for food in the market will increase, in fact, the price value also. With regard to horticulture segment, the unpredictable climate is likely to affect commercial fruit-bearing plants, flowering plants and vegetables with visible outcome of physiological disorders for those crops grown in open field conditions e.g. cracking in pomegranates and spongy tissue in guards. The livestock become more defenseless to the temperature rise or heat waves. The undesirable impact reduces the reproductive capability of the animals and causes impaired milk production (NIRDPR, 2019). Often, the spring and the warm conditions support the growth of parasites and other pathogens to survive. Therefore, the occurrence and emergence of parasite or insect attacks rises tremendously. The extensive application of pesticides leads to pesticide resistance, resulting in bio-magnification affecting human health.

As per the current data, severe drought conditions have reduced the forage quality and availability of feed supplies, inducing livestock migration (IPCC, 2014). An acute water shortage would affect the animal health and production of livestock by-products. Climate change reflects on several natural ecosystems through insect attacks, biodiversity loss, invasive species, severe storms and forest fires. The rise in sea level and ocean acidification, added with storm severity, will affect the primary productivity of the ocean ecosystem (species composition, diseases emergence (Badjeck et al., 2010).

In the fishing sector, inland aquaculture will be under loss due to damage caused to the livelihood, infrastructure and health. In the water sector, the issues will directly interfere with the hydrological cycle. Erratic rainfall and unexpected decline in rainfall is expected to follow by drought condition in different agro ecological climatic condition. Intense short-duration rainfall would be a significant problem (KSAPCC, 2023). Increasing precipitation and melting polar ice caps may lead to surface runoff and reduce groundwater recharge. As a result, water scarcity for irrigation and domestic purposes becomes a challenging concern. The rainfall variation from place to place may be due to the El-Nino effect (Chowdary, 2017). Overall, the effects of climate change hit the marginalized section of society and become an extreme challenge to cope with the exposures.

Rural Development Initiatives for Climate Proofing

The Rural Development sector plays a crucial role in equipping and shaping the rural community to face critical climate impacts that acquire slowly or suddenly. Recognising the range of possibilities and strength of existing RD schemes/strategies, strong and determined transformative action plans can be devised and articulated to meet the targets meticulously for long-term sustainability. Although several schemes and programmes are implemented in rural areas, the disadvantaged population bear a disproportionate share of the consequences concerning climate change (Rudolf et al., 2015; White- Newsome at al., 2018 Younger et al., 2008). The specific objective of the schemes under the Department of Rural Development needs to be clearly focused on climate change for optimal results but currently the planning under the schemes are carried out in isolation, targeting poverty eradication and infrastructure setup.

The Ministry of Panchayat Raj, Government of India, has now started to address the issues holistically under all the welfare schemes obligatory to reflect in the Gram Panchayat Development Plan (GPDP). Through an integrated approach, the climate-related adaptation strategies can be carried out in complementary with mitigation under the RD schemes such as MGNREGS, PMKSY, NRLM, and PMAY-G. Taking into account of the socio-economic indicators such as the population size, gender, basic access to physical infrastructures, livelihood options, and preparedness to the disaster risks etc. for the selection of the most suitable adaptation strategies before implementation, the schemes can focus on the common goal, including direct participation of the people and local governments, elected representatives, private and public sectors, NGOs and the scientific communities. Here a proper, careful planning process incorporating into the GPDP can meet the needs of the people (Economic, Environmental, and Social) and demonstrate reliable, resilient pathways to overcome the adversities confronted by the marginalized sections of society.

It is possible to achieve climate proofing at the community level by using risk based strategy that addresses the risk through adaptation as opposed to mitigation which is effective only for long term goals. Among the four major approaches to Climate Change Risk Management, *viz* 1) Mitigation (activities to reduce GHGs, 2) Adaptation (Increase adaptive capacity to cope with current and anticipated impacts, 3) Geoengineering (deliberate manipulation of the earth system, 4) Knowledge-based expansion (learn and understand about the climate system (AMS, 2014), Climate Change Risk Based Management approach to **adaptation** is more practicable and appropriate as it includes both the likelihood and components of climate impacts for existing and future impacts. It is the most crucial strategy to save the lives and livelihoods of the people.

Depending on the kind of geographical setting, drought-prone or rain-fed areas, various watershed activities like contour trenching, contour bunding, farm bunding, farm ponds, horticulture plantation works, irrigation works, renovation of water can be taken which aid in improving their adaptive capacity. Understanding the situation of the region and considering the sensitivity of the landscape indicators such as the land use and land cover, moisture content of the soil, slope and elevation of the region, adaptive capacity could be built.

One of the interesting features of MGNREGA is the wide range of works and overlapping activities where each activity addresses multiple issues, and several activities address a particular issue. MGNREGA can be considered one of the best platforms for a green global plan to deal with Climate Change (MoEF, 2010) due to its inter-sectoral approach. The scheme aligns with the targets of the Ministry of Agriculture, Ministry of Rural Development, Ministry of Water Resources, Ministry of Environment Forest and Climate Change. The Government of India, in 2019, issued a new guideline recognising MGNREGA as a tool to fight against climate vulnerabilities and has enforced planning activities under the programmes based on the climate- data and the vulnerability assessment report. Guaranteeing 100 days of wage employment in Indian villages, MGNREGA empowers the rural adults with livelihood security and offering environmental (climate) benefits due to its activities taken up under Natural Resource Management (NRM).

The question is, "Whether climate-proofing is possible through MGNREGA support in conjuction with other RD schemes?" If so, then to what extent?

The Mahatma Gandhi National Rural Employment Guarantee Act (MGNREGA) has a direct link with rural livelihood enhancement. It includes the combination of agricultural and livelihood-related works such as tree plantation, horticulture, water conservation, micro and minor irrigation, climate-resilient livelihood options and land-based development activities that aid in tackling climate change issues. These activities centre around adaptation strategies conserving and enhancing the ecosystem and providing sustainable livelihood security to the rural people.

MGNREGS works associated with plantation (Natural Resource Management) aid in minimising the vulnerability and climate variability through carbon offsetting generated due to increased tree growth, biomass production and soil carbon enrichment. Carbon sequestration can be vital in reducing carbon dioxide emissions in the atmosphere (Gavali & Shaik, 2016). The IPCC AR5 reports that most adaptation choices for floods and droughts under the land-based activities for climate action have mitigation power through biomass production and soil carbon enhancement against climate risks like floods and drought. As an emission reduction strategy, plantation activities under the scheme can play a significant role and satisfy the national target of NDC – expansion of 2.5 – 3 billion tonnes of carbon sink of additional forest cover. In India's tropical climate, plantation activities would favour the country by tapping potential carbon and reducing the urban heat island. MGNREGA activities associated with natural

Figure 4. Carbon sequestration trends
Adopted by Ravindranath and Indu, 2018

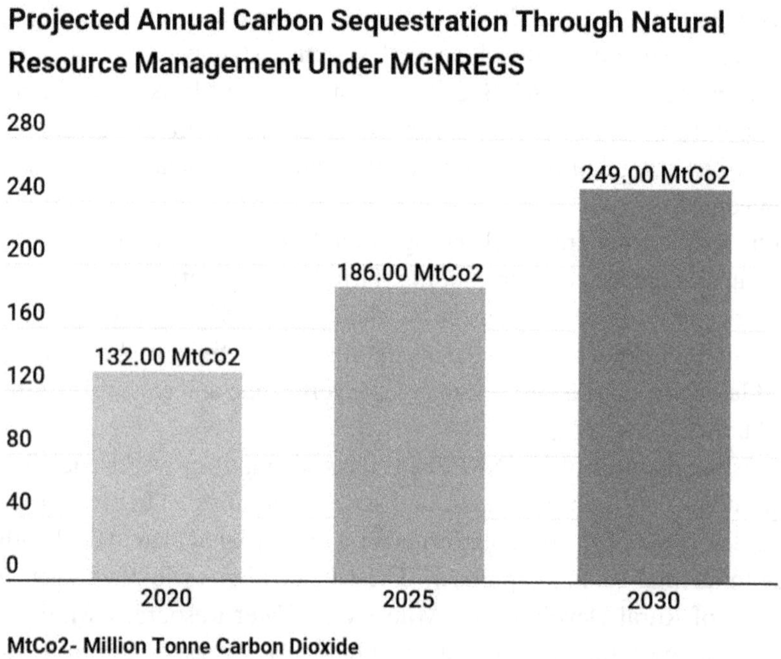

resource management serve as a climate mitigation and adaptation tool. **Figure 4** depicts the average carbon sequestration in the year, projected to surge from 62MtCO_2 (2017-2018) to 132MtCO_2 (2020), 186 MtCO_2 by 2025 and 249 MtCO_2 by 2030(Tiwari et al., 2011; Ravindranath et al., 2018).

Land-based activities of MGNREGA, in association with PMKSY, under the irrigation sector, similarly increase water accessibility. Currently, the country has covered micro irrigation of an area of Drip irrigation – 5.32 Lakh/Ha; Sprinkler 5.70 Lakh/Ha (Total 11.02 Lakh /Ha) and has shown an excellent successful rate to contribute to the adaptation capacity of the local people. This will increase agricultural production and support to ensure food security, ultimately improving the adaptive capacity of a particular region. For instance, wheat production has doubled from 9.68 qt/hh to 19.55 qt/hh (Sinha et al., 2011). With a wide range of designated activities and suitable strength of manpower, MGNREGA, in association with schemes like Pradhan Mantri Krishi Sinchayi Yojana (PMKSY-Watershed Component) under the Department of Land Resources, restore ecological balance through the conservation of soil, water and vegetative cover along with providing sustainable livelihood options to the people in watershed areas. While dealing with water shortage, about 49% of MGNREGA activities are involved in constructing water harvesting and water storage structures. (MoRD, 2009) as they contribute to the conservation of surface water and groundwater in drought-prone areas, it enables groundwater recharge and subsequent rise of water level for drinking and domestic purposes. Construction activities under MGNREGA have also shown major implications for the livelihood options among the rural inhabitants.

The scheme's objectives are directly linked to the watershed models, which are climate resilient and associated with soil and water conservation strategies in drought-prone and flood-affected areas. The initiatives include upscaling the selection of climate-resilient seed varieties/breeds (drought-resistant and short-duration crops), soil nutrient management, weather-based agro-advisories, crop insurance, and climate-neutral farming. Encouraging Zero tillage practices to avoid/reduce GHG emissions and enhance crop productivity, particularly for vulnerable crops like rice (due to erratic monsoon) and wheat (due to heat stress), is also working under a common goal.

As a means to secure livelihood options for the rural population, India, with 6.9 million SHGs with over 75 million members across 783,389 villages, is to endorse sustainable livelihoods through alternative livelihood practices and to diminish poverty by enabling the rural people to access productive, skilled and self-employment. The National Rural Livelihood Mission (NRLM) supports rural livelihoods during climate uncertainties through farm and non-farm livelihood options and value addition aspects for better marketing, which closely align with the climate targets.

Considering the agrarian community, which shares the majority of the population of India, it is possible to opt for alternate livelihood options suitable to the people, such as mushroom cultivation, snail farming, and fish farming. Such options help boost local resilience but also help them gain profitable income and encourage value-added livelihood options, which can be taken up synchronously within the adaptation strategies (Hilson et al., 2009). As mushroom cultivation do not need vast land and can be cultivated year-round, it offers a promising avenue for the local economic market (Wendiro et al., 2019). Other alternate options include the cultivation of Azolla as an alternate feed for cattle to overcome the green fodder shortage (NIRDPR, 2019). A case from Balaram village, Dalai, shows that bout 139 farmers, who cultivated mushrooms at a total expense of Rs. 9000, drew a profit of Rs. 19,952 in six months, an example of adopting a sustainable livelihood option (Dev et al., 2016). On the other hand, through convergence with MGNREGA, the NRLM (National Rural Livelihood Mission) promotes the sustainable harvesting practices of Non-Timber Forest Products (NTFP) species, adding a market value chain (MoEF&CC, 2022).

Moving ahead, under MGNREGA, many disaster-proof houses have been constructed that can help overcome the adverse impacts of natural disasters like floods and cyclones (Subrat & Sukla, 2023). The Government of India, as a poverty alleviation strategy, launched the most extensive social housing programme 'Pradhan Mantri Awas Yojana, intending to reduce the housing shortage among the poor. The housing construction undertaken under the AAWAS Yojana includes disaster-resilient features, including a wide selection of construction design typologies suitable to the place's geo-climatic conditions. The disaster-resilient infrastructures ensure low-carbon building operations at low–cost designs, which are sustainable. The innovative nature-based climate resilient infrastructure is a flexible model that claims to be cheaper than the age-old housing structures. Although climate risks cannot be avoided entirely, experiences have shown that they can manage risk-associated disruptions. Building assets or infrastructures deprived of proper climate-proofing modules flop the economy of the nation. The National Disaster Management Authority (NDMA) reported that between June 01 and July 16, 2018, a total of 54,249 houses were destructed due to severe flood followed by the monsoon; the affected States include Assam (5203), Kerala (8069), Karnataka (8481), West Bengal (6910), Manipur (17846).

Coming to the energy sector, as a part of the climate mitigation strategy, Govt. of India advocates the Pan India project, which is dedicated to establishing energy security for farmers in India, along with upholding India's promise to increase the proportion of electric power produced from non-fossil fuel sources to 40% by the end of 2023 as part of NDC. The project aims to establish 10,000 MW of Decentralised Grid Connected Renewable Energy Power Plants in barren land/wasteland and install 17.50 Lakh stand-alone solar irrigation pumps. Under this component, individual farmers will be supported to install solar irrigation pumps of capacity up to 7.5 HP to substitute existing conventional Agriculture pumps. The third component is to ensure the Solarisation of 10 Lakh Grid Connected Agriculture Pumps. The program can improve the energy efficiency at Gram Panchayats and reduce the emissions and uncertainty in the energy supply to sustain the crops. It was observed that per capita electricity consumption is higher in the southern part of India. A per the REN21, Renewables 2022 Global Status Report, India is the world's fourth-largest producer of solar power and there is good growth in solar generation in Rajasthan, Gujarat, Karnataka, Tamil Nadu, and Maharashtra, over the years since the inception of various incentives in the state solar policy after 2014 (Winter, 2022).

Convergence and Localisation

The adaptation strategies to climate change impacts involve a multi-sectoral approach as the thematic areas of climate action fall under cross-cutting sectors like health, water, forest, fisheries, and sustainability. Different line departments function on the ground with their mandated flagship programmes. However, all the departments plan and implement their programmes in silos without taking consideration of integrated planning at the GP/Block/District level. To take up various schemes, programmes, and initiatives with specific targets that precisely align with the areas, the 'Ministries' mapped to achieve Goal 13 include Ministry of Environment Forest and Climate Change (MoEF&CC), Ministry of New and Renewable Energy (MoNRE), Ministry of Power and Ministry of Home Affairs and partly Ministry of Rural Development and Panchayat Raj. Here, the state plan schemes of many departments can be easily converged with Mahatma Gandhi NREGS like horticulture, agriculture, forest, fisheries, sericulture, and animal husbandry to work in coherence to implement various schemes. Thus, the convergence of schemes and the concerned departments can help achieve common objectives through value addition under targeted and efficient use of financial and human resources for upscaling successful initiatives.

Acknowledging the contributions of MGNREGS, the risks associated with climate change should be combined in the planning and implementation of the program while making the developmental plans. According to the Ministry of Rural Development and Panchayat Raj report in 2022, among the major schemes under the Department of Rural Development (MGNREGA, PMAY, NRLM, and PMKSY), more than half of the government grants are shared by the MGNREGS (54%) followed by Pradhan Mantri Awaas Yojana- Gramin (PMAY – G (15%) and Pradhan Mantri Gram Sadak Yojna (PMGSY – 14%). These figures show ample opportunities to build climate resilience through MGNREGA in a socially acceptable, cost-effective way.

Any scheme can be better delivered if the activities are taken under better coordination within a given timeframe. Moreover, constant feedback, smooth monitoring and evaluation of the activities are necessary for a result-oriented output. Given the technological development, policy frameworks and consistent economic stability, the underlying gaps and challenges strategies inclusive to accomplish the targets should be unravelled to improve the quality of life of every individual. Demonstrative workshops or training, hands-on experience with modern tools or machinery, and its functionalities can support the system for a productive output. Meanwhile, innovative scientific tools like mobile-based data collection and other technical skills can prevent wastage of time and resources and engage farmers to self-operate the device to generate GIS Data, which becomes essential at the planning phase to map the areas exposed to vulnerabilities within or between the regions (Gram Panchayats, blocks, districts).

Currently, the Panchayats Raj and Rural Development under MGNREGS take up convergence activities of plantation works with other departments like horticulture, forest, sericulture, irrigation, National

Highways Authorities of India (NHAI), integrated Tribal Development Agency (ITDA). The combined efforts not only help the government frame policies and align with the Paris Agreement (SDG-13.2) but also help farmers know about their carbon credits, reduce the vulnerability of climate change, and improve quality of Life. Mahatma Gandhi NREGS is a principal platform for converging interventions from agriculture and allied line departments. Climate proofing of the villages is possible by facilitating the farmers under various schemes, particularly agriculture, SHG groups and Mahila Kisan Shashakthirkaran Pariyojana (MKSP).

The District Rural Development Agency, in convergence with the Mission for Integrated Development of Horticulture / National Horticulture Mission, has been implementing programmes for horticulture plantations since the scheme's inception. Construction of check dams, water harvesting structures, percolation tanks, gabion structures, creation and renovation of water bodies, plantation and other activities is converged by different flagship programmes like MGNREGS, WDC-PMKSY, Finance Commission Grant, State Schemes, PVTG Mission, MP & MLA Lad, CSR fund etc. The adverse effects of climate change events like drought, flood and other extreme weather events can be mitigated by utilising the existing resource endowments available to Gram Panchayat. Several SDGs touch on the issue of conservation and sustainable use of agrobiodiversity, such as SDG 2 (Zero hunger), SDG 12 (Responsible consumption and Production, SDG 15 (Life on Earth), a pool of climate-smart technologies under rural development schemes will help farmers to fight against the climate risks. As a result, scaling up upright adaptation practices for climate-proofing villages of different agro ecological regions prevailing on the climatic and geographical conditions is vital. A unified matrix developed that identifies the major climate events, and maps- plan the activities under the RD schemes against the climate risks **(Figure 5; Table 1)** are given belo.

Some Central Schemes that can work in convergence for climate resilience are listed below.

(i) National Mission for Sustainable Agriculture (NMSA) – to enrich agricultural productivity in rain-fed areas
(ii) Mission for Integrated Development of Horticulture (MIDH) – For enhancement of horticulture
(iii) Rain-fed Area Development Programme (NRLP) – To support rural livelihoods and establish an institutional platform for improved rural access.
(iv) Mahila Kisan Sashakthikaran Pariyojanav (MKSP) – To promote sustainable and productive agriculture with the involvement of women
(v) Pradhan Mantri Fasal Bhima Yojana (PMBFY) – For Agricultural Production

Plantation as a Tool for Climate Mitigation

As a case study in Telangana State (Anuradha et al., 2019), carbon offsetting generated by the horticulture plantation works was estimated and shown how rural people have benefitted in managing their livelihood by improving agriculture practices. Carbon sequestration values for selected fruit plants *viz.,* mango (Mangifera indica), sweet orange (Citrus x sinensis), acid lime (citrus aurantifolia) and guava (Psidium guajava) cultivated by farmers have been derived through volume equations and specific gravity values (FSI, 2006; FRI, 2006). Here, perennial horticulture plantations by farmers showcased the positive impacts of sequestering atmospheric CO_2 to about 17164.76MT, which aids in combating climate change. In another study, it is found that by planting more than a half trillion trees, about 205 gigatons

Figure 5. Key drought-proofing activities under Rural Development Initiatives

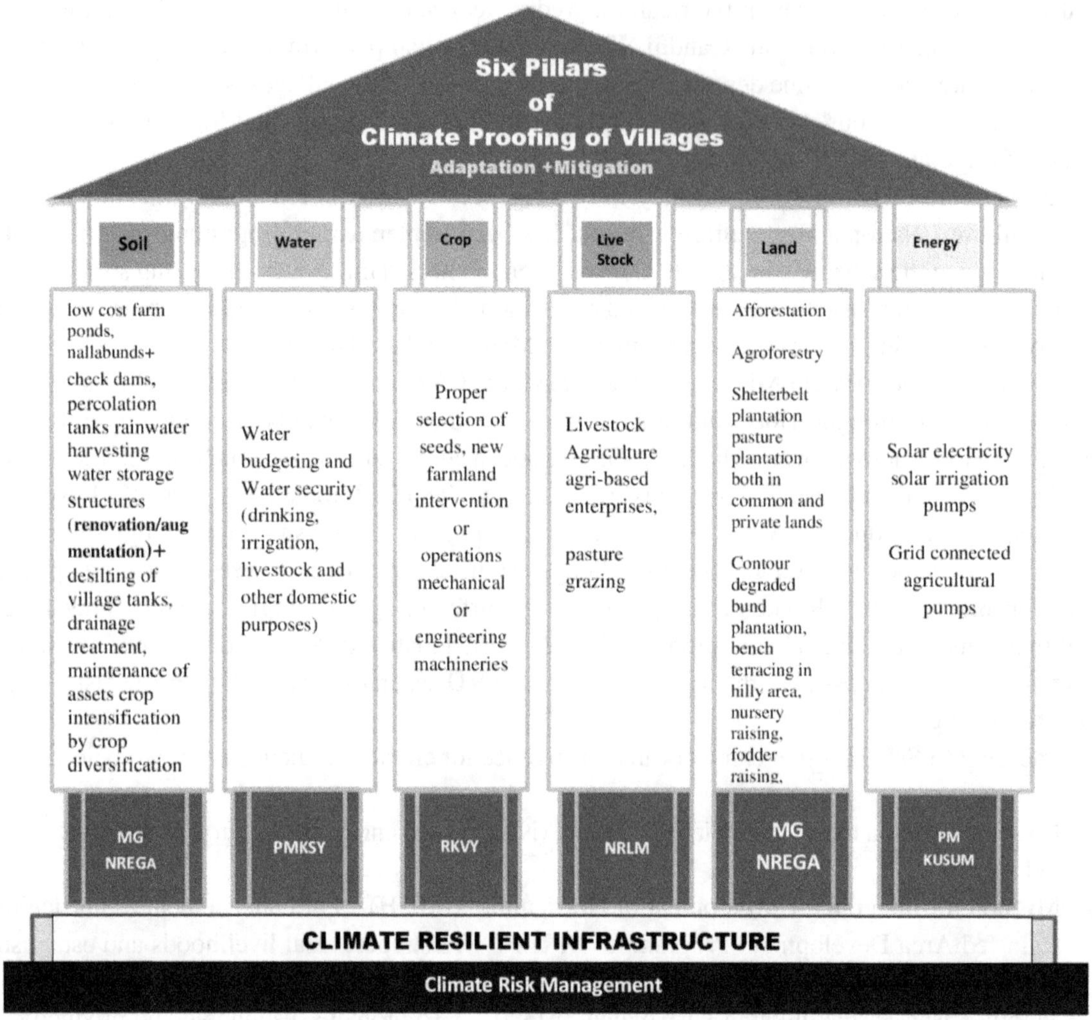

of carbon can be reduced by 25 percent (Alan Buis, 2019). Reported by India Science wire, 2021, the drought proofing activities captures maximum carbon ranging from 0.29 tonne per hectare to 4.50 tonne per hectare annually in Indian villages of selected work sites of different agro ecological zones. Large scale tree planting serves as a climate change solution in terms of carbon reduction, removal of air pollutants, reduction of urban heat (Donovan et al., 2013; Jones et al., 2019; Heo et al., 2021)

Such works scientifically prove that under the MGNREGA scheme, plantation works, afforestation, reforestation and nursery-raising programmes such as Telanganaku Harithaharam, Brahath Palle Prakruthi, Village Nature Parks, Multilayer Avenue Plantations, block plantation, bund plantation, individual plantation, wasteland plantation, canal line plantation, horticulture plantation, avenue plantations, can act as a worthy climate mitigation strategy.

Table 1. Climate resilience through climate proofing activities

Chronic /Acute Climate Risks	Drought Proofing Activities	Resilience to Climate
Heat and Water loss	Construction of check damsWater efficient technologies	water savings
	Direct seeded rice	reduce methane emissionsless soil disturbance
	Integrated Farming Systems Crop diversification	Crop Risk managementCrop efficiency
	-Water efficient technologies (Mechanical to engineering techniques)	Saves water
Drought	-Modified SRI (Machine -Transplantation of rice)	Increase crop efficiency
Vector-borne diseasesEmission of GHGs Water loss	-Alternate Wetting and Drying (AWD)	Reduce methane emissionsLess water usageReduce pest/insect attack
	-Micro Irrigation (Horticulture and commercial Crops)	Increase crop yieldLess water required
water loss	-Community irrigation	Increased water-use efficiency
water loss	-Solar irrigation	sustain economic viability of farm operations
water scarcity/drought	Rainwater harvesting structures	mitigate the impacts of drought and flood
water shortage	Renovation/storage structures	Protects drinking water sourcesCombats water scarcityRestoration of water sources
Drought	Ensure water accessWater budgeting	Improved Water use efficiency
	Groundwater rechargeWater Use and Water Requirements of Gram Panchayat (Drinking, Irrigation, Livestock and other domestic purposes)	increase sustainable yield of an aquifer
Water shortage	Waste water reuse	reduces emission of CO2, CH4, NO2
Increase forest tree cover	-Raising of nurseries-Plantation works(**Individual**)e.g. Homestead, Horticulture,-Teak Bund / Block Plantation,-Pulp wood, Eetha,-Malabar Neem, Mulberry	Improves Carbon offsetting/ removes GHGs from atmosphere
rejuvenate degraded forestsAfforestationReforestation for regeneration of denuded forests	**(Community Plantations)-Institutional**Public Health Centres, Schools, Colleges, Anganwadi centers, Gram Panchayat Buildings, Residential schools and hostels etcCommunity Plantationstank Bunds, Revenue Lands, Canal Bunds, Barren Hills, Degraded forest lands, Tank Foreshore areas**Roadside Avenue Plantation (Single Layer)Multi-layer Avenue Plantation (MLAP):** National Highways,State Highways,	Increase carbon sink 2.5 billion equivalent forest tree cover/More carbon fixation per unit areaEnriched biomassGene pool of local flora and faunaSustainable ecosystemsoil enrichment
	Drought-tolerant species help	supports the rural livelihoods during climate uncertainties
Drought/High temperature/Flood	Low-tech: better practices generate healthier soils and water storage: selective irrigation, mulching, cover crops, crop rotation, reduced ploughing/zero tillage, agroforestry, natural/organic farming.	indirectly helps carbon-dioxide sequestration
	Global Housing Technology Challenge (MoHUA initiative)	low- carbon building operationslow –cost designSustainable, eco-friendly and disaster-resilient infrastructures

Convergence of Rural Development Schemes

Scalable adaptation models using convergence of rural development schemes are available for developing sustainable agriculture livelihoods to address climate change (Gavali et. al., 2023). When sufficient training by the agricultural department offers methods of climate-smart agriculture practices, convergence activities can be taken up and encourage farmers to use drought-resistant seed varieties, cattle sheds, and poultry shed activities for improved milk production and meat production of the locals as on demand. The net benefit is social, economic and environmental benefit (carbon sequestration). Thus, effective utilisation of the barren land and wastelands with plantation works under MGNREGS also helps survive amid social, environmental and economic loss. The negative impacts of drought and flood conditions can be altered, and drinking water scarcity can be reduced through the development of a water security plan and water budgeting under the PMKSY Scheme, which can enhance the water use efficiency of the rural population through adequate awareness on natural resource management (land, water and soil).

Similar is the case with the Fisheries sector, where many schemes can work in convergence with each of the ministries/ departments, starting from the planning stage through monitoring to the implementation level until the desired outcomes are produced. Under the Ministry of Rural Development and the Ministry of Agriculture and Farmers Welfare, the Mahatma Gandhi National Rural Employment Guarantee Act (MGNREGA) and the Rashtriya Vikas Yojana (RKVY) can take up construction of the community ponds and other waterbody development while the Ministry of food processing industries through the Pradhan Manthri Malsya Sampada Yojana (PMMSY) aid in post-harvesting in cold chain facilities. In the meantime, the National Rural Livelihood Mission (NRLM) can support the marketing and admissible activities. The Ministry of Earth Sciences Indian National Centre for Information Services (INCOIS) work for advisories and devices.

Bihar and Madhya Pradesh in India, affected by the delayed rainfall and traditional practice of crop cultivation, struggled to restore the agricultural land. The land was left barren, fields dried, water drained, lost food, no jobs, tended to leave their homeland and were deprived of their basic privileges. The intervention with crop diversification and modern farming techniques laid down by Pradhan Manthri Krishi Sinchayi Yojana Scheme (PMKSY) improved the living conditions of the farmers by earning good agricultural yield, with a steady flow of income and improved adaptive capacity, while the installation of photovoltaic community irrigation system with drip and sprinkler irrigational models in the same affected land by Pradhan Manthri Kisan Urja Suraksha Evam Uttam Mahabiyan Scheme (PMKUSUM) has turned out be economically and socially viable for consistent production of agricultural yield and prevent loss of any energy supply. Accordingly, consistent cultivation of Rabi and Kharif crops gave promising food security results, and the fear of livelihood insecurity prevented them from mass displacement to other places (House, 2019).

Critical demand of water for irrigational purposes happens when there are fewer water storage structures and no plants in the region especially during the rainy season. Mapping the grave impacts and the best-chosen climate-resilient suggested activities, such as the construction of a conservation pond, feeder channel, flood diversion channel, and drainage construction in the net irrigated area, groundwater availability can bring about positive results along with plantation works under the permissible works of MGNREGA contributing to carbon sequestration and water restoration. The convergence enriches the place with ample water supply and, protects the area's biodiversity, and prevents water logging, providing diverse livelihood options.

Another problem is managing floods during the rainy season, which becomes a common issue in a rainfed area. When intense rainfall for an increased number of days and a short period hits the ground, the excess surface runoff may wash away the soil nutrients and, due to waterlogging caused by flash floods, may lead to severe loss of agricultural productivity. Aiming to solve the issue, a combination of the Mahatma Gandhi National Rural Employment Guarantee Act (MGNREGA) and Pradhan Mantri Krishi Sinchayi Yojana Scheme (PMKSY) designing selected climate resilient works like water conservation irrigation channels along with afforestation works can address the issue effectively. Expand the limits of adaptation strategies and widen the scale of work in the face of climate change to avoid the cost of eradicating poverty and mitigating climate risks. In the Indian context, climate-proofing in villages is possible by facilitating the farmers under various schemes, mainly agriculture, through SHG groups and Mahilakisan Shashaktirkaran Pariyojana (MKSP).

CONCLUSION

Risks associated with climate events can be addressed by adopting the best possible nature-based solutions from the success stories mentioned above. However, bearing in mind all the Rural Development Initiatives that primarily focus on the social security of the rural population, it is time to think about the prospects to expand the scope of each scheme and give special attention to those activities that can explicitly target climate action through designing a well-defined unified policy under a common roof. None of the existing rural development schemes are conspicuously focusing on climate action. A new face can be brought about with little financial constraints. The key Ministries of the Government must be prioritized to take up the essential responsibilities. A constant and consistent effort through maximizing the reach and coverage of the schemes should strictly be ensured at every part of the nation, irrespective of geographical barriers. All these combined efforts would help the government frame policies and align with the Paris Agreement (SDG-13).

Way Forward

A toolkit of ideas, activities and approaches that can offer pathways to climate-resilient villages has been showcased here. The chapter discusses that although the goal can be attained without any policy change, strategising area-specific action plans under the RD schemes reviewed above can be brought under a common umbrella to achieve common targets. Learning from previous experiences and identifying the gaps/problems is crucial to bring a change as expected. It is identified that the current recognised problems associated with the implementation of the programmes for climate proofing include delayed disbursal of the fund, lack of proper planning, inadequate capacity building of agency, field practitioners and local community, lack of technical advice and lack of field level demonstration classes, seasonality of the problem etc. Reiterating the need to invigorate the current system of functioning of the departments and their activities, maintaining timely dissemination of the data acquired from each department on climate forecasts can work for effective climate-proofing of the villages. Through scientific support planning and decision-making processes, a clear picture of the critical issues and the said scientific prophecies can be acquired and designed so that proper climate-proof activities can be recommended in affected regions.

In the phase of likely occurrences of floods and droughts, it is time to deliberately think about how community wastelands or land under each landlord or a farmer is liable to manage the damage caused to his/her property and undertakes the responsibility to integrate with new interventions to prepare and adjust before the advent of next climate disasters. A few models and case studies that can be expanded in climate-hit areas are discussed below.

REFERENCES

Anuradha, P., Jayashree, K., Suresh Babu, V., & Sreenivasa Rao, V. (2019) *Addressing Climate Resilience and Livelihood opportunities through MGNREGS Assets in Telangana State. 10th INSEE Biennial Conference on Climate Change and Disaster: Challenges, Opportunities and Responses held at CESS*, Hyderabad.

Bhanumati, P., de Haan, M., & Tebrake, J. W. (2022). *Greenhouse Emissions Rise to Record, Erasing Drop During Pandemic*. IMF Blogs.

Chauhan, A., Singh, R. P., Dash, P., & Kumar, R. (2021). Impact of tropical cyclone "Fani" on land, ocean, atmospheric and meteorological parameters. *Marine Pollution Bulletin*, *162*, 111844. doi:10.1016/j.marpolbul.2020.111844 PMID:33220915

Chowdary, J. S., Harsha, H. S., Gnanaseelan, C., Srinivas, G., Parekh, A., Pillai, P., & Naidu, C. V. (2017). Indian summer monsoon rainfall variability in response to differences in the decay phase of El Niño. *Climate Dynamics*, *48*(7-8), 2707–2727. doi:10.1007/s00382-016-3233-1

Cohen, R., Eames, P. C., Hammond, G. P., Newborough, M., & Norton, B. (2022). Briefing: The 2021 Glasgow Climate Pact: Steps on the transition pathway towards a low carbon world. *Proceedings of the Institution of Civil Engineers-Energy*, *175*(3), 97–102. doi:10.1680/jener.22.00011

Dev, I., Ahlawat, S. P., Palsaniya, D. R., Ram, A., Newaj, R., Tewari, R. K., & Yadav, R. S. (2016). *A sustainable livelihood option for farmers of the semi-arid region: Bamboo+ Chickpea based Agroforestry model.*

Donovan, G. H., Butry, D. T., Michael, Y. L., Prestemon, J. P., Liebhold, A. M., Gatziolis, D., & Mao, M. Y. (2013). The relationship between trees and human health: Evidence from the spread of the emerald ash borer. *American Journal of Preventive Medicine*, *44*(2), 139–145. doi:10.1016/j.amepre.2012.09.066 PMID:23332329

Eckstein, D., Künzel, V., & Schäfer, L. (2021). *The Global Climate Risk Index 2021*. Germanwatch.

Gavali, R. S. (2023). Scalable Adaptation Model for Sustainable Agriculture Livelihoods Under Changing Climate: A Case Study from Bihar and Madhya Pradesh. In S. Nautiyal, A. K. Gupta, M. Goswami, & Y. D. Imran Khan (Eds.), *The Palgrave Handbook of Socio-ecological Resilience in the Face of Climate Change*. doi:10.1007/978-981-99-2206-2_29

Gavali, R. S., & Shaikh, H. M. Y. (2016). Estimation of carbon storage in the tree growth of Solapur University Campus, Maharashtra, India. *International Journal of Scientific Research*, *5*(4), 2364–2367.

Germanwatch. (2022). *Observing, Analysing, Acting - for Global Equity and the Preservation of Livelihoods.* COP 27: U.N. Climate Conference in Egypt.

Heo, S., Chen, C., Kim, H., Sabath, B., Dominici, F., Warren, J. L., Di, Q., Schwartz, J., & Bell, M. L. (2021). Temporal changes in associations between high temperature and hospitalizations by greenspace: Analysis in the Medicare population in 40 US northeast counties. *Environment International*, *156*, 106737. doi:10.1016/j.envint.2021.106737 PMID:34218185

Hilson, G., & Banchirigah, S. M. (2009). Are alternative livelihood projects alleviating poverty in mining communities? Experiences from Ghana. *The Journal of Development Studies*, *45*(2), 172–196. doi:10.1080/00220380802553057

IPCC. (2014). *Climate change 2014: Synthesis Report. Contribution of Working Groups I, II and III to the Fifth Assessment Report of the Intergovernmental Panel on Climate Change* [Core Writing Team, R.K. Pachauri and L.A. Meyer (eds.)]. IPCC.

Jones, J. A., Wei, X., Archer, E., Bishop, K., Blanco, J. A., Ellison, D., & Creed, I. F. (2020). Forest-water interactions under global change. *Forest-water interactions*, 589-624. Research Gate.

Leone, U. (2022). G20 Climate Risk Atlas. Impacts, Policy, Economics. CMCC. *Documenti geografici*, (2), 477-478.

Lin, X., Niu, J., Berndtsson, R., Yu, X., Zhang, L., & Chen, X. (2020). NDVI dynamics and its response to climate change and reforestation in northern China. *Remote Sensing (Basel)*, *12*(24), 4138. doi:10.3390/rs12244138

Masson-Delmotte, V., Zhai, P., Pörtner, H. O., Roberts, D., Skea, J., & Shukla, P. R. (2022). *Global Warming of 1.5 C: IPCC special report on impacts of global warming of 1.5 C above pre-industrial levels in the context of strengthening response to climate change, sustainable Development, and efforts to eradicate poverty*. Cambridge University Press.

Ministry of Rural Development. (2012). *MGNREGS Sameeksha: An Anthology of Research Studies on the Mahatma Gandhi National Rural Employment Guarantee Act, 2005*. Government of India.

Ministry of Rural Development. (2021-2022). *Annual Report*. Government of India.

MOEF. (2010). *India: Greenhouse Gas Emissions 2007*. Indian Network for Climate Change Assessment.

Mohanty, A., & Shreya, W. (2021). *Mapping India's Climate Vulnerability: A District-Level Assessment*. Council on Energy, Environment and Water.

National Determined Contribution. (2022). *India's updated Nationally Determined Contribution under Paris Agreement 2021-2030. Submission to UNFCC*. Government of India.

Ravindrnath, N. H., & Indu, K. M. (2018). Estimation of Carbon Sequestration under MGNREGA: Achievement and Potential. India Indian Institute of Science, Bangalore..

Reddy, K. K., Babu, S. V., & Gavali, R. S. (2019). *Sustainable Livelihood and Adaptation to Climate Change. Training Manual*. National Institute of Rural Development and Panchayath Raj, Ministry of Rural Development, Government of India.

Riondato, E., Pilla, F., Basu, A. S., & Basu, B. (2020). Investigating the effect of trees on urban quality in Dublin by combining air monitoring with i-Tree Eco model. *Sustainable Cities and Society*, *61*, 102356. doi:10.1016/j.scs.2020.102356

Roelfsema, M., van Soest, H. L., Harmsen, M., van Vuuren, D. P., Bertram, C., den Elzen, M., Höhne, N., Iacobuta, G., Krey, V., Kriegler, E., Luderer, G., Riahi, K., Ueckerdt, F., Després, J., Drouet, L., Emmerling, J., Frank, S., Fricko, O., Gidden, M., & Vishwanathan, S. S. (2020). Taking stock of national climate policies to evaluate implementation of the Paris Agreement. *Nature Communications*, *11*(1), 2096. doi:10.1038/s41467-020-15414-6 PMID:32350258

Roome, J. (2015). *Rapid, climate–informed Development is needed to keep climate change from pushing more than 100 million people into poverty by 2030.* World Bank.

Rudolph, L., & Gould, S. (2015). Climate change and health inequities: A framework for action. *Annals of Global Health*, *81*(3), 432–444. doi:10.1016/j.aogh.2015.06.003 PMID:26615079

Sinha, B., Basu, A., & Katiyar, A. S. (2011). *Adapting to Climate Change: Opportunities under MGNREGA*. Presented in the conference on Sustaining Commons: Sustaining Our Future, the Thirteenth Biennial Conference of the International Association for the Study of Common Property, Hyderabad. (https://dlc.dlib.indiana.edu/dlc/handle/10535/7144)

Subrat Kumar Mishra & Firoz Khan. (2023). Guaranteeing employment in Rural India during Covid-19 Pandemic. Economic and Political Weekly.

The Times of India. (2022). *Now, India is the world's largest producer of solar energy*. Ornate Solar.

United Nations. (2017). *Voluntary National Review Report on the implementation of the sustainable development goals*. UN.

United Nations. (2019). *United Nations Sustainable Development Goals: 17 Goals. SDG Progress Report*. Department of Economic and Social Affairs, Government of India.

United Nations. (2020). *High-Level Political Forum on Sustainable Development*. Sustainable Development Goals Knowledge Platform.

United Nations. (2023). *United Nations SDG Progress Report*. Department of Economic and Social Affairs, Government of India.

Wadhawan, S. (2021). *Three steps that India should take to Climate –Proof Infrastructure. Council of Energy Environment and Water*. CEEW.

Waterman, P., Chapman, S., Sargent, S., Laves, G., English, K., & Morvell, G. (2009, October). Climate Proofing as a Regional Development Strategy. In *2009 Sustainable Economic Growth for Regional Australia Conference: Refereed Papers* (pp. 1-22). Management Solutions (QLD) Pty Ltd.

Wendiro, D., Wacoo, A. P., & Wise, G. (2019). Identifying indigenous practices for cultivation of wild saprophytic mushrooms: Responding to the need for sustainable utilization of natural resources. *Journal of Ethnobiology and Ethnomedicine*, *15*(1), 1–15. doi:10.1186/s13002-019-0342-z PMID:31831075

White-Newsome, J. L., Meadows, P., & Kabel, C. (2018). Bridging climate, health, and equity: A growing imperative. *American Journal of Public Health*, *108*(S2), S72–S73. doi:10.2105/AJPH.2017.304133 PMID:29072940

Winter, N. (2022). Renewables 2022. *Global Status Report United States of America Factsheet.*

Younger, M., Morrow-Almeida, H. R., Vindigni, S. M., & Dannenberg, A. L. (2008). The built environment, climate change, and health: Opportunities for co-benefits. *American Journal of Preventive Medicine*, *35*(5), 517–526. doi:10.1016/j.amepre.2008.08.017 PMID:18929978

Ziska, L., Crimmins, A., Auclair, A., DeGrasse, S., Garofalo, J. F., Khan, A. S & Walls, I. (2016). Ch. 7: Food safety, nutrition, and distribution. *The impacts of climate change on human health in the United States: a scientific assessment,* 189-216. Research Gate.

Chapter 9
Industry 5.0, Quality of Life, and Sustainable Adaptation to Climate Change:
Challenges and the Transformational Way Forward

Surabhee Satheesh
Symbiosis International University, India

Kasturi Shukla
Symbiosis Institute of Health Sciences, Symbiosis International University (Deemed), Pune, India

ABSTRACT

Globalization and industrialization have exposed the world to the exacerbating issue of sustainability. With the occurrences of climate related extreme events increasing the world over in the last decade, climate change has come to be recognized as the most profound concern staring in the face of mankind. There have been many separate studies addressing climate change and an equally good number of research in the domain of quality of life. But there is a dearth of research linking climate variability and quality of life. Climate uncertainties have led to loss of lives. Settlements and employments, material living conditions, physical and economic security, natural environment, health, basic rights, and social exchanges among communities stand impacted. All of these metrics are important determinants of quality of life (QoL). Also, alarming is the fact that QoL indicators remain poorly connected to climate change related concerns. This is an important research gap and calls for urgent attention.

INTRODUCTION

The United Nations Framework Convention on Climate Change (UNFCCC) introduced the Paris Agreement in December 2015 at the Conference of Parties (COP21) in Paris to address climate risk by reduction of greenhouse gas emissions globally (UNEP, 2016). Globalization and the ever-escalating

DOI: 10.4018/978-1-6684-9863-7.ch009

consumer demand driven conventional economic model which is the cornerstone for economic upturn, poses a key challenge to sustainability in the global ecosystem. In the post pandemic era, businesses, governments and citizenry are equivocal in their stance towards a resilient economic recovery system based on green finance and green energy framework enhancing the environmental, social and economic sustainability through reformed financial system and a circular economy (Asmelash et al, 2020). Climate change induced ecological instability requires concrete adaptation pathways to be devised and strict policy formulations aiming at individual, community and industry participation for planning, implementation and institutionalization of resilience strategies to reduce climate risk and boost sustainability. Specific industrial sectors and regions battling 'Resource Curse', justice, social, economical and environmental vulnerabilities are other potential challenges staring us in the face due to climate change (Loginova et al, 2019). As geographies around the world have been reeling under severe climate risks in gigantic proportions, there has been a compelling urgency for humanity to embrace sustainability and rein in adaptation as well as mitigation measures posed by the climate hazard (Rizzello, 2022). Prior studies have described the detrimental effects of climate change on human health (Corsi et al, 2020; de Vries, 2019; Ebi et al, 2021; Kotcher et al, 2021; Solomon, & LaRocque, 2019; Xu et al, 2020).

These concerns have led the governments and international monitoring bodies to encourage the formulation of economic models based on sustainability instead of reliance on models based on fossil fuels and greenhouse gas emissions (International Institute for Sustainable Development, 2020; Kasinathan et al, 2022; Maddikunta et al, 2022).

Giving predominance to a phased and gradual shift to clean and green energy model that aid in the development of Industry 5.0 powered by resilient and robust green energy frameworks will help in building a stronger and healthier living conditions for humankind. Under Industry 5.0(I5.0), the new upturns will bring comprehensive changes in the existing value chains from the incubation phase to end of lifecycle phase with embedded systems of waste minimization within (United Nations, 2020). A robust health infrastructure, among other things will need critical power supply and functional water supply, both of which are energy intensive. In terms of normative thinking, a paradigm shift to clean and green energy sources will help in dealing with health risk imposed by climate change. Initiatives as these will lead to mammoth strides in the direction of sustainability. Businesses need to work closely with stakeholder engagement at a greater depth to usher in sustainability hand in glove with I5.0 (Nahavandi, 2019).

Though quality of life is generally associated with wellbeing, satisfaction and happiness, it is a very nebulous concept. There have been numerous definitions given to the term. But here, in this chapter we will be looking at quality of life from the lens of sustainability, meaning: quality of life that is high and also bionomically as well as socially falling under the purview of sustainability. (Constanza et al, 2016) worked out a wellbeing index and linked it to the United Nations Sustainable Development Goals. (Bakar et al, 2015) interlinked economic wellbeing to sustainability. The concept of Sustainable QoL was ushered by (Higgins et al, 2011) in their empirical study and described it as "enjoying life without compromising future generations".

CONCEPTUAL FRAMEWORK: INDUSTRY 5.0 AND BEYOND

Climate change adaptation is analogous to climate risk reduction and hence embedded measures of risk governance and securitization are imperative in the adaptation strategies. Most of the adaptation policies are top down, like those aided by the United Nations, whereas for policy formulations and strategic planning at the local and community level, bottom-up approaches work better with more focussed convergence on the benefits of the local communities. This in turn demands an understanding of the fabric of the region/community where adaptation measures are to be initiated. Dearth of understanding of the local nuclei, will lead to the failure of formulation of region appropriate adaptation strategies. (University of Stavenger, Nov 2020).

Formulation of region specific and appropriate adaptation pathways is therefore imperative. Interdisciplinary research and literature reviews have highlighted three major types of adaptation: incremental, transitional and transformational. (Ng et al, 2021). Incremental adaptation aims towards maintaining the already existing systems, exhibiting resilience with the developmental pathways. (Revi et al, 2014). Transitional adaptation displays a reformist leaning. Tries to bring changes within the existing systems through social interactions, behaviour change etc. Transitional adaptation intertwines socio-economic, political, technical and environmental paradigms as part of dealing with climate vulnerabilities. Transformational adaptation challenges the status quo, tries to break the glass ceiling and address the root cause for long term benefits (Loginova et al, 2019).

The onset of mechanization of industries started with the beginning of industrial revolution more than 200 years ago. Usage of fossil fuels for energy needs both: for domestic and industrial use has been on the rise since then. Ecosystem impacts on an adverse scale have followed. The progress in technology and its related transfer to developing economies has had its share of environmental and social impact (Rahman et al, 2015).

Industry 4.0, had technology, flexibility and productivity as its core values and focussed more on digitization and Artificial Intelligence (AI) driven technologies for production maximization and optimum output generation. The major aim was process automation and reduced human interaction by capitalizing on device interaction using machine learning while Industry 5.0, has three interconnected core principles of human centricity, sustainability and resilience. It aims at leveraging the expertise of humans with powerful machines. Industry 5.0 categorically tries to furnish a fine blend of human cognition and machine intelligence working in tandem. I5.0 thus offers the provision of greener solutions and collaborative robots called cobots. It is thus value based and aims at preserving planetary boundaries and human wellbeing at the core of the production process. It champions the cause of human-machine synergy and vouches for an industrial climate wherein economic, societal and ecological impact of digitization are weighted in the same scale as technological progress.

Transition to green and clean processes and sustainable practices for the benefit of societal values and human wellbeing as well as designing and usage of technological advancements in line with sustainable practices will be at the forefront to make industries human centric, sustainable, resilient and future proof. I5.0, thus supports research and innovation to achieve industry objectives, along with respecting the planetary boundaries and societal social fabric wellbeing, thereby heralding an era of symbiotic reconciliation of humans and machines (Xu et al, 2021). A gradual transition could reduce the ecological impacts caused due to human activities led climate change in alignment with the United Nations Sustainable Development Goals (UNSDG), (Rizzello, 2022; Xu et al, 2021). Such a paradigm shift in the global industrial arena driven by innovation, sustainability, human centricity, resilience and value systems through I5.0 could help in the reduction of climate impacts and aid towards the attainment of SDGs. Cleaner production through the usage of clean fuels, augmented production processes with digitized operating procedures, focus on innovation and research could lead to reduced emissions and sustainability (Atif, 2023).

Figure 1: Core principles of industry 5.0
[Source: Author's Creation]

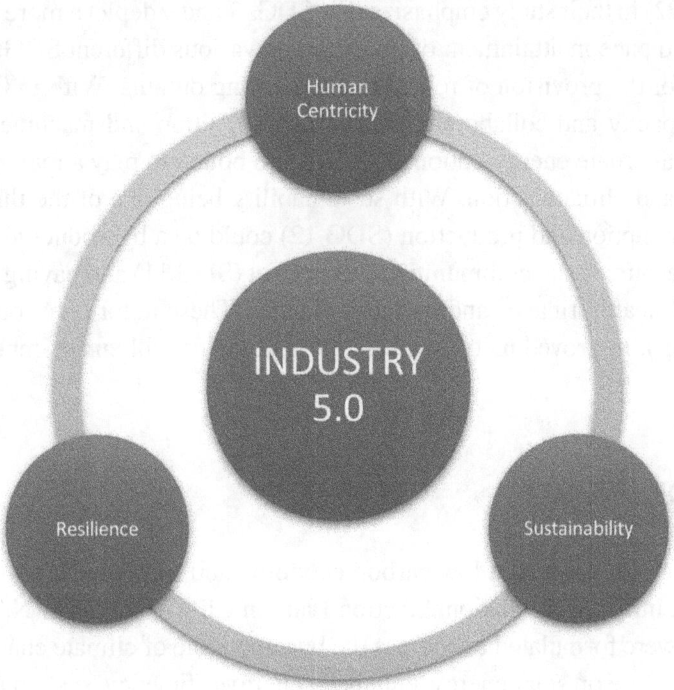

INTERNATIONAL GUIDELINES

As part of climate action and in order to provide I5.0 and beyond with a seamless transition towards green fuels and green economy, various regulations have been introduced by countries worldwide. The European Union introduced the EU Green Deal as part of their policy interventions aimed at facilitating Europe's climate neutrality ambitions. Ecodesign for Sustainable Products Regulation (ESPR) is another European mandate aimed at providing a coherent framework as part of a wider circular economy and sustainability plan. The USA has The National Emission Standards for Hazardous Air Pollutants (NESHAP) which is aimed towards the regulation of hazardous air pollutants in the atmosphere. The Hong Kong Convention, The Basel Convention and the The International Convention for Prevention of Pollution from Ships (MARPOL) are international guidelines directed towards curbing marine pollution (IMO, 2019).

INTERLINKAGES: INDUSTRY 5.0 AND SUSTAINABLE DEVELOPMENT GOALS

The Sustainable Development Goals (SDGs) are an important global agenda encompassing different objectives concerning economic, social, environmental and human development indices with 17 goals, 169 targets and interlinkages between various SDGs. These interlinkages could result either in synergies or trade-offs. This study wishes to draw attention on SDG 3(Good health and wellbeing), SDG 7(Affordable and clean energy), SDG 9(Industry, innovation and infrastructure), SDG 11(Sustainable cities and communities), SDG 12(Responsible consumption and production) and SDG 13(Climate action). (Kasinathan et al, 2022) in their study emphasize that SDGs 7 and 9 depict a more evident synergic effect and aid an accelerated pace in attainment of goals across various different SDGs. Energy and industry work hand in glove for the provision of robust manufacturing outputs. With I5.0(SDG 9) focussing on the double-edged capacity and collaboration of human cognition and machine level automation and precision, green and alternate energy options (SDG 7) are bound to play a major role in the attainment of sustainable goals and climate action. With sustainability being one of the three drivers of industry 5.0, responsible consumption and production (SDG 12) could be a byproduct leading to the protection of the earth's sinks, resources and communities to an extent (SDG 11) and paving the way towards better wellbeing in terms of health, finance and social exchanges. These in turn are some of the determinants of quality of life (QoL). Improved metrics across these domains will ensue improved QoL and human development.

NATIONAL GUIDELINES

With an aim to inch closer towards a low carbon economy and promote sustainable development, the Government of India initiated the National Action Plan on Climate Change (NAPCC) in 2008. Under NAPCC, 8 missions were formulated to address the burning issue of climate change and march towards better resilience. These are on solar energy, enhanced energy efficiency, sustainable habitat, water mission, mission for sustaining the Himalayan Ecosystem, Green India mission, sustainable agriculture and strategic knowledge for climate change. Additionally, the PAT (Perform, Achieve, Trade) scheme is introduced as a regulatory mechanism to reduce specific energy consumption. It caters to emissions trading and energy saving certificates for energy intensive industries. The Recycling of Ships Act came into effect in 2019 with an objective of establishing international standards for the recycling of ships and enforcement mechanisms for the compliance of this mandate.

NET ZERO READINESS: CROSS SECTORAL PERSPECTIVE

Industries account for approximately 40% of the global energy consumption and more than 30% of greenhouse emissions. Dynamic and sustained transformation of these industries is important towards decarbonization goals to reach net zero emissions by 2050. The coming future will further see a manifold increase in demands for products and services spanning multiple industrial sectors. While consumer demand and increased climate awareness has raised the demand for low carbon products, there are challenges galore in transition towards a decarbonized industry. Better value chains and an alternate operating protocol with renewables rather than fossil fuels are needed to strengthen the transitions. Improved regulations to incentivize transformations, economically viable low carbon markets and multi stakeholder collaborations across industrial ecosystems are vital for the implementation of sustainable practices (World Economic Forum, 2022).

We will take a look at the existing scenarios, challenges and transformations required towards sustainability across two different industrial sectors – aluminium and shipping. Though these industries play a major role in maintaining the global economic backbone, these are also major pollution causing ones, inflicting serious environmental harm to the planet (Loginova et al, 2019; Onal et al, 2021; Saevarsdottir et al, 2020).

ALUMINIUM INDUSTRY

Aluminium is a light weight metal with excellent anti corrosive properties, is recyclable, has low density and high thermal conductivity. The malleability and ductility quotient of aluminium is better than most metals. Aluminium's properties make it highly in demand for industries like transport, power, construction, automobile, machinery, packaging among others. Aluminium can thus be counted among the basic industries providing energy for sustaining modern life on our planet.

Manufacturing Process: From Bauxite Ore to Pure Aluminium

The aluminium manufacturing process is long and very technical. We will hence present it here in a very simplistic way. The main raw material in the industry is bauxite, which is a reddish hued clay mineral. It is mined, washed and ground. After grinding, caustic soda solution is mixed with it to produce sodium aluminate. The residue settles and then precipitation of aluminium crystals begins. After filtering impurities, aluminium oxide is obtained, which is reduced. Electric current is passed to break the bond between aluminium and oxygen and 99.8% pure aluminium is the resultant product. We have explained the entire process in the diagram below.

Pure aluminium has to be mixed with other metals and alloyed to make it stronger. Liquid aluminium is processed in three ways to further increase its applicability and make it more usable through extrusion, rolling and casting.

The Culprit: Carbon

As a byproduct of the production process, carbon dioxide is emitted and carbon monoxide is also co evolved. For each tonne of aluminium produced, 1.22 tonnes of carbon dioxide is emitted. Electricity supply is the challenge here with regards to controlling the emissions part (Brough et al, 2020). Alu-

Figure 2: Process flow from bauxite mining to getting pure aluminium
[Source: Author's Creation]

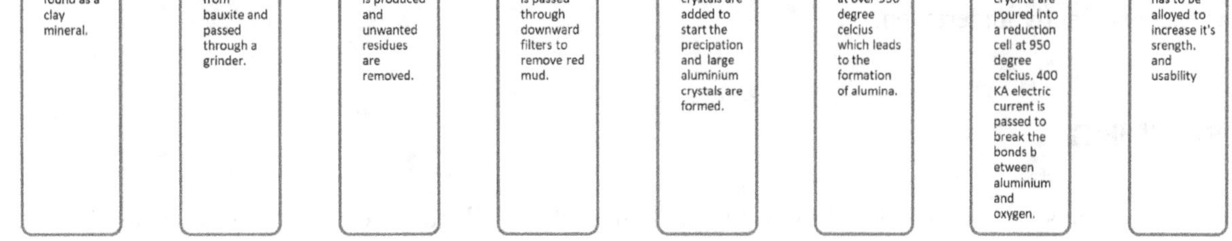

minium industry is also one of the highest emitters of greenhouse gases which are the major causative agents of anthropogenic climate emissions. With increased demand and increased production, increased emissions are also taking place through this sector. Right from mining aluminium ore, to alumina ingot production, the emissions reported rest between 12-17 tonnes of CO2 equivalent emissions per tonne of aluminium. The culprit here is the fuel source used for electrolysis. Fossil fuels like coal, oil and natural gas emit greenhouse gases detrimental to the environment. In addition, the sector has to grapple with the prospect of carbon leakage too. When a production in a country with strict emission mandates shifts its plant to another country with less stringent emission norms, it thereby relocates the process related emissions too. This is referred to as carbon leakage. This can lead to a situation where the country in which the production cessation occurs is relieved of emissions from the plant but the new greenfield plant set up in the alternate country would be quite a good source of emissions. Aluminium industries are at a high risk of carbon leakage, adding to the total global emissions as the new plant set up is added based on the fuel mix which is mostly fossil based. Based on its properties, aluminium is a widely used metal in transport, construction, automobiles, power, consumer goods. With a surge in usage, this highly in demand metal has seen a sharp rise in its production too. A major chunk of aluminium supply comes from China (57% of the 64.3 million metric ton globally produced as per 2018 data) followed by 28 million tonnes from recycling (Saevarsdottir et al, 2020). Most of the electricity requirement for running the plants comes majorly from fossil-based sources like coal, oil and natural gas. Migration to clean energy sources as per demand is still a long way to go considering the demand. This again contributes to increased greenhouse emissions (GHG). The primary process of aluminium production leads to release of GHG emissions due to the usage of carbon electrodes. This apart, solid hazardous wastes like red mud, sulphur and nitrogen are also given out.

Suggested Mitigation Measures for Aluminium Industry

Aluminium industry aiming for reduced emissions should opt for inert electrodes for electrolysis, carbon capture and storage system, carbon leakage risk mitigation, increased recycling efforts and usage of green energy for production by incentivizing transformations in the sector. Continued development and usage of inert electrodes in the electrolysis process to the scale of imminent commercialization within the industry as a mitigation measure towards reduced carbon dioxide emissions are suggested and reduce emissions by 16% (Saevarsdottir et al, 2020). Efficient waste management measures in the treatment and disposal of toxic red mud are imperative here. Going green with hydroelectricity, geothermal energy, wind and solar could be one alternative. As energy efficiency is a global issue and aluminium is an energy intensive sector, effective energy efficiency measures and upgradation to sustainable equipment and technology are the need of the hour. With I5.0, technological sophistication and innovation is spiralling across the industrial domain. Adoption of sustainable measures for mitigation must continue to be introduced and application of the same for further optimization within the aluminium industry has to be persistently undertaken.

SHIPPING

Maritime transport is considered to be the backbone of global trade. The shipping industry plays an important role in the interlinking of various nodes in the global supply and value chains. Though 90%

of global trade occurs over the waves, there are tough challenges that need to be looked into. 2.9% of the global greenhouse gas emissions comes from shipping sector. In the future scenario, the cluster of economic activities through maritime industry is going to spiral to three times the existing volumes (Organization for Economic Cooperation and Development, 2022).

Akin to other sectors, heightened pro environmental mechanisms and green pathways have been initiated for climate mitigation by the shipping industry. During ship building and ship recycling (end of life) phase environmental impacts are grave. The ship recycling industry is mostly centered in the developing countries. India and Bangladesh are major ship breaking centres of the world. They can be termed as the ship breaking capital of the world since a majority of the world's ships come here towards the end of their life phase for recycling. Apart from India and Bangladesh, Pakistan, China, Philippines, Turkey and Vietnam are the other markets as regulations pertaining to health, environment and safety are not very stringent, labour availability is easy and cheaper (The rest are sent to the UK for recycling under stringent health, safety and environment law enforcements (Frey, 2013)).

Ship recycling involves dispossessing the ship of its steel and other metals, when it has reached a no profit stage after an average life span of 25-30 years. Ship building and recycling leads to adverse environmental impacts like contamination of air, water and soil by heavy metals used in ship hull making and antifouling, GHG emissions, acidification, hazardous chemicals, eutrophication, ozone layer depletion and high damage to the workers' health and safety (Onal et al, 2021; Rahman et al, 2015).

Suggested Mitigation Measures for Shipping Industry

To mitigate the environmental impact of this economic activity, green measures are being proposed globally. Marine instruments used should produce less noise and vibration, reduced emissions and lower consumption of energy. Bunker fuel used in shipping is a fossil fuel and contains sulphur and nitrogen. Other options in marine fuels are methanol, low sulphur bunker oil, di methyl ether, liquified natural gas (Rahman et al, 2015) which could be used extensively. Methane slip from marine fuels needs urgent capping measures too. Life Cycle Assessment (LCA) can play a vital role towards establishing a green ship building industry as it can provide an insight into environmental impacts of the vessel by compilation of the raw materials used and emissions released. The greening efforts can further be extended to the improvisations' to be done during the construction, operation and scrapping phase. Minor modifications in the structure can play a vital role in the pro environmental aspect (European Commission, 2021). A gradual shift towards electric or hybrid propulsion systems over the conventional diesel ones would result in reduced GHG emissions, lesser vibration and noise and decreased fuel consumption.

The Green Panorama

Access to Clean Energy to Developed and Developing Economies Globally

The United Nations has announced a set of 17 sustainability goals (SDGs) to be achieved by 2030. For economic models based on sustainability, the UN SDGs could be taken as a base document. SDG 7 advocates access to affordable, reliable, sustainable and modern energy for all. (United Nations, 2020) Presently, many populations across scattered demographics in many developing countries do not have access to electricity. The UNSDGs advocates provision of universal energy solutions to such strata among others in order to maximise the socio-economic benefits of SDG 7 (Nathwani et al, 2019). Energy services are interlinked to a variety of basic life requirements for sustenance like the provisions of health and education (World Health Organization, 2021). Water for agricultural activities, household use and industrial processes, fuels for transport, cooking and industry, different gadgets and machines for various economic activities, the functioning base for all these are energy resources. Going by the United Nations Sustainable Development Goals data, as per the current pace, only about 92% of the world's population will have access to electricity by 2030, leaving about 670 million people unserved. Another UNSDG statistic points out that in the year 2020, about 4 billion people relied on polluting cooking systems.

Realization of the objectives of SDG 7 implies ending energy poverty and achieving energy parity, both in developed and developing countries. Energy is the key propellant for industrial functionalities. The United Nations Industrial Development Organization (hereafter referred to as UNIDO) statistics shows that hunger, poverty, climate change and environmental degradation are the major global challenges. Sustainable industrial development using sustainable energy can be a solution towards creating impactful industries functioning on alternate energy sources through job creation based on SDG 9, to promote, accelerate and energize industrial development well within the realms of climate change variability, QoL and inclusive development by means of economies with reduced GHG, better living standards and preservation of environment for future generations.(UNIDO, 2023). SDG 7 is one of the most powerful enablers of all SDGs as it is delicately linked to many other SDGs like provision of water, education, health and infrastructure among others. Low cost, clean, green and affordable energy for the low income groups and developing economies coupled with synergised basic infrastructure will pave way towards reduction in fossil fuel emissions and carbon neutrality and improved livelihoods (Ng et al, 2021). One of the main metrics of energy parity could be the onboarding of the energy poor into mainstream economic roadmaps through sustained and prevailed allocation and availability of clean energy sources comprehensively (Ng et al, 2021). For universal energy access to all in general and industries in particular, clean diffusion pathways are to be charted and implemented so that energy poverty reduction happens. The International Energy Agency(hereafter referred to as IEA) defined energy access as "*a household having reliable and affordable access to both clean cooking facilities and to electricity, which is enough to supply a basic bundle of energy services initially, and then an increasing level of electricity over time to reach the regional average*"(International Energy Agency, 2022).

To end energy poverty and manage energy deficit, the major enabler is proving energy access to the energy poor. Impending energy disparity can be erased and a stage of energy parity can be reached through affordable energy access to all. A paradigm shift to renewables can provide a major boost to sustainability and quality of life metrics apart from reduced GHG emissions. Magnanimous inputs from the scientific and technological domains and radical amalgamation of I5.0 determinants are required to provide affordable and green energy solutions in scale and volume. With improved energy access, livelihoods, economy and quality of life indicators in third world economies will show improved metrics. The most obvious result of providing affordable energy access to all is drawing the energy poor to the mainstream and ending of energy disparity. Bridging the energy gap thus will remove the barrier posed to economic wellbeing and bring inclusivity within. Energy access disparity among the developed and the developing economies is a raging issue. Reducing this energy disparity by increased energy access can lead to reduced climate pollution by way of decreased GHG emissions and low carbonised and earth sustaining energy systems (Alstone et al, 2015).

To bring together the mammoth scale of energy collation and access to all, convergence of innovation, good governance, research and technology is required. An energy access hub framework mobilizing the existing nodes and networks in the energy access network based on the following objectives can be very useful:

1) Knowledge centres powered by research and innovation for analysis of environmental and human solutions.
2) Financial funding pipeline for resource mobilization, infrastructure boost and provision of economical and green energy access to the energy poor in line with UNFCCC, UNSDG goals for a global extension service linked to energy access hubs.
3) Decentralized energy systems program for universal energy access and climate stabilization.
4) Convergence of I5.0 core values of human centricity, sustainability and resilience aimed at providing earth friendly industrial development to global energy access solutions and bring equity and justice to the energy poor.

Migration to Green Energy as a Feasible Business Strategy

With extreme weather events on the rise, global climate research and climate awareness has increased participation from the scientific community and the population. Intense research over many years have proved the cost efficiency of renewable technologies in contrast to fossil fuel-based energy sources. The The International Renewable Energy Agency's report (hereafter referred to as IRENA) – Renewable energy capacity statistics, 2023 presents renewable energy as the de-facto energy choice for power generation. By the end of 2022, renewables accounted for 40% of global installed power capacities.2022, saw the largest increase in renewable capacity building till date with an addition of 295 GW of renewable energy sources, increasing it by 9.6% and contributing to 83% of global power additions. (IRENA, 2023).

The IRENA report states that the global weighted average levelized cost of electricity (LCOE) of new utility-scale solar PV and hydropower was 11% lower than the cheapest new fossil fuel-fired power generation option in 2021, and 39% lower for onshore wind despite rising material and equipment costs. The global weighted average LCOE of new onshore wind projects in 2021 fell by 15%, year on year to USD 0.33/kWh, and that of utility scale solar PV fell by 13% year on year to USD 0.048/kWh and that of offshore wind declined 13% to USD 0.075/kWh. In contrast, fossil fuel prices are on an upward streak. With rising fossil fuel prices, increased incidences of extreme weather events, heightened climate risk awareness, stringent policy formulations and international synergy for climate adaptation and mitigation, the renewables sector has witnessed an acceleration. Given the fossil fuel price rise crisis, the benefits of renewables have been phenomenal in the year 2022. If we speak for Europe alone for the year 2021, the lifetime cost per kWh for newly added solar and wind facilities would be at least 4 to 6 times lesser than the cost for power generation using fossil fuels in 2022. Global renewables addition in 2021 could reduce electricity generation costs in 2022 by US$ 5 billion. Fossil fuel imports to Europe in the period between January to May 2022 could be reduced to the tune of US$ 50 billion due to new commissions in wind and solar sectors. (IRENA, 2022).

This apart the renewables sector witnessed increased and improved competition. The sector thus is an anchor in phasing out fossil fuels and build a carbon neutral resilient society. The renewable sector's cost competitiveness has thus paved the way for phasing out fossil fuels and reduced emission scenarios in the future.

With the advent in technological and infrastructure advances, transition to renewable energy across economies is on the upswing. Low cost and affordable options in the renewable sector have made climate risk reduction and climate action strategies more feasible ad plausible. Transition to these technologies could result in transformational add ons to the climate adaptation and mitigation efforts initiated worldwide (United Nations, 2020).

Ample studies have focussed on the decreased power generation costs and global transition to energy pathways accommodating the Paris Agreement and pre industrial level temperature threshold of below 2 degrees Celsius (United Nations, 2020).

Green Finance for Risk Reduction

With the renewable sector showing increased cost efficiency and facilitate such transformational economic upheaval towards a robust and sustainable economy helmed by I5.0's core, an efficient green finance pipeline needs to be maintained as pointed out by recent studies (United Nations, 2020). With increased competitiveness in the renewables sector and the heightened cost efficiency, the migration to green and alternate sources in on an acceleration globally. The increased cost factor on fossil fuels further abets the green cause. In keeping with the mandate of international governing bodies about a pertinent shift from fossil fuels to alternate energy models, governments worldwide are in lieu of developing an alternate green financing system in an attempt to gradual phasing out the fossil fuels and open de carbonization pathways to bring in climate stability. To meet the goal of keeping temperature rise to below 1.5-degree Celsius, international environmental bodies have been closely working with governments and financial bodies to create green investment mechanism pipeline to propel the UNSDG delivery. Financial markets and banking institutions provide capital allocation to the green businesses. Investors are provided with financial impetus and encouraged to venture into the renewables sectors. The green finances are necessary to shape production and consumption patterns of the future and preserve the ecosystem from further climate impact. The resource efficiency program of UNEP encourages countries to review their financial ordinances and sustainability roadmaps to assist the nodal bodies to provide a thrust to the domestic financial flow and assist in the formulation of trans national greening collaborations at different tiers of nodal, national, regional and global levels. (Lee et al, 2020). With a focus on inclusive green economy based on the vials of I5.0 and sustainable procurement, financing mechanisms are issuing green bonds and green insurance covers. Facilitating the modelling of a green economic framework necessitates massive investments and fund allocations through structured and systemic financial framework by risk mitigation. Transition of business models, risk reduction, redesigning of debt and equity instruments, micro credit for pilot projects and community enterprises, interactive public private partnerships and dialogues can be the core initiatives of finance sector to achieve sustainability and align with UNSDG. A green financing framework involving governments, businesses and citizens can be pioneered in the march towards sustainability. Global standards for green bonds issuance have already been proposed by international lenders with highlighted attention to climate risks. Regulators are in demand of organizations' performance and their sustainability initiatives. Establishment of The Network of Greening and Financial Systems (NGFS) in 2017 with a view towards sharing of environmental and financial best practices for sustainability by prominent central banks and finance bodies of the world was a pioneering move to mobilize green funds pipeline that are earth friendly and sustainable. NGFS has incorporated blended finance transactions for climate action and formulated a set of best practices to scale up blended finance for climate action (Rode et al, 2019).

Green Bonds are "an increasingly popular way of reaching investors interested in financing sustainable development and the fight against climate change through their investments. Developing countries are increasingly raising money for climate action through green bonds. 19 emerging markets have, since 2016 issued green bonds for climate adaptation funding. In 2016, Fiji became the first emerging market to raise funds through green bonds as way of transition towards a fossil free society. In 2020, Egypt raised $750-dollar green sovereign bonds as part of its climate action steps. Columbia, in 2021 raised green funds through its local currency. $481.9 green bonds finance was heralded by Kuala Lampur in 2021. As of January, 2023, green bonds to the tune of $2.5 trillion have been raised worldwide for sustainable projects. India launched its first green bond in 2023 and raised $2 billion for biodiversity conservation and net zero objectives (Banga, 2019; World Bank, 2023). An active role in green bonds mobilization is being played by the stock exchanges in the last few years the world over. These exchanges facilitate the market development process from market access provision, providing fluidity and regulatory mandates for different stakeholders. Table 1 lists the stock exchanges with an exclusive green bond. This data further accentuates the incursion of many third world economies into sustainable finance model through green bonds.

Table 1. Stock exchanges with an exclusive green bond

Name of Stock Exchange	Type of Dedicated Section	Launch Date
Oslo Stock Exchange	Green bonds	January 2015
Stockholm Stock Exchange	Sustainable bonds	June 2015
London Stock Exchange	Sustainable bonds	July 2015
Shanghai Stock Exchange	Green bonds	March 2016
Mexico Stock Exchange	Green bonds	August 2016
Luxembourg Stock Exchange	Luxembourg Green Exchange	September 2016
Borsa Italiana	Green and Social bonds	March 2017
Taipei Exchange	Sustainable bonds	May 2017
Johannesburg Stock Exchange	Green bonds	October 2017
Japan Exchange Group	Green and Social bonds	January 2018
Vienna Exchange	Green and Social bonds	March 2018
Nasdaq (Multiple stock exchanges)	Sustainable bonds	May 2018
Swiss Stock Exchange	Green and Sustainability bonds	July 2018
The International Stock Exchange	Sustainable bonds	November 2018
Frankfurt Stock Exchange	Green bonds	November 2018
Santiago Stock Exchange	Green and social bonds	July 2019
Euronext (Multiple stock exchanges)	Green, Sustainable and Social bonds	November 2019
Bombay Stock Exchange	Green bonds	June 2019
Argentina Stock Exchange BYMA	Green, Sustainable and Social bonds	September 2019
Brazil Stock Exchange	Green bonds	September 2019
Nigerian Stock Exchange	Sustainable bonds	October 2019
Hong Kong Exchange	Green, Sustainable and Social bonds	June 2020
Korea Exchange	Green, Sustainable and Social bonds	June 2020
Toronto Stock Exchange	Sustainable bonds	November 2020
Singapore Stock Exchange	Green, Sustainable and Social bonds	n.a
Bolsas y Mercados Españoles (BME)	Green, Sustainable and Social bonds	n.a

Adapted from (Climate Bonds Initiative, 2023)

SDG 7 is closely linked to the optimization and realization of many SDGs that are primary life determinants like health. Medical and diagnostic services require uninterrupted supply of power, water for sustenance and health, education and internet technologies for technological seamlessness. Hence investing in sustainable energy solutions for ecological conservation, achieving net zero goals and phasing out fossil fuels with a complementary alternative clean and green energy system is critical.

Industry 5.0 and Quality of Life (QoL) Interlinkages

Industry 4.0 was technology driven with artificial intelligence and internet of things ruling the roost. Industry 5.0 on the other hand is value driven with the three-pronged drivers of sustainability, human centricity and resilience taking centre stage. Machine precision and superlative human cognition function in tandem to produce accurate and efficient outputs. I5.0 has sustainability as one of its drivers. Hence sustainable work systems and their socio ecological impacts are also strong areas of focus of I5.0.

Industry 5.0 also focusses on human centricity and hence human wellbeing is one of its most pivotal metrics. Wellbeing in terms of health, finance, social exchanges, environment etc roughly translate into important indicators of quality of life (QoL). I5.0 can thus contribute to the betterment of QoL metrics through highly improved, efficient and value-based work systems. Though there are many studies showing the interlinkages between I5.0 and Society 5.0, there are very few studies showing interlinkages between I5.0 and QoL. This could hence be an important lacuna for future research.

PROPOSED NEW FRAMEWORK

The authors propose a new framework interlinking I5.0, sustainability and environmentally responsible investment or green finance. The new framework stresses the need for a global green finance model that would aid in scourging and streamlining the requisite funding for sustainable projects through I5.0 and also aimed at climate action. With the crosslinking and meeting the goals of SDGs 3,7,9,11,12 and 13 and giving an impetus to cleaner fuels, the proposed new framework would act as an aid in climate mitigation and improved health, education, environment and socio-economic equities resulting in an improvement in the overall QoL.

Global Green Finance Framework for Sustainability Through Unification of Industry 5.0 and Innovation Exchanging

Climate Bonds Initiative is a registered charity in England and Wales (no. 1154413)

Integration Framework of Industry 5.0, Sustainability and Green Finance

As shown in Figure 3, the rapid increase in the numbers of green bonds, fiscal incentives for clean assets, ESG initiatives and the augmented environmental risk management measures are evidences to the unfolding of a global green finance framework aimed at sustainable practices. The major determinants of Industry 5.0 are human centricity, sustainability and resilience. Clean and fair priced energy sources are therefore integral components of I5.0 which are synergistic to the configuration of framework development for resource development and utilization, responsible production and consumption. Such unified

Figure 3: Global green finance framework for sustainability through unification of Industry 5.0 and innovation exchanging
[Source: Author's Creation]

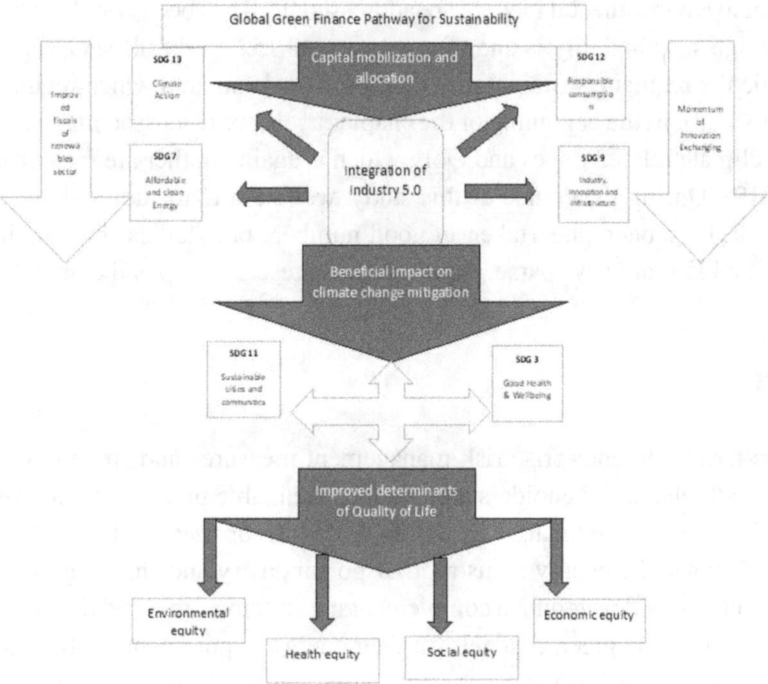

actions on the climate crisis results in improved metrics of social, economic, health, environment and sustainability equities, which are also the determinants of Quality of Life. Green finance thus helps to improve the QoL through climate action. With improved innovation and thrusts in the industry in general due to technological and research initiatives (SDG 9) and a dynamic upswing in the renewables sector and improved measures of energy accessibility and energy parity, industries can avail the usage of clean energy (SDG 7). With a shift towards the renewable energy sources in industries, we are contributing towards SDG 13 (Climate Action) and also SDG 12(Responsible consumption). The outcomes of the shift from fossils to clean energy can be improved metrics of health and wellbeing (SDG 3) and sustainability in living spaces and communities (SDG 11). Along with better managed climate change mitigation, the resultant outcomes rein in better economic, social, environmental and health equities which are all determinants of better quality of life.

The fifth industrial revolution, which is value based is a harbinger of green actions across industrial spaces. Technological augmentation along with human and environmental wellbeing is the basis of Industry 5.0. Intertwined, versatile and multidisciplinary research and innovation are key to its synchronization with climate metrics and embedded metrics of Quality of Life. Complementing the green campaign, world bodies and governments have chosen to work collaboratively to make industrial spaces safer and green through innovation, research, policy formulations and strengthened public private partnerships for increased sustainability and environmental management. I5.0 presents us with intelligent workflows wherein AI algorithms and humans work in tandem to create an active synergy between humans, augmented machines and environmental risk management (Nahavandi, 2019).

Future Scope of Research

I5.0 emerged as an augmentation to Industry 4.0. wherein industry led automation and innovation as well as the aspects of socio environmental metrics are addressed. I5.0 has been widely publicized to stimulate sustainability. Though sustainability is one of the drivers of I5.0, it is still very vague as to how will it deliver and cascade the requisite sustainability outputs. This is an area which requires extensive study. (Ghobakhloo et al, 2022). In the beginning of the chapter itself, we discussed that there is dearth of interlinkages between climate related issues and QoL, which is again another area requiring future research. (Estoque et al, 2019). During the course of this study, we found that studies showing interlinkages of Society 5.0 and QoL have been undertaken in good numbers but studies showing interconnectedness of QoL indicators and I5.0 are vey sparse. Again, this is a research gap and demands future research.

CONCLUSION

The last few years have seen enterprise risk management measures and strategies being incorporated by the industry and all related stakeholders along with sustainable practices whilst keeping in mind the serious climate crisis that we are facing (Ng, 2018).With anthropogenic climate change and the rise of greenhouse gas emissions, the energy industry, transport industry and shipping industry among others can play a catalytic effect by developing a complementary system of reduced dependence on fossil fuels and increasing dependence on green channels for their energy requirements. Infrastructure projects aiding in and important in sustainability measures implementation will require green finance pathways viz a viz green bonds or asset linked bonds which ensure that the proceeds will be diverted to eco-friendly use. Transition to a low carbon and climate resilient economy via green bonds and other linked assets that formulated on a policy of using the funds to finance or re finance environment friendly projects also requires governments and policy makers to have a strategy of mobilization of debt and equity capital for migration to fossil free energy sources. (OECD, 2015).

Sectors and infrastructures aiding in a seamless transition to renewables need to be benchmarked as priority assets as they will be the important aids in the global decarbonization dreams. Entrepreneurial set ups (SMEs and MSMEs) and manufacturing facilities are vital to a resilient economy.

We have argued that climate risk is one of the most delirious issues faced by mankind today. Pandemic instigated economic downturn and charted recovery strategies coupled with a transition to clean energy options are the need of the hour. With lives being affected due to climate disruptions, the overall QoL across various determinants have been impacted. Green financing hubs situated in various countries can accelerate the transition and achieve sustainability targets. Sustainable resource usage to strengthen and build resilience with value chains is vital. Technology transfer to developing economies, cross sectoral economic growth that is decoupled from resource usage, energy equality, investment in training and development initiatives, policy measures in law and governance, compliance mandates and international collaboration among others can help in helming the migration of Industry 4.0 to Industry 5.0 via interlinkages in sustainability and climate adaptation and mitigation. To highlight the case of an urgent need to devise sustainability strategies in the industrial domain, we have presented here, in this chapter, case studies of two major industrial sectors with the current scenarios, the process flows from start to finish, the technologies being in vogue and the technical operations involved, the existing laws and the need to migrate to greener alternatives. The chapter provides comprehensive case studies with respect to two major industrial domains: the aluminium industry and the shipping industry. The authors have chosen these two industries as these are among the highly polluting industries of the world but at the same time, these are also the backbone of the global economy. Aluminium usage is manifested in the smallest of things like pins and foils to automotive parts and huge machines. The coverage of aluminium usage in day to day lives as well as high end machines and medical equipments is well known. Similarly, the waterways are the connecting link of global trade and commerce. We cannot imagine a world without ships and marine trade. Though highly profitable, these two industries are highly polluting. Aluminium is one of the highest emitters of greenhouse gases today and ship building/breaking industry releases effluents that are hazardous and lead to environmental degradation. The compliance mandates too seem ineffective and call for revised regulatory reforms. The writers emphasize the need for asset categorization in industries for mitigation action. The authors have used this platform to highlight the standard operating procedures followed, the problem areas, environmental degradation and challenges posed to the ecosystem, alternative and complementary solutions to mitigate and cope with the triple existential threat of climate change, environmental degradation and quality of life impacts among other challenges faced in the migration to Industry 5.0 goals, green financial framework and transformational solutions in our race towards actualization of SDG 2030.

REFERENCES

Alstone, P., Gershenson, D., & Kammen, D. M. (2015). Decentralized energy systems for clean electricity access. *Nature Climate Change*, *5*(4), 305–314. doi:10.1038/nclimate2512

Asmelash, E., Prakash, G., Gorini, R., & Gielen, D. (2020). Role of IRENA for global transition to 100% renewable energy. *Accelerating the transition to a 100% renewable energy era*, 51-71.

Atif, S. (2023). Analysing the alignment between circular economy and industry 4.0 nexus with industry 5.0 era: An integrative systematic literature review. *Sustainable Development (Bradford)*, *31*(4), 2155–2175. doi:10.1002/sd.2542

Bakar, A. A., Osman, M. M., Bachok, S., Ibrahim, M., & Mohamed, M. Z. (2015). Modelling economic wellbeing and social wellbeing for sustainability: A theoretical concept. *Procedia Environmental Sciences*, *28*, 286–296. doi:10.1016/j.proenv.2015.07.037

Banga, J. (2019). The green bond market: A potential source of climate finance for developing countries. *Journal of Sustainable Finance & Investment*, *9*(1), 17–32. doi:10.1080/20430795.2018.1498617

Brough, D., & Jouhara, H. (2020). The aluminium industry: A review on state-of-the-art technologies, environmental impacts and possibilities for waste heat recovery. *International Journal of Thermofluids*, *1*, 100007. doi:10.1016/j.ijft.2019.100007

Climate Bonds Initiative. (2023). *Green Bonds Stock Exchanges*. Climate Bonds.https://www.climatebonds.net/green-bond-segments-stock-exchanges

Corsi, A., Pagani, R. N., Kovaleski, J. L., & Luiz da Silva, V. (2020). Technology transfer for sustainable development: Social impacts depicted and some other answers to a few questions. *Journal of Cleaner Production*, *245*, 118522. doi:10.1016/j.jclepro.2019.118522

Costanza, R., Daly, L., Fioramonti, L., Giovannini, E., Kubiszewski, I., Mortensen, L. F., Pickett, K. E., Ragnarsdottir, K. V., De Vogli, R., & Wilkinson, R. (2016). Modelling and measuring sustainable wellbeing in connection with the UN Sustainable Development Goals. *Ecological Economics, 130*, 350–355. doi:10.1016/j.ecolecon.2016.07.009

de Vries, B. J. (2019). Inequality, SDG10 and the financial system. *Global Sustainability, 2*, e9. doi:10.1017/sus.2019.6

Ebi, K. L., Vanos, J., Baldwin, J. W., Bell, J. E., Hondula, D. M., Errett, N. A., Hayes, K., Reid, C. E., Saha, S., Spector, J., & Berry, P. (2021). Extreme weather and climate change: Population health and health system implications. *Annual Review of Public Health, 42*(1), 293–315. doi:10.1146/annurev-publhealth-012420-105026 PMID:33406378

Estoque, R. C., Togawa, T., Ooba, M., Gomi, K., Nakamura, S., Hijioka, Y., & Kameyama, Y. (2019). A review of quality of life (QOL) assessments and indicators: Towards a "QOL-Climate" assessment framework. *Ambio, 48*(6), 619–638. doi:10.1007/s13280-018-1090-3 PMID:30206898

Frey, R. S. (2013). *Breaking ships in the world-system: An analysis of two ship breaking capitals*. Alang India and Chittagong.

Ghobakhloo, M., Iranmanesh, M., Mubarak, M. F., Mubarik, M., Rejeb, A., & Nilashi, M. (2022). Identifying industry 5.0 contributions to sustainable development: A strategy roadmap for delivering sustainability values. *Sustainable Production and Consumption, 33*, 716–737. doi:10.1016/j.spc.2022.08.003

Higgins, P., & Campanera, J. M. (2011). (Sustainable) quality of life in English city locations. *Cities (London, England), 28*(4), 290–299. doi:10.1016/j.cities.2011.02.005

International Energy Agency. (2022). *Analysis: Energy-Access*. IEA. https://www.iea.org/topics/energy-access

International Institute for Sustainable Development (IISD). (2020). *Three Ways the Coronavirus is Shaping Sustainable Development*. Winnipeg: IISD. https://www.iisd.org/articles/insight/three-ways-coronavirus-shaping-sustainable-development

International Maritime Organization. (2019). International Convention for the Prevention of Pollution from Ships (MARPOL). https://www.imo.org/en/about/Conventions/Pages/International-Convention-for-the-Prevention-of-Pollution-from-Ships-(MARPOL).aspx#:~:text=The%20MARPOL%20Convention%20was%20adopted,Protocol%20absorbed%20the%20parent%20Convention

International Renewable Energy Agency. (2022). *Renewable power generation costs in 2021*. IREA. https://www.irena.org/publications/2022/Jul/Renewable-Power-Generation-Costs-in-2021

International Renewable Energy Agency. (March 2023). *Renewable-capacity-statistics-2023*. IREA. https://www.irena.org/Publications/2023/Mar/Renewable-capacity-statistics-2023

Kasinathan, P., Pugazhendhi, R., Elavarasan, R. M., Ramachandaramurthy, V. K., Ramanathan, V., Subramanian, S., Kumar, S., Nandhagopal, K., Raghavan, R. R. V., Rangasamy, S., Devendiran, R., & Alsharif, M. H. (2022). Realization of Sustainable Development Goals with Disruptive Technologies by Integrating Industry 5.0, Society 5.0, Smart Cities and Villages. *Sustainability (Basel)*, *14*(22), 15258. doi:10.3390/su142215258

Kotcher, J., Maibach, E., Miller, J., Campbell, E., Alqodmani, L., Maiero, M., & Wyns, A. (2021). Views of health professionals on climate change and health: A multinational survey study. *The Lancet. Planetary Health*, *5*(5), e316–e323. doi:10.1016/S2542-5196(21)00053-X PMID:33838130

Lee, J. W. (2020). Green Finance and Sustainable Development Goals: The Case of China. *Journal of Asian Finance Economics and Business*, *7*(7), 577–586. doi:10.13106/jafeb.2020.vol7.no7.577

Loginova, J., & Batterbury, S. P. (2019). Incremental, transitional and transformational adaptation to climate change in resource extraction regions. *Global Sustainability*, *2*, e17. doi:10.1017/sus.2019.14

Maddikunta, P., Pham, Q.-V., B, P., Deepa, N., Dev, K., Gadekallu, T. R., Ruby, R., & Liyanage, M. (2022). Industry 5.0: A survey on enabling technologies and potential applications. *Journal of Industrial Information Integration*, *26*, 100257. doi:10.1016/j.jii.2021.100257

Nahavandi, S. (2019). Industry 5.0—A human-centric solution. *Sustainability (Basel)*, *11*(16), 4371. doi:10.3390/su11164371

Nathwani, J., & Kammen, D. M. (2019). Affordable energy for humanity: A global movement to support universal clean energy access. *Proceedings of the IEEE*, *107*(9), 1780–1789. doi:10.1109/JPROC.2019.2918758

Ng, A. W. (2018). From sustainability accounting to a green financing system: Institutional legitimacy and market heterogeneity in a global financial centre. *Journal of Cleaner Production*, *195*, 585–592. doi:10.1016/j.jclepro.2018.05.250

Ng, A. W., Nathwani, J., Fu, J., & Zhou, H. (2021). Green financing for global energy sustainability: prospecting transformational adaptation beyond Industry 4.0. *Sustainability: Science. Sustainability*, *17*(1), 377–390. doi:10.1080/15487733.2021.1999079

Önal, M., Neşer, G., & Gürsel, K. T. (2021). Environmental impacts of steel ship hulls building and recycling by life cycle assessment (LCA). *Ships and Offshore Structures*, *16*(10), 1061–1066. doi:10.1080/17445302.2020.1816706

Rahman, A., & Karim, M. M. (2015). Green shipbuilding and recycling: Issues and challenges. *International Journal of Environmental Sciences and Development*, *6*(11), 838–842. doi:10.7763/IJESD.2015.V6.709

Ravindra, K., Rattan, P., Mor, S., & Aggarwal, A. N. (2019). Generalized additive models: Building evidence of air pollution, climate change and human health. *Environment International*, *132*, 104987. doi:10.1016/j.envint.2019.104987 PMID:31398655

Revi, A., Satterthwaite, D., Aragón-Durand, F., Corfee-Morlot, J., Kiunsi, R. B., Pelling, M., Roberts, D., Solecki, W., Gajjar, S. P., & Sverdlik, A. (2014). Towards transformative adaptation in cities: The IPCC's Fifth Assessment. *Environment and Urbanization*, *26*(1), 11–28. doi:10.1177/0956247814523539

Rizzello, A. (2022). Financing the Green Recovery: The New Directions of Finance After the COVID-19 Crisis. In *Green Investing: Changing Paradigms and Future Directions* (pp. 133–158). Springer International Publishing. doi:10.1007/978-3-031-08031-9_6

Rode, J., Pinzon, A., Stabile, M. C., Pirker, J., Bauch, S., Iribarrem, A., Sammon, P., Llerena, C. A., Muniz Alves, L., Orihuela, C. E., & Wittmer, H. (2019). Why 'blended finance'could help transitions to sustainable landscapes: Lessons from the Unlocking Forest Finance project. *Ecosystem Services*, *37*, 100917. doi:10.1016/j.ecoser.2019.100917

Saevarsdottir, G., Kvande, H., & Welch, B. J. (2020). Aluminum production in the times of climate change: The global challenge to reduce the carbon footprint and prevent carbon leakage. *Journal of the Minerals Metals & Materials Society*, *72*(1), 296–308. doi:10.1007/s11837-019-03918-6

Schnurr, R. E., & Walker, T. R. (2019). *Marine transportation and energy use. Reference Module in Earth Systems and Environmental Sciences*. Elsevier.

Solomon, C. G., & LaRocque, R. C. (2019). Climate change—A health emergency. *The New England Journal of Medicine*, *380*(3), 209–211. doi:10.1056/NEJMp1817067 PMID:30650319

The European Commission Director General for Research and Innovation, (2021). *Horizons: New-materials-make-ships-more-sustainable-and-less-noisy-marine-life*. OECD.

The Organisation for Economic Co-operation and Development. (2015). *Green bonds Mobilising the debt capital markets for a low-carbon transition*. OECD. https://www.oecd.org/environment/cc/Green%20bonds%20PP%20%5Bf3%5D%20%5Blr%5D.pdf

The Organisation for Economic Co-operation and Development. (2022). *Shipbuilding-policy-and-market-developments-in-selected-economies-2022*. OECD. https://www.oecd.org/ocean/topics/ocean-shipping/

The World Bank. (2023). *Green Bonds*. The World Bank. https://www.worldbank.org/en/news/feature/2023/04/10/from-india-to-indonesia-green-bonds-help-countries-move-toward-sustainability

The World Economic Forum. (2022). *Net Zero Tracking*. WEF. https://www.weforum.org/reports/the-net-zero-industry-tracker/

The World Health Organization. (2021). *Energy Progress*. WHO. https://www.who.int/news/item/07-06-2021-global-launch-tracking-sdg7-the-energy-progress-report

United Nations. (2020). *Sustainable Development Goals*. New York: United Nations. https://www.un.org/sustainabledevelopment/goal-of-the-month

United Nations. (2020). *Sustainable Development Goals*. New York: United Nations. https://www.un.org/sustainabledevelopment/energy/

United Nations Industrial Development Organization. (2023). *Who We Are*. UN. https://www.unido.org/about-us/who-we-are

University of Stavenger. (2020). *RISKSEC 2.0*. UIS. https://www.uis.no/en/research/climate-change-adaptation-strategies

Ushakov, S., Stenersen, D., & Einang, P. M. (2019). Methane slip from gas fuelled ships: A comprehensive summary based on measurement data. *Journal of Marine Science and Technology*, *24*(4), 1308–1325. doi:10.1007/s00773-018-00622-z

Xu, R., Yu, P., Abramson, M. J., Johnston, F. H., Samet, J. M., Bell, M. L., Haines, A., Ebi, K. L., Li, S., & Guo, Y. (2020). Wildfires, global climate change, and human health. *The New England Journal of Medicine*, *383*(22), 2173–2181. doi:10.1056/NEJMsr2028985 PMID:33034960

Xu, X., Lu, Y., Vogel-Heuser, B., & Wang, L. (2021). Industry 4.0 and Industry 5.0—Inception, conception and perception. *Journal of Manufacturing Systems*, *61*, 530–535. doi:10.1016/j.jmsy.2021.10.006

Chapter 10
Promoting Sustainability:
Mitigating the Water Footprint in AI-Embedded Data Centres

Sheelratan Shashikant Bansode
Walchand Institute of Technology, India

Rahul Hiremath
IIM Raipur, India

Gurudevi Rahul Hiremath
Punyashlok Ahilyadevi Holkar Solapur University, Solapur, India

ABSTRACT

Artificial intelligence (AI) applications are expanding quickly, which has boosted demand for data centers and created serious environmental problems, notably in relation to water use. The main goal of this project is to promote sustainability by reducing the water footprint in data centers with integrated AI. It looks at many techniques and tools that can be used to use less water while still getting the best results and efficiency. This research offers important insights into environmentally friendly practices for AI-embedded data centers by examining water-efficient cooling approaches, hardware optimization, water recycling, and intelligent water management systems. Data center operators will be able to promote the development and expansion of AI technologies while also helping the environment by putting these methods into practice.

INTRODUCTION

Data centers, the foundation of digital infrastructure, have recently seen a spike in demand as a result of the exponential growth of data-driven technologies and artificial intelligence (AI) (Next Move Strategy Consulting, 2023, H.-A. Ounifi et al, 2022). Along with server counts the amount of power demand is also increasing (A. Shehabi et al, 2016). Numerous servers and other pieces of computing hardware are housed in these data centres, which utilize a lot of water and other resources as shown in figure 1.

DOI: 10.4018/978-1-6684-9863-7.ch010

Table 1: Digital and energy indicators for global trends in 2015-2021

Indicators	2015	2021	Change
Internet Users	3 billion	4.9 billion	+60%
Internet Traffic	0.6 ZB	3.4 ZB	+440%
Data Centre workloads	180 million	160 million	+260%
Date Centre energy use (excluding crypto)	200 TWh	220 - 320 TWh	+10-60%
Crypto mining energy use	4 TWh	100-140 TWh	+2300-3300%
Data transmission network energy use	220 TWh	260-340 TWh	+20-60%

Source: Data Centres and Data Transmission Networks – Analysis - IEA

The above table 1 gives the values for the energy consumptions at different ends. Exploring sustainable methods within the context of AI-embedded data centres is vital as the urgent need to address environmental issues like water scarcity and climate change becomes more widely recognized. The amount of water used throughout the construction, operation, and cooling phases of a data centre is referred to as its water footprint. The amount of heat generated in the data centres very high and around 1 megawatt (MW) data centre can consume 18,000 gallons (68 m3) of water a day for cooling. Figure 2 shows the flow of energy in data centre. Particularly cooling, which needs a lot of water for heat dispersion, contributes significantly to the water use in data centres. Concern over the environmental impact and water footprint of these cutting-edge computing systems has increased with the rise of AI and its increasing incorporation into data centres.

Figure 1: Typical data centre infrastructure
Source: Digital Realty

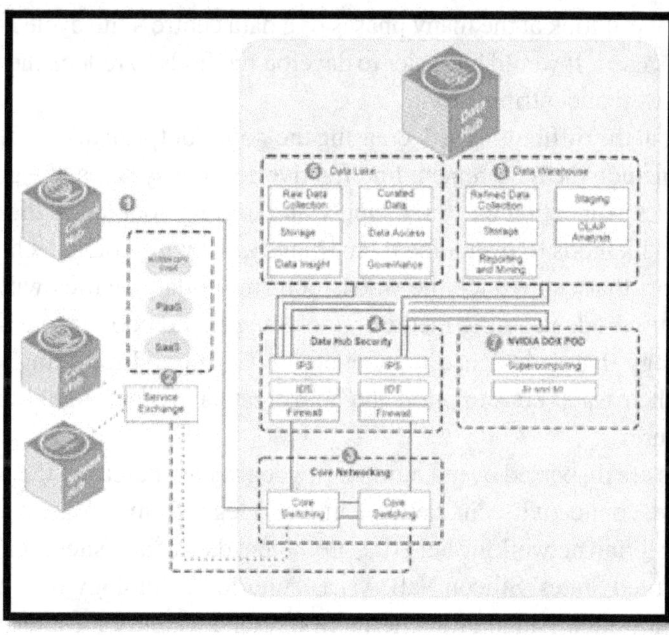

Figure 2: The power flow and heat flow in a data center
Source: Shao, X., et al. (2022)

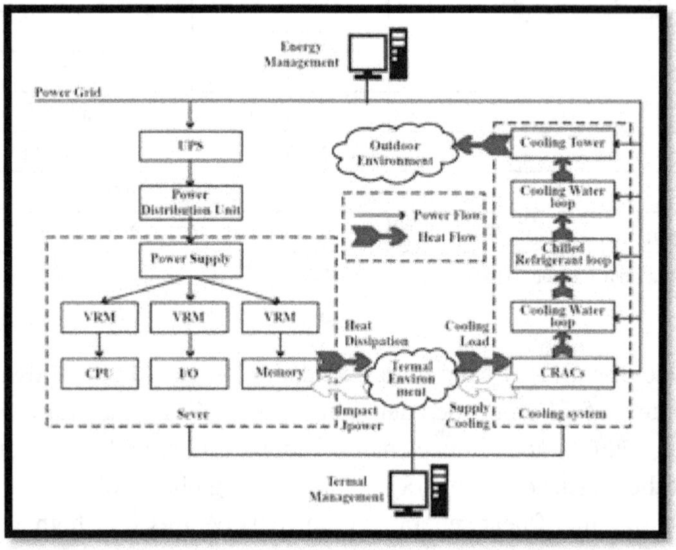

The purpose of this study is to emphasize the significance of reducing the water footprint in AI-integrated data centres and fostering sustainable practices in this field. We may endeavour to lessen the negative effects of data centres on the environment and promote a more sustainable future by being aware of the issues and considering potential solutions. The importance of data centres in the context of the digital revolution and the expansion of AI applications will be discussed first. As a result, a thorough overview of the size and expansion of these centres will be provided, emphasizing the urgent need to address their environmental impact. The idea of the water footprint and its application to data centres will be discussed next. We'll look at the many phases of a data centre's life cycle and pinpoint the major locations where water is used. It would be easier to develop methods to reduce the water footprint if you are aware of the elements that contribute to it.

We will also talk about the difficulties in decreasing the water footprint in data centres with embedded AI. These difficulties include the requirement for effective cooling systems, the growing power density of AI gear, and the effect of location on water supply and use. By looking at these challenges, we can find viable answers and methods to overcome them. Finally, we will look at technology improvements and sustainable practices that can reduce the water footprint in data centres with embedded AI. This section will explore cutting-edge cooling methods, including liquid cooling and direct chip-level cooling, as well as technology that is designed to be energy-efficient and recycling systems. Data centres may drastically lower their water consumption and help create a greener digital infrastructure by using these sustainable measures.

Popular data centres are dispersed over a number of geographical areas in the world because of variables like infrastructure, connectivity, market demand, and legal frameworks. Known as "Data Center Alley" because of its scale and networking benefits, prominent data centre sites include Northern Virginia (United States). The United States' Silicon Valley is a center for technology with a strong infrastructure ecosystem. London (the United Kingdom) benefits from being a centre for international finance and a digital entry point. Frankfurt (Germany), thanks to its strategic position and robust connectivity, serves

Figure 3: Estimated water footprints for training AI in the data centre

Source: P. Li et al. (2023) and G. A. Brady et al. (2013)

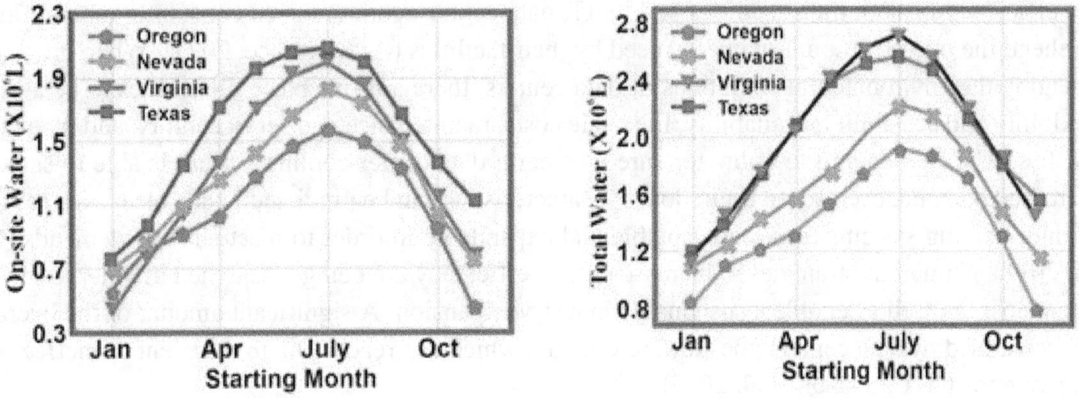

as a significant data centre hub in Europe. Amsterdam (Netherlands) provides connections and network infrastructure with low latency. While Tokyo (Japan) is a notable data centre location with cutting-edge technology infrastructure, Singapore is a significant data centre hub in the Asia-Pacific area. Sydney (Australia) is renowned for its consistent electricity supply and access to international markets. In South America, Sao Paulo (Brazil) has grown to be a key data centre hub. Although popularity may fluctuate over time as new markets and technological advancements take place, certain areas are popular due to reasons like connectivity, infrastructure, market demand, and the regulatory environment. The estimated water footprints for training the AI in the data center is as shown in figure 3 below.

LITERATURE REVIEW

The informational backbone of our increasingly digital world is represented by data centers. Data center infrastructures need to be expanded to a big scale with a larger power capacity, a decrease in energy consumption, and a reduction in carbon emissions due to the internet's quick development. Their services are in high demand, and data-intensive technologies like artificial intelligence, smart and linked energy systems, distributed manufacturing systems, and autonomous vehicles have the potential to drive up demand even higher. These trends have clear consequences for the world's energy demand and need to be thoroughly examined because data centers are energy-intensive businesses that utilize an estimated 1% of the world's electricity (C. Miller, 2011). In a data center, the cooling system uses 38% of the electricity, the IT equipment uses 52%, and the remaining equipment (UPS, electrical power distribution...) uses 10%. For the safety of all the electronic equipment and to maintain ideal air temperature and relative humidity, the cooling system is necessary (C. Nadjahi et al, 2018). Numerous frequently cited but oversimplified analyses assert that the energy consumed by data centers worldwide has doubled over the past ten years and will triple or even quadruple over the following ten years (L. Belkhir, A. Elmeligi, 2018; A.S.G. Andrae, T. Edler, 2015; T. Bawdy, 2016). Such estimations support the belief that as data center services are rapidly in demand, their worldwide energy use must follow suit (E. Masanet et al, 2020). Its ability to lower energy use, capital costs, and carbon emissions while maintaining a secure and dependable data center operation is crucial (Shao, X., et al. 2022). In order to increase the energy efficiency of both their facilities and its IT, data center operators need constantly monitor and comprehend the overall amount of energy used by their facilities (A. Lawrence, 2022). When it comes to managing the environmental conditions of data centers, there are five basic difficulties to be aware of: flexibility and scalability, availability, life cycle costs, maintenance and serviceability, and manageability. It can be challenging to know for sure whether a data center cooling system is able to satisfy the increased demand, even when future loads' characteristics can be predicted in advance. As a result, data center cooling systems must be adaptable and expandable in order to meet shifting demands (C. Miller, 2011). Numerous strategies to increase energy efficiency are being researched in order to lower operating costs as well as enable a sustainable industry expansion. A significant amount of the overall energy consumed by data centers goes toward cooling, which can reach 40% in the event of ineffective cooling systems (A. Capozzoli et al, 2015)

Water is typically used in data centers to help with heat rejection (cooling IT equipment). A water-cooled chilled water system, which delivers cool water to computer room cooling units, is widely used in data centers. IT equipment receives cool, conditioned air that is blown through a chilled water coil by a fan. The water then returns to the chiller where it is chilled again (K. Heslin, 2016).

The total water footprint of a person, a process, a product, a company, a river basin, or a country may all be calculated. They offer insightful information that may be used by firms to assess their water-related business risk, by governments to recognize the importance of water to their economies and their dependence on it, and by consumers to determine how much water is concealed in the goods they use. Most significantly, they support the strategic action needed for equitable, cost-effective, and sustainable water use (Water Footprint Network, 2023).

Public scrutiny has been focused on the growing carbon footprint of artificial intelligence (AI) models, particularly big ones like GPT-3 and GPT-4. But sadly, the vast and equally significant water impact of AI models has gone unnoticed. This is very serious since, in the wake of the fast expanding population, dwindling water supplies, and aging water infrastructures, freshwater shortage has emerged as one of the most urgent concerns that affect us all (P. Li et al, 2023). The low quality of the heat makes it challenging to recover and use this stream from waste heat (K. Ebrahimi et al, 2013). Free cooling technology can significantly reduce the amount of energy needed to cool data centers, but it is highly dependent on regional and climatic factors (F. Zhou et al, 2017).

CONTROVERSIES

Google and Water Footprint

Google has taken steps to minimize its water footprint in data center operations, but it has come under fire and stirred up debate about its water consumption in certain areas. The annual water usage for Google data center is tabulated in the table 2. The revelation that Google has obtained permission to pump groundwater for its data center in South Carolina, USA, caused a noteworthy controversy in 2019 (Google Data Centers, 2023, N. Walsh, 2023).

Table 2: Annual Water Usage for Google data centers (2021)

Data Center	Withdrawals	Discharge	Consumption
Berkeley county, South Carolina	541.3m gallons	65.9m gallons	475.3m gallons
Council Bluffs, Iowa	1,111m gallons	265.2m gallons	845.8m gallons
The Dalles, Oregon	358.3m gallons	83.8m gallons	274.5m gallons
Douglas County, Georgia	421.8m gallons	61.7m gallons	360.1m gallons
Henderson, Nevada	68.9m gallons	28.4m gallons	40.5m gallons
Jackson County, Alabama	92.3m gallons	22.1m gallons	70.2m gallons
Leesburg, Virginia	59.6m gallons	15.0m gallons	44.7m gallons
Lenoir, North Carolina	352.2m gallons	21.5m gallons	330.7m gallons
Mayes County, Oklahoma	863.1m gallons	192.0m gallons	671.2m gallons
Midlothian, Texas	51.3m gallons	11.4m gallons	39.9m gallons
Montgomery County, Tennessee	166.5m gallons	27.3m gallons	139.2m gallons
New Albany, Ohio	27.0m gallons	11.1m gallons	15.9m gallons
Papillion, Nebraska	34.2m gallons	16.7m gallons	17.5m gallons
Sterling, Virginia	51.3m gallons	9.3m gallons	41.9m gallons
Storey County, Nevada	0.6m gallons	0.6m gallons	0.1m gallons
Subtotal – U.S. Locations	***4.20bn gallons***	***832.0m gallons***	***3.37bn gallons***
Additional Global Locations	1.33bn gallons	361.2m gallons	971.0m gallons
Total Locations	**5.53bn gallons**	**1.19bn gallons**	**4.34bn gallons**

Source: Data Center Water Usage: Billions of Gallons Every Year - Dgtl Infra, M. Zhang, 2022

Concerns regarding the possible effect on the region's water supplies, particularly during droughts, were expressed by local residents and environmental organizations. The dispute brought to light the conflict between the local residents' demand for water and the data centers' need for water. In another case, criticism arose over Google's data center project in Quilicura, Chile, because of the facility's heavy water use in an area with a water shortage (A. Mollen, 2023). The use of water by the data center, according to critics, might worsen local water shortages and have a severe influence on community water supply. The figure 4 below gives the water use by Google and Microsoft cloud service providers. Increased scrutiny of Google's water management procedures is a result of these problems. Critics contend that internet corporations should be more accountable for their water use, especially in areas with limited water resources, and put the demands of local populations ahead of those of their data centers.

Google has taken action to solve water-related problems in response to these worries. The business has made a commitment to using sustainable water management techniques and has taken steps to reduce its water use. In order to assist efforts to save water, this involves making investments in water-saving technology, looking into alternate cooling techniques, and forming collaborations with the local community. Google has also promised to restore more water than it uses in certain areas. For instance, Google has pledged to refill 120% of the water consumed in its data center operations in places where water shortage is an issue (M. Zhang, 2022, Efficiency-Data Centers, 2023).

Despite these attempts, the water disputes around Google's data centers underscore the more general difficulty of striking a balance between the water requirements of large-scale technological infrastructure

Figure 4: Water use by Google and Microsoft's cloud service providers in billions of litres (2016-2019)
Source: Mytton, D. (2021)

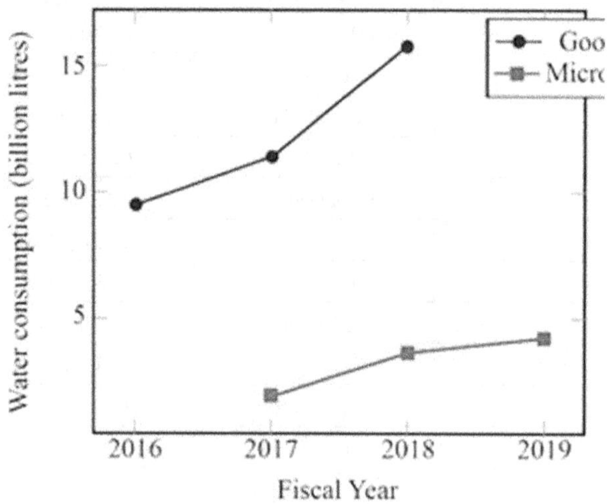

and the demands of local populations and the environment. It emphasizes the need of constant communication, teamwork, and open water management procedures to overcome these issues.

Water for Cooling Data Centers

Google has taken steps to minimize its water footprint in data center operations, but it has come under fire and stirred up debate about its water consumption in certain areas. The revelation that Google has obtained permission to pump groundwater for its data center in South Carolina, USA, caused a noteworthy controversy in 2019. Concerns regarding the possible effect on the region's water supplies, particularly during droughts, were expressed by local residents and environmental organizations (S. McCammon, 2023). The dispute brought to light the conflict between the local residents' demand for water and the data centers' need for water.

In another case, criticism arose over Google's data center project in Quilicura, Chile, because of the facility's heavy water use in an area with a water shortage. The use of water by the data center, according to critics, might worsen local water shortages and have a severe influence on community water supply. Increased scrutiny of Google's water management procedures is a result of these problems. Critics contend that internet corporations should be more accountable for their water use, especially in areas with limited water resources, and put the demands of local populations ahead of those of their data centers. Google has taken action to solve water-related problems in response to these worries. The business has made a commitment to using sustainable water management techniques and has taken steps to reduce its water use. In order to assist efforts to save water, this involves making investments in water-saving technology, looking into alternate cooling techniques, and forming collaborations with the local community.

Google has also promised to restore more water than it uses in certain areas. Despite these attempts, the water disputes around Google's data centers underscore the more general difficulty of striking a balance between the water requirements of large-scale technological infrastructure and the demands of local populations and the environment. It emphasizes the need of constant communication, teamwork, and open water management procedures to overcome these issues.

SUSTAINABLE STRATEGIES

Data centers can efficiently lower their water footprint, encourage sustainability, and contribute to a greener, more environmentally conscious digital infrastructure by putting the following measures into practice.

Implement Efficient Cooling Systems

- To reduce water usage, use modern cooling techniques including liquid cooling and direct chip-level cooling.
- To increase cooling effectiveness and lessen the need for excessive cooling, optimize airflow management and equipment architecture.
- Examine the utilization of air-saving strategies to lessen dependency on water-intensive cooling processes by utilizing ambient air for cooling.

Support water reuse and recycling

- Put in place wastewater treatment and reuse technologies for the data center facility. (D. Azevedo et al, 2011)
- Look at ways to recycle water for uses other than drinking, such as toilet flushing, landscaping, or cleaning equipment as shown in figure 5.
- Work together with local water authorities to find potential uses for recycled or reclaimed water.

Figure 5: Data center water footprint: on-site water use for cooling the data center and off-site water use for generating power
Source: P. Li et al. (2023)

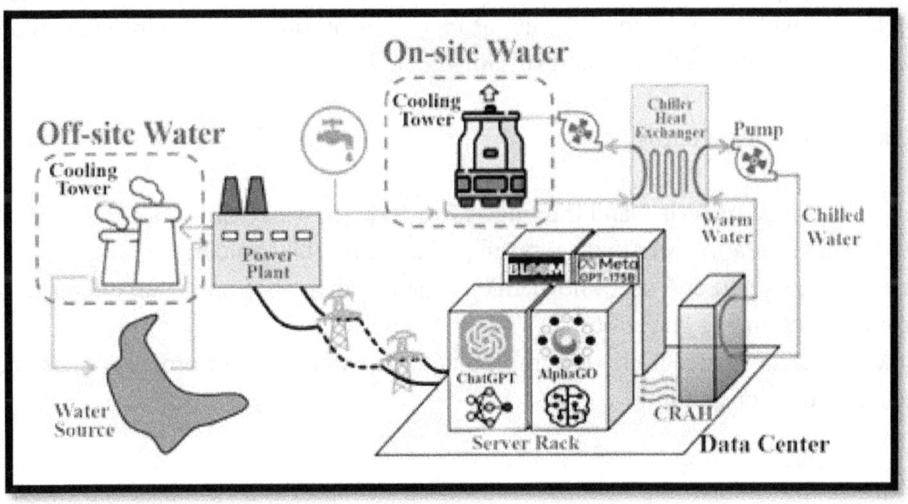

Improve Hardware and electricity Management

- Use water-efficient hardware to generate less electricity overall by using energy-efficient hardware designs and components.
- Put intelligent power management technologies in place to ensure effective computing resource use and reduce energy waste (International Organization for Standardization, 2016).
- To lower computational needs and associated cooling requirements, take into account the usage of low-power processors and energy-efficient AI algorithms.

Give Renewable Energy Sources Priority

- Invest in renewable energy initiatives to power data centers, such solar or wind power.
- Form partnerships with producers of renewable energy or build on-site facilities for their production.
- Look into renewable energy certificates (RECs) or power purchase agreements (PPAs) to aid in the switch to clean energy sources.

Improve Data Center Location Selection

- When choosing a location for a data center, take the local climate and water availability into account.
- Choose locations with easy access to water resources or those where the environment permits effective cooling without using a lot of water.
- Consider the possible effects of data centers on regional water supplies and use sustainable water management techniques.

Encourage Industry Collaboration and Knowledge Sharing

- Participate in industry-wide partnerships to exchange best practices, scientific discoveries, and technology breakthroughs relevant to data centers' water footprint reduction.
- Attend industry gatherings, workshops, and forums devoted to running sustainable data centers.
- Form alliances with government agencies, academic institutions, and environmental organizations to advance coordinated action toward sustainability.

Educate and Raise Awareness

- Create instructional materials and training sessions on the value of sustainability and water footprint reduction for data center managers, employees, and stakeholders.
- Increase consumer and end-user knowledge of the negative environmental effects of data centers and promote sustainable behaviour.
- Implement internal sustainability programs and transparently report progress to promote an environmentally conscious culture.

Track Progress

- Track and measure water use, energy use, and overall water footprint on a regular basis to judge how well adopted methods are working.
- Establish goals and key performance indicators (KPIs) to monitor progress and guarantee ongoing improvement.
- Consistently assess and update sustainability plans in light of cutting-edge technological advancements, market norms, and changing environmental issues.

BEST CASE STUDIES

The water footprint of data centers has been significantly reduced and controlled by a number of businesses. Here are some noteworthy case studies and illustrations:

- **Google:** When it comes to data center sustainability, Google has been at the forefront. They have made large expenditures in water conservation and renewable energy. They have a cutting-edge cooling system that employs recycled wastewater for their cooling towers in one of their data centers in Douglas County, Georgia (Google Data Centers, 2023). Google is able to lessen its dependency on freshwater sources and conserve water by working with local water agencies.
- **Microsoft:** To reduce the water footprint of its data centers, Microsoft has put in place a number of measures. They have installed cutting-edge evaporative cooling technology in their Quincy, Washington, data center, which drastically lowers water usage. This technology reduces the requirement for mechanical cooling by using outside air for cooling, eliminating the need for water-intensive cooling techniques. [M. Zhang, 2022, N. Walsh, 2022, Microsoft, 2021]
- **Facebook:** When operating its data centers, Facebook has given special attention to sustainable procedures. Their plant in Lule, Sweden, uses the area's chilly weather to install an extremely effective cooling system. They have reduced the amount of water necessary for cooling by using outdoor air. In order to further lessen its environmental impact, Facebook is committed to funding renewable energy initiatives to power its data centers.
- **Switch:** Switch, a provider of data centers, has taken significant steps to manage water usage in its facilities. In order to reduce the amount of water needed for cooling, their data centers in Las Vegas, Nevada, combine indirect evaporative cooling and air economization methods. These systems make use of the dry climate and low humidity to reduce water use and maximize energy efficiency.
- **NTT Global Data Centers:** NTT Global Data Centers (formerly RagingWire) has put in place a number of steps to lower water usage in their data centers. Their Ashburn, Virginia, facility uses a closed-loop cooling system to cool the building while also purifying the cooling water to use less water overall. NTT Global Data Centers has successfully decreased its water footprint by enhancing the cooling infrastructure and utilizing cutting-edge cooling technologies.

These case studies show how companies may embrace cutting-edge cooling technologies, work with local water authorities, and make a commitment to renewable energy sources to successfully manage and reduce the water footprint of data centers. These businesses serve as examples of the viability and value of sustainable data center operations in reducing environmental impact.

CONCLUSION

In conclusion, this article underlines the significance of encouraging sustainability and reducing the water footprint in data centers with integrated AI. It draws attention to the rapidly expanding data center industry and the rising demand for artificial intelligence, both of which have increased water use. The study looks at case studies of companies including Google, Microsoft, Facebook, Switch, and NTT Global Data Centers that have successfully reduced their water impact. The study describes the process used to examine this problem, which includes completing a review of the literature, gathering data on water use, assessing cooling systems, looking into sustainable technology, and coming up with water-saving plans. The article places a strong emphasis on the necessity of ongoing field improvement and stakeholder participation.

In the concluding remarks, it is emphasized how important it is to foster sustainability in AI-integrated data centers in order to create a digital infrastructure that is more ecologically friendly. It recognizes the difficulties, including rising power densities, the need for effective cooling, and the scarcity of water. However, it emphasizes the opportunity for major advancement through cutting-edge technologies, teamwork, and knowledge exchange. The data centers can support water conservation, reduce climate change, and build more resilient and effective digital infrastructure by reducing their water footprint and adopting sustainable practices. It underlines the social and economic advantages of such projects and urges stakeholders to drive sustainability efforts. Overall, the study emphasized how critical it is to manage the water footprint in AI-integrated data centers and offers suggestions and tactics to advance sustainability in this area.

REFERENCES

Andrae, A. & Edler, T. (2015). On Global Electricity Usage of Communication Technology: Trends to 2030. *Challenges 6*(1), 117-157. . doi:10.3390/challe6010117

Artificial Intelligence (AI) Bots Market Analysis – 2030. (2023). NextMSC. https://www.nextmsc.com/report/artificial-intelligence-ai-bots-market

Azevedo, S. C. (2011). *Water usage effectiveness (WUETM): A green grid datacenter sustainability metric*. Green Grid.

Bawdy, T. (2016). Global warming: Data centres to consume three times as much energy in next decade, experts warn. *The Independent*.

Belkhir, L., & Elmeligi, A. (2018). Assessing ICT global emissions footprint: Trends to 2040 & recommendations. *Journal of Cleaner Production*, *177*, 448–463. doi:10.1016/j.jclepro.2017.12.239

Brady, G. A., Kapur, N., Summers, J. L., & Thompson, H. M. (2013, December). A case study and critical assessment in calculating power usage effectiveness for a data centre. *Energy Conversion and Management, 76*, 155–161. doi:10.1016/j.enconman.2013.07.035

Google. (n.d.). *Douglas County, Georgia – Data Centers*. Google. https://www.google.com/about/datacenters/locations/douglas-county/

Ebrahimi, K., Jones, G. F., & Fleischer, A. S. (2014, March). A review of data center cooling technology, operating conditions and the corresponding low-grade waste heat recovery opportunities. *Renewable & Sustainable Energy Reviews, 31*, 622–638. doi:10.1016/j.rser.2013.12.007

Google. (n.d.). *Efficiency – Data Centers*. Google. https://www.google.com/about/datacenters/efficiency/

Frizziero, M. (2016). Rethinking Chilled Water Temperatures Can Bring Big Savings in Data Center Cooling. *Schneider Electric Blog*. https://blog.se.com/datacenter/2016/08/17/water-temperatures-data-center-cooling/ .

Google. (n.d.). *Google Environmental Report*. Google Inc. https://www.gstatic.com/gumdrop/sustainability/google-2022-environmental-report.pdf

Heslin, K. (2016). Ignore Data Center Water Consumption at Your Own Peril. *Uptime Institute Blog*. https://journal.uptimeinstitute.com/dont-ignore-water-consumption/

IEA. (2022). *Data Centres and Data Transmission Networks – Analysis*. IEA. https://www.iea.org/reports/data-centres-and-data-transmission-networks.

International Organization for Standardization. (2016). *ISO/IEC 30134–2: 2016 Information Technology—Data Centres—Key Performance Indicators—Part 2: Power Usage Effectiveness (PUE)*. International Organization for Standardization: Geneva Switzerland. https://www.iso.org/standard/63451.html

Lawrence. (2019). Is PUE actually going UP? *Uptime Institute Blog*. https://journal.uptimeinstitute.com/is-pue-actually-going-up/

Li, P., Yang, J., Islam, M. A., & Ren, S. (2023). *Making AI Less 'Thirsty': Uncovering and Addressing the Secret Water Footprint of AI Models.* arxiv. doi:10.48550/ARXIV.2304.03271

Masanet, A., Shehabi, A., Lei, N., Smith, S., & Koomey, J. (2020, February). Shehabi, N. Lei, S. Smith, and J. Koomey, "Recalibrating global data center energy-use estimates,". *Science, 367*(6481), 984–986. doi:10.1126/science.aba3758 PMID:32108103

McCammon, S. (2017). *Google Moves In And Wants To Pump 1.5 Million Gallons Of Water Per Day.* NPR.https://www.npr.org/2017/05/08/527214026/google-moves-in-and-wants-to-pump-1-5-million-gallons-of-water-per-day

Meta. (2022). *2021 Sustainability Report.* Meta Inc. https://sustainability.fb.com/wp-content/uploads/2022/06/Meta-2021-Sustainability-Report.pdf

Microsoft. (n.d.). *Innovative cooling solutions reduce datacenter water consumption*. Microsoft. https://local.microsoft.com/wp-content/uploads/2021/11/Datacenter-water-consumption-fact-sheet.pdf

Mollen, A. (2023). Sustain-AI and the Challenge of Sustainability, a report on The Sustainability Index for Artificial Intelligence. *Algorithm Watch.* https://algorithmwatch.org/en/wp-content/uploads/2023/03/SustAIn-magazine-issue-2.pdf

Mytton, D. (2021). Data centre water consumption. *npj Clean Water, 4*(1), 11.

Nadjahi, H. L., & Lemasson, S. (2018, September). A review of thermal management and innovative cooling strategies for data center. *Sustainable Computing : Informatics and Systems, 19*, 14–28. doi:10.1016/j.suscom.2018.05.002

Ounifi, H.-A., Gherbi, A., & Kara, N. (2022, August). Deep machine learning-based power usage effectiveness prediction for sustainable cloud infrastructures. *Sustainable Energy Technologies and Assessments, 52*, 101967. doi:10.1016/j.seta.2022.101967

Shao, X., Zhang, Z., Song, P., Feng, Y., & Wang, X. (2022). A review of energy efficiency evaluation metrics for data centers. *Energy and Building, 271*, 112308. doi:10.1016/j.enbuild.2022.112308

Shehabi, Smith, S., Sartor, D., Brown, R., Herrlin, M., Koomey, J., Masanet, E., Horner, N., Azevedo, I., & Lintner, W. (2016, June). United States Data Center Energy Usage Report. *LBNL, 1005775*, 1372902. doi:10.2172/1372902

Walsh, N. (2022). Sharing the latest improvements to efficiency in Microsoft's datacenters. *Azure Blog*. https://azure.microsoft.com/en-us/blog/sharing-the-latest-improvements-to-efficiency-in-microsoft-s-datacenters/

Water Footprint. (n.d.). *Water Footprint Network*. Water Footprint. https://www.waterfootprint.org/water-footprint-2/

Zhang, M. (2022). *Data Center Water Usage: Billions of Gallons Every Year*. Dgtl Infra. https://dgtlinfra.com/data-center-water-usage/

Zhou, C., Shen, C., Ma, G., & Yan, X. (2022, August). Power usage effectiveness analysis of a liquid-pump-driven hybrid cooling system for data centers in subclimate zones. *Sustainable Energy Technologies and Assessments, 52*, 102277. doi:10.1016/j.seta.2022.102277

Chapter 11
Factors Impacting Adoption of Climate Change Mitigation Strategies in the Indian Iron and Steel Industry

Yogesh Patil
Symbiosis International University, India

Puja Singh
Symbiosis International University, India

ABSTRACT

This chapter provides an insight on the factors critical to the adoption of climate change mitigation strategies (CCMS) in the Indian iron and steel industry (IISI). The authors identified key factors and the relationship between them to develop a systems approach in prioritizing climate change mitigation strategies in energy intensive (coal dependent) iron and steel plants. The analysis took into account the benefits of a multi-criteria decision making tool, TISM framework, in particular to model the value and extent of cross-sectoral connections, balance conflicting needs, and identify solutions to achieve financial stability in order to test the parameters. To comprehend how institutional and policy processes operate, what shapes they take, and how they regulate the adoption of CCMS in the iron and steel industries of India have been discussed in detail. This, in turn, will assist corporations in getting over roadblocks and receiving encouragement to adopt CCMS.

INTRODUCTION

The process of India's shift towards a diverse and environmentally sustainable energy system is now in progress, facilitated by the implementation of supportive governmental policies and good economic conditions (Dey *et al.*, 2022). The intricate and pivotal matter of the interplay between quality of life, climate change, sustainable adaptation, and social-ecological resilience necessitates careful examination.

DOI: 10.4018/978-1-6684-9863-7.ch011

Climate change has extensive ramifications on ecosystems, cultures, and economies, all of which possess the potential to substantially influence individuals' well-being. Provided is a comprehensive review of the interrelated concepts at hand.

Climate change is responsible for an increased occurrence and heightened intensity of extreme weather phenomena, including hurricanes, heatwaves, floods, and droughts. These occurrences have the potential to lead to fatalities, destruction of assets, and disturbances to routine activities (AghaKouchak *et al.*, 2020). Elevated temperatures and shifting climatic patterns can have adverse effects on human health, which includes an increased susceptibility to heat-related illnesses, the proliferation of vector-borne diseases, and the emergence of mental health concerns stemming from climate-induced stress (Fong and Fong, 2020). It has the potential to disrupt agricultural systems, resulting in crop failures and subsequent scarcity of food resources. This, in turn, can have adverse effects on both nutrition and the accessibility of food (Shahzad *et al.*, 2021). Water resources are subject to the impact of changing precipitation patterns, which can result in the shortage of water and have implications for water quality (Bhatta *et al.*, 2021). These changes have the potential to give rise to conflicts over the allocation and utilization of water resources (Xiang *et al.*, 2021). Disasters and disruptions brought on by climate change can have a negative impact on livelihoods, leading to job losses, and exacerbate economic inequality.

The primary objective of sustainable adaptation is to facilitate the ability of communities and ecosystems to effectively respond to evolving circumstances, all the while upholding or enhancing the general standard of living. Sustainable adaptation encompasses a range of methods and strategies aimed at mitigating the most severe effects of climate change, while concurrently fostering economic, social, and environmental sustainability. Illustrations of sustainable adaptation methods encompass the construction of robust infrastructure, the promotion of sustainable agricultural and forestry practices, and the allocation of resources towards renewable energy sources in order to mitigate greenhouse gas emissions. The correlation between climate change and the overall well-being of individuals is closely intertwined (Schneider *et al.*, 2021; Lawrance *et al.*, 2022). Effectively tackling this matter necessitates a comprehensive strategy that encompasses two key aspects: mitigating climate change through the adoption of sustainable practices, and facilitating the adaptation of people and ecosystems to the consequences of climate change. The preservation of the welfare of current and future generations in the context of climate change is heavily reliant on social-ecological resilience (Fedele *et al.*, 2019; Turner *et al.*, 2022). Communities that exhibit both social and ecological resilience possess enhanced capabilities to effectively address climate-related concerns, safeguard their overall well-being, and maintain the sustainability of their livelihoods.

Steel, a versatile metal used across a diverse array of sectors, spanning infrastructure, transportation, packaging, and renewable energy systems, thereby representing a key component of sustainable solutions (Worldsteel, 2018). Steel possesses the capacity to serve as an environmentally-friendly substance when manufactured and utilized in a conscientious manner, and it remains a prominent contributor to contemporary sustainable progress. It is imperative to acknowledge that the environmental ramifications of steel can fluctuate contingent upon variables such as the manufacturing procedure, energy origin, techniques of transportation, and disposal at the finish of its lifecycle. Hence, in order to optimize the environmental sustainability of steel, it is imperative to embrace sustainable approaches throughout its whole life cycle, encompassing responsible procurement, recycling, and repurposing. The backbone of industrialisation in developing nations will unavoidably continue to be steel, which is also the foundation of contemporary civilization (Shatokha, 2018). Global demand for steel has been rising rapidly in recent years due to population and GDP development and is predicted to keep expanding, particularly due to the economies of Africa, India, and the ASEAN nations (IEA Report, 2021). Steel production on a large scale is vital and closely related to a nation's development but also causes environmental damage results from mass production. Therefore, a comprehensive plan is indispensable for reducing the carbon footprints of this vital yet energy intensive industry. One of the most recyclable materials on earth is steel. The material possesses the ability to undergo several recycling processes without experiencing any degradation in its inherent characteristics. The practice of recycling steel serves to mitigate the necessity of engaging in fresh mining activities, thereby contributing to the conservation of valuable natural resources and diminishing the energy demands associated with the production process.

Under the base case scenario, there is a projected upward trend in the utilization of fossil fuels, with an estimated consumption of 36,543 petajoules (PJ) by the year 2050. This represents a notable increase from the 30,448 PJ consumed in 2021. The utilization of coal is projected to reach its highest point in the year 2037, with a total of 1,098 million metric tons (Mt) of coal being consumed. Subsequently, this figure is anticipated to decline to 970Mt by the year 2050, representing a 20% increase compared to the levels observed in 2021. The steel industry, which is the greatest contributor to India's industrial emissions, is projected to experience a nearly threefold increase in emissions from 351MtCO2 in 2021 to 948MtCO2 by 2050. The utilization of coal in the steel production industry exhibits a comparable upward trend, attaining a volume of 399 million metric tons (Mt) by the middle of the twenty-first century. Although the steel industry in India faces several obstacles, including the absence of domestically sourced metallurgical grade coal, the presence of lower-quality ore, the need for effective solid waste and water waste management, and the issue of water scarcity (Shanmugam *et al.*, 2021). Technological developments have led to increased energy efficiency in steel production. Integrated mills employ waste heat recovery systems, which enable the utilization of residual heat. Additionally, electric arc furnaces have the potential to be fuelled by environmentally friendly energy sources, thereby mitigating the overall carbon emissions (Kieush *et al.*, 2023). There is a growing trend among steel manufacturers and users to embrace a circular economy framework, wherein the focus is on developing goods and implementing processes that aim to limit waste generation while maximizing opportunities for reuse and recycling. The incorporation of technological advancements is vital for India's ongoing economic expansion, environmental viability, and international competitiveness. Collaborative efforts between governmental entities, industrial sectors, and research institutions are being undertaken to foster the advancement and implementation of these technologies in several domains (Yang and Gu, 2021). The integration of renewable energy sources into the electrical grid is being facilitated through the implementation of advanced grid management and storage technology (Oyekale *et al.*, 2020). The presence of a proficient labour force capable of efficiently utilizing and coordinating cutting-edge technology poses a significant obstacle in numerous nations, India being one of them. India possesses a substantial and heterogeneous workforce; however, the integration of modern technologies frequently necessitates specialized competencies and instruction that may not be readily accessible (Singh, 2019; Mallett and Pal, 2022).

India holds substantial reserves of iron ore and coal; however, these resources are frequently characterized by inferior quality, necessitating the use of beneficiation or cleaning techniques to fulfil the requisite criteria for steel manufacturing. India possesses substantial reserves of iron ore in areas such as Odisha, Chhattisgarh, and Jharkhand. Coal reserves are distributed throughout multiple regions, including as Jharkhand, West Bengal, and Chhattisgarh. The geographic location of steel plants and mines in India is influenced by a confluence of historical circumstances, the accessibility of raw materials, the state of infrastructure, market requirements, and governmental regulations (Mallett and Pal, 2022). Certain businesses have intentionally positioned themselves in locations that offer advantageous closeness to sources of raw materials. In contrast, other businesses have chosen to establish their operations in places that provide convenient access to transportation networks and markets. The transfer of raw materials, particularly coal and iron ore, from mining regions to steel mills poses significant challenges due to insufficient transportation infrastructure, including railways and highways.

It is imperative to acknowledge that the steel sector in India is currently undergoing continuous change. Certain steel factories are undertaking initiatives to diversify their sources of raw materials, investigate the utilization of alternative resources, and enhance energy efficiency in order to mitigate their dependence on specific mining locations (Domaracka *et al.*, 2022). The volatility of imported coking coal and iron ore prices can have a significant impact on both the price structure and revenue of the steel industry. Moreover, the consideration of environmental and sustainability factors is exerting an impact on the determination of suitable locations, as there is an increasing focus on the adoption of cleaner technologies and the implementation of responsible mining methods. In the Paris Climate Change Agreement, 2015 India pledged to reduce its emission intensity as 33 -35% of its gross domestic product from prior pledge of 20 to 25% reduction between 2005 and 2020. There are two key articulations regarding this commitment NAPCC (National Action Plan on Climate Change) for domestic focus and INDC (Intended Nationally Determined Commitments) for external affairs. India has pledged to follow the procedure of rules and standards mentioned in the programmes in order to promote sustainability, accountability, transparency and climate change mitigation (Deshpande *et al.*, 2023). The primary objective of these regulatory frameworks and regulations is to establish a favourable setting for the IISI to address its climate change impact through the implementation of energy-efficient measures, emission reduction strategies, and the promotion of environmental sustainability. Adherence to these regulations is crucial for enterprises operating within the industry, as there exist both incentives and punishments to ensure compliance with these measures aimed at mitigating climate change. NAPCC delineates India's comprehensive approach to tackling climate change, encompassing several initiatives such as the National Mission on Enhanced Energy Efficiency (NMEEE). NMEEE aims to facilitate the implementation of energy efficiency measures within energy-intensive sectors, such as the iron and steel industry. It includes the Perform, Achieve, and Trade (PAT) plan, which imposes energy efficiency targets on selected customers, such as steel mills. Excess energy savings can be exchanged through the utilization of Energy Savings Certificates (ESCerts). The issuance of guidance and standards pertaining to energy efficiency and sustainable practices in IISI is carried out by the Bureau of Energy Efficiency (BEE) in conjunction with the Ministry of Steel. The iron and steel industry is required to comply with air pollution control rules and emission guidelines for particulate matter, by the State and Central Pollution Control Board (Ministry of Steel, 2022). In brief, these strategies exert both direct and indirect influence on the climate change mitigation techniques adopted by the IISI. These policies are closely linked to the industry's endeavours to cut emissions, enhance energy efficiency, and make a contribution to India's domestic and global obligations in addressing climate change. The iron and steel sector in India is expected to undergo further development and adapt its strategy in accordance with global climate objectives, taking into account the outcomes of COP 27 meeting.

Considering the preceding circumstances, the aim is to analyse and establish the interconnections and significance of the factors to adopt a comprehensive systems approach. Thereby paving ways for prioritizing measures aimed at mitigating deleterious impacts of climate change from energy-intensive iron and steel industries. Part of this chapter was published by the authors in *Management of Environmental Quality: An International Journal* (Singh *et al.*, 2023).

REVIEW OF LITERATURE

As a result of the adverse consequences of carbon dioxide (CO2) emissions on the natural environment, there is a growing demand from regulatory bodies, customers, and society as a whole to mitigate and minimize the overall levels of CO2 emissions. The institutions that are most frequently called upon to respond are companies in the energy and manufacturing sectors operating with a high carbon footprint. The heightened awareness and apprehension over the environment and the phenomenon of climate change have prompted increased engagement in environmental matters, particularly in relation to climate change, due to mounting institutional pressures and more stringent regulatory measures.(Kumari and Patil, 2018).

A 'climate change mitigation strategy' is defined as a pattern of a firm's action to reduce its CO2 emissions (Cadez and Czerny, 2016). In the realm of business, a strategy commonly pertains to the establishment of enduring corporate objectives, the formulation of initiatives, and the allocation of resources necessary for the organization to attain these objectives (Snow and Hambrick, 1980). These strategic actions illustrating the commitment of an institution can be transition of fossil fuel to other energy resources, cogenerating power and electricity from the same energy source, material substitution, upgrading via recycling, utilisation of carbon capture storage, down streaming supply chain to upstream carbon trading, decarbonization (Büchs *et al.*, 2011; Rissman *et al.*, 2020; Fan and Friedmann, 2021; An et *al.*, 2021). The fundamental motives behind adoption of any strategy passes through the behavioural dimensions of the decision making body of the institution. Any factor favourably affecting a firm's aspirations for innovation and paving a path for adoption of innovative practices acts as a driving force (Sonsale *et al.*, 2023). These factors can be socioeconomic, cultural, geographic, cultural and political. The following subsection presents a deep understanding of such factors, critically involved in the adoption of mitigation strategies from the detrimental effects of climate change in an energy intensive iron and steel industry.

Pragmatic and cost-effective technology (F1): The use of optimal technological advancements has the potential to substantially mitigate emissions originating from steel factories. The cost-effectiveness and environmental friendliness of substituting coke and coal with hydrogen-containing resources, such as steam, natural gas, and waste polymers, have been the subject of debate. The utilization of green hydrogen, derived from sustainable sources such as wind or solar power, as a substitute for carbon-intensive coking coal during the steel production process has the potential to yield substantial reductions in emissions. Advanced technologies like as electric arc furnaces (EAFs) and direct reduction iron (DRI) processes exhibit higher energy efficiency in comparison to conventional blast furnaces. In addition, the use of sensors and automation has the potential to enhance energy efficiency across the entirety of the production process. Enhanced energy efficiency can be achieved by the implementation of a proficient energy management system, hence resulting in reduced operational expenditures (He and Wang, 2017; Brunke *et al.*,2014; Hasan *et al.*, 2018).

Less supply chain complexity (F2): The influence of supply chain complexity in the steel sector on climate change is substantial, since it affects emissions, resource utilization, and resilience. Addressing these issues necessitates a synergistic approach encompassing supply chain efficiency, efforts to promote sustainability, and industry-wide collaboration. The process commences with the identification of asset placement, encompassing facilities such as plants, warehouses, and central marketing centers, and extends to the movement of materials across the supply chain. Additionally, it addresses a diverse range of objectives, such as gaining a competitive edge in the market and providing finished products at reduced costs. The need of timely delivery is emphasized when managing high-value commodities, as it helps to minimize inventory levels and the associated holding costs (Homayouni *et al.*, 2021; Chakraborty *et al.*, 2020).

Robust policy and legal framework (F3): The rules and regulations establish quantifiable emission, energy use, and efficiency objectives that are obligatory for all recipients to fulfill in order to prevent potential financial and/or legal consequences. Climate change mitigation initiatives also have an impact on social policy. The adoption of robust policies has a mutual impact on taxation, infrastructure, and public budget. The low intensity of environmental regulation has a negative impact on enhancing the level of energy-environmental efficiency. Legal policies and frameworks have a significant impact on the implementation of climate change mitigation strategies in steel plants. They achieve this by setting emissions regulations, providing incentives for the use of clean technology, supporting the use of renewable energy sources, and ensuring openness and accountability in these efforts. These policies play a crucial role in ensuring that the actions of the steel sector are in compliance with sustainable development goals and commitments (Büchs *et al.*, 2011; Wu and Lin, 2022).

Public participation (F4): A broad definition of "public engagement with climate change" is "individuals'" evaluation of and reaction to climate change, which includes "cognitive," "emotional," and "behavioral" components. Public action is not just prompted by information campaigns, but also by the need for public engagement and support in the implementation of new innovative technologies and infrastructure adaption. The involvement of the public in the implementation of climate change mitigation measures plays a crucial role in enhancing the efficiency, equity, and social responsibility of these strategies, hence facilitating the reduction of emissions and the attainment of sustainability objectives. By incorporating a wide range of stakeholders, leaders in the business provides a more robust framework for tackling the issue of climate change, while concurrently yielding advantages for local communities and the wider society (Lorenzoni *et al.*, 2007; Nuttall, 2012).

Institutional effectiveness (F5): Diverse factors, such as motivation, participation, and the quality of activities, play a significant role in assessing institutional performance. Various elements, such as culture, ethics, capacity, and complexity, have a significant role in influencing institutional effectiveness. These factors encompass aspects such as standards of peers, social norms, network dynamics, and managerial paradigms (Ahmad *et al.*, 2020; Beccarello and Di Foggia, 2023).

Environmental awareness (F6): The conduct of managers in relation to environmental sustainability is notably impacted, both directly and indirectly, by their level of environmental education and awareness. Additional characteristics such as decision-making attitude, commitment to sustainable practices, and behavioral goals are also influenced. The accumulation and dissemination of environmental information among managers and engineers has been shown to enhance both the environmental and financial performance of companies (Safari *et al.*, 2018; Huang and Shih, 2009).

Improved resource availability (F7): The utilization of scrap steel in secondary and small to medium scale steel factories can serve as a beneficial measure in conserving energy and reducing the consumption of iron resources. The utilization of heat recovery from blast furnace operations has the potential to serve as a highly efficient resource. Furthermore, it is anticipated that this measure will enhance the sustainability prospects of steel manufacturing facilities (Tiejun, 2011; Keplinger *et al.*, 2018).

Financial stability (F8): The adoption of new plans, policies, or practices is thwarted by inadequate capital access and suboptimal market value. The main roadblock in the process of retrofitting is primarily linked with the financial capabilities of the organization. The implementation of adoption mitigation measures within the steel industry frequently results in additional expenses. The cost encompasses the comprehension, acceptance, and recognition of government mitigation policies, the substantial upfront expenses associated with mitigation technologies, the significant overhead costs and the expenses related to information analysis and collection, inconveniences, and disruptions to production. The steel sector encounters challenges in obtaining financial capital, particularly in relation to limited access for implementing emission reduction measures. In addition to the preceding factors, it is worth noting that budgetary systems generally lack incentives that effectively promote the development of low carbon technologies. The steel sector encounters considerable impediments, such as the difficulty in evaluating the economic performance or poor financial performance characterized by low returns and longer payback periods on investments in certain mitigation measures (Brunke *et al.*, 2014; Ahmad *et al.*, 2020; Debiagi *et al.*, 2022; Balsara *et al.*, 2021).

Recycle and reproduction (F9): In addition to facilitating financial recovery, the utilization of recovered waste plastic in reduction processes yields a notable reduction of approximately 30% in the carbon footprint of iron and steel factories. When comparing plastics to coal, it can be shown that polymers possess a higher hydrogen (H2) content and exhibit lower carbon dioxide (CO2) emissions. The utilization of plastic substitution results in a reduction in the process temperature by a minimum of 100–200 °C, hence enhancing the energy efficiency of the process. The utilization of scrap steel as a primary input material exhibits a positive correlation with the reduction in demand for iron ore and the subsequent mitigation of emissions linked to mining activities and ore processing. The implementation of advanced sorting and recycling technologies has the potential to enhance the efficacy of scrap collection and processing. Technologies such as combined heat and power (CHP) systems have the potential to be utilized for the purpose of waste heat recovery (Hodgkinson *et al.*, 2021; Tong *et al.*, 2019; Devasahayam *et al.*, 2019; Yellishetty *et al.*, 2011).

Assertive media and communication (F10): The media has a crucial role in facilitating the widespread implementation of climate change mitigation strategies, not alone by serving as a forum for disclosures and reporting. Investigative journalism and reporting play a crucial role in promoting corporate accountability regarding environmental policies, fostering a culture of transparency in emissions reporting, and encouraging the advancement of sustainability activities. Positive advertising and ethical publication practices can foster the organization of joint seminars and collaborative efforts aimed at protecting the environment. It has the ability to impact public opinion and exert influence over corporate behavior to shift towards more sustainable and ecologically responsible practices within energy intensive industries. There exists a direct correlation between performance and disclosure. Positive media disclosures have the potential to decrease a company's creditor expenses and contribute to the establishment of an environmentally conscious reputation within the market (Meng *et al.*, 2014; Lou *et al.*, 2019).

METHODOLOGY

A thorough review of the literature was conducted in order to ascertain the elements that contribute to the successful adoption of climate change mitigation strategies in Indian iron and steel industries. The identification of variables was achieved by conducting a comprehensive literature review, which encompasses the essential sub-themes associated with the effective implementation of strategies. Sometimes identified variables for a study maybe published globally yet not operational in the context of current study (Sushil, 2018). Consequently, the objective of the present study is to examine the interconnections among these parameters and afterwards establish their relative rankings. The 10 elements are determined by the utilization of the aforementioned sub-themes in conjunction with the expert's perspective. The panel of experts consists top managers and the head of environment control division. They were interviewed to have a deeper understanding of the vertical hierarchy of the respective primary and secondary steel industries, India. The 22 professionals possess a minimum of fifteen years of experience and knowledge regarding the adoption and implementation of climate change mitigation strategies aimed at improving environmental performance of iron and steel industries. The difference between primary and secondary steel plants is the reduction of iron ore to make steel in blast furnaces. The latter does not have a blast furnace and uses melting scrap from primary plants to make final products. Steel plants which are listed in the public sector undertaking, both primary and secondary from six Indian states (Maharashtra, Tamil Nadu, Chhattisgarh, Orissa, Jharkhand, and West Bengal), were selected for the study. The data collection and survey were done in person by the researcher at all the locations. An average of 60 minutes was spent per expert, depending upon their availability.

A semi-structured questionnaire is employed to gather the perspectives of specialists regarding the interrelationships among several aspects. This study follows the Total Interpretive Structural Modeling (TISM) which is an advancement to the Interpretive Structural Modeling (ISM) by giving meaning to the relationships of variables by using the tool of interpretive matrix (Bhavan *et al.*, 2005). It is a system that enables people to make intricate connections between various variables in a complicated scenario (Warfield, 1974). ISM-related research first emerged in the 1970s and TISM is a suitable method to combine quantitative measurement with qualitative semi-structured interviews that capture the context of the unique organizational settings. Researchers have used TISM as an approach to study supply chain, sustainable waste management (Dohale *et al.*, 2022), artificial intelligence in banking (Lamba and Thareja, 2021), food supply chain (Patil *et al.*, 2018), cybersecurity management (Rajan *et al.*, 2021), accelerators of automotive industry (Singhal *et al.*, 2022), cross border joint venture studies (Sharma *et al.*, 2022). It is a 9-step process as follows:

Step I involves the identification and definition of the factors associated with the mitigation strategies. The identification of factors can be accomplished through a review of relevant literature or by employing various idea generation techniques, such as group discussions or grounded theory.

Step II involves the establishment of a contextual link between the drivers and obstacles that have been identified. The establishment of the contextual relationship between the elements for which the model is to be constructed holds significant importance. The contextual relationship is contingent upon the specific structure being considered, such as goal, priority, driving and dependence power. This process leads to the creation of the Structural Self Interaction Matrix (SSIM), which is represented in Table 1. The identification of linkages is facilitated through the utilization of industry expert opinion, as evidenced by Table 3 and Figure 1.

Step III of the process, the initial reachability matrix (IRM) is constructed by converting the alphabetical symbols in the SSIM into binary representation, as shown in Table 2. The elucidation of the connection would be advantageous in order to render the underlying information overt. The perception of the link between objectives would vary depending on the specific pair under consideration.

Step IV, the Final Reachability Matrix (FRM) is constructed by integrating transitive linkages into the Initial Reachability Matrix (IRM), as depicted in Table 4.

In Step V, the division of each FRM into distinct levels is determined by the associations between the driving variables (reachability) and the dependent variables (antecedent), as indicated in Table 5.

Step VI involves assigning Level 1 to drivers/barriers based on the equality of their reachability set (row values) and intersection set (common values between reachability set and antecedent set). Any variable present in the previous level is then eliminated throughout the remainder of the level partitioning process. This process is continued until all the variables under investigation have been assigned a level. The levelof iteration does not necessarily ensure the same level of variables.

In the step VII, digraphs are formed through the identification of different levels of constituents once transitivity has been confirmed.

In step VIII, the digraph that is obtained is transformed into a model by elucidating the connections among the variable nodes. Figures 2 illustrate the use of solid lines to represent direct relationships and dashed lines to represent accepted significant transitive links.

The last step IX involves analysis of the model by experts to mitigate any potential inconsistencies,. The relationship statement yields a structured hierarchy of the factors for CCMS in IISI.

RESULTS

Self Structural Interaction Matrix (SSIM)

This section examines the self-interaction level of the factors. The row of the matrix is denoted by the variable *i*, whereas the column is denoted by the variable *j*. The relationships among the factors are represented by the letters V, A, X, and O. The allocation of these letters inside the matrix is determined by a process of expert interviews. The symbols can be interpreted in a meaningful manner as follows: V where *i* leads to *j* but *j* does not lead to *i*, A when the letter "*j*" precedes the letter "*i*" but the letter "*i*" does not precede the letter "*j*," X, when the variable i leads to the variable j and the variable j leads to the variable *i*, O in all cases where there is no discernible correlation between variables *i* and *j*.

The SSIM values are converted into reachability matrices by transforming the alphabetical symbols into the binary system, thereby defining the driving power variables. This transformation is accomplished by assigning a value of 1 to relationships represented by V, where *i* leads to j, while assigning a value of 0 to relationships where *j* leads to *i*. Symbol A is assigned the value of 0 when the directed relationship is from *i* to *j*, and the value of 1 when the directed relationship is from *j* to *i*. The values of 1 and 0 are assigned to X and O, regardless of the orientation of the relationship matrix.

Table 1. SSIM drivers in IISI

	F10	F9	F8	F7	F6	F5	F4	F3	F2	F1
F1	O	V	O	O	V	O	O	X	X	X
F2	O	O	O	V	V	O	O	O	X	
F3	0	0	V	V	V	O	O	X		
F4	O	V	O	A	A	X	X			
F5	V	V	O	A	A	X				
F6	O	V	O	V	X					
F7	V	V	V	X						
F8	A	A	X							
F9	O	X								
F10	X									

Source: Author's own elaboration and adopted from Singh *et al.*, 2023

Table 2. IRM drivers in IISI

	F1	F2	F3	F4	F5	F6	F7	F8	F9	F10
F1	1	1	1	0	0	1	0	0	1	0
F2	1	1	0	0	0	1	1	0	0	0
F3	1	0	1	0	0	1	1	1	0	0
F4	0	0	0	1	1	0	0	0	1	0
F5	0	0	0	1	1	0	0	0	1	1
F6	0	0	0	1	1	1	1	0	1	0
F7	0	0	0	1	1	0	1	1	1	1
F8	0	0	0	0	0	0	0	1	0	0
F9	0	0	0	0	0	0	0	1	1	0
F10	0	0	0	0	0	0	0	1	0	1

Source: Author's own elaboration and adopted from Singh *et al.*, 2023

Table 3. Interpretive knowledge base of transitive links for factors as per industry experts

Linkage between drivers	Expert score (avg of 5 experts' score)	Interpretation	Relationship
D1-D4	2	Technology might spark awareness and base for knowledge sharing	Weak
D1-D5	2	Relationship of negligible direct impact	Weak
D1-D10	2	Technological advancements attract media attention	Weak
D2-D3	1	Supply chain management is one of the industrial policies	Weak
D2-D4	2	Ground level management of resources	Weak
D2-D5	2	Seamless operations enhance industry's image	Weak
D2-D8	1	Stable operational capacity brings financial stability	Weak
D2-D9	1	Pro-active strategies impact adaptation strategies	Weak
D2-D10	1	Impacts the perception of industry in public eye	Weak
D3-D2	2	Stringency demonstrates better use and demand for resources	Weak
D3-D4	1	Clarity and transparency of projects	Weak
D3-D9	2	Adoption of 3R principle (Recycle, Reduce, Reuse)	Weak
D3-D10	2	Check on publishing sensitive information	Weak
D4-D8	2	Irreflexive relationship	Weak
D5-D8	1	Institutional capacity aids in monetary capabilities	Weak
D6-D10	1	Pro-environment attitude builds a window of green image	Weak

Figure 1. Significant transitive relationships between the factors

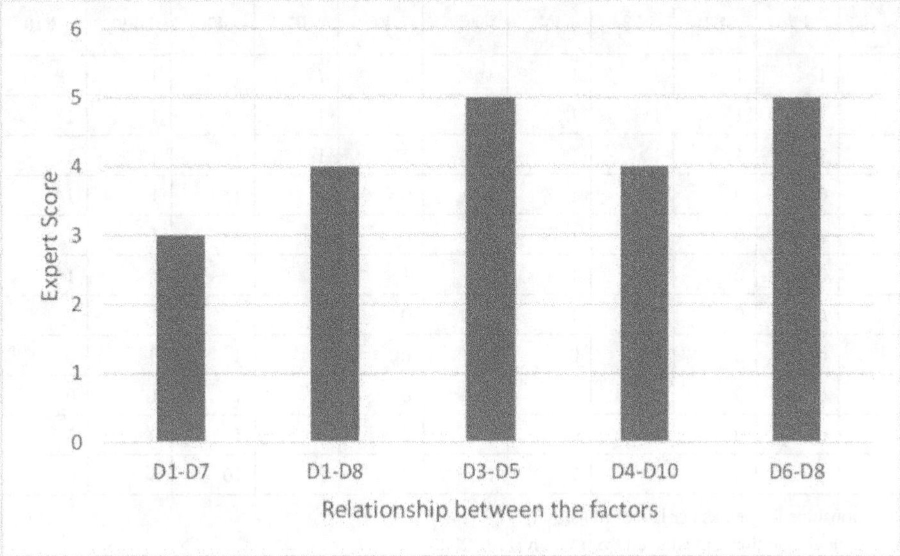

Final Reachability Matrix (FRM)

The binary-valued tables 1 and 2are transformed into the Final Reachability Matrix (FRM) of drivers and obstacles by adding the principle of transitivity. The concept of transitivity can be comprehended as follows: if there exists a relationship between variable A and variable B, and a separate relationship between variable B and variable C, then it follows that there is also a relationship between variable A and variable C. This assumption facilitates the preservation of conceptual coherence (Sushil, 2012; Dubey and Ali, 2014).

In order to ascertain the noteworthy transitive connections, a follow-up interview was conducted with five experts from the pool of previously interviewed experts. The purpose of this interview was to assess the transitive relationships identified by the authors, using a five-point Likert scale (see to Table 3 and Figure 1). A threshold of 3 or greater was established to determine the significance of transitive linkages (Rajesh, 2017). Connections having a score of 2 or lower are eliminated from the ultimate directed graph. However, this does not always indicate that the linkages are non-transitive. This action was undertaken in order to ensure that the ultimate digraph alone encompasses the most significant transitive links. In order to clarify the ambiguity surrounding the various values represented by 1* in terms of their significance as either transitive connections or weak links, we have resolved this issue by assigning the value 1* to designate substantial transitive linkages, as illustrated in the digraph. Additionally, we have assigned the value *1** to represent weak transitive links in Table 4. The process of categorizing the factors into levels is conducted through the hierarchical ranking as depicted in Table 5.

On following Step VI, factors are assigned a Level 1 based on the comparison of their reachability set (row values) and intersection set (common values between reachability set and antecedent set). Each variable that exists in the preceding level is subsequently eradicated during the next stages of the level partitioning procedure. The procedure is iterated until all the variables under scrutiny have been allocated a level. The level of iteration does not necessarily guarantee the same level to factors.

Table 4. FRM drivers in IISI

	F1	F2	F3	F4	F5	F6	F7	F8	F9	F10	Driving power
F1	1	1	1	1*	1*	1	*1**	*1**	1	1*	10
F2	1	1	1*	1*	1*	1	1	1*	1*	1*	10
F3	1	1*	1	1*	*1**	1	1	1	1*	1*	10
F4	0	0	0	1	1	0	0	1*	1	*1**	5
F5	0	0	0	1	1	0	0	1*	1	1	5
F6	0	0	0	1	1	1	1	*1**	1	1*	7
F7	0	0	0	1	1	0	1	1	1	1	6
F8	0	0	0	0	0	0	0	1	0	0	1
F9	0	0	0	0	0	0	0	1	1	0	2
F10	0	0	0	0	0	0	0	1	0	1	2
Dependence	3	3	3	7	7	4	5	10	8	8	

* denotes the relationships formed as per law of transitivity.
Source: Author's own elaboration and adopted from Singh *et al.*, 2023

MICMAC Analysis

The analysis of the hierarchical relationship among factors is conducted by the application of the Cross Impact Matrix-multiplication Applied to Classification (MICMAC) approach, which takes into account their driving force and dependence power (Warfield, 1994). The MICMAC analysis method employs matrix multiplication attributes and considers the dispersion of impacts according to their driving and dependent power in order to classify the components of a complex system into four quadrants, namely autonomous (1), dependent (2), linking (3), and driving (4). The utilization of MICMAC analysis enables firms to effectively allocate their resources by pinpointing critical areas for intervention and comprehending the intricate dynamics of a complex system.

Quadrant 1 consists of variables that exhibit both weak driving strength and dependence, resulting in minimal impact on the whole system. Based on the analysis of Figure 3, it can be inferred that all the drivers examined in the present study exhibit strong driving power and reliance, as none of them demonstrate weak characteristics in these regards. This quadrant signifies variables that exhibit a reasonably high degree of independence from one another. The aforementioned elements exert a substantial influence on the system, however their susceptibility to external forces inside the system is rather limited. These elements are commonly regarded as strategic in nature and necessitate careful consideration. The comprehensive examination of existing scholarly works and the incorporation of specialized expertise have successfully identified the significant factors influencing the subject matter.

Quadrant 2 exhibits variables that have a greater degree of reliance but possess relatively modest driving power. The dependent variables in decision making are positioned at the highest level within the TISM hierarchy. The study revealed that the performance criteria for public engagement (D4), institutional effectiveness (D5), recycle and reproduction (D9), assertive media and communication (D10), and financial stability (D8) are strongly reliant on the regulation of quadrant 4. The quadrant under consideration encompasses factors that are subject to influence from other components within

Table 5. Level partitioning of factors

Drivers	Reachability Set	Intersection Set	Level
Iteration 1			
F1	1,2,3,4,5,6,7,8,9,10	1,2,3	
F2	1,2,3,4,5,6,7,8,9,10	1,2,3	
F3	1,2,3,4,5,6,7,8,9,10	1,2,3	
F4	4,5,8,9,10	4,5	
F5	4,5,8,9,10	4,5	
F6	4,5,6,7,8,9,10	6	
F7	4,5,7,8,9,10	7	
F8	8	8	I
F9	8,9	9	
F10	8,10	10	
Iteration 2			
F1	1,2,3,4,5,6,7,9,10	1,2,3	
F2	1,2,3,4,5,6,7,9,10	1,2,3	
F3	1,2,3,4,5,6,7,9,10	1,2,3	
F4	4,5,9,10	4,5	
F5	4,5,9,10	4,5	
F6	4,5,6,7,9,10	6	
F7	4,5,7,9,10	7	
F9	9	9	II
F10	10	10	II
Iteration 3			
F1	1,2,3,4,5,6,7	1,2,3	
F2	1,2,3,4,5,6,7	1,2,3	
F3	1,2,3,4,5,6,7	1,2,3	
F4	4,5	4,5	III
F5	4,5	4,5	III
F6	4,5,6,7	6	
F7	4,5,7	7	
Iteration 4			
F1	1,2,3,6,7	1,2,3	
F2	1,2,3,6,7	1,2,3	
F3	1,2,3,6,7	1,2,3	
F6	6,7	6	
F7	7	7	IV
Iteration 5			
F1	1,2,3,6	1,2,3	
F2	1,2,3,6	1,2,3	
F3	1,2,3,6	1,2,3	
F6	6	6	V
Iteration 6			
F1	1,2,3	1,2,3	V
F2	1,2,3	1,2,3	VI
F3	1,2,3	1,2,3	VI

Figure 2. TISM of factors

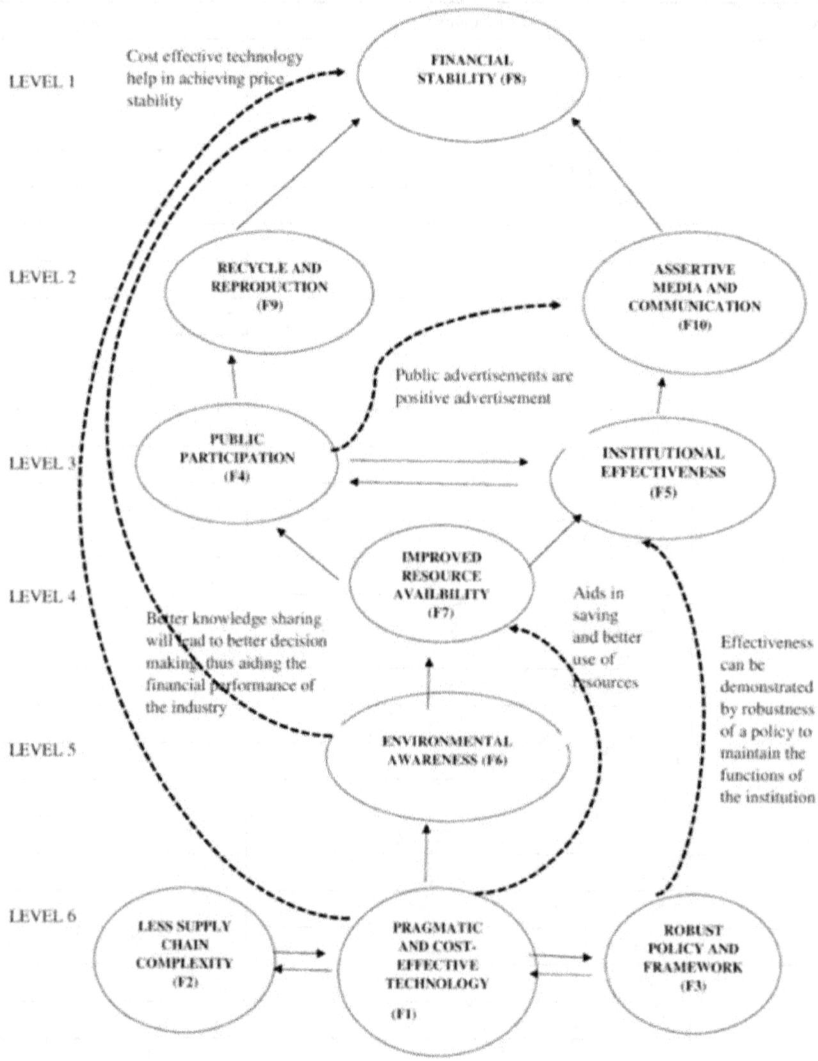

the system, while simultaneously exerting impact on other factors. Operational or tactical factors are frequently represented by them. Therefore, it is imperative that all activities pertaining to CCMS be specifically targeted at these drivers.

Notes: F1, pragmatic and cost-effective technology; F2, less supply chain complexity; F3, robust policy and legal framework; F4, public participation; F5, institutional effectiveness; F6, environmental awareness; F7, improved resource availability; F8, financial stability; F9, recycle and reproduction; F10, assertive media and communication

Quadrant 3 comprises variables that exhibit both a significant driving force and a high level of reliance, while also being characterized by inherent instability. The primary objective of this cluster is to augment the knowledge and skills of decision makers, enabling them to effectively attain the targeted level of performance. This cluster serves as an intermediary, facilitating communication and collaboration between quadrant 2 and 4. In general, they are often positioned at the apex of a TISM hierarchy. This quadrant comprises components that possess a high degree of influence while also being subject to influence from other factors. These factors are deemed strategic due to their capacity to exert a substantial influence on the system and their susceptibility to be affected by strategic measures. Nevertheless, our investigation yielded no evidence of drivers that may account for the observed outcomes, suggesting the possibility that independent factors may have a direct impact on other variables within the study.

Quadrant 4 exhibits crucial predictive variables characterized by their elevated driving strength and little reliance. This quadrant encompasses variables that possess a somewhat diminished level of influence, however has the potential to exert a substantial impact on other variables. Factors necessitating monitoring or modification are frequently seen as such. Hence, it is imperative to address these factors as a priority, as they serve as the fundamental catalyst for influencing all other variables within the system. In the present investigation, a total of five variables were found as exerting influence over the overall system. The factors contributing to the reduction of supply chain complexity (F2), the adoption of pragmatic and cost-effective technology (F1), the establishment of a solid policy and legal framework (F3), the promotion of environmental awareness (F6), and the enhancement of resource availability (F7) are of significant importance.

Essentially, the utilization of MICMAC analysis proves to be a valuable instrument in comprehending the intricacies inherent in complex systems and facilitating the process of making well-informed judgments. The process assists businesses in effectively allocating their efforts and resources by discerning crucial areas for intervention and strategic action, taking into account the complex interaction of multiple components inside the system.

DISCUSSION

The conceptual framework presented for the adoption of climate change mitigation strategies has identified ten factors on a hierarchical priority levels. The most critical factors, less supply chain complexity(F2), pragmatic and cost effective technology(F1) and robust policy and legal framework(F3) are at the bottom of developed TISM. Financial stability(F8) shares a direct transitive link pragmatic and cost effective technology(F1) and environmental awareness(F6).

Based on the framework depicted in Figure 3, it can be observed that the variable with the greatest significance among all the factors is the combination of less supply chain complexity (F2), pragmatic and cost-effective technology (F1), and a robust policy and regulatory framework (F3) at level 6. These elements provide the fundamental framework for any industry and ought to get accorded the appropriate level of emphasis. The findings of the MICMAC analysis has validated the same, as the results have successfully classified them into the dependent group. They possess the greatest power and maintain a bidirectional direct connection with one another, operating at an equivalent level. Additionally, they establish a direct connection with environmental awareness at level 5. An effective assessment of the existing resources inside the industry can lead to a decrease in reliance on the supplier, and conversely, an increase in self-sufficiency (Tiejun, 2011; Chakraborty *et al.*, 2020). The relationship between pragmatic and cost-effective technology (F1) and financial stability (F8) is noteworthy due to the transitive connection they share. This connection arises from the fact that cost-effective technologies contribute to the attainment of price stability. The relationship between environmental awareness (F6) and financial stability (F8) is noteworthy, since it is evident at level 1 that the exchange of knowledge can contribute to improved decision-making concerning the industry's financial performance. The establishment of a major transitive link between pragmatic and cost-effective technology (F1) and increased resource availability (F7) at level 4 is based on the principles of saving and better utilization of resources. The relationship between a robust policy and legal framework (F3) and institutional effectiveness (F5) at level 3 is one of transitive linkage. This is because the efficacy of an institution can be evidenced by the robustness of its policies in maintaining its functions. At level 3 of public participation (F4), there exists a notable transitive connection with assertive media and communication (F10) at level 2. This is due to the fact that any public advertisements, such as workshops or organizational camps, aimed at promoting new innovations within the industry, will be presented as a favourable demonstration of support for CCMS. The presentation of a socially aware portrayal of an industry's tactics through media will generate a need for recycled products, hence improving its financial performance. This assertion is supported by the established correlation between recycling and reproduction (F9) as well as financial stability (F8).

The paradigm employed for the study highlights the importance of attaining financial stability within energy-intensive sectors, such as the IISI. According to Chao et al. (2023), there is empirical evidence supporting the notion that financial integration has a favourable effect on the consumption of renewable energy. Enhancing vertical coordination among the listed ten factors can effectively mitigate supply chain complexity, facilitate the deployment of cost-effective technologies, and ensure adherence to a rigorous policy framework. This highlights the significance of establishing effective vertical coordination among the key stakeholders involved in the study on the deployment of CCMS in IISI.

The interplay between finance and climate change is no secret. One of the key point from COP 27 states "Finance is at the heart of all that the world is doing to combat climate change. Mitigation, adaptation, loss and damage, climate technology – all of it requires sufficient funds to function properly and to yield the desired results". The significance of enhancing the magnitude and accessibility of climate finance entails the mobilization of finance from both the public and private sectors, together with the exploration of novel financing options, in order to address the disparity between existing investment levels and the financial requirements for implementing climate change effectively. The allocation of funds for adaptation measures is expected to garner significant focus, given the escalating severity of climate change effects. The discussions of COP 27 conference is expected to place significant emphasis on the imperative of providing enhanced financial assistance to disadvantaged communities and nations, with the aim of bolstering their capacity to withstand and adapt to climate-induced hazards. The promotion of private sector engagement in climate finance is a perennial topic. Conversations centre on strategies for promoting private investments in renewable energy, sustainable technologies, and infrastructure projects that are robust to climate change. Debt relief and the correlation between debt and climate change are topics of discussion in light of the economic difficulties experienced by developing nations. The complicated matter of simultaneously addressing debt sustainability and accomplishing climate goals may warrant inclusion on the agenda. Reaffirming their pledge to contribute climate money to support developing countries' efforts at adaptation and mitigation is expected of governments of especially developed nations.

Financial stability and climate change has a trilateral relationship involving physical, regulatory and transitional risks. Climate-related emergencies, such as extreme weather events and increasing sea levels, possess the potential to inflict havoc upon infrastructure, disrupt supply networks, and result in financial losses for businesses and insurers. These losses have the potential to propagate across the economic system. The government are promulgating laws and regulations in order to address the issue of climate change. Financial firms that fail to sufficiently consider and effectively address climate-related issues may encounter regulatory sanctions or be compelled to devalue carbon-intensive assets. But there are transition risks encompassing the financial hazards inherently linked to the process of transitioning towards a low-carbon economy. The potential decline in asset values and profitability of enterprises primarily dependent on fossil fuels may arise if governments and businesses undertake measures to mitigate greenhouse gas emissions. These events could potentially have an impact on the stability of financial institutions that have investments in such sectors. Similarly technology plays a crucial role in the mitigation of climate change through facilitating the shift towards renewable energy sources, enhancing energy efficiency, and advancing carbon capture and storage solutions. Technological progressions have the potential to expedite the process of reducing carbon emissions and achieving a state of decarbonization within the global economy. Adaptation technologies are vital in addressing the challenges posed by climate change consequences. Technology bears a significant role in facilitating adaptation efforts. This spans advancements in multiple sectors such as infrastructure, agriculture, water management, and disaster preparedness, which have the potential to bolster resilience against climate-related hazards. The utilization of remote sensing, satellite technology, and data analytics plays a crucial role in monitoring alterations within the environment, forecasting occurrences of severe weather phenomena, and evaluating the susceptibility of assets and areas to the consequences of climate change. The utilization of technology not only facilitates enhanced monitoring and evaluation but gauzing the climate related risks as well.

The presence of sufficient resources and investment in technological research and development has the potential to stimulate innovation in sectors such as clean energy and transportation, which are of utmost importance in tackling the issue of climate change. The utilization of technology can enhance the ability of financial institutions to comprehensively comprehend and effectively manage risks associated with climate change. This can be achieved through the implementation of advanced modelling techniques, scenario analysis, and stress testing methodologies. There is a growing trend among financial institutions to allocate investments towards green technologies and sustainable initiatives. The current trend towards green finance is aimed at promoting the advancement and implementation of environmentally sustainable technology, while simultaneously mitigating the risks associated with carbon-intensive investments.

The global steel industry has been responding to the challenges posed by climate change by implementing a range of methods and technologies aimed at mitigating its environmental impact and promoting sustainable practices. U.S. Steel has expressed its commitment to mitigating greenhouse gas emissions by adopting renewable energy sources, including solar and wind power, as alternatives for its operational activities (Holappa, 2020). In Sweden, it is achieved through the adoption of fossil-free technologies, which serve as a viable alternative to conventional coal-based blast furnaces. The HYBRIT initiative is a noteworthy project that involves a partnership between SSAB, LKAB, and Vattenfall. This collaborative effort seeks to achieve the production of steel without reliance on fossil fuels by employing the method of hydrogen reduction of iron ore (Muslemnai *et al.*, 2021).

In Germany, programs aimed at promoting the use of a circular economy for the steel industry. One method that can be employed to mitigate the demand for virgin raw materials is the recycling of steel scrap, which facilitates the production of new steel products (Ogunmakinde, 2019). Nippon Steel, a prominent steel manufacturer based in Japan, is actively engaged in the research and development of cutting-edge technology aimed at mitigating carbon dioxide (CO2) emissions within the steel production process. The researchers are currently investigating advanced technologies such as the hydrogen direct reduction process and carbon capture and utilization (CCU) approaches (Oda and Akimoto, 2022). POSCO, an established steel corporation in South Korea, has prioritized the implementation of environmentally sustainable practices in its steel production operations, as part of its strategic endeavor known as the

"Green Transformation" campaign. In order to mitigate emissions, the adoption of technology such as hydrogen reduction process and carbon capture is being pursued (Yang, 2021).

Developing countries like China, being the foremost global producer of steel, has established high objectives with regards to achieving carbon neutrality. A growing number of Chinese steel enterprises are embracing environmentally-friendly technology, such as electric arc furnaces, and integrating carbon capture and utilization (CCU) methods in order to mitigate emissions (Wei *et al.*, 2022). In the Indian context, there is a growing interdependence between the economic policies of distinct areas and climate change policies, particularly within the industrial sector. The governmental bodies at both the national and state levels have been engaged in the development and execution of policies aimed at encouraging sustainable behaviors, mitigating carbon emissions, and fostering environmental preservation. India has implemented a range of sustainable practices, encompassing the recycling of steel scrap, the improvement of energy efficiency, and the integration of waste heat recovery systems, with the aim of mitigating emissions and reducing energy consumption (Wang, 2022).

The regional governments have demonstrated a proactive approach by implementing economic incentives, laws, and support systems to encourage industries to embrace sustainable and environmentally-friendly practices. This initiative aims to address the implications of climate change and foster the development of a greener, low-carbon economy. The state of Odisha possesses abundant reserves of iron ore and coal, rendering it a prominent geographical region for the steel industry. In recent years, the state has placed significant emphasis on incorporating energy efficiency, waste heat recovery, and greener technology into its policy framework pertaining to the steel business. The promotion of the use of novel technologies is advocated in order to mitigate carbon emissions and enhance energy efficiency (Rout and Pradhan, 2023).

In essence, the interconnection between financial stability, technology, and climate change is evident. Climate change presents significant financial risks that can be effectively addressed through the implementation of technology-driven solutions and the adoption of sound financial strategies. Furthermore, technology assumes a crucial role in tackling the issue of climate change by enabling and supporting both mitigation and adaptation endeavours. The combination of technology and sustainable finance will remain a crucial catalyst for progress towards a more stable and sustainable future as governments, corporations, and financial institutions navigate the challenges posed by a shifting climate.

ACKNOWLEDGEMENTS

The first author (PS) expresses gratitude towards Symbiosis International (Deemed University), Pune for the provision of a Junior Research Fellowship (JRF). The authors are also thankful to the experts from IISI for their invaluable time and assistance.

REFERENCES

AghaKouchak, A., Chiang, F., Huning, L. S., Love, C. A., Mallakpour, I., Mazdiyasni, O., Moftakhari, H., Papalexiou, S. M., Ragno, E., & Sadegh, M. (2020, May 30). Climate Extremes and Compound Hazards in a Warming World. *Annual Review of Earth and Planetary Sciences*, *48*(1), 519–548. doi:10.1146/annurev-earth-071719-055228

Ahmad, I., Arif, M. S., Cheema, I. I., Thollander, P., & Khan, M. A. (2020). Drivers and barriers for efficient energy management practices in energy-intensive industries: A case-study of iron and steel sector. *Sustainability (Basel)*, *12*(18), 7703. doi:10.3390/su12187703

An, R., Yu, B., Li, R., & Wei, Y. M. (2018). Potential of energy savings and CO2 emission reduction in China's iron and steel industry. *Applied Energy*, *226*, 862–880. doi:10.1016/j.apenergy.2018.06.044

Balsara, S., Jain, P. K., & Ramesh, A. (2021). An integrated methodology to overcome barriers to climate change mitigation strategies: A case of the cement industry in India. *Environmental Science and Pollution Research International*, *28*(16), 20451–20475. doi:10.1007/s11356-020-11566-6 PMID:33410023

Beccarello, M., & Di Foggia, G. (2023). The Strategic Role of Energy Efficiency and Industry 4.0 Interventions in Manufacturing. *Journal of Management and Sustainability*, *13*(1), 31–44. doi:10.5539/jms.v13n1p31

Bhatta, B., Shrestha, S., Shrestha, P. K., & Talchabhadel, R. (2019). Evaluation and application of a SWAT model to assess the climate change impact on the hydrology of the Himalayan River Basin. *Catena*, *181*, 104082. doi:10.1016/j.catena.2019.104082

Brunke, J. C., Johansson, M., & Thollander, P. (2014). Empirical investigation of barriers and drivers to the adoption of energy conservation measures, energy management practices and energy services in the Swedish iron and steel industry. *Journal of Cleaner Production*, *84*, 509–525. doi:10.1016/j.jclepro.2014.04.078

Büchs, M., Bardsley, N., & Duwe, S. (2011). Who bears the brunt? Distributional effects of climate change mitigation policies. *Critical Social Policy*, *31*(2), 285–307. doi:10.1177/0261018310396036

Cadez, S., & Czerny, A. (2016). Climate change mitigation strategies in carbon-intensive firms. *Journal of Cleaner Production*, *112*, 4132–4143. doi:10.1016/j.jclepro.2015.07.099

Chakraborty, S., Chattopadhyay, R., & Chakraborty, S. (2020). An integrated D-MARCOS method for supplier selection in an iron and steel industry. *Decision Making: Applications in Management and Engineering*, *3*(2), 49–69.

Chao, T., Yunbao, X., Chengbo, D., Bo, L., & Ullah, S. (2023). Financial integration and renewable energy consumption in China: Do education and digital economy development matter? *Environmental Science and Pollution Research International*, *30*(5), 12944–12952. doi:10.1007/s11356-022-22852-w PMID:36121627

Debiagi, P., Rocha, R. C., Scholtissek, A., Janicka, J., & Hasse, C. (2022). Iron as a sustainable chemical carrier of renewable energy: Analysis of opportunities and challenges for retrofitting coal-fired power plants. *Renewable & Sustainable Energy Reviews*, *165*, 112579. doi:10.1016/j.rser.2022.112579

Deshpande, T., Mukherji, R., & Sastry, M. (2023). Policy styles and India's national action plan on climate change (NAPCC). *Policy Studies*, 1–18. doi:10.1080/01442872.2023.2261388

Devasahayam, S., Raju, G. B., & Hussain, C. M. (2019). Utilization and recycling of end of life plastics for sustainable and clean industrial processes including the iron and steel industry. *Materials Science for Energy Technologies*, *2*(3), 634–646. doi:10.1016/j.mset.2019.08.002

Dey, S., Sreenivasulu, A., Veerendra, G. T. N., Rao, K. V., & Babu, P. A. (2022). Renewable energy present status and future potentials in India: An overview. *Innovation and Green Development*, 100006.

Dohale, V., Ambilkar, P., Gunasekaran, A., & Verma, P. (2022). Supply chain risk mitigation strategies during COVID-19: Exploratory cases of "make-to-order" handloom saree apparel industries. *International Journal of Physical Distribution & Logistics Management*, *52*(2), 109–129. doi:10.1108/IJPDLM-12-2020-0450

Domaracka, L., Matuskova, S., Tausova, M., Senova, A., & Kowal, B. (2022). Efficient use of critical raw materials for optimal resource management in EU countries. *Sustainability (Basel)*, *14*(11), 6554. doi:10.3390/su14116554

Fan, Z., & Friedmann, S. J. (2021). Low-carbon production of iron and steel: Technology options, economic assessment, and policy. *Joule*, *5*(4), 829–862. doi:10.1016/j.joule.2021.02.018

Fedele, G., Donatti, C. I., Harvey, C. A., Hannah, L., & Hole, D. G. (2019). Transformative adaptation to climate change for sustainable social-ecological systems. *Environmental Science & Policy*, *101*, 116–125. doi:10.1016/j.envsci.2019.07.001

Fong, I. W., & Fong, I. W. (2020). Climate change: Impact on health and infectious diseases globally. *Current trends and concerns in infectious diseases*, 165-190.

Hasan, A. M., Hoq, M. T., & Thollander, P. (2018). Energy management practices in Bangladesh's iron and steel industries. *Energy strategy reviews, 22*, 230-236.

He, K., & Wang, L. (2017). A review of energy use and energy-efficient technologies for the iron and steel industry. *Renewable & Sustainable Energy Reviews*, *70*, 1022–1039. doi:10.1016/j.rser.2016.12.007

Hodgkinson, J. H., & Smith, M. H. (2021). Climate change and sustainability as drivers for the next mining and metals boom: The need for climate-smart mining and recycling. *Resources Policy*, *74*, 101205. doi:10.1016/j.resourpol.2018.05.016

Holappa, L. (2020). A general vision for reduction of energy consumption and CO2 emissions from the steel industry. *Metals*, *10*(9), 1117. doi:10.3390/met10091117

Homayouni, Z., Pishvaee, M. S., Jahani, H., & Ivanov, D. (2021). A robust-heuristic optimization approach to a green supply chain design with consideration of assorted vehicle types and carbon policies under uncertainty. *Annals of Operations Research*, 1–41.

Huang, P. S., & Shih, L. H. (2009). Effective environmental management through environmental knowledge management. *International Journal of Environmental Science and Technology*, *6*(1), 35–50. doi:10.1007/BF03326058

IEA. (2021). *India Energy Outlook 2021.World Energy Outlook Special Report*. Paris: OECD. https://www.iea.org/data-and-statistics/charts/CO2-emissions-from-the-indian-energy-sector.

Keplinger, T., Haider, M., Steinparzer, T., Patrejko, A., Trunner, P., & Haselgrübler, M. (2018). Dynamic simulation of an electric arc furnace waste heat recovery system for steam production. *Applied Thermal Engineering*, *135*, 188–196. doi:10.1016/j.applthermaleng.2018.02.060

Kieush, L., Schenk, J., Koveria, A., Rantitsch, G., Hrubiak, A., & Hopfinger, H. (2023). Utilization of Renewable Carbon in Electric Arc Furnace-Based Steel Production: Comparative Evaluation of Properties of Conventional and Non-Conventional Carbon-Bearing Sources. *Metals*, *13*(4), 722. doi:10.3390/met13040722

Kumari, S., & Patil, Y. B. (2019). Enablers of sustainable industrial ecosystem: Framework and future research directions. *Management of Environmental Quality*, *30*(1), 61–86. doi:10.1108/MEQ-02-2018-0044

Lamba, N., & Thareja, P. (2021). Developing the structural model based on analyzing the relationship between the barriers of green supply chain management using TOPSIS approach. *Materials Today: Proceedings*, *43*, 1–8. doi:10.1016/j.matpr.2020.09.487

Lawrance, E. L., Thompson, R., Newberry Le Vay, J., Page, L., & Jennings, N. (2022). The impact of climate change on mental health and emotional wellbeing: A narrative review of current evidence, and its implications. *International Review of Psychiatry (Abingdon, England)*, *34*(5), 443–498. doi:10.1080/09540261.2022.2128725 PMID:36165756

Lorenzoni, I., Nicholson-Cole, S., & Whitmarsh, L. (2007). Barriers perceived to engaging with climate change among the UK public and their policy implications. *Global Environmental Change*, *17*(3-4), 445–459. doi:10.1016/j.gloenvcha.2007.01.004

Lou, C., Tan, S. S., & Chen, X. (2019). Investigating consumer engagement with influencer-vs. brand-promoted ads: The roles of source and disclosure. *Journal of Interactive Advertising*, *19*(3), 169–186. doi:10.1080/15252019.2019.1667928

Mallett, A., & Pal, P. (2022). Green transformation in the iron and steel industry in India: Rethinking patterns of innovation. *Energy Strategy Reviews*, *44*, 100968. doi:10.1016/j.esr.2022.100968

Meng, X. H., Zeng, S. X., Shi, J. J., Qi, G. Y., & Zhang, Z. B. (2014). The relationship between corporate environmental performance and environmental disclosure: An empirical study in China. *Journal of Environmental Management*, *145*, 357–367. doi:10.1016/j.jenvman.2014.07.009 PMID:25113230

Muslemani, H., Liang, X., Kaesehage, K., Ascui, F., & Wilson, J. (2021). Opportunities and challenges for decarbonizing steel production by creating markets for 'green steel'products. *Journal of Cleaner Production*, *315*, 128127. doi:10.1016/j.jclepro.2021.128127

Nuttall, M. (2012). Imagining and governing the Greenlandic resource frontier. *The Polar Journal*, *2*(1), 113–124. doi:10.1080/2154896X.2012.679563

Oda, J., & Akimoto, K. (2022). Carbon intensity of the Japanese Iron and steel Industry: Analysis of factors from 2000 to 2019. *Journal of Cleaner Production*, *345*, 130920. doi:10.1016/j.jclepro.2022.130920

Ogunmakinde, O. E. (2019). A review of circular economy development models in China, Germany and Japan. *Recycling*, *4*(3), 27. doi:10.3390/recycling4030027

Oyekale, J., Petrollese, M., Tola, V., & Cau, G. (2020). Impacts of renewable energy resources on effectiveness of grid-integrated systems: Succinct review of current challenges and potential solution strategies. *Energies*, *13*(18), 4856. doi:10.3390/en13184856

Rajan, R., Rana, N. P., Parameswar, N., Dhir, S., & Dwivedi, Y. K. (2021). Developing a modified total interpretive structural model (M-TISM) for organizational strategic cybersecurity management. *Technological Forecasting and Social Change*, *170*, 120872. doi:10.1016/j.techfore.2021.120872

Rissman, J., Bataille, C., Masanet, E., Aden, N., Morrow, W. R. III, Zhou, N., Elliott, N., Dell, R., Heeren, N., Huckestein, B., Cresko, J., Miller, S. A., Roy, J., Fennell, P., Cremmins, B., Koch Blank, T., Hone, D., Williams, E. D., de la Rue du Can, S., & Helseth, J. (2020). Technologies and policies to decarbonize global industry: Review and assessment of mitigation drivers through 2070. *Applied Energy*, *266*, 114848. doi:10.1016/j.apenergy.2020.114848

Rout, B., & Pradhan, J. P. (2023). Odisha's Extractive Industries: Present Status and Future Prospects. *Productivity*, *64*(1), 107–114. doi:10.32381/PROD.2023.64.01.10

Safari, A., Salehzadeh, R., Panahi, R., & Abolghasemian, S. (2018). Multiple pathways linking environmental knowledge and awareness to employees' green behavior. *Corporate Governance: The international journal of business in society, 18*(1), 81-103.

Schneider, C. R., Zaval, L., & Markowitz, E. M. (2021). Positive emotions and climate change. *Current Opinion in Behavioral Sciences*, *42*, 114–120. doi:10.1016/j.cobeha.2021.04.009

Shahzad, A., Ullah, S., Dar, A. A., Sardar, M. F., Mehmood, T., Tufail, M. A., Shakoor, A., & Haris, M. (2021). Nexus on climate change: Agriculture and possible solution to cope future climate change stresses. *Environmental Science and Pollution Research International*, *28*(12), 14211–14232. doi:10.1007/s11356-021-12649-8 PMID:33515149

Shanmugam, S. P., Nurni, V. N., Manjini, S., Chandra, S., & Holappa, L. E. (2021). Challenges and outlines of steelmaking toward the year 2030 and beyond—Indian perspective. *Metals*, *11*(10), 1654. doi:10.3390/met11101654

Sharma, R., Kannan, D., Darbari, J. D., & Jha, P. C. (2022). Analysis of Collaborative Sustainable Practices in multi-tier food supply chain using integrated TISM-Fuzzy MICMAC model: A supply chain practice view. *Journal of Cleaner Production*, *354*, 131271. doi:10.1016/j.jclepro.2022.131271

Shatokha, V. (2016). Environmental sustainability of the iron and steel industry: Towards reaching the climate goals. *European Journal of Sustainable Development*, *5*(4), 289–289. doi:10.14207/ejsd.2016.v5n4p289

Singh, D. (2019). Implementation of technology innovation in MSMEs in India: Case study in select firms from Northern region. *Journal of Science and Technology Policy Management*, *10*(3), 769–792. doi:10.1108/JSTPM-06-2018-0065

Singh, P., Pradhan, V. S., & Patil, Y. B. (2024). Modeling drivers and barriers of climate change mitigation strategies in Indian iron and steel industry: A TISM-based approach. *Management of Environmental Quality*, *35*(1), 38–60. doi:10.1108/MEQ-04-2023-0097

Singhal, A., Dube, P., & Jain, V. K. (2022). Modelling barriers of artificial intelligence in banking sectors using total interpretive structural modelling. *International Journal of Applied Decision Sciences*, *15*(3), 311–335. doi:10.1504/IJADS.2022.122649

Snow, C. C., & Hambrick, D. C. (1980). Measuring organizational strategies: Some theoretical and methodological problems. *Academy of Management Review*, *5*(4), 527–538. doi:10.2307/257458

Sonsale, A. N., Pohekar, S. D., & Purohit, J. K. (2023). Drivers to energy efficiency measures in recycled paper and pulp industry in India: An interpretive structural modelling-based framework. *Sustainable Energy Technologies and Assessments*, *55*, 102961. doi:10.1016/j.seta.2022.102961

Sushil. (2018). How to check correctness of total interpretive structural models? *Annals of Operations Research, 270*(1-2), 473-487.

Tiejun, D. (2011). The influence of iron flow on iron resource efficiency in the steel manufacturing process. *Resources, Conservation and Recycling*, *55*(8), 760–771. doi:10.1016/j.resconrec.2011.03.002

Tong, Y., Cai, J., Zhang, Q., Gao, C., Wang, L., Li, P., Hu, S., Liu, C., He, Z., & Yang, J. (2019). Life cycle water use and wastewater discharge of steel production based on material-energy-water flows: A case study in China. *Journal of Cleaner Production*, *241*, 118410. doi:10.1016/j.jclepro.2019.118410

Turner, B., Devisscher, T., Chabaneix, N., Woroniecki, S., Messier, C., & Seddon, N. (2022). The role of nature-based solutions in supporting social-ecological resilience for climate change adaptation. *Annual Review of Environment and Resources*, *47*(1), 123–148. doi:10.1146/annurev-environ-012220-010017

Wang, R. (2022). *Energy saving technologies and optimisation of energy use for decarbonised iron and steel industry*.

Warfield, J. N. (1974). Developing interconnection matrices in structural modeling. *IEEE Transactions on Systems, Man, and Cybernetics*, *SMC-4*(1), 81–87. doi:10.1109/TSMC.1974.5408524

Wei, N., Liu, S., Jiao, Z., & Li, X. C. (2022). A possible contribution of carbon capture, geological utilization, and storage in the Chinese crude steel industry for carbon neutrality. *Journal of Cleaner Production*, *374*, 133793. doi:10.1016/j.jclepro.2022.133793

Worldsteel. (2018). *Steel facts.* World Steel Association, Brussels, Belgium. htt ps://www.worldsteel.org/en/dam/jcr:ab8be93e-1d2f-4215-9143-4eba6808bf0 3/20190207_steelFacts.pdf.

Wu, R., & Lin, B. (2022). Environmental regulation and its influence on energy-environmental performance: Evidence on the Porter Hypothesis from China's iron and steel industry. *Resources, Conservation and Recycling*, *176*, 105954. doi:10.1016/j.resconrec.2021.105954

Xiang, X., Li, Q., Khan, S., & Khalaf, O. I. (2021). Urban water resource management for sustainable environment planning using artificial intelligence techniques. *Environmental Impact Assessment Review*, *86*, 106515. doi:10.1016/j.eiar.2020.106515

Yang, F., & Gu, S. (2021). Industry 4.0, a revolution that requires technology and national strategies. *Complex & Intelligent Systems*, *7*(3), 1311–1325. doi:10.1007/s40747-020-00267-9

Yang, J. (2021). How Does State Policy Shape East Asia's Steel Industry? A Selective Review. *Growth Mechanisms and Sustainability: Economic Analysis of the Steel Industry in East Asia*, 11-42.

Yellishetty, M., Mudd, G. M., Ranjith, P. G., & Tharumarajah, A. (2011). Environmental life-cycle comparisons of steel production and recycling: Sustainability issues, problems and prospects. *Environmental Science & Policy*, *14*(6), 650–663. doi:10.1016/j.envsci.2011.04.008

Chapter 12
Advocating for Indigenous Knowledge Systems for Veld-Fire Reduction in Mitigating Climate Change in Masvingo Province of Zimbabwe

Menard Musendekwa
https://orcid.org/0000-0002-6644-8727
Reformed Church University, Zimbabwe

Liveson Tatara
University of Zimbabwe

ABSTRACT

In Zimbabwe, vast tracks of vegetation are destroyed by veld fire annually, thereby facilitating the occurrence of climate change. According to the environmental management agency in Zimbabwe, veld fire gutted 165 352 94 hectares of land since the onset of the 2020 fire season, which mainly stretches from 31 July to 31 October every year. To reduce uncontrolled veld fires, this chapter makes a case for Indigenous knowledge systems (IKS) as a viable means of addressing the issue. Findings reveal that households employ various strategies such as evoking taboos that prevent destruction of sacred forests, among others, to reduce veld-fire in their community, which in turn enables them to maintain sources of their livelihood annually. Implications of these findings are discussed broadly in terms of the ways in which IKS can contribute towards governing of veld fire in Zimbabwe.

INTRODUCTION

This chapter advocates for Indigenous Knowledge Systems for veld-fire reduction in mitigating climate change in Masvingo Province of Zimbabwe. The chapter contends that through the application of in-

DOI: 10.4018/978-1-6684-9863-7.ch012

digenous knowledge systems, climate change can be arrested. There are certain issues of the environment which can be best tackled by imposing social restrictions through engaging taboos rather than legal restrictions. Legal restrictions can be evaded, sometimes challenged and also their effectiveness can be weakened by corruption. However, the consequences of violating taboos seem deterrent enough since they are spiritual and mystical in nature. In Masvingo Province of Zimbabwe, there are places that are designated as sacred and actions which are prohibited, thereby protecting both the physical and atmospheric environments. The phenomenon of protecting the environment in order to mitigate climate change through things like the protection of graves or the areas where people are buried, prevention of burning forests or cutting them down and other issues to do with the environment would be interrogated in this chapter.

It is pertinent to engage Indigenous knowledge systems when dealing with the environment considering that in Zimbabwe, vast tracks of vegetation are destroyed by veld fire annually, thereby facilitating the occurrence of climate change. Brazier (2015: x) defines climate change as follows:

Climate change is the long-term change in the Earth's climate caused by the release of greenhouse gases – such as carbon dioxide [Co2] and methane [CH4] – which trap heat in the atmosphere, causing the planet to become hotter (global warming). Greenhouse gases are released by human activities which use fossil fuels (coal, oil and natural gas) as well as by large-scale commercial agriculture and deforestation.

Some observed effects of climate change include, but not limited to, reduced reliability of rainfall, increased frequency of floods, prolonged dry spells, droughts and other extreme events (Watson, Zinyowera, Moss and Dokken, 1998 and Brazier, 2015). In Zimbabwe, recorded temperatures have risen by 1 degree Celsius over the past 40 years of the 20th Century, while rainfall and runoff decreased by 20 and 30 percent respectively (Watson et al (1998).

The destruction of the environment, one of such through veld-fire, inevitably causes climate change. According to the environmental management agency in Zimbabwe, veld fire gutted 165 352 94 hectares of land since the onset of the 2020 fire season, which mainly stretches from 31 July to 31 October every year. To reduce uncontrolled veld fires, this chapter makes a case for Indigenous knowledge systems (IKS) as a viable means of addressing the issue. Primary data is obtained using semi-structured interviews with purposively selected participants in Masvingo province in Zimbabwe. Masvingo province has been for decades highly susceptible to veld fires which threaten the livelihoods of community members. Findings reveal that households employ various strategies such as evoking taboos that prevent destruction of sacred forests, among others, to reduce veld-fire in their community, which in turn enables them to maintain sources of their livelihood annually. The implications of these findings are discussed broadly in terms of the ways in which IKS can contribute towards controlling of veld fire in Zimbabwe thereby mitigating climate change.

Climate change has become a very serious concern worldwide and cannot be left to the western methods of mitigation. The indigenous knowledge system should assist regulatory authorities in crafting indigenous policies for mitigating climate change. The continuing environmental degradation, depletion of forests through veld-fire being one aspect that has damaged the environment resulting in global warming. Global warming would result in climate change as earlier shown in this chapter. Recent spate of heat waves across the globe is blamed on various human activities that includes veld fire. Veld - fire has two major problems i.e., destruction of vegetation which forms a great part of human livelihood and ecological support to the environment. Veld - fire emits carbon monoxide into the atmosphere and generates heat that impacts on weather patterns.

Therefore, this chapter is of the opinion that by employing indigenous knowledge system, veld – fire can be curbed in Masvingo Province in Zimbabwe. The Vice President of Zimbabwe, Honourable Kembo Mohadi, is quoted as saying, Zimbabwe loses over one million hectares of forest and grasslands every year. In addition, plantations, orchards, private property, game reserves and areas that support tourism industry were destroyed (Kawadza, 2019). Before the commencement of the fire season, which mainly stretches from 31 July to 31 October, of 2020 the Environmental Management Agency (EMA) in Zimbabwe, recorded 165 352 94 hectares of land destroyed by veld - fire (Guvamombe 2020). While EMA has the mandate to ensure the protection of natural resources, very little can be achieved without the participation of the local community. Mohadi eluded that the local leaders should take responsibility to reverse the trend veld - fire has taken (Kawadza 2019). The way the local government in Zimbabwe is structured makes it easier to engage indigenous governing system in which IKS would play an important role in addressing the fire hazard that has reached such an alarming level. This is because in rural areas and farming area where veld – fires seem rampant, councillors stay with the people and are aware of the indigenous knowledge which protect the environment. Generally, such counsellors are opinion makers and highly respected in their communities.

According to The Herald (2021), the Minister of Environment and Natural Resources Management, Honourable Francis Nhema, strategised on protecting the environment by limiting the damage caused by veld - fires through awareness campaigns on veld - fires and educating communities of their implications on the environment.

In Zimbabwe, there are instruments that help to arrest incidences of veld-fire. As such there is a statutory instrument on Environmental Impact Assessment and Ecosystem protection regulations that have been enacted to provide guidelines regarding fire management. The regulations stipulate measures of pre-suppression, suppression and post suppression measures during the fire season. Though this is undoubtedly commendable, it is regrettable that very little has been done regarding the application of indigenous knowledge system.

Indigenous knowledge is useful in helping in veld- fire reduction in mitigating climate change. It has been observed that communities which have succeeded in managing disasters have made use of indigenous knowledge (Iloka (2016). The chapter seeks to advocate for indigenous knowledge systems for veld fire reduction in mitigating climate change in Zimbabwe, Masvingo province being the case study. The objectives of the chapter is to investigate the contribution of indigenous knowledge systems to veld fire reduction or management, to identify indigenous knowledge systems that deal with veld fire reduction in the communities in Masvingo province and to find ways of further enhancing the existing indigenous knowledge for veld fire reduction. Before we go into further discussion in this chapter, it is useful at this point so put across the National Climate Policy in Zimbabwe. Though comprehensive as it is, it would be seen that the indigenous knowledge aspect is conspicuously absent. In our opinion, this gap needs to be addressed in order to make it more relevant and beneficial to Zimbabwean communities.

THE NATIONAL CLIMATE POLICY OF ZIMBABWE

The government of Zimbabwe has its own National Climate policy. It also ratified the United National Framework Convention in Climate Change (UNFCCC). In its preface of the Climate Change Policy of

The Republic of Zimbabwe, Ministry of Environment, Water and Climate Policy: (2017: i), it spells out that, The National Climate Policy seeks to create a pathway toward a climate resilient and low carbon development economy in which the people have enough adaptive capacity and contribute to develop in harmony with the environment.

In the same document, the principles underpinning the National Climate Policy are underpinned by the following principles:

(i) Base climate related decisions or actions on sound data based, modelling, robust scientific findings including Indigenous Knowledge Systems (IKS) and rigorous analysis but also integrated, participatory approach;
(ii) Ensure Zimbabwe's resilience to climate related challenges by cross-sectoral engagement inclusive of public-private sector environment in implementation;
(iii) Adopt low carbon development pathway that incorporate national development aspirations, vision and national programmes;
(iv) Build resilience to climate challenges through adaptation programmes that shall be pursued concurrently with climate change mitigation programmes;
(v) Ensure that the climate policy is wholly owned by all Zimbabweans and is gender sensitive and responsive;
(vi) Address implications of the past, current and future debates;
(vii) Guide climate change adaptation and mitigation investment programmes in Zimbabwe's priority areas;
(viii) Inform National Climate Change Response Strategies and Action Plans;
(ix) Adopt new and emerging technologies and innovations that are relevant and dynamic in offering adaptation and mitigation solutions;
(x) Support climate action and compliance in line with the Multi-lateral Environment Agreements (The Republic of Zimbabwe, Ministry of Environment, Water and Climate Policy, 2017:14).

The Zimbabwe National Climate policy seems to be an offshoot of The 2030 Agenda for Sustainable Development and The Sustainable Development Goal 13. In this Agenda, member states, to which Zimbabwe is a member, express their commitment to protect the planet from degradation and take urgent action against climate change (Sdgs.UN.Org/topic/climate change). The Sustainable Development Goal 13 aims to "take urgent action to combat climate change and its impact (Zimbabwe National Climate Policy Nov 2017). Therefore, it is abundantly clear that Zimbabwe is prepared to go with the rest of the world in tackling the issue of climate change. However, in this chapter, we advocate that also indigenous knowledge systems might go a long way in mitigating climate change. The same phenomenon has been observed by Muyambo (2018).

RESEARCH AREA

The maps below show the country of Zimbabwe and the province of Masvingo. Masvingo is one of the provinces of Zimbabwe found in the south eastern part of the country. The province is bordered by Matebeleland South Province to the southwest, Midlands Province to the northwest, Manicaland Province to the northwest, and Mozambique to the southeast (https://www.researchgate.net/profile) accessed 11

Figure 1.
Source: commons.wikimedia.org.

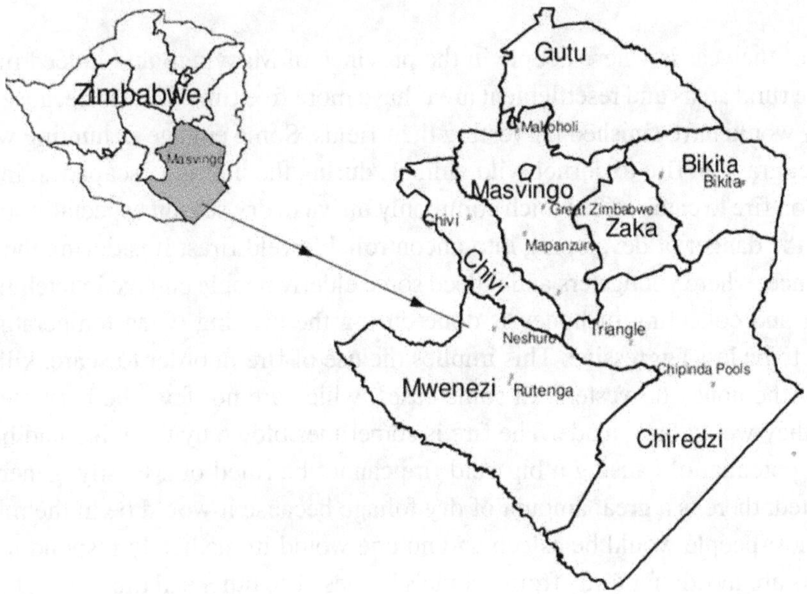

February 2021. When it receives good rainfall during the summer season, most of the province turns green with tall Savannah grass. This put the province at a high risk when it comes to the dry season of August to late November of each year.

The Context of the Problem in Masvingo Province

Masvingo Province, as the other rest of Africa, is prone to veld fire. Africa is considered as the most fire prone continent in the world and so often referred to as fire continent. It is important to see how Masvingo Province becomes essential in the discussion of veld fire. The problem of the province is that the temperatures are very high in the dry season thus the vegetation which would have been green and tall turns dry and easily catches fire. The peak period for the veld fire in Zimbabwe is from 31 July to 31 October. This period constitute Zimbabwe's fire season. Burning is the single largest contributor to greenhouse gases during the dry season in Zimbabwe and in the region (Scholes and Andrea, 2000; Mhlanga, Mashavave and Dzvairo (2017) in their research on Environmental Sub-Committees in Masvingo district found that veld fire contributes 27% of environmental damage.

Nyamadzawo, Gwenzi, Kanda, Kudhlande and Masona (2013:8) inform us of Zimbabwe's Environmental Management Agency Act in the following words:

The Government of Zimbabwe has put in place several legislative provisions with regards to veld fire management and control. It is a statutory requirement under Statutory Instrument of 7 of 2007 (EMA 2007) that any land user, owner or occupier of any land should put in place preventative mechanism … According to Statutory Instrument of 2007 as read with Environmental Management Act (CAP 20:27), no person is allowed to light fire outside residential and commercial premises during the period of 31 July to October, 31 each year, which constitutes Zimbabwe's fire season.

This period of the year is where people in the province of Masvingo and indeed most provinces in Zimbabwe in the rural areas and resettlement areas have more free time. This is because during this time of the year, they would have finished harvesting their fields. Some engage in hunting where they might be tempted to set grass on fire to corner wild animals during the hunting escapades. In certain circumstances they set on fire to catch mice which commonly thrive in dry season especially soon after harvest. Such fires have the danger of developing into uncontrollable veld fires. It is during the same season, in Masvingo province, where youngsters and indeed some elderly people engage in fetching of wild honey.

The fetching and collecting of honey is done during the evening when temperatures are low and bees are known to be less aggressive. This implies the use of fire in order to scare, kill the bees and to provide light for the honey harvesters. In some cases, which are not few, the harvesters leave the fire burning which they would have made. The fire is sometimes blown by the wind and burns the dry environment. The potential of causing a big veld fire cannot be ruled out. Firstly, generally where wild honey is harvested, there is a great amount of dry foliage because it would be in the midst of the forest and secondly, most people would be asleep and no one would immediately respond to the fire. In any case, such forests are mostly far away from people's homes. The other sad thing is that generally no one would like to take responsibility since all happens in the darkness of the night and culprits would like to avoid the penalty which goes with such offence.

In Masvingo Province, the resettlement areas seem to contribute a large number of veld fires compared to the rural areas. This is so because there is more arable land and the foliage is thicker than in rural areas. More to that wild life is much more abundant and people would be tempted to employ even fire to help catch them. The seemly most effective ways of controlling such veld fires is through the use of indigenous knowledge systems. This is so because there are certain taboos and beliefs which prevent the damage of the environment through veld fire. Such taboos are respected since their violations are believed to attract supernatural consequences.

INDIGENOUS KNOWLEDGE SYSTEMS AS A FORM OF MITIGATING VELD FIRE IN MASVINGO

Tatira (2000; 2000b; 2013) has written on taboos and how they help in shaping Shona people's worldview regarding many facets of life, including how they interact with the physical environment. Taboos are used by the people in Masvingo province to discourage the wanton destruction of environment. There are taboos which protect mountains, forests and those that discourage burning of grass and the general playing with fire.

Taboos that Protect Forests and Mountains From Being Destroyed

In the wisdom of the Shona people, and in this particular case, people of Masvingo province, protect forests and mountains from being destroyed by the people by declaring them sacred. People are not allowed to venture into such forests or are prohibited from making fire in the forests or near them. The places are believed to be inhabitancy to the ancestral spirits. There are some mountains where chiefs are buried and such mountains are out of bounce to ordinary people. It is only the elders who belong to the chieftaincy who are allowed to climb up the mountains on specific periods for ritual purposes or for the burial of the chief.

In any case, in Shona culture, it is not only untoward for an individual to be seen around the graves of people one is unrelated to but also an offence that might result in such a person taken to a chief's court for penalty. In most cases, such a visit would be interpreted as an act of witchcraft. The general absence of people from such sacred places and the prohibition of fires allow the places to flourish in vegetation and wild animals. This helps to conserve the environment and mitigate climate change.

Other mountains and forests are declared sacred though they might be no chiefs buried there. The belief would be that the ancestral spirits dwell there or at least would be annoyed by any form of defilement of such places. Defilement might amount to cutting down of trees, defecating in or around the sacred place and worse still setting it on fire. In so doing the environment is conserved.

Graves as Sacred Places

In Shona culture, graves are sacred places. This means that it is not only the chiefs' graves that are sacred. Fire is not allowed to burn close to or over the graves. To do so is taboo and if fire is seen close to the graves every effort is made to extinguish it as quickly as possible and the culprit sought. Taking into account that burial sites are ubiquitous wherever you find people, this helps to curb veld fires as such sites, in form of graves, are guarded against fire. In some cases, people weed around the graves to keep them away from the danger of veld fire. It is the way graves are treated that helps to reduce veld fires in Masvingo province.

Burning a Hut Roof

Culturally it is not permissible to burn the roof of the hut or any other thatched roof. The message behind this prohibition is that fire should not be played with. Children might be tempted to experiment with fire by torching the grass thatched roof or even elders in arson cases. Whenever one sets a hut or house on fire whether by mistake or intentionally, it is an offence that lands a person at the traditional court. Though such issues might end up as criminal cases in the modern courts, here we are interested in how the Indigenous Knowledge Systems play a vital role in preventing fire thus mitigating climate change. For burning a hut or any other dwelling, the culprit would be fined one herd of cattle for the offense by the chief.

The chief would accuse the person of *kuvheneka nyika yaShe* (lightening the Chief's chiefdom). Such an accusation and the subsequent fine help people to be careful with fire thus preventing the burning of the dwellings and the surrounding environment which could also catch fire in the process. In any case, people from childhood are discouraged from playing with fire.

METHODOLOGY

. Primary data has been obtained using semi-structured interviews with purposively selected participants in Masvingo province in Zimbabwe. The selected interviewees were traditional leaders and local authorities. Twenty participants were selected and their names withheld on ethical considerations. Hence where necessary pseudonyms have been used. The outcomes of the interviews can be presented in three ways of reducing veld-fire in the community, which in turn enable people to maintain sources of their livelihood. These are evoking taboos, law enforcement and peer education.

Evoking Taboos That Prevent Destruction of Sacred Forests

Among the Karanga people of Masvingo, there are various taboos that limit the burning of certain tree species. Veld fire would indiscriminately destroy forests if some trees were not regarded as sacred. One local chief in Chivi district of Masvingo confirmed that, "*Muchakata* tree is sacred and should not be used for firewood neither should it be burnt down". There are various beliefs regarding what entails sacredness of *Muchakata* tree. Participants interviewed confirmed the belief that in the traditional times, when one was hungry one could ask for food from under the *Muchakata* tree. Some further indicated that the village could have a specific *Muchakata* tree where they would perform rain - making rituals.

The same tree is also used to predict how a season would turn out to be. On this issue, a chief in Masvingo district of Masvingo province alluded that "kubereka *kwemuchakata kunoratidza kuti gore iroro kune zhara*" (If *muchakata* tree gets too much fruits, it is evident enough that there is going to be famine that particular season). The abundant fruits would provide food for livestock and people. People may eat the fruits or make some porridge or beer out of the fruits. During the drought when people lack food reserves, they can collect fruits, dry them in order to get nuts which can eat raw or roasted. The seeds can also be used to make butter. The *Muchakata* tree is, therefore, considered sacred and highly protected by the community and its leaders from being burnt.

It is apparent to note that the *Muchakata* tree does not require much rainfall in order for it to produce fruits. For this reason, it remains the source of food during such difficult times and hence should not be cut or destroyed through veld fire.

In Masvingo there are also sacred mountains and forests. Such forests are considered sacred and no one is allowed to burn or pollute them in whatever form. One such hill is Chirora of Murairwa village in Gutu district of Masvingo province. One informant confirmed that the hill is sacred and no one is allowed to climb it. Even cattle are not allowed to graze on and around this hill. Even fetching of fire hood is not permitted. Mythical stories circulate that anyone who takes courage to climb up the hill would disappear. This makes the hill sacred and conserved from any form of destruction.

Law Enforcement

Law enforcement is executed at various levels in communities in Masvingo province. These include village courts and the chief's court. In a village, the reports of offense committed against a neighbour are normally first presented before the sub-chief's court. Incidences of veld fire may be caused by fire which might get out of control as villagers prepare their land in preparation for the farming season. In some cases, this fire apart from destroying the environment, it ends up destroying other people's property. The sub-chief in Bikita district of Masvingo province, commenting on the issue of veld fire retorted, "*Takapiwa simba neHurumende rokuranga vanhu vanopisa minda yevamwe*" (We were empowered by the Government to punish people who burn other people's fields). Anyone who intends to clear his/her field by burning should first make a fire guard. Fire guards are made either by clearing right round one's area most commonly using a plough. This exercise is highly recommended and is also regulated by EMA especially in farming areas. According to a chief in Zaka district of Masvingo province, who owns a farm, he notes, "Everyone should be responsible for own area and during the fire season EMA inspects every farm and fines anyone without a fire guard". It is noted that those with farms are bound by the requirements of EMA while those in rural areas though regulated by EMA, they are not expected to make fire guards. In rural areas, no one is allowed to burn the grazing areas especially during the dry season. The penalty is much heavier during the dry season than when towards the rainy season. It should, however, be noted that controlled fire is allowed when one burns one's stalks and the contour ridges. This kind of exercise does not destroy the environment.

Interviewed chiefs across the province, concurred that the rules of the land do not allow anyone to leave fire unattended. There is a general discouragement that people should not leave the fire burning made at bus stops when people wait to catch up early buses. This is because generally people feel cold and make fire which they often leave unattended when they board the bus. People who harvest honey are discouraged from using open fire but rather they are recommended to use smoke from cow dung for the reason that it can be easily controlled than open fire.

Anyone who causes veld fire is expected to seek the help of other villagers to control or rather put off the fire. According to one village head, "the offender must report the case of veld fire to the village-head promptly". In some areas of Masvingo, there are fire teams who help put off the fire. The cultural practice of keeping yards clear of anything, which likely makes it easy to catch fire, helps in reducing incidences of veld fire. The cleared dirty is put in a rubbish pit. Every household generally has two separate pits, one for rubbish and the other one for ashes. This is echoed by one sub- chief in Mwenezi who says, "*Imba yoga-yoga ine durunhuru*" (Every household has a place for depositing dirt and for depositing ashes). When ashes are deposited one is expected to ensure that there is no live charcoal, otherwise wind may blow it and cause veld fire. All fire should be extinguished when not used.

Great care is taken to avoid rubbish pits from causing veld fire. This is because penalty for causing veld fire is heavy considering the paucity of resources in rural areas. The owner of a burnt property would be given a chance to charge the compensation he/she desires. However, in the case of burning of a grazing area, it is the chief who fines the culprit. Another chief talk of such penalty as follows, "The court fees include a goat which is to be slaughtered at the court and beer for the courtiers once the court is over. Further charges would be made depending on the gravity of the offense. Therefore, the chief's courts have the capacity to restrain villagers from making unnecessary starting veld fires. Such penalties are determined by the chief because the grazing area does not belong to particular individuals but to the chief who is the custodian of the land.

In an interview, another chief declared that setting a house on fire is always an offense. He also alluded to the tradition that when one torches a house, the whole village would join and try to put off the fire. The offense for torching a house goes beyond the amount of damage it has caused as it is also regarded to be an offense to the spiritual realm for which the chief represents. The village head takes the matter to the chief who would then send his messengers to summon the culprit. If no culprit is found, the whole village would often be summoned. The whole village may sometimes be fined for failing to bring the culprit to book. The charges are increased if the fire would have gotten out of control and damaged the environment.

It is noted that there are variations in determining penalties related to veld fire from place to place and from time to time considering the damage caused.

Peer Education

The Shona culture has its own education system since time immemorial. In the evening, especially after harvest, is always family time and children are taught ethical conduct through telling of stories. These stories may also help family members to desist from spreading veldfire. This section explores the views given by some local authorities.

In an interview with one counsellor, regarding socialisation on environmental issues, he said, "*Chikoro chinotangira mumba*" (Education begins in the household). Older children take turns in storytelling. These stories have moral value and would equip the children to become responsible citizens. These moral stories cascade to village level as lessons when the village-head inculcates morality among his subjects. The village heads would in turn use those fora as opportunity for conscientising subjects on the dangers of the veld-fire. They also teach their subjects the taboos they do not need to break. Similarly, they develop an understanding on the consequences of violating the laws of the land. They also would include issues related to veld fire especially at the inception of the dry season. This means the significance of peer education in reduction of veld fire has been critical in Masvingo. The same counsellor also heeded that at every funeral, the counsellor is given an opportunity to address mourners and it is on such occasion one grabs the opportunity to conscientise the local communities on the dangers and potential damages caused by fire.

Another counsellor further alluded that Churches and church leaders are also involved in campaigns against veld fire in their communities.

CONCLUSION

The Zimbabwe government is supportive of mitigating climate change as evident by its National Climate Policy of 2017. The Masvingo province which was the case study of this chapter is also very supportive in mitigating climate change with regards to control of veld fire. Communities and local leadership across the province are conscious of the danger of veld fire to the environment. They use their indigenous knowledge systems, local leadership and EMA to control veld fires.

Their use of fire is regulated from family level, village level and to community level in an attempt to protect their environment. Heads of the family, village heads, counsellors, church leaders, sub- chiefs and chiefs all play important roles in protecting the environment against veld fire. The traditional leadership uses multi-pronged approach in dealing with the issue of veld fire. The traditional leadership use persuasion, teaching, taboos and penalties to ensure that people refrain from destroying the environment through veld fire. Related to taboos, they evoke mythical stories of what befall people who disregard sacredness of certain places. These stories are scaring and many people are discouraged from defying taboos. In any case, if one is caught defying such taboos he/she is punished by the chief by being made to pay a heavy fine.

Therefore, in formulating policies on climate, it is advisable that the government recognises and promotes indigenous adaptation strategies. The commendable effort exhibited by the local leadership must be supported by the National Government by way of incentives for those communities that have low statistics in veld fire. Leading communities should be given an opportunity to share their successes with other districts and provinces. Publishing and distributing across the country various indigenous ways of conserving the environment as practised in different communities would go a long way in empowering all communities. This would help in reducing incidences of veld fire thereby mitigating climate change.

In doing so the government would definitely be complementing and further supporting the efforts exhibited by the people of Masvingo province in mitigating climate change through the reduction of veld fire.

This study in no way can it claim to be exhaustive. The authors of this chapter hope that, since climate change, is a reality and an enormous threat to humanity, more comprehensive research would be undertaken to fill the gap left by this study.

REFERENCES

Brazier, A. (2015). *Climate Change in Zimbabwe: Facts for Planners and Decision Makers.* Konrad-Adenaner – Stiftung.

Chenje, D., Sola, L., & Paleczny, D. (1998). *The State of the Zimbabwe's Environment. Ministry of Mines, Environment and Tourism.* Government Printers.

Furusa, M. (2024). *Indigenous Knowledge and Technology in African and Diasporan Communities: Multi-Disciplinary Approaches.* University of Zimbabwe: Southern African Association for Culture and DevelopmentStudies.

Guvamombe, T. (2020). EMA alerts nation on veld fire season" in Spotlight. https://spotlight.co.zw/ema-alerts-nation-on-veld-fire-season/

Gwidi, R. (2019). Masvingo Least Affected by veld Fire. *Herald.* https://www.herald.co.zw/masvingo-least-affected-by-veld-fire/

Kawadza, S. (2021). Veld fires a national concern: VP Mohadi. *The Herald.* https://www.herald.co.zw/veld-fires-a-national-concern-vp-mohadi/

Lloka, N. G. (2016). "Indigenous knowledge for disaster risk resolution: An African perspective." In *Journal of Disaster risk. Studies, 8*(1), 1–7.

Mhlanga, W., Mashavave, T., & Dzvairo, S. (2017). A Study of the roles, opportunities and challenges of environmental sub-committees (ESCs): A case of Masvingo district, Zimbabwe. In South Africa Journal of Education, Science and Technology, 4(1).

Muyambo, T. (2018). *Indigenous religion and environmental challenges: Phenomenological reflections on the role of the Environmental Management Agency in Zimbabwe.* Routledge. doi:10.4324/9781351066501-13

Nyamadzawo, G., Gwenzi, W., Kanda, A., Kundhlande, A., & Masona, C. (2013). Understanding the causes, socio-economic and environmental impacts, and management of veld fires in tropical Zimbabwe. *Fire Science Reviews, 2*(1), 2. http://www.firesciencereviews.com/content/2/1/2. doi:10.1186/2193-0414-2-2

Scholes, M., & Andreae, M. O. (2000). Biogenic and pyrogenic emissions from Africa and their impact on Global atmosphere. *Ambio, 29*(1), 23–26. doi:10.1579/0044-7447-29.1.23

Tatira, L. (2000). *Zviera zvaVaShona*. Mambo Press.

Tatira, L. (2000b). The role of *Zviera* in socialisation. In Chiwome,E. Gambahaya, Z. and Tatira, L. (2013). *Some Aspects of Ubuntu/Unhu as encapsulated in Zviera (Taboos)*. Harare: Bhabhu books.

The Herald. (2021). Veld fire destroys environment. *The Herald.* https://www.herald.co.zw/veld-fires-destroy-environment/

Watson, R. T., Zinyowera, M. C., Moss, R. H., & Dokken, D. J. (1998). *The Regional Impacts of Climate Change*. IPCC.

Chapter 13
Blockchain Technology in Disaster Rescue Wings

Mukesh Kumar D.
Hindusthan College of Arts and Sciences, India

A. V. Senthil Kumar
https://orcid.org/0000-0002-8587-7017
Hindusthan College of Arts and Sciences, India

Indrarini Dyah Irawati
Telkom University, Indonesia

Rohaya Latip
Universiti Putra Malaysia, Malaysia

Ibrahiem M. M. El Emary
King Abdulaziz University, Saudi Arabia

Ismail Bin Musirin
Universiti Teknologi MARA, Malaysia

Hesham Mohammed Ali Abdullah
Jouf University, Saudi Arabia

Manjunatha Rao L.
National Assessment and Accreditation Council, India

S. Chaya
Nagarjuna College of Engineering and Technology, India

Nitin Arya
India Para Power Lifting, India

ABSTRACT

Blockchain, the base of Bitcoin, has recently received broad attention. Blockchain acts as an immutable ledger that allows transactions to take place in a decentralized manner. Blockchain-based applications are growing, covering many sectors, including financial services. The reputation of Blockchain, the base of Bitcoin, has recently received broad attention. Blockchain acts as management, and the internet of things (IoT). During disaster events, timely and targeted information provision and exchange could provide great help to the stricken population in difficult and complicated environments. This chapter reports a service-oriented system, called BlockchainDisaster Rescue Wings, for providing emergency support to sufferers and rescuers in disasters. The authors identify the most frequent request sequence patterns (FRSP) of Rescue Wings, and develop a new application for efficiently scheduling the requests to minimize the response delay. The system has been tested in several disaster rescue drills, and has been successfully applied.

DOI: 10.4018/978-1-6684-9863-7.ch013

INTRODUCTION

Each year, unexpected incidents generate $140 billion in losses in several nations and the deaths of 150,000 individuals globally. Iran is one of the most disaster-prone nations in the world for a variety of reasons, and at the moment, earthquake and flood risks are present in around 70% and 50% of the country, respectively. To avoid unforeseen outcomes, it is crucial today to draw lessons from the painful past while also taking risks and the uncertain future into consideration. Therefore, it is crucial to adopt new crisis management techniques and reduce losses and damages when dealing with such incidents. Moreover, in an emergency, "rescue," "treatment," and "support" organisations must be ready; otherwise,It is clear that the three aforementioned domains (1–3) can all benefit from modern information and communication technology. It is inevitable that during an emergency, new information and communication technologies would be used to get to the accident scene swiftly and enhance the quality and quantity of rescue efforts. Communication and information. (Gao W, Hatcher WG, 2018)

The researchers have also recommended using delay tolerant network (DTN) to build an emergency post disaster communication network using the various communication features of smartphones, such as Wi-Fi Direct or Bluetooth, taking into account the effects of natural disasters on communication infrastructure. Volunteers for disaster relief and drivers of supply vehicles all use these smartphones when operating in DTN mode [1-6]. DTNs, however, frequently break up, lack an end-to-end link, and have inconsistent connectivity, which results in subpar network performance. Establishing DropBoxes, which are fxed relay nodes that assist in message transfer and are commonly found on smartphones and laptops with internet connectivity, is one potential method for enhancing network throughput.A scientific forecasting model that can predict the requests for post-disaster emergency relief items and list their uses is essential for reducing the issue of ad-hoc distribution. Using mathematical modelling, a few works have made an effort to predict the precise need for relief supplies in each emergency shelter [5] and to list the particular uses of such supplies.(Zheng Z, Xie S, Dai H, Chen X, 2017)

On the other hand, it becomes essential to have a document for providing an unquestionable and globally accessible record of relief requirement vis-à-vis allocation for effective relief management in order to solve the challenges of allocation transparency, resource diversion, etc. The incorporation of blockchain technology has been suggested [7– 14] by a number of disaster response groups and researchers in order to produce such a shared and unchallengeable set of records. Nevertheless, the availability of end-to-end internet connectivity, which, as was previously discussed, may not be present in a post- disaster situation, limits the use of blockchain.(Izadi DS, Sharif-Khatibi Z,2020)

In this paper, ReliefChain, a blockchain-based post-disaster relief allocation system using a delay-tolerant network, is proposed. The suggested system gathers the relief needs of various shelters at a specified Location, which in turn verifies these needs to prevent resource divergence, estimates the precise need, and enumerates the particular uses of these relief goods. As is common knowledge, smart contracts are a subset of Ethereum accounts that are used to create and send transactions via the blockchain network. DropBox uses smart contracts to initiate new transactions that record information about requests, services, and allocations on the blockchain. As any alteration to the record would necessitate several validations, logging these particulars into a blockchain renders them tamper-proof.

The coordination of crisis management during an emergency and the rescue process depends heavily on communication and information exchange. As a result, one of the key elements in ensuring an adequate reaction to emergencies will be the development of a strong, complete, and consistent communication infrastructure. It is necessary to continually assess the usage of new technologies that appear to assure trustworthy communication and information exchange.(Siemon C, Rueckel D, Krumay B,2020)

Distributed Ledger

User Information

This module deals with the authentication reports of user or admin. In user authentication module, it stored the verification details of the user. The type of user and their usage, identification details like useid, name, address etc. In admin authentication the verification of the user occurs. The detail of this module helps to verify the user information.\

Disaster Information

A natural disaster is a major adverse event resulting from natural processes of the Earth; examples include floods, hurricanes, tornadoes, volcanic eruptions, earthquakes, tsunamis, and other geologic processes. A natural disaster can cause loss of life or property damage, and typically leaves some economic damage in its wake. the disasters information stored and share the information about damage to other users.

Emergency Request

In this module request to public. In this module disaster damage and other information details update this web page its user for affected people. people can help them and intimate to others.

Rescue Details

Rescue Wings deploys its client-side applications to webpage of end users, who can upload and maintain their profiles on the servers. The servants are responsible for actively collecting real-time information and monitoring the states of the clients on the spot, and providing required services to assist the users in self protection, escape, search and rescue (S&R). When detecting abnormal states, the servants can also remotely request ad hoc operations of the clients. Rescue Wings has access to a number of public services for obtaining rescue information, and provides a set of intelligent services for supporting the responders in different stages of the rescue operation.

Search Disaster

This module is about users' input data and its analysis. At first, the user enters the disaster and views that disaster-affected area. Users can search the text. This text will be used to find relevant data in the database. Then the user can find the location and loss of disaster.(Ali MS, Vecchio M, Pincheira M, Dolui K, Antonelli F,2017)

Introduction to the Blockchain Technology Concept

Technology known as "blockchain" has recently received a lot of interest and is currently the subject of extensive media campaigns. This technology has several different definitions. It is a collection of protocols and encryptions that inventors claim may be used to store data securely across distant networks. It serves as a platform for creating digital currencies in the world of commerce and economy, and for technologists, it influences how the future iteration of the Internet will be. It is also seen as a way to change society and the economy in order to create a decentralised world. The breadth of this science's application is one of the reasons why blockchain technology has overtaken the Internet as the greatest human innovation.

The words "Block" and "Chain" are combined to form the word "blockchain." This technology is built up of various pieces. Data is kept in batches in carefully constructed blocks using "blockchain" technology, which uses codes to connect each block to the one before it. When a block is joined to a chain, the veracity and accuracy of its information are established, and a connected and legitimate chain is produced. Consequently, blockchain is more than simply a technology; it is also a culture and a society, whose ultimate purpose is to establish decentralisation and a more just world. Blockchain is also a type of reporting and information system.

All users of the network can access the data saved in this system, and utilising encryption makes it difficult to remove or change the data collected. In general, blockchain technology is regarded as a huge technological advancement that has the potential to profoundly alter corporate models, operational procedures, and the whole economic system. Also, this ground-breaking technology is widely used in a wide range of industries, from specialised areas like fraud and financial abuse to routine information systems tasks like organisational modelling. This suggests that blockchain technology is a type of innovation in both technology and business (16). The results of earlier research support the claim that one of the key advantages of employing blockchain technology is its decentralised organisation and distribution, which may boost objectivity and confidence (15).(Düdder, B. and Ross, O. 2017)

Benefits of Blockchain Technology

The informants cited a wide range of benefits that come with blockchain technology. They have been combined under the headings of decentralisation, immutability, transparency, and trust. Decentralization and trust were highlighted as advantages by every informant, whereas immutability and transparency were mentioned by 11 people and openness by 4 people.

All of the sources agree that decentralisation is the primary advantage of blockchain technology. According to the informants, there are a number of advantages to dispersing information among users as opposed to keeping it in one location. First off, sharing information also means sharing control and authority. Blockchain technology has the potential to include parties in processes that are typically disregarded in the humanitarian sector, where ownership of all parties involved, including the beneficiaries, is respected and sought for. Currently, creating a blockchain wallet for someone does not grant them the same access that a bank card would, but it does connect them to a network.Future advancements could cause coins on blockchains to be accepted as currency in many nations, allowing those with blockchain wallets access to a variety of services they would typically be denied. In any event, decentralisation implies that the end user has more power and that they are the only proprietors of their data draws attention to this aspect.

"The cooperation aspect of blockchain will be the advantage, not the blockchain itself." Thirdly, by exposing operations that do not bring value to a project, system, or operation, decentralisation removes intermediaries that do not add value (Informant 9). The local banks, who often operate as a go-between for a foreign relief organisation and the recipients, are the disintermediated actors in the case of CBT (Informant 4). Taking out the local banks reduces practically all transaction costs, some previously present dangers, such as the financial risk of sending significant sums of money to a bank in an unstable climate, and eliminates the requirement for advanced transfers (Informant 4).(Engelenburg, S.V.S. van et al. 2017)

Traceabilityis another aspect of blockchain technology that 11 out of 12 survey participants saw as advantageous. Transparency and immutability interact to produce it (Informant 8). This makes it simpler to track the whereabouts of donated funds while emergency help is being provided Fraud and corruption are significantly harder to commit and much simpler to spot. The likelihood of individuals who are most in need of assistance obtaining it may increase if inclusion and exclusion errors were greatly reduced If they are no longer dependent on the information provided by the banks but can trace the data themselves, assistance organisations might also use blockchain technology to enhance their monitoring and evaluation processes.This could imply that, in the right setting, supermarkets in refugee camps would consider and stock food and other things more in line with the needs of the refugees (Informant 4). As the procedures are automated, additional use cases, such supply chain management and traceability of in-kind aid, would also be feasible (Informant 4). All of these proposed procedures would make organisations more accountable, both to their supporters and to the people they are helping

Trustis the result of the previously mentioned blockchain technology qualities and a valuable asset in a distrustless world (1F2). To some extent, trust has always been the cornerstone of human relationship Because of globalisation, trust is becoming rarer and more difficult to build. provides an explanation as to the reason for the case.

Opennessis another advantage of blockchain technology that only four respondents cited. It refers to the technology's open availability, transparency in how it works and how its operations are carried out, and its independence from any one corporation or organisation This means that in the case of help organisations, they can efficiently create or buy a blockchain protocol Additionally, once an aid organisation has fulfilled its mission in a country, this may allow it to transfer a specific blockchain to another, local help organisation or the government . As a result, the beneficiaries' ongoing help could be ensured, and greater long-term results could be tracked and assessed.

Additional advantages, such as security resilience and anonymity (Informant 10), are only stated by a small number of people and should be viewed as debatable, as will be shown in the next section. (Gaetani, E. et al. 2017)

Challenges With Blockchain Technology

Each participant shared their experiences with the difficulties that came with using blockchain technology. Nonetheless, they can all be broadly divided into four categories: immaturity, ecosystem, scepticism, and technical hurdles. Some of these were particular to the project or use case at hand. All of the informants discussed ecological immaturity in a variety of ways. Ten and six informants, respectively, spoke about scepticism and technical difficulties.

According to the informants, blockchain technology's obvious issue is immaturity. Many organisations and businesses are still waiting to see how the technology will evolve in the future with only a decade under its belt and a few years of actual use cases All of the informants consider the technology to be in its infancy and point out the scant amount of research that has been done on the difficulties it presents and the potential effects it could have on people and processes. Other informants claim that there hasn't been enough time for a successful use case of the technology to be discovered.The existing use cases, such the WFP's "Building Blocks" initiative, might have been carried out with other, more conventional databases, according to informants 1 and 4.

The entire potential of the technology has not been explored, and its benefits over other technologies are constrained, as long as its trump card, the capacity for cooperation, has not been utilised.According to the informants, the ecosystem in which blockchain technology is intended to be deployed frequently poses the initial issue. Many conditions must be met for blockchain technology to function in a humanitarian context. The first need for processing transactions based on the blockchain is internet connectivity. Such an internet connection can be established during long-lasting crises, like the Syrian civil war and the ensuing refugee camps. Aid organisations are unable to use a blockchain in some situations, such as those where the camps are farther away or if a natural disaster has destroyed the networks. Second, the intended recipients must be reached digitally.

In this situation, the beneficiary must first develop a digital identity before they can access any blockchain-based aid. Thirdly, many of the previously described benefits do not occur if the coins or tokens used to represent a currency on the blockchain are not widely accepted . Because of the magnitude of the camp, the WFP's "Building Blocks" programme in the Zaatari refugee camp might be compared to a city.The beneficiaries can be identified in the camp's supermarkets using iris scanners, and they may make purchases there using their blockchain wallets Yet, the camp is the only one to benefit from the blockchain. It does not provide access to a bank or other options for refugees outside of the camp, resulting in no systemic change. Additionally, as summarised by the advantages of traceability, monitoring, and evaluation by the organisations are also constrained in such a situation.

One obstacle to get past is scepticism from all sides, which may be done by effectively raising awareness and managing change. Informant 3 emphasises how crucial change management is to the technological success.

"Blockchain is merely a technical option. It is about managing the entire collaboration as well as change management, not about picking blockchain or not.

Independent of blockchain technology, for any new technology to be successful, decision-makers must be flexible and understand that maintaining the status quo will neither increase productivity or promote growth. Yet, as stated by Informant 4: "Creating such a consensus can be challenging in conventional, hierarchical structures and bureaucratic processes.

The first problem that arises when trying to convey blockchain potential is frequently its connection to Bitcoin and other cryptocurrencies, as well as the volatility such currencies have seen recently. As a result, blockchain has a negative image for many people who only have a cursory understanding of the technology. To combat this scepticism, one must fight the lack of specialists and information surrounding blockchain technology and place faith in a still-evolving technology. As a result of the ongoing scepticism, businesses are currently not adopting the technology, and governments are not passing any regulations or laws.Both of these issues pose difficulties for those who are now working on blockchain solutions because they are navigating an uncertain future and unsure of whether their protocol will satisfy potential future regulations and market demand.

Despite only comprising a small portion of the difficulties highlighted, those researching blockchain technology also encounter technical difficulties. Data security and immutability are the two most important of these. Concerns concerning blockchain technology' compliance with the EU GDPR have been raised by certain individuals working in the humanitarian sector; nevertheless, some people do not see this as a problem. The consensus among all interviewees is that in order to reduce the chance of sensitive data being misused, no sensitive data should ever be housed on a blockchain. The fact that the data cannot be changed and that no organisation can be sure that it will always have control over its systems necessitates a discussion of these concerns and an examination of their ramifications.Currently, the UNHCR identifying code, location, family size, phone number, and iris scan are saved onto the blockchain in the WFP's "Building Blocks" project. Date of birth, gender, ethnicity, and other details are purposefully omitted.

In addition to this risk to data security, Informant 10 argues that a so-called 51% attack could weaken the system's resiliency and compromise data security. Decentralization in this case makes a cyberattack more difficult and time-consuming, but it does not render it impossible Informant 10 even goes so far as to claim that because there is no centralised, trustworthy authority guarding the data, decentralisation makes fraud easier.Some potential technological issues are viewed as less urgent, including interoperability, scalability, and electricity consumption. Scalability is an issue for people who want to gradually extend their blockchain protocol, but it has not been encountered thus far because the majority of private blockchains only have a small number of nodes. Thus, energy consumption does not appear to be a worry for businesses and would only become a problem once large-scale blockchains are installed. (Balcik, B., Beamon, B. M., Krejci, C. C., Muramatsu, K. M., & Ramirez, M.2010).

DISASTER RESCUE WINGS

Disaster response is the second stage of the disaster management cycle. It consists of a number of components, including as warning/evacuation, search and rescue, immediate help, assessing damage, continuous assistance, and the quick construction of infrastructure or restoration of infrastructure (i.e. provisional storm drains or diversion dams). To maintain life, enhance health, and boost the morale of the affected community, emergency response seeks to offer rapid aid. Such aid could take the form of building semi-permanent settlements in camps and other places, as well as more focused but limited assistance such providing transportation, temporary housing, and food for refugees. Initial fixes to broken items or a shift in infrastructure may also be necessary.

The goal of the reaction phase is to keep people safe, stop further disasters, and provide for their basic needs while more long-term and sustainable solutions are sought. The government or governments whose territory the disaster has occurred are primarily responsible for attending to these demands and responding to a disaster. Also, during this stage of the disaster management cycle, humanitarian organisations are frequently quite active, especially in nations where the government has the resources to adequately address the requirements.

Disaster response is described as decisions and activities conducted in accordance with the strategic, tactical, and operational objectives established by emergency responders in the Emergency Response and Recovery guidance published by the British Government. They will, at a high level, aim to protect life, control and lessen the effects of the emergency, and establish the circumstances for a return to normalcy. The decisions and steps made to address an emergency's immediate repercussions are collectively referred to as a response. Rapid deployment of collaboration, coordination, and communication arrangements are crucial because it is often likely to be brief, lasting only a few hours or days. Reaction includes making an effort to address issues beyond the immediate consequences of the catastrophe (e.g. fighting fires, rescuing individuals). (Becker, P.2014).

Disaster Rescue Technologies Communications

The Government Emergency Telecommunications Service provides emergency access and priority handling for local and long distance calls over the public switched telephone network in order to support federal, state, local, and tribal government personnel, business, and non-governmental organisations during a crisis or emergency. A customer can use the Nationwide Wireless Priority Service to wait for available cellular bandwidth.

Wireless mesh networks can be quickly set up to offer Internet connectivity, replace downed mobile phone networks, and enable emergency and post-disaster communication, notably for coordinating emergency calls and disaster response. Often times, volunteer communities with limited resources design and operate open-source mesh networks like B.A.T.M.A.N.

Electricity

Emergency power systems, such as mobile microgeneration units, mobile charging and power supply stations, or specially designed or extended smart grids, could restore power to small regions whose connections to the main power grid were severed or support critical electrical systems in the event of a loss of normal power supply.

Transportation

A disaster may have rendered the transportation infrastructure impassable, hampering logistics, evacuation, and disaster response. Technology might make it possible to quickly and adequately replace or restore some of the transportation network. One example is the quick construction of stable bridges using lightweight, portable, and/or locally sourced materials or components, which military have done.

Waste Management

Disaster waste is frequently handled haphazardly. A disaster's trash production can overwhelm current solid waste management infrastructure and interfere with other response efforts. Conventional garbage may need to be managed in a similar manner depending on the sort of disaster, its scale, and the length of time it takes to recover, and both may be connected to the restoration of the transportation network.

Emergency Accommodation

Infrastructure is occasionally thought of as including housing as it relates to emergency housing. It is a problem to house humans and animals temporarily following calamities. Occasionally the infrastructure and logistics of private accommodations are repurposed for the disaster response.

Water Supply

Disasters may affect the infrastructure for the supply of water, drainage and sewage, and the operation of wastewater treatment facilities.

Response Coordination Websites

Websites and other similar ICTs can be used to coordinate volunteers as well as other participants in a disaster response, such as locals and civil organisations like the TechnischeHilfswerk, to prevent traffic jams, "disaster tourists," and other obstructions of the transportation network, to allocate various forms of aid to areas in need, to report missing people, and to increase efficiency. Following the European floods, such webpages have been created for the many devastated regions.(Brinkmann, S., &Kvale, S. 2015).

Uses of Blockchain Technology in Crisis Management and Rescue Operations

Emergencies are unplanned situations brought on by accidents or destructive disasters that endanger the environment, property, and human lives. Emergency situations can only be controlled if all the actors demonstrate more effective collaboration and coordination, regardless of the severity of the effect or whether they are wholly unplanned or predictable. Since the decisions of any decision-maker can have far-reaching effects, accurate and optimal information availability is a crucial prerequisite for crisis management and disaster response.Managing the issue requires a range of information, skills, and abilities, all of which the role-maker offers. In a real emergency, however, all efforts must be focused on a single objective, therefore this diversity is also seen as one of the impediments. Although any planner may find use for this strategy, it may cause serious problems with the command and control center's network and cooperation.

Since the shared infrastructures may be temporarily or permanently unavailable or unreliable in local, national, and international emergencies, communication and information exchange are of utmost importance. Designing and assessing ideas for the use of cutting-edge technologies seems necessary because communication and information exchange are so important to emergency and rescue processes (such as alerting, facility deployment and distribution, and localization of resources required for rescue processes).

With regard to the nature of emergencies, using new technology is crucial to enhancing the logistics and necessary assistance to respond and lowering costs.When used in crisis management during crises cutting-edge blockchain technology can encourage collaboration among numerous stakeholders participating in the response process, resulting in transparency. Also, it is crucial to identify the deceased in mass disasters with international victims for legal, moral, religious, and human rights reasons as well as to give surviving family members the opportunity to finish the grieving process (27).

Disaster Victim Identification (DVI) is a difficult process that involves collecting post-mortem identification (PM) data, primarily fingerprints, DNA, and teeth, to compare with ante-mortem data (AM) in the list of missing people (26). Despite the fact that there are methods for identifying people, they all have shortcomings and frequently fail to produce a prompt identification of a person's identity. It is extremely difficult to handle the list of missing people, which spans years and occasionally decades, as well as keep the AM family's data sources up to date. Moreover, there is currently no all-inclusive technical solution for managing PM and AM to find people who can assist international and intergovernmental participatory procedures (27, 28).

Blockchain technology offers resources to make it easier to build robust ecosystems that can publish AM and PM data across systems and guard against data breaches, duplications, incompatibilities, and errors. Additionally, this technology has the potential to completely transform how people are identified throughout the world in terms of managing missing persons lists, AM information resources for living individuals, and PM information resources for anonymously discovered victims. It also compares compatible biological specifications for unmistakable identification.In order to demonstrate how blockchain technology addresses these difficulties and shares information across other fields, the researchers offer a variety of relevant uses based on real-world scenarios. (Carminati, B., Ferrari, E., &Rondanini, C.2018).

Blockchain in Government Healthcare

The healthcare industry can greatly assistance from blockchain technology. Over the past 30 years, the healthcare industry has been affected by the advent of constitutional data systems, health data adjustment, and a mandate to focus on digitizing medical data in organization with different computerized medical record (EMR) service providers. The bulk of archive processing information owned by healthcare providers, medicine community, and other stakeholders in the health and medical environs, do not interact with one another. The lack of interoperability among most systems containing lengthwise health data at both the original (patient) and population (public health) levels explains systemic barriers often anticipated in the following situations [2].

- When patients wish to consider or seek medical services from other healthcare providers
- When the administration of a clinical trial wants to validate the vast medical data of its competition
- When medicine community seek to ensure the accuracy of drugs circulating global markets.

Because of the inability to securely share data and the siloed administration of medical records, patients spend precious time and resources seeking constantly medical care (e.g. executing duplicate blood tests or physicals). In emergencies, physicians and other health specialist providing care may not have full brightness over a patient's medical history (e.g. explicit documentation highlighting patient allergies, prior or withstanding medical conditions, authority of controlled substances, etc.) in which case they risk improper treatment. n drug traceability, secure supply chain tracking is necessary to prevent the circulation of counterfeit or illicit drugs. Counterfeit drugs pose risks when active quantity is altered or in illegal amount. Such risks may compromise the bond and success of treatment or lend to addiction and drug confidence. They also carry the exposure of unknown side-effects, some of which may be destructive[15].

In order to address illegal transport or illegal construction of controlled material, blockchain technology presents the convenience to securely track, announce chronology, and highlight definitive data (makeup and composition) of material in an immutable record. Enterprise Ethereum provides smart contracts and systems interoperability that will change the industry.

Enterprise Ethereum enables secure data administration, consent management, medical device and drug hunting, and other blockchain use case. (Casino, F., Dasaklis, T. K., &Patsakis, C.2019).

Blockchain Vaccination:

Blockchain is a decentralised technology that is excellent for transparency and accountability. It uses tamper-proof digital ledger technology to address problems in the real world [29]. Since its creation, blockchain has been used as a smart contract and transaction system in several industries. Through a few cryptocurrencies like Bitcoin, Ethereum, Litecoin, and Zero cash, blockchain enables the distribution of financial operations [30]. Blockchain has introduced the idea of decentralisation, in which a system would not be controlled by a single central authority. The network's nodes get these transaction records while transactions are taking place. These won't be uploaded to the network as a block until current nodes accept them, which is known as verification

We don't even have exact estimations of the rates of waste in our existing situation. In the lack of data from individual countries, WHO publishes indicative vaccine wastage rates and an estimation tool. Although the tool itself acknowledges a lack of precise and pertinent data at the national level, this is partly due to overworked systems providing data that is late or incomplete, as well as the challenges associated with determining a vaccine's target demographic. A report by the WHO on immunisation information systems under Project Optimize, a WHO-PATH project to facilitate an effective vaccine supply chain, suggests that even the Gavi Alliance has seen its attempts to make financial support proportional to performance targets get limited by questions regarding data variances. (Ellram, L. M. (1991).

Blockchain Technology Impact in Vaccination Tracking

Recording inoculation data on the blockchain enables schools, allowance, and medical providers to validate vaccinations quickly. This process naturally triggers corresponding micropayments and delegates access to benefits based on medical status.

With COVID-19 on the rise, information technology (IT) is becoming increasingly important. It aids in analysing the COVID-19 pandemic epidemic, coronavirus statistical breakdown, diagnosing COVID-19 using a variety of symptoms, and vaccine development [1]. Around the world, contract tracing makes extensive use of it. These systems, however, are created and maintained separately. They are unable to converse with one another as a result. Getting a comprehensive understanding of transmission, testing, and immunisation is difficult for policymakers. Furthermore, there was some doubt regarding the veracity of the testing and vaccination statistics, particularly in underdeveloped nations. For instance, test fraud occurred in Bangladesh. Hospitals in Bangladesh with the names Regent, JKG, and Shahabuddin were found to be defrauding patients by fabricating false COVID tests, administering incorrect care. (Etemadi, N., Borbon-Galvez, Y., Strozzi, F., & Etemadi, T.2021).

BLOCKCHAIN IN DISASTER RESCUE WINGS

The recent flood tragedy in Sylhet and Sunamgonj has revealed a lack of coordinated attempts to recover despite the enormous support response from the many ends. People from different backgrounds contributed, offered assistance, and offered support. The inability of all stakeholders to interact on a single platform, however, diminished the impact of their efforts. Aid did not always get to the places where it was most needed in time as a result. Simply put, the demand for resources was too great for aid workers to meet.

Several parties must act quickly and in concert to provide an effective emergency response to natural catastrophes. Sadly, it suffices to remark that many relief initiatives up to this point have fallen short of this standard of effectiveness. Before a tragedy occurs and help organisations are on hand, days or even weeks may pass. The community must safeguard itself in the interim.

The aforementioned problems could be resolved with blockchain technology. Blockchain technology implementation would produce a shared, distributed ledger of transactions to which all pertinent parties have immediate access. The implementation of a blockchain-based disaster management strategy can reduce corruption, promote and facilitate the creation of partnerships between various authorities, provide verified and timely disaster communications, and more.improve the distribution of essential resources, and permit consistent access to the important information and restoration efforts. The resulting platform offers the confidentiality and privacy required for effective stakeholder communication and coordination.

The blockchain solution is the best way to handle the difficulties associated with catastrophe management. Thankfully, Hyperledger Foundation provides a succinct recommendation for the answer. By having a single trusted ledger, blockchain fosters collaboration and coordination among all parties involved in rescue operations. The data cannot be easily changed after it has been recorded, resulting in an audit trail and a trustworthy system for the acquisition and distribution of products and services. Data may be quickly shared with important stakeholders and managed securely. (Dorri, A., Kanhere, S. S., &Jurdak, R. 2018).

Demands (needs) from several fields can be gathered in a single network of dependable participants and satisfied by various institutions. This prevents wastage and duplication. Third parties can validate their requirements to foster greater confidence. Inventory of consumables can be traced from warehouse receipt to delivery to the final recipient. More transparency, simpler logistics, quicker reaction times, and the avoidance of labour duplication amongst agencies are all benefits of this. Volunteer time can be tracked, logged, acknowledged, and audited. Increased community involvement might result from incentives or even just increased awareness.

Blockchain has transformed from a murky niche technology to a popular one during the past ten years. The technical foundation of blockchain includes features like distributed storage, peer-to-peer networks, decentralisation, consensus process, traceability, information sharing, resource reserve, and distribution, as well as power and responsibility structures, monitoring, and rewards.

When responding to natural disasters, coordination of activities has always been the top priority, but it has also historically been the weakest link. Fraud, response delays, a failure to plan or respond appropriately, and an ineffective use of resources are just a few of the issues that are putting our society's foundation of trust under more and more pressure. Natural calamities in the future are beyond our control, but the inherent trust and transparency of the blockchain can help us all prepare. (de la Torre, L. E., Dolinskaya, I. S., &Smilowitz, K. R. 2012).

Disaster Wings Supply Chain Model

As a result of their high sensitivity and dangers, supply chains during disasters are complex networks. The system remains dynamic, effective, and robust by having a thorough awareness of supply chain risks and management techniques. Collaboration serves as the glue that keeps businesses together in the event of a tragedy (Glenn Richey Jr, 2009). Regardless of the circumstance, organisations pursue a number of various and, in some cases, contradictory strategic goals.

This study suggests a decision-management framework for a supply chain for disaster assistance that takes the complexity of disasters into account. The suggested model was created using a survey of the literature, Blockchain technology for decentralised network brainstorming (mind mapping), and a simulated data stream.The distribution of supplies to the afflicted people swiftly and in sufficient numbers is one of the major issues in the catastrophe supply chain, according to Thomas and Kopczak's 2005 argument. If there is an oversupply of non-essential items, the logistical response may become delayed. Hence, it is necessary to prioritise the demands in a methodical manner, involving all players and establishing quick trust to provide channels for collaboration.

Each stage of a disaster can be supported by information technology by offering real- time monitoring, online dashboards, interactive communication, and cooperation. Decentralization is one strategy that supports acting rapidly.In the literature, the DLT has been used in the field of disaster management for transparency, efficiency, scale, and sustainability. Several data centres with independent hardware, infrastructure, implementation, and resource-sharing capabilities are needed to service and monitor the catastrophe network, and users can send requests to various data centres inside a cloud datacenter.

One way to serve a large number of consumers at a lower cost is through the utilisation of the cloud. Given that the catastrophe network requires dispersed computing, networking, and services, virtual machines can improve performance while transferring to other nodes without interruption and dynamically alter the resource quantities allotted to a client. Cloud data centres can save operating costs and boost service quality.Using a load balancing strategy prevents overloaded processing on one of the servers and optimises the load distribution over different hops. As employed by C. Xu et al. (2017) for the effective blockchain-based decentralised resource management framework to lower the cost of energy, DVFS is one of the techniques that may be put to use. As a result, this study creates a blockchain-based disaster management model that uses the DVFS algorithm to avoid overloading on a single hop and cut down on the energy and processing time required for the entire system. (Dorri, A., Kanhere, S. S., &Jurdak, R. 2018).

Disaster Supply Chain Critical Demands

In Figure [1] Disaster Supply chain is the literature, the DLT has been used in the field of disaster management for transparency, efficiency, scale, and sustainability. Several data centres with independent hardware, infrastructure, implementation, and resource-sharing capabilities are needed to service and monitor the catastrophe network, and users can send requests to various data centres inside a cloud datacenter. One way to serve a large number of consumers at a lower cost is through the utilisation of the cloud.

Blockchain-Enabled DVFS-Based Model

The supply chain for disaster relief must be organised by a network authority, according to a rigorous analysis of various recent disasters. The system and preparation of a thorough communication strategy can aid in coordinating the actions and responses of all teams. Every party concerned is given the chance to be in charge of their area of expertise and contribute to a disaster relief operation successfully for an extended period. To improve management and coordination of relief activities, the network may involve government agencies, NGOs, the military, and business organisations. Building an effective disaster management system is a challenging process that involves several obstacles, such as gathering victim demand requests, setting up a framework to handle time-sensitive needs, allocating resources for decision-making, etc.The possibility to address such issues lies in new data-driven methodologies. Using modern technology has made it possible for parties to communicate in multiple directions and offers a modern method of interacting. (Patsakis, C. 2019).

In the proposed models, a group of transactions (C1, C2, etc.) are applied from a group of clients in the first portion. The IoT nodes that reflect the forced flow of the calamity are the clients. A network consists of the nodes and a few servers. The round-robin (RR) method is executed and randomly assigned to the virtual machines in the IoT format. The goal of the study is to determine how to manage the resources using the algorithms. To reduce task failures, the online data streams can be sorted and assigned to the server that is best prepared by incorporating the DVFS algorithm.

Figure 1. Disaster supply chain
[Source: Author's Creation]

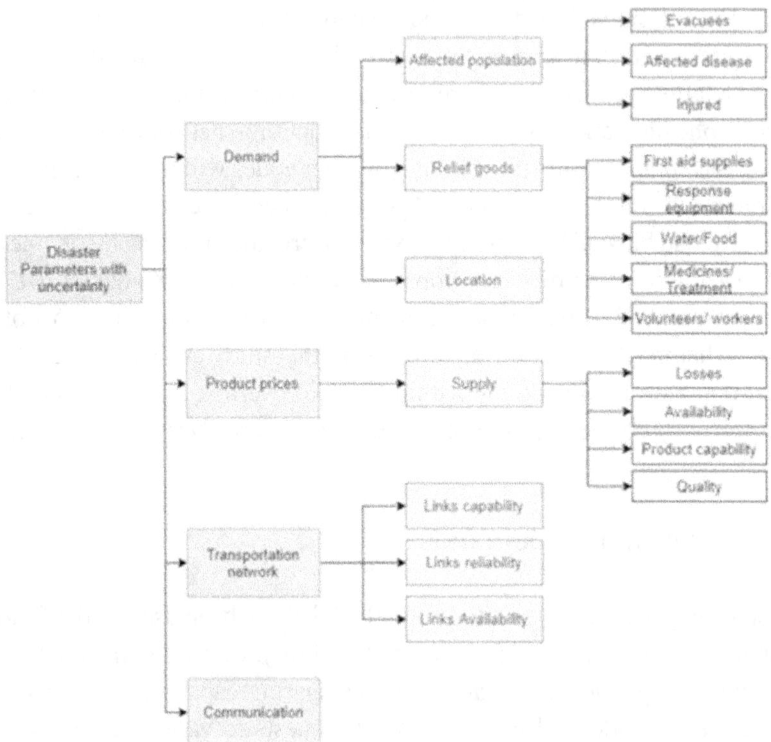

The disaster is simulated using a matrix that can generate flows under various circumstances. The supply chain procedure for the group of clients is taken into account in the matrix. The approach seeks to manage catastrophe data quickly and effectively while allocating client inputs in a chain system to increase efficiency.

The tasks include a heavy load of stream data entering the network. The model strives to handle those duties effectively. The jobs will be added to blocks and move past the chain stage. The model's key feature is figuring out how to organise the virtual machines and which server in each virtual machine is appropriate for running the chain.

The second part describes how the DVFS algorithm is used on each chain in the blockchain to examine and organise the queue of processes that are waiting to be executed. The queue is used to update the calculations using various real and virtual machines. DVFS is used to pinpoint the exact location inside the network where the catastrophe happened as a result of the extremely high volume of client requests in the transaction sector. By employing DVFS, one can execute a queueing algorithm, find out where the demands are kept, and enable monitoring of those requests. As a result, the entire network is transparent and is therefore watchable. The variables response time, delay, throughput, and other outcomes of this tracking and monitoring. Theresultsofthistrackingandmonitoringwillberepresentedinthevariables ofresponsetime, delay, throughput,and successful immigration.

The Network is thought of as having many inputs, and it is necessary to update and monitor the network at the same time. The agents review and store the data in the storage component that has been

Figure 2: Disaster preparedness and relief

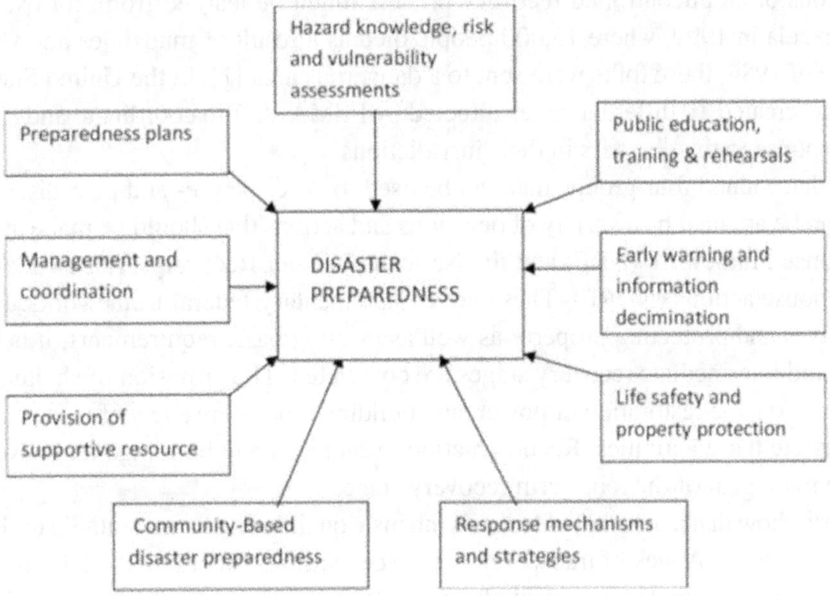

activated by a trigger. The relationship between blockchains and DVFS is taken into account by the binary Agent-based condition.

DVFS is used to sort and evaluate smart contracts. A relationship between update tracking and action management is taken into account when the agent is engaged. The network's entities can communicate with one another thanks to the MAS. The disaster management stage, which can keep an eye on the impacted area and the response team, would be the culmination of these three processes. (Fogli, D., Greppi, C., &Guida, G. 2017).

3.2.1 Blockchain-centricResourceSystem for Disaster Response and Relief

The Figure [2] Disaster Preparedness and Relief is the Method of producing encrypted data through cryptography is crucial. It is based on encryption that the science of employing mathematical formulas for the cryptography of existing data was developed. Following the safe node's determination of the shortest path, blockchain encryption is used to gather the middle layer nodes' public keys. Using private keys on all the intermediary nodes, a package is encrypted and distributed via blockchain technology. The model is implemented using two equations. (Papadopoulos, T. 2020).

Disaster Relief Applications of Blockchain Technologies

The Ineffective management of the permitting procedure, the post-disaster recovery process can be slowed considerably. Repairs and new construction permits can be issued more quickly by incorporating fast-tracked methods and improving engagement with other stakeholders [59]. When creating a post-disaster management plan, insufficient attempts are made to draw attention to these issues. The objectives of post-disaster recovery are to rebuild a constructed environment that is resilient to future disaster events. The repercussions of an uncontrolled recovery process might be learned from, for example, what occurred in Venezuela in 1999, where 15,000 people died as a result of mudslides and floods. After the major landslide of 1984, these folks were sent to a dangerous area [1]. In the United States, the Federal governments has created 10 different zones directed by FEMA staff to coordinate and support the local government response to the disasters in their jurisdictions.

FEMA has determined four groups that can be used to describe pre- and post-disaster planning. A full recovery can be attained by a variety of decisions and actions that should be made at each stage.The National Response Framework (NRF) and the National Disaster Recovery Framework both addressed the various response actions (NDRF). These are complementary federal frameworks. For the purpose of preserving lives and protecting property as well as meeting basic requirements, it is imperative that the short-term and intermediate recovery stages are completed. The provision of shelter, rescue efforts, the removal of debris, the restoration of power, and building repairs are a few of these tasks. It may take months to complete these activities. Reconstruction of facilities and buildings that have been severely damaged is the main goal of the long-term recovery stage.

Recent efforts show that a distributed ledger's intrinsic qualities make it a suitable tool to use in disaster management systems. A lack of transparency between various functionality, laborious coordination between several parties, complex logistical planning, delayed deliveries, shortages, and resource waste frequently hinder connecting services responsible for transporting food, water, and other relief.claims that only number of diverse can give the exact transparency that crisis scenarios demand. By linking all parties and improving communication ease while sustaining a secure protocol over the network, BCT can therefore serve as the central system of all operations.A proposal put forth by [63] proposes an integrative model over a blockchain network that links governmental entities, healthcare providers, housing providers, organisations that give humanitarian help, telecom service providers, citizens, and organisations that provide transportation.

In addition to eliminating the need for laborious paperwork and human involvement in granting approvals [8], the transparency and automated dissemination of recorded information that are features of BCT can also further streamline other managerial processes.

The World Food Program's "Building Blocks" blockchain programme was introduced in January 2017. (WFP). This programme aims to improve the overall effectiveness of the cash-based transfer scheme across a range of activities, including refugee registration and financial accounting, and is focused on the Azrap refugee camp in Hordan [63]. In less than five months, initial development and deployment were completed.Building Blocks was designed to supplement and improve the effectiveness of the WFP's existing digital infrastructure for beneficiary information registration and payment mechanisms by improving back-end operations while keeping the beneficiaries' cash system unaltered [63]. (Fogli, D., Greppi, C., &Guida, G. 2017).

INFORMATIONAL DIFFICULTIES FOR INTERNATIONAL DISASTER MANAGEMENT

Identification of Disaster Victims

Mass disasters rarely affect a single country in our global society (INTERPOL, 2018). The DVI process is the facilities management of a disaster or incident that results in a death toll greater than the resources and response capacity of a community. They encompass both unintentional and intentional man-made disasters, as well as natural disasters (such as hurricanes, floods, tornadoes, volcanoes, wildfires, earthquakes, and tsunamis) (Nuzzolese and Di Vella, 2007; Senn and Weems, 2013). DVI is crucial to law enforcement investigations into crimes since many nations utilise placeholder names (such as "John Doe" or "John Smith") for people whose genuine identities are unknown and/or cannot be confirmed.

This is true from both a judicial and human rights standpoint (Nuzzolese, 2012).The Disaster Victim Identification (DVI) process is a complicated process where postmortem (PM), the after death, primary identifying data, essentially fingerprints, DNA, and dental data, are collected from the deceased or human remains to be compared with equivalent antemortem (AM), the before death, data, related to the reported missing persons involved in the disaster. On the one hand, these primary identifiers are the most accurate tools for identifying individuals because each one may identify or rule out a specific person. As a backup for the identification procedure, secondary forms of identification including scars, jewellery, and tattoos are also gathered.

There are predetermined parameters for the DVI process that must be followed based on Interpol protocol and regional regulations. According to Interpol DVI, managing a disaster and identifying the victims involves five steps, which are described in further detail (AlQahtani et al., 2019) and shown in Figure 1 below. Aside from the debriefing phase, they are phase 1 accident or scene site: recovery of human remains and properties; phase 2 temporary mortuary or PM:

Where forensic DVI specialists examine human remains to determine the cause of death and collect all primary identification data and forensic evidence relevant to the case; and phase 3 AM data collection: recovery of all primary identification data and forensic evidence; and phase 4 debriefing.The gathering of the missing person's medical and dental records, physical traits like build, hair and eye colour, scars, tattoos, as well as DNA from their home or family, are all included in this phase. But the most crucial component is phase

4 data reconciliation, in which the data from phases 2 and 3 are compared to find correspondences, matches, but also exclusions, in order to achieve a positive identification. Phase 5 debriefing: At this phase, all DVI activity is analysed and examined for quality control. This process can take a long time depending on the complexity of human remains and the availability of AM data (Morgan et al., 2006).

Classification of Human Remains in a Disaster

The creation of a PM profile begins with the recovery of human remains from the tragedy scene and their examination in a mortuary. Before the reconciliation procedure, the AM team gathers the police-provided AM data, analyses it, and then aggregates it. By comparing the AM primary identifiers tracked in a time-stamped ledger with the gathered PM data and the blockchain-based PM data, a blockchain-based solution would be able to manage both AM and PM data on one platform and support the human identification process.Regardless of the matching techniques, whether done manually or with the aid of artificial intelligence algorithms, this should compile all the data required for the DVI process quickly, leaving the reconciliation team in charge of making the final determination of whether an identification is accurate. This will conserve time and resources.Creating a blockchain would allow for the tracking and distribution of specific disaster specifics among various DVI teams, facilitating the sharing of disaster-related information across several countries.(Abolhasan, M., Abbas, H., & Ni, W. 2019).

CONCLUSION

Over the years, organisations supplying emergency help to people affected have faced difficulties due to the opaque nature of disasters, including corruption and a lack of coordination among humanitarian aid agencies. The creation of a more efficient and transparent system for the organisations themselves and the beneficiaries has been thwarted by numerous attempts, but none have been very effective. Less traditional approaches, such as technical ones like blockchain, should be carefully studied in order to clear the air, making sure that the available resources are used in accordance with the fundamentals of humanitarian aid.

The goal of this thesis was to critically evaluate the conditions under which blockchain technology may be utilised to distribute emergency aid, thereby resolving the ongoing problems. To do this, a scoping research and an interview study with 12 informants were conducted. To better grasp the potential impact of the technology, an overview of the knowledge already available about the advantages of blockchain technology and the issues it poses was established. Both study formats discovered that the technology's unique benefits of decentralisation, transparency, and traceability set it apart from other products on the market.

By enabling greater cooperation between aid organisations, raising the transparency of the financial trail, and improving beneficiary ownership, these advantages can help to develop a more effective payment service for cash-based transfers in the humanitarian relief sector.facilitating cash-based transfers by enabling increased collaboration between aid organisations, raising the transparency of the financial trail, extending beneficiary ownership, and significantly lowering transaction costs. The openness of the technology and the confidence it might foster in an otherwise distrustful atmosphere, according to insiders in the humanitarian sector, can be advantageous to assistance organisations and donors. Notwithstanding these positive aspects, there are still certain difficulties with blockchain technology, and if it is not used properly, there are significant hazards.

Some of these, including issues with scalability and philosophy of the technology, are not relevant to the humanitarian context, while others have implications for emergency relief and must therefore be addressed. A few of these include the scepticism surrounding the technology, particularly ina relative lack of maturity with regard to data security. With investments in research as well as pilot projects applying blockchain technology in practise and evaluating its impact, scepticism and immaturity can both be overcome. Good change management can also help clear up misconceptions. It is true that responsible consideration will need to be given to both the current risks associated with data security and any additional possible dangers that may be found in the future.

Second, it was discovered that blockchain technology can be a useful instrument for the distribution of emergency relief based on the knowledge that has been collected about it. Due to its special qualities, blockchain technology has the ability to address problems like a lack of collaboration and enduring corruption, developing a system that serves recipients, assistance organisations, and funders more effectively and efficiently. This decision is not without limitations, though. There are four prerequisites that have been established for a successful implementation of blockchain technology. The fundamental technical prerequisite for transactions to be recorded on the blockchain is internet connectivity. Actors on the blockchain must acquire a digital identity in order to function in the system once this is formed. Both political will and an initial investment are then necessary to build a blockchain project.

Lastly, by connecting the prerequisites for a successful implementation with crisis scenarios, the breadth of blockchain technology's application for urgent aid was reduced. Above all else, the environment must be appropriate for transferring money rather than offering in-kind aid. A stable internet connection must also be possible given the connectedness of the area before the tragedy and the extent of the event's destruction, which must have left the telecommunications network substantially unharmed. In an environment where organisations have established a long-term presence and are serving the demands of numerous customers, it might be extremely beneficial to test the technology's performance.

Blockchain technology has the potential to be especially useful in the future when dealing with risks that present a high risk of onset quickly since it can be implemented more quickly than conventional solutions. Before implementing blockchain technology in these settings, anthropogenic dangers like terrorist attacks or bloody wars should be thoroughly assessed to make sure some of the technology's risks aren't being taken advantage

Future use cases for blockchain technology may include more than just providing transparent and effective emergency help to people in need. It may also be applied to other situations. With the help of technology, it may be possible to store refugees' digital identities, including their diplomas and other official papers, and to provide change management, such as for tracking in-kind aid. Block by block, blockchain technology enables the humanitarian industry to reinvent itself and emphasise ideals like transparency and cooperation, lifting the murky atmosphere that surrounds the post-disaster environment.

Since the decisions of any decision-maker can have broad repercussions, accurate and effective information is a necessary prerequisite for crisis management and disaster response. Because to its decentralised and distributed organisation and management, cutting-edge blockchain technology is widely used in many industries and is typically highly well-liked (26–23). The primary goal of this study was to look into the framework for using blockchain technology in an emergency. The actual world can make use of this technology and the knowledge that goes with it. The results of this study demonstrated how blockchain technology satisfies all standards for trust and information security.

REFERENCES

Balcik, B., Beamon, B. M., Krejci, C. C., Muramatsu, K. M., & Ramirez, M. (2010). Coordination in humanitarian relief chains: Practices, challenges and opportunities. *International Journal of Production Economics*, *126*(1), 22–34. doi:10.1016/j.ijpe.2009.09.008

Becker, P. (2014). *Sustainability Science: Managing Risk and Resilience for Sustainable Development*. Elsevier.

Brinkmann, S., & Kvale, S. (2015). *Interviews: Learning the Craft of Qualitative Research Interviewing*. Sage Publication.

Carminati, B., Ferrari, E., & Rondanini, C. (2018). Blockchain as a platform for secure interorganizational business processes. *2018 IEEE 4th International Conference on Collaboration and Internet Computing* (122–129). IEEE.

Casino, F., Dasaklis, T. K., & Patsakis, C. (2019). A systematic literature review of blockchain-based applications: Current status, classification and open issues. *Telematics and Informatics, 36*, 55–58. doi:10.1016/j.tele.2018.11.006

Dakhil, A., & Alshawi, M. (2014). Client's Role in Building Disaster Management through Building Information Modelling. [CrossRef]. *Procedia Economics and Finance, 18*, 47–54. doi:10.1016/S2212-5671(14)00912-5

Day, J. M., Melnyk, S. A., Larson, P. D., Davis, E. W., & Whybark, D. C. (2012). Humanitarian and disaster relief supply chains: A matter of life and death. *The Journal of Supply Chain Management, 48*(2), 21–36. doi:10.1111/j.1745-493X.2012.03267.x

de la Torre, L. E., Dolinskaya, I. S., & Smilowitz, K. R. (2012). Disaster relief routing: Integrating research and practice. *Socio-Economic Planning Sciences, 46*(1), 88–97. doi:10.1016/j.seps.2011.06.001

Dorri, A., Kanhere, S. S., & Jurdak, R. (2018). Multi-agent systems: A survey. *IEEE Access : Practical Innovations, Open Solutions, 6*, 28573–28593. doi:10.1109/ACCESS.2018.2831228

Dubey, R., Gunasekaran, A., Bryde, D. J., Dwivedi, Y. K., & Papadopoulos, T. (2020). Blockchain technology for enhancing swift-trust, collaboration and resilience within a humanitarian supply chain setting. *International Journal of Production Research, 58*(11), 1–18. doi:10.1080/00207543.2020.1722860

Düdder, B. & Ross, O. (2017). Timber tracking: Reducing complexity of due diligence by using blockchain technology (position paper). *CEUR Workshop Proceedings*. Research Gate.

Ellram, L. M. (1991). Supply-chain management: The industrial organisation perspective. *International Journal of Physical Distribution & Logistics Management, 21*(1), 13–22. doi:10.1108/09600039110137082

Engelenburg, S.V.S. van et al. (2017, July). Design of a software architecture supporting business-to-government information sharing to improve public safety and security. *Journal of Intelligent Information Systems*. doi:10.1007/s10844-017- 0478z

Etemadi, N., Borbon-Galvez, Y., Strozzi, F., & Etemadi, T. (2021). Supply chain disruption risk management with blockchain: A dynamic literature review. *Information (Basel), 12*(2), 70. doi:10.3390/info12020070

Fogli, D., Greppi, C., & Guida, G. (2017). Design patterns for emergency management: An exercise in reflective practice. *Information & Management, 54*(7), 971–986. doi:10.1016/j.im.2017.02.002

Gaetani, E. (2017). *Blockchain-based database to ensure data integrity in cloud computing environments*. CEUR Workshop Proceedings.

Gao, W., Hatcher, W. G., & Yu, W. (2018). *A survey of blockchain: techniques, applications, andchallenges. 27th International Conference on Computer Communication and Networks (ICCCN)*, Hangzhou, China.

Gunes, E., & Kovel, J. P. (2000). Using GIS in Emergency Management Operations. [CrossRef]. *Journal of Urban Planning and Development*, *126*(3), 136–149. doi:10.1061/(ASCE)0733-9488(2000)126:3(136)

Izadi, D. S., & Sharif-Khatibi, Z. (2020). Blockchain and its usage in storing information as a secure distributed database. *J Informat Communication TechnolPolic*, *1*(2), 85–106.

Lin, I. C., & Liao, T. C. (2017). A Survey of Blockchain Security Issues and Challenges. *International Journal of Network Security*, *19*(5), 653–659.

Lindman, J., Tuunainen, V. K., & Rossi, M. (2017). Opportunities and risks of Blockchain Technologies in payments - a research agenda. Presented at *50th Hawaii International Conference on System Sciences*, Hawaii, US.

Link, L. E. (2010). The anatomy of a disaster, an overview of Hurricane Katrina and New Orleans. [CrossRef]. *Ocean Engineering*, *37*(1), 4–12. doi:10.1016/j.oceaneng.2009.09.002

Lo, S. K., Xu, X., Chiam, Y. K., & Lu, Q. (2017). *Evaluating Suitability of Applying Blockchain,* presented at International Conference on Engineering of Complex Computer Systems, Fukuoka, Japan.

Lu, Y. (2018). Blockchain and the related issues: A review of current research topics. *Journal of Management Analytics*, *5*(4), 231–255. doi:10.1080/23270012.2018.1516523

Madory, D. (2017). Puerto Rico's Slow Internet Recovery. Oracle. Retrieved from.) A secure incentive scheme for vehicular delay tolerant networks using cryptocurrency. *Security and Communication Networks*, 2018.

Makhdoom, I., Abolhasan, M., Abbas, H., & Ni, W. (2019). Blockchain's adoption in IoT: The challenges, and a way forward. *Journal of Network and Computer Applications*, *125*, 251–279. doi:10.1016/j.jnca.2018.10.019

Maxmen, A. (2018). Cape Town scientists prepare for "Day Zero". *Nature*, 13–14.

Min, H. (2019). Blockchain technology for enhancing supply chain resilience. *Business Horizons*, *62*(1), 35–45. doi:10.1016/j.bushor.2018.08.012

Minear, L. (2002). *The Humanitarian Enterprise: Dilemmas and Discoveries*. Kumarian Press.

Niranjanamurthy, M., Nithya, B. N., & Jagannatha, S. (2018). Analysis of Blockchain technology: Pros, cons and SWOT. *Cluster Computing*, *5*(2), 1–15.

Purwanti, S. H. (2013). From Bali Bombing to Sukhoi Tragedy. Jakarta Timu, 17–25. Rayyana Komunikasindo.

REUTERS. (2019). *Accenture Microsoft Team up on Blockchain-Based Digital ID Network*. Reuters. https://www.reuters.com/article/us-microsoftaccenture- digitalid/accenture-microsoft-team-up-on-blockchain-baseddigital-id-network- idUSKBN19A22B.

See Khoo, L., Soon Lai, P., Hafizam Hasmi, A., & Shah Mahmood, M. (2016). Secondary identifier for positive identification in DVI. *Forensic Science and Criminology*, *1*(1), 1–3. doi:10.15761/FSC.1000102

Senn, R. D., & Weems, R. A. (2013). *Manual of Forensic Odontology* (5th ed.). CRC Press.

Siemon, C., Rueckel, D., & Krumay, B. (2020). *Blockchain technology for emergency response. Proceedings of the 53rd Hawaii International Conference on System Sciences*, Maui, Hawaii.

Zheng, Z., Xie, S., Dai, H., Chen, X., & Wang, H. (2017). *An overview of blockchain technology: Architecture, consensus, and future trends*. International Congress on Big Data (Big Data Congress), Honolulu, HI, USA.

Chapter 14
How Caste Comes Into Play When Climate Changes:
Re-Thinking the Environmental Paradigms

Bianca Cherechés
https://orcid.org/0000-0002-6083-6235
University of Zaragoza, Spain

ABSTRACT

Climate change has diverse impacts that disproportionately affect marginalised communities, exacerbating existing inequalities. In the Indian context, climate change has been experienced for decades, particularly affecting Dalits and Adivasi communities who rely on nature for sustenance but have limited resources to adapt. Thus, recognition of caste as a significant factor in climate action policies is crucial. In the same vein, environmental literature in India has often neglected the ecological perspectives of marginalised castes. This oversight stems from the assumption that Dalits are subsumed within broader categories of poverty and vulnerability. This chapter aims to examine Dalit literature, often viewed as protest literature, as a valuable insight into the interactions between Dalits and their environments. By exploring Baby Kamble's The Prisons We Broke and Bama's Karukku the intricate relationship between Dalits and nature can be illuminated, contributing to ecological, literary, and cultural theories.

INTRODUCTION

Climate change takes on various forms, including abnormal rains, floods, and extreme temperatures, among other phenomena. Its effects are not uniform and disproportionately impact individuals and communities. Within the prevailing neo-capitalist socio-economic framework, the disparity between the privileged and marginalised continues to grow, intensifying inequality. Paradoxically, attempts by dominant factions to safeguard the environment often reinforce existing social inequities instead of mitigating them (Bullard, 2001). Consequently, it becomes imperative to address the unique circumstances of diverse communities, particularly those on the margins.

DOI: 10.4018/978-1-6684-9863-7.ch014

In the Indian context, although the term 'climate change' is gradually gaining recognition, its effects have been felt for decades, manifesting in rainfall fluctuation, changing monsoon patterns, increasing floods, heat waves, unpredictable weather, and coastal erosion. Dalits—along with Adivasi or tribal communities[1]—have an inherent and intricate relationship with nature, relying on it for sustenance at various levels. However, due to ongoing socio-economic and political deprivation, coupled with systemic discrimination, caste-oppressed communities have limited resources to adapt to the damage caused by climate change (Sharma, 2017). As a result, they not only bear a disproportionate burden from the impacts of climate change but also encounter discrimination during rescue, rehabilitation, and recovery from climate-related calamities. Therefore, it is crucial for both the United Nations Framework Convention on Climate Change and national climate action policies to recognise caste as a significant factor, in addition to class, ethnicity, and gender.

Similarly, in recent decades, the field of environmental literature has expanded its scope to include the intersections of ecology, society, and culture. In the realm of Indian 'nature writing', there has been a longstanding tradition of incorporating ecological, geographical, and sacral elements. However, the broader canon often fails to acknowledge the ecological foundations present in the writings of marginalised castes. Thus, this chapter seeks to address two key aspects: firstly, the oversight of environmental perspectives on caste-related issues, and secondly, the importance of viewing the environment through the lens of caste, with a particular focus on Dalits. By analysing two Dalit autobiographies, Bama's *Karukku* (2012) and Baby Kamble's *The Prisons We Broke* (2009), written in different regions, languages, time periods, and with different intentions, this chapter aims to illuminate an unexplored dimension of Dalit writings.[2]

MAPPING INDIAN ENVIRONMENTALISM

Initially, discussions surrounding the environment primarily centred around saving 'wild nature or wilderness' in the United States and 'safeguarding the countryside' in Europe (Elliott, 2023). These discussions gradually expanded to encompass a wider range of concerns, including climate change, global warming, toxicity, and their intersections with ethnicity and gender. In the Indian context, conflicts over natural resources were already prevalent during the colonial era, as the struggles for resources fuelled conflicts between the colonisers and the colonised. Consequently, environmental movements in India emerged with the goal of advocating for social justice for disenfranchised communities. As stated by Ramachandra Guha, "Indian environmentalism was an environmentalism of the poor, which married the concerns of social justice on the one hand with sustainability on the other" (2013, n.p.).

The Chipko movement, which originated in a remote Himalayan village in 1973, stands as one of the most significant environmental movements in India. This movement ended up encompassing a broad range of conflicts regarding natural resources in the 70s and 80s, including disputes over land, forests, and water, with a focus on protecting natural resources rather than engaging in a fight for the right to exploit them (Guha, 2013). Alongside Chipko, several other Indian environmental movements, such as Sulabh International (1970),[3] Ralegan Siddhi (1975),[4] and the Narmada Bachao Andolan (1985),[5] arose to safeguard the rights of the disenfranchised and marginal communities to natural resources (Kumar & Mishra, 2022, p. 581). However, these movements failed to adequately address the specific ecological concerns faced by a particular section of marginalised communities: Dalits (Aiyadurai & Ingole, 2021; Omvedt, 1997; Sharma, 2017). Similarly, mainstream environmental movements and ecological discourses have also neglected the ecological crisis experienced by Dalits in terms of access to and exclusion from various natural resources (Kumar & Mishra, 2022, p. 581). This oversight may partly stem from the assumption that Dalits are subsumed within broader categories of poor, marginalised, vulnerable, displaced, and environmental refugees and migrants (Sharma, 2012, p. 46). Another reason could be that, unlike Adivasis, Dalits have not received recognition for their contributions to the ecological knowledge base (Krishnan & Naga, 2017). It becomes necessary then to examine and delve into the environmental insights of Dalits and recognise their undeniable connection with nature.

Bhimrao Ramji Ambedkar, a prominent Dalit leader who spearheaded one of the earliest organised Dalit environmental movements called the Mahad Satyagraha in 1927 (Omvedt, 2008; Teltumbde, 2016), holds a foundational position in anti-caste ecological scholarship.[6] Jyotirao Phule, another iconic figure among Dalit environmentalists, acknowledged and highlighted the significance of nature and ecology for marginalised sectors of society (Sharma, 2017, p. 97). Both Ambedkar and Phule stand as initial inspirations for Dalit environmentalists, motivating their struggle and resistance against the socio-economic and politico-ecological injustice that has afflicted them for generations.

Dalit perspectives challenge the dominant discourse of Indian environmentalism in two ways. On one hand, prominent environmental discourses in India have often attributed ecological degradation to the imposition of Western colonial civilisation and globalised neoliberal economic reforms over indigenous and green Indian culture, thus idealising the Indian past (Sharma, 2012, p. 47).[7] For Dalits, such discourses glorify and romanticise a tradition and past that have directly contributed to their marginalisation and oppression. On the other hand, Indian eco-feminists attribute the exploitation of nature—alongside the oppression of women—to patriarchy and male dominance, as expressed in the following quote: "[Men] want to be creators themselves. But for that they must first rob women and nature of their subjectiveness, that is, of their own dignity, their spirituality, and turn them into lifeless, controllable matter" (Shiva & Mies, 2014, p. 25). While this framework helps highlight particular gender-related environmental injustices, it often fails to recognise the intersecting three-axis discrimination faced by Dalit women on a daily basis, namely, caste, class, and gender (Agarwal, 2002; Jackson, 2003; Kings, 2017; Mies & Shiva, 2014).

Hence, despite acknowledging various influencing factors such as ethnicity, class, or gender, environmental studies overlook the significance of caste. To rectify this, incorporating Dalit insights may provide ecological sensibilities to shape the future of environmentalism, such as diverse perspectives on labour, natural resources, rural communities, public spaces, and development. Even though these issues may not have been explicitly articulated in a traditional 'environmental language' by Dalits, they undoubtedly provide a fresh and necessary nuance to the field.

(RE)MAPPING THE ENVIRONMENTAL LITERARY SPACE

Mapping environmental representations and meanings in a country as regionally, linguistically, and culturally varied as India poses a challenging task. Nonetheless, we can identify three overarching trends in this domain. Firstly, there are nature writings that intimately engage with the natural world, offering positive observations and reflective connections.[8] Secondly, religious writings celebrate the beauty, divinity, and sanctity of nature, often exploring concepts such as nature worship and mysticism.[9] Lastly, eco-literature underscores the intersection of nature, culture, identity, and politics, with a focus on fostering ecological awareness.[10] However, Dalit eco-literary traditions have been overlooked within these

Indian environmental writings. Similarly, while Dalit struggles against inequality and injustice have lately gained recognition in global literary circles, they have received limited attention within the field of global environmental humanities. This exclusion can be attributed, in part, to the failure to acknowledge or recognise the pervasive presence of nature, ecological symbols, and environmental thought within Dalit narratives. Even when these elements are acknowledged, there is often a lack of reflection on their broader implications for environmental histories and scholarly discourse. To rectify this blind spot, it is essential to explore the intricate relationship between caste, nature, and environmental imagination and shed light on the imaginative and ecological prowess of Dalit writers in navigating the intersections of casteism with eco-criticism and the pursuit of eco-justice.

Eco-criticism explores the complex connections between literature and the physical environment (Glotfelty, 1996, p. xviii). The emergence of 'eco-criticism of the Global South' challenges the dominance of political, economic, and social power by highlighting the intersections of nature and culture across diverse locations, beliefs, and practices.[11] Scholars such as Frantz Fanon have contributed to postcolonial ecology by emphasising the humanistic bond between life, land, and resistance (Clare, 2013). Similarly, Edward Saïd's reinterpretation of land, dispossession, displacement, identity, and memory questions power dynamics and control over people and places, transcending traditional geographical notions (Wainwright, 2005). Drawing from these intellectual foundations, ideas of environmental justice have taken various theoretical and practical forms, often revolving around resource ownership and the equitable distribution of benefits and burdens within a community (Dobson, 1998). Within the domain of eco-criticism, there exists a valuable opportunity to situate Dalit literature within the natural world. This integration helps illuminate Dalit cultural representations and socio-political struggles within a 'natural' space and place. Simultaneously, Dalit texts broaden the scope of eco-criticism by narrating their fractured connections with the earth, where histories of caste injustices and existence within hierarchies position them as the disenfranchised 'other' in their everyday environment.

Environmental justice, closely intertwined with eco-criticism, provides a framework for addressing environmental discrimination based on race, ethnicity, income, or cultural differences, among other factors. The convergence of human rights and environmental justice movements has further underscored the importance of social justice, equality, and diversity within environmental spheres. Literature, according to Julie Sze, offers a unique lens for exploring environmental justice by employing visual imagery and metaphors that go beyond mere statistics. This approach enables a deeper comprehension of the challenges faced by communities battling against environmental racism (2006, p. 163). In a similar vein, Dalit texts can be seen as a form of literature embodying environmental justice. As Mukul Sharma highlights,

Caste has shaped the history of the environment in India in some prominent ways. [...] caste has created a concept of natural and social order where people, places, occupation, and knowledge are characterised by pollution and ritual cleanliness; where bodies, behaviours, situations, and actions are isolated, 'out of place', and 'untouched', because of deep down hierarchical boundaries. (2017, p. xix)

The ideology of caste continues to deny subjectivity and personhood to Dalits while still determining who holds control over means of production and access to essential resources,[12] putting forth the need to incorporate the concept of 'eco-casteism' (Sharma, 2017, p. xix) into discussion.[13]

The literature often overlooks the environmental challenges faced by Dalits, despite their significant ties with nature. Part of the role and significance of Dalit literature lies precisely in its ability to question and broaden the discourse on eco-criticism and environmental justice, and to raise awareness and confront the inherent casteism in mainstream environmental activism in India. In the following section, Bama's *Karukku* and Baby Kamble's *The Prisons We Broke* will be explored as examples that illustrate the intertwined relationship between Dalits and nature, highlighting how Dalit writings can contribute to our understanding of ecological, literary, and cultural theories.

Ecological Ethnographies

Dalit literature—including poems, autobiographies, short stories, and novels—has traditionally been viewed as a form of protest literature, representing the denunciation, agency, and resistance of Dalits against oppression. Lately, some scholars have re-evaluated these works to explore the interactions between Dalits and their environment, address their ecological vulnerabilities, and recognise their valuable contributions to the environmental discourse. Notably, Mukul Sharma's ground-breaking research has played a pivotal role in integrating caste into Indian environmental literature. Sharma observes that "In different and creative ways [Dalit writers] explore the multi-textured relationship between land and labour, water and wounds, place and culture" (2017, p. xxix). Thus, although Dalit texts may not be "explicitly environmental", they abound with ecological symbols and motifs (p. xxvii).

This chapter brings together two texts that may not explicitly be ecological, but nonetheless provide valuable insights into the ecological implications embedded within the dominant ideology in India. Baby—Babytai—Kamble, born 1929 in the state of Maharashtra in central western India, wrote *The Prisons We Broke* (original written in Marathi as 'Jina Amucha') in 2009, largely drawing upon her personal experiences as a Dalit child, young woman, and adult. Kamble guides her readers through the challenges of living in a society deeply rooted in oppressive caste norms and illustrates the gradual journey towards emancipation. While the book touches on the otherness of Dalit women within the Dalit community, its focus is on the socio-economic disparities and oppression faced by Dalits, painting a raw imagery of the harsh realities of their world. Growing up in a Maharwada—a Dalit 'neighbourhood' on the outskirts of a village—Kamble shares her first-hand experiences, exemplifying the prejudices of the Hindu caste system prevalent in and around Maharashtra.

In line with Kamble's text, *Karukku* chronicles Bama's life, from childhood to early adulthood as a Dalit Christian nun, shedding light on socio-ecological injustices, albeit from a different regional and religious perspective—Kerala and Dalit Christianity. Originally written in Tamil in 1992 and later translated into the English by Lakshmi Holmström, *Karukku* serves as a tribute to the community in which Bama grew up. She portrays the vibrant and colourful aspects of Dalit life, avoiding a narrative that reduces it to self-pity and contempt, although caste identity is an ever-present aspect that they cannot forget or escape. Despite the differences in religion, region, and landscape depicted in these texts, they are interconnected through the pervasive dimensions of socio-political and environmental injustices perpetuated by the caste system across India. As highlighted by Baviskar and Gidwani,

How we regard those who deal with waste and pollution is influenced by enduring ideas about racial and ethnic difference as marked on bodies, rendering them susceptible or resistant to environmental ills in differing degrees. While the caste system is the most deep-seated and obdurate of such systems of classification, colonial ideologies of race and climate also continue to shape the epistemology of risk and resilience. (2019, n.p.)

Dalit writings employ specific narrative techniques characterised by unique historical and spatial characteristics, which Rob Nixon (2005) refers to as 'bioregional writing' and Pramod K. Nayar labels as 'eco-tropics' (2013, p. 2). Nayar argues that these tropes are "rooted in the land and local culture" (p. 2) and condensed in themes of suffering, fear, loss, labour, and community resulting from hardships, anger, anxiety, physical trauma, and exploitation tied to their location and land. Eco-tropics encompass both a mode of expression and a material context, functioning as "an ecology of protest by proposing a 'moral economy', of land use, conservation, and way of life […] in the place of the existing unjust economy" (p. 2). The moral economy presented in Dalit texts offers an alternative to India's neoliberal, caste-based, and exploitative development model, where market forces and social inequalities perpetuate suffering among communities closely connected to nature (Nayar, 2012, p. 105). Similarly, Indulata Prasad introduces the concept of 'Dalit ecologies' as a framework to understand Dalit interactions with their environment and their experiences of eco-casteism (2022, p. 98). Drawing on the radical—even utopian—visions of activists such as Ambedkar and Phule, Dalit ecologies explore caste-based segregation, dispossession, dehumanisation, appropriation of Dalit labour and ecological knowledge, and the impact of the global ecological crisis on Dalit communities, including deforestation and water scarcity, among other ecological concerns. Both Nayar's notion of eco-tropics and Prasad's framework of Dalit ecologies provide relevant frameworks for this analysis.

Dalit life narratives convey a profound and intimate connection with nature, revealing how ecological spaces and diverse forms of liveability are ingrained in their experiences. The landscape holds immense significance, often serving as the very foundation of their imagination. Bama skilfully portrays the natural beauty of her village, describing mountain ranges, woods, fields, rains, lakes and ponds coexisting harmoniously. In her storytelling, nature emerges as a dynamic presence, constantly evolving and attaining consciousness. She writes:

When we used to go out in the early morning to relieve ourselves, a bright red sun, huge and round, would wake up in the east and climb into the sky. It would make its way, peering between the trees, glowing, its light spilling and sparkling. And in the same way, at evening time, when it went and dropped through the mountains, all the fields around would be luminous with a yellow light. (2012, p. 4)

Such vivid and intricate memories of nature imbue her text, capturing the essence of her everyday village life in pre-industrialised India, where the land was revered, celebrated, and shared. Similarly, Kamble acknowledges that nature is their only wealth: "All the wealth of the family lay around […]. One big clay pot, one mud bowl to eat from and a cracked coconut shell with a piece of wood nailed to it which functioned as a spoon" (2009, p. 39). Alongside this modest wealth, their sustenance came from the organic world, as Bama narrates: "We'd go to the lakeside in the evening, pluck some wild greens like Kuppaikira or thoyilukkira, and eat it with a quickly stirred and thickened ragi dough. Sometimes we boiled and ate drumstick leaves" (2012, p. 72). Kamble vividly recalls her childhood experiences of frolicking upon harsh terrain:

The ground would be covered with various creepers and thorns. But we were completely oblivious to this. We danced as if drunk, sweat running down our faces, our hair in tangles, and our bodies covered in dust. […] We are the children of this mother earth […]. (2009, pp. 43-44)

Nature holds immense significance for Dalits, who cherish it as their 'Mother Earth', as movingly expressed by Kamble: "You have always been treacherous to us, but we never deceived our mother. We ate dry husk and told ourselves we are eating rich food; we considered our huts great mansions" (2009, p. 38). She further states: "We smeared our foreheads with earth and finally the same earth took us into her arms; in the end we merged with mother earth" (2009, pp. 43-44). These words unequivocally demonstrate the deep connection Dalits have with the land, illustrating how natural spaces offer not only liberation and emotional refuge but also enable a transformation from rootlessness to rootedness for them.

In their coexistence with nature, Dalits have developed a range of techniques and crafted agricultural tools that embody organic and eco-friendly principles, standing in stark contrast to the detrimental effects of industrialisation and capitalism on environmental degradation and labour exploitation. Bama remembers how her family and neighbours would plough, fertilise, water, sow seeds, separate and replant them, weed, and harvest the crops (2012, p. 48). Thus, in contrast to the leisurely and exploitative lifestyle of upper castes, Dalits have nurtured an organic and respectful connection with nature through their way of life. The expertise of Dalits in creating eco-friendly tools, producing organic fertilisers, and their commitment to sustainable agricultural practices deserve recognition and commendation:

Dalit-Bahujans, with an intimate and everyday connection to nature, are constructors of the science of leather technology, builders of scientific use of manure, innovators of tools of productions, which not only improve our production but also keep our environment green and clean. (Ilaiah, 2019, p. 16)

This underscores the significant role played by Dalits, who have historically been socially and culturally marginalised due to their perceived 'pollution', in upholding the cleanliness and purity of nature. Kamble expresses this sentiment as follows: "like [the] ocean that covers mountains of sin under its huge expanse, [Dalits] covered the sins of the high castes. That is why, like the ocean, [they] deserve the admiration of the whole world" (2009, p. 49).

But, although nature is inherently abundant and shared by all, it can also symbolise oppression and exploitation, especially when it comes to shared resources. For Dalits, the interplay between nature and caste reflects their marginalised status, as resources such as land, forests, and water have historically been domains where caste-based hierarchical control has been imposed. Dalit literature consistently emphasises the daily ecological burdens they endure, revealing a distinct hierarchy of experiences: nature often becomes entwined with fear, violence, hardship, bondage, bloodshed, and conflict for them. For instance, water carries a significance beyond its pure and spiritual nature; it serves as a reminder to Dalits of their discrimination, thirst, physical abuse, and struggles for survival. Both Bama and Kamble recount the challenges Dalits had to face in accessing water. Going to the village well, for instance, meant not only traversing long distances but also depending on the generosity of upper-caste villagers for access. As a result, they were often compelled to carry out their daily tasks in unhygienic ditches often filled with rainwater or sewage from the village houses. Regarding forests, Bama recalls how Dalits were prohibited from entering easily accessible forests, forcing them to rely on remote mountainous forests to collect firewood. She vividly describes the ordeal involved in this endeavour, including the need to pay the forest guard for permission and the overall harshness of the environment, such as branches entangling in their hair and twigs and thorns scratching and tearing their skin.

Hence, land holds immense significance for Dalits. Bama and Kamble delve into the complexities surrounding land ownership, Dalit labour, and the inequalities in labour and land distribution. In their writings, most Dalits are exploited labourers, both physically and economically, on lands owned by upper-caste landlords. Bama depicts the relentless toil carried out by her mother, grandmother, community members, and herself from sunrise to sunset. Apart from working the land, their responsibilities encompassed street sweeping, drain cleaning, firewood and cow-dung collection, among many others (2012, pp. 41-42). Due to their socio-economic position, most Dalits working as agricultural labourers in rural areas are still landless today (Bhimraj, 2020). According to Ishan Anand (2016), recent land reforms in India have only provided small residential plots to Dalits, failing to allocate them agricultural land necessary to sustain their livelihoods. Thus, even after more than 50 years, their basic living conditions remain largely unchanged, as they continue to face limited access to agricultural land and resources, further impacting their identity.

The recurring theme of casteist dominance over environmental resources and space prominently emerges in both narratives, illustrating how the mistreatment of Dalits in the social realm often extends to their experiences in the natural world, thus prompting connections between ecological exploitation and social exploitation. As Huggan and Tiffin put it, this discriminatory treatment can be "best understood as a sociological phenomenon, exemplified in the environmentally discriminatory treatment of socially marginalised or economically disadvantaged peoples" (2010, p. 4). Considered environmentally polluting and inferior, Dalits have been "thrown [...] into their garbage pits, on the outskirts of the village [living] in the filthiest conditions possible" (Kamble, 2009, p. 18). In *Karukku*, the stark divisions are also revealed through access and prohibition. Initially, readers are captivated by the beauty of mountains, overflowing ponds, and lush green fields. However, the segregationist force of casteism plunges readers into an oppressive atmosphere of caste consciousness. Bama describes the division of the village into socio-ecological spaces marked by competition and conflict, where upper and lower castes occupy separate parts of the village: "they kept themselves to their part of the village, and we stayed in ours. The post office, the panchayat board, the milk-depot, the big shops, the church, the schools—all stood in [upper-caste] streets" (2012, p. 7). Bama articulates the interconnectedness between social ostracism, economic deprivation, and ecological vulnerability as follows, elevating the plight of Dalits to a broader discourse of human—and ecological—rights:

Karukku was written out of a specific experience, the experience of a Tamil Dalit Christian woman. Yet it has universality at its core which questions all oppressions, disturbs all complacencies, and reaching out, empowers all those who have suffered different oppressions. (2012, p. xiv)

Even today, 90 years after the historic Mahad Satyagraha, Dalits continue to confront the enduring challenges of caste-based discrimination, encompassing issues of access vs. prohibition and purity vs. pollution, and their ecological rights remain suppressed within the confines of a hegemonic centrism (Plumwood, 2002, p. 138). The distressing rise in deaths of manual scavengers and the atrocities committed against Dalits over common grazing lands and forests (Desai, 2020) underscore the pressing need for Dalit writers and activists to amplify their socio-ecological concerns.

Despite the promise of equality offered by the Green Revolution and industrialisation, Maya Pandit, in her introduction to *The Prisons We Broke*, highlights how these reforms have "pushed the already oppressed communities further into desolation, depriving them even of basic resources" (Kamble 2009, p. xii). The reality is that modernity and industrialisation have placed Dalits in the same social position, only to subject them to harmful chemicals and toxic gases. When addressing the present situation of Dalits, Bama contrasts the idyllic and pristine landscapes of her childhood village with the harsh reality of factory premises where even young Dalit children are compelled to toil: "They work at sticking on matchbox labels; they make firecrackers and use chemicals; and they return home exhausted, at seven in the evening" (2012, p. 55).

A 2021 study conducted by Picciariello et al. reveals that marginalised communities are disproportionately vulnerable to the impacts of climate change:

Sustained high temperatures take a disproportionate toll on those who depend on manual outdoor work or live in crowded, poorly ventilated homes. Floods, storm surge and cyclones wreak the most havoc on densely settled, low-income communities not served by risk-reducing infrastructure. One study suggests that declining agricultural productivity and rising cereal prices could increase India's national poverty rate by 3.5% by 2040 compared to a zero-warming scenario; this equates to around 50 million more poor people that year. (2021, p. 31)

In recent years, India, situated in a tropical region, has witnessed increasingly severe and intolerable heatwaves surpassing previous records. As a result, access to water has become a valuable commodity, with clean water sources becoming scarcer and more contaminated each day. This situation has adverse effects on communities that rely directly on nature, including farm labourers, many of whom are Dalits, who struggle to get the necessary resources to cope with these challenges. The recent deaths of nearly 100 people in Uttar Pradesh and Bihar due to scorching heatwaves have shocked the global community (The Hindu, 2023). This serves as a stark reminder that while climate change indeed affects everyone, certain individuals and communities bear the brunt of its consequences earlier and more intensely than others. Moreover, the systemic oppression faced by Dalits is no longer solely a socio-economic concern but has become a troubling environmental issue as well.

CONCLUSION

In conclusion, the intersection of caste, nature, and environmental imagination is a crucial aspect that needs to be explored and acknowledged within the realm of Indian environmentalism and eco-criticism. The experiences of Dalits, who have long been marginalised and oppressed, provide valuable insights into the ecological vulnerabilities faced by caste-oppressed communities and their relationship with the environment. By incorporating Dalit perspectives, the discourse on environmental justice can be expanded to address the specific challenges faced by Dalits in terms of access to natural resources, socio-economic disparities, and discrimination within environmental spheres.

Indian environmentalism has historically focused on social justice and sustainability, with movements such as the Chipko movement advocating for the rights of disenfranchised communities. However, these movements have often failed to adequately address the ecological concerns of Dalits. The oversight of caste in environmental studies can be attributed to the assumption that Dalits are subsumed within broader categories of marginalised groups and the lack of recognition for their contributions to ecological knowledge. Recognising the unique circumstances of Dalits and their intricate relationship with nature is essential for mitigating the disproportionate burden they bear from the impacts of climate change and addressing discrimination during climate-related calamities.

Similarly, within the domain of environmental literature, the perspectives of Dalit writers have been largely overlooked or marginalised. Indian nature writing has traditionally focused on positive observations of the natural world, religious writings that celebrated the sanctity of nature, and eco-literature that has explored the intersection of nature, culture, identity, and politics. However, Dalit eco-literary traditions have been excluded from these narratives. By integrating Dalit literature into the discourse of eco-criticism, the field can be enriched with new perspectives on labour, natural resources, rural communities, public spaces, and development. Dalit texts challenge the dominant discourse by exposing the inherent casteism in mainstream environmental activism and providing ecological sensibilities that shape the future of environmentalism.

The integration of Dalit literature and environmental justice is particularly relevant in the context of eco-criticism. Eco-criticism explores the connections between literature and the physical environment, and the emergence of eco-criticism of the Global South has highlighted the intersections of nature and culture across diverse locations and practices. Dalit literature embodies environmental justice by narrating the fractured connections between Dalits and the land, exposing the histories of caste injustices and hierarchies that position them as the disenfranchised 'other'. By situating Dalit literature within the natural world, eco-criticism can broaden its scope and contribute to a deeper understanding of the challenges faced by marginalised communities battling against environmental racism.

Dalit texts such as Bama's *Karukku* and Baby Kamble's *The Prisons We Broke* offer valuable insights into the ecological implications embedded within the dominant ideology in India. While these texts may not be explicitly environmental, they are rich in ecological symbols and motifs, providing a nuanced understanding of the multi-textured relationship between land, labour, water, place, and culture. Through these texts, the socio-ecological injustices faced by Dalits are brought to the forefront, challenging the romanticisation of the Indian past and highlighting the intersections of caste, gender, and class.

Incorporating Dalit perspectives into environmental studies and eco-criticism is not only crucial for understanding the unique challenges faced by marginalised communities but also for rectifying the historical oversight of caste in environmental discourse. By recognising the contributions of Dalit writers to environmental knowledge, a more comprehensive and inclusive understanding of the environment can be achieved. This recognition can also lead to more effective and equitable environmental policies and practices that address the diverse needs and experiences of marginalised communities.

Ultimately, the intertwining of caste, nature, and environmentalism in India highlights the urgent need for inclusive and intersectional approaches to environmental justice. By amplifying Dalit voices, scholars, policymakers, and activists can develop a more comprehensive understanding of the environmental challenges faced by marginalised communities and work towards more equitable and sustainable solutions. Only through such inclusive and intersectional approaches can we foster a just and harmonious relationship between people, nature, and the environment, where no community is left behind or subjected to discrimination and environmental degradation.

REFERENCES

Agarwal, B. (2002). *Are we not peasants too? Land rights and women's claims in India*. The Populations Council, Inc. doi:10.31899/pgy2.1005

Aiyadurai, A., & Ingole, P. (2021, November 29). Invisibility of caste in environmental studies. *The Indian Express*. https://indianexpress.com/article/opinion/columns/invisibilityof-caste-in-environmental-studies-7644987/

Anand, I. (2016). Dalit emancipation and the land question. *Economic and Political Weekly*, *51* (47), 12–14.

Bama (2012). Karukku (L. Holmström Trans.). Oxford University Press.

Baviskar, A., & Gidwani, V. (2019). The lives of waste and pollution. *Economic and Political Weekly*, *54*(47), 7–8.

Bhimraj, M. (2020). A Dalit critique of environmental justice in India. [Rajiv Gandhi National University of Law.]. In R. Soni, S. Mukhopadhyay, B. Agrawal, S. Ghosh, & A. Hargava (Eds.), Contemporary environmental concerns: Multi-disciplinary aspects of environmental law (pp. 89-120).

Bindra, P. S. (2017). *The vanishing: India's wildlife crisis*. Penguin Books.

Bond, R. (2008). *The book of nature*. Penguin India.

Bullard, R. D. (2001). Environmental justice in the 21st century: Race still matters. *Phylon*, *49*(3/4), 151–171. https://www.jstor.org/stable/3132626?origin=crossref. doi:10.2307/3132626

Clare, S. (2013). Geopower: The politics of life and land in Frantz Fanon's writing. *Diacritics*, *41*(4), 60–80. doi:10.1353/dia.2013.0026

Colopy, C. (2012). *Dirty, sacred rivers: Confronting south Asia's water crisis*. Oxford University Press. doi:10.1093/oso/9780199845019.001.0001

Desai, D. (2020, January 27). 282 Deaths in last 4 years: How Swachh Bharat Mission failed India's manual scavengers. *The Print*. https://theprint.in/india/282-deaths-in-last-4-years-how-swachh-bharat-mission-failed-indias-manual-scavengers/354116/

Dobson, A. (1998). *Justice and the environment: Conceptions of environmental sustainability and theories of distributive justice*. Oxford University Press. doi:10.1093/0198294956.001.0001

Eck, D. L. (2012). India: A sacred geography. Three Rivers Press.

Elliott, L. (2023, December 26). Environmentalism. *Encyclopedia Britannica*. https://www.britannica.com/topic/environmentalism

Gadgil, M., & Guha, R. (2004). *The use and abuse of nature*. Oxford India Paperbacks.

Gandhi, T. (2006). *A bird's eye view: Collected essays and shorter writings of Salim Ali* (Vol. I & II). Permanent Black.

Glotfelty, C. (1996). Introduction: Literary studies in an age of environmental crisis. In C. Glotfelty & H. Fromm (Eds.), *The ecocriticism reader: Landmarks in literary ecology* (pp. xv–xxxvii). University of Georgia Press.

Goyal, S., & Gupta, V. (2009). Sulabh International: Social transformation through sanitation. *Vikalpa*, *34*(1), 89–105. https://journals.sagepub.com/doi/pdf/10.1177/0256090920090107#:~:text=The%20Sulabh%20movement%20rehabilitated%20more,regular%20flush%20toilets%20by%20households.&text=Sulabh%20was%20the%20pioneering%20private%20initiative%20in%20the%20area%20of%20sanitation. doi:10.1177/0256090920090107

Guha, R. (1997). *Nature's spokesman: M. Krishnan and Indian wildlife*. Oxford University Press.

Guha, R. (2013, March 27). The past and present of Indian environmentalism. *The Hindu.* https://www.thehindu.com/opinion/lead/the-past-present-of-indian-environmentalism/article4551665.ece accessed 11 Jan. 2021

Huggan, G., & Tiffin, H. (2010). *Postcolonial ecocriticism: Literature, animals, environment*. Routledge.

Ilaiah, K. (2019). *Why I am not a Hindu: A Sudra critique of Hindutva philosophy, culture and political economy*. Sage Publications.

Jackson, C. (2003). Gender analysis of land: Beyond land rights for women? *Journal of Agrarian Change, 3*(4), 453–480. doi:10.1111/1471-0366.00062

Jairam, R. (2016, May). Ecological concerns in Mahabharata. *IOSR Journal of Humanities and Social Sciences, 21*(5), 63–65.

Kamble, B. (2009). *The prisons we broke* (M. Pandit, Trans.). Orient BlackSwan. (Original work published 1986)

Kela, S. (2018, April). Where the wild things are not: The curious absence of contemporary nature writing in India. *The Caravan.* http://www.caravanmagazine.in/reviews-essays/absence-contemporary-nature-writingindia

Kings, A. E. (2017). Intersectionality and the changing face of ecofeminism. *Ethics and the Environment, 22*(1), 63–87. doi:10.2979/ethicsenviro.22.1.04

Kodiveri, A. (2021). The delicate task of including different voices in environmental law making in India. *Journal of Environmental Law, 33*(3), 495–500. doi:10.1093/jel/eqab030

Krishnan, R., & Naga, R. (2017). 'Ecological warriors' versus 'Indigenous performers': Understanding state responses to resistance movements in Jagatsinghpur and Niyamgiri in Odisha. *South Asia, 40*(4), 878–894. doi:10.1080/00856401.2017.1375730

Kumar, A. (2020). B. R. Ambedkar on caste and land relations in India. *Review of Agrarian Studies, 10*(1), 37–56.

Kumar, V., & Mishra, B. (2022). Environmental casteism and the democratisation of natural resources: Reimagining Dalit testimonies. *South Asia, 45*(3), 577–595. doi:10.1080/00856401.2022.2014157

Mcquade, J. (2019, April 18). Earth day: Colonialism's role in the overexploitation of natural resources. *The Conversation*. https://theconversation.com/earth-day-colonialisms-role-in-the-overexploitation-of-natural-resources-113995

Moosvi, S. (2010). Environmental concerns in Mughal era. *Journal of History & Social Sciences, 1*(1), 1–4.

Mosse, D. (1999). Colonial and contemporary ideologies of 'community management': The case of tank irrigation development in South India. *Modern Asian Studies, 33*(2), 303–338. doi:10.1017/S0026749X99003285 PMID:22026039

Mukherjee, U. P. (2010). *Post-colonial environments: Nature, culture and the contemporary Indian novel in English*. Palgrave. doi:10.1057/9780230251328

Nayar, P. K. (2012). *Writing wrongs: The cultural construction of human rights in India*. Routledge.

Nayar, P. K. (2013). Indigenous cultures and the ecology of protest: Moral economy and 'knowing subalternity' in Dalit and Tribal writing from India. *Journal of Postcolonial Writing, 50*(3), 291–303. doi:10.1080/17449855.2013.815127

Nixon, R. (2005). Environmentalism and postcolonialism. In A. Loomba, S. Kaul, M. Bunzl, A. Burton, & J. Esty (Eds.), *Postcolonial studies and beyond* (pp. 233–251). Permanent Black. doi:10.1215/9780822386650-011

Omvedt, G. (1997, June 24). Why Dalits dislike environmentalists. *The Hindu*. http://ces.iisc.ernet.in/hpg/envis/doc97html/envenv627.html

Omvedt, G. (2008). *Seeking Begumpura: The social vision of anticaste intellectuals*. Navayana Pub.

Picciariello, A., Colenbrander, S., Bazaz, A., & Roy, R. (2021, June). *The costs of climate change in India: A review of the climate-related risks facing India, and their economic and social costs*. Overseas Development Institute. https://cdn.odi.org/media/documents/ODI-JR-costclimatechangeindia-final.pdf

Plumwood, V. (2002). *Environmental culture: The ecological crisis of reason*. Routledge.

Prasad, I. (2022). Towards Dalit ecologies. *Environment and Society, 13*(1), 98–120. doi:10.3167/ares.2022.130107

Ravikumar, V. M. (2014). Green democracy: Relevance of Ambedkar's ideas for Indian environmentalism. *The Journal of Tribal Intellectual Collective India, 2*(1), 24–34.

Sharma, M. (2012). Dalits and Indian environmental politics. *Economic and Political Weekly, 47*(23), 46–52.

Sharma, M. (2017). *Caste and nature: Dalits and Indian environmental politics*. Oxford University Press.

Shiva, V., & Mies, M. (2014). *Ecofeminism*. Zed Books.

Slovic, S., Rangarajan, S., & Sarveswaran, V. (2015). *Ecocriticism of the Global South*. Lexington Books.

Sze, J. (2006). From environmental justice literature to the literature of environmental justice. In J. Adamson, M. M. Evans, & R. Stein (Eds.), *The environmental justice reader: Politics, poetics and pedagogy*. University of Arizona Press.

The Hindu. (2023, June 18). Heatwave in U.P. and Bihar kill nearly 100. *The Hindu*. https://www.thehindu.com/news/national/other-states/heatwave-up-bihar-kill-100-india-heatwave/article66982736.ece

Wainwright, J. (2005). The geographies of political ecology: After Edward Saïd. *Environment & Planning A, 37*(6), 1033–1043. doi:10.1068/a37166

ADDITIONAL READING

Bhaduri, A. (2013). *Impact of climate change on life & livelihood of Dalits: An exploratory study from disaster risk reduction lens*. National Campaign on Dalit Human Rights (NCDHR).

Hasan, Z., Huq, A. Z., Nussbaum, M. C., & Verma, V. (2018). *The empire of disgust: Prejudice, discrimination, and policy in India and the US*. Academic., doi:10.1093/oso/9780199487837.001.0001

Popescu, C. R. (2021). Measuring progress towards the sustainable development goals. In C. Popescu (Ed.), *Handbook of research on novel practices and current successes in achieving the sustainable development goals* (pp. 125–136). IGI Global., doi:10.4018/978-1-7998-8426-2.ch006

Sharma, M. (2017). *Caste and nature: Dalits and Indian environmental politics*. Oxford University Press.

Sharma, M. (2019). My world is a different world: Caste and Dalit eco-literary traditions. *South Asia, 42*(6), 1013–1030. doi:10.1080/00856401.2019.1667057

KEY TERMS AND DEFINITIONS

Caste: It denotes a social and occupational grouping of individuals within a society, typically tied to specific professions or social classes and passed down through generations. While the concept of caste has been historically present in several societies, it has particularly played a significant role in India.

Climate Change: This refers to long-term changes in the planet's climate, primarily caused by human activities and leading to global warming and significant impacts on weather patterns, sea levels, biodiversity, and ecosystems.

Dalit: This term, previously referred to as 'untouchable', is used to advocate for social justice and equal opportunities for a community that has historically faced marginalisation and oppression in South Asia, particularly in India, as a result of their position at the lowest rung of the Hindu caste system.

Eco-Casteism: A connection between ecological exploitation and degradation and caste-based discrimination.

Ecocriticism: A field of study that analyses the relationships between literature, ecology, and the environmental discourse.

Ecology: The study of the functioning, evolution, and responses of ecosystems to different factors, as well as the relationships between living organisms and their environments.

Protest Literature: Written works created with the aim of expressing dissent, criticism, or resistance against various social, political, economic, or cultural injustices, either implicitly or explicitly.

ENDNOTES

1 The two groups are not mutually exclusive, as the experiences of Dalits and Adivasis are often intertwined, and they frequently encounter comparable patterns of environmental marginalisation and deprivation (Krishnan & Naga, 2017). Nevertheless, it is important to acknowledge that terms

such as 'indigenous' in the realm of environmental politics are administrative constructs created to address governmental strategies (Mosse, 1999, p. 62).

2 It is important to note that while the two texts chosen for analysis were authored by Dalit women, this analysis will not adopt a Dalit ecofeminist standpoint since the focus remains on highlighting the socio-ecological concerns of Dalits within the broader framework of environmentalism.

3 'Sulabh Shauchalaya Sansthan'—now known as Sulabh International Social Service Organization, SISSO—was established in 1970 as a prominent non-governmental organisation engaged in the field of sanitation at both national and international levels. The inception of 'Sulabh' can be attributed to Bindeshwar Pathak whose exposal to the challenges faced by scavengers sensitised him to their plight. Motivated by Gandhian ideals, Pathak made a conscious decision to combat the social stigma of untouchability attached to this marginalised group of individuals (Goyal & Gupta, 2009, p. 88).

4 Ralegan Siddhi is a village in Maharashtra (India) considered nowadays a model of environmental conservations. Back in 1975, the village suffered from severe drought and environmental activist Anna Hazare initiated a development programme in which she urged villagers to contribute towards repairing the village tank. Due to their collective efforts, the repaired tank allowed to hold water for the first time in generations.

5 The Narmada Bachao Andolan (NBA) emerged as a prominent movement in 1985, aiming to protest the lack of a proper policy for resettlement and rehabilitation (R&R) for over 250,000 individuals who faced displacement due to the construction of large dams along the Narmada river in the Indian state of Madhya Pradesh. Originally called the 'Narmada Dharangrast Samiti' or 'Committee for Narmada Dam-affected people', the movement was later renamed 'NBA' in 1989. The Narmada Valley project was initially conceived in 1946, but actual work on it began in 1978 following the final orders of the Narmada Water Disputes Tribunal (NWDT), which included plans for R&R. The project's objective was to build 30 major dams, 135 medium dams, and 3,000 small dams across the Narmada river, spanning from Madhya Pradesh to Gujarat. To this day, the NBA persists in its pursuit of justice regarding resettlement and rehabilitation. However, forced relocation of villages has disrupted the social fabric and many of the displaced individuals have been coerced into accepting unfavourable labour and living conditions in their new lives. Furthermore, the compensatory agricultural land provided often turned out to be rocky and infertile. As a result, many families have been forced to migrate to cities and settle in slum areas (Frontline, 2022).

6 B. R. Ambedkar believed that democratising natural resources held the key to addressing various environmental challenges. His ecological perspective is of great significance to Dalit ecological visions. In addition to the Mahad Satyagraha, his contributions to Indian environmentalism can be observed in the Temple Entry Movements, Central Water Commission, River Valley Authority and Central Electricity Authority. His and the generalised Dalit objective in these movements was not to seize land but to attain self-respect and dignity. Unfortunately, these policies and mass movements went largely unnoticed by the media and the wider public (Kumar, 2020; Ravikumar, 2014, pp. 24-34; Sharma, 2017, p. 14).

7 According to numerous environmental thinkers and historians, colonialism bears the responsibility for deforestation and the environmental crisis (McQuade, 2019, n.p.).

8 British and independent India witnessed the emergence of numerous nature writers (Moosvi, 2010; Kela, 2018). Notably, figures such as Salim Ali (Gandhi, 2006), M. Krishnan (Guha, 1997), Ruskin

Bond (2008), Prerna Singh Bindra (2017), and Cheryl Colopy (2012) have made significant contributions to the field of environmental literature in more recent times.

9 The study of nature writing in ancient India delves into the portrayal of nature found in texts such as the *Rig Veda*, the *Ramayana* and the *Mahabharata*, and in the works of the Sanskrit poet Kalidasa (Jairam, 2016), as well as in other religious and spiritual commentaries (Eck, 2012).

10 Ecocritical writing aims to bridge the divide between the scientific and intellectual understanding of nature and the broader comprehension of landscapes and physical environments. By delving into 'eco-ethics', these narratives frequently portray rural and ethnic communities enduring the brunt of exploitation within their natural surroundings, while the urban bourgeoisie often occupy positions of privilege and power (Mukherjee, 2010).

11 See Scott Slovic, Swarnalatha Rangarajan and Vidya Sarveswaran's *Ecocriticism of the Global South* (2015).

12 The exclusion of Dalits from the progressive Forest Rights Act (FRA) of 2006 highlighted the state's failure to acknowledge them as stakeholders in the environment (Kodiveri, 2021, p. 1151).

13 Sharma added 'eco-organicism' (2017, p. 16) and 'eco-naturalism' (p. 20) to this framework, suggesting that these ideologies align caste with Brahminical thinking and overshadow the everyday environmental experiences of Dalits (p. xvii). According to Sharma, eco-organicism combines the protection of nature with the protection of Indian society against Western influences such as modernity and consumerism. Similarly, eco-naturalism associates environmental protection with preserving life in its natural order (p. xxi).

Chapter 15
Climate Change and Gender:
An Appraisal From the Public Health Perspective

Rudrajit Paul
Ruby General Hospital, Kolkata, India

Meera Suresh
https://orcid.org/0000-0002-8566-8784
Lincoln University College, Malaysia

ABSTRACT

One of the biggest threats facing mankind in the immediate future is climate change. However, climate change doesn't affect all humans equally; the adverse effects are accentuated for vulnerable groups like women and gender minorities. The SDG goal 13 of the United Nations is about Climate Action and for that goal to be achieved, academic research and societal application must go in concord. While the gender disparities of climate change are starkly manifest in multiple areas like economic opportunity, housing or political action, this chapter will concentrate on the public health perspective only. The authors of this chapter found that there is a dearth of comprehensive academic discourse on the nuanced health effects of climate disasters in different gender groups. While women and LGBTQ+ people are particularly victimized by climate disasters, the men are also not without their own challenges. The chapter discusses its contents within the framework of a conceptual model. Suitable real-life examples from prominent historical events and current events have been depicted. The impacts of climate change on mental health, sexual and reproductive health, gender violence and LGBTQ+ community have also been explored with due diligence. A special section on fragile ecosystems like the coastal areas and the situation in South Asia, the most climate sensitive region of the world, has also been added. Online data resources like PubMed, PLoS, WHO and UNDP were used along with relevant information from several government bodies and NGOs, as available online. In conclusion, the chapter reveals the dual role of males both as perpetrators of climate change and its victims, the situation of females as disproportionate sufferers and LGBTQ+ community facing additional adversities due to the consequential effects of climate change.

DOI: 10.4018/978-1-6684-9863-7.ch015

INTRODUCTION

Climate change is perhaps the biggest threat facing mankind in the near future. According to the World Health Organization (WHO), climate change has the potential to affect all aspects of good health (like food supply, clean air etc.) and undermine years of progress achieved in public health (World Health Organization, 2023). By 2050, the global GDP is expected to quadruple (OECD Environmental outlook, n.d.), and this will put a severe strain on natural resources, leading to massive exploitation of the environment. There will be around 70% increase in energy-related CO_2 emissions. Also, over the next two decades, most of the world population will be shifting to urban areas. Unless proper actions are taken, air pollution is expected to become the leading cause of premature mortality by 2050. Thus, overall, over the next two decades, climate change will cause a lot of financial damage and lead to excess mortality and morbidity (WHO, 2023).

However, although climate change is a threat for everyone, it does not affect all humans equally (UNDP, 2023). Rather, such events accentuate the already existing inequalities in the society and disproportionately increase the suffering of vulnerable and deprived groups (UNDP, 2023). Gender is one such factor which creates inequalities in the human society. Climate changes, like any other humanitarian crisis, have greater negative impact on women and other gender minorities. This is not a gender-neutral phenomenon and a single birds' eye view of the devastation is likely to gloss over the nuanced suffering of different gender groups. Thus, there needs to be a separate discussion on the impact of climate change on health from the gender perspective.

GENDER: THE DEFINITION

But before we go any further, let's discuss the meaning of "gender".

Gender, according to the WHO (2011) is defined as

"the characteristics of women, men, girls and boys that are socially constructed. This includes norms, behaviours and roles associated with being a woman, man, girl or boy, as well as relationships with others. As a social construct, gender varies from society to society and can change over time".

Or in simpler terms, "Gender determines what is expected, permitted and valued in a woman or a man in a pre-determined context" (WHO, 2014). So, the term "gender" includes both biological and social contexts. This social construct has significant effects on the life-path of the person and determines both their strengths and vulnerabilities.

To state an example; a woman in South Asia is supposed to feed her family and in quest for drinking water, she may have to travel considerable distance in challenging weather conditions. Similarly, in the time of emergencies, males are expected to be at the forefront risking their lives more than women. Thus, each gender group has its own challenges in the ensuing climate crisis. Neumayer & Plümper (2007) in their study have substantiated the fact that this disparity is more pronounced in those countries where the socioeconomic status of women is lesser to men. Whereas, in countries, where women share a status similar to men, the impact of natural disasters is nearly equal for both genders. (Neumayer & Plümper, 2007). Thus, the nuanced experiences of all genders deserve equitable attention.

The impact of natural disasters on males, females, and LGBTQ+ individuals vary over time and across societies. Vulnerability to climate-induced catastrophes is experienced by both genders, shaped not only by biological differences but also by societal roles and responsibilities assigned to them. (Easterling, 2000; Wisner et al., 2004). However, there are a few gender groups which deserve special attention from a public health point of view. This will be introduced in the next section.

THE LGBTQ COMMUNITY

Climate breakdown will affect all aspects of society like housing, medical care and food security. But pre-existing inequalities of the society will mean that the net impact will be disproportionately severe for those already at a disadvantage. This is where the plight of the LGBTQ community raises its head. LGBTIQ+, are group of people who identify themselves as lesbian, gay, bisexual, transgender, intersex or queer (WHO, n.d.$_a$). The plus sign (+) signifies vast range of individuals with different sexual orientation, gender identity, expression and sex characteristics (SOGIESC).

People who are outside the traditional family-centric binary gender norms of the society face multiple hurdles in all fields. While it is true that the LGBTQ community has come a long way in fighting for basic rights, their voice is still missing from important discourse areas. For example, in India, the most populous country of the world, there is still lack of usable data on the health status of LGBTQ community (Nikarthil and Kejriwal, 2021). Thus, this lack of data, combined with lack of awareness, make it very difficult to have gender-minority specific issues in the health policy. According to unofficial estimates, more than 100 million people in India belong to the LGBTQ community (Nikarthil and Kejriwal, 2021). So, this means that, the unmet needs of this large population are basically invisible to the policy-makers. However, experiences from other parts of the world have shown that sustained efforts can bring meaningful change. For example, in the USA, there is the LGBTQ health caucus which advocates for various aspects of health of the gender minorities with a rights-based approach (APHA, (n.d.)).

PROBLEM STATEMENT

Climate change is going to usher in an era of existential crisis for humanity. But different sections of the society will be impacted differently by the disasters. Gender is one factor which will determine the final impact of the disaster to a large extent. However, there are very few comprehensive academic discourses on the nuanced effects of gender in climate disasters.

ORGANIZATION OF THE CHAPTER

This chapter is arranged according to a conceptual framework as discussed later. The framework is divided into four sections describing the inter-relation of climate change and gender. In each section, suitable recent real-life examples will be discussed. Other topics like the impact on the LGBTQ community, the effect of climate change on mental health and gender violence exacerbated by climate change will be discussed according to their relevance for each section of this framework. Towards the end, there will be two special sections: the public health in coastal areas and on the situation in south Asia, one of the most populous and climate-sensitive areas of the world where millions of people are facing climate disasters as we speak. We will also explore climate impact specifically on sexual and reproductive health across genders. The chapter will end with some suggestions for the policy-makers and monitoring and evaluation strategies.

MATERIALS AND METHODS

This review is based on online search of biomedical literature available in English. Online databases like PubMed, Digital Commons Network, PLoS, Cochrane open access etc. were searched using terms like "Climate change", "gender", "LGBTQ", "Environmental disaster", "floods", "wildfire", "tropical storm", "heatwave", "air pollution", "female death", "sexual and reproductive health" etc in various combinations. In a typical Google search on the topic "gender and climate change", the results that emerge mostly are women-oriented. When conducted a targeted search by replacing the term "gender" to "males", the outcomes yielded are mostly indicating males as both perpetrators and sufferers in the issues created by climate change.

Further, the online resources of organizations like the WHO, UNDP and NGOs (Non-government organization) like Greenpeace and Environmental Defence Fund were scoured for relevant data. Any organization working in the field of climate change, which has an English language website, was searched for relevant information (e.g. Climate Action Network). Health data (in English) from several government bodies, as available in online, were also used. The open-source reporting of several national and international news organizations on various aspects of climate change were also consulted for relevant material.

Before going to the conceptual framework explanation, let's first take the example of air pollution, one of the most visible effects of climate change, to provide the reader with a primer on the current state of affairs.

AIR POLLUTION: THE NEW KILLER

As mentioned above, it has been seen that air pollution, both indoor and outdoor, affects female health disproportionately. Most household kitchens in the Indian subcontinent lack proper ventilation and the females, who are the cooks in almost all households, are exposed to the toxic fumes (both fossil fuel and vegetable oil) daily. This not only affects their respiratory system but may also cause reproductive health problems or a decrease in immunity. This domestic harm gets multiplied when the same women are exposed to outdoor air pollution due to environmental calamities. The following diagram (figure 1) depicts the effects of air pollution on health.

Though the hydra-headed effects of air pollution on health is vividly outlined in this diagram, very little is done both at micro- and macro-level to address these harms. Scientists say that this inertia is not due to lack of evidence, but due to lack of political will at all levels. Women are generally invisible in the climate discourse and this silent effect often stays under the radar (Shetty, 2021). However, this does not mean that men are exempt from indoor air pollution. A study by Kankaria et al. (2014) showed that in India, men also bear the burden of indoor air pollution and this health effect starts from a very young age for both sexes.

Figure 1. The health effects of air pollution
(Figure 1: Author's own creation. Information taken from: "WHO Air quality and health", (n.d.$_b$) https://www.who.int/teams/environment-climate-change-and-health/air-quality-and-health/health-impacts)

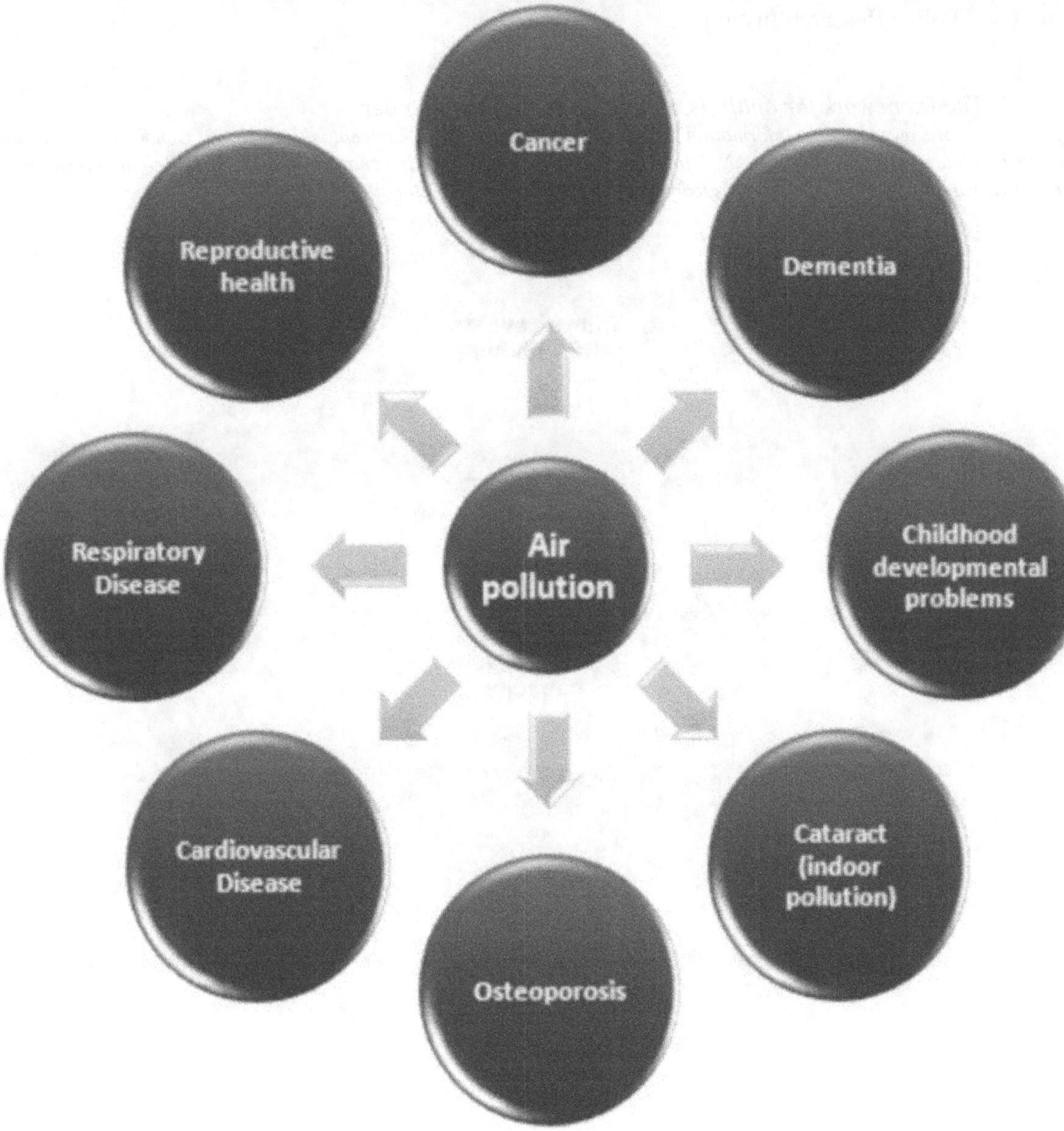

The conventional body of knowledge regarding climate change is male-dominated and the publications and policies usually stress only the economic, scientific and structural aspects of the problem (Oudraat & Brown, 2022). But this male dominance is one of the root causes of the problem. Discussion about the contribution of a patriarchal social structure in this humanitarian crisis and, by extension, on the peace and security of the society at large, is often neglected (Oudraat & Brown, 2022). But men are not only the perpetrators; they are also the sufferers. The effect of this societal system is not limited to women's health but affects that of all genders.

CONCEPTUAL FRAMEWORK

The conceptual framework (figure 2) for this discussion is adapted from that formulated by the International Food Policy Research Institute.

Figure 2. The framework for analysis of climate change and gender
(Figure 2: Author's own creation. Information taken from: Bryan et al., 2017. Conceptual framework: Gender, climate change, and nutrition integration initiative. GCAN Policy Note 7. Washington, D.C.: International Food Policy Research Institute (IFPRI). Retrieved from http://ebrary.ifpri.org/cdm/singleitem/collection/p15738coll2/id/132300)

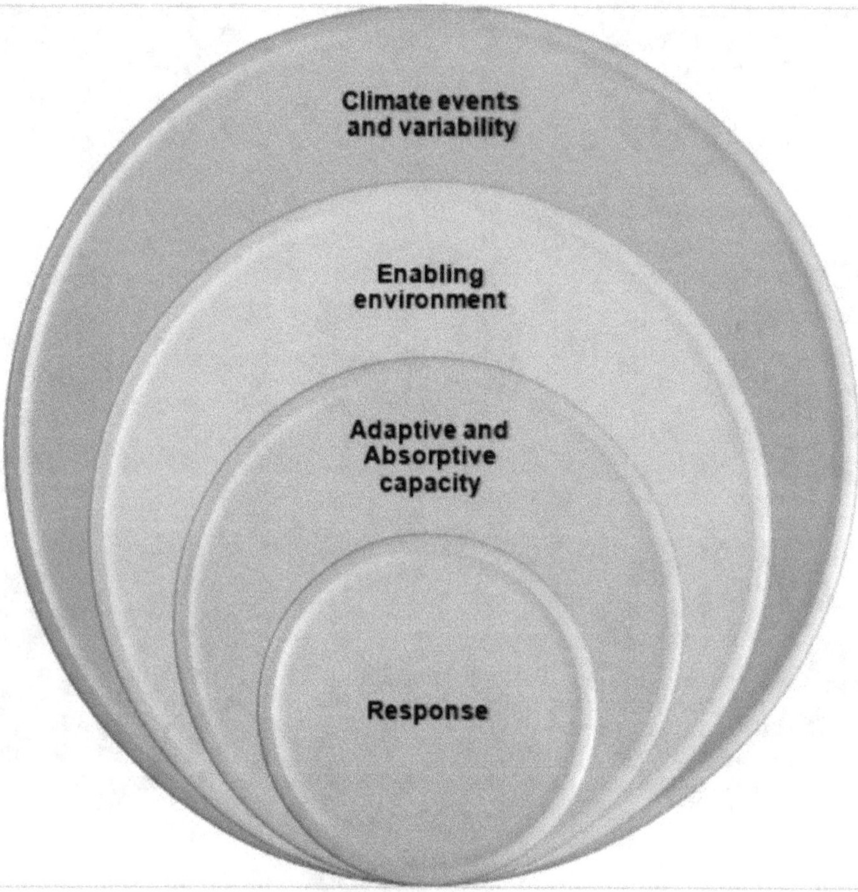

According to this framework, climate catastrophes like floods or heatwaves are further modified by the micro-environment in a society. The final effect further depends on the adaptive and absorptive capacity of the different social groups. All this chain of influencing factors finally determines the response of a particular group to the climate event and the final impact of the event on their lives. We will next discuss some examples before going into the framework analysis.

The Male Gender and Climate Crisis

Nagel and Lies (2022) effectively illustrated how males are exclusively assigned responsibility for climate change across various domains, encompassing the definition of climate change and resolution of the issue. Predominantly, men play a leading role in the significant carbon-intensive sectors, including the military (Nagel & Lies, 2022), as well as the automobile, fuel and energy production, agriculture, and food production industries ("Most polluting industries in 2023 revealed," n.d.), among others.

The above paragraph though mentions about males as the main contributors to climate change, also the main decision and policy makers, it discusses only about a limited percentage of males who are affluent, influential, and relatively shielded from the adverse impacts of climate change. However, focusing on males solely through this lens provides a distorted perspective, overshadowing the vulnerability of millions of poor, impoverished males worldwide who lack such privileged status or protection. Hence, the authors aim to underscore the viewpoint that males are also adversely affected by the consequences of climate change.

Doing a critical analysis of various researches conducted worldwide it could be seen that in some of the hostile weather catastrophes the mortality or morbidity rates among males were higher to females. For example, during the heatwaves in Chicago, in July 1995, elderly males were more prone to health risks than females (Semenza, 1996; Whitman et al., 1997) and Klinenberg (2002) ascertained the cause might be due to the social exclusion of geriatric men. CDC (2006) has also stated that it is mostly the male population who are at risk of heatstroke mortality as they are more likely to be active in hot weather. Similarly, Kovats and Allen (2008) have drawn out the conclusion in their study that during flooding, men are more at risk of drowning as they would take up the rescuer role to uphold their "heroic" behaviour entitled upon them by the society.

Long continued draughts, especially in the countries where agriculture is the main source of livelihood, issues of reduced income and consequential indebtedness take a toll on mental health of males especially leading them to depression and related suicides. This was particularly witnessed in Australia (Nicholls et al., 2006) and in semi-arid regions of India (Behere & Behere, 2008; Nagaraj, 2008). Now, let's look at the situation for the female gender.

EXAMPLES OF THE PREDICAMENT OF WOMEN IN CLIMATE DISASTERS

In 2018, Kerala faced catastrophic floods, the most severe in over a century for the state. Hundreds of people died, many went missing. However, two heavily pregnant women were airlifted by the Indian navy and the expectant mothers successfully delivered babies (Hindustan times, 2018). However, in disaster zones, women may not be as fortunate, facing adverse health consequences during childbirth in relief camps.

In the 2022 Uttarakhand flood, a woman's body was found in her house's sludge. Hindered by the sudden flash flood, she couldn't escape (The Indian Express, 2023). Similarly, on July 8, 2023, in New Delhi, another woman was crushed under a collapsed apartment ceiling during heavy rains (India.com, 2023). In both the scenarios, women faced perilous situations while salvaging household items, emphasizing the vulnerability they experience in such crises.

In the 2022 Pakistan flood, a young girl died while collecting leaves for her goats near a river bank. Women and girls in these areas undertake tasks like gathering cattle food and firewood, exposing them to risks (BBC News, 2022).

On June 25, 2023, in New Delhi, a woman was electrocuted while trying to maintain balance on an electric pole due to flooded roads from the previous night's rain (India Today, 2023).

In these incidents, we highlighted specific cases illustrating the differential impact of climate change on females, considering ecological determinants. While two pregnant women were rescued in one instance, women often faced fatalities due to home-care responsibilities, addressing family and animal needs, and systemic failures. This chapter examines climate change's population impact through a gender perspective, emphasizing that women in low-carbon-emitting countries bear a disproportionately severe brunt of its detrimental effects.

DIFFERENT SECTIONS OF THE CONCEPTUAL FRAMEWORK

Now, let's examine each section of the conceptual framework.

1. Climate events and variability

Climate change is affecting all aspects of human life on this planet. Due to various human activities, average global temperature is rising significantly. For example, since 1880, the average temperature is rising 0.14°F per decade and after 1981, the rate of this rise has almost doubled (Lindsey & Dahlman, 2023). The ten warmest years on record have all occurred after 2010 (Lindsey & Dahlman, 2023).

This rapid change of weather is giving rise to frequent extreme climate events like floods, heatwaves and hurricanes. This, in turn, is causing newer health emergencies like spread of zoonoses, increase in pollution related morbidity and mental health problems. Most of these events are occurring in low- and middle-income countries (LMIC); places which have the least contribution to climate change but which suffer the most.

Recent examples of climate catastrophe include the Cyclone Gabrielle in Feb, 2023 in New Zealand, causing damages to the tune of USD 8.4 billion. Also, in 2022, Pakistan had an unprecedented flood (mentioned earlier) which caused massive financial loss and mortality. In 2019-20, Australia had a "mega fire" event where millions of hectares of forest were gutted by bushfire. All of these events were either the direct or indirect results of climate change and its consequences. That is why, the United Nations climate action panel has started the "Race to zero" campaign to galvanize all stakeholders and mitigate the harm (Race to Zero, n.d.).

However, a macro-level statistic of these natural disasters often hides the fact that there are marked differences in impact it has on different genders. For example, in the aforementioned Pakistan floods, though there were a lot of human displacement and deaths however, the NGO, *Rozan*, mentioned that women and girls faced additional harms in the form of domestic violence, sexual exploitation and disruption of reproductive health services (UN Women, 2022a). The impact of such disasters on women does not stop even after the disaster is over. For example, going back a decade here, after the 2010 floods in Pakistan, there was a sustained rise in maternal mortality, teenage pregnancy and premature deaths among women over the next few years (Bukhari & Rizvi, 2015). Villages once destroyed by flood took a long time to rebuild and a lot of women were forced to give birth in makeshift roadside camps with little medical support and privacy. Rigid gender norms in a patriarchal society already reduce resilience of women in normal times and natural disasters tend to push them over the edge. In addition to physical health problems, the mental health of women also suffers a lot post to disasters. The rates of self-harm and suicide are higher after such catastrophes, especially among women (Kolves et al., 2013).

Climate Change and Mental Health

Historically, the health response or disaster-preparedness responses have focussed only on the injury, infection or exposure -related illness. Although clinicians have documented mental health declines following disasters, this topic has not received enough spotlight or policy inclusion. Thus, like indoor air pollution affecting women, the problem is again, not lack of evidence but a stark lack of policy initiatives. Further studies are needed to understand the expected behavioural changes following calamities and to identify the communities at risk of psychological perturbations. Lessons from previous events can help us to be better prepared for these adverse effects and include mental health as a component of the public health response (Yun et al., 2010).

In fact, a completely new term is now doing the rounds: *eco-psychiatry*. This term encompasses a lot of different issues. Let's take an example to clarify better. Sundarbans, a mangrove forest area in Eastern India is often affected by cyclones and floods. This lead to the destruction of home and agricultural land which causes mental stress in the short term. But the impact extends far beyond that. As forests are destroyed, the man-eating tigers often come out into the locality to hunt and that causes loss of lives. Moreover, as agricultural lands are destroyed, the villagers living on the fringes of the forest are forced to foray inside the jungle in search of honey or firewood. This increases man-animal contact and leads to further mental stress in the long term. Usually, it is the men who are attacked and killed by tigers. Then, their widows are left without financial support which affects their overall health. Also,

Table 1. Mental health effects of natural disasters

Climate change and psychological problems
Mild stress and insomnia
High-risk coping behaviour such as increased alcohol use
Depression
Anxiety
Post-traumatic stress
Increased risk of suicide
Long term adverse effects on children
Stress and depression in first responders and relief workers
Increase in aggressive behaviour in heat induced stress
Worsening of pre-existing mental illness
Autism
Malnutrition like iron deficiency may increase bipolar disorder
Propensity for increased domestic violence

(Table 1: Author's own creation. Information taken from: American Psychiatric Association. (n.d.). Climate Change and Mental Health. Retrieved from https://www.psychiatry.org/patients-families/climate-change-and-mental-health-connections/affects-on-mental-health)

after every natural disaster in the Sundarbans, the rate of human trafficking of young girls rises manifold (Jalais, 2019). Following table (table 1), gives guideline on various psychological factors to be explored aftermath to climatic catastrophes.

Effect on LGBTQ People

Climate disasters also have marked adverse effect on the LGBTQ community. This community is already uniquely vulnerable to exclusion and exploitation due to social stigma and discrimination (Behal, 2021). In most scenarios, normal social institutions like healthcare facilities are inapproachable for them. They are economically deprived, are forced to work in dangerous professions and often suffer from homelessness (Behal, 2021). These conditions make them especially vulnerable to climate hazards. The LGBTQ people often lack information about natural disasters and are likely to be marginalized during relief efforts (Tabary, 2018). Thus, the LGBTQ liberation movement must be linked with climate justice.

2. Enabling environment

The effects of climate change occur within certain contexts, where enabling factors determine the ability of individuals and groups to cope with the disasters. These enabling factors include international cooperation and commitments as also local and regional policies or laws.

It is now well recognized that people who are more dependent on natural resources and have the least capacity to respond to natural hazards are the ones worst affected by climate change (United Nations climate change, n.d.). Globally, this vulnerable group is predominantly composed of women. The United Nations Framework Convention on Climate Change (UNFCCC) recognizes this fact and the parties to the convention have pledged to make the climate policies more gender-responsive (United Nations climate change, n.d.). However, there is still much progress to be made. The UNDP policy document, *"Gender, Climate and Security"* (2020) has rung the alarm bell about this "critical gap" in gender-sensitive policy making for climatic action. As recently as in 2019, an international debate on *"Addressing the Impacts of Climate Related Disasters on International Peace and Security"* in the UN Security Council witnessed a marked lack of awareness about gender issues among the participating member states (United Nations Environment Programme et al., 2020).

Climate Change and Female Empowerment

While the previous paragraph highlights women's heightened vulnerability to climate change, it is concurrently creating new spaces for female empowerment. For example, droughts in Sudan compelled the menfolk of many pastoral communities to move in search of new grassland, leaving the women in charge of the household. This shift, supported by local NGOs like PENHA, enabled women to acquire new skills, such as livestock raising, enhancing their economic productivity (Van Dijk, 2012). These changing social norms can empower women and promote their participation in decision-making.

Experiences elsewhere have also shown that if women are empowered, they not only mitigate the effects of climate change, but also, act as active drivers for environmental improvement (Gurung, n.d.). The United Nations Convention to Combat Desertification (UNCCD) has recognized the importance of involving women in dry land management (Gurung, n.d.). Inclusion of women in the decision-making process of agricultural and pastoral activities in dry lands like Niger and the highland deserts of Tibet have shown to revamp crop biodiversity, increase in household income and better land utilization (Gurung, n.d.). This resulted in improved family nutrition and better spending for health. It is now recognized that indigenous women in climate-vulnerable regions possess valuable knowledge on sustainable environmental practices. Gender mainstreaming can harness this knowledge for enhanced environmental security. The Andhra Pradesh Tribal Development Project in India has also shown that the folk knowledge passed down among women can be effective in improving land management of a region.

The UN Womenwatch ("Women, Gender Equality and Climate Change," n.d.), from a policy perspective, emphasizes specific financing mechanisms to boost female participation in various activities, advocating for the formation of women's groups and networks. It stresses the need for definite female representation in climate governance.

Experiences show that enhancing women's access to natural resources and the decision-making process enriches the community as a whole and enable them to mitigate climate change effects. Yet legal support is crucial for female representation in environmental policy making and recognition of female leadership by patriarchal authorities.

LGBTQ Movement and Climate

As the LGBTQ liberation movement is gathering pace all over the world, it's been realized that the existing environmental movements will need to integrate the demands of gender minorities. There is currently very limited research on the potential impact of climate change on the gender minority groups. Most conversations and policy responses to climate change follow the heteronormative practice (Whitley & Bowers, 2023). In addition to the impact of a changing climate, the plight of LGBTQ community is likely to be further aggravated by loco-regional policies which perpetuate the structural and institutional barriers (Whitley & Bowers, 2023). For example, recently in the USA, a lot of anti-LGBTQ bills have been filed, which aim to further restrict access of these people to public utilities (Whitley & Bowers, 2023). These legislations, if enacted, are likely to increase the sufferings of gender minorities during disasters.

However, surveys have found that LGBTQ people are more likely to participate in movements about climate change. According to the social identity theory, people often bond together due to a shared experience (Goldberg et al., 2019). For LGBTQ people, this shared experience is that of quintessential social ostracization. As they realize the compounding effect of climate change on their predicament, they are likely to raise their voice for policy change. In its policy brief in July, 2022 the UNDP has also stressed the importance of gender and social inclusion (GSI) in public climate finance (UNDP Thailand, 2023). A recent pilot project in Thailand has shown the importance of GSI in local infrastructure budgets. With help from UNDP, there was local capacity building and leadership support to create the enabling environment for climate finance in an inclusive manner (UNDP Thailand, 2023).

3. Adaptive and absorptive capacity

Absorptive capacity is defined as the sensitivity of individuals or groups to stressor events (like climate disaster). Adaptive capacity is defined as the ability of these groups to respond to stressors (Bryan et al., 2017). Both of these are determined by pre-existing personal, environmental and social factors. According to Bryan et al. (2017) there are three main factors that determine these capacities in individuals and groups: Rights, Information and Empowerment.

At the personal level, absorptive capacity is determined by health status of individuals. Natural disasters are times of extreme physical stress in the form of injuries, infectious diseases and lack of availability of care for chronic conditions. Also, in these times, individuals bear the added responsibility of caring for the sick and procuring food and resources for the family. Often, these duties fall entirely on women. Therefore, women's health status not only influences their own resilience but also shapes the fate of the entire family. As aforementioned, female empowerment in disaster risk reduction decision-making offers numerous advantages. Empowering the women not only enable them to address their specific needs and vulnerabilities, but also could bring their valuable experiences into the disaster management process.

Health Status of Women in Disaster-Prone Areas

In many countries, women already suffer from malnutrition and lack of proper health services access. For example, the National Family Health Survey in India found that almost one in four women are underweight. In some impoverished regions of the country, the level of female malnutrition is even higher than 30%. Incidentally, these are the regions of the country (Bihar, Jharkhand) that are also more prone to climate events. The repeated exposure to climate disasters is also one of the reasons for this background malnutrition. Thus, the impact of climate change on the female population in these regions will be higher (Ministry of Women and Child Development, Government of India, 2020).

This pre-existing status reduces their absorptive capacity further. In addition to these, Bryan et al. (2017) states that the adaptive capacity is also determined by factors like access to information, level of participation in economic activities and policy priorities. Women and gender minorities are usually at a disadvantage in this regard. For example, in South Asia, women still have unequal access to household finances and less power in household decision making. Even after the introduction of government programs targeting women, this inequality has not vanished. Women, although being given access to finances and skills, still lacked the social environment to exercise their choices. In this background, a natural disaster is further likely to marginalize women as they lack the power to decide on priorities for their own health and well-being (Kochar et al., 2022).

Women constitute majority of the world's poor and are more dependent on natural resources for livelihood (UN Womenwatch, n.d.). A UN estimate says that seventy percent of the world's poorest humans are women. Despite being the primary workforce in food production worldwide, women own only a tiny fraction of the land and assets. Hence, though they are involved in the utilization of natural resources, they have very little say in the environmental management.

Gender Violence and Climate Change

Women also have less access to transformative technology. Hence, during natural disasters, women are the first to endure shortages of food and shelter. Additionally, they face considerable social, economic, and political barriers that impede their coping mechanisms. Many studies have found that violence against women increases after disasters (Thurston et al., 2021). This is probably due to creation of an enabling environment that promotes such violence and accentuation of pre-existing drivers for violence (Thurston et al., 2021). According to the Staff Correspondent of *The Daily Star* (2008) after a flood in Bangladesh, more than 50% of women reported facing physical, sexual and psychological abuse at the relief camps. The correspondent stated that this reduces adaptive capacity of women and deters them from seeking healthcare. However, this phenomenon is not limited to LMIC and studies across the world have shown that violence against women and girls spike significantly after disasters. Anastario et al. (2009) reported a significant increase in gender-based violence against women in the USA after Hurricane Katrina, persisting for two years beyond the baseline. This underscores the necessity for specific policies addressing these issues.

The palpable impact of these gender inequalities are starkly brought out by the mortality data following natural disasters. For example, after the 1991 Cyclone Gorky in Bangladesh, 91% of those who died were women. During the 2003 Heat wave in Europe, 75% of the dead were females. In 2016, after Cyclone Winston in Fiji, many gender minorities were denied shelter at the relief camps. This led to increased negative health consequences ("Climate Change is Not Gender Neutral," 2021).

Post-disasters, women often face homelessness and loss of employment. Hence, urgent action is required for gender-sensitive disaster risk reduction policies that address women's concerns during relief operations. UN Women conducted a gender analysis in flood-affected areas in Bangladesh after the 2022 floods. After situation analysis, their recommendations included provision of gender protection in disaster planning, gender-specific training for volunteers and health workers, financial aid for women, and advocacy with policy makers ensuring safe spaces for women and girls (UN Women, 2022b).

Hurdles Faced by the LGBTQ Community

Transitioning from our previous discussion on women, let's now explore the distinct challenges faced by the LGBTQ community. They face unique barriers during natural disasters. For example, after the 2004 Tsunami in Tamil Nadu, a gender minority group in the area were excluded from food relief as they were denied the requisite government papers (Blanchard, 2021). Likewise, after the Hurricane Katrina in USA, gender minorities faced inhuman conditions which accentuated their sufferings. For example, they were turned back from relief centres, the same-gender families were denied insurance claims and, in some cases, people faced legal problems even for the use of restrooms (D'Ooge, n.d.). According to the minority stress theory, these events will increase the mental health issues in this population. A recent and similar study from Philippines found that following typhoons, LGBTQ people often find it difficult to access resources. A transwoman was forced to use the name at birth instead of her chosen name to get the essential supplies. In many cases, same-sex couples were denied supplies designated for families as their "family" status was not recognized. In some interviews, young LGBTQ people of Philippines expressed their apprehension about the possible lack of social support during future climate events too (Santos, 2022).

While there are entrenched social stigma and "othering" of this group, there is also failure at policy level. For example, the Sendai Framework for Disaster Risk Reduction 2015-2030, although calling out for the formulation of an inclusive disaster relief policy, does not explicitly mention gender minorities (Blanchard, 2021). Thus, there is very little incentive for policy makers and governments to stress on this issue. The UNDRR identifies this omission as a policy blind spot, hampering the adaptive capacity of LGBTQ individuals during disasters. Addressing this requires centralized guidance and engagement with grassroots workers and community leaders. On a positive note, a gender-inclusive approach has demonstrated more meaningful climate action. For example, countries with more number of female parliamentarians are more likely to ratify environmental treaties. As a tangible result, it has been seen that increased female parliamentary representation leads to lower carbon dioxide emission from the country. Comparing Bahrain (2% female parliamentarians) to Denmark (37%), it is seen that the climate change policies of the latter are about 8 times as stringent as the former (Mavisakalyan & Tarverdi, 2019).

4. Response

The response of different individuals and social groups to climate change can be short term or long term. In the short term, they try to minimize damages and overcome the problems of homelessness, food scarcity or security. In the long term, there may be transformative changes that try to increase coping capacity and minimising environmental damage. However, these responses often involve complex decision-making processes and people with more bargaining power usually get the net results skewed in their favour. This is where gender advocacy becomes paramount because women and other gender minorities usually lack the capacity to place their demands.

Improvement in health after a climate shock usually follows one or more of six pathways: food production, income, asset dynamics, labour, natural resources and cooperation (Bryan et al., 2017). A few of these will be discussed next.

Climate Responsive Agriculture and Gender

Climate change is causing the disruption of food production due to unpredictable weather. This affects the nutritional status of the families and leads to ill-health in the long term. As seen earlier, the inclusion of women and their traditional knowledge, in the decision-making process can lead to better climate-responsive agricultural practices. The World Bank calls this Climate Smart Agriculture (Nieuwkoop et al., 2022). It has been seen that if women are given the decision making power, they are more likely to plant subsistence crops, thereby leading to better food security and better health outcomes (UN Women, n.d.). Thus, the formulation of a gender-sensitive food policy is essential to enable women to become agents of positive change.

The WHO report on gender and climate change highlights the fact that climate disasters tend to kill more women and at a younger age (WHO, 2014). Hence, natural disasters exacerbate the gender gap in life expectancy and climate-sensitive health events like chronic malnutrition and malaria also exhibit marked gender differentials (WHO, 2014). A study in Europe has shown that women are more likely to die of heat waves than men (WHO, 2014).

In many countries like India, women lack access to healthcare because of distance, lack of a male companion to accompany them and financial constraints. During natural disasters, this is likely to be augmented. For example, a recent study in Bangladesh found that floods often reduce the utilization of ante-natal services (Orderud et al., 2022). Even the UNFPA have declared that essential health needs and rights of women are often neglected during disaster relief (UNFPA, 2015). Thus, a fit-for-all humanitarian aid policy is likely to leave a significant number of people behind and there is need of a nuanced approach in battling disasters. The policy changes brought about by climate disasters can go a long way in achieving gender equality and battling the systemic deprivations.

Hitherto, the analysis has focused on the impact of climate change on genders in general. In the next section, let's delve into more specific subjects, such as the effects of climate change on women living in coastal areas and of women living in South Asia as they are more risk-prone and fragile populations. Additionally, the exploration will extend to the specific ways in which climate change affects the sexual and reproductive health across genders.

Coastal regions:

The coastal regions are some of the most threatened ecosystems of the world in the era of climate change (Prakash et al., 2022). These are regions with a fragile and shifting human settlement and far from mainstream economic activities. The aquatic food systems, ocean-dependent livelihood chains, coastal land and water utilization activities are likely to be severely disrupted in the near future as climate change makes weather patterns hostile (Prakash et al., 2022). Hence, while comprehensive climate policy action is essential, special attention is needed for coastal climate action. As noted earlier, like other spheres of climate policy, coastal climate change action plans must also be gender-sensitive.

Females have historically played a major role in fish and seafood farming, trading and processing. But their contributions have been traditionally kept invisible and there is a wide pay gap in the sector. There is also a lack of data about the gender roles and specific gender contributions in this seafood sector. Trading in coastal areas is largely unorganized and outside the formal economy in countries like India. This opens the doors to the continuance of exploitation, especially of female workers. As climate change destroys this fragile economy, the downtrodden female workers will be the first to bear the brunt of its impact. Nonetheless, coastal climate policies often neglect the specific needs of the female cohort.

As coastal regions are more vulnerable to extreme events like Tsunami, cyclones and flooding, there is need for training the inhabitants in disaster management and alternative livelihood adaptation. However, women, who are historically deprived of access to information and decision making, are often left out of the loop. The IPCC report, drawing on studies from Bangladesh, has shown that women are often left out of micro- and macro-level economic and societal decision-making processes. Thus, when climate events disrupt the livelihood of females in these regions, there is very little coping mechanism for them. One example is the Sundarbans region of West Bengal, India. Cyclones often destroy homes and make agricultural lands saline in this region. But women are often not trained or equipped to court alternative occupations. Thus, they are forced to be employed as prawn-seed catchers in the coasts. This job involves standing in the murky water for hours to trap the prawn seeds with makeshift hand-nets. Adult women and school-drop out young girls are mainly employed in this business. This causes tremendous physical harm in the long term and leads to overall morbidity. This ill-health further reduces resilience of these women during future climate events (Verma et al., 2011).

Gender-Responsive Coastal Area Planning

Despite the stark scenarios mentioned above, there are also positive reports from some studies. For example, in Africa, women of some coastal communities have been encouraged to practice seaweed farming. This has resulted in a much sustainable livelihood in the face of climate change and has led to overall improvement of the family financial condition (Msuya et al., 2017). This led to female empowerment and women were able to assert themselves in the social hierarchy. Experiences from Bangladesh have taught us that, in addition to government level policy action, NGOs play a very important role in creating a sustainable, gender-responsive economy in coastal regions (Roy et al., 2022). Bangladesh, the largest deltaic region in the world, is virtually sitting on a time bomb as sea levels are rising due to climate change. Thus, this region is running out of time as extreme weather events continue to batter down on the vulnerable population.

In coastal Bangladesh, women have been shown to have valuable input in keeping food dry, devising innovative ways to store food and ensuring fresh water supply for the family during disasters (Roy et al., 2022). The traditional knowledge possessed by women, for example, kitchen gardening and preservation of fresh-water sources in coastal regions can be utilised in the mainstream economy to boost micro-level income and overall health of the family and community. Involving women in disaster planning, supported by education and livelihood security at the policy level, has shown improved overall outcomes, emphasizing the crucial role of gender-inclusive approaches.

The Situation in South Asia

Climate change is a global emergency that transcends national barriers. But still, some regions of the world are more vulnerable than others. One such fragile region is South Asia (CANSA, 2022). South Asia has many fragile ecosystems like newly formed slopes of the Himalayas, Ramsar water bodies and mangrove forests which can be extremely vulnerable regions. Around 40% of all global climate-related disasters are experienced in this region and events like landslides, glacier melt, droughts and flooding are frequent occurrences. Many of the countries here are losing GDP due to climate events and the resulting poverty further widens the social inequalities, including the gender gap.

The following table (table 2) shows the relation between climate change and gender inequality in three South Asian countries.

The wide gender gap contributes to issues such as childhood malnutrition, domestic violence, and early marriages. This creates a vicious cycle where climate-induced damage accentuates gender inequality, and vice versa. Extreme weather events often lead to male migration in search of livelihood, leaving women responsible for households, particularly from regions like West Bengal, India where cyclones and flooding are common to the Middle East or other industrialised zones of India. This adds to the burden on women, limiting their economic opportunities and education. Coupled with existing social and systemic barriers, this hinders female education and meaningful economic participation, consequently reducing their resilience to further natural disasters. This is demonstrated in following figure (figure 3).

Thus, it is clear that climate action policies should not only be gender-sensitive but also, gender-transformative. Special action is needed in areas like agriculture, forest and water management: sectors which are likely to be worst affected by climate change and which directly affect women to a greater extent.

Table 2. Statistics of South Asian countries related to climate risk

Name of country	Climate risk data	Gender inequality index/ Ranking
Nepal	16th in fatalities among all countries; 18th in fatalities per 100,000 inhabitants; 56th in losses; 40th in losses per unit GDP.	0.452/ (113/170)
Bangladesh	Seventh most climate change vulnerable country (Germanwatch);Average annual losses to disaster at around $3 billion, or around 1% to 2% of GDP (UNISDR)	0.53/ (131/170)
Sri Lanka	Sri Lanka ranks 35th in fatalities among all countries ;42nd in fatalities per 100000 inhabitants; 28th in losses; 45th in losses per unit GDP	0.383/ (92/170)

(Table 2: Author's own creation. Information taken from: Climate Action Network South Asia (CANSA). (2022, December). Gender And Child Inclusion In Policies On Climate Change In South Asia: A study from Nepal, Sri Lanka and Bangladesh. Retrieved from https://cansouthasia.net/gender-and-child-inclusion-in-policies-on-climate-change-in-south-asia/)

Figure 3. Effect of climate disasters on women
(Figure 3: Author's own creation)

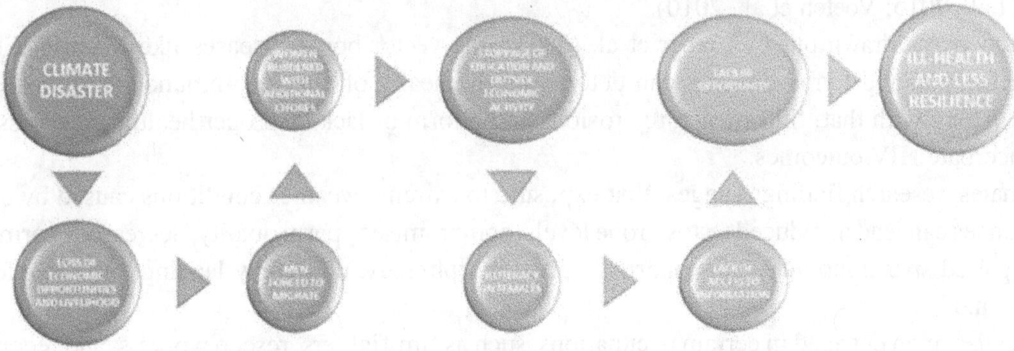

One such attempt has recently been started in Kandhmal District of Odissa, India. As climate change is making food production uncertain, the local tribal people of this area are engaging in deforestation to increase agricultural land. This, in turn, is causing more environmental problems. As a solution, the tribal women of the region, the main agricultural workforce, has been trained in millet farming. Millet farming is sustainable, requires less use of water and can be done in barren land too. This farming system is thus aimed at empowering the local women and increasing their food security (Barik, 2022). Thus, from being victims of climate change, these tribal women are being transformed into "change agents".

Some action has already been taken in this regard in some of the countries in South Asia. For example, the Bangladesh Water Act, 2013 has mandated that at least one-third of local water-body management groups be female (CANSA, 2022). The Coastal Zone Policy, 2005 of Bangladesh includes special training programs for women of coastal regions, in order to increase their economic opportunities and improve their coping capacity. However, in many countries of this region, in spite of an attempt at gender mainstreaming at policy level, the corresponding budgetary allocations have not been commensurate with the promises. Thus, although the theme of gender-equality remains on paper, its practical application gets hindered by lack of resources.

CLIMATE CHANGE AND IMPACT ON SEXUAL AND REPRODUCTIVE HEALTH ACROSS GENDERS

Recent research indicates that climate change affects sexual health outcomes, notably in regions like sub-Saharan Africa and Haiti, where severe drought and extreme weather events pose health risks, including those related to HIV (Logie et al., 2021). Watts et al. (2019) reported in the 2020 Lancet Countdown that climate change will lead to a rise in human migration and displacement. Both these phenomena can result in sexual health risks (Singh et al., 2018) especially in marginalized groups; such as migrants, sexually and gender diverse individuals, those living in poverty, adolescent girls, and in young women (Logie et al., 2021). Studies from sub-Saharan Africa (Baker, 2020) involving 400,000 individuals across 25 sub-Saharan countries revealed a positive correlation between prolonged temperature rise and HIV prevalence, particularly in rural areas and among younger age groups possibly connected to men's migration and involvement in transactional sex. Extreme weather events, including drought and flooding, are associated with risky sexual behaviours like condomless and transactional sex and early sexual debut, increasing the risk of HIV and other STIs in young people (Low et al., 2019; Robinson & Yeh, 2011; Burke et al., 2015; Voeten et al., 2010).

Moreover, as drawn out by Lieber et al. (2021), the vector borne-diseases like Malaria, Chikunguniya, Dengue and Lyme diseases can deteriorate the health of already immunocompromised HIV patient. Along with that, 'infrastructure erosion' in the form of lack of proper healthcare accessibility can exacerbate HIV outcomes.

In males, research findings suggest that exposure to extreme weather conditions caused by climate disturbances can lead to reduced testosterone levels, compromised sperm quality, decreased sperm quantity, impaired sperm motility, and abnormal sperm morphology, ultimately leading to male infertility (Legacy, n.d.)

Likewise, men engaged in certain occupations, such as fire-fighters, rescue workers, and reconstruction workers involved in responding to natural disasters caused by climate change, face an enhanced risk of sustaining physical disabilities or even fatalities (Nagel & Lies, 2022).

The impact of climate change on sexual and reproductive health (SRH) of women especially in rural areas of LMICs is substantially significant and has been examined at different levels as following figure (figure 4).

According to the WHO/UNICEF Joint Monitoring Programme (WHO/UNICEF JMP, 2023) most of the women in rural areas of LMIC use reusable materials for menstrual purposes. Natural disasters disrupt the availability of clean water, soap, and private places for menstrual hygiene, leading to increased cases of reproductive tract diseases (Ipas, 2022). In the same Ipas study (2022), during an in-depth interview one of the adolescent victim was vocal about her unintended pregnancy as she couldn't take the birth control pills on time due to a sudden storm hit in a residential area and was transferred to an emergency camp with no access to oral contraceptive pills to avoid her unintended pregnancy. The unmet needs of SRH care in the temporary humanitarian shelters are profound (UNFPA ESARO, 2021).

Figure 4. Impact of natural disasters on SRH
(Figure 4: Author's own creation)

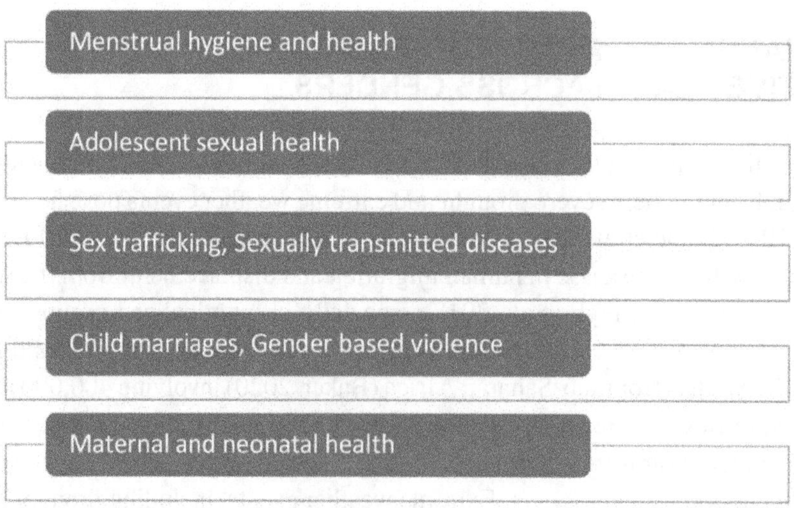

Post-climatic catastrophes, human displacement triggers an economic crisis, contributing to increased child marriages, sexual trafficking, sexually transmitted diseases, and gender-based violence (Alston et al., 2014; HRW, 2015; Ahmed et al., 2019; McLeod, 2019; Thurston et al., 2021; UNFPA ESARO, 2021; Atmos, 2023,). Instances of domestic violence, sexual abuse, and female genital mutilation rise during droughts in Uganda (Le Masson et al., 2019). Tropical cyclones in Vanuatu in 2011 led to a 300 percent surge in new domestic violence cases (CARE, 2015). Climate change and its impact of gender based violence has been detailed earlier. Although not exclusively falling under the category of SRH, factors like "school absenteeism" is also accentuated by climate-related events, affecting approximately four million young girls in LMICs (Malala Fund, 2022).

Heat waves pose threats to maternal health, causing complications like oligohydramnios and increasing stillbirths, preterm births, and low birth weight babies (Davison, C., 2022). This has also been established by a meta-analysis study (Chersich et al., 2020) comprising 70 studies, which found that for every 1°C rise in environmental temperature, the risk of stillbirths and premature births escalated by 5%. Similarly, rising sea levels and hurricanes expose expectant women to contaminated water and barriers to healthcare facilities, resulting in health complications (The Guardian, 2023; Karl et al., 2009). The Zika virus epidemic in Brazil in 2015, linked to conducive climatic conditions, led to a surge in microcephaly cases among new-borns (Nolen & Galdieri, 2022). Likewise, floods in Pakistan in 2022 increased neonatal mortality due to diseases like cholera, dengue fever, and malaria (Saifi et al., 2022). Selby, D. (2015) reported that aforementioned mosquito-borne diseases impacted pregnant women also significantly because of their increased engagement in domestic tasks like cooking, washing etc. due to close proximity to stagnant water.

Thus, we conclude this chapter with the final section. As all readers are aware, public health policies which look excellent on paper, often fizzle out in want of proper planning framework. One of the essential components of public health planning is a monitoring system. We will end the chapter with proposed monitoring system for gender action in climate crisis.

MONITORING AND EVALUATION

Monitoring and evaluation (M&E) plays an important role in understanding the complex interplay between climate change and gender dynamics, and provides valuable insights into the effectiveness and impact of programmes and policies aimed at addressing these issues. Climate change affects men and women differently, with distinct gender roles, power dynamics, and vulnerabilities shaping their experiences and responses. M&E furnishes a comprehensive framework which enables researchers and policymakers to identify gaps, track progress, and design evidence-based strategies that promote gender equality and resilience in the face of environmental challenges. The next section of the article deals with some important M/E indicators which have been adapted from Gardner (2021).

Policy related:

1. Number and proportion of female members and LGBTQ members in disaster-management and disaster-policy making groups.
2. Number of national programs directly linked to female empowerment.

3. Number of government reports concerning natural disasters containing sex-disaggregated data.
4. Number and proportion of climate policy documents containing at least one mention of LGBTQ issues.
5. Number of climate policy documents where gender equality is mentioned as one of the objectives.

Diverse representation, including women and LGBTQ individuals, in disaster-management and policy-making groups is highlighted as an essential metric, emphasizing the importance of varied perspectives. The number of national programs directly linked to female empowerment underscores the significance of targeted initiatives in addressing gender inequalities amid climate challenges. The inclusion of sex-disaggregated data in government reports on natural disasters is emphasized for its role in identifying specific vulnerabilities and tailoring interventions. Acknowledging LGBTQ issues in climate policy documents and integrating gender equality as an objective are crucial components of comprehensive strategies that M&E facilitates.

Exposure related:

1. Sex-disaggregated statistics about number of people affected in natural disasters.
2. Mortality data after natural disasters: disaggregated by gender.
3. Number of people attending health care facilities after natural disasters: disaggregated by gender.
4. Number of out-of-hospital births one week, one month and six months after a natural disaster.
5. Number of reported physical abuse cases against women and LGBTQ as a function of time after a disaster.

Others: -

1. Number of female students who dropped out of school after a natural disaster.
2. Number of missing females in an area one month and six months after a natural disaster.
3. Number of teenage pregnancies one month and six months after a natural disaster.
4. Number of suicides one month and six months after a disaster: disaggregated by gender.
5. Number and proportion of relief workers receiving training in gender sensitivity.
6. Number of visits to mental health facilities after a disaster.

These M&E indicators offer essential insights post-natural disasters. Tracking the number of female students dropping out and missing females assesses the educational and safety impacts. Monitoring teen-age pregnancies provides data on reproductive health vulnerabilities. Suicides, disaggregated by gender, reveal mental health strains. Training relief workers in gender sensitivity ensures equitable assistance, and tracking visits to mental health facilities gauge psychological impacts. These indicators guide targeted interventions, shaping policies that address the diverse needs of affected populations.

RECOMMENDATIONS

As explored in the article, the interlinking of climate changes to gender is nuanced. Hence the recommendations suggested would also have to be multidimensional.

- Gather gender-disaggregated data. This will help in introducing targeted interventions and gender-responsive policies. Additionally, this data will also aid in monitoring and evaluation of the effectiveness of interventions and/or policies.
- Scale up women's and LGBTQ people's financial empowerment along with decision making power. Former can be done by ensuring their formal education and also providing awareness about the climate change and their vulnerability. Latter by providing them skill-based-trainings especially on climate-resilient livelihood opportunities.
- Use Women self-help groups to train women, adolescent girls and other weaker sections of the society, the basic life-saving skills especially during and post disasters.
- Give equal opportunities to weaker and marginalised genders of the society a role in policy making and also to be a part at grass root levels to ensure its execution.
- Ensure that sexual and reproductive health services are available for women affected by climate-induced disasters. Safer and private spaces should be provided for weaker genders so that their vulnerability to any form of violence is adequately addressed.
- Promote researches and advocacies in climate and gender issues. Assure everyone in the society is aware of the climate changes happening around, hence, include awareness raising education and activities from school level education.
- Recognize and include the valuable traditional knowledge possessed by indigenous women in climate change efforts. Respect their role as protectors of biodiversity and natural resources.
- Collaboration of various governments, NGO's at local, national and international levels as climate disaster is a global threat. There should be a two-way feedback and communication loop between these stakeholders, and not a top-down approach.
- Lastly, proper monitoring and evaluation of existing climate programs to ensure coverage and outreach.

CONCLUSION

In summary, our examination has elucidated the intricate dynamics of gender in the context of climate change, revealing the dual role played by males as perpetrators and victims of its repercussions. Females, in particular, bear a disproportionate brunt, facing heightened vulnerabilities in terms of health, mortality, morbidity, and disruptions in sexual and reproductive health care. The LGBTQ+ community, already marginalized, confronts additional adversities. The discernment of air pollution, intricately linked to climate change, has accentuated the pressing need for gender-sensitive environmental policies, especially given its disparate impact on women. Further exploration delved into the adaptive capacities exhibited by different genders, the transformative shifts in livelihood techniques, and the indispensable role played by women's traditional knowledge. The interplay of climate change with domestic violence and mental health has been expounded upon, with specific emphasis on coastal areas, South Asia. Throughout the discourse, the chapter underscored the indispensable role of monitoring and evaluation in comprehending and alleviating these intricate gendered impacts, culminating in recommendations for fostering more resilient, inclusive, and equitable approaches to confront the challenges posed by climate change.

REFERENCES

Ahmed, K., Atiqul Haq, S. M., & Bartiaux, F. (2019). The nexus between extreme weather events, sexual violence, and early marriage: A study of vulnerable populations in Bangladesh. *Population and Environment, 40*(3), 303–324. doi:10.1007/s11111-019-0312-3

Alston, M., Whittenbury, K., et al. (2014). Are climate challenges reinforcing child and forced marriage and dowry as adaptation strategies in the context of Bangladesh? *Women's Studies International Forum, 47*(Part A), 137-144. doi:10.1016/j.wsif.2014.08.001

American Psychiatric Association. (n.d.). *Climate Change and Mental Health.* APA https://www.psychiatry.org/patients-families/climate-change-and-mental-health-connections/affects-on-mental-health

American Public Health Association. (n.d). *LGBTQ Health Caucus.* APHA. https://www.apha.org/apha-communities/caucuses/lgbtq-health-caucus)

Anastario, M., Shehab, N., & Lawry, L. (2009). Increased gender-based violence among women internally displaced in Mississippi 2 years post-Hurricane Katrina. *Disaster Medicine and Public Health Preparedness, 3*(1), 18–26. doi:10.1097/DMP.0b013e3181979c32 PMID:19293740

Atmos. (2023, March 27). *Pakistan's climate crisis of Child marriage.* Atmos. https://atmos.earth/pakistan-floods-child-marriage

Baker, R. E. (2020). Climate change drives increase in modeled HIV prevalence. *Climatic Change, 163*(1), 237–252. doi:10.1007/s10584-020-02753-y

Barik, S. (2022, November 25). Millets in the mainstream: How Odisha's Kutia Kondh tribe rediscovered a palate for the "poor man's food." *The Hindu.* https://www.thehindu.com/news/national/other-states/a-tribes-efforts-in-odisha-to-bring-millets-back-on-their-platters/article66178864.ece

BBC News. (2022, August 28). *Pakistan floods: Appeals for aid as 119 more die in a day. BBC News.* BBC News. https://www.bbc.com/news/world-asia-62704004.amp

Behal, A. (2021, January 11). How climate change is affecting the LGBTQIA+ community. DownToEarth. Retrieved May 25, 2023, from https://www.downtoearth.org.in/blog/environment/how-climate-change-is-affecting-the-lgbtqia-community-74988

Behere, P. B., & Behere, A. P. (2008). Farmers' suicide in Vidarbha region of Maharashtra state: A myth or reality? *Indian Journal of Psychiatry, 50*(2), 124–127. doi:10.4103/0019-5545.42401 PMID:19742218

Blanchard, K. (2021, September 16). *LGBTQIA+ people and disasters: A need for policy guidance.* UNDRR Prevention Web. https://www.preventionweb.net/blog/lgbtqia-people-and-disasters-need-policy-guidance

Bryan, E., Theis, S., Choufani, J., De Pinto, A., Meinzen-Dick, R. S., & Ringler, C. (2017). *Conceptual framework: Gender, climate change, and nutrition integration initiative*. (GCAN Policy Note 7). Washington, D.C.: International Food Policy Research Institute (IFPRI). http://ebrary.ifpri.org/cdm/singleitem/collection/p15738coll2/id/132300

Bukhari, S. I. A., & Rizvi, S. S. (2015). Impact of Floods on Women: With Special Reference to Flooding Experience of 2010 Flood in Pakistan. *Journal of Geography & Natural Disasters*, *5*(2).

Burke, M., Gong, E., & Jones, K. (2015). Income shocks and HIV in Africa. *Economic Journal (London)*, *125*(585), 1157–1189. doi:10.1111/ecoj.12149

CARE. (2015). *Rapid Gender Analysis Cyclone Pam Vanuatu*. CARE. https://care.ca/wp-content/uploads/2018/12/RGA-Cyclone-Pam-Vanuatu-7-April-2015.pdf

CDC. (2006). Heat-related deaths: United States, 1999–2003. *Morbidity and Mortality Weekly Report*, *55*, 769–798. PMID:16874294

Chersich, M. F., Pham, M. D., Areal, A., Haghighi, M. M., Manyuchi, A., Swift, C. P., Wernecke, B., Robinson, M., Hetem, R., Boeckmann, M., & Hajat, S. (2020). Associations between high temperatures in pregnancy and risk of preterm birth, low birth weight, and stillbirths: Systematic Review and meta-analysis. *BMJ (Clinical Research Ed.)*, *m3811*, m3811. doi:10.1136/bmj.m3811 PMID:33148618

Climate Action Network South Asia (CANSA). (2022, December). *Gender And Child Inclusion In Policies On Climate Change In South Asia: A study from Nepal, Sri Lanka and Bangladesh*. CANSA. https://cansouthasia.net/gender-and-child-inclusion-in-policies-on-climate-change-in-south-asia/

Climate Change is Not Gender Neutral [Blog]. (2021, June). IUCN. https://www.iucn.org/news/gender/202106/climate-change-not-gender-neutral

D'Ooge, C. (n.d.). *Chapter II. Queer Katrina: Gender and sexual orientation matters in the aftermath of the disaster*. Newcomb Institute. https://issuu.com/newcombcollegeinstitute/docs/nccrowreport08/s/11775539

Davison, C. (2022, June 1). Indian heatwaves threaten maternal health. *Nature and Environment India*. [https://www.dw.com/en/climate-crisis-heat-baby-prenatal-maternal-health-mortality/a-61952364]

Easterling, D. R., Evans, J. L., Groisman, P. Y., Karl, T. R., Kunkel, K. E., & Ambenje, P. (2000). Observed variability and trends in extreme climate events: A brief review. *Bulletin of the American Meteorological Society*, *81*(3), 417–425. doi:10.1175/1520-0477(2000)081<0417:OVATIE>2.3.CO;2

Gardner, J. (2021). Integrating Gender. In *Climate Change And Disaster-Related Statistics In Asia & The Pacific*. UN Women.

Goldberg, M. H., Gustafson, A., Ballew, M. T., Rosenthal, S. A., & Leiserowitz, A. (2019). A Social Identity Approach to Engaging Christians in the Issue of Climate Change. *Science Communication*, *41*(4), 442–463. doi:10.1177/1075547019860847

Gurung, J. D. (n.d.). *Gender and Desertification: Expanding roles for women to restore dryland areas*. IFAD; Rome. https://www.ifad.org/documents/38714170/39148759/Gender+and+desertification+-+Expanding+roles+for+women+to+restore+drylands.pdf/76b3905e-f1ed-428b-8673-822fb923dcd8

Hindustan Times. (2018, August 18). Pregnant woman airlifted by navy from flooded Kerala village, mother and baby fine. *Hindustan Times*. https://www.hindustantimes.com/kerala/pregnant-woman-airlifted-by-navy-from-flooded-kerala-village-mother-and-baby-fine/story-S4FRiwEmfkivJFrGfUNuXP.html

Human Rights Watch. (2015). *Marry Before Your House is Swept Away - Child Marriage in Bangladesh*. HRW.

India Today. (2023, June 25). Delhi woman dies of electrocution after coming in contact with electric pole during rain floods. *India Today*. https://www.indiatoday.in/cities/delhi/story/delhi-woman-dies-electrocution-electric-pole-rain-floods-2397666-2023-06-25

India.com. (2023, July 8). Delhi Rains: Woman Crushed to Death Under Collapsed Ceiling in Karol Bagh. *India.com*. https://www.india.com/news/delhi/delhi-rains-woman-crushed-to-death-under-collapsed-ceiling-in-karol-bagh-6156125/

Ipas study. (2022, September 23). *New research is in: Climate change impacts women's sexual and Reproductive Health*. IPAS. [https://www.ipas.org/our-work/climate-justice/climate-change-impacts-womens-sexual-and-reproductive-health/]

Jalais, A. (2019). Do human-wildlife 'interactions' affect mental health in the Sundarbans. In A. Ghosh (Ed.), *The Tides of Life: Surviving Between the Margins*.

Kankaria, A. (2014). Indoor Air Pollution in India: Implications on Health and its Control. *Indian J Community Med. Oct-Dec; 39*(4): 203–207.

Karl, T. R., Melillo, J. M., & Peterson, T. C. (Eds.). (2009). *Global Climate Change Impacts in the United States*. Cambridge University Press.

Klinenberg, E. (2002). *Heat wave: A social autopsy of disaster in Chicago*. University of Chicago Press. doi:10.7208/chicago/9780226026718.001.0001

Kochar, A., Nagabhushana, C., Sarkar, R., Shah, R., & Singh, G. (2022). Financial access and women's role in household decisions: Empirical evidence from India's National Rural Livelihoods project. *Journal of Development Economics*, *155*, 102821. doi:10.1016/j.jdeveco.2022.102821 PMID:35241869

Kolves, K., Kolves, K. E., & de Leo, D. (2013). Natural disasters and suicidal behaviours: A systematic literature review. *Journal of Affective Disorders*, *146*(1), 1–14. doi:10.1016/j.jad.2012.07.037 PMID:22917940

Kovats, S., & Allen, M. (2008). Flooding, windstorms and climate change. In *Department of Health and Health Protection Agency of the UK. Health effects of climate change in the UK 2008: An update of the Department of Health report 2001–2002*. Department of Health and Health Protection Agency of the UK., http://www.bvsde.paho.org/bvsacd/cd68/UK2008/cap2.pdf

Le Masson, V., Benoudji, C., Sotelo Reyes, S., & Bernard, G. (2019). How violence against women and girls undermines resilience to climate risks in Chad. *Disasters*, *43*(S3), S245–S257. doi:10.1111/disa.12343 PMID:30945771

Legacy. (n.d.). *Climate Change and Male Fertility*. Give Legacy. https://www.givelegacy.com/resources/climate-change-and-male-fertility/

Lieber, M., Chin-Hong, P., Whittle, H. J., Hogg, R., & Weiser, S. D. (2021). The synergistic relationship between climate change and the HIV/AIDS epidemic: A conceptual framework. *AIDS and Behavior, 25*(7), 2266–2277. doi:10.1007/s10461-020-03155-y PMID:33452659

Lindsey, R., & Dahlman, L. (2023, January 18). Climate Change: Global Temperature. *Climate.gov*. https://www.climate.gov/news-features/understanding-climate/climate-change-global-temperature#:~:text=Earth's%20temperature%20has%20risen%20by,0.18%C2%B0%20C)%20per%20decade

Logie, C. H., Toccalino, D., Reed, A. C., Malama, K., Newman, P. A., Weiser, S., Harris, O., Berry, I., & Adedimeji, A. (2021). Exploring linkages between climate change and sexual health: A scoping review protocol. *BMJ Open, 11*(1), e054720. doi:10.1136/bmjopen-2021-054720 PMID:34663670

Low, A. J., Frederix, K., McCracken, S., Manyau, S., Gummerson, E., Radin, E., Davia, S., Longwe, H., Ahmed, N., Parekh, B., Findley, S., & Schwitters, A. (2019). Association between severe drought and HIV prevention and care behaviors in Lesotho: A population-based survey 2016-2017. *PLoS Medicine, 16*(1), e1002727. doi:10.1371/journal.pmed.1002727 PMID:30640916

Malala Fund. (2022, March 29). *Malala Fund publishes report on climate change and girls' education*. Malala Fund. https://malala.org/newsroom/malala-fund-publishes-report-on-climate-change-and-girls-education#:~:text=Malala%20Fund%E2%80%99s%20new%20report%2C%20A%20Tipping%20Point%20for%20Climate%20Action%20and%20Girls%E2%80%99%20Education%2C%20reveals%20that%20climate%20change%20will%20prevent%20an%20estimated%20four%20million%20girls%20in%20low%2D%20and%20lower%2Dmiddle%20income%20countries%20from%20completing%20their%20education

Maviskalyan, A., & Tarverdi, Y. (2019). Gender and climate change: Do female parliamentarians make a difference? *European Journal of Political Economy, 56*, 151–164. doi:10.1016/j.ejpoleco.2018.08.001

McLed, C., Barr, H., & Rall, K. (2019). Does climate change increase the risk of child marriage: A look at what we know and what we don't with lessons from Bangladesh and Mozambique. *Columbia Journal of Gender and Law, 38*(1), 96–146.

Ministy of Women and Child Development, Government of India. (2020). Malnutrition among Women. Ministry of Women and Child Development.

Most polluting industries in 2023 revealed. (n.d.). Heatable. https://heatable.co.uk/boiler-advice/most-polluting-industries

Msuya, F. E., & Hurtado, A. Q. (2017). The role of women in seaweed aquaculture in the Western Indian Ocean and South-East Asia. *European Journal of Phycology, 52*(4), 482–494. doi:10.1080/09670262.2017.1357084

Nagaraj, K. (2008). *Farmers' suicides in India: Magnitudes, trends and spatial patterns*. Madras Institute of Development Studies. https://www.macroscan.org/anl/mar08/pdf/Farmers_Suicides.pdf

Nagel, J., & Lies, T. S. (2022). Re-gendering Climate Change: Men and Masculinity in Climate Research, Policy, and Practice. *Frontiers in Climate, 4*, 856869. doi:10.3389/fclim.2022.856869

Neumayer, E., & Plümper, T. (2007). The gendered nature of natural disasters: The impact of catastrophic events on the gender gap in life expectancy, 1981–2002. *Annals of the Association of American Geographers*, *97*(3), 551–566. doi:10.1111/j.1467-8306.2007.00563.x

Nicholls, N., Butler, C. D., & Hanigan, I. (2006). Inter-annual rainfall variations and suicide in New South Wales, Australia, 1964–2001. *International Journal of Biometeorology*, *50*(3), 139–143. doi:10.1007/s00484-005-0002-y PMID:16235091

Nieuwkoop, M. V., Van de Velde, P., Huyer, S., & Freeman, K. K. (2022, March 8). Gender-smart agriculture: The only way forward for women and climate. *World Bank Blogs*. https://blogs.worldbank.org/climatechange/gender-smart-agriculture-only-way-forward-women-and-climate

Nikarthil, D. S., & Kejriwal, S. (2021). *India's Healthcare Systems Persistently Exclude LGBTQ+ People. This Needs To Change.* The Wire.

Nolen, S., & Galdieri, D. (2022, August 16). The forgotten virus: Zika families and researchers struggle for support. *The New York Times*. https://www.nytimes.com/2022/08/16/health/zika-children-research.html

OECD. (n.d.). *Environmental Outlook to 2050: The Consequences of Inaction - Key Facts and Figures.* OECD. https://www.oecd.org/env/indicators-modelling-outlooks/oecdenvironmentaloutlookto2050theconsequencesofinaction-keyfactsandfigures.htm

Orderud, H., Harkonen, J., Harsaker, C. T., & Bogren, M. (2022). Floods and maternal healthcare utilization in Bangladesh. *Population and Environment*, *44*(3-4), 193–225. doi:10.1007/s11111-022-00410-3

Oudraat, C. J., & Brown, M. E. (2022, January 25). *Gender, Climate Change, and Security: Making the Connections.* Wilson Center. Wilson Center. https://www.wilsoncenter.org/article/gender-climate-change-and-security-making-connections

Prakash, A., McGlade, K., Roxy, M. K., Roy, J., Some, S., & Rao, N. (2022). Climate Adaptation Interventions in Coastal Areas: A Rapid Review of Social and Gender Dimensions. *Frontiers in Climate*, *4*, 785212. doi:10.3389/fclim.2022.785212

Race to Zero. (n.d.). *Global Climate action*. Race to Zero. https://racetozero.unfccc.int/system/race-to-zero/

Robinson, J., & Yeh, E. (2011). Transactional sex as a response to risk in Western Kenya. *American Economic Journal. Applied Economics*, *3*(1), 35–64. doi:10.1257/app.3.1.35

Roy, S., Tandukar, S., & Bhattarai, U. (2022). Gender, Climate Change Adaptation, and Cultural Sustainability: Insights From Bangladesh. *Frontiers in Climate*, *4*, 2022. doi:10.3389/fclim.2022.841488

Saifi, S., Mogul, R., Coren, A., Sidhu, S., Iqbal, J., & Mandhro, S. (2022, September 26). *First came the floods. Now, Pakistan's children face a new disaster.* CNN. https://edition.cnn.com/2022/09/25/asia/pakistan-floods-children-water-borne-disease-intl-hnk-dst/index.html

Santos, E. (2022, June 28). *[Opinion] Why climate justice is queer justice*. Rappler. https://www.rappler.com/voices/ispeak/opinion-why-climate-justice-is-queer-justice/

Selby, D. (2015). Climate Change: Reorienting the Development Agenda. In G. McCann & S. McCloskey (Eds.), From the Local to the Global (3rd edition): Key Issues in Development Studies (pp. 113–131). London: Pluto Press. doi:10.2307/j.ctt183p5fn.11

Semenza, J. C. (1996). Deaths in the Chicago heat wave. *The New England Journal of Medicine*, *335*(24), 1848–1849. doi:10.1056/NEJM199612123352418 PMID:8649494

Shetty, D. (2021). *Why air pollution conversations leave out women's health.* Devex. https://www.devex.com/news/why-air-pollution-conversations-leave-out-women-s-health-102011

Singh, N. S., Aryasinghe, S., Smith, J., Khosla, R., Say, L., & Blanchet, K. (2018). A long way to go: A systematic review to assess the utilization of sexual and reproductive health services during humanitarian crises. *BMJ Global Health*, *3*(2), e000682. doi:10.1136/bmjgh-2017-000682 PMID:29736272

Tabary, Z. (2018, March 8). *Women, LGBT people 'invisible' victims of disasters – experts.* Thomson Reuters Foundation. https://news.trust.org/item/20180308114259-otaio/

Thailand U. N. D. P. (2023). Policy Brief: Gender Responsive And Socially Inclusive Public Climate Finance. UN. https://www.undp.org/thailand/publications/policy-brief-gender-responsive-and-socially-inclusive-public-climate-finance

The Daily Star. (2008, April 9). Staff Correspondent. Women more vulnerable to violence during floods. *The Daily Star.* https://www.thedailystar.net/news-detail-31477

The Guardian. (2023, April 10). Sea-level rise poses existential threat to Miami and New Orleans, studies find. *The Guardian.* https://www.theguardian.com/environment/2023/apr/10/sea-level-rise-climate-crisis-miami-new-orleans

The Indian Express. (2023, July 14). Woman missing, 28 villages inundated after cloudburst in Uttarakhand's Pithoragarh. *The Indian Express.* https://indianexpress.com/article/india/woman-missing-28-villages-inundated-cloudburst-uttarakhands-pithoragarh-8142486/

Thurston, A. M., Stockl, H., & Ranganathan, M. (2021). Natural hazards, disasters and violence against women and girls: A global mixed-methods systematic review. *BMJ Global Health*, *6*(4), e004377. doi:10.1136/bmjgh-2020-004377 PMID:33958379

UNDP. (2023, February 28). *What does gender equality have to do with climate change?* [Blog post]. UN. https://climatepromise.undp.org/news-and-stories/what-does-gender-equality-have-do-climate-change#:~:text=Giving%20women%20increased%20access%20to,benefit%20the%20needs%20of%20women

UNFPA. (2015, December 3). *Essential health needs of women often neglected in assistance after natural disasters, conflicts* [Press release]. UN. https://www.unfpa.org/press/essential-health-needs-women-often-neglected-assistance-after-natural-disasters-conflicts

UNFPA East and Southern Africa Regional Office. (2021). *The impact of the climate crisis on sexual and reproductive health and rights (SRHR).* UN. https://esaro.unfpa.org/sites/default/files/pub-pdf/advocacy_-_impact_of_climate_crisis_on_srhr.pdf

United Nations climate change. (n.d.). *Gender & Climate Change: an important connection*. UN. https://unfccc.int/gender

United Nations Environment Programme, UN Women, UNDP, & UNDPPA/PBSO. (2020). *Gender, Climate and Security: Sustaining inclusive peace on the frontiers of climate change*. UN.

Van Dijk, K. (2012). *Empowerment of Pastoralist Women*. PENHA narrative report. https://oxfamilibrary.openrepository.com/bitstream/handle/10546/222733/cs-empowerment-pastoralist-women-sudan-penha-2012-en.pdf;jsessionid=A74E516013AED6518687A3C1C29F8874?sequence=1

Verma, R., Nellemann, C., & Hislop, L. (2011). *Women at the Frontline of Climate Change: Gender Risks and Hopes*.

Voeten, H. A. C. M., Vissers, D. C. J., Gregson, S., Zaba, B., White, R. G., de Vlas, S. J., & Habbema, J. D. F. (2010). Strong association between in-migration and HIV prevalence in urban sub-Saharan Africa. *Sexually Transmitted Diseases*, *37*(4), 240–243. doi:10.1097/OLQ.0b013e3181c3f2d0 PMID:19959971

Watts, N., Amann, M., Arnell, N., Ayeb-Karlsson, S., Belesova, K., Boykoff, M., Byass, P., Cai, W., Campbell-Lendrum, D., Capstick, S., Chambers, J., Dalin, C., Daly, M., Dasandi, N., Davies, M., Drummond, P., Dubrow, R., Ebi, K. L., Eckelman, M., & Montgomery, H. (2019). The 2019 report of the Lancet Countdown on health and climate change: Ensuring that the health of a child born today is not defined by a changing climate. *Lancet*, *394*(10211), 1836–1878. doi:10.1016/S0140-6736(19)32596-6 PMID:31733928

Whitley, C. T., & Bowers, M. M. (2023). Queering Climate Change: Exploring the Influence of LGBTQ+ Identity on Climate Change Belief and Risk Perceptions. *Sociological Inquiry*, *93*(3), 1–27. doi:10.1111/soin.12534

Whitman, S., Good, G., Donoghue, E. R., Benbow, N., Shou, W., & Mou, S. (1997). Mortality in Chicago attributed to the July 1995 heat wave. *American Journal of Public Health*, *87*(9), 1515–1518. doi:10.2105/AJPH.87.9.1515 PMID:9314806

WHO/UNICEF Joint Monitoring Programme (JMP). (2023). *Progress on household drinking water, sanitation and hygiene 2000-2022: Special focus on gender* WHO. [https://cdn.who.int/media/docs/default-source/wash-documents/jmp-2023_layout_v3launch_5july_lowreswhowebsite.pdf?sfvrsn=c52136f5_3&download=true]

Wisner, B. (2004). *At risk: Natural hazards, people's vulnerability and disasters* (2nd ed.). Routledge.

Women U. N. (2022a, November 18). *Stepping up for women and girls during the flood crisis in Pakistan*. UN Women. https://www.unwomen.org/en/news-stories/feature-story/2022/11/stepping-up-for-women-and-girls-during-the-flood-crisis-in-pakistan

Women U. N. (2022b). Rapid Gender Analysis of Flood Situation in North and North-Eastern Bangladesh: Gender in Humanitarian Action Working Group Bangladesh. *Asia Pacific*. https://asiapacific.unwomen.org/sites/default/files/2022-07/bd-Rapid-gender-analysis-north-northeaster-flood-2022.pdf

Women U. N. (n.d.). Cl*imate Change and Women in Agriculture. EmPower project*. UN. https://www.uncclearn.org/wp-content/uploads/library/02-issue-brief_climate-change-and-women-in-agriculture.pdf

Womenwatch, U. N. (n.d.). *Women, gender equality and climate change: Fact Sheet*. UN. https://www.un.org/womenwatch/feature/climate_change/

World Health Organization. (2011). *Gender and health*. World Health Organization. https://www.who.int/health-topics/gender#tab=tab_1.

World Health Organization. (2014). *Gender, climate change and health*. WHO. https://apps.who.int/iris/bitstream/handle/10665/144781/9789241508186_eng.pdf

World Health Organization. (2023, May 24). *Climate Change*. WHO. https://www.who.int/health-topics/climate-change#tab=tab_1

World Health Organization. (n.d.a). *Improving LGBTIQ+ health and well-being with consideration for SOGIESC*. WHO. https://www.who.int/activities/improving-lgbtiq-health-and-well--being-with-consideration-for-sogiesc

World Health Organization. (n.d.,b). *Air quality and health*. WHO. https://www.who.int/teams/environment-climate-change-and-health/air-quality-and-health/health-impacts

Yun, K., Lurie, N., & Hyde, P. S. (2010). Moving Mental Health into the Disaster-Preparedness Spotlight. *The New England Journal of Medicine*, *363*(13), 1193–1195. doi:10.1056/NEJMp1008304 PMID:20860498

Chapter 16
Sustainable and Resilient Future Hospital Models in the Light of Climate Change and Resultant Quality of Life

Kasturi Shukla
https://orcid.org/0000-0001-7006-9001
Symbiosis Institute of Health Sciences, Symbiosis International University (Deemed), Pune, India

Sayali Shilewant
https://orcid.org/0009-0008-7537-3967
National Health Mission, India

Beauty Sinha
Fitterfly Health tech Pvt. Ltd., India

ABSTRACT

Climate change is a long-term shift in temperature and weather patterns. The United Nations states that since the late 18th century, human activities - primarily the combustion of fossil fuels like coal, oil, and gas - have been the primary cause of climate change. Across the different sectors, the greenhouse effect is most prominent which forms a blanket wrapped around the Earth, trapping the sun's heat and raising temperatures. According to the UN, clearing of land and deforestation is responsible for increased carbon dioxide levels in the environment. The main sectors responsible for greenhouse gas emission are energy, transport, construction, agriculture, healthcare, and high land use. The healthcare industry is the largest source of greenhouse gasses, accounting for almost 5% of global greenhouse emissions. This chapter reviews the theories and models proposed by national and international organizations like United Nations and World Health Organization to explore and explain the impact of hospitals on climate change and the short-term and long-term impact on sustainability.

DOI: 10.4018/978-1-6684-9863-7.ch016

INTRODUCTION

Climate change refers to long-term shifts in temperatures and weather patterns. Since the late 18th century, human activities - primarily the combustion of fossil fuels like coal, oil, and gas - have been the primary cause of climate change (UN, n.d.). The greenhouse effect is most prominent which forms a blanket wrapped around the Earth, trapping the sun's heat and raising temperatures. The gasses which are linked with climate change include carbon dioxide and methane which are emitted due to gasoline combustion in vehicles or coal burning for multiple purposes. According to the UN, clearing of land and deforestation is responsible for increased carbon dioxide levels in the environment. The main sectors responsible for greenhouse gas emission are energy, transport, construction, agriculture, healthcare and high land use (UN, n.d.).

The climatic changes are very gradual but have a long lasting impact on living beings, environment and business. Many diseases are linked to climate change especially in vulnerable populations in low- and middle- income countries. The rise in temperature has a long term impact on the sustainable aspects of the environment and society. Sustainable life refers to a lifestyle that aims to reduce the negative impact on the environment while maintaining a balance with social and economic well-being (Sustainable lifestyles, 2023). A focus on sustainability refers to the conscious choices with a long-term effect on the planet and future generations.

Figure 1: Impact of climate change on health
(WHO, 2023)

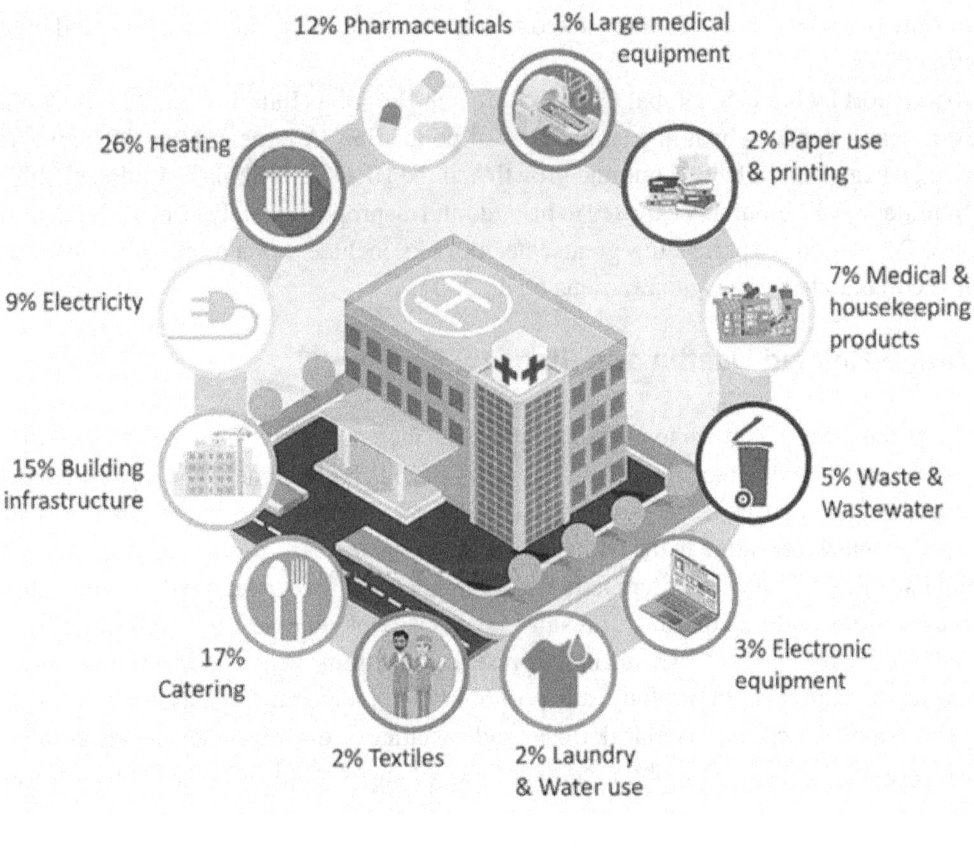

As shown in figure 1, climate change affects human health by altering exposures to heat waves, floods, droughts, and other extreme events; vector-, food- and waterborne infectious diseases; changes in the quality and safety of air, food, and water; and stresses to mental health and well-being. (WHO, 2021). Climate change also affects human health by changing the severity, duration, or frequency of health problems and by creating unprecedented or unanticipated health problems or health threats in places or populations where they have not previously occurred (Balbus et al., 2016a). There is a tremendous opportunity for economic growth by supporting sustainable businesses and necessitates a shift in mindset towards responsible consumption and responsible production. It requires individuals to rethink their choices and make decisions that align with sustainability goals.

WHO defines Quality of Life as an individual's perception of their position in life in the context of the culture and value systems in which they live and in relation to their goals, expectations, standards and concerns (WHO, 2012) and Health as a state of complete physical, mental and social well-being and not merely the absence of disease or infirmity. The concept of Health-Related Quality of Life (HRQoL) and its determinants have evolved since the 1980s to encompass those aspects of overall quality of life that can be clearly shown to affect health—either physical or mental. On the individual level, HRQOL includes physical and mental health perceptions (e.g., energy level, mood) and their correlates - including health risks and conditions, functional status, social support, and socioeconomic status (CDC, 2001). At the community level, HRQOL includes community-level resources, conditions, policies, and practices that influence a population's health perceptions and functional status (Rodríguez-Artalejo et al., 2005). Adopting a sustainable lifestyle can have a significant impact on the HRQOL (WHO, 2023). Sustainable practices involve 3Rs that includes Reduce, Reuse and Recycle like, reducing waste, conserving water and energy, and using environmentally friendly products can contribute to a healthier and more sustainable environment. This is further linked to improvement of social well-being and creating balanced communities.

The 2016 report by the U.S. Global Change Research Program (Balbus et al, 2016b) describes several types of health threats which impact the QoL of population. A large number of studies prove that climate change negatively affects economic growth and the HRQOL (Nikola & Radoica, 2022). People and communities are differentially exposed to hazards and disproportionately affected by climate-related health risks. Populations experiencing greater health risks include women, children, older adults and low-income communities (Abeygunawardena et al., 2004).

Temperature Related Deaths and Illness

Increasing concentrations of greenhouse gasses lead to an increase of both average and extreme temperatures. This is expected to lead to an increase in deaths and illness from heat and a potential decrease in deaths from cold. Days that are hotter than the average seasonal temperature in the summer or colder than the average seasonal temperature in the winter cause increased levels of illness and death by compromising the body's ability to regulate its temperature or by inducing direct or indirect health complications. Loss of internal temperature control can result in a cascade of illnesses, including heat cramps, heat exhaustion, heatstroke, and hyperthermia in the presence of extreme heat, and hypothermia and frostbite in the presence of extreme cold. Prolonged exposure to high temperatures is associated with increased hospital admissions for cardiovascular, kidney, and respiratory disorders. An increase in population tolerance to extreme heat has been observed over time. Changes in this tolerance have been associated with increased use of air conditioning, improved social responses, and/or physiological acclimatization, among other factors (Balbus et al, 2016b).

Climate change is expected to alter the geographic and seasonal distributions of existing vectors and vector-borne diseases. Vector-borne diseases are illnesses that are transmitted by *vectors*, which include mosquitoes, ticks, and fleas. These vectors can carry infective pathogens such as viruses, bacteria, and protozoa, which can be transferred from one host (carrier) to another seasonality, distribution, and prevalence of vector-borne diseases are influenced significantly by climate factors, primarily high and low temperature extremes and precipitation patterns. Climate change is likely to have both short- and long-term effects on vector-borne disease transmission and infection patterns, affecting both seasonal risk and broad geographic changes in disease occurrence over decades. A study from Nepal showed that a 1-degree Celsius increase in mean temperature increased the incidence of malaria by 25% (Balbus et al, 2016b; WHO, 2023).

Air Quality

Changes in the climate affect the air we breathe, both indoors and outdoors. The changing climate has modified weather patterns, which in turn have influenced the levels and location of outdoor air pollutants such as ground-level ozone (O3) and fine particulate matter. Increasing carbon dioxide (CO2) levels also promote the growth of plants that release airborne allergens (aeroallergens). Climate change is projected to increase the number and severity of naturally occurring wildfires emit fine particles and ozone precursors that in turn increase the risk of premature death and adverse chronic and acute cardiovascular and respiratory health outcomes. Poor air quality, whether outdoors or indoors, can negatively affect the human respiratory and cardiovascular systems. Higher pollen concentrations and longer pollen seasons can increase allergic sensitization and asthma episodes and thereby limit productivity of the humans (Balbus et al, 2016b).

Extreme Events

Climate change will increase exposure risk due to projected increases in the frequency and/or intensity of drought, wildfires, and flooding related to extreme precipitation and hurricanes. Health impacts associated with climate-related changes in exposure to extreme events include death, injury, or illness; exacerbation of underlying medical conditions; and adverse effects on mental health. Many types of extreme events related to climate change cause disruption of infrastructure, including power, water, transportation, and communication systems, that are essential to maintaining access to health care and emergency response services and safeguarding human health (Balbus et al, 2016b).

Other Morbidities

Climate change is expected to affect fresh and marine water resources in ways that will increase people's exposure to water-related contaminants that cause illness. Water-related illnesses include waterborne diseases caused by pathogens, such as bacteria, viruses, and protozoa. These are also caused by toxins produced by certain harmful algae and cyanobacteria and by chemicals introduced into the environment by human activities.

The effects of global climate change on mental health and well-being are integral parts of the overall climate-related human health impacts. Mental health consequences of climate change range from minimal stress and distress symptoms to clinical disorders, such as anxiety, depression, post-traumatic stress, and suicidality. Media and popular culture representations of climate change influence stress responses and mental health and well-being. The interactive and cumulative nature of climate change effects on health, mental health, and well-being are critical factors in understanding the overall consequences of climate change on human health (Balbus et al, 2016b).

Increases in some extreme weather events and storm surges can increase the risk that infrastructure for drinking water, wastewater, and storm water will fail due to either damage or crossing of system capacity, especially in areas with aging infrastructure. As a result, the risk of exposure to water-related pathogens, chemicals, and algal toxins will increase in recreational and fish harvesting waters and in drinking water where treatment barriers break down.

A safe and nutritious food supply is a vital component of food security. Climate change is very likely to affect global, regional, and local food security by disrupting food availability, decreasing access to food, and making utilization more difficult. There are two overarching means by which increasing carbon dioxide (CO2) and climate change alter safety, nutrition, and distribution of food. The first is associated with rising global temperatures and the subsequent changes in weather patterns and extreme climate events. The second pathway is through the direct CO2 "fertilization" effect on plant photosynthesis.

Elevated sea surface temperatures will lead to greater accumulation of mercury in seafood while increases in extreme weather events will introduce contaminants into the food chain. Rising carbon dioxide concentrations and climate change will alter incidence and distribution of pests, parasites, and microbes leading to increases in the use of pesticides and veterinary drugs. Increases in the frequency or intensity of some extreme weather events associated with climate change will increase disruptions of food distribution by damaging existing infrastructure or slowing food shipments. These impediments lead to increased risk for food damage, spoilage, or contamination, which will limit availability of and access to safe and nutritious food depending on the extent of disruption and the resilience of food distribution infrastructure (Balbus et al, 2016b).

Challenges in Designing Sustainable Hospitals

Healthcare industry is a surprisingly large source of greenhouse gasses, accounting for almost 5% of global greenhouse emissions (Smith, 2022). In the indoor environment, thermal comfort is a crucial design factor for the well-being of patients as well as for staff (Yuan et al., 2022). The American Society of Heating, Refrigerating and Air-Conditioning Engineers (ASHRAE), a hospital indoor environment can be classified as: critical care, nursing, ancillary services, administration, diagnostic and treatment, sterilizing and supply, service, etc. The requirements of the indoor environment are different for each functional zone (Yuan et al., 2022). Hospitals have a huge role to play in instilling sustainability in the ecosystem and preventing damage to the environment.

Carbon Footprint Associated With Different Categories of Hospitals

Hospitals exhibit diverse characteristics in terms of types and services, leading to significant variations in their environmental impact. The following considerations shed light on the carbon footprint associated with different categories of hospitals:

(a) *Primary Health Centers and Specialized Units:* Primary Health Centers (PHCs): These smaller healthcare facilities primarily concentrate on delivering fundamental healthcare services, preventive care, and health education within specific communities. Due to their limited scale and service scope, PHCs typically maintain a lower carbon footprint compared to specialized units.

(b) *Specialized Units (e.g., Bone Marrow Transplant Unit):* Specialized units, dealing with intricate procedures and advanced medical technologies, often incur a higher environmental impact. The energy-intensive nature of certain medical equipment and the need for specialized resources contribute significantly to this heightened footprint.

(c) *Tertiary Care Centers and PHCs:* These expansive, advanced medical facilities provide specialized and highly complex services. Tertiary care centers frequently register a heightened environmental impact owing to their size, advanced medical equipment, and the energy demands associated with maintaining sophisticated infrastructure.

(d) *Multi-Specialty Hospitals and Community Care Centers:* These large-scale medical facilities offer an extensive range of medical services and specialties under a single roof. While their overall energy consumption may be higher, the efficiency of multi-specialty hospitals can fluctuate based on factors such as building design, energy management practices, and utilization of renewable energy sources.

(e) *Community Care Centers:* These community-based facilities prioritize providing accessible healthcare services to local populations. Their environmental impact tends to be lower compared to multi-specialty hospitals, particularly when emphasizing sustainable practices and energy-efficient technologies.

Hospitals play a crucial role in the healthcare industry, but they are also significant contributors to climate change. The healthcare sector is responsible for a significant amount (about 4.5% of worldwide emissions) of greenhouse gas emissions, and hospitals are a significant contributor to this problem [Berniak-Woźny and Rataj, 2023]. One of the primary ways hospitals contribute to climate change is through energy consumption. Hospitals require a vast amount of energy to operate, including heating, cooling, and lighting systems, as well as medical equipment. The energy used in hospitals is often generated through the burning of fossil fuels, which releases carbon dioxide into the atmosphere. Figure 2 shows the linkages of environmental impact due to energy consumption in different areas of hospitals.

Therefore, one area where sustainability can be brought with minimum effort is energy conservation. Hospitals consume a vast amount of energy, and reducing energy consumption can significantly reduce the carbon footprint of the facility (Sampath et al., 2022).

Figure 3 shows that for different parameters a hospital typically consumes more energy than any other office (Srishti, 2021). One good example of compounding efficiencies is lighting in the OR (operating room). Directed lighting that provides high contrast is very crucial during a surgery and is usually provided by 2–4 overhead task lights in conjunction with headband lights worn by the surgeons. The halogen task lights are about 2–3.5 ft (0.6–1m) in diameter and hold a tungsten filament that is incandescent at a

Figure 2: Environmental key spots of hospitals
(Source: Keller et al., 2021)

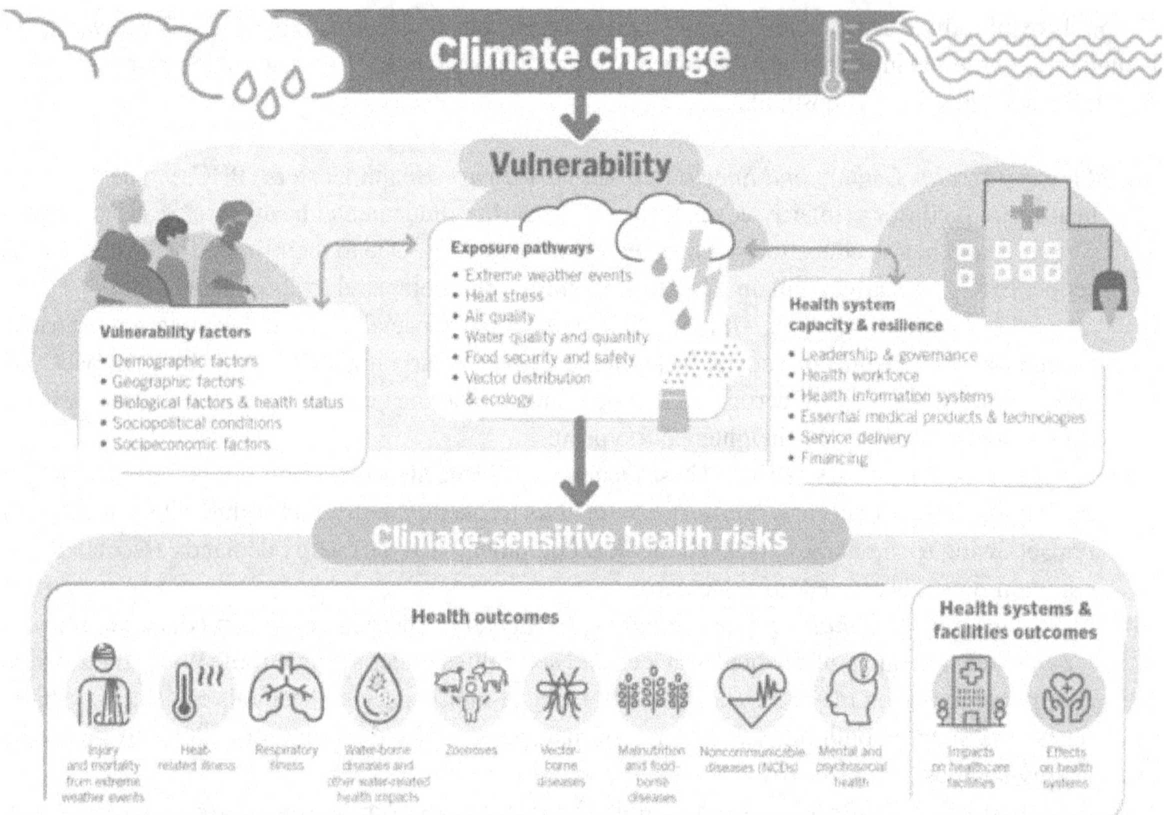

temperature over 3600° F (2000° C), radiating tremendous amounts of heat and making the doctors, who wear full personal protective equipment (masks, gowns, etc.), overheat quickly. To combat the effect, they turn up the air conditioning. By code, the temperature is not supposed to drop below 68° F, though most hospitals apply for an exception and keep their operating theaters at much cooler temperatures for surgeons' comfort. Such low ambient air temperatures would make patients undergoing surgery essentially hypothermic. This condition causes blood to drain away from the skin, which in turn makes them much more vulnerable to postoperative infection. To mitigate this risk, doctors wrap the patient in an electric blanket, which is responsible for the largest electric draw in the room.

Simple solutions such as these are the low hanging fruits for hospitals. As such acute care hospitals are particularly difficult structures to retrofit for numerous reasons, though two of the largest hurdles that must be faced are logistical and structural. First, hospitals cannot simply go offline for renovations. Almost no facilities have a temporary space to which they can transfer operations while the main building undergoes a deep retrofit. This means that an entire hospital structure cannot be shut down all at once without seriously jeopardizing the financial security of the hospital as well as the health of the people and community it serves. As such, capital improvements are often done piecemeal, with only certain divisions shutting down for a short period of time. Even these renovation projects are limited by the

Figure 3: Energy consumption of a hospital compared to any other typical office in Mumbai, India (Srishti, 2021)

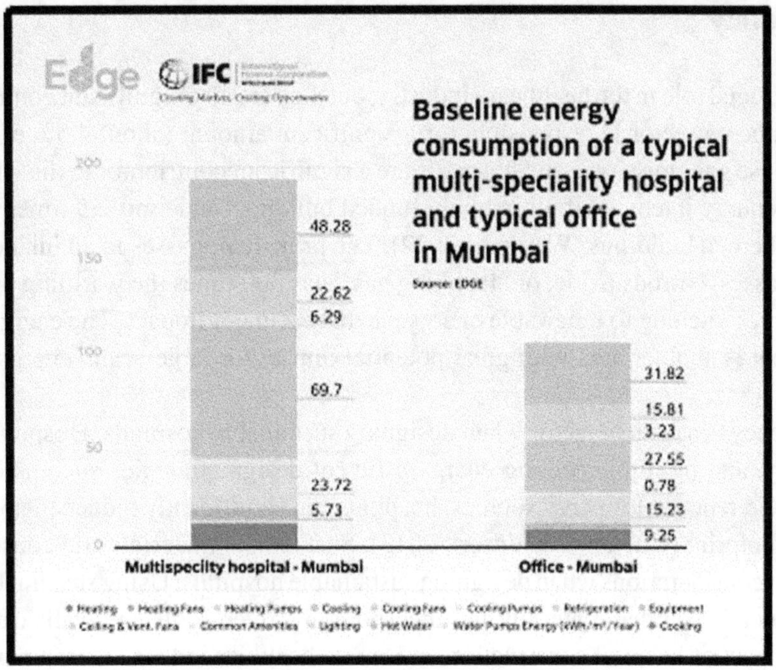

necessity of continuing critical care operations and protecting the well-being of patients and staff that remain in the building throughout the construction period.

Hospital staff and patients often travel to and from hospitals, which can result in significant emissions from vehicles. Additionally, hospitals contribute to climate change through transportation emissions. Hospitals are exploring alternative transportation methods, such as public transportation or electric vehicles, to reduce emissions. However, many hospitals are taking steps to reduce their environmental impact and contribute to the fight against climate change. Some hospitals are implementing energy-efficient systems, using renewable energy sources, and promoting sustainable waste management practices.

Waste Management and Water Conservation

Another way hospitals can better contribute to sustainability is through efficient waste management. Hospitals generate a significant amount of hazardous waste, including medical equipment, expired medications, and biological waste, which can have negative environmental impacts if not disposed of correctly [Berniak-Woźny and Małgorzata, 2023]. Hospitals generate a large amount of waste, and proper management of this waste is crucial to minimize its impact on the environment. Hospitals use a significant amount of water for cleaning, sanitation, and other purposes. By using water-efficient fixtures and water conservation strategies needs to be implemented. While there are areas where sustainability can be brought with minimum effort in various industries, some areas require significant investments to yield long-term returns. In the context of sustainability, investments in these areas are critical as they help healthcare organizations to become more sustainable, reduce their environmental impact, and improve their social responsibility.

POSSIBLE SOLUTIONS AND RECOMMENDATIONS

Energy Efficiency

Hospitals play a crucial role in the healthcare industry, but they are also significant contributors to climate change. The healthcare sector is responsible for a significant amount (about 4.5% of worldwide emissions) of greenhouse gas emissions, and hospitals are a significant contributor to this problem. Hospitals have the highest energy intensity of all publicly funded buildings and emit 2.5 times more greenhouse gasses than commercial buildings (Wesdrop, 2022). The propellant gasses in an inhaler or in anesthetic are greenhouse gasses. Nitrous oxide, or "laughing gas," has 300 times the warming potential of carbon dioxide. Therefore, switching to renewable energy can have a major impact. The energy consumption of medical equipment is another area with great potential impact for large-scale, energy-intensive operations like hospitals.

Energy efficiency is a critical factor when designing sustainable hospitals. Hospitals consume a vast amount of energy, and by implementing energy-efficient design strategies such as efficient lighting, HVAC systems, and renewable energy sources, hospitals can significantly reduce their energy consumption and carbon footprint (Guenther & Vittori, 2013). Sustainable materials and construction practices are also important considerations when designing sustainable hospitals. Using sustainable materials such as recycled and locally sourced materials and incorporating green roofs and walls can help reduce the environmental impact of hospital construction and create healthier indoor environments.

Certain hospitals have successfully implemented energy-saving policies to curtail their carbon footprint. Examples include the adoption of energy-efficient lighting, HVAC systems, medical equipment, and the incorporation of renewable energy sources, all contributing to enhanced sustainability. For instance, the Cleveland Clinic in the United States has implemented robust energy-efficient measures. This includes the deployment of combined heat and power systems and energy-efficient lighting, resulting in substantial energy savings. Further, energy star-rated solar water heaters are energy-efficient alternatives for water heating in hospitals. Hospitals, especially public hospitals, have a large potential for adoption of solar thermal or electric hybrid water heating systems (NCDC, 2023).

Hospitals are fitted with computers, machinery and lighting that run 24/7, and are more than twice as energy intensive as other commercial buildings (Choi-Schagrin, 2021). Simple measures such as installing energy-efficient light bulbs and equipment, utilizing natural light sources, and turning off equipment when not in use can lead to significant energy savings. As per a project case study by Cleaveland Clinic published by Rocky Mountain Institute (RMI), using the measures of energy efficient lighting such as replacing the thousands of incandescent bulbs with LEDs, installation of variable-frequency drives (VFDs), UV lights on cooling coils and locked thermostats, has multiple benefits and some measures even created compounding efficiencies that led to remarkable results (RMI, 2012). By replacing traditional halogen lights with LED surgical lamps the hospital reaps multiple benefits. First and foremost, the LEDs provide superior performance: their color rendering and temperature (light quality) and lumen output (light quantity) can be finely tuned to meet the needs of the situation, delivering excellent visual acuity for less energy. They also generate only a small amount of heat that dissipates upward and doesn't overheat the doctors who are performing surgeries. As a result, surgeons don't need to cool the operating room and patients don't require heating blankets. The simple change to LEDs saves 60% of the energy used for lighting while significantly reducing the energy demands of cooling the surgeons and warming their patients.

Renewable energy is one of the areas where organizations can invest to become more sustainable as it gives returns in the long term. By transitioning to renewable energy sources such as solar, wind, and hydropower, organizations can significantly reduce their carbon footprint and contribute to the fight against climate change. Although the initial investment may be high, the returns can be significant in the long term in the form of energy cost savings and environmental benefits.

National Centre for Disease Control (NCDC, 2023) has recently published relevant recommendation for more energy efficient HVAC systems, like, hospitals should promote the adoption of high-efficiency BEE star-labelled Room Air Conditioners (RACs). The lower-rated RACs should be upgraded to 3-5 star labelled as early as possible. Among the administrators and operators, the awareness about BEE star ratings can be helpful for better decision-making. The variable frequency drives (VFD) for chillers enable better capacity control and energy savings. Hospitals should aim to transition from High-Global Warming Potential (GWP) refrigerants to Low-GWP alternatives to reduce greenhouse gas emissions. State and national level health departments should promote the use of solar refrigerators in sub-centers and primary health centres in areas where the grid connectivity is unreliable (NCDC, 2023).

Green building practices are critical for long term sustainability. By designing and constructing buildings that use energy-efficient materials and technologies, have efficient heating and cooling systems, and incorporate green roofs and walls, organizations can reduce their environmental impact and create healthier and more comfortable spaces for their occupants. Green Roofs will not only provide a natural habitat for birds and insects but also regulate temperature, reduce noise pollution and help reduce storm water runoff (O'Hara et al, 2022). These practices may require higher investments upfront, but they can result in significant cost savings over time and improve the overall value of the building.

Waste Management and Water Conservation

Waste management is a critical factor to consider when designing sustainable hospitals. Hospitals generate a significant amount of waste, and proper management of this waste is crucial to minimize its impact on the environment. To promote hygiene, hospitals rely heavily on single-use items. Hospitals can produce up to 43 pounds of waste per patient per day. Some items can only be used once, but the F.D.A. does allow certain devices to be reused with proper cleaning (Choi-Schagrin, 2021). By implementing recycling programs and reducing the use of disposable products such as plastic bags and paper cups, hospitals can significantly reduce the amount of waste generated. Additionally, hospitals can promote sustainable food practices by serving locally sourced and organic food in their cafeterias, reducing food waste by composting, and promoting plant-based options.

Water conservation is also an important factor to consider when designing sustainable hospitals. Hospitals use a significant amount of water for cleaning, sanitation, and other purposes. By installing water-efficient fixtures such as low-flow toilets and faucets hospitals can significantly reduce their water consumption and conserve this precious resource (Guenther & Vittori, 2013). Implementing water-saving measures such as reusing water for irrigation and other non-potable uses, hospitals can reduce their water consumption and conserve water. By implementing water-saving technologies such as rainwater harvesting, organizations can reduce their water consumption. These investments may require significant capital upfront, but they can yield significant returns in the form of reduced water bills and improved water security (O'Hara et al, 2022).

Technology and Sustainability

Technology has played a crucial role in bringing environment sustainability in hospitals. With the growing awareness of the negative impacts of healthcare on the environment, hospitals are increasingly adopting technology to reduce their environmental impact, improve their energy efficiency, and promote sustainable practices. There are several ways in which technology is playing a role in bringing environment sustainability via hospitals (Cripps & Scarbrough, 2022). Smart building automation systems are one of the technologies that hospitals are using to reduce their energy consumption and promote sustainable practices. These systems use sensors and algorithms to optimize the heating, ventilation, and air conditioning (HVAC) systems in hospitals, reducing energy usage and improving energy efficiency.

Telemedicine and Telehealth are essential modalities to create sustainable models. Telehealth services support in reducing carbon footprints by reducing the need for travel and transportation to appointments. This can also improve access to healthcare for those who are unable to travel [Thomas & Suresh, 2022]. Telemedicine has played a big role in bringing environment sustainability in hospitals. With telemedicine, patients can receive medical consultations and treatment remotely, reducing the need for travel to hospitals and medical facilities, and thereby reducing carbon emissions from transportation.

Electronic Health Records (EHRs) are also playing a significant role in bringing environment sustainability via hospitals. EHRs allow healthcare providers to store, share, and access patient medical records electronically, reducing the need for paper-based records, and thereby reducing the environmental impact of healthcare (Cripps & Scarbrough, 2022).

Sustainable medical equipment is another significant area where technology is playing a role in bringing environment sustainability via hospitals. Manufacturers are developing medical equipment that uses less energy, produces less waste, and is more durable, thereby reducing the environmental impact of healthcare equipment. During business hours, medical imaging equipment should be kept on energy-saving "low power" mode when not in use (i.e., not scanning) which results in tremendous power saving (NCDC, 2023).

Recent studies have recommended that national and sub-national governments should establish action plans to decarbonize their health systems, build resilience, and improve health outcomes. By reducing the carbon footprint, preserving natural resources, and minimizing waste, we can create a more sustainable future for ourselves and future generations (Tennison et al., 2021). Further research is necessary to better understand trends in the interplay of health care and climate change, so as to facilitate the transition of the health sector to a climate-smart future (Karliner et al, 2019).

CONCLUSION

Designing sustainable hospitals is essential for reducing the environmental impact of healthcare facilities and promoting sustainable practices. Sustainable hospitals not only help reduce the carbon footprint of healthcare but also create healthier environments for patients, staff, and visitors. Decisions regarding the construction and operation of healthcare facilities should carefully consider factors such as population healthcare needs, geographic distribution, and environmental sustainability goals. While larger hospitals and specialized units may exhibit higher energy demands, the implementation of energy-saving policies can effectively mitigate their environmental impact. Local conditions, healthcare priorities, and sustainable practices should serve as guiding principles in these decisions.

REFERENCES

Abeygunawardena, P., Vyas, Y., Knill, P., Foy, T., Harrold, M., Steele, P., Tanner, T., Hirsch, D., Oosterman, M., & Rooimans, J. (2004). *Poverty and Climate Change: Reducing the Vulnerability of the Poor through Adaptation*. World Bank. https://documents1.worldbank.org/curated/en/534871468155709473/pdf/521760WP0pover1e0Box35554B01PUBLIC1.pdf. Retrieved: 20 May 2023.

Aydın, D., Yaldız, E., & Buyuksahin, S. (2017, December). *Sustainable hospital design for sustainable development*. In *Proceedings of the 8th International Conference on Urban Planning, Architecture, Civil and Environment Engineering*, Dubai, UAE (pp. 21-22).

Balbus, J., Beard, C. B., Bell, J. E., Crimmins, A., Dodgen, D., Easterling, D. R., Eisen, R. J., Fann, N., Gamble, J. L., Hawkins, M. D., Herring, S. C., Jantarasami, L., Kunkel, K. E., Mills, D. M., Saha, S., Sarofim, M. C., Trtanj, J., & Ziska, L. (2016b). *The Impacts of Climate Change on Human Health in the United States: A Scientific Assessment*. U.S. Global Change Research Program. doi:10.7930/J0R49N

Balbus, J., Crimmins, A., Easterling, D. R., Gamble, J. L., Kunkel, K. E., Saha, S., & Sarofim, M. C. (2016a). Chapter 1: Introduction: Climate Change and Human Health. The Impacts of Climate Change on Human Health in the United States: A Scientific Assessment U.S. Global Change Research Program, 25-42.

Berniak-Woźny, J., & Rataj, M. (2023). Towards Green and Sustainable Healthcare: A Literature Review and Research Agenda for Green Leadership in the Healthcare Sector. *International Journal of Environmental Research and Public Health*, *20*(2), 908. doi:10.3390/ijerph20020908 PMID:36673663

CDC. (2001). *Measuring Healthy Days: Population Assessment of Health-related Quality of Life*. CDC. https://www.cdc.gov/hrqol/pdfs/mhd.pdf.

Choi-Schagrin, W. (Nov. 6, 2021). How Hospitals Fuel Climate Change. *The New York Times*. https://www.nytimes.com/interactive/2021/11/05/climate/climate-change-health.html

Cripps, M., & Scarbrough, H. (2022). Making Digital Health "Solutions" Sustainable in Healthcare Systems: A Practitioner Perspective. *Frontiers in Digital Health*, *4*, 727421. doi:10.3389/fdgth.2022.727421 PMID:35434699

Fathy, R., Nelson, C. A., & Barbieri, J. S. (2021). Combating climate change in the clinic: Cost-effective strategies to decrease the carbon footprint of outpatient dermatologic practice. *International Journal of Women's Dermatology*, *7*(1), 107–111. doi:10.1016/j.ijwd.2020.05.015 PMID:33537400

Guenther, R., & Vittori, G. (2013). *Sustainable Healthcare Architecture*. John Wiley & Sons.

Karliner, J., Slotterback, S., Boyd, R., Ashby, B., & Steele, K. (2019). Health care's climate footprint. Health Care Without Harm Climate-smart health care series. *Green Paper Number One*.

Keller, R. L., Muir, K., Roth, F., Jattke, M., & Stucki, M. (2021). From bandages to buildings: Identifying the environmental hotspots of hospitals. *Journal of Cleaner Production*, *319*, 128479. doi:10.1016/j.jclepro.2021.128479

Larriva, M. T. B., Mendes, A. S., & Forcada, N. (2022). The effect of climatic conditions on occupants' thermal comfort in naturally ventilated nursing homes. *Building and Environment*, *214*, 108930. doi:10.1016/j.buildenv.2022.108930

NCDC. (2023). *Towards Climate-smart Hospitals: Insights from a National Hospital Energy Consumption Survey (85/NCDC/NPCCHH/2022-23/NatEnergySurvey)*. National Centre for Disease Control, Directorate General of Health Services, Ministry of Health & Family Welfare, Government of India.

Nikola, F., & Radoica, L. (2022). *Climate Change & Quality of Life*. Research Gate. https://www.researchgate.net/publication/360996574_CLIMATE_CHANGE_AND_QUALITY_OF_LIFE.

O'Hara, A. C., Miller, A. C., Spinks, H., Seifert, A., Mills, T., & Tuininga, A. R. (2022). The Sustainable Prescription: Benefits of Green Roof Implementation for Urban Hospitals. *Frontiers in Sustainable Cities*, *4*, 798012. doi:10.3389/frsc.2022.798012

RMI. (2012). *Project Case Study*. Rocky Mountain Institute. https://rmi.org/wp-content/uploads/2017/04/Pathways-to-Zero_Cleveland-Clinic-Retrofit_Case-Study_2012.pdf

Rodríguez-Artalejo, F., Guallar-Castillón, P., Pascual, C. R., Otero, C. M., Montes, A. O., García, A. N., Conthe, P., Chiva, M. O., Banegas, J. R., & Herrera, M. C. (2005). Health-related quality of life as a predictor of hospital readmission and death among patients with heart failure. *Archives of Internal Medicine*, *165*(11), 1274–1279. doi:10.1001/archinte.165.11.1274 PMID:15956007

Rus, T., Cruciat, G., Nemeti, G., Mare, R., & Muresan, D. (2022). Thermal comfort in maternity wards: Summer vs. winter conditions. *Journal of Building Engineering*, *51*, 104356. doi:10.1016/j.jobe.2022.104356

Sampath, B., Jensen, M., Lenoci-Edwards, J., Little, K., Singh, H., & Sherman, J. D. (2022). *Reducing Healthcare Carbon Emissions: A Primer on Measures and Actions for Healthcare Organizations to Mitigate Climate Change*. (Prepared by Institute for Healthcare Improvement under Contract No. 75Q80122P00007.) AHRQ Publication No. 22-M011. Rockville, MD: Agency for Healthcare Research and Quality.

Smith, L. (2022). Healthcare and Climate Change. The Nexus between Climate Change and Healthcare. *Health Policy Partnership*. https://www.healthpolicypartnership.com/the-nexus-between-climate-change-and-healthcare/.

Srishti. (2021). *Green Hospitals – Key to a healthy future*. Srishti. https://www.architectandinteriorsindia.com/insights/green-hospitals-key-to-a-healthy-future.

Sustainable lifestyles. (2023). UNEP - UN Environment Programme. https://www.unep.org/explore-topics/resource-efficiency/what-we-do/sustainable-lifestyles.

Tennison, I., Roschnik, S., Ashby, B., Boyd, R., Hamilton, I., Oreszczyn, T., Owen, A., Romanello, M., Ruyssevelt, P., Sherman, J. D., Smith, A. Z. P., Steele, K., Watts, N., & Eckelman, M. J. (2021). Health care's response to climate change: A carbon footprint assessment of the NHS in England. *The Lancet. Planetary Health*, *5*(2), e84–e92. doi:10.1016/S2542-5196(20)30271-0 PMID:33581070

Thomas, A., & Suresh, M. (2022). Readiness for sustainable-resilience in healthcare organisations during Covid-19 era. *The International Journal of Organizational Analysis*, *31*(1), 91–123. doi:10.1108/IJOA-09-2021-2960

United Nations (UN). (n.d.). *What Is Climate Change?* United Nations. https://www.un.org/en/climatechange/what-is-climate-change.

Wesdrop, K. (2022). Health Care has a Carbon Problem. Health Tech can Help Fix It. *State News*. https://www.statnews.com/2022/06/22/health-tech-help-solve-health-care-carbon-emissions-problem/.

WHO. (2012). *WHOQOL User Manual* (Vol. 03). WHO. https://apps.who.int/iris/handle/10665/77932. Retrieved: 20 June 2023.

WHO. (2017). *Framework for action in building health systems resilience to climate change in South-East Asia Region 2017-2022*. WHO. https://apps.who.int/iris/handle/10665/258953.

WHO. (2021). *Climate Change and Health*. WHO. https://www.who.int/news-room/fact-sheets/detail/climate-change-and-health.

WHO. (2023). *Climate-change-and-health, June 2023*. WHO. https://www.who.int/news-room/fact-sheets/detail/climate-change-and-health..

Yoon Yi, C. (2022). Thermal comfort modelling of older people living in care homes: An evaluation of heat balance, adaptive comfort, and thermographic methods. *Building and Environment, 207*, 108550.

Yuan, F., Yao, R., Sadrizadeh, S., Li, B., Cao, G., Zhang, S., & Short, C. A. (2022). Thermal comfort in hospital buildings–A literature review. *Journal of Building Engineering*, *45*, 103463. https://www.sciencedirect.com/science/article/pii/S2352710221013218. doi:10.1016/j.jobe.2021.103463

Chapter 17
Climate Change Resilience of Single Female Headed Families in Mongu District, Western Zambia:
Adult Educators Applying a Feminist Lens

Daniel L. Mpolomoka
https://orcid.org/0000-0002-2479-2693
UNICAF University, Zambia

Sandra C. Maputa
Zambian Open University, Zambia

Christine Mushibwe
https://orcid.org/0000-0002-7569-1665
UNICAF University, Zambia

Selina Banda
Copperbelt University, Zambia

ABSTRACT

This chapter shares findings of a study which assessed knowledge, while ascertaining effects and describing responses to climate change of single female headed households. It used a descriptive research design with 150 participants. Cluster, simple rotary, and purposive sampling methods were used. Data was collected using semi-structured interviews and questionnaires, analysed using descriptive statistics and thematic analysis. Findings revealed that the single female headed households rated their knowledge on climate change to be low (32%), very low (25%), good (19), very good (13%), and not sure (11). Knowledge on the causes of climate change where it is caused by God (28%), deforestation (23%), emission of gases (18%), while 26% and 5% were not sure and had no response respectively. Effects include loss of farming productivity (35%), loss of fishing productivity (27%), reduced availability of water and wood fuel (20%), rise in temperature (10%), and displacement of households (8%). Climate change is impacting livelihoods of single female headed households, but traces of mitigation abound.

DOI: 10.4018/978-1-6684-9863-7.ch017

OVERVIEW

This chapter is based on an excerpt of findings from a study investigating the climate change resilience of single female-headed households in Mongu District, Western Zambia. The chapter covers the following: introduction, context, objectives, theoretical framework, review of related literature, findings, discussions, conclusion, recommendations, and references.

INTRODUCTION

The role of a single parent is challenging, especially when the family is headed by a woman. Challenges faced by single mothers include problems related to the upbringing of children and their future. The majority find it difficult to maintain discipline among children due to the absence of male figures to assist in the home (Chant, 2007; Safa, 1998). Other challenges include social exclusion and isolation, especially when attempting to settle down (Lebni et al., 2020). Socially, they tend to isolate themselves to avoid attending social gatherings. Furthermore, they may change their dressing style infrequently due to depression and low self-esteem, leading to indifferent attitudes from the community. As a result, the emotional well-being of single mothers is affected, and many tend to feel lonely, helpless, and hopeless, lacking confidence and a sense of identity.

Culturally, single female parents face discrimination because of their status and are often not respected in their communities. They are neither allowed to make decisions in community meetings nor hold high positions, particularly in rural areas. While married women present their views through their husbands, single female parents are not accorded this privilege, making it challenging for their community challenges to be addressed. Unbalanced power relations between women and men in the domestic home, community, and public domain hinder the advancement of women in Zambia. Zambian women have fewer decision-making positions compared to men at all levels and are disproportionately affected by the country's high unemployment and poverty (GRZ, 2017). This raises the rhetorical question: *If urban women with access to education still face discrimination, how much more do single female parents in rural settings, who may have limited access to education, no skills, and live in communities that uphold traditional cultural traits?*

Economically, single female-headed households in these communities tend to be impoverished and experience poor health due to food insecurity at the household level (Samuel, 2010). They often lack opportunities to engage in more productive activities and primarily resort to hand-to-mouth businesses, such as selling garden produce, fish mongering, and selling charcoal. Many perceive these businesses as unsustainable.

Men, traditionally seen as breadwinners, hold the responsibility of making decisions in the home as the head of the household. In case of any challenges at home or in the community, men make decisions or decide together with their wives on the course of action. Even in situations where there is no food in the house, the man takes the responsibility of finding ways to provide for the family, while women take care of the home. For example, during floods in the plains of Mongu rural, challenges for married women are generally less as their husbands lead in ensuring all resources are in place to evacuate their family to higher ground and construct temporary houses. This is not the case for single female parents.

Women in Africa face significant obstacles, hardships, and discrimination within the family, community, state, and international context compared to males, a situation exacerbated for single female parents (Goebel et al., 2010). Undoubtedly, single female-headed households face compounded challenges. These challenges, combined with the effects of climate change, could be detrimental to the lives of single female-headed households, impeding their capacity to adapt or mitigate the effects of climate change. Even in situations where the government provides relief food as a mitigation measure to climate change effects, it may not be a sustainable means to help single female-headed households. Against this backdrop, this chapter shares an excerpt of a study that investigated the effects of climate change on single female-headed households in Mongu District, Western Zambia.

Context

Climate change poses a significant global threat, impacting ecosystems and various aspects of human life (IPCC, 2007). Zambia, characterized by low adaptive capacity, is particularly vulnerable to the adverse effects of climate change. Women, whose livelihoods heavily rely on natural resources for food, energy, fuel, and water, are disproportionately affected (GRZ and UNDP, 2014). The consequences of climate change are a cause for concern in Zambia, especially in Mongu District, Western Province. The economic vulnerability of certain societal groups, coupled with limited resources to address climate change impacts, exacerbates the challenges. Single female-headed households, identified by Goebel et al. (2010), are among the most vulnerable to the effects of climate change due to their specific circumstances. Reports indicate higher levels of poverty among Zambian females compared to males (GRZ, 2014; Samuel, 2010). Despite the potential serious consequences of climate change on economically challenged female-headed households, the extent of their vulnerability, especially in the context of Mongu District, remains unclear.

A study by Banda, Namafe, and Chakanika (2015) highlighted the reality of climate change, emphasizing threats and dangers to the survival, culture, and livelihoods of Lozi communities in the Barotse Plains. These findings align with those of the World Food Programme (WFP, 2017), which conducted an assessment and mapping of vulnerability in Mongu District. In light of these observations, the key question arises: *How resilient are single female-headed households to the effects of climate change in Mongu District?*

Objectives

These objectives aim to provide a comprehensive understanding of the knowledge, impacts, and responses of single female-headed households to climate change in Mongu District, contributing valuable insights for targeted interventions and policy formulation.

1. Evaluate the knowledge level of single female-headed households in Mongu District regarding climate change.
2. Determine the specific effects of climate change on single female-headed households in Mongu District.
3. Ascertain the resilient measures adopted by single female-headed households in response to the effects of climate change.

4. Examine the impact of climate change through a feminist lens, considering the gendered aspects of vulnerability and resilience among single female-headed households.
5. Explore and document the responses of single female-headed households to the effects of climate change in Mongu District.

THEORETICAL FRAMEWORK

The study is grounded in the Vulnerability Theory by Fineman (2008), which originates from the notion that all human beings are inherently vulnerable and prone to dependency (Kohn, 2014). This theory emphasizes the significance of formal equality as an alternative framework for comprehending substantive equality. It posits that vulnerability is inherent to the human condition, and governments have a responsibility to respond affirmatively by ensuring equal access to societal institutions that distribute resources (Kohn, 2014; Wiseman, 2008). The theory offers an alternative basis for defining the role of government and justifying expansive social welfare policies (Kohn, 2014). According to Wiseman (2008), the Vulnerability Theory has been effective in advancing equality among those disadvantaged by law or policy. The relevance of the theory to this study lies in the fact that women, particularly those heading households, are among the disadvantaged groups in communities. Therefore, state-induced strategies are needed to reduce the vulnerability of women-headed households to the effects of climate change.

This study adopts a feminist perspective on climate change, understanding that knowledge is situated and embodied in Climate Justice, which frames the climate crisis as a social justice concern. This perspective places the most vulnerable, such as single female-headed families, at the center of the discourse.

LITERATURE REVIEW

Knowledge and Awareness of Climate Change in Communities

Numerous studies have explored community knowledge of climate change (Hope, 2016; Issaharou-Matchi, 2019; Wei et al., 2014). For example, Issaharou-Matchi's study (2019:790) on Rural Communities' Perceptions of climate change in South East Niger indicated that rural communities possessed knowledge of climate change, viewing it as changes in temperature, rainfall, and extreme events. It is for this reason that Bomuhangi et al. (2016) and Slegers et al (2008) put emphasis on community knowledge and perceptions of change and changes in weather to formulating community specific adaptation and responses to climate change. Similarly, studies in Guyana (Hope, 2016) and China's Shanxi Province (Wei et al., 2014) found respondents to be relatively aware and knowledgeable about climate change.

However, in Africa, studies reveal a general lack of awareness and knowledge of climate change among vulnerable populations. Refugees in East Africa attributed climate change issues to bad governance and conflicts (Afifi et al., 2012), and farmers in Sudan associated climate change with economic constraints (Mertz, 2009). Even in Kenya, young people demonstrated inadequate knowledge of climate change, perceiving it as affecting only farmers (Huho, 2015). Despite climate change being included in school curricula in East African countries, there still appears to be insufficient knowledge and awareness among vulnerable African populations (Apollo and Mbah, 2021; Silvestri et al., 2012).

The Effects of Climate Change on Communities

Climate change is a complex phenomenon with multifaceted impacts on weather patterns and ecosystems. In the context of Mongu District, Western Province, Zambia, as well as in other regions globally, climate change has resulted in observable alterations in rainfall patterns and temperatures. Climate change has global implications, affecting economies, lives, and communities worldwide. Zambia, like other nations, faces challenges from climate change, disrupting national economies and hindering development goals (Mweemba, 2014; GRZ, 2020; GRZ, 2017). Environmental effects include abnormal rainfall patterns leading to droughts or floods, rising temperatures contributing to glacial melting, and prolonged dry spells (Bwalya, 2010; GIZ, 2020; IPCC, 2001; Lukas, 2013; UNFPA, 2009; WFP, 2016). The impact of climate change on extreme weather conditions was observed in a Zambian vulnerability assessment report, indicating an increased risk due to more frequent and severe droughts, floods, and high temperatures (Riche, 2007). The study focuses on the effects of climate change on single female-headed households in Mongu District, Western Zambia.

The Effects of Climate Change on the Socio-Economy of Communities

The literature reviewed indicates that Zambia, like many other countries (Bwalya, 2010; Jain, 2006; WFP, 2016), is grappling with the adverse effects of climate change, particularly in the agricultural sector. These effects have direct implications for food security and livelihoods, often leading to crop failure, loss of income, and subsequently, food insecurity (ADB, 2024; OPM, 2012; WFP, 2016). Diminished farming productivity results in reduced income for farming communities, escalating poverty levels (Kalantary, 2020; OPM, 2012). Studies also highlight the increasing loss of income and prevalence of poverty among farming communities due to unfavourable climatic variations resulting from climate change (Kalantary, 2020; OPM, 2012).

In Sub-Saharan Africa and many developing countries, climate change poses serious and long-term risks to fisheries (Mohammed and Uraguchi, 2013; World Fish, 2007). Zambia's fishery resources are under threat due to predictions of lower water levels in rivers and lakes, adversely affecting the fishing industry and impacting communities dependent on fishing for their livelihoods (Kalantary, 2010: 93).

Climate change also contributes to the occurrence of diseases, creating an environment conducive to disease outbreaks (Kalantary, 2010; WFP, 2006; WHO, 2003; Mpolomoka, 2021). The physical and economic impact of diseases on communities is significant, hindering individuals' ability to work and earn a livelihood.

Despite the predominantly negative impact of climate change on the environment, communities, and economies, some studies (Lal et al., 2011; Howden et al., 2007) suggest potential benefits, such as improved wheat yields and enhanced fruit production in specific regions. The argument is that increased warming, if within certain thresholds, could positively affect agricultural productivity. A good is example is that of increased fruit production in various regions like the Great Lakes Region as well as the Mid-west region (Field, et al., 2007; Karl, et al., 2009) provided the temperatures do not go further beyond the threshold for various fruit crops. Although climate change devastates the environment, communities and families, humans can still establish productive ways of resilience.

The Response of Communities to Climate Change

Literature reveals diverse community responses to climate change, ranging from feelings of vulnerability, hopefulness for the future through coping and adaptation, fear, sadness (Hope, 2016), to a sense of personal responsibility to mitigate the effects of climate change (Poortinga et al., 2018; Mpolomoka, 2021). Studies in Britain, such as Fisher et al. (2018), show that people generally exhibit moderate concern about climate change and do not feel an extreme responsibility to address it. Various mitigation measures have been proposed in response to climate change. Changes in farming practices, such as transitioning from rice to alternative crops, have been identified (ADB, 2014). Communities are urged to adapt their farming methods to align with the sensitivity of farming to climate conditions (ADB, 2014; Kalantary, 2020; UNFCCC, 2007; Hope, 2016; Banda & Mpolomoka, 2018). These adaptations aim to address the adverse effects of climate change on farming productivity.

Women and Climate Change

Women play a crucial role in environmental stewardship, being active participants in natural resource management and environmental conservation. Empowering women with knowledge about climate change and awareness of their surroundings can contribute positively to their attitudes toward addressing climate change challenges (Aditya, 2016).

Literature presents mixed findings on women's awareness and knowledge of climate change, with some studies indicating that women may have both knowledge and unique experiences related to climate change. For instance, Hope's study in Guyana showed that out of the total respondents who responded "yes" to understanding climate change, there were more females than males (Hope, 2016; Mayesha, et al., 2015). Another study by the Government of Canada (2019) on women and climate change revealed that women have both knowledge and unique experiences with climate change. It is not surprising that female headed households and wives of male farmers in Uganda were significantly less likely to adopt drought tolerant maize to curb climate change, induced crop failure than their male counterparts due to lack of access to resources and less knowledge (Global Gender and Climate Alliance, 2017; Brody et al., 2008).

Women interact more with the surrounding environment in their daily lives (Pasqualino et al., 2014; Osman-Elasha, 2010; Aditya, 2016; Escalante and Maisonnave, 2020) than men. This corroborates with findings by Fuhrman et al (2020) that women are producers and providers of food and caretakers of their family; hold the key to addressing hunger and malnutrition through environmental related livelihoods. Interesting, this was said three decades ago (in the 90s and reaffirmed in the early 2000s), that women produce 60 to 80 percent of food in developing countries as well as half of the food produced in the world (FAO, 1997; Mehra and Rojas, 2008).

In addition, climate change has been seen to affect women as they manage not only agricultural terrain but water, sources of fuel as well as forests in communities around the world (Mago and Gunwal, 2019). This forces woman headed households to face challenges of walking long distances to look for safe drinking water, fetch for wood fuel, take children to hospitals in case of sicknesses and find food for the family (Demetriades and Esplen, 2010; Zulu and Sichikwenkwe, 2011; Paudyal et al., 2019). The impact of environmental destruction is profoundly felt by impoverished rural women, a sentiment echoed by Oxfam (2002). In the face of ecological degradation, especially in impoverished villages, women find themselves at the forefront of the struggle for survival. Each day, irrespective of age or pregnancy status, they embark on arduous journeys to secure essential resources such as fuel, fodder, and water. As environmental conditions deteriorate, this daily struggle becomes increasingly strenuous, pushing poor rural women to the brink of physical endurance. The excerpt underlines the harsh reality faced by these women, caught in a relentless cycle of poverty and environmental devastation, emphasizing the urgent need for sustainable solutions.

Various reasons have been brought forward as to why women seem to be the most adversely affected by the effects of climate change. According to Osman-Elasha (2010), female headed households are most susceptible to poverty because they have fewer revenue streams to supplement their household income (Flato and Pelser, 2016; Gammage, 2010). Horrell and Krishnan (2006: 6) underscore the economic disparities within female-headed households in Zimbabwe, revealing stark contrasts in income levels. Their study asserts that households led by widows experience a total income slightly exceeding half of that garnered by male-headed households. Similarly, de facto female-headed households, those led by women without a legal marriage, command approximately three-quarters of the income generated by their male-headed counterparts. This economic discrepancy highlights the financial challenges faced by female-headed households, particularly those headed by widows, underscoring the need for targeted interventions to address these inequalities.

The World Health Organization (WHO, 2014) draws attention to the challenges faced by women in adapting to climate change, emphasizing that factors such as early childbearing and high fertility rates can impede their ability to earn, save money, and effectively respond to the impacts of climate change. The early onset of childbearing and elevated fertility rates are noted as contributors to poor health and lower education levels among women. This observation aligns with emerging literature that highlights the correlation between poverty levels among single mothers and factors such as educational attainment, limited opportunities for gainful employment, and a lack of involvement in sustainable entrepreneurial activities (Brown and Lichter, 2004; Banda & Mpolomoka, 2018). These intertwined challenges underscore the multifaceted nature of barriers faced by women, emphasizing the importance of comprehensive strategies to address the intersectionality of gender, reproductive health, education, and economic opportunities in the context of climate change adaptation.

The vulnerability of women-headed households to the impacts of climate change is further attributed to differences in social roles and responsibilities compared to men (Easterling, 2000; Wisner et al., 2004). The expectation that women fulfil their roles as caretakers of families places an additional burden on them during extreme climate events, as noted by WHO (2014: 5). This burden is reinforced by cultural norms that assign women specific household chores and roles, contributing to their vulnerability during climate-related challenges (Demetriades and Esplen, 2010; Paudyal et al., 2019).

Moreover, the lack of independence and decision-making power is a significant constraint on women's ability to adapt to climate change (Government of Canada, 2019). Many women have limited control or no control over family finances and assets, impeding their participation in effective climate change action strategies (UNFPA, 2009; Yavinsky, 2012; Mpolomoka, 2021). Environmental transparency, decision-making, accountability, and good environmental governance are highlighted as essential for effective climate change strategies, emphasizing the importance of democratic involvement that includes women in affected communities.

Cultural restrictions can also hinder women's adaptation to the impacts of climate change. In countries like India, women often have less access to critical information on weather alerts and crop patterns compared to men, affecting their ability to respond effectively to climate variability (UNFPA, 2009; Yavinsky, 2012). Additionally, traditional cultural clothing may pose constraints on women engaged in climate change adaptation, potentially limiting their participation in activities such as running and swimming. This limitation makes it harder for women to escape natural disasters associated with climate change, such as floods and extreme winds (Neumayer and Plumper, 2007).

Gaps in the Reviewed Literature

Lack of Specific Study on Single Female Headed Households in Mongu District: The existing literature lacks a specific study on how climate change has affected the living standards of single female-headed households in Mongu District of Western Province, Zambia. This gap highlights the need for a targeted investigation into the unique challenges faced by this specific demographic group in the context of climate change.

Insufficient Specifics in Government Policies: While the Government of Zambia has developed policies for women's empowerment, including the National Gender Policy and the Eighth National Development Plan, the content of these policies does not specifically address the challenges faced by single female-headed households. There is a need for more detailed and targeted policy measures to address the distinct vulnerabilities of this demographic group.

Limited Visibility of Project Outcomes: Despite the implementation of projects in Western Province aimed at empowering women, the results of these programs are not visibly improving the lives of single female parents. The literature points out that women benefitting from these projects are often economically sound and possess knowledge and skills, potentially excluding those who need assistance the most.

Paucity of Information on Single Female Headed Households: The literature lacks sufficient information specifically focusing on single female-headed households in the study area. While there is information on women and their livelihoods, such as farming, there is a lack of disaggregated data that distinguishes between single and married women. This gap underscores the importance of understanding the specific circumstances of single female-headed households.

METHODOLOGY

The study adopted a qualitative research paradigm, specifically employing a descriptive research design. This choice is in alignment with established literature (Aggarwal, 2008; Kerlinger, 1986; Banda, Mpolomoka, Mbono & Sampa, 2017). Kombo and Tromp (2006) emphasize that descriptive studies go beyond fact-finding, often leading to the formulation of essential principles of knowledge and solutions to significant problems. The design facilitated the collection, description, and interpretation of in-depth perspectives, feelings, experiences, attitudes, opinions, beliefs, habits, or any variety of educational or social issues related to the effects of climate change on single female-headed households in Mongu District, including changes in their lifestyles.

The population included officials from government institutions, NGOs operating within Mongu District, and single female-headed households in rural areas. The sample comprised 150 participants: 30 NGO representatives, 30 government representatives, and 90 single female-headed households.

For single female-headed households, cluster random sampling was used, segmenting the district into wards and randomly selecting one for participation. A rotary method was then employed to randomly select households (Msabila and Nalaila, 2013; Salaria, 2012) within the chosen ward (Kothari and Garg, 2014). NGO and government representatives were selected using non-random purposive sampling (White,

2005), considering their privileged positions and specific knowledge related to the study. Structured questionnaires and semi-structured interview schedules were used to collect both narrative and statistical data. Questionnaires were distributed to NGO and government representatives due to their busy schedules, allowing them to complete them at their convenience. Semi-structured interviews were conducted with single female-headed households, enabling them to express their views openly.

Descriptive statistics were used to code and present quantitative data in the form of percentages, figures, and frequencies. Thematic analysis involved arranging qualitative data into themes based on opinions, coding, and presenting in tables to illustrate trends.

Permission and consent were obtained from NGOs, government institutions, and single female-headed households before data collection. Participants were informed about the study's purpose, and their participation was voluntary without coercion.

FINDINGS

Demographic Information

The study included a total of 150 respondents, with 67% (n=101) being female, and 33% (n=49) being male. Among the females, 90 (60%) were identified as single female household heads, while 7% of the total females held leadership positions in NGOs and government departments.

The majority of respondents (47%) fell within the age group of 41-50 years. Age groups 36-40 and below 25 years were the second major groups, each representing 17%. Age groups 31-35 and 25-30 years followed with 12% and 7%, respectively.

43% of the respondents had tertiary education, mainly officials from NGOs and government departments. 28% had primary education. 13% had secondary education. 16% had no formal education, with most being single female household heads.

According to the findings of the study, it was established that the majority of the single female household heads had never married before and they accounted for 46% of single female household heads had never married, 35% were widows and 19% were divorcees.

The findings of the study have revealed that the majority 61% were peasant farmers, 17% were small-scale businesspeople, mainly in fish mongering. 12% were not involved in any occupation and 6% were agricultural officers, as well as shop attendants and ward councillors who represented 2% each of the total 90 respondents.

Level of Knowledge of Single Headed Households on Climate Change in Mongu District

Knowledge Rating on Climate Change

Among the single female household heads (90 respondents), 68% rated their knowledge on climate change as very low (25%), low (32%), or not sure (11%). Only 32% rated their knowledge as good (19%) or very good (13%).

Knowledge on the Definition of Climate Change

Four themes emerged that included climate change being abnormal change in weather due to deforestation and emission of gases; change in climate as planned by God; Climate change as being the end of the world as well as the theme of not knowing the definition of climate change. Using these four themes, data demonstrated that:

- 62% of single female household heads did not know the definition of climate change.
- change in climate as planned by God (27%),
- climate change as the end of the world (21%), and
- having no idea at all (14%).

Knowledge on the Causes of Climate Change

The results of the study revealed that out of 90 respondents, 59% of single female household heads did not know the causes of climate change. Causes cited included God (28%), not sure (26%), and no response (5%). Those expressing knowledge mentioned deforestation (23%) and emission of gases (18%).

Effects of Climate Change on Single Female Household Heads

According to 60 NGO and government officials, major effects included loss of farming productivity (35%), loss of fishing productivity (27%), reduced availability of water and wood fuel (20%), rise in temperatures (10%), and displacement of households (8%).

Responses of Single Female Headed Households to the Effects of Climate Change

Attitudes of Single Female Household Heads

Negative attitudes were dominant, including complaints of drought, floods, hunger, and heat (34%), devastation and frustration (26%), and giving up by doing nothing (12%). Positive attitudes of hopefulness in coping and adapting to climate change were indicated by 28% of respondents.

Mitigation Measures of Single Female Household Heads

The majority (38%) employed new farming practices (crop diversification, drought-resistant crops). Other responses included moving to one meal a day (17%), increased food gathering (15%), joining business and women's business clubs (13%), waiting for external aid (10%), and engaging in prostitution and theft (7%).

DISCUSSION

This section discusses the findings, categorised in thematic topic areas.

Level of Knowledge on Climate Change

The respondents' conceptualization of climate change as a change in weather, while a minority perspective, aligns with the research by Grummell et al. (2020), which found that local communities often perceive climate change as a simple 'change in weather.' Grummell et al. (2020) note that, from the local community perspective, climate change is distinct from the scientific understanding and encompasses variations in weather, seasons, and climate within various tribal groups in Zambia. This suggests a local understanding that differs from the scientific definition.

The study's findings indicate that single female-headed households exhibited a low level of knowledge about climate change, as reflected in their low climate change knowledge ratings and their inability to define climate change or identify its causes. These results contrast with those of Hope (2016), who found in Guyana that student respondents demonstrated knowledge of climate change, citing causes such as carbon emissions, deforestation, burning of fossil fuels, and methane from livestock farming. However, the current findings align with the observations of Mulenga and Wineman (2014), who noted a general lack of adequate knowledge about climate change among most people.

The observed lack of knowledge among single female-headed households about climate change can be attributed to the high levels of illiteracy within this demographic. The study revealed that a significant majority of respondents had only received primary education (42%) or had no formal education at all (16%), totalling 66% of single female-headed households with primary education or lower. This suggests a prevalence of school dropouts among this group, which aligns with broader trends indicating higher dropout rates among girls compared to boys, particularly in rural areas like Mongu District (McConnell and Mupuwaliywa, 2015).

The high dropout rates and low educational attainment may impede these women's ability to comprehend information about climate change, especially considering that climate change information is often disseminated through literature, media, and everyday conversations (Hope, 2016). Furthermore, the study highlighted a lack of serious awareness activities about climate change in the Barotse Plains, where the study was conducted. This lack of awareness, both from government and NGOs, suggests that communities, particularly female-headed households dependent on farming, may not have access to crucial information on climate change, its impact on their livelihoods, and potential mitigation strategies.

The study also emphasized the challenges of sensitizing rural communities, especially women, about climate change. The demanding nature of women's lives as heads of households, responsible for various chores, childcare, and earning a livelihood, hinders their participation in awareness activities. This aligns with broader findings that women face numerous responsibilities that impede their learning at different levels, whether in community sensitization programs or higher learning institutions (Mpolomoka et al., 2018). The study underscores the need for targeted and accessible climate change education and awareness programs tailored to the specific challenges faced by single female-headed households in rural areas.

Effects of Climate Change on Single Female Headed House in Mongu District

The study revealed that climate change has significant adverse effects on single female-headed households in Mongu District. The identified effects include a loss of farming productivity (35%), loss of fishing productivity (27%), reduced availability of water and wood fuel (20%), a rise in temperature (10%), and displacement of households (8%). These findings underscore the profound impact of climate change on the livelihoods of these households. The majority of single female-headed households reported experiencing a loss of farming and fishing productivity, which is particularly significant given that 61% and 17% of these households depend on peasant farming and fish mongering, respectively, based on their occupations. The study attributes the loss in productivity to the occurrence of floods and droughts, both of which are consequences of climate change. Zambia has been witnessing these effects annually since 2000 (GRZ, 2004). Floods have become more recurrent, affecting even areas not historically prone to such disasters (Kalantary, 2010). Prolonged dry spells and floods, intensified by climate change, can lead to crop failures and disruptions in fishing activities, ultimately increasing vulnerability among households. Osman-Elasha (2010) supports this perspective, suggesting that the increased frequency and intensity of droughts and floods can result in crop failure, food insecurity, and heightened risks of malnutrition, famine, and starvation. The disruption of farming and fishing, which serve as essential livelihood activities for rural populations, may contribute to food shortages, hunger, and overall food insecurity among single female-headed households. Similar studies, such as Lewis, Monem, and Impiglia's (2018) research on the impacts of climate change on farming systems and livelihoods in the Near East and North Africa, have highlighted the vulnerability of agricultural yields to climate-induced changes in rainfall patterns.

It is clear that the effects of climate change on single female-headed households in Mongu District are multifaceted, ranging from disruptions in farming and fishing activities to potential food shortages and heightened vulnerability. These findings underscore the urgent need for targeted interventions and adaptive strategies to mitigate the impact of climate change on this vulnerable demographic. The disruptive impact of climate change on the livelihoods of single female-headed households in Mongu District extends beyond immediate agricultural and fishing challenges. The study identifies several secondary effects, including potential increases in poverty levels, displacement of households, heat-related issues, and reduced availability of drinking water and wood fuel. The disruption of livelihoods can lead to a significant reduction in income flow, thereby intensifying poverty levels among the already economically vulnerable rural populations of Mongu District, particularly among single female-headed households. Given that poverty is already more prevalent among women than men (GRZ, 2010), the adverse effects on women's livelihoods can exacerbate poverty levels in their households. Women, who are more dependent on natural resources threatened by climate change, are particularly vulnerable to its impact, not only in Mongu District but throughout the country (Mwanangombe, et al., 2021).

The displacement of households is highlighted as another effect of climate change, with flush floods being a contributing factor. Located in the Barotse plains, some households are prone to flooding, necessitating migration to higher ground for safety. This displacement, especially challenging for single female-headed households, can result in the loss of property and significant disruptions to their lives. Moreover, the study notes excessive heat, reduced availability of drinking water, and wood fuel scarcity as additional effects of climate change on single female-headed households. Excessive heat is attributed to the warming of the earth, leading to discomfort from heatwaves and increased evaporation rates. This, in turn, exacerbates drought conditions, affecting humans, animals, and plants due to water scarcity. Similar observations were made in Peru, where climate change-induced water shortages became a significant problem (IFAD, 2010).

In Mongu District, the drying up of water sources, such as streams and rivers, forces women to travel long distances in search of water. The availability of wood fuel, crucial for these communities, is also impacted as climate change affects plant growth due to insufficient water and dry spells. This results in longer travels for women to collect wood fuel and access water. These challenges align with the observation that climate change adversely affects women whose livelihoods depend on natural resources for food, wood fuel, and water (GRZ, 2014). The study underscores the far-reaching consequences of climate change on single female-headed households in Mongu District, encompassing economic vulnerabilities, displacement, and challenges related to heat, water, and fuel availability. Addressing these issues requires a comprehensive and gender-sensitive approach to climate change adaptation and mitigation strategies.

Response to Climate Change Among the Single Female Headed Households

The responses of single female-headed households to the effects of climate change are varied, reflecting both positive and negative attitudes and adaptive strategies. The study notes that attitudes among these households are predominantly negative, with complaints about droughts, floods, hunger, and heat being common. This negative outlook may be linked to the low levels of knowledge about climate change among the respondents, suggesting that understanding and awareness play a crucial role in shaping attitudes toward climate change. Contrary to findings in other contexts, where respondents demonstrated a more hopeful attitude (Hope, 2016), the study suggests that the vulnerability of single female-headed households, coupled with limited knowledge about climate change, contributes to a prevalent negative mindset. The findings align with the argument that women's vulnerability to climate change is heightened by their roles and responsibilities, which can become overwhelming during extreme climate events (Wisner et al., 2004). In terms of response strategies, the study reveals a range of actions taken by single female-headed households to mitigate the impact of climate change on their livelihoods. Notably, there is a focus on adapting farming practices, including crop diversification, crop rotation, and planting drought-resistant crops. Given that the majority of these households rely on farming as their primary livelihood, these adaptive measures align with the need to protect and enhance agricultural productivity in the face of climate change challenges.

The findings underscore the profound impact of climate change on the agricultural livelihoods of single female-headed households, emphasizing the need for adaptive strategies to cope with the challenges posed. The primary reliance on farming and fishing activities makes these households particularly susceptible to the adverse effects of climate change, such as disruptions in productivity, increased temperatures, and altered weather patterns. The observed responses, including new farming practices like crop diversification, crop rotation, and planting drought-resistant crops, indicate a proactive effort to adapt agricultural methods to the changing climate. These strategies align with broader global discussions on climate-smart agriculture, emphasizing sustainable practices that enhance resilience to climate change impacts.

However, the study also highlights concerning responses such as engaging in prostitution and theft, potentially driven by the desperation to secure alternative income sources in the face of climate-induced livelihood challenges. These activities pose risks not only to the economic well-being of these households but also to the health and safety of individuals, as indicated by the potential danger of contracting HIV and other sexually transmitted diseases. The response of waiting for external aid suggests a reliance on external assistance, indicating a perception that individual adaptive capacity may be limited. This emphasizes the importance of effective and targeted support mechanisms, both from government agencies and non-governmental organizations, to build resilience and empower these households in the face of climate change.

Joining business and women's business clubs, like SILK, indicates a recognition of the value of collective efforts and support networks in navigating climate-related challenges. Such initiatives can play a crucial role in providing financial assistance, sharing knowledge, and fostering a sense of community resilience. While some responses reflect attempts at climate change resilience, it is essential to address the root causes of vulnerability, including limited knowledge and awareness about climate change. Educational programs, awareness campaigns, and capacity-building initiatives can empower these households to make informed decisions and adopt sustainable practices that enhance their adaptive capacity. The study underscores the complex interplay between climate change impacts, livelihood strategies, and the adaptive capacity of single female-headed households. Efforts to address these challenges should prioritize a holistic approach that combines sustainable agricultural practices, community-based support networks, and targeted interventions to enhance knowledge and awareness.

CONCLUSION

The study's findings on climate change knowledge among single female-headed households highlight a significant gap in awareness and understanding. A majority of these households rated their knowledge on climate change as low or very low, with some expressing uncertainty about their level of understanding. This lack of awareness is further evident in the varied and sometimes incorrect definitions of climate change provided by the respondents, ranging from perceiving it as God's plan to being the end of the world. The study also reveals a notable lack of knowledge about the causes of climate change among single female-headed households. A substantial proportion either attributed climate change to God or expressed uncertainty about its causes. Only a minority correctly identified human activities such as deforestation and the emission of gases as contributing factors to climate change.

Moving on to the effects of climate change, the study emphasizes that loss of farming productivity is the most widely reported impact among single female-headed households. Loss of fishing productivity, reduced availability of essential resources like water and wood fuel, the rise in temperatures, and displacement of households are also identified as significant effects. These findings emphasize the vulnerability of these households, particularly those heavily dependent on agriculture and natural resources for their livelihoods.

In response to the challenges posed by climate change, the study identifies various coping strategies adopted by single female-headed households. These include engaging in new farming practices, adjusting meal frequency, increasing food gathering efforts, participating in business and women's clubs, waiting for external aid, and, notably, resorting to activities like prostitution and theft. While some responses reflect efforts to adapt to climate change, others, such as involvement in illegal activities, underscore the desperation and limited options available to these households in the face of environmental challenges.

In conclusion, the study highpoints the urgent need for targeted educational programs and awareness campaigns to enhance climate change knowledge among single female-headed households. Additionally, the findings emphasize the importance of developing and implementing sustainable and climate-resilient livelihood strategies for these households. Addressing the underlying causes of vulnerability, including limited knowledge and socio-economic constraints, is crucial for empowering these households to navigate the impacts of climate change more effectively.

RECOMMENDATIONS

These recommendations aim to address the identified gaps in knowledge, mitigate the impacts of climate change, and promote sustainable and resilient livelihoods for single female-headed households in Mongu

District. Implementing these measures will contribute to building a more climate-resilient community and ensuring the well-being of vulnerable households in the face of environmental challenges. Implementing the recommendations to enhance climate change knowledge and resilience among single female-headed households indeed comes with cost implications. Overall, a collaborative and multi-stakeholder approach is essential to address the challenges posed by climate change among single female-headed households. Commitment at the governmental, community, and individual levels, along with support from NGOs and international entities, will be critical for the successful implementation of strategies to enhance resilience and knowledge in the face of climate change.

Education and Awareness Programs: NGOs and the Zambian government, particularly the Ministry of Agriculture, should collaborate to implement targeted education and awareness programs on climate change. These initiatives should aim to enhance the knowledge and understanding of climate change among single female-headed households.

Inclusion in Agricultural Programs: The government, through the Ministry of Agriculture, should consider increasing the participation of single female parents in programs like the Farmer Input Support Program (FISP). This could assist in boosting agricultural productivity and resilience in the face of climate change impacts.

Redesigned Relief Strategies: The government, in collaboration with the Ministry of Community Development and the Disaster Management and Mitigation Unit, should reassess and redesign relief strategies. Special consideration and assistance should be provided to single female-headed households in rural areas affected by climate change.

Fossil Fuel Limitation: The government, through the Zambia Environmental Management Agency (ZEMA), should explore and implement mechanisms to limit the use of fossil fuels. This step is crucial for mitigating climate change impacts, as fossil fuel emissions contribute to environmental degradation.

Women Economic Empowerment: The Ministry of Community Development should rejuvenate and strengthen women's economic empowerment programs. Empowering women economically can enhance their capacity to adapt to and cope with the impacts of climate change, reducing vulnerabilities.

Multi-stakeholder Commitment: The commitment of various stakeholders, including the government, communities, and individuals, is crucial for the success of these efforts.

Government Commitment:

Financial Support: The government may need to allocate funds for educational programs, awareness campaigns, and training initiatives focused on climate change for single female-headed households. This includes developing and implementing sustainable livelihood strategies.

Policy Support: Governments can play a crucial role by creating policies that promote climate resilience, gender equality, and sustainable practices. This may involve integrating climate change education into formal and informal education systems.

Community Commitment:

Participation: Communities should actively engage in awareness programs and educational activities. This might involve participating in workshops, training sessions, and community-led initiatives to share knowledge about climate change.

Implementation: Communities can implement sustainable practices and resilience-building measures. This might include adopting climate-smart agriculture, water conservation, and forest management practices.

Individual Commitment:

Learning: Individuals, especially single female-headed households, should commit to learning about climate change and its impacts. This can involve attending training sessions, accessing educational resources, and staying informed about climate-related issues.

Adapting Livelihoods: Individuals may need to adapt their livelihood strategies based on the knowledge acquired. This could include implementing new farming techniques, exploring alternative income-generating activities, and incorporating climate-resilient practices.

NGOs and Other Organizations:

Support Programs: Non-governmental organizations (NGOs) and other relevant entities can develop and implement programs specifically targeting single female-headed households. These programs might include skill-building workshops, access to resources, and support for sustainable livelihoods.

Advocacy: NGOs can advocate for policies that address the unique challenges faced by single female-headed households in the context of climate change. This involves working closely with governments and local communities.

International Support:

Funding: International organizations and donor agencies can contribute by providing financial support for climate change adaptation and resilience-building projects in vulnerable communities.

Capacity Building: Supporting capacity-building initiatives, including training programs and knowledge-sharing platforms, can empower local communities to effectively respond to climate change.

REFERENCES

ADB. (2014). *Climate Change and Rural Communities in the Greater Mekong Sub-Region: A Framework for Assessing Vulnerability and Adaptation Options*. Mandaluyong City: ADB.

Aditya, S. K. (2016). Role of Women in Environmental Conservation. *International Journal of Political Science and Development*, *4*(4), 140–145.

Afifi, T., Govil, R., Sakdapolrak, P., & Warner, K. (2012). *Climate Change, Vulnerability and Human Mobility: Perspectives of Refugees from the East and Horn of Africa*. UNHCR.

Aggarwal, Y. P. (2008). *Science of Educational Research*. Nirmal Book Agency.

Apollo, A., & Mbah, M. F. (2021). 'Challenges and Opportunities for Climate Change Education (CCE) in East Africa. A Critical Review. *Climate (Basel)*, *9*(93), 1–16. doi:10.3390/cli9060093

Banda, S., & Mpolomoka, D. L. (2018). Culturally relevant education and training for communities: A review. *American Educational Research Journal*, *6*(2), 88–93. doi:10.30918/AERJ.62.18.019

Banda, S., Mpolomoka, D. L., Mbono, D., & Sampa, R. L. (2017). Use of questions in qualitative research: How questions guided our study. *International Journal of Developmental Research*, *7*(12), 17895–17898.

Banda, S., Namafe, C. M., & Chakanika, W. W. (2015). Traditional Environmental Knowledge among Lozi Adults in Mitigating Climate Change in the Barotse Plains of Western Zambia. *International Journal of Humanities, Social. Sciences and Education*, *2*(9), 222–239.

Bomuhangi, A., Nabanoga, G., Namaalwa, J. J., Jacobson, M. G., Abwoli, B., & Wich, S. (2016). Local Communities' Perceptions of Climate Variability in the Mt. Elgon Region, Eastern Uganda. *Cogent Environmental Science*, *2*(1168276), 1168276. doi:10.1080/23311843.2016.1168276

Braun, V., & Clarke, V. M. (2012). Thematic Analysis. In H. Cooper, P. M. Camic, D. L. Long, A. T. Panter, D. Rindskopf, & K. J. Sher (Eds.), *Research Designs: Quantitative, Qualitative, Neuropsychological and Biological.' APA Handbook of Research Methods in Psychology, 2, 57-71*. American Psychological Association. doi:10.1037/13620-004

Brody, A., Demetriades, J., & Esplen, E. (2008). *Gender and Climate Change: Mapping the Linkages – A scoping Study on Knowledge and Gaps*. BRIDGE Institute of Development Studies.

Brown, J. B., & Lichter, D. T. (2004). Poverty, Welfare and the Livelihood Strategies of Non-Metropolitan Single Mothers. *Rural Sociology*, *69*(2), 282–301. doi:10.1526/003601104323087615

Bwalya, S. M. (2010). *Climate Change in Zambia: Opportunities for Adaptation and Mitigation through Africa Bio-Carbon Initiative*. Centre for International Forest Research, Southern Africa Regional Office.

Cavanagh, S. (1997). Content Analysis: Concepts, Methods and Citations. *Nurse Researcher*, *4*, 5–16. PMID:27285770

Chant, S. (2007). *Children in Female Headed Households: Interrogating the Concept of an 'Inter-Generational Transmission of Disadvantage' with Particular Reference to the Gambia, Philippines and Costa Rica*. New Working Paper Series, London: London School of Economics, Gender Institute.

Demetriades, J. & Esplen, E. (2010). The Gender Dimension of Poverty and Climate Change Adaptation: Social Dimensions of Climate Change. *Equity and Vulnerability in a Warming World,* 133-143.

Devi, P. B. (2008). *Content Analysis: A Method of Social Science Research*. Sage.

Easterling, D. R., Evans, J. L., Groisman, P. Y., Karl, T. R., Kunkel, K. E., & Ambenje, P. (2000). Observed Variability and Trends in Extreme Climate Events: A Brief Review. *Bulletin of the American Meteorological Society*, *81*(3), 417–425. doi:10.1175/1520-0477(2000)081<0417:OVATIE>2.3.CO;2

Escalante, L., & Maisonnave, H. (2020). *Assessing the Impacts of Climate Change on Women's Poverty and Domestic Burdens: A Bolivian Case Study*. HAL Archives-Ouvertes. https://hal-normandie-univ-archives-ouvertes.fr/hal-02970249

FAO. (1997). *FAO Focus on Women and Food Security*. FAO.

Feikner. (2009). Impact of Climate Change on Rice Production in Thailand. *The American Economic Review*, *99*(2), 205–210. doi:10.1257/aer.99.2.205 PMID:20161598

Field, C. B., Mortsch, L. D., & Brklacich, M. (2007). North America. In M. L. Parry, O. F. Canziani, & J. P. Palutikof (Eds.), *Climate change: Impacts, Adaptation and Vulnerability. Contribution of Working Group II to the Fourth Assessment Report of the Intergovernmental Panel on Climate Change* (pp. 617–652). Cambridge University Press.

Fineman, M. A. (2008). The Vulnerable Subject Anchoring Equality in the Human Condition. *Yale Journal of Law and Feminism*, *1*, 9–15.

Fisher, S., Fitzgerald, R., & Poortinga, W. (2018). *Climate Change: Social Divisions in Beliefs and Behaviour – British Social Attitudes. Climate Change, 35*. National Centre for Social Research.

Flato, M., & Pelser, A. (2017). Women, Weather and Woes: The Triangular Dynamics of Female-Headed Households, Economic Variability, and Climate Variability in South Africa. *World Development, 90*, 41–62. doi:10.1016/j.worlddev.2016.08.015

Fuhrman, S., Janoch, E., Koch, R., Oo, K., Parra, V., & Rawe, T. (2020). *Left out and Left Behind: Ignoring Women will Prevent us from Solving the Hunger Crisis: Policy Report*. Care International.

Gammage, S. (2010). Gender, Time Poverty and Amartya Sen's Capability Approach: Evidence from Guatemala. In S. Chant (ed). The International Handbook of Gender and Poverty: Concepts, Research, Policy, 26, 71-76.

GIZ. (2020). *Climate Change and Rural Development*. Rural Development and Agriculture Division – GIZ.

Global Gender and Climate Alliance. (2017). *Gender and Climate Change in Africa – Facts from Gender and Climate Change: A Closer look at Existing Evidence*. Global Gender and Climate Alliance.

Goebel, A., Dodson, B., & Hill, T. (2010). Urban Advantage or Urban Penalty? A Case Study of Female Headed Households in a South African City. *Health & Place, 16*(3), 573–580. doi:10.1016/j.healthplace.2010.01.002 PMID:20116325

Government of Canada. (2019). *Women and Climate Change*. Government of Canada.

Grummell, B., Daura, M., Chipeta, G. C., Daura, M., Downes, M., Kasulo, V., Lichilo, I., Mataya, B., Mpolomoka, D., Murphy, C., Mwale, J., Noone, M., Sibalwa, D., Ryan, A., Tembo, M., & Yerokun, O. (2020). 'TEN Team,' *Concept: The Journal of Contemporary Community Education Practice Theory, 11*(1), 1-11. http://concept.lib.ed.ac.uk/article/view/4294

GRZ. (2004). *Zambia Vulnerability Assessment Survey*. GRZ.

GRZ. (2014). *Zambia National Gender Policy*. GRZ.

GRZ (2014). *Ministry of Child and Development: Living Conditions Monitoring Survey*. Lusaka 2010: GRZ.

GRZ. (2017). *Seventh National Development Plan 2017-2021*. GRZ.

GRZ (2020). *Presidential Speech on Pilot Programmes for Climate Sites in Lusaka District by Edgar Chagwa Lungu. Lusaka*: GRZ.

GRZ & UNDP. (2010). *Project Document: Adaptation to the Effects of Climate Change in Agro-Ecological Regions I and II in Zambia*. GRZ and UNDP.

GRZ & UNDP. (2013). *Millennium Development Goals Progress Report: Zambia*. GRZ and UNDP.

Harnett, B. M. (2015). *The Value of Content Analysis as a Qualitative Research Method: Research Proposal*. The British University in Dubai.

Hope, S. A. A. (2016). *Knowledge, Attitudes & Practices Study on Climate Change Adaptation & Mitigation in Guyana*. Climate Change Partnership Japan-Caribbean and UNDP.

Horrell, S., & Krishnan, P. (2006). Poverty and Productivity in Female Headed Households in Zimbabwe. *The Journal of Development Studies*, *43*(8), 1–37.

Howden, S. M., Sousanna, J. F., & Tubiello, F. N. (2007). Adapting Agriculture to Climate Change: Proceedings of National academy of Sciences. *Proceedings of the National Academy of Sciences of the United States of America*, *104*(50), 19691–19696. doi:10.1073/pnas.0701890104 PMID:18077402

Huho, J. M. (2015). Climate Change Knowledge Gap in Education System in Kenya. *Int. J. Innov. Res. Educ. Sci*, *2*, 2349–5219.

IFAD. (2010). *Climate Change: Building the Resilience of the Poor Rural Communities*. IFAD.

IPCC. (2001). *Third Report of the Working Group of the Intergovernmental Panel on Climate Change*. IPCC.

IPCC. (2007). *Climate Change 2007: Synthesis Report – Summary for Policy Makers*. Cambridge University Press.

Issaharou-Matchi1, I., Rabiou, H., Mahamane1, M., Idrissa, B., Mahamane, A. andSaadou, M. (2019). Rural Communities' Perceptions of Climate Change and its Impacts on their Livelihood in Southeast Niger. *International Journal of Environment and Climate Change*, *9*(12), 789-800.

Jain, S. (2006). An Empirical Economic Assessment of Impacts of Climate Change on Agriculture in Zambia. *Centre for Environmental Economics and Policy in Africa (CEEPA) Discussion Paper, No. 27*. Pretoria: CEEPA.

Kalantary, C. (2010). Climate Change in Zambia: Impacts and Adaptation. *Global Majority E-Journal*, *1*(2), 85–96.

Karl, T. R. Melillo, J.M., Peterson, T.C., eds., (2009). Global Climate Change Impacts in the United States. Cambridge: Cambridge University Press.

Kerlinger, F. N. (Ed.). (1986). *Foundations ofBehavioural Research* (3rd ed.). Holt, Rinehart and Winston.

Kohn, N. A. (2014). Vulnerability Theory and the role of Government. *Yale Journal of Law and Feminism*, *26*(1), 1–27.

Kombo, D. K., & Tromp, D. L. A. (2006). *Proposal and Thesis Writing: An Introduction.* Nairobi: Pauline Publication Africa.

Kothari, C. R. (2004). Research Methodology: Methods and Techniques. New Delhi: New Age International (P) Limited, Publishers.

Kumar, S., Kumar, V., Singh, H., Kumar, S., & Meena, R. M. (2016). *Climate Change: Challenges and Opportunities for the 21st Century*. Institute of Agricultural Sciences, Banaras Hindu University.

Lal, P., Alavalapati, J. R. R., & Mercer, E. D. (2011). *Social Economic Impacts of Climate Change on Rural United States*. USDA Forest Service/UNL Faculty Publications.

Lebni, J. Y., Gharehghani, M. A. M., Soofizad, G., & Khosravi, B. (2020). Challenges and Opportunities Confronting Female-Headed Households in Iran: A Qualitative Study. *BMC Women's Health*, *20*(1), 183. doi:10.1186/s12905-020-01046-x PMID:32807144

Lewis, P., Monem, M. A., & Impiglia, A. (2018). *Impacts of Climate change on Farming Systems and Livelihoods in the near East and North Africa-with a Special Focus on Small Scale Family Farming*. FAO.

Lukas, K. S. (2013). *Achieving Growth through Constant Implementation of New Professional Life Standard, Techniques and Innovative Ideas*. Nyambari Nyangwine Publishers.

Mago, P., & Gunwal, I. (2019). *Role of Women in Environmental Conservation*. University of New Delhi.

Mayesha, A., Bhatia, R., & Mawby, B. (2015). *Women and Climate Change: Impact and Agency in Human Rights, Security, and Economic Development*. Georgetown Institute for Women, Peace and Security.

McConnell, C., & Mupuwaliywa, M. (2015). *Policy Brief Zambia – Keeping Girls in School: Situation Analysis for Zambia*. World Bank. doi:10.1596/23865

Mehra, R., & Rojas, M. H. (2008). *A Significant Shift: Women, Food Security and Agriculture in a Global Marketplace. International Centre for Research on Women, citing FAO Focus on Women and Food Security*. FAO.

Mertz, O., Mbow, C., Reenberg, A., & Diouf, A. (2009). A Farmers' Perceptions of Climate Change and Agricultural Adaptation Strategies in Rural Sahel. *Environmental Management*, *43*(5), 804–816. doi:10.1007/s00267-008-9197-0 PMID:18810526

Mohammed, E. Y., & Uraguchi, Z. B. (2013). Impacts of Climate Change on Fisheries: Implications for Food Security in Sub-Saharan Africa. In M. A. Hanja (Ed.), *Global Food Security*. Nova Science Publishers, Inc.

Mpolomoka, D. L. (2021). Using Participatory Approaches to Improve Environmental Sustainability in selected parts of Zambia. *Mulungushi University Multidisciplinary Journal*, *2*(1), 1–11.

Mpolomoka, D. L. (2021). Using Participatory Approaches to Improve Environmental Sustainability in selected parts of Zambia. *Mulungushi University Multidisciplinary Journal*, *2*(1), 1–11.

Mpolomoka, D. L., Ikabongo, M. K., & Mushibwe, C. (2018). *Women's Education a Tool of Social Transformation: Experiences of female students in ODeL institutions of higher learning in Zambia*. Knowledge Solutions for the Society, the Economy and the Environment hosted by the Faculty of Education Complex, Egerton University, Njoro, Kenya.

Msabila, D. T., & Nalaila, S. G. (2013). *Research Proposal and Dissertation Writing: Principles and Practice*. Dar-es-Salaam: Nyambari Nyangwine Publishers.

Mulenga, B., & Wineman, A. (2014). Climate Trends and Farmers' Perceptions of Climate Change in Zambia. (Working Paper, Number 8). Lusaka: IAPRI.

Mwanangombe, N. S., Mpolomoka, D. L., & Banda, S. (2021). *An exploration of Financial Literacy of Women Marketeers Participating in Micro Finance Programmes in selected Communities in Lusaka, Zambia.* Paper presented at the 12th Bi-Annual Pan-African Literacy for All (PALFA) conference held at Mulungushi International Conference Centre (MICC), Lusaka, Zambia.

Mweemba, L. (2014). Climate Change in the Zambian Mind: Communicating Risk Perception of Climate Change and Variability in Zambia. *Zambia Social Science Journal*, *5*(1), 71–93.

Neumayer, E., & Plumper, T. (2007). The Gendered Nature of Natural Disasters: The Impact of Catastrophic Events on the Gender Gap in Life Expectancy, 1981-2002. *Annals of the Association of American Geographers*, *97*(1), 551–566. doi:10.1111/j.1467-8306.2007.00563.x

OPM. (2012). *The 2010-2011 Integrated Rainfall Variability Impacts, Needs Assessment and Drought Risk Management Strategy*. Department of Disaster Management.

Osman-Elasha, B. (2010). *Women ... in the Shadow of Climate Change*. United Nations.

Oxfam (Great Britain). (2002). *Natural Resource Management and Gender: A global Source Book*. KIT Publishers.

Pasqualino, M., Kennedy, G., Longley, K., & Thilsted, S. H. (2014). *Food and Nutrition Security in the Barotse Floodplain System*. GGIAR Research, WorldFish, Research Program on Aquatic Agricultural Systems and Biodiversity International.

Paudyal, B. R., Chanana, N., Khatri-Chhetri, A., Sherpa, L., Kadariya, I., & Aggarwal, P. (2019). Gender Integration in Climate Change and Agricultural Policies: The Case of Nepal. *Frontiers in Sustainable Food Systems*, *3*, 3. doi:10.3389/fsufs.2019.00066

doi:10.3389/fsufs.2019.00066

Poortinga, W., Fisher, S., Bohm, G., Steg, L., Whitmarsh, L. and Ogunbode, C. (2018). *European Attitudes to Climate Change and Energy: Topline Results from 8 of the European Social Survey*. London: European Social Survey.

Riche, B. (2007). *Climate Change Vulnerability Assessment in Zambia: IUCN, Climate Change and Development Project, Pilot Phase*. Geneva: IUCN. http://cmsdata.iucn.org/downloads/climate_change_vulnerability_assessment_zambia.pdf

Rochdane, S., Reichert, B., Messouli, M., Babqiqi, A., & Khebiza, M. Y. (2012). Climate Change Impacts on Water Supply and Demand in Rheraya Watershed (Morocco) with Potential Adaptation Strategies. *Water (Basel)*, *4*(1), 28–44. doi:10.3390/w4010028

Safa, H. (1998). Female Headed Households in the Caribbean: Sign of Pathology of Alternative Form of Family Organization? *The Brown Journal of World Affairs*, *5*(2), 203–214.

Salaria, N. (2012). Meaning of the Term-Descriptive Survey Research Method. *International Journal of Transformations in Business Management.*, *1*(6), 1–7.

Samuel, P. (2010). *Final report: Psychometric evaluation of the Family Quality of Life Survey for low-income, minority families*. Wayne State University, Detroit, MI.

Silvestri, S., Bryan, E., Ringler, C., Herrero, M., & Okoba, B. (2012). Climate Change Perception and Adaptation of Agro-Pastoral Communities in Kenya. *Regional Environmental Change*, *12*(4), 791–802. doi:10.1007/s10113-012-0293-6

Slegers, M. F. W. (2008). If only it would rain: Farmers' perceptions of rainfall and drought in semi-arid central Tanzania. *Journal of Arid Environments*, *72*(11), 2106–2123. doi:10.1016/j.jaridenv.2008.06.011

UNFCCC. (2007). *Climate Change: Impacts, Vulnerability and Adaptations in Developing Countries*. UNFCCC.

UNFPA. (2009). *State of the World Population: Women, Population and Climate Change*. UNFPA.

Wei, J., Zhang, Y., Hansen, A., Liu, Q., Sun, Y., & Bi, P. (2014). Perception, Attitude and Behaviour in Relation to Climate Change: A Survey among CDC Health Professionals in Shanxi Province, China. *Environmental Research*, *134*, 301–308. doi:10.1016/j.envres.2014.08.006 PMID:25199970

WFP. (2016). *The Impacts of Climate Change on Food Security and Livelihoods in Karamoja, Uganda*. WFP.

WFP. (2017). *Vulnerability Assessment and Mapping (VAM). VAM Bulletin Number 8*. WFP.

WHO. (2003). *Climate Change and Human Health: Risks and Responses*. WHO.

WHO. (2014). *Gender, Climate Change and Health*. WHO.

Wiseman, A. W. (2008). A Culture of (in)Equality?: A Cross-National Study of Gender Parity and Gender Segregation in National School Systems. *Research in Comparative and International Education*, *3*(2), 179–201. doi:10.2304/rcie.2008.3.2.179

doi:10.2304/rcie.2008.3.2.179

Wisner, B. (2004). *At Risk: Natural Hazards, Peoples' Vulnerability and Disasters* (2nd ed.). Routledge.

World Fish Center (2007). *The Threat to Fish and Aquaculture from Climate Change. Policy Brief.* Penang: WorldFishCenter.

Yavinsky, R. (2012). *Women more Vulnerable than Men to Climate Change*. Population Reference Bureau.

Zulu, D. and Sichikwenkwe, P. (2011). Zambia: The Impact of Climate Change on Women. *Times of Zambia Newspaper*. Ndola: Times of Zambia.

Chapter 18

Bamboo Production and Value Chain for Quality Life in the Era of Climate Change:
A Circular Way for Social Impact

Sneha Kumari
Symbiosis School of Economics, Symbiosis International University (Deemed), Pune, India

Havisha Agarwal
Symbiosis School of Economics, Symbiosis International University (Deemed), Pune, India

V. G. Venkatesh
https://orcid.org/0000-0002-5340-4637
E.M. Normandie Business School, France

Nandini Jain
Symbiosis School of Economics, Symbiosis International University (Deemed), Pune, India

Yangyan Shi
Macquarie University, Australia

ABSTRACT

Bamboo can lead to value-added products like furniture, baskets, containers, fishing materials, medicines, etc. The study answers: How does bamboo forestry impact the livelihood of the rural population, and How does bamboo production impact carbon capture and carbon trading? The aim of the study is to explore the impact of bamboo production on climate change, to study the impact of bamboo production on carbon captures, and to study the impact of bamboo cultivation on the livelihood of the rural population. The study is exploratory and based on an extensive literature review. Many carbon accounting records omit bamboo or do not include it inside forestry because bamboo is botanically a grass rather than a tree. Therefore, neither the Kyoto Protocol, the Marrakech Accords, nor the IPCC's definition of a "forest" sufficiently fits bamboo. There are several definitions of a forest. Threshold parameters, such as the minimum forest area, tree height, and degree of crown coverage, are, however, shared by most definitions. Motivated by the benefits of bamboo, the chapter is an attempt to understand the impact of bamboo forestry on the livelihood of the rural population, and the impact of bamboo production on carbon capture and carbon trading in the era of climate change.

DOI: 10.4018/978-1-6684-9863-7.ch018

INTRODUCTION AND BACKGROUND

Global warming is increasing at such a rate, that in the next 20-25 years, the Earth and the atmosphere will suffer to an extent where recovery would be difficult. Carbon Dioxide is one of the major factors that contribute to climate change and global warming (Salzer et al., 2016). There is a dire need to look into this matter and come up with solutions that effectively work without causing any other harm to the environment. Here's where bamboo production comes into the picture. Bamboo production allows carbon sequestration (Zhou et la., 2019). This means that it stores atmospheric carbon dioxide and helps reduce climate change (Lobovikov et al., 2012). This process happens naturally as well as by using technology. Apart from that its cultivation can be an optimal response to the current devastation of tropical forests (Molari & Garcia, 2021). Bamboo production can also help in carbon trading which will not only help in earning revenues for the carbon credit seller but also promote sustainable living practices. Carbon Trading (trading of emissions of six major greenhouse gases primarily CO2) is a market-based practice aimed at mitigating climate change (Perdan & Azapagic, 2011). India is the second-largest producer of bamboo after China. It covers nearly 13% of India's total forest area. Around 5 million tons of bamboo is produced in India per annum. According to a recent development, India is going to have its first bamboo-based bio-refinery which will start its operations in March 2024. It is going to incorporate bamboo as feedstock and is expected to produce 16,000 tons of ethanol, 16,000 tons of furfural, and 11,000 tons of acetic acid annually (Verma, 2023). India has introduced the *Carbon Credit Trading Scheme* to motivate industrial entities to bring about reductions in their carbon emissions. It is an effective carbon sink. About 17 tons of carbon can be absorbed per annum by one hectare of bamboo. A ton of carbon dioxide is equal to a single carbon credit and the value of a credit in the international market is about 22 Euros. An acre of bamboo can produce approximately 80 carbon credits which can have a value of up to Rs. 60,000. India can thus exploit the opportunity of selling carbon credits to industries and nations and can generate significant revenues. The amalgamation of all the aforementioned factors led to the selection of this specific topic.

The most significant constraints for climate change are uneven terrain, shallow soil depth, poor soil fertility, low water retention capacity, fragmented holdings, high-intensity rainfall (that causes severe soil erosion), frequently erratic distribution (prolonged dry spells), and minimal irrigation potential (10–12%). Deforestation significantly impacts rising atmospheric carbon stocks. The potential for forests to remove carbon from the atmosphere. As averted, deforestation is an expensive carbon abatement option, and there is rising interest in reducing deforestation to combat climate change (Gullison et al., 2007). The extensive carbon storage among terrestrial plant populations is found in forests, and the loss of this store through deforestation is responsible for about 17% of world emissions. (Eliasch, 2008). This is next to coal and oil combustion as the third most significant source of greenhouse gas (GHG) emissions (Brickell, 2009). Bamboo plantations can contribute significantly to climate change (Xu et al., 2023), a significant carbon-storing contributor owing to its rapid expansion and biomass addition both above and below ground (Gu et al., 2019). They are a valuable biomass resource for various reasons, including their ability to adapt to different soils and climates, short harvesting period, and wide range of applications. India produces approximately 11.4 million ha of bamboo which makes up around half of all of Asia's bamboo land (Sudhakara & Jijeesh, 2015). The increasing number of global development projects has led to an increase in the use of bamboo. Owing to the great potential for sustainable production, bamboo could be a significant net sink for Carbon Dioxide (CO2) sequestration (Sudhakara & Jijeesh, 2015). A life cycle analysis of commercial bamboo commodities suggests that they could be considered carbon-free with efficient technology. The study found a negative correlation between daily carbon fixation and atmospheric carbon dioxide levels, indicating that as carbon dioxide concentration decreases, daily carbon fixation increases. Third-party certification is necessary for bamboo products to receive any green designation, as these products need to be thoroughly reevaluated. Bamboo is a versatile plant used for furniture, papermaking, fuel, cloth production, tableware, packaging, medicines, and healthcare products (Flynn et al., 2017). Less than ten suppliers of bamboo flooring existed in the United States in the late 1990s; by 2005, there were around 200, and imports totalled roughly 45 million sq. feet (Bowyer, et al., 2014). Culms take approximately 40 days to reach their maximum heights. During this time, significant gains in biomass and carbon storage were found. Moso bamboo accumulated three-fourths of its biomass in just forty days (Yen, 2016).

One of the characteristics of bamboo that makes it a valuable resource for people is its rapid growth (Dwivedi et al., 2019; Chen et al., 2021. It is also sometimes known for its capacity to absorb and store atmospheric carbon, which helps to ameliorate climate change. Indian Forest Act of 1927 considered bamboo a tree and was removed from that category in 2017. Bamboo considered a grass, can be grown in forests for livelihood opportunities for tribal women. Looking at the benefits of bamboo, there is a need to focus on bamboo production and its opportunities. Bamboo forest ecosystems can be used to mitigate climate change while also offering other services for livelihood. Not only can bamboo production help adapt to climate change, but also, the byproducts of bamboo can be a source of livelihood for tribal people. Bamboo can lead to value-added products like furniture, baskets, containers, fishing materials, medicines, etc. Training provided to rural women led to the production of bamboo products. This has led to the creation of business opportunities in bamboo. Most industries rely on natural materials like bamboo, which has harmful carbon emissions. Bamboo, sometimes referred to as "Green Gold," is a great biomass resource for carbon sequestration because of its fast growth and yearly addition of biomass both above and below ground. As a result, it can operate as a substantial carbon sink. India, which accounts for around half of all bamboo land in Asia, is a major producer of bamboo and a key contributor to reducing climate change and global warming. When it comes to carbon-storing capacity and long-term viability bamboo has significant benefits over tree species due to its rapid growth patterns, flexibility to a variety of soil and climatic conditions, and sustainable production. Motivated by the benefits of bamboo, the chapter is an attempt to understand the impact of bamboo forestry on the livelihood of the rural population, and the impact of bamboo production on carbon capture and carbon trading in the era of climate change

This chapter is followed by sections that explore climate change, bamboo production, the role of bamboo as a carbon footprint, bamboo: a key source of bioenergy and environmental management, bamboo: a foundation for socio-economic development, the value chain for quality life, livelihood importance for chain actors, the impact of bamboo value chain on economic, social, and environment for quality life and different sources of livelihood created that lead to sustainable development.

RESEARCH METHODOLOGY

In the era of climate change, the present chapter helps to address the research question viz. How does bamboo forestry impact the livelihood of the rural population, and How does bamboo production impact carbon capture and carbon trading in the era of climate change? The study aims to understand the impact of bamboo forestry on the livelihoods of the rural population and carbon trading. The study is based on an extensive literature review. The study captures secondary sources to understand the bamboo value chain and its impact on sustainability. Bamboo forestry impacts sustainable development goals and contributes to social, economic, and environmental development. It leads to climate change adaptation and mitigation once recognized as a part of forestry (Lobovikov et al., 2009).

CLIMATE CHANGE

One of the most pressing issues facing the globe is climate change. Due to the concern among leaders from different countries, the average global temperature increase from 1880 to 2012 of 0.85 degrees Celsius, the rise in CO2 emissions, and the average global sea level rise of 19 cm between 1901 and 2010 have earned their place in the list of global SDGs. One of the main contributors to global climate change is the alteration of the atmospheric composition brought about by people (Karl & Trenberth, 2003). One of the critical actions made in 2016 to encourage sustainable growth and lower emission levels in the area was the creation of the Pan-Canadian Framework on Clean Growth and Climate Change (Shaw et al., 2014). The European Union has made tremendous efforts to address the issue of climate change in European countries. The European Climate Law established a framework for combating climate change in European nations, which came into effect on 29th July 2021. This strategy aims to attain climate neutrality by 2050 and a net reduction of 55% in greenhouse gas emissions by 2030 (UNESCAP, n.d.). With stricter goals and standards for energy, transportation, taxes, and climate policy, the European Commission has proposed a package of freshly modified EU climate and energy laws. China recently declared that it would achieve carbon neutrality by 2060. Korea and Japan have each promised to achieve the same goal by 2050. To speed up the transition to a low-carbon economy, changes must be made to land use, energy use, and consumer behavior. In low-income nations, a degree increase in temperature is expected to cause a 1.4% decrease in per capita income (Dell et al., 2014).

KYOTO PROTOCOL

This report uses modelling and comparison studies to decipher that even though bamboo in botanical terms is a grass and not a tree, it can still sequester carbon faster than rapidly growing trees if actively managed (Yiping et al., 2010). The report also mentions the "Kyoto Protocol" which gives mechanisms to decrease the cost of attaining emission targets and has also created two categories of countries with different obligations for trading of emission credits. Since bamboo is not considered as a tree, it is also not included in the Kyoto Protocol. If it had been included, it could have acted as a leading factor in climate change mitigation (Yiping et al., 2010). The Kyoto Protocol (1970) enforces the United Nations Framework Convention on Climate Change and requires industrialized nations and transitioning economies to limit and decrease GHG emissions based on agreed targets. By incorporating village bamboo into traditional home garden schemes, farmers in Northeast India have the possibility of obtaining carbon credits via the Kyoto Protocol's Clean Development Mechanism (CDM). In fact, these bamboo gardens have been shown to store a total of 50.1 t ha−1 of carbon, providing a valuable opportunity for local communities (Pan, et al., 2023). Markets both domestically and internationally that have made a commitment to lessen their carbon footprint are able to exchange certified emission reductions (Nath, Lal, & Das, 2015).

CARBON TRADING

Carbon trading incentivizes industrial houses to reduce their carbon emissions and take advantage of selling those excess credits (Paul, 2011). It has been believed that India can be one of the biggest sellers of carbon credits in the near future. Currently, there are two commodity exchanges that are trading in carbon markets. India is a big player in this market and has great future potential in this sector, but it cannot be achieved without planning and implementation. The role of bamboo in carbon credits and as a carbon sequestration component is great. Bamboo forests can play a crucial role in sequestering carbon from the atmosphere more effectively than other methods (Pan, et al., 2023).

BAMBOO: AN OPPORTUNITY

Climate mitigation policies promoting Nature-based Solutions (NbS) with low value of biodiversity, like non-native monoculture afforestation, may lead to maladaptation, particularly in a world that is changing quickly where biodiversity-based resilience is critical (Seddon, et al., 2020). The development of bamboo production in Italy on environmental as well as ecological analysis has been explored through a case study. In the ecological analysis, the study found that since bamboo doesn't require much use of pesticides and also purifies the air, it can represent the future of agriculture. Economic analysis led to the finding that bamboo represents a great opportunity for the farmers as well as afforestation projects. Its production can also promote carbon trading or the sale or purchase of carbon credits in the international markets. These credits can act as compensation for the opportunity costs of alternative land uses (Lombardo, 2022). The demand for forest products is increasing rapidly, and reforestation with the help of fast-growing species is important. Bamboo is one of the fastest-growing plant species. A study is thus done using five species of bamboo and their growth performance is assessed. Bamboo has the potential to rehabilitate degraded land and it produces more biomass than many other tree species (Getahun et al., 2023). In addition to providing many people with flexible resources for their livelihoods, bamboo helps people adapt to and lessen the effects of climate change. It would be possible to offer low-interest loans and include carbon credits, techniques for encouraging bamboo management with financial rewards. Global population change is affecting everyone, but in the end, the poorest will suffer. Additional policy thought should be devoted to the ways that carbon systems of credit can take the place of conventional financial assistance in reducing poverty (Wu et al., 2014). bamboo cultivation can be helpful in absorbing carbon in the soil, bamboo biochar is an effective carbon sink and it reduces the emission of carbon (Gupta, 2013). One of the major reasons why bamboo is set apart from other forest trees is that it is a fast-growing grass as compared to the slow-growing forest trees. This makes it a great source in helping reduce the carbon footprints. It was found that the aboveground biomass for the 2 species, namely Bambusa tulda and Dendrocalamus longispathus have a great potential in sequestering carbon. Under the Kyoto Protocol, carbon sequestration by bamboo forests can be considered for Clean Development Mechanism Projects (Nath et al., 2009). This is helpful in eliminating poverty and environmental degradation. Bamboo has potential in Nigeria as a sustainable crop and cheap substitute for various supplies, but lack of awareness among citizens and lawmakers hinders its growth. Addressing this could help resolve issues like soil erosion; which is a matter of concern in many regions of Nigeria. Nigeria has a high deforestation rate, making it crucial to explore various methods to address the issue. One promising solution is bamboo forests, which have high biomass productivity and are capable of sequestering substantial amounts of carbon (Isukuru et al., 2023). Bamboo offers a unique opportunity to restore and rejuvenate land tarnished by crude oil pollution in the Niger Delta. Artisanal refining of crude oil, oil bunkering and spillage have left the region's lands barren and contaminated with crude oil. However, the use of bamboo presents a promising solution for remediation (Isukuru et al., 2023).

BAMBOO PRODUCTION FOR QUALITY LIFE IN THE ERA OF CLIMATE CHANGE

One of the key characteristics of bamboo that makes it a valuable resource for people is its rapid growth (Figure 1). It is also known for its capacity to absorb and store atmospheric carbon, which helps to ameliorate climate change. The amount of greenhouse gases released into the atmosphere can be managed by extending the life of resources and increasing biodiversity. Bamboo can make a significant contribution in a world with rising demand for timber and adaptation to climate change. Global regulations must change to favor more practical and environment-friendly manufacturing techniques to meet the growing commodity demand. To accommodate this demand, it will be crucial to manage forests sustainably. The use of bamboo can significantly reduce the demand for forestry resources. For instance, bamboo has become more widely accepted as a potential timber substitute in many industries historically dominated by lumber in China since national logging prohibitions on forests took effect in 1998. Applying bamboo to paper, packaging, furniture, and flooring demonstrates how adaptable bamboo is as a more sustainable alternative material. Given the rising atmospheric carbon dioxide levels, the population has relied on forestry to quantify different forest species performing as sinks (the carbon absorbed and emitted) and as carbon stores. Bamboo cultivation has resulted in income-generating activities like handicrafts (Phimmachanh et al., 2015).

Figure 1: Bamboo production
Source: Taken by Authors

ROLE OF BAMBOO AS A CARBON FOOTPRINT

Bamboo has unique features that significantly impact CDM (*Clean Development Mechanism*) and REDD+ (*Reducing Emissions from Deforestation and Forest Degradation*) projects. They are naturally found in tropical and sub-tropical regions across the globe, except Europe, and are abundant in Asia-Pacific and parts of South and Central America. Bamboos have extensive growth potential in these regions, with a potential forest area estimated to be around 4.06 billion hectares (FAO, 2020). The morphology and growth pattern of bamboo differentiates them from other trees. They comprise aboveground culms, underground rhizomes, and dense, fibrous roots. The two main types of rhizomes, monopodial and sympodial, determine the growth behavior of bamboo species. Culms, the aboveground shoots, elongate rapidly and reach their maximum height within a few months. Unlike trees, bamboo culms continue to acquire biomass and increase in density even after their initial growth phase.

Additionally, the underground biomass of bamboo can increase significantly, making them advantageous for carbon sequestration in afforestation and reforestation projects. Physiologically, bamboos are C3 plants, similar to most timber trees. This characteristic makes them capable of increasing biomass productivity and carbon sequestration in response to a CO2-enriched atmosphere. However, research in this area is limited, and other factors beyond biomass productivity influence carbon stocks and flows in ecosystems. Bamboos have ecological significance beyond their carbon services. They provide habitats for various animal species, including the giant panda, red panda, and numerous bird species. Bamboo protects against natural disasters like earthquakes, floods, and tsunamis. However, they can be invasive and cause slope failures in some instances. Bamboo also measures carbon. Carbon sequestration and green electricity production compensate for transport, energy, and glue emissions. Bamboo plants include 1500 species worldwide (Gu et al., 2019).

BAMBOO: A KEY SOURCE OF BIOENERGY AND ENVIRONMENTAL MANAGEMENT

Bamboo is a primary source of electricity, wherein woody biomass around the world accounts for 13% of the world's primary energy. It has an annual biomass growth that is way higher than pine forests. It is a crucial source for electricity generation given its fuel properties, like low ash content and alkali index. It has a low heating value, and due to its growth being substantially high in density and global abundance, it is being used for pyrolytic production. Its excellent fuel qualities, like high caloric values, volatile content, and low moisture, make it worthwhile for biofuel generation. Another major area where bamboo has the potential to serve as a highly efficient, natural, and low-cost product is carbon sequestration due to its high carbon collecting and storage potential. Its decay-resistant litter allows it to be a carbon sink in the future which would aid it in increasing its forest cover and, at the same time, capture high levels of CO2, which is almost 2-4 times the rate of tropical rainforests and pine plantations. Bamboo cultivation also has substantial implications for reducing land degradation, which has positive externalities as it is beneficial for the environment and the people relying on land-based activities, hence reducing poverty. The restoration cost is timely and also financially burdening as well as the potential returns after restorations are slow, hampering investments.

BAMBOO: A FOUNDATION FOR SOCIO-ECONOMIC DEVELOPMENT

Bamboo also comes in as a source of socio-economic development because of its faster growth rate, lesser management, and requirement of less expertise in production, allowing it to provide a livelihood to several rural low-income groups (Rathour et al., 2022). The waste produced from the sheath and basal section is helpful in the production of pulp during the processing of bioproducts. The shoot tips and young shoots are a great source of nutrient content and have health benefits from bioactive compounds. Bamboo is highly significant in producing various value-added products, such as for medicinal applications, biochar, and composite materials. Its medicinal properties make it useful in pharmaceutical preparations to regulate diabetes, cholesterol, and hypertension. It serves inflammatory, antioxidant, and antiulcer properties, along with its burned root and bark for treating arthritis, bleeding gums, and skin eruptions. It also aids as an antimicrobial, antifungal and helps in weight loss.

Interestingly, a natural flavonoid found in black bamboo also serves against the Covid-19 virus, SARS-CoV-2. Its anti-cancer properties and regulation of high glucose levels make it a perfect food crop for cultivation as a medicinal plant. Further, it can also be used as a soil supplement for crop production, enhancing soil fertility and reducing microbial activities. It has improved water holding capacity and moisture retention and balances the Ph. Bamboo has made a place in the construction markets of India, South America, Africa, and parts of Asia due to its elasticity, wood-like strength, elegance, and lightweight. It is of specific use in earthquake-prone areas. The high amount of silica present in it allows for preventing fire. Moreover, its low cost, biodegradable and long-lasting substance makes it environmentally sustainable. Chairs, tables, and sofas, besides the handicraft items such as baskets, lampshades, and other household utilities made from bamboo, provide the rural economy with long-term benefits. Bamboo also serves as an improvement over the problem of food scarcity for both human and livestock diets. The shoot of the bamboo tree is considered a very healthy food in many Asian countries, including India, because of its high vitamin, mineral, carbohydrates, and low-fat content. It also has the advantage that it can help solve the problem of obesity as it is a rich source of high fibre regulating cholesterol and vitamin requirements. It has been found through studies in countries like Kenya that the use of bamboo for food storage helps to alleviate poverty and nutritional deficiencies (Rathour, 2022). Its healthy content makes it a significant ingredient in the sweetener, sugar, and pharmaceutical industry for preparing antibacterial products.

BAMBOO VALUE CHAIN FOR QUALITY LIFE

The value chain of bamboo involves multiple stages, including bamboo harvesting, processing, manufacturing, and marketing. Harvesting bamboo requires proper management to ensure its sustainable use and conservation—the bamboo value chain results in the strengthening and durability of bamboo products. The processed products help in furniture, flooring, shoots, handicraft, biofuels, and other construction work. It helps improve rural livelihood, employment creation, and sustainable development. Bamboo helps in watershed management, reducing soil erosion and carbon sequestration. Their widespread growth, distinct morphology, physiological traits, environmental contributions, and socio-economic benefits highlight their potential for sustainable development and carbon mitigation efforts. Bamboo Produce can be used to make medicines, paper, fish nets, beehives, containers, furniture, etc. The study will help adopt mitigation practices for climate change and the value chain for bamboo. Bamboo is underutilized in the upcoming *Reducing Emissions from Deforestation and Forest Degradation* (REDD+) instrument and the *Clean Development Mechanism* (CDM) of the *Kyoto Protocol*. In both CDM and REDD+ projects, bamboo has the potential to be used in addition to wood trees, benefiting smallholder livelihoods, rural development, and climate adaptation in addition to carbon sequestration. Bamboo can be harvested regularly with a long shelf life. Bamboo can also be used as a source of renewable energy. It can be used for ceilings, floorings, gardens, and outdoor use like decking. Bamboo strips laminated can produce panels and high-end products. Bamboo is allocated in mills and pressed together. This leads to the storage of biomass. Laminated strips, compressed strips, and stranded woven bamboo are the byproducts.

LIVELIHOOD IMPORTANCE FOR CHAIN ACTORS

The bamboo value chain is a significant source of income for many households, particularly in areas with geography-specific availability of bamboo. The bamboo chain is heavily male-dominated, reflecting the physical labor involved in harvesting, transporting, and processing bamboo. In contrast, other *non-timber forest product* (NTFP) chains have more female participation. The dependence on bamboo varies among actor groups. Harvesters have more diversified incomes, with bamboo contributing a smaller proportion. However, for craftspeople, bamboo forms a significant part of their income, reflecting their specialization and investment in the trade (Ingram & Tieguhong, 2013). Bamboo is consciously included as one of the multiple livelihood capitals. There are many reasons for the acceptance of bamboo as trees within the *United Nations Framework Convention on Climate Change* (UNFCCC). While bamboos are technically grasses, some species exhibit tree-like characteristics, reaching heights and diameters comparable to traditional trees. Their woody biomass, density, and carbon concentrations resemble those of timber trees.

Furthermore, forests that include bamboo provide diverse ecological and social services in addition to carbon storage. The *Intergovernmental Panel on Climate Change* (IPCC) and the *Food and Agriculture Organization* (FAO) already classify bamboo as trees in specific contexts and guidelines. The *Clean Development Mechanism* (CDM) Executive Board recognizes the equivalence of palm trees and bamboo to trees in afforestation/reforestation projects. However, current negotiations on REDD+ do not include references to bamboo, potentially leading to the neglect of significant carbon storage, viable livelihoods, and climate change mitigation and adaptation opportunities. Accepting bamboo as trees under UNFCCC and REDD+ is argued to be crucial for sustainable development and climate action.

IMPACT OF BAMBOO VALUE CHAIN ON ECONOMIC, SOCIAL, AND ENVIRONMENT FOR QUALITY LIFE

Bamboo processing positively impacts the economy by generating business activities for small and medium enterprises. Bamboo benefits in generating electricity, roads, and building other community development projects to improve livelihoods. Bamboo farms have helped in the migrations of plants and species. Infrastructure, livelihood, and income generation have been supported by bamboo production. Bamboo Value Chain has helped rural communities and has been considered a poor man's timber. It helps in the transformation of the livelihood patterns of rural communities. Bamboo is easy to manage by individual farmers with special assistance during harvesting. It offers an ecosystem for holding the soil, reducing soil erosion. The soil is held in place even after harvesting. Bamboo helps in reducing water runoff (Akwada & Akinlabi, 2016). Bamboo has impacted local forests, resource conservation, and biodiversity. Bamboo production has led to regulating water levels and biodiversity. Bamboo is a fast-growing species and can sequester carbon. It also releases 35 percent more oxygen as compared to other trees.

FUTURE OF BAMBOO

Bamboo is a significant crop in India due to its potential as a bio-energy source, and its ability to aid in ecological restoration. The article explores potential avenues for future research and other benefits of bamboo plantations for the economy, society, and the environment. In its 2015 Nationally Determined Contribution (NDC) to the United Nations Framework Convention on Climate Change (UNFCCC), India set a goal to reduce greenhouse gas emissions by 33–35% by 2030 (Rathour, et al., 2022). India's demand for bamboo could lead to significant employment opportunities in rural areas. Studies suggest that 28 million metric tons' demand could produce 98,000 million man-days of work. However, a case study in Purnia, Bihar, shows that only 15% of people in bamboo-related industries earn more than 143.14 USD (Rathour, et al., 2022). The bamboo industry needs to be encouraged through incentives to expedite its expansion and development. By encouraging bamboo-based land-use systems that have a greater rate of carbon storage, Southeast Asia's underdeveloped and degraded land can be exploited for carbon investment. Bamboo agroforestry systems are a suitable model for CDM projects under the Kyoto Protocol since they can convert and conserve these areas successfully (Nath & Das, 2012). The paper found that as soil depth increased, the soil carbon pool showed a decline. Notably, there was a significant difference in carbon sequestration capacity among different species. In the case of the vegetation pool, a higher capacity for carbon sequestration was observed (Kashung, et al., 2022). According to the UN Food and Agriculture Organization (FAO), China had a bamboo resource of just over 5.71 million hectares in 2010. The majority of this bamboo was grown on farmer-managed plantations (Buckingham, Wu, & Lou, 2013).

Research explores the difficulties associated with generating alternative natural resources through an analysis with a focus on bamboo. The study focuses on the feasibility of implementing certification from the Forest Stewardship Council, confirming the legality of timber, and lowering emissions from deforestation and damage to forests in the bamboo sector.

Bamboo was classified as a 'tree', which handed the forest bureaucracy power over the lucrative $225 million trade. As a result, harvesting, trading, or selling of bamboo became a government monopoly (Buckingham, Wu, & Lou, 2013). The advancement of future bamboo carbon projects can be aided by a precise estimation of the product carbon pool (Pan, et al., 2023).

CONCLUSION

Over the past few years, there has been a move towards an advanced industrial society, where developments such as population rise, technological breakthroughs, mechanization, and an upgradation towards a world where countries have been experiencing high levels of economic growth have become a reality. While economic growth and a higher standard of living are being enjoyed, it is not hidden that humans are overutilizing the resources available to them, and the harsh consequences of such exploitation are rising pollution, depletion of natural and renewable resources, and inability to meet basic needs by people with low incomes. Thus, there is a need for bold steps to be taken in order to achieve sustainable development goals for future generations. The research emphasizes that among the many critical areas that need to be addressed, increasing energy requirements is one of the most crucial ones, especially when there has been continuous depletion of fossil fuels, highlighting the role of bioenergy production. It focuses on how important it is to produce bioenergy without compromising on the food reserves available to ordinary people and, at the same time, shifting towards sustainability. It stresses the multifunctionality of bamboo crops and their potential to serve not only as a source of environment management but also as an economically and socially appropriate crop. Diversification in energy production to upsurge the share of renewable energy resources for clean and efficient energy has been targeted in different countries, with a special focus on renewable energy. It has been projected that such a target can be achieved only with bioenergy contributing substantially as a renewable energy source. Forest land resources provide the opportunity for the development of bioenergy, but such an expanding bioenergy plantation can be harmful to hydrosphere and lithosphere resources, posing a threat to food production. The focus is on finding such sources of energy which have a balanced effect on biodiversity. The potential benefits of bamboo for bioenergy production, environmental management, land restoration, its value addition in the production of a variety of commodities, medicinal uses, utility as a source of food, and economic benefits allow India the second-largest producer of bamboo to call it as an energy crop and using it for developing biomass feedstock.

REFERENCES

Akwada, D. R., & Akinlabi, E. T. (2016, July). Economic, social, and environmental assessment of bamboo for infrastructure development. In *International conference on infrastructure development in Africa* (pp. 1-15).

Bowyer, J., Fernholz, K., Frank, M., Howe, J. E. F. F., Bratkovich, S. T. E. V. E., & Pepke, E. (2014). *Bamboo products and their environmental impacts: revisited*. Dovetail Partners, Incorporated.

Brickell, C. D., Alexander, C., David, J. C., Hetterscheid, W. L., Leslie, A. C., Male'cot, V., Jin, X., & Cubey, J. J. 2009. International code of nomenclature for cultivated plants, 8th ed. Scripta Horticulturae.

Buckingham, K. C., Wu, L., & Lou, Y. (2013). *Can't See the (Bamboo) Forest for the Trees: Examining Bamboo's Fit Within International Forestry Institutions*. Springer.

Chen, M., Weng, Y., Semple, K., Zhang, S., Jiang, X., Ma, J., & Dai, C. (2021). Sustainability and innovation of bamboo winding composite pipe products. *Renewable & Sustainable Energy Reviews*, *144*, 110976. doi:10.1016/j.rser.2021.110976

Dell, M., Jones, B. F., & Olken, B. A. (2012). Temperature shocks and economic growth: Evidence from the last half century. *American Economic Journal. Macroeconomics*, *4*(3), 66–95. doi:10.1257/mac.4.3.66

Dwivedi, A. K., Kumar, A., Baredar, P., & Prakash, O. (2019). Bamboo as a complementary crop to address climate change and livelihoods–Insights from India. *Forest Policy and Economics*, *102*, 66–74. doi:10.1016/j.forpol.2019.02.007

Eliasch, J. (2008). *Climate change: financing global forests: the Eliasch review*. Earthscan.

FAO. (2020). *The State of the World's Forests*. FAO. https://www.fao.org/state-of-forests/2020/en/,.

Flynn, A., Chan, K. W., Zhu, Z. H., & Yu, L. (2017). Sustainability, space and supply chains: The role of bamboo in Anji County, China. *Journal of Rural Studies*, *49*, 128–139. doi:10.1016/j.jrurstud.2016.11.012

Getahun, A., Kebede, Y., Mulatu, Y., & Adamu, F. (2023). *Growth and biomass production of five exotic bamboo species in North-western Ethiopia. Bahir Dar*. Elsevier.

Gu, L., Zhou, Y., Mei, T., Zhou, G., & Xu, L. (2019). Carbon footprint analysis of bamboo scrimber flooring—Implications for carbon sequestration of bamboo forests and its products. *Forests*, *10*(1), 51. doi:10.3390/f10010051

Gullison, R. E., Frumhoff, P. C., Canadell, J. G., Field, C. B., Nepstad, D. C., Hayhoe, K., Avissar, R., Curran, L. M., Friedlingstein, P., Jones, C. D., & Nobre, C. (2007). Tropical forests and climate policy. *Science*, *316*(5827), 985–986. doi:10.1126/science.1136163 PMID:17495140

Gupta, A. (2013). *Bamboos in Mitigating Climate Change. International Union for conservation of Nature*. International Union for Conservation of Nature. https://www.iucn.org/our-work/nature-based-solutions

Ingram, V., & Tieguhong, J. C. (2013). Bars to jars: Bamboo value chains in Cameroon. *Ambio*, *42*(3), 320–333. doi:10.1007/s13280-012-0347-5 PMID:23015325

Isukuru, E. J., Ogunkeyede, A. O., Adebayo, A. A., & Uruejoma, M. F. (2023). Potentials of bamboo and its ecological benefits in Nigeria. *Elsevier B.V.*, 01-11.

Karl, T. R., & Trenberth, K. E. (2003). Modern global climate change. *science, 302* (5651), 1719-1723.

Kashung, Y., Gap, N., Debbarma, D., Das, B., Paul, A., Singha, L. B., & Tripathi, O. P. (2022). Carbon stock and pool assessment in tropical and subtropical bamboo brakes of Arunachal Pradesh, India. *Vegetos*, *36*(2), 668–674. doi:10.1007/s42535-022-00436-w

Lobovikov, M., Lou, Y., Schoene, D., & Widenoja, R. (2009). The poor man's carbon sink: bamboo in climate change and poverty alleviation. *Non-Wood Forest Products Working Document*, (8).

Lobovikov, M., Schoene, D., & Yping, L. (2012). Bamboo in climate change and rural livelihoods. *Mitigation and Adaptation Strategies for Global Change*, *17*(3), 261–276. doi:10.1007/s11027-011-9324-8

Lombardo, E. (2022). An overview of bamboo cultivation in Southern Italy. *Advances in Bamboo Science, 100002*.

Molari, L., & Garcia, J. (2021). On the radial variation of the transverse mechanical properties of bamboo. *Build. Eng.*, 33-34.

Nath, A. J., & Das, A. K. (2012). Carbon pool and sequestration potential of village bamboos in the agroforestry system of northeast India. *Tropical Ecology*, 287–293.

Nath, A. J., Lal, R., & Das, A. K. (2015). Managing woody bamboos for carbon farming and carbon trading. *Global Ecology and Conservation*, *III*, 654–663. doi:10.1016/j.gecco.2015.03.002

Nath, J. A., Gitasree, G. D., & Ashesh, D. K. (2009). *Aboveground standing biomass and carbon storage in village bamboos in North East India*. Elsevier.

Pan, C., Zhou, G., Shrestha, A. K., Chen, J., Kozak, R., Li, N., & Wang, G. (2023). *Bamboo as a Nuture-Based Solution (NbS) for Climate Change Mitigation: Biomass, Products, and Carbon Credits*. MDPI.

Patel, B., Gami, B., & Patel, P. (2017). Carbon sequestration by bamboo farming on marginal land and sustainable use of wood waste for bioenergy: case studies from Abellon clean energy. In *Wood is Good: current trends and future prospects in wood utilization* (pp. 451–467). Springer Singapore. doi:10.1007/978-981-10-3115-1_43

Paul, A. (2011). Carbon Credit and Carbon Trading: An Overview. *Business Studies*, 136-146.

Perdan, S., & Azapagic, A. (2011). Carbon trading: Current schemes and future developments. *Energy Policy*, *39*(10), 6040–6054. doi:10.1016/j.enpol.2011.07.003

Phimmachanh, S., Ying, Z., & Beckline, M. (2015). Bamboo resources utilization: A potential source of income to support rural livelihoods. *Applied Ecology and Environmental Sciences*, *3*(6), 176–183.

Rathour, R., Kumar, H., Prasad, K., Anerao, P., Kumar, M., Kapley, A., Pandey, A., Kumar Awasthi, M., & Singh, L. (2022). Multifunctional applications of bamboo crop beyond environmental management: An Indian prospective. *Bioengineered*, *13*(4), 8893–8914. doi:10.1080/21655979.2022.2056689 PMID:35333141

Salzer, C., Wallbaum, H., Lopez, L. F., & Kouyoumji, J. L. (2016). Sustainability of social housing in Asia: A holistic multi-perspective development process for bamboo-based construction in the Philippines. *Sustainability (Basel)*, *8*(2), 151. doi:10.3390/su8020151

Seddon, N., Chausson, A., Berry, P., Girardin, C. A., Smith, A., & Turner, B. (2020). *Understanding the value and limits of nature-based solutions to climate change and other gloabl challenges*. The Royal Society Publisher.

Shaw, A., Burch, S., Kristensen, F., Robinson, J., & Dale, A. (2014). Accelerating the sustainability transition: Exploring synergies between adaptation and mitigation in British Columbian communities. *Global Environmental Change*, *25*, 41–51. doi:10.1016/j.gloenvcha.2014.01.002

Sudhakara, K., & Jijeesh, C. (2015). *Bamboos: Emerging carbon sink for global climate change mitigation. Thrissur*. Department of Forestry.

UNESCAP. (n.d.). *Climate Action*. ESCAP. https://www.unescap.org/sites/default/d8files/event- documents/SDG%2013%20Goal%20Profile_Final%20Web_EDIT%20v3.pdf, as retrieved on 18th March 2023.

Verma, N. (2023). *India's first bamboo-based biorefinery to start operations in March*. Reuters.

Wu, W., Liu, Q., Zhu, Z., & Shen, Y. (2014). *Managing Bamboo for Carbon Sequestration Bamboo*. Springer.

Xu, P., Zhu, J., Li, H., Xiong, Z., & Xu, X. (2023). Coupling analysis between cost and carbon emission of bamboo building materials: A perspective of supply chain. *Energy and Building*, *280*, 112718. doi:10.1016/j.enbuild.2022.112718

Yen, T.-M. (2016). Culm height development, biomass accumulation and carbon storage in an initial growth stage for a fast-growing moso bamboo (Phyllostachy pubescens). *Botanical Studies*, *57*(1), 10. doi:10.1186/s40529-016-0126-x PMID:28597419

Yiping, L., Yanxia, L., Henley, K., & Zhou, G. (2010). *Bamboo and Climate Change Mitigation:a comparative analysis of carbon sequestration.* International Network for Bamboo and Rattan.

Zhou, G., & Xu, L. (2019). Carbon footprint analysis of bamboo scrimber flooring—Implications for carbon sequestration of bamboo forests and its products. *Forests*, *10*(1), 51. doi:10.3390/f10010051

Chapter 19
Climate Change and Humans With a Cyborg Scientific Theory Approach

Fatma Aydogmus
https://orcid.org/0000-0003-1434-2143
Faculty of Science, Istanbul University, Turkey

ABSTRACT

The signs of climate change are increasing rapidly. There is no doubt that an irreversible transformation that may occur in climate change will have important consequences for our world's habitat. The effects that may occur from the global economy to social and human structures are closely followed by scientists, and theories and models are tried to be produced on what may happen. In addition, it is tried to predict the effects of climate change on the quality of life (QoL) of individuals and communities. In this book chapter, the signs of global climate change and their causes will be summarized, and whether the temporal evolution of climate change that has taken place throughout world history is chaotic or not will be examined. The conditions for the emergence of this evolution process after the chaos caused by global warming will be discussed. Climate change and its possible effects will be examined within the framework of a cyborg science theory, which is put forward as the postmodern science of the future.

INTRODUCTION

The effects of climate change have become a priority that cannot be ignored in every platform that concerns life. Especially in recent years, it was determined that excessive consumption of fossil fuels and other sources due to rapid energy consumption have caused significant increases in the amount of greenhouse gases in the atmosphere, there has been an increase in the frequency and impact of extreme weather events such as drought, floods, severe hurricanes, the melting of glaciers and the retreat of the poles, glaciers and snow cover on the tops of high mountains are decreasing, the sea water levels increased, the tropical plants and fish that love warm air and water spread towards the poles, the delicate bird species that are sensitive to airborne pollutants decrease, and age rings on trees grow faster (NOAA

DOI: 10.4018/978-1-6684-9863-7.ch019

Report, 2023; NASA, 2023). These and similar symptoms are considered as the most important signs of climate change in the scientific world (Kadıoğlu, 2001; NOAA Report, 2023; NASA, 2023).). The effects of climate change have become a priority that cannot be ignored in every platform that concerns life.

From the results of geological and stratigraphic research, it has been understood that there are significant climate changes in our world until today. However, even if the global warming in the last 1400 years is considered as a normal development of the periodic background, it can be thought that the rapid increase in recent years has reached a speed that can disrupt the periodic background of the temporal evolution of global warming. Considering that even an average of 6° Celsius change in temperature in the northern hemisphere causes serious changes in the climate, this situation, which proves how serious the sensitization in climate change can be, also shows us how much the life of the world and the Quality of Life (QoL) of living things are under threat (Sumer, 2022).

The importance of complying with the temperature change value determined with 1.5°C in the Paris Agreement in order to keep global warming under control is also discussed in the report issued by the IPCC (IPCC, 2018). Limiting warming to 1.5°C would require halving emissions by 2030 and achieving net zero emissions by 2050 (Rogelj et al., 2018). However, studies show that this value can be exceeded. If an increase is observed above this value, it is predicted that it may cause undesirable natural events.

Especially after Lorenz's studies, it was understood that the atmosphere is an autonomous system. From the periodic background characteristic of the change in global warming, it can be concluded that the temporal evolution of the future temperature change will also exhibit periodic behavior (Berkmen, 2007). As can be seen from the logistic map and many chaotic maps, there is a periodic motion before every chaotic behavior (Valle et al., 2022). In other words, global warming accelerated by the increase of greenhouse gases in the atmosphere outside the natural cycle in recent years may turn this periodic background into a chaotic structure. Considering the temporal evolution in global warming, as well as all other factors that cause climate change, it is possible that a complex structure that may emerge will reach a critical value (the threshold of chaos) that will trigger a leap in climate change, which is in a sensitive relationship with the environment. Irregular fluctuations in climate change symptom data will herald the potential leap in climate change. Such a chaotic leap will seriously affect the QoL of individuals and communities in rural and urban environments around the world, and may transform the earth's atmosphere into a new structure with a different global balance, and life on earth may evolve according to this structure (Aydogmus et al., 2007).

Since we cannot know when and how the conditions for reaching the threshold of chaos will occur due to the temporal evolution of climate change and its multi-parameter structure, it is difficult to predict the leap and its consequences in climate change, which will turn into a complex structure after the threshold of chaos. However, the irregularities in the climate change symptom data, which are necessary for the emergence of such a dangerous situation, have not been observed until now (Patterson et al.,1995; Aydogmus et al., 2017). The linear appearance of symptoms against the rapid increase in global warming, which is one of the most important factors triggering climate change, shows that it will be possible to develop predictive theories on the effects of climate change on various natural structures and systems in the future (Jifan, 1989). Even in the periodic situation that chaotic leap is not possible, humanity will face another danger due to climate change; In addition to digital technology and communication (media) mechanisms, climate change will further facilitate the control of the machine-human hybrid (cyborg) body (Haraway, 1991; Ghan, 2019). taking on new states and shaping. However, in addition to mental differentiation, geographical imbalances due to climate change in the production of energy may cause irreversible breaks between cyborgized human societies. One of the most important reasons for man's quest to shape himself with a machine (cyborgization) is to balance bodily behavior against the increase in energy use. Since our world has endless energy resources such as nuclear, solar and wind energy, it will be inevitable that the rate of machines in the cyborg will increase rapidly as the increase in energy production continues. For this reason, the projects of shaping human cyborgization and developing technologies and methods to control cyborg relations have been transferred to a new science, cyborg science, which accepts the computer as a paradigm (Mirowski, 2002; Kaswan et al., 2023).

The signs of climate change and its effects on the QoL of living things are followed with great concern. In this framework, mathematical theories and simulation models are tried to be developed that can predict future global as well as local negative developments in the scientific world (Kaper et al., 2013). In this book chapter, the important signs of climate change that have been detected so far will be briefly given and the dynamic of climate change and chaos will be summarized, considering that the rapid increase in global warming, which behaves as if it has a periodic background, in recent years has reached a speed that can disrupt the periodic background of the temporal evolution of global climate change. With the effect of climate change, small mental differences in the bodily behaviors of the machine-human hybrid (cyborg) can cause unpredictable emergences and unstable fluctuations that may occur around these emergences will be criticized with a scientific theoretical approach (Kaswan et al., 2023).

SOME IMPORTANT SIGNS AND EFFECTS OF CLIMATE CHANGE

Some signs and effects of global climate change (IPCC, 2001;2007;2013;2014; Koçak, 2005; NOAA, 2023):

Rising Sea Water Levels

Two-thirds of cities with populations over 5 million are located on low-lying coastlines. Flooding combined with rising sea levels and storm swell could pose a threat to commerce, jobs and livelihoods, with far-reaching effects on populations, property, ecosystems and the QoL of individuals and communities.

Rising Temperatures and Global Warming

Increasing and excessive amounts of gas have been released into the atmosphere from fossil fuels and other sources consumed in the last 150 years, and there have been significant increases in the amount of greenhouse gases (Olivier, 2017, NOAA, 2021;2022;2023, NASA, 2023). The greenhouse process, which has become stronger in the atmosphere due to these increases, has also revealed the current global warming problem. Global warming is a term that expresses only one of the potential effects of increasing greenhouse gases in the atmosphere. In other words, one of the most obvious symptoms of current climate change is warming. Average temperature rise in some cities may be more than 4°C by 2100, with high seasonal upper temperatures. Also, according to NOAA's report, the highest monthly sea surface temperature anomaly (+1.03°C or +1.85°F) was observed in August 2023 (NOAA, September 2023). Warmer days can lead to more heat-related health problems and possibly air pollution, worsening urban heat island effects and QoL.

Extreme Weather Events

As a result of the increase in heavy rains, it can cause sewer flooding leaks and increase the sewage volume by 40% in some cities. Uncontrolled urban development can often make inland flooding worse (NOAA, 2021;2022).

Water Scarcity

Risks to freshwater resources, such as drought, can lead to water shortages in urban areas, electricity shortages related to water power and facility cooling, an increase in water-related diseases, and adverse development such as increased food prices due to reduced supply and food security issues. These may lead to negative economic effects and an increase in rural-urban migration.

Decrease in Food Production

Decreased food production and loss of productive arable land are likely in many regions. All aspects of food security, including food access, use and price stability, are potentially affected by climate change. Climate change is expected to cause a decline in food production in some regions due to warming and acidification (NOAA, 2023).

Melting Glaciers

Glaciers have been retreating worldwide since the end of the Little Ice Age, but in recent years they have begun to melt at rates that cannot be explained by historical trends. Climate change predicted in the next century will further affect the rate of melting of glaciers. Average global temperatures are expected to rise by 1.4-5.8°C by the end of the 21st century. Simulations predict that a 4°C increase in temperature would remove nearly all of the world's glaciers. Also, according to the NOAA report, the melting of glaciers has increased by a record amount in 2023. Sea ice extent in August 2023 has decreased by approximately 550,000 square miles from the previous record low in August 2019 (NOAA, August 2023). Continued significant melting of glaciers will seriously affect QoL, threatening coastal communities and habitats, leading to floods, water shortages for millions of people, and sea level rise.

The Effect of Climate Change on Some Animal and Plant Species

All scientific studies and observations show that the world is experiencing a remarkable and rapid decline in biodiversity (WWWF Report, 2015). As a result of monitoring more than 10,000 vertebrate populations, it was determined that these populations decreased by 52 percent between 1970 and 2010. The International Union for Conservation of Nature (IUCN) estimates that 35 percent of bird species, 52 percent of amphibians and 72 percent of reef corals will be vulnerable to the effects of climate change in the next century. In the 5th assessment report prepared by the Intergovernmental Panel on Climate Change (IPCC), it was stated that climate change as a result of human activities intensifies the current pressure of humans on ecosystems and wildlife. The spread of some plant and fish species that love hot air and water and live in tropical climates towards the poles and reduction of delicate bird species sensitive to airborne pollutants are important signs of climate change (IPCC, 2001;2007;2013;2014, NOAA, 2021;2022;2023, NASA, 2023).

Faster Growth of Age Rings in Trees

Trees are very sensitive to local climatic conditions such as rain and temperature. The unusualness in the growth rate of age rings in trees is another important sign of climate change.

ATMOSPHERE AND CHAOS THEORY

Systems whose properties change over time in accordance with certain rules are called dynamic systems. The atmosphere is also a dynamic system. A table with summary information about the history of dynamics is given below (Aydogmus, 2013; Rizaoglu, 2011):

In cases where time is continuous, we can express a dynamic system with systems of ordinary differential equations. During the evolution of a dynamic system, it is very important that the rules to which it is subject are nonlinear. We know that linear models can give limited information about the structure of nature. The real structure of nature conforms to nonlinear features (Scott, 1999; Aydogmus, 2013; 2022; Rizaoglu, 2011).

The analytical solution of nonlinear differential equations is often impossible. Since nonlinear equations do not have exact solutions, phase diagrams and attractors give information about nonlinear dynamic structure; Numerical methods and the phase spaces of the solutions obtained from these methods are applied to have a view on the dynamics and evolution of nonlinear equation solutions. The variables of the system construct the phase space. We represent each state of the system as a point in the phase space of the system. As the system evolves over time, the points representing the system in phase space define a curve. This curve is called the phase curve (trajectory) (Aydogmus, 2013;2022). Every point in the phase space can be taken as an initial condition for the system. As such, the phase curves of the system fill the phase space. The collection of phase space points on a given phase curve is known as an orbit. The collection of phase trajectories starting from different phase points and reflecting the characteristic behavior of the system is called the phase portrait of the system. Since analytically obtaining the phase

Table 1. Dynamic: A historical overview

1600-1700s	Newton	Invention of analysis and explanation of planetary motion, Classical Mechanics
1800s		Studies on planetary motion
1890s	Pioncare	Sprouting of chaos
1920-1960s	BirkhoffKolmogorovArnol'dMoser	Complex behaviors in mechanics
1963s	Lorenz	Chaos studies
1970-2000s	Ruelle&Takens, May, Feigenbaum, Winfree, Mandelbrot	Widespread interest in turbulence, the logistics map, phase transitions, fractals, oscillators and their applications

trajectories for a nonlinear dynamic system is out of the question, it is tried to obtain the phase portrait directly from the properties of the system's variables. Equilibrium points with different properties, periodic phase trajectories and their stable/unstable manifolds play an important role in drawing phase portraits. The phase portrait gives us information about the structure of the phase space as a whole. Phase portraits are constructed in the framework of x^* fixed points defined by $f(x^*)=0$ and these represent equilibrium solutions (Aydogmus, 2022). The stability of these solutions, that is, whether they maintain their condition after a small perturbation, is important. The stability of these solutions is related to the behavior of the phase curves in their immediate vicinity (Strogatz, 1994; Tel, 2006; Aydogmus, 2013; 2022; Rizaoglu, 2011).

It is called bifurcation when the system changes its behavior depending on the change of control parameters and shows another long-time behavior. In the analysis of nonlinear systems, there is a simple differential equation system that simulates the behavior of the system around a bifurcation point. There are different types of bifurcations (Saddle-Node Bifurcation, Transcritical Bifurcation, Pitchfork Bifurcation, Hopf Bifurcation..) (Strogatz, 1994).

The phase curves of a system can look very complex. In this case, in order to talk about the behavior of the system, simplifications are made so that the characteristics of the behavior of the system continue to be reflected. If there is no periodic external effect in the system, the phase trajectory is not tangent to the line/plane, and the points where it intersects a certain line or a plane while always passing in the same direction are taken into account. All of these points are called the Poincaré section of the phase orbit (Strogatz, 1994; Tel, 2006).

The numbers that characterize the behavior of any dynamic system, obtained as a result of experiment or by iteration or in any way, are called the time series of the system. These are the values that the variables we use to determine the system can take. We can see the change of the time series by plotting these numbers according to time or iteration step. If the change in question does not occur according to any order, but in a completely random and irregular manner, the behavior of the system determined by these numbers may show chaos. There are a very large number of numerical values in the time series. Therefore, it is not easy to understand whether the time series contains any order by looking at the above-mentioned graph. To understand the existence of order, if any, in the time series, a dimensionless quantity called the autocorrelation coefficient of the system is defined (Szemplinska, 1988; Strogatz, 1994; Tel, 2006, Aydogmus, 2013;2022).

Chaos is defined as the finite and random behavior of a deterministic dynamical system, which shows extreme sensitivity to initial conditions and whose physical parameters such as amplitude and period are not predicted to evolve over time. Systems that are not deterministic and have random external effects are stochastic systems. Chaotic behavior is non-periodic, unpredictable, and sensitive to initial conditions. According to chaos theory, a simple, predictable set of entities and laws; can have a complex and unpredictable outcome. It is an important feature of chaos that very small changes lead to larger changes. Examples of this; Chaos occurs in many places, be it the weather, the stock market, a dripping faucet, or sometimes the beating of the human heart (Strogatz, 1994; Aydogmus, 2013;2022).

Understanding the mechanisms behind the chaos is not just the behavior of a floating leaf, irregular heartbeats and a dripping faucet; It will also help us understand many aspects of the universe on both the small and large scales. Because of this feature, chaos theory has taken its place in all branches of science. It is possible to find beautiful examples of chaos in mathematics. Solving seemingly simple problems reveals extremely complex behavior. Although simple, the solution of such nonlinear problems has only become easier with the introduction of computers. Therefore, chaos can be called the science of the computer age. Not all nonlinear problems are chaotic, but all chaotic problems are nonlinear (Strogatz, 1994; Koçak, 2005; Aydogmus, 2013).

A system in which a completely deterministic and completely random process exists in the same place is a complex system. A big reason for complexity is uncertainty. This claim is based on the assumption that the definition of a more complex system should be more detailed. It means that it contains more variables and may not be known all at once; therefore information is lacking and there is uncertainty. Complexity is not an intrinsic property of the system. Complex systems are not in equilibrium. Complex problems invariably involve nonlinearity: nonlinear events can be regular or irregular. Simple equations can exhibit very complex dynamics. If simple things can seem so coiled up, it's possible to understand why real situations are so complex. The complexity expresses itself as follows (Cambel, 1993):

1) Complexity can exist in natural and man-made systems and social structures.
2) Complex dynamical systems can be very large, very small, or both.
3) Its physical shape can be regular or irregular.
4) It can be in energy-conserving or dispersive systems.
5) The system is neither completely deterministic nor completely ordinary. (Both can happen at the same time).
6) The causes and effects of events in the system are not proportional
7) The different parts of these systems are chained and affect each other synergistically.
8) There is positive or negative pushback.
9) Level of complexity; It depends on the character of the system, its environment and the natural interaction between them.
10) Complex systems are open so; They exchange matter, energy and information with those around them.
11) They tend to be irreversible (irreversible) processes.
12) It is dynamic, not balanced. They are similar to travel, not to the destination, and can track the moving target.
13) Most complex systems are not well mobile and undergo frequent changes, indicating that their functional relationships are not differentiated.
14) There are paradoxes, fast and slow events, regular and irregular forms, organic and inorganic structures.

Not all complex systems are chaotic. The dynamics of the complex system can change. Depending on the dynamic stability: it can be fixed, temporary or chaotic. A system that is nonlinear, unstable, deterministic, dynamic, sensitive to initial conditions, has strange attractors and positive Lyapunov coefficients is called chaotic (Cambel, 1993; Strogatz, 1994). In order for chaos to occur, the nonlinear equations must have the properties of being sensitive to the initial conditions of the phase trajectories, as well as the properties of staying inside a finite phase volume. Two phase trajectories starting from two extremely close points will diverge exponentially from each other after a certain point due to the sensitive dependence on the initial conditions. This is achieved by the rapid extension of these two curves. If the curves did not lengthen, they would not be able to diverge rapidly from each other. In this case, they have nothing to do but to return to stay in the finite phase volume, that is, to follow the points close to the points they came from. It should be noted here that, due to determinism, one phase trajectory cannot intersect itself and two different phase trajectories do not intersect each other. Under these conditions, the phase trajectories will fit into the finite phase region as the nonlinearity of the system allows. Thus, it cannot be predicted how the internal structure of the ball will be formed. Since the phase points represent the states of the system, unexpected situations may follow each other. The extension of the phase curve is known as stretching and its reversal in a finite phase volume is known as folding (Cambel, 1993; Strogatz, 1994, Aydogmus, 2022).

The disorder in chaos due to the presence of the attractor is not random disorder. The existence of the attractor means that the phase trajectories representing the long-term behavior of the system are locked onto low-dimensional phase subspaces. In the most general case, there are three types of attractors: point attractors, periodic attractors, and strange attractors. If the attractors are fractal, they are called strange attractors. The structure of a fractal formation is independent of scale. In other words, there are similarities of the same motif in every scale in the formation in question. Mandelbrot, one of the founders of chaos theory, discovered fractals, the geometry of the chaos landscape, in the 1960s, emphasizing the contradiction of the regular forms we use in geometry with the real World (Cambel, 1993; Strogatz, 1994).

For a system to be chaotic (Strogatz, 1994; Tel, 2006);

- The system should be nonlinear and the time series should be irregular.
- The behavior of the system should be sensitive to initial conditions.
- The system must have a strange attractor; usually this means it will have fractal dimension.
- Lypanov coefficients must be positive.
- Kolmogrov entropy must be positive in dispersive systems.

The theory and models, which are in linear structure, do not fully comply with the structure of nature. Nature is nonlinear. Therefore, theories and models in nonlinear structure are of great importance in understanding and examining the real structure of nature. Biologists see chaos in, for example, changing insect and bird populations, the spread of epidemics, and cell metabolism. Physicists again encounter chaos in the motion of the electron, in the atoms of molecules, in gases, and in the theory of elementary particles. On the other hand, engineers working in different fields now take chaos into account in their designs. Atmospheric scientists are faced with the existence of chaos in atmosphere and climate events. The study of the movements of the atmosphere, which determines the weather, constitutes one of the most exciting application areas of chaos theory. Meteorologists try to predict what the weather will be like in the future using highly complex mathematical models that explain the motion of the atmosphere. The atmosphere has an unpredictable character by nature.

Weather regimes can be determined numerically from the records of past weather conditions. Such problems aroused the interest of meteorologist Edward Lorenz. Lorenz, in his studies at the Massachusetts Institute of Technology (MIT) (Lorenz, 1963), saw that the atmosphere behaves like a turbulent fluid. He expressed the atmosphere with a differential equation system that is nonlinear and sensitive to initial conditions.

Lorenz also saw that sensitive dependence on initial conditions would make weather forecasting a difficult problem. To make sure his ideas were correct, he tried to make the equations simpler, provided the basic features remained the same. As a result of these studies, he reduced the complex system of equations governing the behavior of the fluid to a simple model with only three variables x, y and z. In this model, air at a given moment is represented by a point in three-dimensional phase space, and the course of air in time is represented by a trajectory passing through these points. The set of possible weather conditions obtained as a result of running the model is called the Lorenz attractor. The Lorenz dynamical system has only three variables, namely three degrees of freedom (Lorenz, 1963). Therefore, it cannot be expected to fully explain the behavior of the real atmosphere. However, increasing the degree of freedom will increase the quality of the predictions to be made. Today, the degrees of freedom of the models used in weather forecasting are expressed in millions.

Chaos theory is an important theory that shapes the prediction of climate change, which is a very sensitive subject, in the coming years. In this theory, the aim is not to predict an individual trajectory on the climate attractor, but to determine the shape of an entire climate attractor and its position in phase space, for example, in the event of an increase in greenhouse gases. In the case of a very small perturbation, it is important to predict whether or how the climate attractor will be affected (Palmer, 1998;2000; Aydogmus, 2013;2022).

Considering the hammer as a whole, questions such as whether there will be a significant change in its shape and location, indicating a devastating weather situation that has never been seen in today's climate, are critical questions that need to be answered. Chaos theory plays an active role in the decisions taken for climate change. An interdisciplinary theory of chaos, understanding the behavior of the earth's atmosphere that surrounds and protects us, with the help of computer technology, will continue to play an important role in understanding the dynamics and evolution of the climate change process that threatens the earth and its QoL.

DYNAMIC OF CLIMATE CHANGE

The results of the geological and stratigraphic researches show that the ice age ended 16,000 years ago, and our world experienced a rapid warming period of 2,000 years between 16,000 and 14,000 years ago. Afterwards, after a period of periodic re-cooling and warming, there was a very rapid cooling of 1,000 years between 11,000 and 10,000 years ago (Patterson, 1995).

Climate is the meteorological situation that emerges as a result of effects such as temperature, precipitation, precipitation pattern, humidity, air pressure and wind observed in a region for many years. Weather is a term for shorter-term data, while climate is obtained by averaging long-term data. Climate remains among the extremes in this long run. Climate change is related to the change in average conditions, and more importantly, the increase and continuity of the renewal and renewal frequency of extreme values. Weather forecasting is one of the first value problems. The climate change problem focuses on the statistics of long-term weather forecasts and explores how certain climate forces affect these statistics (Palmer, 1998;2000). Extreme situations that have an impact on the climate have existed throughout history. However, the occurrence and frequency of extreme situations have changed dimensions with the emergence of developments in the field of industry. After the industrial development, a rapid increase in the emission of greenhouse gases has been observed. CO_2 emissions, one of the most important greenhouse gases (Olivier, 2017), have reached 410 ppm levels, and the global temperature has increased by 1.01°C since 1880 (IPCC 5 2014; NOAA, 2023).

However; Even if the global warming in the last 1400 years is considered as a normal development of the periodic background, it can be thought that the rapid increase in recent years has reached a speed that can disrupt the periodic background of the temporal evolution of global warming. Considering that even an average of 6 degrees Celsius change in temperature in the northern hemisphere causes serious changes in the climate, this situation, which proves how serious the sensitization in climate change can become, also shows us how much of a threat the world's life and QoL are under.

Studies show that the temporal evolution of global warming is periodic, and it can be interpreted that future changes will also exhibit periodic behavior. From the periodic background characteristic of global temperature change, it is normal to conclude that the temporal evolution of the future temperature change will also exhibit periodic behavior (Berkmen, 2007). The fact that the atmosphere is understood to be an autonomous system also indicates that this result may be valid for today. However, as is known, the climate has a dynamic that is affected by many external factors other than warming. Also, as can be seen from the logistic map, there is a periodic movement before every chaotic behavior (Lorenz, 1963; Strogatz, 1994). In other words, climate change accelerated by the increase of greenhouse gases in the atmosphere outside (Olivier, 2017) the natural cycle in recent years may turn this periodic background into a chaotic structure. The important thing here is whether a complex structure that may emerge can catch a critical value (ridge of chaos) that will trigger a jump in climate change, which is in a sensitive relationship with the environment. Signs of climate change need to be monitored very closely. Because the irregular fluctuations that will appear in these symptom data will herald the leap in climate change. Such a chaotic leap may cause the end of life on earth or transform the earth's atmosphere into a new structure with a different global balance, and life on earth will evolve according to this structure (Berkmen, 2007; Aydogmus et al., 2007; 2008; 2015; 2017; 2022). In IPCC reports, it is seen how this situation makes the QoL difficult and measures that can be taken in this regard are indicated (IPCC, 2013;2014). However, if the measures are not fully implemented, the future effects of climate change are a matter of curiosity. The extent of the climate change in the next century is still not known exactly, and this uncertainty is also related to the errors in the weather forecasts (Palmer, 1998;2000).

CYBORG SCIENCE THEORY

Since we do not know when and how the conditions for reaching the ridge of chaos we mentioned in the previous section will occur, it is impossible to predict the leap and its consequences in the climate change that will turn into a complex structure after the ridge of chaos. However, the irregularities in the climate change symptom data, which are necessary for the emergence of such a dangerous situation, have not been observed so far (Patterson, 1995). Despite the rapid increase in global warming, the linear appearance of its symptoms will show that it will be possible to develop predictive theories on the effects of climate change on various natural structures and systems in the future. Even in the periodic situation where chaotic leap is not possible, humanity will face another danger due to climate change; In addition to digital technology and communication (media) mechanisms, climate change will further facilitate the control of the machine-human hybrid (cyborg) body (Haraway, 1991; Kaswan et.al.,2023) taking on new states and shaping. However, in addition to mental differentiation, geographical imbalances due to climate change in the production of energy will cause irreversible breaks between cyborgized human societies.

One of the most important reasons for man's quest to shape himself with a machine, that is, cyborgization, is to balance the bodily behavior against the increase in energy use. Since our world has endless energy resources such as nuclear, solar and wind energy, it will be inevitable that the rate of machines in the cyborg will increase rapidly as the increase in energy production continues. For this reason, the projects of shaping human cyborgization and developing technologies and methods to control cyborg relations have been transferred to a new science that accepts the computer as a paradigm, namely cyborg science (Mirowski, 2002; Philip, 2002). What is expected from cyborg science here is to remove people from the rapid money flow control in the economy, especially in military studies, and from space and high energy experiments where big money is spent, and to search for ways to replace a computer-captive human-machine hybrid, that is, a cyborg.

However, we know that it is not enough to evaluate the machinic shaping to be imposed on the human body in this century as the alienation of man's relationship with himself and nature and the cunning control of his body. For example, if complexity is accepted as the paradigm instead of reductionism, as the human body becomes mechanized, small mental differences may cause unpredictable manifestations (Akdeniz, 2007). For example, according to the DDIDD theory, the symptoms of the cyborgized body as a simulator will be more closely tied to the reality principles of modernity. This addiction will increase the importance of the incomprehensibility of human spiritual disorders, the mind will cease to be a dimension and become a complex situation. Elements of this complexity will be structures such as education, culture, tradition, spatial harmony. The mental differences that arise from these will give unpredictable results in the simulacrums that will be formed on the cyborgized human body. DDIDD simulation theory (Akdeniz, 2007), against the conclusion that simulacra that may arise in the bodily behavior of the cyborg cannot be excluded from the simulation theory of Baudrillard (Baudrillard, 2004), which accepts the discourses of modernity as the reality principle; He says that mental differences can produce emergences in the cyborg that modernity cannot predict.

More energy consumption will increase the symptoms of climate change in parallel, and accordingly, human cyborgization process will naturally fluctuate unstable due to climate change. To give a concrete example of the anxiety that may occur with the effect of climate change:

Observations show that sea water level is increasing continuously and regularly with global warming. This increase in sea water level indicates that global warming is out of chaotic structuring and that the periodic structure is preserved (Aydogmus, 2007). However, the rapid increase in sea water level will try to balance the human cyborg body. This, in turn, will destruct the human body, which has become cyborg, in the direction of living in water. The human body will evolve to continue living within the framework of the Cyborg science theory. Thinking that this process will accelerate with our world's ability to contain infinite energy, cyborg scientists have started to work to take this shaping and formation under control. Apart from these impositions, we know that the first cyborg hero in science fiction was the man living under the sea (Jean de La Hire, 1909). However, there will be significant differences in the behavior of these cyborg fish-men due to mental differences. The problem for cyborg science is the unforeseen cyborg fish men and their oppositions that can arise in these differences.

CLIMATE CHANGE, QUALITY OF LIFE (QoL), AND CYBORG

The geography of the world has changed many times in the period until humans appeared on the stage of history. Today, it is accepted in the entire scientific community that there is a deterioration in the world climate system. Changes observed in the climate cause changes that will affect people's health and the entire Quality of Life (QoL), such as the places where people can live, the suitable living environments of plants, other ecosystems, the development environments of new types of businesses, the condition of buildings and infrastructure.

According to the scenarios put forward by the IPCC, it is predicted that there will be an average increase of 1 to 3,5 degrees in global temperature by 2100 (IPCC, 2018). As a result; there will be significant consequences reaching disaster dimensions that will affect the QoL resulting from sea level rise, temperature and precipitation anomalies. Today, we are exposed to many of these consequences, such as floods, overflows, drought and desertification, storms, and epidemics of biological origin that affect the QoL.

As is known, the atmosphere has a dynamic structure that exhibits chaotic behavior and is sensitive to initial conditions. The formation of a periodic movement in front of chaotic behavior and the periodic background feature of global climate change prepare the ground for a potential leap in climate change. This potential jump will also affect some physical behaviors. In addition, technological developments emphasize machines and hardware operating with artificial intelligence as means of existence. Along with these developments, climate change will play a role in controlling the evolution of the machine-human cyborg body (Haraway, 1991; Kaswan et al., 2023) into new situations. Geographic imbalances due to climate change will also cause significant divisions among cyborgized human societies, and geographical factors will play an important role in the QoL. As a result of all these developments, our world is living in a new era that has not yet been named. In this new era, the effects of climate change on the QoL in global, regional and urban areas are felt as much as possible. In addition, with the cyborgization process accelerating as a result of increasing energy consumption, classifications such as male-female, old-young, good-bad, living-inanimate are also disappearing. People of the modern world are metaphorically transgendered (Baudrillard, 2004). Donna Haraway says that this metaphor will obscure separate categories and identities, thus completing the machine-human hybrid, that is, the cyborg ((Haraway, 1991).

CONCLUSION

In summary; the rapid increase in global warming, which has been known to last for 1400 years, may over-sensitize the irregular but autonomous feature of the atmosphere. An unforeseen emergence in climate change, which will become irregular with this sensitization, may integrate two different scales after a ridge of chaos. This new complex whole; it could even cause a global catastrophe, that is, the end of life on our earth. Or, a new structure that will form in the atmosphere, which will favor survival after this ridge of chaos, may enable living things to evolve by leaps and bounds. However, for these to occur, irregularities must be observed in the signs of climate change (Aydogmus et al., 2007; 2013; 2017). Due to the rapid global warming in recent years, no deterioration was observed in the linear and regular increases in the symptoms. This shows that climate change can cause deconstruction in the human body. Cyborg scientists continue to work in order to predict the effects of climate change to the process of human cyborgization and QoL.

It is understood that the possible effects of climate change in the process of possible cyborgization may be outside the prediction limits of cyborg science. With the influence of developing technology, cyborgized bodies in birth, life control and normal life processes can bring the QoL to a completely different dimension. Because with these effects, the cyborg may have mental differences that cannot be covered by reality principles and cannot adapt to rapid energy consumption. These differences may create unforeseen manifestations not in the shape of the body, but in its behavior.

Whether it is the end of life with a chaotic ridge that may arise in climate change or a cyborgization process of human beings, the only solution that can prevent their emergence is to minimize all negative human-induced factors that cause climate change. Studies show us that in addition to the measures to be taken to stop climate change, it is also aimed to meet the increasing energy need in the world with productions such as solar and wind energy, which are among the endless energy sources. Discussions about climate change and its effects on the Quality of Life have become a priority that cannot be ignored on every platform of life, and the symptoms of climate change need to be closely followed with great seriousness.

REFERENCES

Akdeniz, G. (2007). *DDİDD Simulation Theory*. Gedizak deniz. www.gedizakdeniz.com

Aydogmus, F. (2013). *Phase Space Behaviors of Spinor Type Instanton Attractors*. Istanbul University, Institute of Science and Technology.

Aydogmus F. (2022). Synchronization in Chaotic Dynamics of Spinor-type Fermionic Instantons for Different Dimension. *International Journal of Dynamics and Control, 2022*.

Aydogmus, F., & Akdeniz, G. (2007). Is Global Climate Change Chaotic? *Nuclear Energy Symposium*. Istanbul University Faculty of Forestry.

Aydogmus, F., & Akdeniz, G. (2008). Global Climate Change And Human With A Syborg Science Doctrin, IV. *International Symposium on Sciences of Atmosphere*, Istanbul Teknik University, pp.437-443.

Aydogmus, F., & Tosyali, E. (2015). *Global Climate Change in Consideration of Chaos Theory*. 9th International Physics Conference of the Balkan Physical Union – BPU9, Istanbul.

Aydogmus, F., & Tosyali, E. (2017). The Effect of Chaos Theory on Predictability of the Atmosphere. *The Middle East Journal*, 26–31.

Baudrillard, J. (2004). *Transparency of Evil*. Ayrıntı Publications.

Berkmen, H. (2007). From Cosmos to Chaos and From Chaos to Cosmos, Irregular Systems: Theory and Practices Working Group VII. *National Symposium*.

Cambel, A. B. (1993). *Applied Chaos Theory a Paradigm for Complexity*. Academic Press.

Ghan, B. (2019). *Hybrids of Mind and Body: The Forms and Freedoms of the Cyborg in Posthumanist Science Fiction*. University of Toronto.

Haraway, D. (1991). A Cyborg Manifesto: Science, Technology, and Socialist-Feminism in the Late Twentieth Century. In *Simians, Cyborgs and Women: The Reinvention of Nature* (pp. 149–181). Routledge.

Jean de La Hire. (1909). *L'Homme Qui Peut Vivre dans l'Eau*. The Man Who Could Live Underwater.

Jifan, C. (1989). Predictability of the Atmosphere. *Advances in Atmospheric Sciences*, *6*(3), 335–346. doi:10.1007/BF02661539

Kadıoğlu, M. (2001). *The End of the Weather You Know - Global Climate Change and Turkey*. Current Publishing.

Kaper, H., & Engler, H. (2013). *Mathematics and Climate*. doi:10.1137/1.9781611972610

Koçak, K. (2005). *Human Factor in Climate Change*. ITU. https://web.itu.edu.tr/~kkocak/iklimpdf.pdf

Lorenz, E. (1963). Deterministic nonperiodic flow. *Journal of the Atmospheric Sciences*, *20*(2), 130–141. doi:10.1175/1520-0469(1963)020<0130:DNF>2.0.CO;2

Mirowski, P. (2002). *Machine Dreams: Economics Becomes a Cyborg Science*. Cambridge University Press.

Olivier, J. G. J., Schure, K. M., & Peters, J. A. H. W. (2017). *Trends in global CO2 and total greenhouse gas emissions. 2017 Report*. PBL.

Palmer, T. N. (1998). Climate change from a nonlinear dynamical perspective. *Journal of Climate*, *12*, 575–591. doi:10.1175/1520-0442(1999)012<0575:ANDPOC>2.0.CO;2

Palmer, T. N. (2000). Uncertainty in Forecast of Weather and Climate. *Reports on Progress in Physics*, *63*(2), 71–116. doi:10.1088/0034-4885/63/2/201

Patterson. (1995). Foraminiferal Evidence of Younger Dryas Age Cooling on the British Colombia Shelf. *Geography et Quaternaries*, *49*(3), 409.

Philip, M. (2002). *Machine Dreams: Economics Becomes a Cyborg Science*. Cambridge University Press.

Qo, S. K., Dhatterwal, J. S., Baliyan, A., & Rani, S. (2023). *Cyborg Human and Machine Communication Paradigm*. CRC press.

Report, I. P. C. C. (2001). *Intergovernmental Panel on Climate Change*. Swiss.

Report I. P. C. C. (2007). Synthesis Report. Contribution of Working Groups I, II and III to the Fourth Assessment Report of the Intergovernmental Panel on Climate Change. Switzerland.

Report I. P. C. C. (2013). The Physical Science Basis. Contribution of Working Group I to the Fifth Assessment Report of the Intergovernmental Panel on Climate. Cambridge University Press, Cambridge, United Kingdom and New York.

Report, I. P. C. C. (2014). *Intergovernmental Panel on Climate Change*.

Report, I. P. C. C. (2018). *Intergovernmental Panel on Climate Change*.

Rizaoglu, E. (2011). *Classical Mechanics*. Seçkin Press.

Rogelj, J., Shindell, D., Jiang, K., & Beşta, S. (2018). Mitigation Pathways Compatible with 1.5°C in the Context of Sustainable Development. *IPCC SR1 5*, ss. 93-174.

Scott, A. (1999). Nonlinear Science. *Oxford University Pres, Oxford, 01985017*, 2.

Strogatz, S. H. (1994). *Nonlinear Dynamics and Chaos: with applications to physics, biology, chemistry, and engineering*. Perseus Books.

Sumer, G. (2022). The Problem of Climate Change in Terms of Different Disciplines, Kahramanmaraş Sütçü İmam University. *Journal of Social Sciences*.

Szemplinska Stupnicka, W., & Iooss, G. ve Moon F. C. (1988). Chaotic Motions in Nonlinear Dynamical Systems. Springer Verlag Wien.

Tel, T., & Gruiz, M. (2006). *Chaotic Dynamics*. Cmabridge University Press. doi:10.1017/CBO9780511803277

Chapter 20
Home Gardening:
Enabling Grassroots Efforts in an Urban Socialist Society, Penang

Pavitira Manogaran
https://orcid.org/0000-0002-3434-9505
Universiti Malaysia Kelantan, Malaysia

Nur Aliah Mansor
https://orcid.org/0000-0001-7485-594X
Universiti Tunku Abdul Rahman, Malaysia

ABSTRACT

The COVID-19 pandemic brought about social distancing, self-isolation, and movement control orders, leading to an increased demand for food and disruptions in the supply chain. Consequently, home gardening gained popularity as a productive and environmentally friendly activity, inspiring Penang residents to grow their greens. The state government and organisations like Think City and the Consumers' Association of Penang recognised the importance of urban gardening initiatives. This research investigates the impact of home gardening on society, highlighting its potential as an approach to a socialist society and its alignment with sustainable development goals, including zero hunger and climate action. Home gardening promotes self-sufficiency, food security, and sustainable practices, especially in urban areas, contributing to food system resilience. It fosters community involvement, and cohesiveness, enhancing well-being and happiness among gardeners. The study underscores the importance of promoting home gardening to achieve a more resilient and sustainable future for Penang.

INTRODUCTION

Amid the 2019 pandemic, societies grappled with unprecedented challenges, social distancing, isolation, and mobility restrictions in the fight against COVID-19. These measures, while crucial, strained food supply chains, exacerbating unemployment and triggering heightened food demand (Wilkins, 2020). Amid the prolonged homebound phase, a compelling home gardening trend emerged worldwide, notably

DOI: 10.4018/978-1-6684-9863-7.ch020

in Malaysia. Home gardening gained prominence as solace and sustenance (The Borneo Post, 2021; Walljasper & Polansek, 2020). This phenomenon, spanning urban and rural landscapes, symbolized the potency of urban agriculture (El Khateeb, Saber & Shawket, 2023). At its core, home gardening transformed unused living spaces into vibrant sources of sustenance (Nova, Pinto, Chaves & Silva, 2020). It took root in residential confines, whether indoors or outdoors, adapting to diverse scales and settings (Zasada, Weltin, Zoll & Benninger, 2020). Amid leisure pursuits and experimentation, home gardening surfaced as an ecologically conscious endeavour, uniting living spaces with nature (Nova et al., 2020).

The pandemic sparked a surge in demand for plants, soil, manure and seeds, with home gardening transcending borders, fostering gardening tool sales in nations like Brazil, China, Germany, Italy, the Republic of Korea, the Russian Federation, South Africa, Switzerland, and Turkey (The Star, 2020 ; UNCTAD, 2020). This trend, uninhibited by gender, offered respite and a source of joy, serenity, and communal cohesion (El Khateeb et al., 2023; Ambrose, Das, Fan & Ramaswami, 2023). The ripple effects of home gardening transcended individual gratification, spurring societal transformation (Galhena, Freed & Maredia, 2013). In Penang, Malaysia, the upswing in urban gardening courses mirrored governmental initiatives, such as Think City's autonomous urban farm, fostering resilience, nutrition education and employment (The Citymaker, 2020). Consumer Association of Penang also played a role, guiding Penangites in converting urban agriculture into income streams, underscoring the social potency of the movement (McIntyre, 2021).

As a lens into a socialist society, home gardening reveals diverse virtues. Self-sufficiency and food security are central to its ethos, particularly poignant during the pandemic's peak (Turnšek et al., 2022). However, its impact stretches beyond tangible yields, embodying sustainable narratives like composting and organic waste management in sync with broader environmental stewardship (Phooi et al., 2022). Moreover, high-rises embrace vertical farming, a testament to the resilience of urban food systems (Song et al., 2022). Simultaneously, home gardening stands as an emblem of community engagement and unity. However, barriers persist, including limited resources and misconceptions, warranting targeted interventions (Kanasvamhira, 2023). Beyond personal realms, home gardening extends into schools, nurturing healthy eating habits and fostering community involvement (Orighoye, Apekey & Maynard, 2023).

In Penang's context, the interplay of environmental conservation and sustainable agriculture charts a trajectory towards a resilient future (Chan et al., 2021). Through organic farming techniques and water management, Penang navigates climate uncertainties (Kh'ng et al., 2021). Amid urban dynamism, sustainable agriculture emerges as an economic stimulant, harmonizing prosperity with ecological preservation (Stiepani et al., 2021). Beyond economics, home gardening weaves wellness, fostering connections and camaraderie (Machida, 2019; Eng et al., 2019). Penang's narrative of urbanization and agricultural decline finds a counterbalance in sustainable practices, enabling prosperity while safeguarding ecosystems (Elhadary, Samat & Hasni, 2011). Home gardening catalyzes societal upheaval, promoting holistic transformation, resilience, and shared prosperity.

Policies and Deliberate Ramifications

Policies are instrumental in steering the trajectory of home gardening, especially in response to the challenges introduced by the COVID-19 pandemic. In regions like Penang, Malaysia, where deliberate governmental actions have been taken, policies extend beyond simple encouragement, embracing a holistic approach that spans resilience, nutrition education, and employment generation. These policies are not confined to a singular scale; they accommodate large-scale cultivation initiatives and cater to the needs of small-scale farmers and growers, ensuring inclusivity in their approach.

Furthermore, the deliberate ramifications of these policies are accentuated through funding initiatives from both public and private sectors. Government-backed programs and financial support empower small-scale farmers, providing the necessary resources for cultivation and gardening (McIntyre, 2021 ; The Citymaker, 2020). Internationally, these initiatives are aligned with SDGs, reflecting a concerted effort to address global challenges through localized actions. The nexus between local and international policies becomes evident as these initiatives contribute to regional food security and the overarching goals of poverty reduction, health improvement, and environmental sustainability. Climate change policies and associated funding mechanisms further bolster these initiatives, recognizing the role of home gardening in building resilience against the impacts of a changing climate (Rahaman et al., 2022). As we delve into the intricate web of policies and their deliberate ramifications, the interconnectedness of local and global agendas and their symbiotic relationship with climate change policies becomes a focal point of analysis.

Significance of the Chapter

One noteworthy benefit is the increased availability of locally grown produce. By growing fruits, vegetables, and herbs in their backyards or balconies, urban populations gain direct access to nutritious and flavoursome foods. This reduces their dependence on the food supply chain and potentially lowers carbon emissions associated with transportation.

Home gardeners also play a role in preserving biodiversity. Unlike large-scale farming focusing on various crops, home gardens allow for cultivating heirlooms and diverse plant species. Embracing variety in their gardens ensures a range of nutrients in their diets and safeguards rare plant varieties that may be at risk of disappearing within industrialized agricultural systems. By selecting and saving seeds from plants, home gardeners actively contribute to preserving traditional and local plant varieties while promoting biodiversity conservation.

Furthermore, practising home gardening contributes to sustainability. Large-scale farming can lead to soil degradation, water pollution, and increased reliance on chemical inputs like pesticides and fertilizers. In contrast, home gardening allows for practices prioritizing soil health while minimizing negative impacts on water quality. On the other hand, people tend to adopt environmentally friendly practices regarding home gardening. They focus on composting, using methods to control pests, and conserving water. By embracing these gardening techniques, they can reduce the impact of agriculture on the environment and promote a healthier ecosystem.

Apart from the benefits, home gardening also brings about a sense of empowerment and self-sufficiency. Those who grow their food gain an understanding of how it is produced, which leads to a stronger connection with nature and an appreciation for seasonal changes. This knowledge empowers individuals to make food choices, potentially leading to healthier and more sustainable lifestyles.

Furthermore, grassroots initiatives within the society of Penang can effectively promote development practices. When local communities actively participate in the planning and implementation stages, they prioritize projects and conservation efforts. This approach ensures that development takes place with a long-term perspective, safeguarding resources and preserving the environment for generations.

Grassroots efforts hold significance in Penang's society as they embody the values of community empowerment, social justice, and sustainable progress. By engaging communities in decision-making processes addressing local challenges, advocating for fair resource allocation, and promoting sustainable development practices, grassroots initiatives create an inclusive and prosperous society that benefits everyone.

The primary objective of this chapter is to comprehensively examine the phenomenon of home gardening in the context of the COVID-19 pandemic, elucidating its underlying drivers and the resultant societal transformations. Through a multi-faceted exploration, the chapter seeks to contribute nuanced insights into the role of home gardening in response to crises, particularly its resonance within urban environments such as Penang. The rationale for investigating home gardening amidst the pandemic is rooted in its transformative potential across ecological, social, and economic domains. Understanding the surge and its manifold impacts is crucial for informing sustainable policies community interventions, and shaping narratives of resilience in the face of global challenges.

This chapter is structured as follows:

LITERATURE REVIEW

Understanding the Socialist Society

Amidst the diverse societies, the concept of a socialist society unfolds as a socio-economic of collective ownership and democratic control over production and distribution. In fostering the well-being of all, this ideology aims to dismantle disparities and uplift the marginalized. This egalitarian pursuit finds expression through equitable redistribution of wealth, fortified by progressive taxation systems and robust welfare programs. The aspiration to create a just milieu, where collective prosperity through shared resources, resonates with socialism's core tenets (Giordano, 2020; McCarthy, 2019).

Central to this ethos is social justice, a guiding light for socialist societies, illuminating the path towards eliminating the dual shadows of social and economic inequality. This commitment galvanizes provisions for universal access to education, healthcare, housing, and employment. Such societies strive to transcend social divisions, embracing an inclusive paradigm where every individual's potential thrives, regardless of socioeconomic background (Schrand, 2020).

In investments, socialist societies promote a democratic transformation, fusing financial decisions with collective consensus. This participatory model aligns investments with diverse aspirations, amplifying the essence of democracy inherent in socialism. Investment mechanisms change into responsive tools tailored to the community's welfare rather than the narrow gains of capitalism (Benanav, 2022).

The horizon of benefits in a socialist society stretches far. It is with hues of sustainability, as renewable energy, eco-friendly transportation, and conservation weave against environmental degradation and climate change (Tran, 2020). Guided by community empowerment, these societies nurture engagement in decision-making and encourage shared ownership. Grassroots efforts and cooperative endeavours embolden individuals to shape their destinies, enhancing their lives within a collective embrace (de Laforcade & Springer, 2019).

The integration of sustainable agriculture emerges as a beacon of hope for food security within socialist frameworks. Whether inspired by the Cuban journey or reshaping regions like Penang, community-driven urban farming and home gardening initiatives foster ecological harmony and accessible nourishment (Hanon, 2019).

Nevertheless, challenges persist. As landscapes change towards socialism, formidable economic overhauls beckon, necessitating resource redistribution. Overcoming resistance and minimizing disruptions mandate astute strategies for navigating the changes (Dinu, 2022). Fusing equitable distribution, efficient allocation, and responsible utilization is a delicate waltz between community well-being and environmental sustenance (Salimah et al., 2023).

In dissecting labour's transformation within the embrace of socialism, Baecker (2020) reveals its changes under communal systems. Medicine, art, education, and even domestic spheres reshape through productive labour's lens. Gendered nuances underscore shifts aiming to dismantle labour hierarchies, while governance and participation embolden these dynamics, shaping societal progress.

A quintessential tightrope walk defines socialist societies, balancing individual dreams and collective prosperity. Amidst equity and social welfare, this equilibrium hinges on preserving individual liberties while nurturing the common good. A symphony of political dynamics and social policies harmonizes personal and societal aspirations, unfurling a paradigm that epitomizes the essence of socialism (Sen, 2022).

Philosophical Underpinning

Ecological modernization theory proposes that societies can attain environmental sustainability by transforming crucial institutions, encompassing technology, markets, state interventions, and social movements (Frouws & Mol, 1997). This theory asserts that ecological rationality, which places environmental concerns at the forefront, can evolve within the agricultural sector. Recognizing agriculture as a steward of natural resources, the theory contends that it is pivotal for applying ecological rationality (Thompson, 2017). However, the theory underscores the necessity to curtail the prevailing economic rationality in agriculture, particularly evident in intensive farming regions like the Netherlands, where ecological restructuring is imperative due to challenges such as diseases, water pollution, biodiversity loss, and soil dehydration (Pal, Gopal & Ramkumar, 2023).

While recent changes in agriculture may suggest ecological modernization marked by agro-environmental policies, environmental agreements, and the promotion of eco-friendly technologies, the theory acknowledges challenges and counterfactual developments (Kais & Islam, 2023). Delays, legitimacy issues in manure policy, and obstacles in reducing pesticide pollution underscore the complexities (Helming et al., 2023). In essence, ecological modernization theory in agriculture contends that despite positive shifts towards ecological rationality, substantial challenges persist, necessitating a nuanced analysis of institutional domains for a comprehensive understanding of the extent of ecological restructuring in agriculture.

The Concept of Grassroots Movements and Citizen Involvement in Penang

In the bustling Penang landscape, the vibrant strokes of grassroots movements and the unwavering participation of its citizens' transformation address the intricate challenges encompassing food security and sustainable agriculture. These movements echo far beyond their grassroots origins, propelling individuals and local neighbourhoods to the forefront of decision-making and collective action (Institute for Societal Leadership, 2015). In Penang's intricate narrative, these initiatives become more than mere undertakings; they symbolize the essence of community ownership, ushering in resilient and self-reliant neighbourhoods (Ismail, 2021). Here, the fertile ground of transformative change is cultivated, where residents are not just growing crops but are nurturing a profound connection to their environment, reshaping it in harmony with their aspirations (Rulia, 2023). These movements promise a future intricately shaped by the collective wisdom and aspirations of the grassroots, nurturing a sustainable legacy that transcends time.

Within these grassroots movements, rich knowledge-sharing and capacity-building flourishes, fostering a communal reservoir of expertise (Engelbutzeder et al., 2023). These movements transform into fertile grounds for exchanging experiences, where best practices in sustainable agriculture are painted across community gatherings, workshops, and training initiatives (Zollet, 2022). Each interaction becomes a brushstroke of empowerment, equipping citizens with the tools to embark on endeavours such as home gardening and urban farming and cultivate a future woven with resilience and ecological mindfulness.

In the heart of Penang's diverse landscape, grassroots movements craft an intricate mosaic of unity and camaraderie, weaving together individuals under the banner of sustainable agriculture (Penang2030, 2023). Common interests include shared endeavours as the rhythm of community cohesion and social networks draws residents into a harmonious collaboration. Here, the gardens of collaboration flourish, where community gardens are not merely plots of land but fertile grounds where expertise, resources, and aspirations intertwine (Malaysiakini, 2022). Through this collaborative ecosystem, challenges change into opportunities, fostering a community of resilience and the promise of a secure food future.

In the dynamic symphony of Penang's aspirations, grassroots movements crescendo as potent advocates for change, channelling citizen voices into the corridors of local and regional decision-making (Ahmed et al., 2016). A harmony of advocacy resonates, steering the course towards policies that embrace urban agriculture, nurture small-scale farmers, and honour ecologically conscious practices. In this symphony, the aspirations of a community intertwined with the evolution of policies, crafting a future where sustenance and sustainability stand as intertwined pillars (Penang Institute, 2021).

Krutilin et al. (2022) posit that grassroots movements are not just agents of change but also champions of social equity and environmental harmony within the food system. These movements spotlight the promise of sustainable agriculture, a formidable tool against climate change and a guardian of Penang's biodiversity and natural resources. The narrative weaves social justice and ecological stewardship together in their endeavours, creating nourishment and enabling nature to thrive.

In Penang's destiny, grassroots movements and the dynamic engagement of its citizens stand as the vibrant threads that intricately weave a solution to the complexities of food security and sustainable agriculture. These movements serve as beacons of empowerment, kindling change within individuals and communities. Through the symphony of shared wisdom and skill-building, citizens become the architects of transformation, wielding the knowledge to sow the seeds of sustainability. As the ripples of their efforts spread through the community, the unity grows stronger, fostering bonds that enhance resilience and nurture collective progress. With unity in purpose and concerted determination, grassroots movements can enact policy changes that promote an era where urban agriculture flourishes, small-scale farmers thrive, and the heartbeat of the environment remains unwavering. In this journey, the duality of justice and harmony intertwine as equitable nourishment and ecological preservation align to pave the path for a flourishing Penang that echoes through generations to come.

Home Gardening as a Catalyst for Societal Transformation

In Penang, home gardening emerges as a compelling catalyst for societal transformation, addressing intricate food security concerns, sustainable agriculture, and empowered communities. This practice beckons individuals and communities to reclaim agency over their sustenance, curtailing dependency on external sources. This self-reliance fortifies food security and nurtures healthier dietary habits while reducing the carbon footprint entwined with food transportation (Feola et al., 2023). Furthermore, home gardening intricately interlaces with environmental sustainability. As it cultivates communion with the land, it upholds organic practices that shield soil vitality, preserve water purity, and safeguard ecosystems (Dwi Laksmana., 2023). This harmony with nature simultaneously aspires to enhance Penang's urban biodiversity.

Amidst this transformative narrative, grassroots movements and civic participation emerge as pivotal brushstrokes. Through these vibrant channels, a culture of proactive involvement, wisdom-sharing, and policy advocacy burgeons. These movements are not solitary endeavours but synergistic collaborations, birthing resilient communities knitted by shared aspirations. However, a terrain marked by spatial constraints and resource accessibility necessitates careful tending. In this context, policymakers wield the mantle of nurturing an enabling ecosystem characterized by policies endorsing community gardens, educational initiatives, and workshops (Sou, Ooi & Zhang, 2023). Education, amplified through awareness campaigns, emerges as a beacon illuminating the path to sustainable gardening methods and the bounty they yield.

Within the socialist of Penang, home gardening unfurls as a linchpin fostering socio-economic transformations that empower individuals and galvanize communities. At its core, this practice weaves a food security narrative by sowing the seeds of self-sufficiency. Individuals brace against disruptions and secure their nourishment. This burgeoning resilience finds its partner in economic empowerment (Kodoh et al., 2023). The bounty reaped from home gardens can foster local economies through surplus sales, farmers' markets, or the spin-offs of food processing and agro-tourism. In this embrace of home gardening, the ripple effect extends to financial parity, unfurling equitable opportunities within the community.

This practice, in its essence, transcends garden boundaries to sculpt landscapes of collective growth and social harmony. The shared spaces burgeon into sanctuaries for connection, bridging generational gaps and echoing the exchange of wisdom (Ortiz Cuadra, 2023). This interplay of ages nurtures plants and relationships, bolstering a sense of belonging and shared guardianship. Nevertheless, the Earth cultivates belonging and devotion to environmental stewardship. The organic methods championed by home gardening resonate with the ethos of a sustainable Penang. This symphony of nature-conscious practices not only preserves resources but also becomes a testament to a community nurturing its environment.

Rooted in soil yet branching into empowerment, home gardening's transformative essence stems from intricately woven strands. The narrative unfurls through food sovereignty as individuals command their nutrition destiny. This intricate autonomy spurs skill acquisition and wisdom sharing, fostering a rich mastery that flows between generations. A parallel narrative unwinds towards environmental stewardship, where these gardens become schools of conservation. Their organic cultivation sync with the rhythm of Penang's sustainability ethos. Each water-wise technique and composting is a verse in this symphony of responsibility (Nasrida et al., 2023).

Moreover, health and well-being are enhanced in this garden of transformation. Individuals are granted access to a healthier lifestyle through active engagement and access to homegrown produce (Kersten et al., 2023). The nurturing aura of these gardens extends beyond the physical; each stroke of the hoe serves as therapy, nurturing plants and human well-being. Essentially, these gardens sow plants and connective threads, weaving communities tighter. This intertwined embrace extends towards economic opportunities, where individuals can reap financial rewards from surplus sales or entrepreneurial ventures (Shahidullah, Islam & Kendall, 2023). This nurturing of self and society finds harmony within the diverse yet unified home gardening landscape.

Home gardening emerges as a resilient thread within Penang's landscape that binds communities and forges social unity. It transcends mere cultivation, birthing shared spaces where knowledge is sown, connections flourish, and a sense of shared responsibility is nurtured (Moeti, 2023). These gardens echo intergenerational stories, where wisdom is between hands and hearts. Akin to seeds sprouting, this practice cultivates a sense of belonging and a shared commitment to nurturing the community.

Through home gardening, resilience is seeded at both personal and communal levels. As individuals cultivate sustenance, they become architects of their food security, poised against adversity (Chadha, 2023). In times of crisis, these gardens offer more than produce; they become symbols of stability, nourishing both body and spirit. Nevertheless, this practice does not bloom in isolation; it flourishes within a network of social support and mutual aid (Korpelainen, 2023). Gardens become spaces of reciprocity, where watering a neighbour's plants embodies collective care. Penang can weave this potent thread of resilience by fostering collaborative initiatives, bridging generations, and embracing sustainable practices that strengthen the community.

Home Gardening as a Solution for Food Security and Sustainable Agriculture

The main discussion about home gardening is its role as a solution to solving food security issues. In a study conducted by Turnšek et al. (2022), the fear surrounding general food security and the anxiety related to food security during the COVID-19 pandemic has shown an upsurge in home gardening activities. Food security-related issues have been around even before the pandemic. However, the food source was almost paralyzed due to the limited agriculture activities locally and globally, making global food security a major crisis. This has led individuals to find ways to enhance their food ecosystem to be self-sufficient and take more proactive measures to mitigate food availability at the household and local community levels.

This uncertainty has led to the emergence of home gardening trends among urban populations (Sofo & Sofo, 2020). Traditionally, home gardening is more common in rural areas. Pandemic lockdowns have caused urban people to start home gardening proactively. This is because the urban populations first have increased leisure time due to pandemic lockdowns. This is evident in Malaysia, where Google Trends' top searches since the beginning of the movement control order in March 2020 were horticulture and home gardening (The Borneo Post, 2021). Easy-to-grow potted plants like Brazilian spinach and chilis were among the popular choices among budding home gardeners.

Due to these circumstances, many people have developed an inclination to decrease their reliance on food delivery systems, which have faced disruptions due to lockdowns and restrictions. One method individuals adopt is cultivating their crops, allowing them to ensure a healthy food supply while bypassing distribution challenges. As a result, many residents have started creating home gardens to secure food for their families (Lal, 2020).

During the pandemic, there was a significant increase in the number of people gardening at home, which had repercussions that extended beyond maintaining enough food supplies. People also started to recognize the significance of food production and the benefits of decreasing their carbon footprint by producing food in their backyards. This event launched a movement towards sustainability and resilience (Schanbacher & Cavendish, 2023).

However, despite substantial support from this point of view, research conducted by Du Toit et al. (2022) found that more than having a garden is needed to tackle the issue of food insecurity among people living in cities effectively. While the study explicitly examines suburban areas in South Africa, where food security is a pressing issue, it also highlights the cultural importance of home gardens, where decorative plants are preferred over vegetable gardens.

Growing gardens on one's property allows households to enhance their self-sufficiency in food production and achieve autonomy. Any extra harvest can be sold within the community, promoting a sense of self-reliance. While this may not eliminate the need to purchase groceries from supermarkets and retailers, it does contribute to an empowered lifestyle.

It is important to note, however, that a study conducted by Turnšek et al. (2022) suggests that home gardening may not always result in cost savings, as some might assume. Nevertheless, it is crucial to acknowledge the advantages that home gardening offers, including improved food quality, reduced environmental impact and enhanced community unity. It plays a role in bolstering economies and strengthening bonds within the community. Engaging in gardening encourages interactions and mutual support, benefiting both parties involved. Thus, home gardening has played a role that extends beyond households, especially during the pandemic.

Getting involved in gardening influences building connections and encouraging community engagement. When people embrace gardening, they actively share advice, seeds, and the fruits of their labour with their neighbours and friends, enhancing relationships. It also strengthens the sense of community even when physical distancing is necessary. The supportive network formed through gardening goes beyond growing food; it fosters a shared understanding and reliance on each other's knowledge and resources. Ultimately, shared interests in home gardening have the potential to create a knit and interdependent community (Veen, 2015).

Gardening techniques can have effects on the environment. A study conducted in Italy by Sanyé Mengual et al. (2018) found that residential gardens consume an amount of electricity during the fertigation process. However, there are ways to make home gardening eco-friendly by adopting sustainable practices. These include using low-automation gardening methods, recycling compost, and optimizing rainwater collection. These measures promote farming practices, especially when it comes to harvesting rainwater. It is recommended that home gardeners prioritize using resources like rainwater instead of relying solely on freshwater sources or clean water from the distribution system. Home gardeners can help conserve water resources and reduce their impact by doing this.

A study conducted by Turnšek et al. (2022) indicates that the outbreak of COVID-19 has impacted individuals' involvement in home gardening. Those residing in regions have increased their gardening activities due to concerns regarding food accessibility in general and during the pandemic. Engaging in home gardening alleviated food scarcity anxieties, fostered community bonds, and instilled a greater appreciation for supporting local farmers. Although the world has changed since then, the lessons learned are bound to influence for years.

Despite the advantages mentioned, there are limitations and misunderstandings surrounding the connection between home gardening and food security. Some believe that relying on home gardening may have little impact in tackling larger-scale food security challenges. While it is true that home gardening alone cannot be considered the solution to food security issues, it is an integral part that works with other strategies, like sustainable agricultural practices, efficient food distribution networks, and policies designed to alleviate poverty and hunger.

CONCLUSION

In creating Penang's socialist society, it is crucial to emphasize the significance of endeavours in promoting community empowerment and sustainable progress. In the socialist society framework, the primary focus lies on ownership, social justice, and individuals' well-being, thus making grassroots initiatives to achieve these goals.

In Penang, grassroots efforts play a role in involving local communities in decision-making processes and development projects. By engaging citizens from the ground level, their needs, aspirations, and concerns are given importance. This leads to responsive policies and programs that take into account their perspectives. As a result, it fosters a sense of ownership and participation among residents while strengthening their connection with society. It also promotes a commitment to the good.

Moreover, grassroots initiatives have the potential to address challenges effectively. They are better equipped to understand the context and nuances of communities in their local community. This allows for tailor-made solutions that align with their economic and environmental circumstances. Consequently, these efforts have a lasting impact on creating an equitable and just society.

In a society where wealth and resources are shared for everyone's welfare, grassroots endeavours can catalyze policies. By giving a voice to marginalized communities and ensuring their involvement in decision-making processes, grassroots movements advocate for the distribution of resources and opportunities. This helps to reduce inequalities and foster social cohesion.

REFERENCES

Ahmed, F., Al-Amin, A. Q., Mohamad, Z. F., & Chenayah, S. (2016). Agriculture and food security challenge of climate change: A dynamic analysis for policy selection. *Scientia Agrícola*, *73*(4), 311–321. doi:10.1590/0103-9016-2015-0141

Ambrose, G., Das, K., Fan, Y., & Ramaswami, A. (2023). Comparing happiness associated with household and community gardening: Implications for Food Action Planning. *Landscape and Urban Planning*, *230*, 104593. doi:10.1016/j.landurbplan.2022.104593

Baecker, A. (2020). Reimagining Revolutionary Labor in the People's Commune: Amateurism and Social Reproduction in the Maoist Countryside (Doctoral dissertation).

Benanav, A. (2022). Socialist investment, dynamic planning, and the politics of human need. *Rethinking Marxism*, *34*(2), 193–204. doi:10.1080/08935696.2022.2051375

Chadha, M. L. (2023). Home Gardening: The Way Forward to Be Safe and Healthy. In Vegetables for Nutrition and Entrepreneurship (pp. 217-239). Springer Nature.

Chalmin-Pui, L. S., Griffiths, A., Roe, J., Heaton, T., & Cameron, R. (2021). Why garden?–Attitudes and the perceived health benefits of home gardening. *Cities (London, England)*, *112*, 103118. doi:10.1016/j.cities.2021.103118

Chan, N. W., Ghani, A. A., Samat, N., Roy, R., Tan, M. L., & Rahman, H. A. (2021). Addressing water resources shortfalls due to climate change in Penang, Malaysia. In Water Security in Asia: Opportunities and Challenges in the Context of Climate Change (pp. 239-249). Springer International Publishing.

de Laforcade, G., & Springer, D. (2019). The Red Barrial Afrodescendiente: A Cuban Experiment in Black Community Empowerment. *Souls*, *21*(4), 339–346. doi:10.1080/10999949.2019.1736904

Dinu, R. H. (2022). Recent Historiographical Trends in Scholarship on Disability and Socialism in Eastern Europe. *Scandinavian Journal of Disability Research*, *24*(1), 42–53. doi:10.16993/sjdr.829

Du Toit, M. J., Rendón, O., Cologna, V., Cilliers, S. S., & Dallimer, M. (2022). Why home gardens fail in enhancing food security and dietary diversity. *Frontiers in Ecology and Evolution*, *10*, 38. doi:10.3389/fevo.2022.804523

Dwi Laksmana, D. (2023). Knowledge in the making: embodying transdisciplinary moments on organic agriculture in Yogyakarta, Indonesia [Doctoral dissertation, Universität Passau].

El Khateeb, S., Saber, M., & Shawket, I. M. (2023). Urban reflections through home gardening; Does Gender Matter? *Ain Shams Engineering Journal*, *14*(3), 101885. doi:10.1016/j.asej.2022.101885

Elhadary, Y. A. E., Samat, N., & Hasni, R. (2011). Land use changes impact local communities at Seberang Perai Tengah of Penang, Malaysia. In Issues in peri-urban regions and ways towards sustainable peri-urban futures. Conference Room, School of Humanities. Universiti Sains Malaysia.

Eng, S., Khun, T., Jower, S., & Murro, M. J. (2019). Healthy lifestyle through home gardening: The art of sharing. [PubMed]. *American Journal of Lifestyle Medicine*, *13*(4), 347–350. doi:10.1177/1559827619842068 PMID:31285714

Engelbutzeder, P., Randell, D., Landwehr, M., Aal, K., Stevens, G., & Wulf, V. (2023). From surplus and scarcity towards abundance: Understanding the use of ICT in food resource sharing practices: "Give a man a fish, and you feed him for a day; teach a man to fish, and you feed him for a lifetime."–Lao Tsu. *ACM Transactions on Computer-Human Interaction*. Advance online publication. doi:10.1145/3589957

Feola, G., Goodman, M. K., Suzunaga, J., & Soler, J. (2023). Collective memories, place-framing and the politics of imaginary futures in sustainability transitions and transformation. *Geoforum*, *138*, 103668. doi:10.1016/j.geoforum.2022.103668

Frouws, J., & Mol, A. P. (1997). Ecological modernization theory and agricultural reform. In Images and Realities of Rural Life. Wageningen Perspectives on Rural Transformations (pp. 269-286). Van Gorcum.

Galhena, D. H., Freed, R., & Maredia, K. M. (2013). Home gardens: A promising approach to enhance household food security and well-being. *Agriculture & Food Security*, *2*(1), 1–13. doi:10.1186/2048-7010-2-8

Giordano, C. (2020). From Sicily to Malaysia via Bulgaria: Finding the Common Thread of One's Fieldwork. Recalling Fieldwork: People. *Places and Encounters*, *51*, 19.

Hanon, I. (2019). Workers' participation: The challenge of Cuban socialism. *Socialism and Democracy*, *33*(1), 88–111. doi:10.1080/08854300.2018.1510671

Hansen, L. S., Sorgho, R., Mank, I., Nayna Schwerdtle, P., Agure, E., Bärnighausen, T., & Danquah, I. (2022). Home gardening in sub-Saharan Africa: A scoping review on practices and nutrition outcomes in rural Burkina Faso and Kenya. *Food and Energy Security*, *11*(3), e388. doi:10.1002/fes3.388

Helming, J., Daatselaar, C., van Dijk, W., Mollenhorst, H., & Pishgar-Komleh, S. H. (2023). Model Collaboration between Farm Level Models with Application on Dutch Dairy and Arable Farms Regarding Circular Agricultural Policy. *Sustainability (Basel), 15*(6), 5020. doi:10.3390/su15065020

Institute for Societal Leadership. (2015). *Societal Leadership in Southeast Asia: National Landscapes, Challenges and Opportunities to Enhance Societal Impact.* Institute of Societal Leadership Research Collection.

Ismail, W. A. A. Z. W. (2021). *Development Approaches for the Orang Asli Laut As An Indicator of 'Leaving No One Behind.* Research Gate.

Kais, S. M., & Islam, M. S. (2023). Climate Change, Ecological Modernization, and Disaster Management: The Coastal Embankment Project in Southwestern Bangladesh. [PubMed]. *International Journal of Environmental Research and Public Health, 20*(12), 6086. doi:10.3390/ijerph20126086 PMID:37372673

Kanosvamhira, T. P. (2023). How do we get the community gardening?: Grassroots perspectives from urban gardeners in Cape Town, South Africa. *Journal of Cultural Geography, 40*(1), 47–63. doi:10.1080/08873631.2023.2187509

Kersten, M., Carrazco, L., Rosing, H., Swenski, T., Russell, D., Idrovo, J., & Lofton, S. (2023). Evaluation of the Grow Your Groceries Home Gardening Program in Chicago, Illinois. [PubMed]. *Journal of Community Health, 48*(2), 179–188. doi:10.1007/s10900-022-01152-x PMID:36336753

Kh'ng, X. Y., Teh, S. Y., Koh, H. L., & Shuib, S. (2021). Sea level rise undermines SDG2 and SDG6 in Pantai Acheh, Penang, Malaysia. *Journal of Coastal Conservation, 25*(1), 1–14. doi:10.1007/s11852-021-00797-5

Kodoh, J., Jaunis, O., Maid, M., Lintangah, W., & Mojiol, A. R. (2023, February). Urban Forestry in Kota Kinabalu: Home Garden practices by urban villages for biodiversity, socio-economic and environmental services. [). IOP Publishing.]. *IOP Conference Series. Earth and Environmental Science, 1145*(1), 012012. doi:10.1088/1755-1315/1145/1/012012

Korpelainen, H. (2023). The Role of Home Gardens in Promoting Biodiversity and Food Security. [PubMed]. *Plants, 12*(13), 2473. doi:10.3390/plants12132473 PMID:37447034

Krutilin, A. A., Bazieva, A. M., Dugina, T. A., & Giyazov, A. T. (2022). Sustainable Agriculture for Food Security: Conceptual Framework and Benefits of Digitalization. In Sustainable Agriculture: Circular to Reconstructive, Volume 1 (pp. 35-42). Springer.

Lal, R. (2020). Home gardening and urban agriculture for advancing food and nutritional security in response to the COVID-19 pandemic. [PubMed]. *Food Security, 12*(4), 871–876. doi:10.1007/s12571-020-01058-3 PMID:32837634

Machida, D. (2019). Relationship between community or home gardening and health of the elderly: A web-based cross-sectional survey in Japan. [PubMed]. *International Journal of Environmental Research and Public Health, 16*(8), 1389. doi:10.3390/ijerph16081389 PMID:30999687

Macs, B., & GGdeLW, S. (2020). The concept of environmental leadership: The case of Solid Waste Management (SWM) in Balangoda municipal council in Sri Lanka. *Open Access J Environ Soil Sci.*, *5*(5), 728–734.

Malaysiakini. (2022). Gardening fans, let's gather at New World Park, Penang. The largest Plant Market, City Gardening KL is coming to town! Malaysiakini. https://www.malaysiakini.com/announcement/637177

McCarthy, M. A. (2019). *The Socialist Journey*. Jacobin.

McIntyre, I. (2020). Pandemic sees more Penangites keen to grow own greens. The Vibes. https://www.thevibes.com/articles/news/12180/pandemic-sees-more-penangites-keen-to-grow-owngreens

Moeti, I. (2023). Sustainability and Consumer Education on Food Sustainability [Doctoral dissertation, London South Bank University.

Morgan, J. (2023). Sustainable Contradictions? The Prospects for an Eco-Socialist School Geography. In Understanding Sustainability with Pedagogical Practice: A Contribution from Geography Education (pp. 67-77). Springer Nature. 10.1007/978-981-99-2687-9_6

Nasrida, N., Saediman, H., Rianse, I. S., & Hidrawati, H. (2023). Contribution of Home Gardening to Household Vegetable Consumption During COVID-19 Pandemic. International Journal of Research in Engineering. *Science and Management*, *6*(5), 5–9.

Ngarava, S., Zhou, L., Ningi, T., & Chari, M. M. (2022). Differentiated Intra-Household Food Utilisation in Raymond Mhlaba Local Municipality, South Africa. In Sustainable Agriculture and Food Security (pp. 87–106). Springer International Publishing. 10.1007/978-3-030-98617-9_6

Nova, P., Pinto, E., Chaves, B., & Silva, M. (2020). Urban organic community gardening to promote environmental sustainability practices and increase fruit, vegetables and organic food consumption. [PubMed]. *Gaceta Sanitaria*, *34*(1), 4–9. doi:10.1016/j.gaceta.2018.09.001 PMID:30471837

Orighoye, O., Apekey, T. A., & Maynard, M. J. (2023). Informing Diet and Physical Activity Interventions with Family Involvement in an Urban Setting: Views of Children and Adults in Lagos, Nigeria. *Sustainability (Basel)*, *15*(10), 7850. doi:10.3390/su15107850

Ortiz Cuadra, J. J. (2023). *Learning for Crisis: Improving food security in Uppsala County through participative localized food production*. Research Gate.

Pal, P., Gopal, P. R. C., & Ramkumar, M. (2023). Impact of transportation on climate change: An ecological modernization theoretical perspective. *Transport Policy*, *130*, 167–183. doi:10.1016/j.tranpol.2022.11.008

Penang2030. (2023). Unity in Diversity: Penang Experience and Approach. Penang 2030. https://penang2030.com/2023/03/unity-in-diversity-penang-experience-and-approach

Penang Institute. (2021). Strengthening Food Security in Light of Covid-19. Penang Institute. https://penanginstitute.org/publications/monographs/strengthening-food-security-in-light-of-covid-19

Phooi, C. L., Azman, E. A., Ismail, R., & Tongkaemkaew, U. (2022). *Call home gardening for enhancing food in the urban area*. Research Gate.

Rahaman, Z. A., Kafy, A. A., Saha, M., Rahim, A. A., Almulhim, A. I., Rahaman, S. N., Fattah, M. A., & Rahman, M. T., S, K., Faisal, A.-A., & Al Rakib, A. (2022). Assessing the impacts of vegetation cover loss on surface temperature, urban heat island and carbon emission in Penang City, Malaysia. *Building and Environment*, *222*, 109335. doi:10.1016/j.buildenv.2022.109335

Rao, N., Patil, S., Singh, C., Roy, P., Pryor, C., Poonacha, P., & Genes, M. (2022). Cultivating sustainable and healthy cities: A systematic literature review of the outcomes of urban and peri-urban agriculture. *Sustainable Cities and Society*, *85*, 104063. doi:10.1016/j.scs.2022.104063

Robinson, L. (2018). *Sustainable Agriculture and Farming*. Larsen and Keller Education.

Rulia, A. (2023). The Importance Of Sustainable Agriculture. Bernama. https://www.bernama.com/en/thoughts/news.php?id=2199261

Salimah, N., Siallagan, M. A., Banjarnahor, H. E., & Matondang, K. (2023). Comparison of Capitalist and Socialist Economic Models in International Trade. *Journal of Business Management and Economic Development*, *1*(01), 63–69. doi:10.59653/jbmed.v1i01.39

Santos, M., Moreira, H., Cabral, J. A., Gabriel, R., Teixeira, A., Bastos, R., & Aires, A. (2022). Contribution of home gardens to sustainable development: Perspectives from a supported opinion essay. [PubMed]. *International Journal of Environmental Research and Public Health*, *19*(20), 13715. doi:10.3390/ijerph192013715 PMID:36294295

Sanyé-Mengual, E., Gasperi, D., Michelon, N., Orsini, F., Ponchia, G., & Gianquinto, G. (2018). Eco-efficiency assessment and food security potential of home gardening: A case study in Padua, Italy. *Sustainability (Basel)*, *10*(7), 2124. doi:10.3390/su10072124

Schanbacher, W. D., & Cavendish, J. C. (2023). The effects of COVID-19 on Central Florida's community gardens: Lessons for promoting food security and overall community well-being. [PubMed]. *Frontiers in Public Health*, *11*, 11. doi:10.3389/fpubh.2023.1147967 PMID:37483929

Schrand, T. G. (2020). Utopianism and the equity path to sustainability. *Journal of Environmental Studies and Sciences*, *10*(4), 457–466. doi:10.1007/s13412-020-00619-4

Sen, S. (2022). Law and social policy in India: From growth-based welfare to welfare entitlement. 102-131.

Shahidullah, A. K. M., Islam, A., & Kendall, L. (2023, July). Women Agro-Entrepreneurship Promoting Vegetables at a Family Level: A Healthcare Approach towards Non-Communicable Disease Risk Reduction. [). MDPI.] [PubMed]. *Health Care*, *11*(15), 2165. PMID:37570405

Sofo, A., & Sofo, A. (2020). Converting home spaces into food gardens at the time of Covid-19 quarantine: All the benefits of plants in this difficult and unprecedented period. [PubMed]. *Human Ecology*, *48*(2), 131–139. doi:10.1007/s10745-020-00147-3 PMID:32317823

Song, S., Cheong, J. C., Lee, J. S., Tan, J. K., Chiam, Z., Arora, S., Png, K. J. Q., Seow, J. W. C., Leong, F. W. S., Palliwal, A., Biljecki, F., Tablada, A., & Tan, H. T. (2022). Home Gardening in Singapore: A feasibility study on the utilization of the vertical space of retrofitted high-rise public housing apartment buildings to increase urban vegetable self-sufficiency. *Urban Forestry & Urban Greening*, *78*, 127755. doi:10.1016/j.ufug.2022.127755

Sou, G., Ooi, C. S., & Zhang, Y. (2023). Resilience and Resistance. In Islands and Resilience: Experiences from the Pandemic Era (pp. 53-64). Springer Nature Singapore. 10.1007/978-981-19-9964-2_4

Stiepani, J., Gillis, L. G., Chee, S. Y., Pfeiffer, M., & Nordhaus, I. (2021). Impacts of urbanization on mangrove forests and brachyuran crabs in Penang, Malaysia. *Regional Environmental Change, 21*(3), 69. doi:10.1007/s10113-021-01800-3

The Borneo Post. (2021). Covid-19 pandemic sees more Malaysians getting into home gardening. The Borneo Post. https://www.theborneopost.com/2021/12/22/covid-19-pandemic-sees-more-malaysians-gettinginto home-gardening/

The CityMaker. (2020). Kebun Kita(r): Penang's 1st Self-Sustaining Community Urban Farm Opens at Penang Digital Library. The City Maker. https://thecitymaker.com.my/kebun-kitar-penang-urbancommunity-digital/

The Star. (2020). MCO boosts demand for seeds, gardening tools. The Star. https://www.thestar.com.my/news/nation/2020/04/09/mco-boosts-demand-for-seeds-gardening-tools

Thompson, P. B. (2017). *The spirit of the soil: Agriculture and environmental ethics*. Taylor & Francis., doi:10.4324/9781315559971

Tran, T. P. (2020). Ho Chi Minh thought on solidarity and the social sustainable development in Vietnam. In *E3S Web of Conferences* (Vol. 164, p. 12004). EDP Sciences., doi:10.1051/e3sconf/202016412004

Turnšek, M., Gangenes Skar, S. L., Piirman, M., Thorarinsdottir, R. I., Bavec, M., & Junge, R. (2022). Home Gardening and Food Security Concerns during the COVID-19 Pandemic. *Horticulturae, 8*(9), 778. doi:10.3390/horticulturae8090778

UNCTAD. (2020). COVID-19 has changed online shopping forever, survey shows. UN. https://unctad.org/news/covid-19-has-changed-online-shopping-forever-survey-shows

United Nations. (2022). Sustainable Development Goal 2 Zero Hunger. UN https://malaysia.un.org/en/sdgs/2

Veen, E. J. (2015). *Community gardens in urban areas: a critical reflection on the extent to which they strenghten social cohesion and provide alternative food*. Wageningen University and Research.

Walljasper, C., & Polansek, T. (2020). Home gardening blooms around the world during coronavirus lockdowns. Reuters. https://www.reuters.com/article/us-health-coronavirus-gardensidUSKBN2220D3

Wilkins, J. L. (2020). Challenges and opportunities created by the COVID-19 pandemic. [PubMed]. *Journal of Nutrition Education and Behavior, 52*(7), 669–670. doi:10.1016/j.jneb.2020.05.005 PMID:32646596

Wong, T. A., Bustami, M. R., & Salleh, S. M. (2022). Applying Social Capital Theory on Conservation Policies of Protected Areas: A Case Study of Penang, Malaysia. *International Journal of Business and Society, 23*(1), 172–187. doi:10.33736/ijbs.4607.2022

Yuniarsih, P. D. S. E. T., Hanifa, P. D. A. P., Haryati, M. S. Y., Lestari, I. P., Saleh, M. S. Y., Nurmalinda, M. S. N., & Lesmayati, M. S. S. (2023). *Perspective Chapter: How Important Is Urban Farming in Indonesia to Support Food Sovereignty?* Research Gate.

Zasada, I., Weltin, M., Zoll, F., & Benninger, S. L. (2020). Home gardening practice in Pune (India), the role of communities, urban environment and the contribution to urban sustainability. *Urban Ecosystems*, *23*(2), 403–417. doi:10.1007/s11252-019-00921-2

Zollet, S. (2023). Hybrid food networks and sustainability transitions: Shared and contested values and practices in food relocalization and resocialization. *Sociologia Ruralis*, *63*(1), 117–139. doi:10.1111/soru.12391

KEY TERMS AND DEFINITIONS

Climate Change: Climate change refers to the long-term alterations in Earth's climate, including changes in temperature, precipitation patterns, and weather events. These changes are primarily caused by human activities, such as releasing greenhouse gases from burning fossil fuels and deforestation. The consequences of climate change are extensive and include global warming, rising sea levels, and ecosystem disruptions.

Food Security: Food security refers to the condition where people have constant access to sufficient and safe food resources to support their living conditions.

Grassroots Agriculture: Grassroots agriculture involves local communities or individuals actively participating in small-scale farming practices. This community-driven approach often centers on sustainable and environmentally friendly practices.

Home Gardening: Home gardening refers to cultivating a diverse range of plants, encompassing vegetables, fruits, herbs, and ornamental plants, within the confines of residential areas. This activity serves multiple objectives, including food production, aesthetic appreciation, and promoting sustainable lifestyles.

Socialist Society: A socialist society is characterized by the collective ownership of resources and the means of production to reduce economic inequalities, improve social welfare, and achieve equitable distribution of wealth and assets among its constituents.

Compilation of References

Aasmoe, L., Bang, B., Andorsen, G. S., Evans, R., Gram, I. T., & Lochen, M. L. (2005). Skin symptoms in the seafood-processing industry in north Norway. *Contact Dermatitis, 52*(2), 102–107. doi:10.1111/j.0105-1873.2005.00515.x

Abbasnia, M., & Toros, H. (2019). Analysis of long-term changes in extreme climatic indices: A case study of the Mediterranean climate, Marmara Region, Turkey. İçinde Meteorology and climatology of the Mediterranean and Black Seas (ss. 141-153). Springer.

Abbass, K., Qasim, M. Z., Song, H., Murshed, M., Mahmood, H., & Younis, I. (2022). A review of the global climate change impacts, adaptation, and sustainable mitigation measures. *Environmental Science and Pollution Research International, 29*(28), 42539–42559. doi:10.1007/s11356-022-19718-6 PMID:35378646

Abbriano, R. M., Carranza, M., Hogle, S., Levin, R., Netburn, A., Seto, K., Snyder, S., & Franks, P. (2011). Deepwater Horizon oil spill: A review of the planktonic response. *Oceanography (Washington, D.C.), 24*(3), 294–301. doi:10.5670/oceanog.2011.80

Abeygunawardena, P., Vyas, Y., Knill, P., Foy, T., Harrold, M., Steele, P., Tanner, T., Hirsch, D., Oosterman, M., & Rooimans, J. (2004). *Poverty and Climate Change: Reducing the Vulnerability of the Poor through Adaptation*. World Bank. https://documents1.worldbank.org/curated/en/534871468155709473/pdf/521760WP0pover1e0Box35554B01PUBLIC1.pdf. Retrieved: 20 May 2023.

Achieving Cross-Border Government Innovation. (2021, October 18). OECD-OPSI. https://cross-border.oecd-opsi.org/

Adam, I., Walker, T. R., Bezerra, J. C., & Clayton, A. (2020). Policies to reduce single-use plastic marine pollution in West Africa. *Marine Policy, 116*, 103928. doi:10.1016/j.marpol.2020.103928

Adams, S. M. (2005). Assessing cause and effect of multiple stressors on marine systems. *Mar. Pollut. Bull., 51*, 8–12. https://www.sciencedirect.com/science/article/pii/ S0025326X04004667.

ADB. (2014). *Climate Change and Rural Communities in the Greater Mekong Sub-Region: A Framework for Assessing Vulnerability and Adaptation Options*. Mandaluyong City: ADB.

Aditya, S. K. (2016). Role of Women in Environmental Conservation. *International Journal of Political Science and Development, 4*(4), 140–145.

Adnan, M., Badi, W., Dereczynski, C., Luca, A. D., Ghosh, S., Iskandar, I., Kossin, J., Lewis, S., Otto, F., Pinto, I., Satoh, M., Vicente-Serrano, S. M., Wehner, M., & Zhou, B. (2021). Weather and Climate Extreme Events in a Changing Climate. In V. Masson-Delmotte P, A. Zhai, S. L. Pirani, C. Connors, S. Pean, N. Berger, Y. Caud, L. Chen, M. I. Goldfarb, M. Gomis, K. Huang, E. Leitzell, J. B. R. Lonnoy, T. K. Matthews, T. Maycock, O. Waterfield, R. Y. Yelekci, & B. Zhou (Eds.), *Climate Change 2021: The physical science basis. Contribution of working group I to the sixth assessment report of the Intergovernmental Panel on Climate Change* (pp. 1513–1766). Cambridge University Press. doi:10.1017/9781009157896.013

Afifi, T., Govil, R., Sakdapolrak, P., & Warner, K. (2012). *Climate Change, Vulnerability and Human Mobility: Perspectives of Refugees from the East and Horn of Africa*. UNHCR.

Agarwal, B. (2002). *Are we not peasants too? Land rights and women's claims in India*. The Populations Council, Inc. doi:10.31899/pgy2.1005

Aggarwal, A., Mittal, M., & Battineni, G. (2021). Generative Adversarial Network: An overview of theory and applications. *International Journal of Information Management Data Insights*, *1*(1), 100004. doi:10.1016/j.jjimei.2020.100004

Aggarwal, Y. P. (2008). *Science of Educational Research*. Nirmal Book Agency.

AghaKouchak, A., Chiang, F., Huning, L. S., Love, C. A., Mallakpour, I., Mazdiyasni, O., Moftakhari, H., Papalexiou, S. M., Ragno, E., & Sadegh, M. (2020, May 30). Climate Extremes and Compound Hazards in a Warming World. *Annual Review of Earth and Planetary Sciences*, *48*(1), 519–548. doi:10.1146/annurev-earth-071719-055228

Ahangari, H., Tavassoli, M., Khezerlou, A., Salmi, N. K., Ehsani, A., & Reza, M. (2023). A review on the determination of biogenic amines in fresh and processed fish products using HPLC, LC-MS/MS and other chromatographic methods. *Combinatorial Chemistry & High Throughput Screening*, *26*(15), 1386–2073. doi:10.2174/1386207326666230316141040

Ahmad, I., Arif, M. S., Cheema, I. I., Thollander, P., & Khan, M. A. (2020). Drivers and barriers for efficient energy management practices in energy-intensive industries: A case-study of iron and steel sector. *Sustainability (Basel)*, *12*(18), 7703. doi:10.3390/su12187703

Ahmed Ali, K., Ahmad, M. I., & Yusup, Y. (2020). Issues, impacts, and mitigations of carbon dioxide emissions in the building sector. *Sustainability (Basel)*, *12*(18), 7427. doi:10.3390/su12187427

Ahmed, A. A. (2020). Glimpse of fish as perishable staple. *Al-Qadisiyah Journal for Agriculture Sciences*, *10*(2), 349–375. doi:10.33794/qjas.2020.167497

Ahmed, F., Al-Amin, A. Q., Mohamad, Z. F., & Chenayah, S. (2016). Agriculture and food security challenge of climate change: A dynamic analysis for policy selection. *Scientia Agrícola*, *73*(4), 311–321. doi:10.1590/0103-9016-2015-0141

Ahmed, K., Atiqul Haq, S. M., & Bartiaux, F. (2019). The nexus between extreme weather events, sexual violence, and early marriage: A study of vulnerable populations in Bangladesh. *Population and Environment*, *40*(3), 303–324. doi:10.1007/s11111-019-0312-3

Ahmed, Z. A. M., Dosoki, M. I., & Nasr, S. A. A. (2012). Review article; occupational hazards in fish industry. *World Journal of Fish and Marine Sciences*, *4*(2), 201–210. doi:10.5829/idosi.wjfms.2012.04.02.61203

Aiming higher: Elevating meaningful youth engagement for climate action: United nations development programme. (2022). UNDP. https://www.undp.org/publications/aiming-higher-elevating-meaningful-youth-engagement-climate-action

Aiyadurai, A., & Ingole, P. (2021, November 29). Invisibility of caste in environmental studies. *The Indian Express*. https://indianexpress.com/article/opinion/columns/invisibilityof-caste-in-environmental-studies-7644987/

Akbari, N., Failler, P., Pan, H., Drakeford, B. M., & Forse, A. (2023). The impact of fisheries on the economy: A systematic review on the application of general equilibrium and input–output methods. *Sustainability*, *15*(7), 6089–6089. doi:10.3390/su15076089

Akdeniz, G. (2007). *DDİDD Simulation Theory*. Gedizak deniz. www.gedizakdeniz.com

Akhiljith, P. J., Liya, V. B., Rojith, G., Zacharia, P. U., Grinson, G., Ajith, S., Lakshmi, P. M., Sajna, V. H., & Sathianandan, T. V. (2019). Climatic projections of Indian ocean during 2030, 2050, 2080 with implications on fisheries sector. *Journal of Coastal Research*, *86*(sp1), 198. doi:10.2112/si86-030.1

Akinsanola, A. A., Ogunjobi, K. O., Abolude, A. T., & Salack, S. (2021). Projected changes in wind speed and wind energy potential over West Africa in CMIP6 models. *Environmental Research Letters*, *16*(4), 044033. doi:10.1088/1748-9326/abed7a

Aktan, Y., Dede, A., & Ciftci, P.S. (2008). *Mucilage event associated with diatoms and dinoflagellates in Sea of Marmara, Turkey*. Harmful Algae News.

Akwada, D. R., & Akinlabi, E. T. (2016, July). Economic, social, and environmental assessment of bamboo for infrastructure development. In *International conference on infrastructure development in Africa* (pp. 1-15).

Al-Abri, M., Al-Ghafri, B., Bora, T., Dobretsov, S., Dutta, J., Castelletto, S., Rosa, L., & Boretti, A. (2019). Author correction: Chlorination disadvantages and alternative routes for biofouling control in reverse osmosis desalination. *npj Clean Water*, *2*(1). doi:10.1038/s41545-019-0033-2

Alfonso, S., Gesto, M., & Sadoul, B. (2020). Temperature increase and its effects on fish stress physiology in the context of global warming. *Journal of Fish Biology*, *98*(6), 1496–1508. doi:10.1111/jfb.14599

Ali, N. A. M., Mohamad, M., Juahir, H., Kamarudin, M. K. A., Muhammad, N., & Lee, M. S. M. R. (2018). Source apportionment of social support and quality of life index among drug abuse inmates using multiple linear regression. *Journal of Fundamental and Applied Sciences*, *9*(2S), 46. doi:10.4314/jfas.v9i2s.4

Allen, M. R., Dube, O. P., Solecki, W. A., Aragon-Durand, F., Cramer, W., Humphreys, S., Kainuma, M., Kala, J., Mahowald, N., Mulugetta, Y., Perez, R., Wairiu, M., & Zickfeld, K. (2019). Framing and context. In Masson-Delmotte, V., Zhai, P., Portner, H. O., Roberts, D., Skea, J., Shukla, P. R., Pirani, A., Moufouma-Okia, W., Pean, C., Pidcock, R., Connors, S., Matthews, J. B. R., Chen, Y., Zhou, X., Gomis, M. I., Lonnoy, E., Maycock, T., Tignor, M. and Waterfield, T. (Eds.). *Global Warming of 1.5°C. An IPCC Special Report on the impacts of global warming of 1.5°C above pre-industrial levels and related global greenhouse gas emission pathways, in the context of strengthening the global response to the threat of climate change, sustainable development, and efforts to eradicate poverty* (pp. 49-92). Cambridge University Press. doi:10.1017/9781009157940.003

Allison, E. H., Perry, A. L., Badjeck, M. C., Adger, W. N., Brown, K., Conway, D., Halls, A. S., Pilling, G. M., Reynolds, J. D., Andrew, N. L., & Dulvy, N. K. (2009). Vulnerability of national economies to the impacts of climate change on fisheries. *Fish and Fisheries*, *10*(2), 173–196. doi:10.1111/j.1467-2979.2008.00310.x

Allouzi, M. M. A., Tang, D. Y. Y., Chew, K. W., Rinklebe, J., Bolan, N., Allouzi, S. M. A., & Show, P. L. (2021). Micro (nano) plastic pollution: The ecological influence on soil-plant system and human health. *The Science of the Total Environment*, *788*, 147815. doi:10.1016/j.scitotenv.2021.147815 PMID:34034191

Al-Saidi, M., Das, P., & Saadaoui, I. (2021). Circular economy in basic supply: Framing the approach for the water and food sectors of the Gulf cooperation council countries. *Sustainable Production and Consumption*, *27*, 1273–1285. doi:10.1016/j.spc.2021.03.004

Alston, M., Whittenbury, K., et al. (2014). Are climate challenges reinforcing child and forced marriage and dowry as adaptation strategies in the context of Bangladesh? *Women's Studies International Forum, 47*(Part A), 137-144. doi:10.1016/j.wsif.2014.08.001

Alstone, P., Gershenson, D., & Kammen, D. M. (2015). Decentralized energy systems for clean electricity access. *Nature Climate Change, 5*(4), 305–314. doi:10.1038/nclimate2512

Amaravathi, T., Parimalam, P., Premalatha, M., Hemalatha, G., & Ganguli, A. (2016). Health hazard of women employed in small scale seafood processing units. *Indian Journal of Geo-Marine Sciences, 45*(4), 574–582. https://nopr.niscpr.res.in/bitstream/123456789/35094/1/IJMS%2045%284%29%20574-582.pdf

Ambrose, G., Das, K., Fan, Y., & Ramaswami, A. (2023). Comparing happiness associated with household and community gardening: Implications for Food Action Planning. *Landscape and Urban Planning, 230*, 104593. doi:10.1016/j.landurbplan.2022.104593

American Psychiatric Association. (n.d.). *Climate Change and Mental Health*. APA https://www.psychiatry.org/patients-families/climate-change-and-mental-health-connections/affects-on-mental-health

American Public Health Association. (n.d). *LGBTQ Health Caucus*. APHA. https://www.apha.org/apha-communities/caucuses/lgbtq-health-caucus)

Amorim-Maia, A. T., Anguelovski, I., Chu, E., & Connolly, J. (2022). Intersectional climate justice: A conceptual pathway for bridging adaptation planning, transformative action, and social equity. *Urban Climate, 41*, 101053. doi:10.1016/j.uclim.2021.101053

Anand, I. (2016). Dalit emancipation and the land question. *Economic and Political Weekly, 51* (47), 12–14.

Anastario, M., Shehab, N., & Lawry, L. (2009). Increased gender-based violence among women internally displaced in Mississippi 2 years post-Hurricane Katrina. *Disaster Medicine and Public Health Preparedness, 3*(1), 18–26. doi:10.1097/DMP.0b013e3181979c32 PMID:19293740

Andrae, A. & Edler, T. (2015). On Global Electricity Usage of Communication Technology: Trends to 2030. *Challenges 6*(1), 117-157. . doi:10.3390/challe6010117

Andrews, F. M. (1974). Social indicators of perceived life quality. *Social Indicators Research, 1*(3), 279–299. doi:10.1007/BF00303860

Andrijevic, M., Crespo Cuaresma, J., Muttarak, R., & Schleussner, C. F. (2020). Governance in socioeconomic pathways and its role for future adaptive capacity. *Nature Sustainability, 3*(1), 35–41. doi:10.1038/s41893-019-0405-0

Anisha, P., Athira, P. S., Anagha, B., Charles, P. E., Prabakaran, K., & Rajaram, R. (2023). First report on occurrence of heavy metals in dried fishes from major fishing villages in Kerala coast, Southwest India. *Hygiene and Environmental Health Advances, 6*, 100051. doi:10.1016/j.heha.2023.100051

Anonymous. (2019) *The ins and outs of carbon dioxide*. Sequestration. http://www.sequestration.org/science/carbondioxide.html

An, R., Yu, B., Li, R., & Wei, Y. M. (2018). Potential of energy savings and CO2 emission reduction in China's iron and steel industry. *Applied Energy, 226*, 862–880. doi:10.1016/j.apenergy.2018.06.044

Anuradha, P., Jayashree, K., Suresh Babu, V., & Sreenivasa Rao, V. (2019) *Addressing Climate Resilience and Livelihood opportunities through MGNREGS Assets in Telangana State. 10th INSEE Biennial Conference on Climate Change and Disaster: Challenges, Opportunities and Responses held at CESS*, Hyderabad.

Apollo, A., & Mbah, M. F. (2021). 'Challenges and Opportunities for Climate Change Education (CCE) in East Africa. A Critical Review. *Climate (Basel)*, *9*(93), 1–16. doi:10.3390/cli9060093

Arteaga, L. A., & Rousseaux, C. S. (2023). Impact of Pacific Ocean heatwaves on phytoplankton community composition. *Communications Biology*, *6*(1), 263. doi:10.1038/s42003-023-04645-0

Artificial Intelligence (AI) Bots Market Analysis – 2030. (2023). NextMSC. https://www.nextmsc.com/report/artificial-intelligence-ai-bots-market

Asia-Pacific Fishery Commission (APFIC). (2010). *Best practices to support and improve livelihoods of small-scale fisheries and aquaculture households* (pp. 140). RAP Publication 2010/21. ISBN: 978-92-5-106739-0. https://www.fao.org/3/i1966e/i1966e00.pdf

Asmelash, E., Prakash, G., Gorini, R., & Gielen, D. (2020). Role of IRENA for global transition to 100% renewable energy. *Accelerating the transition to a 100% renewable energy era*, 51-71.

Atif, S. (2023). Analysing the alignment between circular economy and industry 4.0 nexus with industry 5.0 era: An integrative systematic literature review. *Sustainable Development (Bradford)*, *31*(4), 2155–2175. doi:10.1002/sd.2542

Atmos. (2023, March 27). *Pakistan's climate crisis of Child marriage.* Atmos. https://atmos.earth/pakistan-floods-child-marriage

Austin, B. (1998). The effects of pollution on fish health. *Journal of Applied Microbiology, 85*(S1), 234S242S. doi:10.1111/j.1365-2672.1998.tb05303.x

Aydın, M. E. (2021). Mucilage formation in Marmara Sea possible causes and recommendations. I. Ozturk & M. Şeker (Eds.). In Ecology of the Marmara Sea: Formation and Interactions of Marine Mucilage, and Recommendations for Solutions (pp.49-68). Tek Ses Ofset Matbaacılık. Ankara.

Aydın, D., Yaldız, E., & Buyuksahin, S. (2017, December). *Sustainable hospital design for sustainable development.* In *Proceedings of the 8th International Conference on Urban Planning, Architecture, Civil and Environment Engineering*, Dubai, UAE (pp. 21-22).

Aydogmus F. (2022). Synchronization in Chaotic Dynamics of Spinor-type Fermionic Instantons for Different Dimension. *International Journal of Dynamics and Control, 2022.*

Aydogmus, F., & Tosyali, E. (2015). *Global Climate Change in Consideration of Chaos Theory.* 9th International Physics Conference of the Balkan Physical Union – BPU9, Istanbul.

Aydogmus, F. (2013). *Phase Space Behaviors of Spinor Type Instanton Attractors.* Istanbul University, Institute of Science and Technology.

Aydogmus, F., & Akdeniz, G. (2007). Is Global Climate Change Chaotic? *Nuclear Energy Symposium.* Istanbul University Faculty of Forestry.

Aydogmus, F., & Akdeniz, G. (2008). Global Climate Change And Human With A Syborg Science Doctrin, IV. *International Symposium on Sciences of Atmosphere*, Istanbul Teknik University, pp.437-443.

Aydogmus, F., & Tosyali, E. (2017). The Effect of Chaos Theory on Predictability of the Atmosphere. *The Middle East Journal*, 26–31.

Azevedo, S. C. (2011). *Water usage effectiveness (WUETM): A green grid datacenter sustainability metric.* Green Grid.

Baecker, A. (2020). Reimagining Revolutionary Labor in the People's Commune: Amateurism and Social Reproduction in the Maoist Countryside (Doctoral dissertation).

Bai, M., & Li, D. (2020). Quantity of plastic waste input into the ocean from China based on a material fow analysis model. *Anthropoc Coasts, 3*(1), 1–5. doi:10.1139/anc-2018-0028

Baird, R. W., & Hooker, S. K. (2000). Ingestion of plastic and unusual prey by a juvenile harbour porpoise. *Marine Pollution Bulletin, 40*(8), 719–720. doi:10.1016/S0025-326X(00)00051-5

Bakar, A. A., Osman, M. M., Bachok, S., Ibrahim, M., & Mohamed, M. Z. (2015). Modelling economic wellbeing and social wellbeing for sustainability: A theoretical concept. *Procedia Environmental Sciences, 28*, 286–296. doi:10.1016/j.proenv.2015.07.037

Baker, R. E. (2020). Climate change drives increase in modeled HIV prevalence. *Climatic Change, 163*(1), 237–252. doi:10.1007/s10584-020-02753-y

Balbus, J., Crimmins, A., Easterling, D. R., Gamble, J. L., Kunkel, K. E., Saha, S., & Sarofim, M. C. (2016a). Chapter 1: Introduction: Climate Change and Human Health. The Impacts of Climate Change on Human Health in the United States: A Scientific Assessment U.S. Global Change Research Program, 25-42.

Balbus, J., Beard, C. B., Bell, J. E., Crimmins, A., Dodgen, D., Easterling, D. R., Eisen, R. J., Fann, N., Gamble, J. L., Hawkins, M. D., Herring, S. C., Jantarasami, L., Kunkel, K. E., Mills, D. M., Saha, S., Sarofim, M. C., Trtanj, J., & Ziska, L. (2016b). *The Impacts of Climate Change on Human Health in the United States: A Scientific Assessment*. U.S. Global Change Research Program. doi:10.7930/J0R49N

Balcik, B., Beamon, B. M., Krejci, C. C., Muramatsu, K. M., & Ramirez, M. (2010). Coordination in humanitarian relief chains: Practices, challenges and opportunities. *International Journal of Production Economics, 126*(1), 22–34. doi:10.1016/j.ijpe.2009.09.008

Balsara, S., Jain, P. K., & Ramesh, A. (2021). An integrated methodology to overcome barriers to climate change mitigation strategies: A case of the cement industry in India. *Environmental Science and Pollution Research International, 28*(16), 20451–20475. doi:10.1007/s11356-020-11566-6 PMID:33410023

Balwan, W. K., Saba, N., & Rasool, N. (2021). An overview of climate change and food security in India. *Annals of the Romanian Society for Cell Biology, 25*(4), 20124–20137. http://annalsofrscb.ro/index.php/journal/article/view/9066/6625

Bama (2012). Karukku (L. Holmström Trans.). Oxford University Press.

Banda, S., & Mpolomoka, D. L. (2018). Culturally relevant education and training for communities: A review. *American Educational Research Journal, 6*(2), 88–93. doi:10.30918/AERJ.62.18.019

Banda, S., Mpolomoka, D. L., Mbono, D., & Sampa, R. L. (2017). Use of questions in qualitative research: How questions guided our study. *International Journal of Developmental Research, 7*(12), 17895–17898.

Banda, S., Namafe, C. M., & Chakanika, W. W. (2015). Traditional Environmental Knowledge among Lozi Adults in Mitigating Climate Change in the Barotse Plains of Western Zambia. *International Journal of Humanities, Social. Sciences and Education, 2*(9), 222–239.

Banga, J. (2019). The green bond market: A potential source of climate finance for developing countries. *Journal of Sustainable Finance & Investment, 9*(1), 17–32. doi:10.1080/20430795.2018.1498617

Barik, S. (2022, November 25). Millets in the mainstream: How Odisha's Kutia Kondh tribe rediscovered a palate for the "poor man's food." *The Hindu*. https://www.thehindu.com/news/national/other-states/a-tribes-efforts-in-odisha-to-bring-millets-back-on-their-platters/article66178864.ece

Barnes, D. K., Galgani, F., Thompson, R. C., & Barlaz, M. (2009). Accumulation and fragmentation of plastic debris in global environments. *Philosophical Transactions of the Royal Society of London. Series B, Biological Sciences*, *364*(1526), 1985–1998. doi:10.1098/rstb.2008.0205 PMID:19528051

Barreiros, J. P., & Barcelos, J. (2001). Plastic ingestion by a leatherback turtle Dermochelys coriacea from the Azores (NE Atlantic). *Marine Pollution Bulletin*, *42*(11), 1196–1197. doi:10.1016/S0025-326X(01)00215-6 PMID:11763235

Barreto Torres, L. D., Asmus, G. F., & da Cal Seixas, S. R. (2019). Quality of Life and Sustainable Development. Encyclopedia of Sustainability in Higher Education, 1333-1340. Taylor & Francis.

Bastardie, F., Hornborg, S., Ziegler, F., Gislason, H., & Eigaard, O. R. (2022). Reducing the fuel use intensity of fisheries: Through efficient fishing techniques and recovered fish stocks. *Frontiers in Marine Science, 9.* doi:10.3389/fmars.2022.817335

Baudrillard, J. (2004). *Transparency of Evil.* Ayrıntı Publications.

Baviskar, A., & Gidwani, V. (2019). The lives of waste and pollution. *Economic and Political Weekly*, *54*(47), 7–8.

Bawdy, T. (2016). Global warming: Data centres to consume three times as much energy in next decade, experts warn. *The Independent.*

Bax, N., Novaglio, C., Maxwell, K. H., Meyers, K., Mccann, J., Jennings, S., Frusher, S., Fulton, E. A., Nursey-Bray, M., Fischer, M., Anderson, K., Layton, C., Emad, R., Alexander, K. A., Rousseau, Y., Lunn, Z., & Carter, C. G. (2022). Ocean resource use: Building the coastal blue economy. *Reviews in Fish Biology and Fisheries*, *32*, 189–207. doi:10.1007/s11160-021-09636-0

BBC News. (2022, August 28). *Pakistan floods: Appeals for aid as 119 more die in a day. BBC News.* BBC News. https://www.bbc.com/news/world-asia-62704004.amp

Beccarello, M., & Di Foggia, G. (2023). The Strategic Role of Energy Efficiency and Industry 4.0 Interventions in Manufacturing. *Journal of Management and Sustainability*, *13*(1), 31–44. doi:10.5539/jms.v13n1p31

Becker, P. (2014). *Sustainability Science: Managing Risk and Resilience for Sustainable Development*. Elsevier.

Begum, M., Masud, M. M., Alam, L., Mokhtar, M. B., & Amir, A. A. (2022). The impact of climate variables on marine fish production: An empirical evidence from Bangladesh based on autoregressive distributed lag (ARDL) approach. *Environmental Science and Pollution Research International*, *29*(58), 87923–87937. doi:10.1007/s11356-022-21845-z

Behal, A. (2021, January 11). How climate change is affecting the LGBTQIA+ community. DownToEarth. Retrieved May 25, 2023, from https://www.downtoearth.org.in/blog/environment/how-climate-change-is-affecting-the-lgbtqia-community-74988

Behere, P. B., & Behere, A. P. (2008). Farmers' suicide in Vidarbha region of Maharashtra state: A myth or reality? *Indian Journal of Psychiatry*, *50*(2), 124–127. doi:10.4103/0019-5545.42401 PMID:19742218

Belkhir, L., & Elmeligi, A. (2018). Assessing ICT global emissions footprint: Trends to 2040 & recommendations. *Journal of Cleaner Production*, *177*, 448–463. doi:10.1016/j.jclepro.2017.12.239

Benanav, A. (2022). Socialist investment, dynamic planning, and the politics of human need. *Rethinking Marxism*, *34*(2), 193–204. doi:10.1080/08935696.2022.2051375

Berkmen, H. (2007). From Cosmos to Chaos and From Chaos to Cosmos, Irregular Systems: Theory and Practices Working Group VII. *National Symposium.*

Berniak-Woźny, J., & Rataj, M. (2023). Towards Green and Sustainable Healthcare: A Literature Review and Research Agenda for Green Leadership in the Healthcare Sector. *International Journal of Environmental Research and Public Health, 20*(2), 908. doi:10.3390/ijerph20020908 PMID:36673663

Bezerra, J. C., Walker, T. R., Clayton, C. A., & Adam, I. (2021). Single-Use Plastic Bag Policies in the Southern African Development Community. *Environmental Challenges, 3*, 100029. doi:10.1016/j.envc.2021.100029

Bhanumati, P., de Haan, M., & Tebrake, J. W. (2022). *Greenhouse Emissions Rise to Record, Erasing Drop During Pandemic*. IMF Blogs.

Bhatia, S., & Albarrak, A. S. (2023). A blockchain-driven food supply chain management using QR code and XAI-faster RCNN architecture. *Sustainability (Basel), 15*(3), 2579. doi:10.3390/su15032579

Bhatta, B., Shrestha, S., Shrestha, P. K., & Talchabhadel, R. (2019). Evaluation and application of a SWAT model to assess the climate change impact on the hydrology of the Himalayan River Basin. *Catena, 181*, 104082. doi:10.1016/j.catena.2019.104082

Bhatti, S. S., Tripathi, N. K., Nagai, M., & Nitivattananon, V. (2017). Spatial interrelationships of quality of life with land use/land cover, demography and urbanization. *Social Indicators Research, 132*(3), 1193–1216. doi:10.1007/s11205-016-1336-z

Bhimraj, M. (2020). A Dalit critique of environmental justice in India. [Rajiv Gandhi National University of Law.]. In R. Soni, S. Mukhopadhyay, B. Agrawal, S. Ghosh, & A. Hargava (Eds.), Contemporary environmental concerns: Multidisciplinary aspects of environmental law (pp. 89-120).

Bindra, P. S. (2017). *The vanishing: India's wildlife crisis*. Penguin Books.

Birkmann, J., Jamshed, A., McMillan, J. M., Feldmeyer, D., Totin, E., Solecki, W., Ibrahim, Z. Z., Roberts, D., Kerr, R. B., Poertner, H.-O., Pelling, M., Djalante, R., Garschagen, M., Leal Filho, W., Guha-Sapir, D., & Alegría, A. (2022). Understanding human vulnerability to climate change: A global perspective on index validation for adaptation planning. *The Science of the Total Environment, 803*, 150065. doi:10.1016/j.scitotenv.2021.150065 PMID:34525713

Black, R. (2021). *Taking Stock: A Global Assessment of Net Zero Targets* (ECIU and Oxford Net Zero, 2021).

Black, M., Karki, J., Lee, A. C. K., Makai, P., Baral, Y. R., Kritsotakis, E. I., Bernier, A., & Heckmann, A. F. (2019). The health risks of informal waste workers in the Kathmandu Valley: A cross-sectional survey. *Public Health, 166*, 10–18. doi:10.1016/j.puhe.2018.09.026

Blair, R. M., Waldron, S., Phoenix, V., & Gauchotte-Lindsay, C. (2017). Micro- and nanoplastic pollution of freshwater and wastewater treatment systems. *Springer Science Reviews, 5*(1-2), 19–30. doi:10.1007/s40362-017-0044-7

Blanchard, K. (2021, September 16). *LGBTQIA+ people and disasters: A need for policy guidance.* UNDRR Prevention Web. https://www.preventionweb.net/blog/lgbtqia-people-and-disasters-need-policy-guidance

Bleeker, E. A. J., & Verbruggen, E. (2009). *Bioaccumulation of polycyclic aromatic hydrocarbons in aquatic organisms.* National Institute for Human Health and the Environment (RIVM). https://www.researchgate.net/publication/281107718_Bioaccumulation_of_Polycyclic_Aromatic_Hydrocarbons_in_Aquatic_Organisms

Bochenek, B., & Ustrnul, Z. (2022). Machine learning in weather prediction and climate analyses—Applications and perspectives. *Atmosphere (Basel), 13*(2), 180. doi:10.3390/atmos13020180

Bolte, G., Dandolo, L., Gepp, S., Hornberg, C., & Lumbi, S. L. (2023). Climate change and health equity: A public health perspective on climate justice. *Journal of Health Monitoring, 8*(Suppl 6), 3. PMID:38105794

Bomuhangi, A., Nabanoga, G., Namaalwa, J. J., Jacobson, M. G., Abwoli, B., & Wich, S. (2016). Local Communities' Perceptions of Climate Variability in the Mt. Elgon Region, Eastern Uganda. *Cogent Environmental Science*, *2*(1168276), 1168276. doi:10.1080/23311843.2016.1168276

Bond, R. (2008). *The book of nature*. Penguin India.

Bond, T., Ferrandiz-Mas, V., Felipe-Sotelo, M., & van Sebille, E. (2018). The occurrence and degradation of aquatic plastic litter based on polymer physicochemical properties: A review. *Critical Reviews in Environmental Science and Technology*, *48*(7-9), 685–722. doi:10.1080/10643389.2018.1483155

Bowling, A. (1995). What things are important in people's lives? A survey of the public's judgements to inform scales of health related quality of life. *Social Science & Medicine*, *41*(10), 1447–1462. doi:10.1016/0277-9536(95)00113-L PMID:8560313

Bowyer, J., Fernholz, K., Frank, M., Howe, J. E. F. F., Bratkovich, S. T. E. V. E., & Pepke, E. (2014). *Bamboo products and their environmental impacts: revisited*. Dovetail Partners, Incorporated.

Brady, G. A., Kapur, N., Summers, J. L., & Thompson, H. M. (2013, December). A case study and critical assessment in calculating power usage effectiveness for a data centre. *Energy Conversion and Management*, *76*, 155–161. doi:10.1016/j.enconman.2013.07.035

Brander, K. M. (2007). Global fish production and climate change. *Proceedings of the National Academy of Sciences of the United States of America*, *104*(50), 19709–19714. doi:10.1073/pnas.0702059104

Braun, V., & Clarke, V. M. (2012). Thematic Analysis. In H. Cooper, P. M. Camic, D. L. Long, A. T. Panter, D. Rindskopf, & K. J. Sher (Eds.), *Research Designs: Quantitative, Qualitative, Neuropsychological and Biological.' APA Handbook of Research Methods in Psychology, 2, 57-71*. American Psychological Association. doi:10.1037/13620-004

Bravo-Olivas, M., Chavez-Dagostino, R., Malcolm, C., & Espinoza-Sanchez, R. (2015). Notes on the quality of life of artisanal small-scale fishermen along the pacific coast of Jalisco, Mexico. *Sustainability*, *7*(5), 6046–6068. doi:10.3390/su7056046

Brazier, A. (2015). *Climate Change in Zimbabwe: Facts for Planners and Decision Makers*. Konrad-Adenaner – Stiftung.

Brennecke, D., Duarte, B., Paiva, F., Caçador, I., & Canning-Clode, J. (2016). Microplastics as vector for heavy metal contamination from the marine environment. *Estuarine, Coastal and Shelf Science*, *178*, 189–195. doi:10.1016/j.ecss.2015.12.003

Briassoulis, D., Babou, E., Hiskakis, M., Scarascia, G., Picuno, P., Guarde, D., & Dejean, C. (2013). Review, mapping and analysis of the agricultural plastic waste generation and consolidation in Europe. *Waste Management & Research*, *31*(12), 1262–1278. doi:10.1177/0734242X13507968 PMID:24293230

Brickell, C. D., Alexander, C., David, J. C., Hetterscheid, W. L., Leslie, A. C., Male'cot, V., Jin, X., & Cubey, J. J. 2009. International code of nomenclature for cultivated plants, 8th ed. Scripta Horticulturae.

Brinkmann, S., & Kvale, S. (2015). *Interviews: Learning the Craft of Qualitative Research Interviewing*. Sage Publication.

Brody, A., Demetriades, J., & Esplen, E. (2008). *Gender and Climate Change: Mapping the Linkages – A scoping Study on Knowledge and Gaps*. BRIDGE Institute of Development Studies.

Brough, D., & Jouhara, H. (2020). The aluminium industry: A review on state-of-the-art technologies, environmental impacts and possibilities for waste heat recovery. *International Journal of Thermofluids*, *1*, 100007. doi:10.1016/j.ijft.2019.100007

Brown, J. B., & Lichter, D. T. (2004). Poverty, Welfare and the Livelihood Strategies of Non-Metropolitan Single Mothers. *Rural Sociology, 69*(2), 282–301. doi:10.1526/003601104323087615

Brunke, J. C., Johansson, M., & Thollander, P. (2014). Empirical investigation of barriers and drivers to the adoption of energy conservation measures, energy management practices and energy services in the Swedish iron and steel industry. *Journal of Cleaner Production, 84*, 509–525. doi:10.1016/j.jclepro.2014.04.078

Bryan, E., Theis, S., Choufani, J., De Pinto, A., Meinzen-Dick, R. S., & Ringler, C. (2017). *Conceptual framework: Gender, climate change, and nutrition integration initiative*. (GCAN Policy Note 7). Washington, D.C.: International Food Policy Research Institute (IFPRI). http://ebrary.ifpri.org/cdm/singleitem/collection/p15738coll2/id/132300

Büchs, M., Bardsley, N., & Duwe, S. (2011). Who bears the brunt? Distributional effects of climate change mitigation policies. *Critical Social Policy, 31*(2), 285–307. doi:10.1177/0261018310396036

Buckingham, K. C., Wu, L., & Lou, Y. (2013). *Can't See the (Bamboo) Forest for the Trees: Examining Bamboo's Fit Within International Forestry Institutions*. Springer.

Bukhari, S. I. A., & Rizvi, S. S. (2015). Impact of Floods on Women: With Special Reference to Flooding Experience of 2010 Flood in Pakistan. *Journal of Geography & Natural Disasters, 5*(2).

Bullard, R. D. (2001). Environmental justice in the 21st century: Race still matters. *Phylon, 49*(3/4), 151–171. https://www.jstor.org/stable/3132626?origin=crossref. doi:10.2307/3132626

Bull, N., Riise, T., & Moen, B. E. (2002). Work-related injuries and occupational health and safety factors in smaller enterprises—a prospective study. *Occupational Medicine, 52*(2), 70–74. doi:10.1093/occmed/52.2.70

Burke, M., Gong, E., & Jones, K. (2015). Income shocks and HIV in Africa. *Economic Journal (London), 125*(585), 1157–1189. doi:10.1111/ecoj.12149

Burridge, L., Weis, J. S., Cabello, F., Pizarro, J., & Bostick, K. (2010). Chemical use in salmon aquaculture: A review of current practices and possible environmental effects. *Aquaculture (Amsterdam, Netherlands), 306*(1-4), 7–23. doi:10.1016/j.aquaculture.2010.05.020

Bwalya, S. M. (2010). *Climate Change in Zambia: Opportunities for Adaptation and Mitigation through Africa Bio-Carbon Initiative*. Centre for International Forest Research, Southern Africa Regional Office.

Byrne, D., Garcia-Soto, C., Hamilton, G., Leuliette, E., Yu, L., Campos, E., Durack, P. J., Manzella, G. M. R., Tadokoro, K., Schmitt, R. W., Arkin, P., Bryden, H., Nurse, L., Milliman, J., Inniss, L., & Bernal, P. (2017). The Ocean's role in the hydrological cycle. In United Nations (Ed.), *The first global integrated marine assessment world ocean assessment I* (pp. 91–104). Cambridge University Press., doi:10.1017/9781108186148.007

Cadez, S., & Czerny, A. (2016). Climate change mitigation strategies in carbon-intensive firms. *Journal of Cleaner Production, 112*, 4132–4143. doi:10.1016/j.jclepro.2015.07.099

Cambel, A. B. (1993). *Applied Chaos Theory a Paradigm for Complexity*. Academic Press.

Carbon credit (2021), Cop26-Key-Outcomes-Agreed-At-The-Un-Climate-Talks-In-Glasgow https://www.carbonbrief.org/cop26-key-outcomes-agreed-at-the-un-climate-talks-in-glasgow/

CARE. (2015). *Rapid Gender Analysis Cyclone Pam Vanuatu*. CARE. https://care.ca/wp-content/uploads/2018/12/RGA-Cyclone-Pam-Vanuatu-7-April-2015.pdf

Carminati, B., Ferrari, E., & Rondanini, C. (2018). Blockchain as a platform for secure interorganizational business processes. *2018 IEEE 4th International Conference on Collaboration and Internet Computing* (122–129). IEEE.

Carmona, R. (2022). Resilience requires change: Assessing Pehuenche responses to climate change impacts in Southern Chile. *Environmental Justice*, *15*(3), 185–195. doi:10.1089/env.2021.0044 PMID:35686287

Carpenter, E. J., & Smith, K. L. Jr. (1972). Plastics on the Sargasso Sea surface. *Science*, *175*(4027), 1240–1241. doi:10.1126/science.175.4027.1240 PMID:5061243

Casino, F., Dasaklis, T. K., & Patsakis, C. (2019). A systematic literature review of blockchain-based applications: Current status, classification and open issues. *Telematics and Informatics*, *36*, 55–58. doi:10.1016/j.tele.2018.11.006

Cassou, E. M. (2018). *Plastics. Agricultural Pollution: World Bank.* World Bank. https://openknowledge.worldbank.org/handle/10986/29505.

Cavanagh, S. (1997). Content Analysis: Concepts, Methods and Citations. *Nurse Researcher*, *4*, 5–16. PMID:27285770

CDC. (2001). *Measuring Healthy Days: Population Assessment of Health-related Quality of Life*. CDC. https://www.cdc.gov/hrqol/pdfs/mhd.pdf.

CDC. (2006). Heat-related deaths: United States, 1999–2003. *Morbidity and Mortality Weekly Report*, *55*, 769–798. PMID:16874294

Central Marine Fisheries Research Institute (CMFRI)-National Innovations on Climate Resilient Agriculture (NICRA). (2013). *Marine fisheries*. Report of work done at CMFRI submitted to CRIDA, (pp. 14). http://eprints.cmfri.org.in/9465/1/CMFRI_Annual_Report_2012-13.pdf

Central Marine Fisheries Research Institute (CMFRI)-National Innovations on Climate Resilient Agriculture (NICRA). (2014). Marine fisheries. Report of work done at CMFRI submitted to CRIDA, (pp. 32). http://eprints.cmfri.org.in/10164/1/CMFRI%20Annual%20Report%202013-14.pdf

Central Marine Fisheries Research Institute (CMFRI)-National Innovations on Climate Resilient Agriculture (NICRA). (2016). Marine fisheries. Report of work done at CMFRI submitted to CRIDA (pp. 27. http://eprints.cmfri.org.in/10897/1/CMFRI%20ANNUAL%20REPORT%202015-16.pdf

Chadha, M. L. (2023). Home Gardening: The Way Forward to Be Safe and Healthy. In Vegetables for Nutrition and Entrepreneurship (pp. 217-239). Springer Nature.

Chae, Y., & An, Y. J. (2018). Current research trends on plastic pollution and ecological impacts on the soil ecosystem: A review. *Environmental Pollution*, *240*, 387–395. doi:10.1016/j.envpol.2018.05.008 PMID:29753246

Chakraborty, L., Rus, H., Henstra, D., Thistlethwaite, J., & Scott, D. (2020). A place-based socioeconomic status index: Measuring social vulnerability to flood hazards in the context of environmental justice. *International Journal of Disaster Risk Reduction*, *43*, 101394. doi:10.1016/j.ijdrr.2019.101394

Chakraborty, S., Chattopadhyay, R., & Chakraborty, S. (2020). An integrated D-MARCOS method for supplier selection in an iron and steel industry. *Decision Making: Applications in Management and Engineering*, *3*(2), 49–69.

Chalmin-Pui, L. S., Griffiths, A., Roe, J., Heaton, T., & Cameron, R. (2021). Why garden?–Attitudes and the perceived health benefits of home gardening. *Cities (London, England)*, *112*, 103118. doi:10.1016/j.cities.2021.103118

Chan, N. W., Ghani, A. A., Samat, N., Roy, R., Tan, M. L., & Rahman, H. A. (2021). Addressing water resources shortfalls due to climate change in Penang, Malaysia. In Water Security in Asia: Opportunities and Challenges in the Context of Climate Change (pp. 239-249). Springer International Publishing.

Chant, S. (2007). *Children in Female Headed Households: Interrogating the Concept of an 'Inter-Generational Transmission of Disadvantage' with Particular Reference to the Gambia, Philippines and Costa Rica.* New Working Paper Series, London: London School of Economics, Gender Institute.

Chao, T., Yunbao, X., Chengbo, D., Bo, L., & Ullah, S. (2023). Financial integration and renewable energy consumption in China: Do education and digital economy development matter? *Environmental Science and Pollution Research International*, *30*(5), 12944–12952. doi:10.1007/s11356-022-22852-w PMID:36121627

Chauhan, A., Singh, R. P., Dash, P., & Kumar, R. (2021). Impact of tropical cyclone "Fani" on land, ocean, atmospheric and meteorological parameters. *Marine Pollution Bulletin*, *162*, 111844. doi:10.1016/j.marpolbul.2020.111844 PMID:33220915

Chenje, D., Sola, L., & Paleczny, D. (1998). *The State of the Zimbabwe's Environment. Ministry of Mines, Environment and Tourism.* Government Printers.

Chen, L., Chen, Z., Zhang, Y., Liu, Y., Osman, A. I., Farghali, M., Hua, J., Al-Fatesh, A., Ihara, I., Rooney, D. W., & Yap, P.-S. (2023, October). Yap, P.-S. (2023). *Artificial Intelligence-based solutions for climate change: A Review. Environmental Chemistry Letters*, *21*(5), 2525–2557. doi:10.1007/s10311-023-01617-y

Chen, M., Weng, Y., Semple, K., Zhang, S., Jiang, X., Ma, J., & Dai, C. (2021). Sustainability and innovation of bamboo winding composite pipe products. *Renewable & Sustainable Energy Reviews*, *144*, 110976. doi:10.1016/j.rser.2021.110976

Chen, Q., Allgeier, A., Yin, D., & Hollert, H. (2019). Leaching of endocrine disrupting chemicals from marine microplastics and mesoplastics under common life stress conditions. *Environment International*, *130*, 104938. doi:10.1016/j.envint.2019.104938 PMID:31234003

Chen, Q., Su, H., Yu, X., & Hu, Q. (2020). Livelihood vulnerability of marine fishermen to multi-stresses under the vessel buyback and fishermen transfer programs in China: The case of Zhoushan city, Zhejiang Province. *International Journal of Environmental Research and Public Health*, *17*(3), 765. doi:10.3390/ijerph17030765

Chepkoech, W., Mungai, N. W., Stöber, S., & Lotze-Campen, H. (2020). Understanding adaptive capacity of smallholder African indigenous vegetable farmers to climate change in Kenya. *Climate Risk Management*, *27*, 100204. doi:10.1016/j.crm.2019.100204

Chersich, M. F., Pham, M. D., Areal, A., Haghighi, M. M., Manyuchi, A., Swift, C. P., Wernecke, B., Robinson, M., Hetem, R., Boeckmann, M., & Hajat, S. (2020). Associations between high temperatures in pregnancy and risk of preterm birth, low birth weight, and stillbirths: Systematic Review and meta-analysis. *BMJ (Clinical Research Ed.)*, *m3811*, m3811. doi:10.1136/bmj.m3811 PMID:33148618

Chiang, H. C., Ko, Y. C., Chen, S. S., Yu, H. S., Wu, T. N., & Chang, P. Y. (1993). Prevalence of shoulder and upper-limb disorders among workers in the fish-processing industry. *Scandinavian Journal of Work, Environment & Health*, *19*(2), 126–131. doi:10.5271/sjweh.1496

Cho, R., Wiseman, J., Amblee, R., Amaan, A., Green, J., & Sharma, Y. Hakim, Basak, S., Singh, R., (2022, September 13). Artificial Intelligence-a game changer for climate change and the environment. *State of the Planet.* https://news.climate.columbia.edu/2018/06/05/artificial-intelligence-climate-environment/

Choi-Schagrin, W. (Nov. 6, 2021). How Hospitals Fuel Climate Change. *The New York Times.* https://www.nytimes.com/interactive/2021/11/05/climate/climate-change-health.html

Chowdary, J. S., Harsha, H. S., Gnanaseelan, C., Srinivas, G., Parekh, A., Pillai, P., & Naidu, C. V. (2017). Indian summer monsoon rainfall variability in response to differences in the decay phase of El Niño. *Climate Dynamics*, *48*(7-8), 2707–2727. doi:10.1007/s00382-016-3233-1

Chung, S. Y., Yettella, R. R., Kim, J. S., Kwon, K., Kim, M. C., & Min, D. B. (2011). Effects of grilling and roasting on the levels of polycyclic aromatic hydrocarbons in beef and pork. *Food Chemistry, 129*(4), 1420–1426. doi:10.1016/j.foodchem.2011.05.092

Cianconi, P., Betrò, S., & Janiri, L. (2020). The impact of climate change on mental health: A systematic descriptive review. *Frontiers in Psychiatry, 11*, 74. doi:10.3389/fpsyt.2020.00074 PMID:32210846

Clare, S. (2013). Geopower: The politics of life and land in Frantz Fanon's writing. *Diacritics, 41*(4), 60–80. doi:10.1353/dia.2013.0026

Clayton, C. A., Walker, T. R., Bezerra, J. C., & Adam, I. (2020). Policy responses to reduce single-use plastic marine pollution in the Caribbean. *Marine Pollution Bulletin, 162*, 111833. doi:10.1016/j.marpolbul.2020.111833 PMID:33213855

Climate Action Network South Asia (CANSA). (2022, December). *Gender And Child Inclusion In Policies On Climate Change In South Asia: A study from Nepal, Sri Lanka and Bangladesh*. CANSA. https://cansouthasia.net/gender-and-child-inclusion-in-policies-on-climate-change-in-south-asia/

Climate Bonds Initiative. (2023). *Green Bonds Stock Exchanges*. Climate Bonds.https://www.climatebonds.net/green-bond-segments-stock-exchanges

Climate Change is Not Gender Neutral [Blog]. (2021, June). IUCN. https://www.iucn.org/news/gender/202106/climate-change-not-gender-neutral

Climate Change Litigation – insights into the evolving Global Landscape. (2021, July 5). Grantham Research Institute on climate change and the environment. https://www.lse.ac.uk/granthaminstitute/publication/climate-change-litigation-insights-into-the-evolving-global-landscape/

Cochrane, K. L. (2002). Fisheries management. In K. L. Cochrane (Eds.), *A Fishery manager's guidebook - Management measures and their application*. FAO Fisheries Technical Paper. No. 424. ISBN: 92-5-10473204. https://www.fao.org/3/y3427e/y3427e00.htm#Contents

Cohen, R., Eames, P. C., Hammond, G. P., Newborough, M., & Norton, B. (2022). Briefing: The 2021 Glasgow Climate Pact: Steps on the transition pathway towards a low carbon world. *Proceedings of the Institution of Civil Engineers-Energy, 175*(3), 97–102. doi:10.1680/jener.22.00011

Coleman, N., Esmalian, A., & Mostafavi, A. (2020). Equitable resilience in infrastructure systems: Empirical assessment of disparities in hardship experiences of vulnerable populations during service disruptions. *Natural Hazards Review, 21*(4), 04020034. doi:10.1061/(ASCE)NH.1527-6996.0000401

Collins, M., Ashok, K., Barreiro, M., Roxy, M. K., Kang, S. M., Frölicher, T. L., Wang, G., & Tedeschi, R. G. (2021). Editorial: New techniques for improving climate models, predictions and projections. *Frontiers in Climate, 3*, 811205. doi:10.3389/fclim.2021.811205

Colopy, C. (2012). *Dirty, sacred rivers: Confronting south Asia's water crisis*. Oxford University Press. doi:10.1093/oso/9780199845019.001.0001

Cooper, P., Ernst, P., Kiewell, D., & Pinner, D. (2022, May 19). *Quantum computing just might save the planet*. McKinsey & Company. https://www.mckinsey.com/capabilities/mckinsey-digital/our-insights/quantum-computing-just-might-save-the-planet

Corbin, J., & Strauss, A. (2008). *Basics of qualitative research: Techniques and procedures for developing grounded theory* (3rd ed.). Sage. doi:10.4135/9781452230153

Corsi, A., Pagani, R. N., Kovaleski, J. L., & Luiz da Silva, V. (2020). Technology transfer for sustainable development: Social impacts depicted and some other answers to a few questions. *Journal of Cleaner Production, 245*, 118522. doi:10.1016/j.jclepro.2019.118522

Costanza, R., Daly, L., Fioramonti, L., Giovannini, E., Kubiszewski, I., Mortensen, L. F., Pickett, K. E., Ragnarsdottir, K. V., De Vogli, R., & Wilkinson, R. (2016). Modelling and measuring sustainable wellbeing in connection with the UN Sustainable Development Goals. *Ecological Economics, 130*, 350–355. doi:10.1016/j.ecolecon.2016.07.009

Costanza, R., Fisher, B., Ali, S., Beer, C., Bond, L., Boumans, R., Danigelis, N. L., Dickinson, J., Elliott, C., Farley, J., Gayer, D. E., Glenn, L. M. D., Hudspeth, T., Mahoney, D., McCahill, L., McIntosh, B., Reed, B., Rizvi, S. A. T., Rizzo, D. M., & Snapp, R. (2007). Quality of life: An approach integrating opportunities, human needs, and subjective well-being. *Ecological Economics, 61*(2-3), 267–276. doi:10.1016/j.ecolecon.2006.02.023

Cowls, J., Tsamados, A., Taddeo, M., & Floridi, L. (2021). The AI Gambit: Leveraging artificial intelligence to combat climate change—opportunities, challenges, and recommendations. *AI & Society, 38*(1), 283–307. doi:10.1007/s00146-021-01294-x PMID:34690449

CPCB. (2021). *Annual Report 2020-21 on Implementation of PWM Rules 2016*. CPCB. https://www.cpcb.nic.in/

Cripps, M., & Scarbrough, H. (2022). Making Digital Health "Solutions" Sustainable in Healthcare Systems: A Practitioner Perspective. *Frontiers in Digital Health, 4*, 727421. doi:10.3389/fdgth.2022.727421 PMID:35434699

D'Ooge, C. (n.d.). *Chapter II. Queer Katrina: Gender and sexual orientation matters in the aftermath of the disaster.* Newcomb Institute. https://issuu.com/newcombcollegeinstitute/docs/nccrowreport08/s/11775539

Daanen, H. A. M., Wammes, L. J. A., & Vrijkotte, T. J. M. (1993). Wind chill and dexterity. Report IZF A-7, TNO Institute for Perception.

Dagnachew, A. G., Hof, A., Soest, H. V., & Vuuren, D. V. (2021). Climate change measures and sustainable development goals. *PBL Netherlands Environmenta l Assessment Agency.* https://sdgs.un.org/sites/default/files/2023-01/pbl-2021-climate-change-measures-and-sustainable-development-goals_4639.pdf

Dakhil, A., & Alshawi, M. (2014). Client's Role in Building Disaster Management through Building Information Modelling. [CrossRef]. *Procedia Economics and Finance, 18*, 47–54. doi:10.1016/S2212-5671(14)00912-5

Dal, T., Ramli, I., & Garaizar, J. (2023). Effect of climate change on nature and human health with a special focus on infectious diseases in the Mediterranean region. *Journal of Infection in Developing Countries, 17*(11), 1501–1510. doi:10.3855/jidc.17995 PMID:38064385

Danovaro, R., Fonda Umani, S., & Pusceddu, A. (2009). Climate change and the potential spreading of marine mucilage and microbial pathogens in the Mediterranean Sea. *PLoS One, 4*(9), e7006. doi:10.1371/journal.pone.0007006 PMID:19759910

Davison, C. (2022, June 1). Indian heatwaves threaten maternal health. *Nature and Environment India.* [https://www.dw.com/en/climate-crisis-heat-baby-prenatal-maternal-health-mortality/a-61952364]

Daw, T., Adger, W. N., Brown, K., & Badjeck, M. C. (2009). Climate change and capture fisheries: potential impacts, adaptation and mitigation. In K. Cochrane, C. De Young, D. Soto and T. Bahri (Eds.). *Climate change implications for fisheries and aquaculture: overview of current scientific knowledge.* FAO Fisheries and Aquaculture Technical Paper. No. 530 (pp. 107-150). https://digitalarchive.worldfishcenter.org/bitstream/handle/20.500.12348/1339/WF_2545.pdf?sequence=1

Day, J. M., Melnyk, S. A., Larson, P. D., Davis, E. W., & Whybark, D. C. (2012). Humanitarian and disaster relief supply chains: A matter of life and death. *The Journal of Supply Chain Management*, *48*(2), 21–36. doi:10.1111/j.1745-493X.2012.03267.x

de la Torre, L. E., Dolinskaya, I. S., & Smilowitz, K. R. (2012). Disaster relief routing: Integrating research and practice. *Socio-Economic Planning Sciences*, *46*(1), 88–97. doi:10.1016/j.seps.2011.06.001

de Laforcade, G., & Springer, D. (2019). The Red Barrial Afrodescendiente: A Cuban Experiment in Black Community Empowerment. *Souls*, *21*(4), 339–346. doi:10.1080/10999949.2019.1736904

De Murieta, E. S., Galarraga, I., & Olazabal, M. (2021). How well do climate adaptation policies align with risk-based approaches? An assessment framework for cities. *Cities (London, England)*, *109*, 103018. doi:10.1016/j.cities.2020.103018

De Silva, S. S., & Soto, D. (2009). Climate change and aquaculture: potential impacts, adaptation and mitigation. In K. Cochrane, C. De Young, D. Soto, & T. Bahri (Eds.), Climate change implications for fisheries and aquaculture: overview of current scientific knowledge (pp. 151–212). Food and Agriculture Organization (FAO) of the United Nations. ISBN: 978-92-5-106347-7

de Vries, B. J. (2019). Inequality, SDG10 and the financial system. *Global Sustainability*, *2*, e9. doi:10.1017/sus.2019.6

Debiagi, P., Rocha, R. C., Scholtissek, A., Janicka, J., & Hasse, C. (2022). Iron as a sustainable chemical carrier of renewable energy: Analysis of opportunities and challenges for retrofitting coal-fired power plants. *Renewable & Sustainable Energy Reviews*, *165*, 112579. doi:10.1016/j.rser.2022.112579

Del Bene, D., &; Ávila, S. (2023). Critical mapping for researching and acting upon environmental conflicts – the case of the EJAtlas. *Studies in Ecological Economics*, 195–210. doi:10.1007/978-3-031-22566-6_17

Delbeke, J., Runge-Metzger, A., Slingenberg, Y., & Werksman, J. (2019). The paris agreement. In *Towards a climate-neutral Europe* (pp. 24–45). Routledge. doi:10.4324/9789276082569-2

Delgado, C. L., Wada, N., Rosegrant, M. W., Meijer, S., & Ahmed, M. (2003). *Outlook for fish to 2020: Meeting global demand* (pp. 28). International Food Policy Research Institute (IFPRI) and the World Fish Centre. https://ebrary.ifpri.org/utils/getfile/collection/p15738coll2/id/62912/filename/62913.pdf

Dell, M., Jones, B. F., & Olken, B. A. (2012). Temperature shocks and economic growth: Evidence from the last half century. *American Economic Journal. Macroeconomics*, *4*(3), 66–95. doi:10.1257/mac.4.3.66

Demeke, A., & Tassew, A. (2016). A review on water quality and its impact on fish health. *International Journal of Fauna and Biological Studies*, *3*(1), 21–31. https://api.semanticscholar.org/CorpusID:54571218

Demetriades, J. & Esplen, E. (2010). The Gender Dimension of Poverty and Climate Change Adaptation: Social Dimensions of Climate Change. *Equity and Vulnerability in a Warming World,* 133-143.

Dendooven, L., Patiño-Zúñiga, L., Verhulst, N., Luna-Guido, M., Marsch, R., & Govaerts, B. (2012). Global warming potential of agricultural systems with contrasting tillage and residue management in the central highlands of Mexico. *Agriculture, Ecosystems & Environment*, *152*, 50–58. doi:10.1016/j.agee.2012.02.010

Derraik, J. G. B. (2002). The pollution of the marine environment by plastic debris: A review. *Marine Pollution Bulletin*, *44*(9), 842–852. doi:10.1016/S0025-326X(02)00220-5 PMID:12405208

Desai, D. (2020, January 27). 282 Deaths in last 4 years: How Swachh Bharat Mission failed India's manual scavengers. *The Print.* https://theprint.in/india/282-deaths-in-last-4-years-how-swachh-bharat-mission-failed-indias-manual-scavengers/354116/

Deshpande, T., Mukherji, R., & Sastry, M. (2023). Policy styles and India's national action plan on climate change (NAPCC). *Policy Studies*, 1–18. doi:10.1080/01442872.2023.2261388

Dev, I., Ahlawat, S. P., Palsaniya, D. R., Ram, A., Newaj, R., Tewari, R. K., & Yadav, R. S. (2016). *A sustainable livelihood option for farmers of the semi-arid region: Bamboo+ Chickpea based Agroforestry model.*

Devasahayam, S., Raju, G. B., & Hussain, C. M. (2019). Utilization and recycling of end of life plastics for sustainable and clean industrial processes including the iron and steel industry. *Materials Science for Energy Technologies*, *2*(3), 634–646. doi:10.1016/j.mset.2019.08.002

Devi, P. B. (2008). *Content Analysis: A Method of Social Science Research.* Sage.

Dey, S., Sreenivasulu, A., Veerendra, G. T. N., Rao, K. V., & Babu, P. A. (2022). Renewable energy present status and future potentials in India: An overview. *Innovation and Green Development*, 100006.

Dhanya, P. R., & Mathai, A. (2017). *Occupational hazards in seafood industry: A case study in Ernakulam district, Kerala.* 6th International conference on recent trends in engineering, science and management, Punjab, India. ISBN: 978-93-86171-21-4

Diezmartínez, C. V., & Short Gianotti, A. G. (2022). US cities increasingly integrate justice into climate planning and create policy tools for climate justice. *Nature Communications*, *13*(1), 5763. Advance online publication. doi:10.1038/s41467-022-33392-9 PMID:36180440

Dinu, R. H. (2022). Recent Historiographical Trends in Scholarship on Disability and Socialism in Eastern Europe. *Scandinavian Journal of Disability Research*, *24*(1), 42–53. doi:10.16993/sjdr.829

Do Sul, J. A. I., & Costa, M. F. (2014). The present and future of microplastic pollution in the marine environment. *Environmental Pollution*, *185*, 352–364. doi:10.1016/j.envpol.2013.10.036 PMID:24275078

Dobson, A. (1998). *Justice and the environment: Conceptions of environmental sustainability and theories of distributive justice*. Oxford University Press. doi:10.1093/0198294956.001.0001

Dohale, V., Ambilkar, P., Gunasekaran, A., & Verma, P. (2022). Supply chain risk mitigation strategies during COVID-19: Exploratory cases of "make-to-order" handloom saree apparel industries. *International Journal of Physical Distribution & Logistics Management*, *52*(2), 109–129. doi:10.1108/IJPDLM-12-2020-0450

Domaracka, L., Matuskova, S., Tausova, M., Senova, A., & Kowal, B. (2022). Efficient use of critical raw materials for optimal resource management in EU countries. *Sustainability (Basel)*, *14*(11), 6554. doi:10.3390/su14116554

Donovan, G. H., Butry, D. T., Michael, Y. L., Prestemon, J. P., Liebhold, A. M., Gatziolis, D., & Mao, M. Y. (2013). The relationship between trees and human health: Evidence from the spread of the emerald ash borer. *American Journal of Preventive Medicine*, *44*(2), 139–145. doi:10.1016/j.amepre.2012.09.066 PMID:23332329

Dorri, A., Kanhere, S. S., & Jurdak, R. (2018). Multi-agent systems: A survey. *IEEE Access : Practical Innovations, Open Solutions*, *6*, 28573–28593. doi:10.1109/ACCESS.2018.2831228

Du Toit, M. J., Rendón, O., Cologna, V., Cilliers, S. S., & Dallimer, M. (2022). Why home gardens fail in enhancing food security and dietary diversity. *Frontiers in Ecology and Evolution*, *10*, 38. doi:10.3389/fevo.2022.804523

Dube, K., Nhamo, G., & Chikodzi, D. (2022). Flooding trends and their impacts on coastal communities of Western Cape Province, South Africa. *GeoJournal*, *87*(S4, Suppl 4), 453–468. doi:10.1007/s10708-021-10460-z PMID:34188356

Dubey, R., Gunasekaran, A., Bryde, D. J., Dwivedi, Y. K., & Papadopoulos, T. (2020). Blockchain technology for enhancing swift-trust, collaboration and resilience within a humanitarian supply chain setting. *International Journal of Production Research*, *58*(11), 1–18. doi:10.1080/00207543.2020.1722860

Düdder, B. & Ross, O. (2017). Timber tracking: Reducing complexity of due diligence by using blockchain technology (position paper). *CEUR Workshop Proceedings*. Research Gate.

Durán-Romero, G., López, A. M., Beliaeva, T., Ferasso, M., Garonne, C., & Jones, P. (2020). Bridging the gap between circular economy and climate change mitigation policies through eco-innovations and Quintuple Helix Model. *Technological Forecasting and Social Change*, *160*, 120246. doi:10.1016/j.techfore.2020.120246

Dwi Laksmana, D. (2023). Knowledge in the making: embodying transdisciplinary moments on organic agriculture in Yogyakarta, Indonesia [Doctoral dissertation, Universität Passau].

Dwivedi, A. K., Kumar, A., Baredar, P., & Prakash, O. (2019). Bamboo as a complementary crop to address climate change and livelihoods–Insights from India. *Forest Policy and Economics*, *102*, 66–74. doi:10.1016/j.forpol.2019.02.007

Dwivedi, B., & Banerjee, S. (2017). Fish disease as a monitor of lake water pollution. *International Journal of Research*, *5*(3), 209–212. doi:10.5281/zenodo.439573

Dwivedi, Y. K., Hughes, L., Kar, A. K., Baabdullah, A. M., Grover, P., Abbas, R., Andreini, D., Abumoghli, I., Barlette, Y., Bunker, D., Chandra Kruse, L., Constantiou, I., Davison, R. M., De', R., Dubey, R., Fenby-Taylor, H., Gupta, B., He, W., Kodama, M., & Wade, M. (2022). Climate change and COP26: Are Digital Technologies and information management part of the problem or the solution? an editorial reflection and call to action. *International Journal of Information Management*, *63*, 102456. doi:10.1016/j.ijinfomgt.2021.102456

Dzoga, M., Simatele, D., & Munga, C. (2018). Assessment of ecological vulnerability to climate variability on coastal fishing communities: A study of Ungwana Bay and Lower Tana Estuary, Kenya. *Ocean and Coastal Management*, *163*, 437–444. doi:10.1016/j.ocecoaman.2018.07.015

Easterling, D. R., Evans, J. L., Groisman, P. Y., Karl, T. R., Kunkel, K. E., & Ambenje, P. (2000). Observed variability and trends in extreme climate events: A brief review. *Bulletin of the American Meteorological Society*, *81*(3), 417–425. doi:10.1175/1520-0477(2000)081<0417:OVATIE>2.3.CO;2

Ebi, K. L., & Hess, J. J. (2020). Health Risks Due To Climate Change: Inequity In Causes And Consequences: Study examines health risks due to climate change. *Health Affairs*, *39*(12), 2056–2062. doi:10.1377/hlthaff.2020.01125 PMID:33284705

Ebi, K. L., Vanos, J., Baldwin, J. W., Bell, J. E., Hondula, D. M., Errett, N. A., Hayes, K., Reid, C. E., Saha, S., Spector, J., & Berry, P. (2021). Extreme weather and climate change: Population health and health system implications. *Annual Review of Public Health*, *42*(1), 293–315. doi:10.1146/annurev-publhealth-012420-105026 PMID:33406378

Ebrahimi, K., Jones, G. F., & Fleischer, A. S. (2014, March). A review of data center cooling technology, operating conditions and the corresponding low-grade waste heat recovery opportunities. *Renewable & Sustainable Energy Reviews*, *31*, 622–638. doi:10.1016/j.rser.2013.12.007

Eck, D. L. (2012). India: A sacred geography. Three Rivers Press.

Eckstein, D., Künzel, V., & Schäfer, L. (2021). *The Global Climate Risk Index 2021*. Germanwatch.

EIP-AGRIE Focus Group. (2021). *Reducing the plastic footprint of agriculture, Final Report*. EIP-AGRIE Focus Group.

El Khateeb, S., Saber, M., & Shawket, I. M. (2023). Urban reflections through home gardening; Does Gender Matter? *Ain Shams Engineering Journal*, *14*(3), 101885. doi:10.1016/j.asej.2022.101885

Electronics forum. (2021). P*lastic Waste To Renewable Energy–Implementing A Circular Economy*. Electronics Forum. https://www.electronicsforu.com/market-verticals/power-electronics/plastic-waste-renewable-energy-implementing-circular-economy

Elhadary, Y. A. E., Samat, N., & Hasni, R. (2011). Land use changes impact local communities at Seberang Perai Tengah of Penang, Malaysia. In Issues in peri-urban regions and ways towards sustainable peri-urban futures. Conference Room, School of Humanities. Universiti Sains Malaysia.

Eliasch, J. (2008). *Climate change: financing global forests: the Eliasch review*. Earthscan.

Elliott, L. (2023, December 26). Environmentalism. *Encyclopedia Britannica*. https://www.britannica.com/topic/environmentalism

Ellram, L. M. (1991). Supply-chain management: The industrial organisation perspective. *International Journal of Physical Distribution & Logistics Management, 21*(1), 13–22. doi:10.1108/09600039110137082

El-Saadawy, M., Soliman, N., El Tayeb, I., & Hammouda, M. (2014). Some occupational health hazards among fishermen in Alexandria city. *Gaziantep Medical Journal, 20*(1), 71. doi:10.5455/gmj-30-44689

El-Sayed, Y. Y., & Khalil, R. H. (2009). Toxicity, biochemical effects and residue of aflatoxin B1 in marine water-reared sea bass (Dicentrarchuslabrax L.). *Food and Chemical Toxicology, 47*(7), 1606–1609. doi:10.1016/j.fct.2009.04.008

Engelbutzeder, P., Randell, D., Landwehr, M., Aal, K., Stevens, G., & Wulf, V. (2023). From surplus and scarcity towards abundance: Understanding the use of ICT in food resource sharing practices: "Give a man a fish, and you feed him for a day; teach a man to fish, and you feed him for a lifetime."–Lao Tsu. *ACM Transactions on Computer-Human Interaction*. Advance online publication. doi:10.1145/3589957

Engelenburg, S.V.S. van et al. (2017, July). Design of a software architecture supporting business-to- government information sharing to improve public safety and security. *Journal of Intelligent Information Systems*. doi:10.1007/s10844-017- 0478z

Engelhard, G. H., Howes, E. L., Pinnegar, J. K., & Le Quesne, W. J. F. (2022).Assessing the risk of climate change to aquaculture: A national-scale case study for the Sultanate of Oman. *Climate Risk Management 35*. 100416. https://doi.org/10.1016/j.crm.2022.100416

Engler, R. E. (2012). The complex interaction between marine debris and toxic chemicals in the ocean. *Environmental Science & Technology, 46*(22), 12302–12315. doi:10.1021/es3027105 PMID:23088563

Eng, S., Khun, T., Jower, S., & Murro, M. J. (2019). Healthy lifestyle through home gardening: The art of sharing. [PubMed]. *American Journal of Lifestyle Medicine, 13*(4), 347–350. doi:10.1177/1559827619842068 PMID:31285714

Eriksen, C., Simon, G. L., Roth, F., Lakhina, S. J., Wisner, B., Adler, C., Thomalla, F., Scolobig, A., Brady, K., Bründl, M., Neisser, F., Grenfell, M., Maduz, L., & Prior, T. (2020). Rethinking the interplay between affluence and vulnerability to aid climate change adaptive capacity. *Climatic Change, 162*(1), 25–39. doi:10.1007/s10584-020-02819-x PMID:33184523

Eriksen, M., Lebreton, L. C. M., Carson, H. S., Thiel, M., Moore, C. J., Borerro, J. C., Galgani, F., Ryan, P. G., & Reisser, J. (2014). Plastic pollution in the world's oceans: More than 5 trillion plastic pieces weighing over 250,000 tons afloat at sea. *PLoS One, 9*(12), e111913. doi:10.1371/journal.pone.0111913 PMID:25494041

Erkic, A. (2022, July 7). *How to break down team silos and improve collaboration*, Pumble. https://pumble.com/blog/break-down-team-silos/

Escalante, L., & Maisonnave, H. (2020). *Assessing the Impacts of Climate Change on Women's Poverty and Domestic Burdens: A Bolivian Case Study*. HAL Archives-Ouvertes. https://hal-normandie-univ-archives-ouvertes.fr/hal-02970249

Estoque, R. C. (2022). Complexity and diversity of nexuses: A review of the nexus approach in the sustainability context. *The Science of the Total Environment*, 158612. PMID:36089051

Estoque, R. C., Togawa, T., Ooba, M., Gomi, K., Nakamura, S., Hijioka, Y., & Kameyama, Y. (2019). A review of quality of life (QOL) assessments and indicators: Towards a "QOL-Climate" assessment framework. *Ambio*, *48*(6), 619–638. doi:10.1007/s13280-018-1090-3 PMID:30206898

Etemadi, N., Borbon-Galvez, Y., Strozzi, F., & Etemadi, T. (2021). Supply chain disruption risk management with blockchain: A dynamic literature review. *Information (Basel)*, *12*(2), 70. doi:10.3390/info12020070

European Commission. (2023). *Consequences of climate change*. Climate Action. https://climate.ec.europa.eu/climate-change/consequences-climate-change_en

Eurostat. (2020). *Packaging waste statistics*. Europea. https://ec.europa.eu/ eurostat/statistics-explained/index.php?title=Packaging_waste_ statistics.

Fankhauser, S., Smith, S. M., Allen, M., Axelsson, K., Hale, T., Hepburn, C., Kendall, J. M., Khosla, R., Lezaun, J., Mitchell-Larson, E., Obersteiner, M., Rajamani, L., Rickaby, R., Seddon, N., & Wetzer, T. (2022). The meaning of net zero and how to get it right. *Nature Climate Change*, *12*(1), 15–21. doi:10.1038/s41558-021-01245-w

Fan, Z., & Friedmann, S. J. (2021). Low-carbon production of iron and steel: Technology options, economic assessment, and policy. *Joule*, *5*(4), 829–862. doi:10.1016/j.joule.2021.02.018

FAO (2020). *The State of World Fisheries and Aquaculture 2020. Sustainability in action*. Rome. . doi:10.4060/ca9229en

FAO, ISSCFG. (2013). *International standard statistical classification of fishing gear*. FAO.

FAO. (1997). *FAO Focus on Women and Food Security*. FAO.

FAO. (2020). *The State of the World's Forests*. FAO. https://www.fao.org/state-of-forests/2020/en/,.

Fathy, R., Nelson, C. A., & Barbieri, J. S. (2021). Combating climate change in the clinic: Cost-effective strategies to decrease the carbon footprint of outpatient dermatologic practice. *International Journal of Women's Dermatology*, *7*(1), 107–111. doi:10.1016/j.ijwd.2020.05.015 PMID:33537400

Fedele, G., Donatti, C. I., Harvey, C. A., Hannah, L., & Hole, D. G. (2019). Transformative adaptation to climate change for sustainable social-ecological systems. *Environmental Science & Policy*, *101*, 116–125. doi:10.1016/j.envsci.2019.07.001

Federation of Indian Chambers of Commerce & Industry (FICCI) - PricewaterhouseCoopers (PwC) (2022). *Championing the blue economy: Promoting sustainable growth of the fisheries sector in India* https://www.pwc.in/assets/pdfs/grid/agriculture/championing-the-blue-economy-promoting-sustainable-growth-of-the-fisheries-sector-in-india.pdf

Feikner. (2009). Impact of Climate Change on Rice Production in Thailand. *The American Economic Review*, *99*(2), 205–210. doi:10.1257/aer.99.2.205 PMID:20161598

Fekete, H., Kuramochi, T., Roelfsema, M., den Elzen, M., Forsell, N., Höhne, N., & Gusti, M. (2021). A review of successful climate change mitigation policies in major emitting economies and the potential of global replication. *Renewable & Sustainable Energy Reviews*, *137*, 110602. doi:10.1016/j.rser.2020.110602

Feola, G., Goodman, M. K., Suzunaga, J., & Soler, J. (2023). Collective memories, place-framing and the politics of imaginary futures in sustainability transitions and transformation. *Geoforum*, *138*, 103668. doi:10.1016/j.geoforum.2022.103668

Ficke, A. D., Myrick, C. A., & Hansen, L. J. (2007). Potential impacts of global climate change on freshwater fisheries. *Reviews in Fish Biology and Fisheries*, *17*, 581–613. doi:10.1007/s11160-007-9059-5

Field, C. B., Mortsch, L. D., & Brklacich, M. (2007). North America. In M. L. Parry, O. F. Canziani, & J. P. Palutikof (Eds.), *Climate change: Impacts, Adaptation and Vulnerability. Contribution of Working Group II to the Fourth Assessment Report of the Intergovernmental Panel on Climate Change* (pp. 617–652). Cambridge University Press.

Fineman, M. A. (2008). The Vulnerable Subject Anchoring Equality in the Human Condition. *Yale Journal of Law and Feminism, 1*, 9–15.

Fisher, S., Fitzgerald, R., & Poortinga, W. (2018). *Climate Change: Social Divisions in Beliefs and Behaviour – British Social Attitudes. Climate Change, 35*. National Centre for Social Research.

Flato, M., & Pelser, A. (2017). Women, Weather and Woes: The Triangular Dynamics of Female-Headed Households, Economic Variability, and Climate Variability in South Africa. *World Development, 90*, 41–62. doi:10.1016/j.worlddev.2016.08.015

Flynn, A., Chan, K. W., Zhu, Z. H., & Yu, L. (2017). Sustainability, space and supply chains: The role of bamboo in Anji County, China. *Journal of Rural Studies, 49*, 128–139. doi:10.1016/j.jrurstud.2016.11.012

Fogli, D., Greppi, C., & Guida, G. (2017). Design patterns for emergency management: An exercise in reflective practice. *Information & Management, 54*(7), 971–986. doi:10.1016/j.im.2017.02.002

Fong, I. W., & Fong, I. W. (2020). Climate change: Impact on health and infectious diseases globally. *Current trends and concerns in infectious diseases*, 165-190.

Food and Agriculture Organization (FAO) of the United Nations and World Health Organization. (WHO). 2020. *Code of practice for fish and fishery products.* doi:10.4060/cb0658en

Food and Agriculture Organization (FAO) of the United Nations. (2012). *The prevention of losses in cured fish.* FAO.

Food and Agriculture Organization (FAO) of the United Nations. (2016). The State of World Fisheries and Aquaculture 2016 - Contributing to food security and nutrition for all (pp. 200). ISBN: 978-92-5-109185-2

Food and Agriculture Organization (FAO) of the United Nations. (2018a). The State of World Fisheries and Aquaculture 2018 - Meeting the sustainable development goals (pp. 200). ISBN: 978-92-5-130562-1

Food and Agriculture Organization (FAO) of the United Nations. (2018b). *Impacts of climate change on fisheries and aquaculture - Synthesis of current knowledge, adaptation and mitigation options.* ISBN: 978-92-5-130607-9. https://www.fao.org/3/i9705en/i9705en.pdf

Food and Agriculture Organization (FAO) of the United Nations. (2022). *The State of World Fisheries and Aquaculture 2022. Towards Blue Transformation* (pp. 266). http://www.fao.org/state-of-fisheries-aquaculture

Food and Agriculture Organization of the United Nations (FAO). (2015). *Climate change and food security: risks and responses.* ISBN: 978-92-5-108998-9. https://www.fao.org/3/i5188e/I5188E.pdf

Food and Agriculture Organization of the United Nations (FAO). (2021). *Recent advances in climate change vulnerability/risk assessments in the fisheries and aquaculture sector.* ISBN: ISBN 978-92-5-134374-6. . https://doi.org/10.4060/cb4585en.pdf

Food and Agriculture Organization of the United Nations (FAO). (n.d.). *Disaster risk management in fisheries.* CC4 Fias Policy Brief, 1-4. https://www.fao.org/3/cb2729en/cb2729en.pdf

Fox, M., Zuidema, C., Bauman, B., Burke, T., & Sheehan, M. (2019). Integrating public health into climate change policy and planning: State of practice update. *International Journal of Environmental Research and Public Health, 16*(18), 3232. doi:10.3390/ijerph16183232

Frey, R. S. (2013). *Breaking ships in the world-system: An analysis of two ship breaking capitals*. Alang India and Chittagong.

Friedman, N., & Ormiston, J. (2022). Blockchain as a sustainability-oriented innovation?: Opportunities for and resistance to blockchain technology as a driver of sustainability in global food supply chains. *Technological Forecasting and Social Change*, *175*, 121403. doi:10.1016/j.techfore.2021.121403

Frizziero, M. (2016). Rethinking Chilled Water Temperatures Can Bring Big Savings in Data Center Cooling. *Schneider Electric Blog*. https://blog.se.com/datacenter/2016/08/17/water-temperatures-data-center-cooling/ .

Frouws, J., & Mol, A. P. (1997). Ecological modernization theory and agricultural reform. In Images and Realities of Rural Life. Wageningen Perspectives on Rural Transformations (pp. 269-286). Van Gorcum.

Fuhrman, S., Janoch, E., Koch, R., Oo, K., Parra, V., & Rawe, T. (2020). *Left out and Left Behind: Ignoring Women will Prevent us from Solving the Hunger Crisis: Policy Report*. Care International.

Furusa, M. (2024). *Indigenous Knowledge and Technology in African and Diasporan Communities: Multi-Disciplinary Approaches*. University of Zimbabwe: Southern African Association for Culture and DevelopmentStudies.

Gadgil, M., & Guha, R. (2004). *The use and abuse of nature*. Oxford India Paperbacks.

Gaetani, E. (2017). *Blockchain-based database to ensure data integrity in cloud computing environments*. CEUR Workshop Proceedings.

Galhena, D. H., Freed, R., & Maredia, K. M. (2013). Home gardens: A promising approach to enhance household food security and well-being. *Agriculture & Food Security*, *2*(1), 1–13. doi:10.1186/2048-7010-2-8

Galloway, T., & Lewis, C. (2016). Marine microplastics spell big problems for future generations. *Proceedings of the National Academy of Sciences of the United States of America*, *113*(9), 2331–2333. doi:10.1073/pnas.1600715113 PMID:26903632

Gammage, S. (2010). Gender, Time Poverty and Amartya Sen's Capability Approach: Evidence from Guatemala. In S. Chant (ed). The International Handbook of Gender and Poverty: Concepts, Research, Policy, 26, 71-76.

Gandhi, T. (2006). *A bird's eye view: Collected essays and shorter writings of Salim Ali* (Vol. I & II). Permanent Black.

Gao, W., Hatcher, W. G., & Yu, W. (2018). *A survey of blockchain: techniques, applications, andchallenges. 27th International Conference on Computer Communication and Networks (ICCCN)*, Hangzhou, China.

Gardner, J. (2021). Integrating Gender. In *Climate Change And Disaster-Related Statistics In Asia & The Pacific*. UN Women.

Gasalla, M. A., Abdallah, P. R., & Lemos, D. (2018). Potential impacts of climate change in Brazilian marine fisheries and aquaculture. In B. F. Phillips & M. Perez-Ramirez (Eds.), *Climate change impacts on fisheries and aquaculture: a global analysis* (pp. 455–472). Wiley Blackwell., doi:10.1002/9781119154051.ch14

Gavali, R. S. (2023). Scalable Adaptation Model for Sustainable Agriculture Livelihoods Under Changing Climate: A Case Study from Bihar and Madhya Pradesh. In S. Nautiyal, A. K. Gupta, M. Goswami, & Y. D. Imran Khan (Eds.), *The Palgrave Handbook of Socio-ecological Resilience in the Face of Climate Change*. doi:10.1007/978-981-99-2206-2_29

Gavali, R. S., & Shaikh, H. M. Y. (2016). Estimation of carbon storage in the tree growth of Solapur University Campus, Maharashtra, India. *International Journal of Scientific Research*, *5*(4), 2364–2367.

George, G., Vethamony, P., Sudheesh, K., & Babu, M. T. (2011). Fish larval transport in a macro-tidal regime: Gulf of Kachchh, West Coast of India. *Fisheries Research*, *110*(1), 160–169. doi:10.1016/j.fishres.2011.04.002

Georgian, S. E., Anderson, O. F., & Rowden, A. A. (2019). Ensemble habitat suitability modeling of vulnerable marine ecosystem indicator taxa to inform deep-sea fisheries management in the South Pacific Ocean. *Fisheries Research, 211*, 256–274. doi:10.1016/j.fishres.2018.11.020

Germanwatch. (2022). *Observing, Analysing, Acting - for Global Equity and the Preservation of Livelihoods.* COP 27: U.N. Climate Conference in Egypt.

GESAMP. (2020). *Sea-Based Sources of Marine Litter – A Review of Current Knowledge and Assessment of Data Gaps.* Second Interim Report of GESAMP Working Group 43.

Getahun, A., Kebede, Y., Mulatu, Y., & Adamu, F. (2023). *Growth and biomass production of five exotic bamboo species in North-western Ethiopia. Bahir Dar*. Elsevier.

Geyer, R. (2020). Production, Chapter 2- Production, Use and Fate of Synthetic Polymers. In Letcher, T.M. (ed.). Plastic Waste and Recycling. Cambridge, MA: Academic Press.

Ghan, B. (2019). *Hybrids of Mind and Body: The Forms and Freedoms of the Cyborg in Posthumanist Science Fiction.* University of Toronto.

Ghani, N. A., Raub, M. A., Adam, F., Abdullah, B., Eusoff, Y. A., & Puteh, D. A. H. M. A. (2017). Quality of Life (QoL) of fishermen in the west coast states of Peninsular Malaysia. *International Journal of Academic Research in Business & Social Sciences*, *7*(4), 313–330. doi:10.6007/ijarbss/v7-i4/2808

Ghobakhloo, M., Iranmanesh, M., Mubarak, M. F., Mubarik, M., Rejeb, A., & Nilashi, M. (2022). Identifying industry 5.0 contributions to sustainable development: A strategy roadmap for delivering sustainability values. *Sustainable Production and Consumption*, *33*, 716–737. doi:10.1016/j.spc.2022.08.003

Ghosh, S., Rao, M. V. H., Kumar, M. S., Mahesh, V. U., Menon, M., & Zacharia, P. U. (2014). Carbon footprint of marine fisheries: Life cycle analysis from Visakhapatnam. *Current Science*, *107*(3), 515–521. http://eprints.cmfri.org.in/10166/1/Ghosh_Current_Science.pdf

Ghosh, S., Rao, M. V. H., Kumar, M. S., Mahesh, V. U., Menon, M., & Zacharia, P. U. (2015). Carbon footprint in life cycle of marine fisheries at major fishing harbours of Andhra Pradesh, India. *Indian Journal of Fisheries*, *62*(4), 37–44. http://eprints.cmfri.org.in/10605/1/IJF_62.4_Ghosh_4.pdf

Gillanders, B., & Kingsford, M. (2002). Impact of Changes in Flow of Freshwater on Estuarine and Open Coastal Habitats and the Associated Organisms. *Oceanography and Marine Biology - an Annual Review*, *40*, 233–309. doi:10.1201/9780203180594.ch5

Giordano, C. (2020). From Sicily to Malaysia via Bulgaria: Finding the Common Thread of One's Fieldwork. Recalling Fieldwork: People. *Places and Encounters*, *51*, 19.

Gissi, E., Manea, E., Mazaris, A. D., Fraschetti, S., Almpanidou, V., Bevilacqua, S., Coll, M., Guarnieri, G., Lloret-Lloret, E., Pascual, M., Petza, D., Rilov, G., Schonwald, M., Stelzenmüller, V., & Katsanevakis, S. (2020). A review of the combined effects of climate change and other local human stressors on the marine environment. *The Science of the Total Environment*, *755*(1), 142564. doi:10.1016/j.scitotenv.2020.142564

GIZ. (2020). *Climate Change and Rural Development*. Rural Development and Agriculture Division – GIZ.

Glasser, R., Johnstone, C., & Kapetas, A. (Eds.). (2022). *The geopolitics of climate and security in the Indo-Pacific.* Australian Strategic Policy Institute.

Global Gender and Climate Alliance. (2017). *Gender and Climate Change in Africa – Facts from Gender and Climate Change: A Closer look at Existing Evidence*. Global Gender and Climate Alliance.

Global trends in climate change litigation: 2022 snapshot. (2023, June 7). Grantham Research Institute on climate change and the environment. https://www.lse.ac.uk/granthaminstitute/publication/global-trends-in-climate-change-litigation-2022/

Glotfelty, C. (1996). Introduction: Literary studies in an age of environmental crisis. In C. Glotfelty & H. Fromm (Eds.), *The ecocriticism reader: Landmarks in literary ecology* (pp. xv–xxxvii). University of Georgia Press.

Glynn, P. W. (1991). Coral reef bleaching in the 1980s and possible connections with global warming. *Trends in Ecology & Evolution*, *6*(6), 175–179. doi:10.1016/0169-5347(91)90208-f

Godvin, V. S., Kumar, D. M., Pugazhendi, A., Bajhaiya, A. K., Gugulothu, P., & Banu, R. J. (2021). Biofuel production from Macroalgae: Present scenario and future scope. *Bioengineered*, *12*(2), 9216–9238. doi:10.1080/21655979.2021.1996019

Goebel, A., Dodson, B., & Hill, T. (2010). Urban Advantage or Urban Penalty? A Case Study of Female Headed Households in a South African City. *Health & Place*, *16*(3), 573–580. doi:10.1016/j.healthplace.2010.01.002 PMID:20116325

Goggin, C., & Lester, R. (1995). Perkinsus, a protistan parasite of abalone in Australia: A review. *Marine and Freshwater Research*, *46*(3), 639–646. doi:10.1071/mf9950639

Goldberg, M. H., Gustafson, A., Ballew, M. T., Rosenthal, S. A., & Leiserowitz, A. (2019). A Social Identity Approach to Engaging Christians in the Issue of Climate Change. *Science Communication*, *41*(4), 442–463. doi:10.1177/1075547019860847

Gomez-Zavaglia, A., Mejuto, J. C., & Simal-Gandara, J. (2020). Mitigation of emerging implications of climate change on food production systems. *Food Research International*, *134*, 109256. doi:10.1016/j.foodres.2020.109256 PMID:32517948

Google. (n.d.). *Douglas County, Georgia – Data Centers*. Google. https://www.google.com/about/datacenters/locations/douglas-county/

Google. (n.d.). *Efficiency – Data Centers*. Google. https://www.google.com/about/datacenters/efficiency/

Google. (n.d.). *Google Environmental Report*. Google Inc. https://www.gstatic.com/gumdrop/sustainability/google-2022-environmental-report.pdf

Gopalakrishnan, A., Zacharia, P. U., & Grinson, G. (2020). *Impact, vulnerability and adaptation strategies for marine fisheries of India: IVAS-MFI final report. CMFRI special Publication No.135*. ICAR-Central Marine Fisheries Research Institute., http://eprints.cmfri.org.in/14771/1/CMFRI%20Special%20Publication_135_IVAS.pdf

Gopal, N. (2021). Labour challenges in fishing and fish processing industries. In C. N. Ravishankar, K. A. Kumar, L. Edwin, S. Mathew, A. K. Mohanty, A. A. Zynudheen, G. Ninan, T. C. Joseph, V. Chandrasekar, V. K. Sajesh, K. K. Prajith, V. Renuka, S. J. Laly, K. R. Sreelakshmi, K. Ranjit, R. K. Nadella, & S. Murali (Eds.), *Recent advances in harvest and post-harvest technologies in fisheries (Training Manual)* (pp. 407–418). ICAR-Central Institute of Fisheries Technology., http://krishi.icar.gov.in/jspui/handle/123456789/70437

Government of Canada. (2019). *Women and Climate Change*. Government of Canada.

Goyal, S., & Gupta, V. (2009). Sulabh International: Social transformation through sanitation. *Vikalpa*, *34*(1), 89–105. https://journals.sagepub.com/doi/pdf/10.1177/0256090920090107#:~:text=The%20Sulabh%20movement%20rehabilitated%20more,regular%20flush%20toilets%20by%20households.&text=Sulabh%20was%20the%20pioneering%20private%20initiative%20in%20the%20area%20of%20sanitation. doi:10.1177/0256090920090107

Greentumble. (2023). *What Causes Carbon Dioxide Emissions in the Air?* Greentumble. https://greentumble.com/what-causes-carbon-emissions

Gregory, M. R. (2009). Environmental implications of plastic debris in marine settings—Entanglement, ingestion, smothering, hangers-on, hitch-hiking and alien invasions. *Philosophical Transactions of the Royal Society of London. Series B, Biological Sciences*, *364*(1526), 2013–2025. doi:10.1098/rstb.2008.0265 PMID:19528053

Grummell, B., Daura, M., Chipeta, G. C., Daura, M., Downes, M., Kasulo, V., Lichilo, I., Mataya, B., Mpolomoka, D., Murphy, C., Mwale, J., Noone, M., Sibalwa, D., Ryan, A., Tembo, M., & Yerokun, O. (2020). 'TEN Team,' *Concept: The Journal of Contemporary Community Education Practice Theory*, *11*(1), 1-11. http://concept.lib.ed.ac.uk/article/view/4294

GRZ & UNDP. (2010). *Project Document: Adaptation to the Effects of Climate Change in Agro-Ecological Regions I and II in Zambia*. GRZ and UNDP.

GRZ & UNDP. (2013). *Millennium Development Goals Progress Report: Zambia*. GRZ and UNDP.

GRZ (2014). *Ministry of Child and Development: Living Conditions Monitoring Survey*. Lusaka 2010: GRZ.

GRZ (2020). *Presidential Speech on Pilot Programmes for Climate Sites in Lusaka District by Edgar Chagwa Lungu. Lusaka*: GRZ.

GRZ. (2004). *Zambia Vulnerability Assessment Survey*. GRZ.

GRZ. (2014). *Zambia National Gender Policy*. GRZ.

GRZ. (2017). *Seventh National Development Plan 2017-2021*. GRZ.

Guenther, R., & Vittori, G. (2013). *Sustainable Healthcare Architecture*. John Wiley & Sons.

Guha, R. (2013, March 27). The past and present of Indian environmentalism. *The Hindu*. https://www.thehindu.com/opinion/lead/the-past-present-of-indian-environmentalism/article4551665.ece accessed 11 Jan. 2021

Guha, R. (1997). *Nature's spokesman: M. Krishnan and Indian wildlife*. Oxford University Press.

Guillard, V., Gaucel, S., Fornaciari, C., Angellier-Coussy, H., Buche, P., & Gontard, N. (2018). The next generation of sustainable food packaging to preserve our environment in a circular economy context. *Frontiers in Nutrition*, *5*, 121. doi:10.3389/fnut.2018.00121 PMID:30564581

Gu, L., Zhou, Y., Mei, T., Zhou, G., & Xu, L. (2019). Carbon footprint analysis of bamboo scrimber flooring—Implications for carbon sequestration of bamboo forests and its products. *Forests*, *10*(1), 51. doi:10.3390/f10010051

Gullison, R. E., Frumhoff, P. C., Canadell, J. G., Field, C. B., Nepstad, D. C., Hayhoe, K., Avissar, R., Curran, L. M., Friedlingstein, P., Jones, C. D., & Nobre, C. (2007). Tropical forests and climate policy. *Science*, *316*(5827), 985–986. doi:10.1126/science.1136163 PMID:17495140

Gunes, E., & Kovel, J. P. (2000). Using GIS in Emergency Management Operations. [CrossRef]. *Journal of Urban Planning and Development*, *126*(3), 136–149. doi:10.1061/(ASCE)0733-9488(2000)126:3(136)

Gupta, A. (2013). *Bamboos in Mitigating Climate Change. International Union for conservation of Nature*. International Union for Conservation of Nature. https://www.iucn.org/our-work/nature-based-solutions

Gupta, A., & Ghosh, J. (2023). *India's climate research agenda: 2030 and beyond*. https://dst.gov.in/sites/default/files/India%27s%20Climate%20Research%20Agenda%202030%20and%20beyond.pdf

Gupta, Y. (2011). Carbon credit: A step towards green environment. *Global Journal of Management and Business Research*, *11*(5), 16–19. https://www.statista.com/statistics/1284586/india-methane-emission/

Gurung, J. D. (n.d.). *Gender and Desertification: Expanding roles for women to restore dryland areas.* IFAD; Rome. https://www.ifad.org/documents/38714170/39148759/Gender+and+desertification+-+Expanding+roles+for+women+to+restore+drylands.pdf/76b3905e-f1ed-428b-8673-822fb923dcd8

Guvamombe, T. (2020). EMA alerts nation on veld fire season" in Spotlight. https://spotlight.co.zw/ema-alerts-nation-on-veld-fire-season/

Gwidi, R. (2019). Masvingo Least Affected by veld Fire. *Herald.* https://www.herald.co.zw/masvingo-least-affected-by-veld-fire/

Ha, H. D., Tan, N. Q., Thang, T. N., Que, H. T. H., & Van Tuyen, T. (2023). Towards Inclusive Resilience: Understanding the Intersectionality of Gender, Ethnicity, and Livelihoods in Climate-prone Upland Areas of Central Vietnam. *Environment and Ecology Research, 11*(4), 630–646. doi:10.13189/eer.2023.110410

Hallerman, E. M., & Kapuscinski, A. R. (1992). Ecological implications of using transgenic fishes in aquaculture. *International Council for the Exploration of the Seas Marine Science Symposium, 194*, 56–66.

Handbook on fisheries statistics (2020). Department of Fisheries, Ministry of Fisheries, Animal Husbandry and Dairying, *Government of India*, 99-100. https://dof.gov.in/sites/default/files/2021-02/Final_Book.pdf

Hanon, I. (2019). Workers' participation: The challenge of Cuban socialism. *Socialism and Democracy, 33*(1), 88–111. doi:10.1080/08854300.2018.1510671

Hansen, L. S., Sorgho, R., Mank, I., Nayna Schwerdtle, P., Agure, E., Bärnighausen, T., & Danquah, I. (2022). Home gardening in sub-Saharan Africa: A scoping review on practices and nutrition outcomes in rural Burkina Faso and Kenya. *Food and Energy Security, 11*(3), e388. doi:10.1002/fes3.388

Haraway, D. (1991). A Cyborg Manifesto: Science, Technology, and Socialist-Feminism in the Late Twentieth Century. In *Simians, Cyborgs and Women: The Reinvention of Nature* (pp. 149–181). Routledge.

Harnett, B. M. (2015). *The Value of Content Analysis as a Qualitative Research Method: Research Proposal.* The British University in Dubai.

Harvell, C. D., Mitchell, C. E., Ward, J. R., Altizer, S., Dobson, A. P., Ostfeld, R. S., & Samuel, M. D. (2002). Climate warming and disease risks for terrestrial and marine biota. *Science, 296*(5576), 2158–2162. doi:10.1126/science.1063699

Hasan, A. M., Hoq, M. T., & Thollander, P. (2018). Energy management practices in Bangladesh's iron and steel industries. *Energy strategy reviews, 22*, 230-236.

Hauer, M. E., Fussell, E., Mueller, V., Burkett, M., Call, M., Abel, K., McLeman, R., & Wrathall, D. (2020). Sea-level rise and human migration. *Nature Reviews. Earth & Environment, 1*(1), 28–39. doi:10.1038/s43017-019-0002-9

He, K., & Wang, L. (2017). A review of energy use and energy-efficient technologies for the iron and steel industry. *Renewable & Sustainable Energy Reviews, 70*, 1022–1039. doi:10.1016/j.rser.2016.12.007

Helming, J., Daatselaar, C., van Dijk, W., Mollenhorst, H., & Pishgar-Komleh, S. H. (2023). Model Collaboration between Farm Level Models with Application on Dutch Dairy and Arable Farms Regarding Circular Agricultural Policy. *Sustainability (Basel), 15*(6), 5020. doi:10.3390/su15065020

Hendricks, M. D., & Van Zandt, S. (2021). Unequal protection revisited: Planning for environmental justice, hazard vulnerability, and critical infrastructure in communities of color. *Environmental Justice, 14*(2), 87–97. doi:10.1089/env.2020.0054

Heo, S., Chen, C., Kim, H., Sabath, B., Dominici, F., Warren, J. L., Di, Q., Schwartz, J., & Bell, M. L. (2021). Temporal changes in associations between high temperature and hospitalizations by greenspace: Analysis in the Medicare population in 40 US northeast counties. *Environment International*, *156*, 106737. doi:10.1016/j.envint.2021.106737 PMID:34218185

Hermabessiere, L., Dehaut, A., Paul-Pont, I., Lacroix, C., Jezequel, R., Soudant, P., & Duflos, G. (2017). Occurrence and efects of plastic additives on marine environments and organisms: A review. *Chemosphere*, *182*, 781–793. doi:10.1016/j.chemosphere.2017.05.096 PMID:28545000

Hersher, R. (2023, January 9), *Climate change makes heat waves, storms and droughts worse, climate report confirms.* NPR. https://www.npr.org/2023/01/09/1147805696/climate-change-makes-heat-waves-storms-and-droughts-worse-climate-report-confirm

Heslin, K. (2016). Ignore Data Center Water Consumption at Your Own Peril. *Uptime Institute Blog*. https://journal.uptimeinstitute.com/dont-ignore-water-consumption/

Heus, R., Daanen, H. A., & Havenith, G. (1995). Physiological criteria for functioning of hands in the cold: A review. *Applied Ergonomics*, *26*(1), 5–13. doi:10.1016/0003-6870(94)00004-i

Higgins, P., & Campanera, J. M. (2011). (Sustainable) quality of life in English city locations. *Cities (London, England)*, *28*(4), 290–299. doi:10.1016/j.cities.2011.02.005

Hilborn, R., Amoroso, R. O., Anderson, C. M., Baum, J. K., Branch, T. A., Costello, C., de Moor, C. L., Faraj, A., Hively, D., Jensen, O. P., Kurota, H., Little, L. R., Mace, P., McClanahan, T., Melnychuk, M. C., Minto, C., Osio, G. C., Parma, A. M., Pons, M., & Segurado, S. (2020). Effective fisheries management instrumental in improving fish stock status. *Proceedings of the National Academy of Sciences of the United States of America*, *117*(4), 2218–2224. doi:10.1073/pnas.1909726116

Hilson, G., & Banchirigah, S. M. (2009). Are alternative livelihood projects alleviating poverty in mining communities? Experiences from Ghana. *The Journal of Development Studies*, *45*(2), 172–196. doi:10.1080/00220380802553057

Hindustan Times. (2018, August 18). Pregnant woman airlifted by navy from flooded Kerala village, mother and baby fine. *Hindustan Times*. https://www.hindustantimes.com/kerala/pregnant-woman-airlifted-by-navy-from-flooded-kerala-village-mother-and-baby-fine/story-S4FRiwEmfkivJFrGfUNuXP.html

Hirai, H., Takada, H., Ogata, Y., Yamashita, R., Mizukawa, K., Saha, M., Kwan, C., Moore, C., Gray, H., Laursen, D., Zettler, E. R., Farrington, J. W., Reddy, C. M., Peacock, E. E., & Ward, M. W. (2011). Organic micropollutants in marine plastic debris from the open ocean and remote and urban beaches. *Marine Pollution Bulletin*, *62*(8), 1683–1692. doi:10.1016/j.marpolbul.2011.06.004 PMID:21719036

Hodgkinson, J. H., & Smith, M. H. (2021). Climate change and sustainability as drivers for the next mining and metals boom: The need for climate-smart mining and recycling. *Resources Policy*, *74*, 101205. doi:10.1016/j.resourpol.2018.05.016

Hoffman, J., & Lott, C. (April 20, 2022). *The intersection of economics, environment and equity*. The Crimson White. https://thecrimsonwhite.com/98316/culture/the-intersection-of-economics-environment-and-equity/

Holappa, L. (2020). A general vision for reduction of energy consumption and CO2 emissions from the steel industry. *Metals*, *10*(9), 1117. doi:10.3390/met10091117

Homayouni, Z., Pishvaee, M. S., Jahani, H., & Ivanov, D. (2021). A robust-heuristic optimization approach to a green supply chain design with consideration of assorted vehicle types and carbon policies under uncertainty. *Annals of Operations Research*, 1–41.

Hope, S. A. A. (2016). *Knowledge, Attitudes & Practices Study on Climate Change Adaptation & Mitigation in Guyana.* Climate Change Partnership Japan-Caribbean and UNDP.

Horrell, S., & Krishnan, P. (2006). Poverty and Productivity in Female Headed Households in Zimbabwe. *The Journal of Development Studies*, *43*(8), 1–37.

How Walmart Canada uses blockchain to solve supply-chain challenges. (2022, January 5). Harvard Business Review. https://hbr.org/2022/01/how-walmart-canada-uses-blockchain-to-solve-supply-chain-challenges

Howden, S. M., Sousanna, J. F., & Tubiello, F. N. (2007). Adapting Agriculture to Climate Change: Proceedings of National academy of Sciences. *Proceedings of the National Academy of Sciences of the United States of America*, *104*(50), 19691–19696. doi:10.1073/pnas.0701890104 PMID:18077402

Huang, P. S., & Shih, L. H. (2009). Effective environmental management through environmental knowledge management. *International Journal of Environmental Science and Technology*, *6*(1), 35–50. doi:10.1007/BF03326058

Huang, Y., Liu, Q., Jia, W., Yan, C., & Wang, J. (2020). Agricultural plastic mulching as a source of microplastics in the terrestrial environment. *Environmental Pollution*, *260*, 114096. doi:10.1016/j.envpol.2020.114096 PMID:32041035

Huang, Y., Ma, Y., Liu, T., & Luo, M. (2020). Climate change impacts on extreme flows under IPCC RCP scenarios in the mountainous Kaiduwatershed, Tarimriver basin. *Sustainability*, *12*(5), 2090. https://doi.org/10.3390/su12052090

Huggan, G., & Tiffin, H. (2010). *Postcolonial ecocriticism: Literature, animals, environment.* Routledge.

Huho, J. M. (2015). Climate Change Knowledge Gap in Education System in Kenya. *Int. J. Innov. Res. Educ. Sci*, *2*, 2349–5219.

Human Rights Watch. (2015). *Marry Before Your House is Swept Away - Child Marriage in Bangladesh.* HRW.

Huntington, T. (2019). *Marine Litter and Aquaculture Gear – White Paper.* Report produced by Poseidon Aquatic Resources Management Ltd for the Aquaculture Stewardship Council.

Hurlbert, M., Krishnaswamy, J., Davin, E., Johnson, F. X., Mena, C. F., Morton, J., Myeong, S., Viner, D., Warner, K., Wreford, A., Zakieldeen, S., & Zommers, Z. (2019). Risk Management and Decision making in Relation to Sustainable Development. In P. R. Shukla, J. Skea, E. C. Buendia, V. Masson-Delmotte, H. O. Pörtner, D. C. Roberts, P. Zhai, R. Slade, S. Connors, R. V. Diemen, M. Ferrat, E. Haughey, S. Luz, S. Neogi, M. Pathak, J. Petzold, J. P. Pereira, P. Vyas, E. Huntley, & K. Kissick (Eds.), Climate Change and Land: an IPCC special report on climate change, desertification, land degradation, sustainable land management, food security, and greenhouse gas fluxes in terrestrial ecosystems (pp. 673–800). IPCC. https://doi.org/ doi:10.1017/9781009157988.009

Hurley, R., Horton, A., Lusher, A., & Nizzetto, L. (2020). Plastic waste in the terrestrial environment. Chaper 7. *Plastic Waste and Recycling Environmental Impact, Societal Issues, Prevention, and Solutions.*, *2020*, 163–193. doi:10.1016/B978-0-12-817880-5.00007-4

Hu, W., Zheng, X., Li, Y., Du, J., Lv, Y., Su, S., Xiao, B., Ye, X., Jiang, Q., Tan, H., Liao, B., & Chen, B. (2022). High vulnerability and a big conservation gap: Mapping the vulnerability of coastal scleractinian corals in South China. *The Science of the Total Environment*, *847*, 157363. doi:10.1016/j.scitotenv.2022.157363

Idowu, T. A., & Sogbesan, O. A. (2017). Synthetic chemical application in aquaculture production. *International Journal of Environment & Agricultural Science*, *1*(2), 010.https://www.researchgate.net/publication/322675339_Synthetic_Chemical_Application_in_Aquaculture_Production

IEA. (2021). *India Energy Outlook 2021.World Energy Outlook Special Report.* Paris: OECD. https://www.iea.org/data-and-statistics/charts/CO2-emissions-from-the-indian-energy-sector.

IEA. (2022). *Data Centres and Data Transmission Networks – Analysis*. IEA. https://www.iea.org/reports/data-centres-and-data-transmission-networks.

IEA. (2023). *Carbon capture, utilisation and storage*. IEA. https://www.iea.org/energy-system/carbon-capture-utilisation-and-storage

IEA. (2023). *CO_2 Emissions in 2022*. IEA, Paris https://www.iea.org/reports/co2-emissions-in-2022, License: CC BY 4.0

IFAD. (2010). *Climate Change: Building the Resilience of the Poor Rural Communities*. IFAD.

Ilaiah, K. (2019). *Why I am not a Hindu: A Sudra critique of Hindutva philosophy, culture and political economy*. Sage Publications.

India Today. (2023, June 25). Delhi woman dies of electrocution after coming in contact with electric pole during rain floods. *India Today*. https://www.indiatoday.in/cities/delhi/story/delhi-woman-dies-electrocution-electric-pole-rain-floods-2397666-2023-06-25

India.com. (2023, July 8). Delhi Rains: Woman Crushed to Death Under Collapsed Ceiling in Karol Bagh. *India.com*. https://www.india.com/news/delhi/delhi-rains-woman-crushed-to-death-under-collapsed-ceiling-in-karol-bagh-6156125/

India's blue economy net getting bigger! Country ranks third in fisheries and second in aquaculture. (2020, February 14). Financial express.https://www.financialexpress.com/opinion/indias-blue-economy-net-getting-biggercountry-ranks-third-in-fisheries

Indian Council of Agricultural Research (ICAR)-Central Institute of Fisheries Technology (CIFT). (2016). *Annual Report 2015-16* (pp. 160). https://krishi.icar.gov.in/jspui/bitstream/123456789/8266/1/CIFT%20Annual%20Report%20 2015-16.pdf

Indian Meteorological Department (IMD). 2021. *Standard operation procedure weather forecasting and warning*. https://mausam.imd.gov.in/imd_latest/contents/pdf/forecasting_sop.pdf

Ingram, V., & Tieguhong, J. C. (2013). Bars to jars: Bamboo value chains in Cameroon. *Ambio*, *42*(3), 320–333. doi:10.1007/s13280-012-0347-5 PMID:23015325

Institute for Societal Leadership. (2015). *Societal Leadership in Southeast Asia: National Landscapes, Challenges and Opportunities to Enhance Societal Impact*. Institute of Societal Leadership Research Collection.

Intergovernmental Panel on Climate Change (IPCC). (2007). Climate change 2007: impacts, adaptation and vulnerability: Working Group II contribution to the fourth assessment report of the IPCC Intergovernmental Panel on Climate Change. In M. Parry, O. Canziani, J. Palutikof, P. van der Linden, & C. Hanson (Eds.), (p. 976). Cambridge University Press., https://www.ipcc.ch/site/assets/uploads/2018/03/ar4_wg2_full_report.pdf

Intergovernmental Panel on Climate Change (IPCC). (2022a). Sea Level Rise and Implications for Low-Lying Islands, Coasts and Communities. In *The Ocean and Cryosphere in a Changing Climate: Special Report of the Intergovernmental Panel on Climate Change* (pp. 321–446). Cambridge University Press., doi:10.1017/9781009157964.006

Intergovernmental Panel on Climate Change. (IPCC, 2022b). *Climate change 2022: Impacts, adaptation, and vulnerability. Contribution of working group II to the sixth assessment report of the Intergovernmental Panel on Climate Change.* Cambridge University Press. https://www.ipcc.ch/report/ar6/wg2/

Intergovernmental Panel on Climate Change. (IPCC, 2022c). *Climate change 2022: Mitigation of climate change. Contribution of working group iii to the sixth assessment report of the Intergovernmental Panel on Climate Change.* Cambridge University Press. https://doi.org/10.1017/9781009157926

International Energy Agency. (2022). *Analysis: Energy-Access*. IEA. https://www.iea.org/topics/energy-access

International Institute for Sustainable Development (IISD). (2020). *Three Ways the Coronavirus is Shaping Sustainable Development*. Winnipeg: IISD. https://www.iisd.org/articles/insight/three-ways-coronavirus-shaping-sustainable-development

International Labour Organization (ILO). (1999). *Fishing among the most dangerous of all professions, says ILO. International Labour Office Report*. https://www.ilo.org/global/about-the-ilo/newsroom/news/WCMS_071324/lang--en/index.htm

International Labour Organization (ILO). (2004). *Danger at sea - Working in the fishing sector*. https://www.ilo.org/global/about-the-ilo/mission-and-objectives/features/WCMS_075579/lang-en/index.html

International Labour Organization (ILO). (n.d.). *Psychosocial risks and stress at work*.https://www.ilo.org/global/topics/safety-and-health-at-work/areasofwork/occupational-safety-and-health-management-systems/WCMS_861077/lang--en/index.htm

International Labour Organization. (ILO, 2016). Psychosocial risks, stress and violence in the world of work.*International Journal of Labour Research*. *8*(1-2). https://www.ilo.org/wcmsp5/groups/public/---ed_dialogue/---actrav/documents/publication/wcms_551796.pdf

International Maritime Organization. (2019). International Convention for the Prevention of Pollution from Ships (MARPOL). https://www.imo.org/en/about/Conventions/Pages/International-Convention-for-the-Prevention-of-Pollution-from-Ships-(MARPOL).aspx#:~:text=The%20MARPOL%20Convention%20was%20adopted,Protocol%20absorbed%20the%20parent%20Convention

International Organization for Standardization. (2016). *ISO/IEC 30134–2: 2016 Information Technology—Data Centres—Key Performance Indicators—Part 2: Power Usage Effectiveness (PUE)*. International Organization for Standardization: Geneva Switzerland. https://www.iso.org/standard/63451.html

International Renewable Energy Agency. (2022). *Renewable power generation costs in 2021*. IREA. https://www.irena.org/publications/2022/Jul/Renewable-Power-Generation-Costs-in-2021

International Renewable Energy Agency. (March 2023). *Renewable-capacity-statistics-2023*. IREA. https://www.irena.org/Publications/2023/Mar/Renewable-capacity-statistics-2023

Ipas study. (2022, September 23). *New research is in: Climate change impacts women's sexual and Reproductive Health*. IPAS. [https://www.ipas.org/our-work/climate-justice/climate-change-impacts-womens-sexual-and-reproductive-health/]

IPCC. (2001). *Third Report of the Working Group of the Intergovernmental Panel on Climate Change*. IPCC.

IPCC. (2007). *Climate Change 2007: Synthesis Report – Summary for Policy Makers*. Cambridge University Press.

IPCC. (2014). *Climate change 2014: Synthesis Report. Contribution of Working Groups I, II and III to the Fifth Assessment Report of the Intergovernmental Panel on Climate Change* [Core Writing Team, R.K. Pachauri and L.A. Meyer (eds.)]. IPCC.

Iqbal, S., Xu, J., Allen, S. D., Khan, S., Nadir, S., Arif, M. S., & Yasmeen, T. (2020). Unraveling consequences of soil micro- and nano-plastic pollution on soil-plant system: Implications for nitrogen (N) cycling and soil microbial activity. *Chemosphere*, *260*, 127578. doi:10.1016/j.chemosphere.2020.127578 PMID:32683024

Ismail, W. A. A. Z. W. (2021). *Development Approaches for the Orang Asli Laut As An Indicator of 'Leaving No One Behind*. Research Gate.

Issaharou-Matchi1, I., Rabiou, H., Mahamane1, M., Idrissa, B., Mahamane, A. andSaadou, M. (2019). Rural Communities' Perceptions of Climate Change and its Impacts on their Livelihood in Southeast Niger. *International Journal of Environment and Climate Change*, *9*(12), 789-800.

Isukuru, E. J., Ogunkeyede, A. O., Adebayo, A. A., & Uruejoma, M. F. (2023). Potentials of bamboo and its ecological benefits in Nigeria. *Elsevier B.V.*, 01-11.

Izadi, D. S., & Sharif-Khatibi, Z. (2020). Blockchain and its usage in storing information as a secure distributed database. *J Informat Communication TechnolPolic*, *1*(2), 85–106.

Jackson, C. (2003). Gender analysis of land: Beyond land rights for women? *Journal ofAgrarian Change*, *3*(4), 453–480. doi:10.1111/1471-0366.00062

Jain, S. (2006). An Empirical Economic Assessment of Impacts of Climate Change on Agriculture in Zambia. *Centre for Environmental Economics and Policy in Africa (CEEPA) Discussion Paper, No. 27*. Pretoria: CEEPA.

Jairam, R. (2016, May). Ecological concerns in Mahabharata. *IOSR Journal of Humanities and Social Sciences*, *21*(5), 63–65.

Jalais, A. (2019). Do human-wildlife 'interactions' affect mental health in the Sundarbans. In A. Ghosh (Ed.), *The Tides of Life: Surviving Between the Margins.*

Jamal, Md. R., Kristiansen, P., Kabir, Md. J., & de Bruyn, L. L. (2023). Risks and adaptation dynamics in shrimp and prawn-based farming systems in southwest coastal Bangladesh. *Aquaculture (Amsterdam, Netherlands)*, *562*, 738819. doi:10.1016/j.aquaculture.2022.738819

Jambeck, J. R., Geyer, R., Wilcox, C., Siegler, T. R., Perryman, M., Andrady, A., Narayan, R., & Law, K. L. (2015). Plastic waste inputs from land into the ocean. *Science*, *347*(6223), 768–771. doi:10.1126/science.1260352 PMID:25678662

Jayakumar, R., Sakthivel, M., Abdul Nazar, A. K., Tamilmani, G., Rameshkumar, P., Samal, A. K., Anikuttan, K. K., Anbarasu, M., Balamurugan, V., Thiagu, R., Sirajudeen, S., & Gopakumar, G. (2018). Impact of increase in temperature and light intensity on development and metamorphosis of hatchery reared silver pompano Trachinotusblochii (Lacepede, 1801) larvae. *Indian Journal of Fisheries*, *65*. Advance online publication. doi:10.21077/ijf.2018.65.2.37091-18

Jayasingh, S. I., Janani, M. S., & Sowndharya, R. (2022). Bacterial diseases in fish with relation to pollution and their consequences—a global scenario. In *Bacterial Fish Diseases* (pp. 113–131). Academic Press., https://doi.org/, 10.1016/B978-0-323-85624-9.00022-1

Jean de La Hire. (1909). *L'Homme Qui Peut Vivre dans l'Eau*. The Man Who Could Live Underwater.

Jeebhay, M. F., Robins, T., & Lopata, A. (2004). World at work: Fish processing workers. *Occupational and Environmental Medicine*, *61*(5), 471–474. doi:10.1136/oem.2002.001099

Jifan, C. (1989). Predictability of the Atmosphere. *Advances in Atmospheric Sciences*, *6*(3), 335–346. doi:10.1007/BF02661539

Johns, D.G., & Reid, P.C. (2001). *An overview of plankton ecology in the North Sea* (TR_005; Strategic Environmental Assessment – SEA2). SAHFOS.

Jones, J. A., Wei, X., Archer, E., Bishop, K., Blanco, J. A., Ellison, D., & Creed, I. F. (2020). Forest-water interactions under global change. *Forest-water interactions*, 589-624. Research Gate.

Kadıoğlu, M. (2001). *The End of the Weather You Know - Global Climate Change and Turkey*. Current Publishing.

Kais, S. M., & Islam, M. S. (2023). Climate Change, Ecological Modernization, and Disaster Management: The Coastal Embankment Project in Southwestern Bangladesh. [PubMed]. *International Journal of Environmental Research and Public Health*, *20*(12), 6086. doi:10.3390/ijerph20126086 PMID:37372673

Kalantary, C. (2010). Climate Change in Zambia: Impacts and Adaptation. *Global Majority E-Journal*, *1*(2), 85–96.

Kamble, B. (2009). *The prisons we broke* (M. Pandit, Trans.). Orient BlackSwan. (Original work published 1986)

Kamel Boulos, M. N., & Wilson, J. P. (2023). Geospatial techniques for monitoring and mitigating climate change and its effects on human health. *International Journal of Health Geographics*, *22*(1), 2. doi:10.1186/s12942-023-00324-9 PMID:36707823

Kankaria, A. (2014). Indoor Air Pollution in India: Implications on Health and its Control. *Indian J Community Med. Oct-Dec; 39*(4): 203–207.

Kanosvamhira, T. P. (2023). How do we get the community gardening?: Grassroots perspectives from urban gardeners in Cape Town, South Africa. *Journal of Cultural Geography*, *40*(1), 47–63. doi:10.1080/08873631.2023.2187509

Kaper, H., & Engler, H. (2013). *Mathematics and Climate*. doi:10.1137/1.9781611972610

Karadurmus, U., & Sari, M. (2022). Marine mucilage in the Sea of Marmara and its effects on the marine ecosystem: Mass deaths. *Turkish Journal of Zoology*, *46*, 93–102.

Karbalaei, S., Hanachi, P., Walker, T. R., & Cole, M. (2018). Occurrence, sources, human health impacts and mitigation of microplastic pollution. *Environmental Science and Pollution Research International*, *25*(36), 36046–36063. doi:10.1007/s11356-018-3508-7 PMID:30382517

Karl, T. R. Melillo, J.M., Peterson, T.C., eds., (2009). Global Climate Change Impacts in the United States. Cambridge: Cambridge University Press.

Karl, T. R., & Trenberth, K. E. (2003). Modern global climate change. *science, 302* (5651), 1719-1723.

Karliner, J., Slotterback, S., Boyd, R., Ashby, B., & Steele, K. (2019). Health care's climate footprint. Health Care Without Harm Climate-smart health care series. *Green Paper Number One*.

Karl, T. R., Melillo, J. M., & Peterson, T. C. (Eds.). (2009). *Global Climate Change Impacts in the United States*. Cambridge University Press.

Karunarathna, H., Maduwantha, P., Kamranzad, B., Rathnasooriya, H., & De Silva, K. (2020). Impacts of global climate change on the future ocean wave power potential: A case study from the Indian Ocean. *Energies*, *13*(11), 3028. doi:10.3390/en13113028

Kashung, Y., Gap, N., Debbarma, D., Das, B., Paul, A., Singha, L. B., & Tripathi, O. P. (2022). Carbon stock and pool assessment in tropical and subtropical bamboo brakes of Arunachal Pradesh, India. *Vegetos*, *36*(2), 668–674. doi:10.1007/s42535-022-00436-w

Kasinathan, P., Pugazhendhi, R., Elavarasan, R. M., Ramachandaramurthy, V. K., Ramanathan, V., Subramanian, S., Kumar, S., Nandhagopal, K., Raghavan, R. R. V., Rangasamy, S., Devendiran, R., & Alsharif, M. H. (2022). Realization of Sustainable Development Goals with Disruptive Technologies by Integrating Industry 5.0, Society 5.0, Smart Cities and Villages. *Sustainability (Basel)*, *14*(22), 15258. doi:10.3390/su142215258

Kasirajan, S., & Ngouajio, M. (2012). Polyethylene and biodegradable mulches for agricultural applications: A review. *Agronomy for Sustainable Development*, *32*(2), 501–529. doi:10.1007/s13593-011-0068-3

Kathleen, A., & Feris, L. (2022). South Africa's small-scale fishers have been marginalised since apartheid – what needs to change. *The Conversation.*: https://www.researchgate.net/publication/365743989

Kawadza, S. (2021). Veld fires a national concern: VP Mohadi. *The Herald.* https://www.herald.co.zw/veld-fires-a-national-concern-vp-mohadi/

Kawamiya, M., Hajima, T., Tachiiri, K., Watanabe, S., & Yokohata, T. (2020). Two decades of Earth system modeling with an emphasis on Model for Interdisciplinary Research on Climate (MIROC). *Progress in Earth and Planetary Science*, *7*(1), 1–13. doi:10.1186/s40645-020-00369-5

Keese, M., & Marcolin, L. (2023). *Labour and social policies for the green transition: A conceptual framework. OECD Social, Employment and Migration Working Papers, No. 295.* OECD Publishing., doi:10.1787/028ffbeb-en

Kela, S. (2018, April). Where the wild things are not: The curious absence of contemporary nature writing in India. *The Caravan.* http://www.caravanmagazine.in/reviews-essays/absence-contemporary-nature-writingindia

Keleş, G., Yılmaz, S., & Zengin, M. (2020). Possible economic effects of mucilage on Sea of Marmara fisheries. *International Journal of Agriculture Forestry and Life Sciences*, *4*(2), 173–177.

Kelkar, N., Arthur, R., Dey, S., & Krishnaswamy, J. (2022). Flood-Pulse variability and climate change effects increase uncertainty in fish yields: Revisiting narratives of declining fish catches in India's Ganga river. *Hydrology*, *9*(4), 53. doi:10.3390/hydrology9040053

Keller, R. L., Muir, K., Roth, F., Jattke, M., & Stucki, M. (2021). From bandages to buildings: Identifying the environmental hotspots of hospitals. *Journal of Cleaner Production*, *319*, 128479. doi:10.1016/j.jclepro.2021.128479

Kemarau, R. A., & Eboy, O. V. (2022). How do El Nino Southern Oscillation (ENSO) events impact fish catch in Sarawak water? *Journal of Physics: Conference Series*, *2314*(1), 012013. doi:10.1088/1742-6596/2314/1/012013

Keplinger, T., Haider, M., Steinparzer, T., Patrejko, A., Trunner, P., & Haselgrübler, M. (2018). Dynamic simulation of an electric arc furnace waste heat recovery system for steam production. *Applied Thermal Engineering*, *135*, 188–196. doi:10.1016/j.applthermaleng.2018.02.060

Kerlinger, F. N. (Ed.). (1986). *Foundations of Behavioural Research* (3rd ed.). Holt, Rinehart and Winston.

Kersten, M., Carrazco, L., Rosing, H., Swenski, T., Russell, D., Idrovo, J., & Lofton, S. (2023). Evaluation of the Grow Your Groceries Home Gardening Program in Chicago, Illinois. [PubMed]. *Journal of Community Health*, *48*(2), 179–188. doi:10.1007/s10900-022-01152-x PMID:36336753

Kh'ng, X. Y., Teh, S. Y., Koh, H. L., & Shuib, S. (2021). Sea level rise undermines SDG2 and SDG6 in Pantai Acheh, Penang, Malaysia. *Journal of Coastal Conservation*, *25*(1), 1–14. doi:10.1007/s11852-021-00797-5

Khalid, N., Aqeel, M., & Noman, A. (2020). Microplastics could be a threat to plants in terrestrial systems directly or indirectly. *Environmental Pollution*, *267*, 115653. doi:10.1016/j.envpol.2020.115653 PMID:33254725

Kieush, L., Schenk, J., Koveria, A., Rantitsch, G., Hrubiak, A., & Hopfinger, H. (2023). Utilization of Renewable Carbon in Electric Arc Furnace-Based Steel Production: Comparative Evaluation of Properties of Conventional and Non-Conventional Carbon-Bearing Sources. *Metals*, *13*(4), 722. doi:10.3390/met13040722

Kings, A. E. (2017). Intersectionality and the changing face of ecofeminism. *Ethics and the Environment*, *22*(1), 63–87. doi:10.2979/ethicsenviro.22.1.04

King, T. J., Turner, R., Versace, V., Abernethy, K., Kilpatrick, S., & Brumby, S. (2021). Mental health in the commercial fishing industry: Modern uncertainties and traditional risks. *Fish and Fisheries*, *22*(5), 1136–1149. doi:10.1111/faf.12572

Kirelli, Y., & Arslankaya, S. (2020). Sentiment analysis of shared tweets on global warming on Twitter with data mining methods: A case study on Turkish language. *Computational Intelligence and Neuroscience*, *2020*, 1–9. doi:10.1155/2020/1904172 PMID:32963511

Klinenberg, E. (2002). *Heat wave: A social autopsy of disaster in Chicago*. University of Chicago Press. doi:10.7208/chicago/9780226026718.001.0001

Koçak, K. (2005). *Human Factor in Climate Change*. ITU. https://web.itu.edu.tr/~kkocak/iklimpdf.pdf

Kochar, A., Nagabhushana, C., Sarkar, R., Shah, R., & Singh, G. (2022). Financial access and women's role in household decisions: Empirical evidence from India's National Rural Livelihoods project. *Journal of Development Economics*, *155*, 102821. doi:10.1016/j.jdeveco.2022.102821 PMID:35241869

Kodiveri, A. (2021). The delicate task of including different voices in environmental law making in India. *Journal of Environmental Law*, *33*(3), 495–500. doi:10.1093/jel/eqab030

Kodoh, J., Jaunis, O., Maid, M., Lintangah, W., & Mojiol, A. R. (2023, February). Urban Forestry in Kota Kinabalu: Home Garden practices by urban villages for biodiversity, socio-economic and environmental services. []. IOP Publishing.]. *IOP Conference Series. Earth and Environmental Science*, *1145*(1), 012012. doi:10.1088/1755-1315/1145/1/012012

Kohn, N. A. (2014). Vulnerability Theory and the role of Government. *Yale Journal of Law and Feminism*, *26*(1), 1–27.

Kolawole, O. D., & Bolobilwe, K. (2018). Survival at a cost: How artisanal fishers perceive occupational hazards in the Okavango Delta, Botswana. *The South African Geographical Journal*, *101*(1), 51–71. doi:10.1080/03736245.2018.1541019

Kolves, K., Kolves, K. E., & de Leo, D. (2013). Natural disasters and suicidal behaviours: A systematic literature review. *Journal of Affective Disorders*, *146*(1), 1–14. doi:10.1016/j.jad.2012.07.037 PMID:22917940

Kombo, D. K., & Tromp, D. L. A. (2006). *Proposal and Thesis Writing: An Introduction*. Nairobi: Pauline Publication Africa.

Koncagul, M., Dulger, N. E., & Yinanc, A. (2022). Formation and the effect of musilage in Marmara Sea and in the world. *European Journal of Engineering and Applied Sciences*, *5*(2), 73–79.

Korpelainen, H. (2023). The Role of Home Gardens in Promoting Biodiversity and Food Security. [PubMed]. *Plants*, *12*(13), 2473. doi:10.3390/plants12132473 PMID:37447034

Kosanic, A., Petzold, J., Martín-López, B., & Razanajatovo, M. (2022). An inclusive future: Disabled populations in the context of climate and environmental change. *Current Opinion in Environmental Sustainability*, *55*, 101159. doi:10.1016/j.cosust.2022.101159

Kotcher, J., Maibach, E., Miller, J., Campbell, E., Alqodmani, L., Maiero, M., & Wyns, A. (2021). Views of health professionals on climate change and health: A multinational survey study. *The Lancet. Planetary Health*, *5*(5), e316–e323. doi:10.1016/S2542-5196(21)00053-X PMID:33838130

Kothari, C. R. (2004). Research Methodology: Methods and Techniques. New Delhi: New Age International (P) Limited, Publishers.

Kourtzanidis, K., Angelakoglou, K., Apostolopoulos, V., Giourka, P., & Nikolopoulos, N. (2021). Assessing impact, performance and sustainability potential of smart city projects: Towards a case agnostic evaluation framework. *Sustainability (Basel)*, *13*(13), 7395. doi:10.3390/su13137395

Kovats, S., & Allen, M. (2008). Flooding, windstorms and climate change. In *Department of Health and Health Protection Agency of the UK. Health effects of climate change in the UK 2008: An update of the Department of Health report 2001–2002*. Department of Health and Health Protection Agency of the UK., http://www.bvsde.paho.org/bvsacd/cd68/UK2008/cap2.pdf

Krishnan, R., & Naga, R. (2017). 'Ecological warriors' versus 'Indigenous performers': Understanding state responses to resistance movements in Jagatsinghpur and Niyamgiri in Odisha. *South Asia*, *40*(4), 878–894. doi:10.1080/00856401.2017.1375730

Kroeker, K. J., Kordas, R. L., Crim, R., Hendriks, I. E., Ramajo, L., Singh, G. S., Duarte, C. M., & Gattuso, J. P. (2013). Impacts of ocean acidification on marine organisms: Quantifying sensitivities and interaction with warming. *Global Change Biology*, *19*(6), 1884–1896. doi:10.1111/gcb.12179

Krutilin, A. A., Bazieva, A. M., Dugina, T. A., & Giyazov, A. T. (2022). Sustainable Agriculture for Food Security: Conceptual Framework and Benefits of Digitalization. In Sustainable Agriculture: Circular to Reconstructive, Volume 1 (pp. 35-42). Springer.

Kulkarni, M. A., Duguay, C., & Ost, K. (2022). Charting the evidence for climate change impacts on the global spread of malaria and dengue and adaptive responses: A scoping review of reviews. *Globalization and Health*, *18*(1), 1–18. doi:10.1186/s12992-021-00793-2 PMID:34980187

Kumar, A. (2020). B. R. Ambedkar on caste and land relations in India. *Review of Agrarian Studies*, *10*(1), 37–56.

Kumari, S., & Patil, Y. B. (2019). Enablers of sustainable industrial ecosystem: Framework and future research directions. *Management of Environmental Quality*, *30*(1), 61–86. doi:10.1108/MEQ-02-2018-0044

Kumari, S., Zacharia, P. U., Kripa, V., Sreenath, K. R., & George, G. (2015). Distribution pattern and community structure of zoanthids (Zoantharia) along the coast of Saurashtra, Gujarat, India. *Journal of the Marine Biological Association of the United Kingdom*, *96*(8), 1577–1584. doi:10.1017/s0025315415001885

Kumar, J. A., Mrralidhar, M., Jayanthi, M., & Kumaran, M. (2013). Trend analysis of weather data in shrimp farming areas of Nagapattinam district of Tamil Nadu. *Journal of Agrometeorology*, *15*(2), 129–134. https://doi.org/10.54386/jam.v15i2.1459

Kumar, S., Kumar, V., Singh, H., Kumar, S., & Meena, R. M. (2016). *Climate Change: Challenges and Opportunities for the 21st Century*. Institute of Agricultural Sciences, Banaras Hindu University.

Kumar, V., & Mishra, B. (2022). Environmental casteism and the democratisation of natural resources: Reimagining Dalit testimonies. *South Asia*, *45*(3), 577–595. doi:10.1080/00856401.2022.2014157

Kuran, C. H. A., Morsut, C., Kruke, B. I., Krüger, M., Segnestam, L., Orru, K., Nævestad, T. O., Airola, M., Keränen, J., Gabel, F., Hansson, S., & Torpan, S. (2020). Vulnerability and vulnerable groups from an intersectionality perspective. *International Journal of Disaster Risk Reduction*, *50*, 101826. doi:10.1016/j.ijdrr.2020.101826

Laffoley, D., & Baxter, J. M. (Eds.). (2016). Explaining ocean warming: Causes, scale, effects and consequences. International Union for Conservation of Nature (IUCN) (pp. 456). http://dx.doi.org/ doi:10.2305/IUCN.CH.2016.08.en

Lakra, W. S., & Gopalakrishnan, A. (2021). Blue revolution in India: Status and future perspectives. *Indian Journal of Fisheries*, *68*(1), 137–150. doi:10.21077/ijf.2021.68.1.109283-19

Lal, P., Alavalapati, J. R. R., & Mercer, E. D. (2011). *Social Economic Impacts of Climate Change on Rural United States*. USDA Forest Service/UNL Faculty Publications.

Lal, R. (2020). Home gardening and urban agriculture for advancing food and nutritional security in response to the COVID-19 pandemic. [PubMed]. *Food Security*, *12*(4), 871–876. doi:10.1007/s12571-020-01058-3 PMID:32837634

Lamba, N., & Thareja, P. (2021). Developing the structural model based on analyzing the relationship between the barriers of green supply chain management using TOPSIS approach. *Materials Today: Proceedings*, *43*, 1–8. doi:10.1016/j.matpr.2020.09.487

Lan, K.-W., Chang, Y.-J., & Wu, Y.-C. (2020). Influence of oceanographic and climatic variability on the catch rate of yellowfin tuna (Thunnusalbacares) cohorts in the Indian Ocean. *Deep-sea Research. Part II, Topical Studies in Oceanography*, *175*, 104681–104681. doi:10.1016/j.dsr2.2019.104681

Larriva, M. T. B., Mendes, A. S., & Forcada, N. (2022). The effect of climatic conditions on occupants' thermal comfort in naturally ventilated nursing homes. *Building and Environment*, *214*, 108930. doi:10.1016/j.buildenv.2022.108930

Lavers, J. L., Bond, A. L., & Rolsky, C. (2022). Far from a distraction: Plastic pollution and the planetary emergency. *Biological Conservation*, *272*, 109655. doi:10.1016/j.biocon.2022.109655

Lawrance, E. L., Thompson, R., Newberry Le Vay, J., Page, L., & Jennings, N. (2022). The impact of climate change on mental health and emotional wellbeing: A narrative review of current evidence, and its implications. *International Review of Psychiatry (Abingdon, England)*, *34*(5), 443–498. doi:10.1080/09540261.2022.2128725 PMID:36165756

Lawrence. (2019). Is PUE actually going UP? *Uptime Institute Blog*. https://journal.uptimeinstitute.com/is-pue-actually-going-up/

Le Masson, V., Benoudji, C., Sotelo Reyes, S., & Bernard, G. (2019). How violence against women and girls undermines resilience to climate risks in Chad. *Disasters*, *43*(S3), S245–S257. doi:10.1111/disa.12343 PMID:30945771

Leal Filho, W., Perry, P., Heim, H., Dinis, M. A., Moda, H., Ebhuoma, E., & Paço, A. (2022). An overview of the contribution of the textiles sector to climate change. *Frontiers in Environmental Science*, *10*, 973102. doi:10.3389/fenvs.2022.973102

Lebbe, T. B., Rey-Valette, H., Chaumillon, E., Camus, G., Almar, R., Cazenave, A., Claudet, J., Rocle, N., Meur-Ferec, C., Viard, F., Mercier, D., Dupuy, C., Menard, F., Rossel, B. A., Mullineaux, L., Sicre, M.-A., Zivian, A., Gaill, F., & Euzen, A. (2021). Designing Coastal Adaptation Strategies to Tackle Sea Level Rise. *Frontiers in Marine Science*, *8*, 740602. doi:10.3389/fmars.2021.740602

Lebni, J. Y., Gharehghani, M. A. M., Soofizad, G., & Khosravi, B. (2020). Challenges and Opportunities Confronting Female-Headed Households in Iran: A Qualitative Study. *BMC Women's Health*, *20*(1), 183. doi:10.1186/s12905-020-01046-x PMID:32807144

Lebreton, L. C. M., Van der Zwet, J., Damsteeg, J. W., Slat, B., Andrady, A., & Reisser, J. (2017). River plastic emissions to the world's oceans. *Nature Communications*, *8*(1), 15611. doi:10.1038/ncomms15611 PMID:28589961

Lee, J. W. (2020). Green Finance and Sustainable Development Goals: The Case of China. *Journal of Asian Finance Economics and Business*, *7*(7), 577–586. doi:10.13106/jafeb.2020.vol7.no7.577

Lefevre, J., Le Gallic, T., Fragkos, P., Mercure, J. F., Simsek, Y., & Paroussos, L. (2022). Global socio-economic and climate change mitigation scenarios through the lens of structural change. *Global Environmental Change*, *74*, 102510. doi:10.1016/j.gloenvcha.2022.102510

Legacy. (n.d.). *Climate Change and Male Fertility*. Give Legacy. https://www.givelegacy.com/resources/climate-change-and-male-fertility/

Lehmuskallio, E., Hassi, J., & Kettunen, P. (2002). The skin in the cold. *International Journal of Circumpolar Health, 61*(3), 277–286. doi:10.3402/ijch.v61i3.17475

Lehtonen, H. (1996). Potential effects of global warming on northern European freshwater fish and fisheries. *Fisheries Management and Ecology, 3*(1), 59–71. doi:10.1111/j.1365-2400.1996.tb00130.x

Lentisco, A., & Robert, U. L. (2015). *A review of women's access to fish in small-scale fisheries.* FAO Fisheries and Aquaculture Circular No. 1098. ISBN: 978-92-5-108857-9. https://www.fao.org/3/i4884e/i4884e.pdf

Leone, U. (2022). G20 Climate Risk Atlas. Impacts, Policy, Economics. CMCC. *Documenti geografici*, (2), 477-478.

Lewis, P., Monem, M. A., & Impiglia, A. (2018). *Impacts of Climate change on Farming Systems and Livelihoods in the near East and North Africa-with a Special Focus on Small Scale Family Farming*. FAO.

Li, J., Liu, H., & Paul Chen, J. (2018). Microplastics in freshwater systems: A review on occurrence, environmental effects, and methods for microplastics detection. *Water Research, 137*, 362-374. https://www. sciencedirect.com/science/article.

Li, P., Yang, J., Islam, M. A., & Ren, S. (2023). *Making AI Less 'Thirsty': Uncovering and Addressing the Secret Water Footprint of AI Models.* arxiv. doi:10.48550/ARXIV.2304.03271

Liang, L., Deng, X., Wang, P., Wang, Z., & Wang, L. (2020). Assessment of the impact of climate change on cities livability in China. *The Science of the Total Environment, 726*, 138339. doi:10.1016/j.scitotenv.2020.138339 PMID:32325298

Lieber, M., Chin-Hong, P., Whittle, H. J., Hogg, R., & Weiser, S. D. (2021). The synergistic relationship between climate change and the HIV/AIDS epidemic: A conceptual framework. *AIDS and Behavior, 25*(7), 2266–2277. doi:10.1007/s10461-020-03155-y PMID:33452659

Liebers, V., Brüning, T., & Raulf, M. (2020). Occupational endotoxin exposure and health effects. *Archives of Toxicology, 94*(11), 3629–3644. https://doi.org/10.1007/s00204-020-02905-0

Lincoln, J. M., Carruth, A., Cherry, D., Kincl, L., & Syron, L. N. (2021). Occupational health research in the commercial fishing industry. *Journal of Agromedicine, 26*(1), 28–30. doi:10.1080/1059924x.2021.1849494

Lindman, J., Tuunainen, V. K., & Rossi, M. (2017). Opportunities and risks of Blockchain Technologies in payments - a research agenda. Presented at *50th Hawaii International Conference on System Sciences*, Hawaii, US.

Lindsey, R., & Dahlman, L. (2023, January 18). Climate Change: Global Temperature. *Climate.gov*. https://www.climate.gov/news-features/understanding-climate/climate-change-global-temperature#:~:text=Earth's%20temperature%20has%20risen%20by,0.18%C2%B0%20C)%20per%20decade

Lin, I. C., & Liao, T. C. (2017). A Survey of Blockchain Security Issues and Challenges. *International Journal of Network Security, 19*(5), 653–659.

Link, L. E. (2010). The anatomy of a disaster, an overview of Hurricane Katrina and New Orleans. [CrossRef]. *Ocean Engineering, 37*(1), 4–12. doi:10.1016/j.oceaneng.2009.09.002

Lin, X., Niu, J., Berndtsson, R., Yu, X., Zhang, L., & Chen, X. (2020). NDVI dynamics and its response to climate change and reforestation in northern China. *Remote Sensing (Basel), 12*(24), 4138. doi:10.3390/rs12244138

Li, P., Wang, X., Su, M., Zou, X., Duan, L., & Zhang, H. (2021). Characteristics of Plastic Pollution in the Environment: A Review. *Bulletin of Environmental Contamination and Toxicology, 107*(4), 577–584. doi:10.1007/s00128-020-02820-1 PMID:32166334

Liu, E. K., He, W. Q., & Yan, C. R. (2014). White revolution to white pollution—Agricultural plastic film mulch in China. *Environmental Research Letters, 9*(9), 091001. doi:10.1088/1748-9326/9/9/091001

Liu, Z., Deng, Z., Davis, S., & Ciais, P. (2023). Monitoring global carbon emissions in 2022. *Nature Reviews. Earth & Environment*, *4*(4), 205–206. doi:10.1038/s43017-023-00406-z PMID:37065615

Li, W. C., Tse, H. F., & Fok, L. (2016). Plastic waste in the marine environment: A review of sources, occurrence and efects. *The Science of the Total Environment*, *566*, 333–349. doi:10.1016/j.scitotenv.2016.05.084 PMID:27232963

Lix, J. K., Venkatesan, R., Grinson, G., Rao, R. R., Jineesh, V. K., Arul, M. M., Vengatesan, G., Ramasundaram, S., Sundar, R., & Atmanand, M. A. (2016). Differential bleaching of corals based on El Niño type and intensity in the Andaman Sea, southeast Bay of Bengal. *Environmental Monitoring and Assessment*, *188*(3), 175. doi:10.1007/s10661-016-5176-8

Lloka, N. G. (2016). "Indigenous knowledge for disaster risk resolution: An African perspective." In *Journal of Disaster risk. Studies*, *8*(1), 1–7.

Lo, S. K., Xu, X., Chiam, Y. K., & Lu, Q. (2017). *Evaluating Suitability of Applying Blockchain,* presented at International Conference on Engineering of Complex Computer Systems, Fukuoka, Japan.

Lobovikov, M., Lou, Y., Schoene, D., & Widenoja, R. (2009). The poor man's carbon sink: bamboo in climate change and poverty alleviation. *Non-Wood Forest Products Working Document*, (8).

Lobovikov, M., Schoene, D., & Yping, L. (2012). Bamboo in climate change and rural livelihoods. *Mitigation and Adaptation Strategies for Global Change*, *17*(3), 261–276. doi:10.1007/s11027-011-9324-8

Logie, C. H., Toccalino, D., Reed, A. C., Malama, K., Newman, P. A., Weiser, S., Harris, O., Berry, I., & Adedimeji, A. (2021). Exploring linkages between climate change and sexual health: A scoping review protocol. *BMJ Open*, *11*(1), e054720. doi:10.1136/bmjopen-2021-054720 PMID:34663670

Loginova, J., & Batterbury, S. P. (2019). Incremental, transitional and transformational adaptation to climate change in resource extraction regions. *Global Sustainability*, *2*, e17. doi:10.1017/sus.2019.14

Lombardo, E. (2022). An overview of bamboo cultivation in Southern Italy. *Advances in Bamboo Science, 100002.*

Long, M. C., Deutsch, C., & Ito, T. (2016). Finding forced trends in oceanic oxygen. *Global Biogeochemical Cycles*, *30*(2), 381–397. doi:10.1002/2015gb005310

Lopata, A. L., Baatjies, R., Thrower, S. J., & Jeebhay, M. F. (2004). Occupational allergies in the seafood industry-a comparative study of Australian and South African workplaces. *International Maritime Health*, *55*(1-4), 61–73.https://pubmed.ncbi.nlm.nih.gov/15881544/

Lorenz, E. (1963). Deterministic nonperiodic flow. *Journal of the Atmospheric Sciences*, *20*(2), 130–141. doi:10.1175/1520-0469(1963)020<0130:DNF>2.0.CO;2

Lorenzoni, I., Nicholson-Cole, S., & Whitmarsh, L. (2007). Barriers perceived to engaging with climate change among the UK public and their policy implications. *Global Environmental Change*, *17*(3-4), 445–459. doi:10.1016/j.gloenvcha.2007.01.004

Lou, C., Tan, S. S., & Chen, X. (2019). Investigating consumer engagement with influencer-vs. brand-promoted ads: The roles of source and disclosure. *Journal of Interactive Advertising*, *19*(3), 169–186. doi:10.1080/15252019.2019.1667928

Louw, A. S., Fu, J., Raut, A., Zulhilmi, A., Yao, S., McAlinn, M., Fujikawa, A., Siddique, M. T., Wang, X., Yu, X., Mandvikar, K., & Avtar, R. (2022). The role of remote sensing during a global disaster: Covid-19 pandemic as case study. *Remote Sensing Applications: Society and Environment*, *27*, 100789. doi:10.1016/j.rsase.2022.100789 PMID:35774725

Low, A. J., Frederix, K., McCracken, S., Manyau, S., Gummerson, E., Radin, E., Davia, S., Longwe, H., Ahmed, N., Parekh, B., Findley, S., & Schwitters, A. (2019). Association between severe drought and HIV prevention and care behaviors in Lesotho: A population-based survey 2016-2017. *PLoS Medicine*, *16*(1), e1002727. doi:10.1371/journal.pmed.1002727 PMID:30640916

Lucas, D., Lucas, R., Boniface, K., Jegaden, D., Lodde, B., & Dewitte, J. A. (2010). Occupational asthma in the commercial fishing industry: A case series and review of the literature. *International Maritime Health*, *61*(1), 13–17. https://pubmed.ncbi.nlm.nih.gov/20496322/

Lukas, K. S. (2013). *Achieving Growth through Constant Implementation of New Professional Life Standard, Techniques and Innovative Ideas*. Nyambari Nyangwine Publishers.

Lu, Y. (2018). Blockchain and the related issues: A review of current research topics. *Journal of Management Analytics*, *5*(4), 231–255. doi:10.1080/23270012.2018.1516523

Machida, D. (2019). Relationship between community or home gardening and health of the elderly: A web-based cross-sectional survey in Japan. [PubMed]. *International Journal of Environmental Research and Public Health*, *16*(8), 1389. doi:10.3390/ijerph16081389 PMID:30999687

Macs, B., & GGdeLW, S. (2020). The concept of environmental leadership: The case of Solid Waste Management (SWM) in Balangoda municipal council in Sri Lanka. *Open Access J Environ Soil Sci.*, *5*(5), 728–734.

Maddikunta, P., Pham, Q.-V., B, P., Deepa, N., Dev, K., Gadekallu, T. R., Ruby, R., & Liyanage, M. (2022). Industry 5.0: A survey on enabling technologies and potential applications. *Journal of Industrial Information Integration*, *26*, 100257. doi:10.1016/j.jii.2021.100257

Madory, D. (2017). Puerto Rico's Slow Internet Recovery. Oracle. Retrieved from.) A secure incentive scheme for vehicular delay tolerant networks using cryptocurrency. *Security and Communication Networks*, 2018.

Maersk and IBM introduce TradeLens blockchain shipping solution. (n.d.). IBM Newsroom. https://newsroom.ibm.com/2018-08-09-Maersk-and-IBM-Introduce-TradeLens-Blockchain-Shipping-Solution

Mago, P., & Gunwal, I. (2019). *Role of Women in Environmental Conservation*. University of New Delhi.

Makhdoom, I., Abolhasan, M., Abbas, H., & Ni, W. (2019). Blockchain's adoption in IoT: The challenges, and a way forward. *Journal of Network and Computer Applications*, *125*, 251–279. doi:10.1016/j.jnca.2018.10.019

Malala Fund. (2022, March 29). *Malala Fund publishes report on climate change and girls' education*. Malala Fund. https://malala.org/newsroom/malala-fund-publishes-report-on-climate-change-and-girls-education#:~:text=Malala%20Fund%E2%80%99s%20new%20report%2C%20A%20Tipping%20Point%20for%20Climate%20Action%20and%20Girls%E2%80%99%20Education%2C%20reveals%20that%20climate%20change%20will%20prevent%20an%20estimated%20four%20million%20girls%20in%20low%2D%20and%20lower%2Dmiddle%20income%20countries%20from%20completing%20their%20education

Malaysiakini. (2022). Gardening fans, let's gather at New World Park, Penang. The largest Plant Market, City Gardening KL is coming to town! Malaysiakini. https://www.malaysiakini.com/announcement/637177

Mallela, J., & Fox, R. J. (2018). The role of parrotfishes in the destruction and construction of coral reefs. In A. S. Hoey& R. M. Bonaldo (Eds.), Biology of Parrotfishes (pp. 161–196). CRC Press. isbn: 978036778140

Mallett, A., & Pal, P. (2022). Green transformation in the iron and steel industry in India: Rethinking patterns of innovation. *Energy Strategy Reviews*, *44*, 100968. doi:10.1016/j.esr.2022.100968

Manikfan, G. H. (2007). Importance of human resource development for the fishing industry. In S. Girija (Ed.), *Spearheading quality fish processing, Integrated Fisheries Project Souvenir* (pp. 13–16). Integrated Fisheries Project.

Marmara Sea Integrated Strategic Plan. (2021). Republic of Türkiye Ministry of Environment, Urbanization, and Climate Change. Ankara.

Masanet, A., Shehabi, A., Lei, N., Smith, S., & Koomey, J. (2020, February). Shehabi, N. Lei, S. Smith, and J. Koomey, "Recalibrating global data center energy-use estimates,". *Science*, *367*(6481), 984–986. doi:10.1126/science.aba3758 PMID:32108103

Masson-Delmotte, V., Zhai, P., Pörtner, H. O., Roberts, D., Skea, J., & Shukla, P. R. (2022). *Global Warming of 1.5 C: IPCC special report on impacts of global warming of 1.5 C above pre-industrial levels in the context of strengthening response to climate change, sustainable Development, and efforts to eradicate poverty*. Cambridge University Press.

Massos, A., & Turner, A. (2017). Cadmium, lead and bromine in beached microplastics. *Environmental Pollution*, *227*, 139–145. doi:10.1016/j.envpol.2017.04.034 PMID:28458244

Mathis, J. T., Cooley, S. R., Lucey, N., Colt, S., Ekstrom, J., Hurst, T., Hauri, C., Evans, W., Cross, J. N., & Feely, R. A. (2015). Ocean acidification risk assessment for Alaska's fishery sector. *Progress in Oceanography*, *136*, 71–91. doi:10.1016/j.pocean.2014.07.001

Mato, Y., Isobe, T., Takada, H., Kanehiro, H., Ohtake, C., & Kaminuma, T. (2001). Plastic resin pellets as a transport medium for toxic chemicals in the marine environment. *Environmental Science & Technology*, *35*(2), 318–324. doi:10.1021/es0010498 PMID:11347604

Maulu, S., Hasimuna, O. J., Haambiya, L. H., Monde, C., Musuka, C. G., Makorwa, T. H., Munganga, B. P., Phiri, K. J., & Nsekanabo, J. D. (2021). Climate change effects on aquaculture production: Sustainability implications, mitigation, and adaptations. *Frontiers in Sustainable Food Systems*, *5*, 609097. doi:10.3389/fsufs.2021.609097

Maviskalyan, A., & Tarverdi, Y. (2019). Gender and climate change: Do female parliamentarians make a difference? *European Journal of Political Economy*, *56*, 151–164. doi:10.1016/j.ejpoleco.2018.08.001

Maxmen, A. (2018). Cape Town scientists prepare for "Day Zero". *Nature*, 13–14.

Mayesha, A., Bhatia, R., & Mawby, B. (2015). *Women and Climate Change: Impact and Agency in Human Rights, Security, and Economic Development*. Georgetown Institute for Women, Peace and Security.

McAdam, J. (2020). Protecting people displaced by the impacts of climate change: The UN Human Rights Committee and the principle of non-refoulement. *The American Journal of International Law*, *114*(4), 708–725. doi:10.1017/ajil.2020.31

McCammon, S. (2017). *Google Moves In And Wants To Pump 1.5 Million Gallons Of Water Per Day*. NPR.https://www.npr.org/2017/05/08/527214026/google-moves-in-and-wants-to-pump-1-5-million-gallons-of-water-per-day

McCarthy, M. A. (2019). *The Socialist Journey*. Jacobin.

McConnell, C., & Mupuwaliywa, M. (2015). *Policy Brief Zambia – Keeping Girls in School: Situation Analysis for Zambia*. World Bank. doi:10.1596/23865

McDonald, P. (2021, March 31). *Resolving climate change disputes through arbitration*. Pinsent Masons. https://www.pinsentmasons.com/out-law/analysis/resolving-climate-change-disputes-through-arbitration

McIntyre, I. (2020). Pandemic sees more Penangites keen to grow own greens. The Vibes. https://www.thevibes.com/articles/news/12180/pandemic-sees-more-penangites-keen-to-grow-owngreens

McLed, C., Barr, H., & Rall, K. (2019). Does climate change increase the risk of child marriage: A look at what we know and what we don't with lessons from Bangladesh and Mozambique. *Columbia Journal of Gender and Law, 38*(1), 96–146.

Mcquade, J. (2019, April 18). Earth day: Colonialism's role in the overexploitation of natural resources. *The Conversation.* https://theconversation.com/earth-day-colonialisms-role-in-the-overexploitation-of-natural-resources-113995

McRae, H. (2022). *The world in 2050.* Bloomsbury Publishing.

Meaden, G. J., & Aguilar-Manjarrez, J. (Eds.). 2013. *Advances in geographic information systems and remote sensing for fisheries and aquaculture.* Summary version (pp. 425). FAO Fisheries and Aquaculture Technical Paper No. 552. https://doi.org/ doi:10.13140/RG.2.1.4037.7682

Mecozzi, M., Acquistucci, R., Noto, V. D., Pietrantonio, E., Amici, M., & Cardarilli, D. (2001). Characterization of mucilage aggreagtes in Adriatic and Tyrrhenian Sea: Structure similarities between mucilage samples and the insoluble fractions of marine humic substances. *Chemosphere, 44*(4), 709–720. doi:10.1016/S0045-6535(00)00375-1 PMID:11482660

Mecozzi, M., Pietrantonio, E., Di Noto, V., & Pápai, Z. (2005). The humin structure of mucilage aggregates in the Adriatic and Tyrrhenian seas: Hypothesis about the reasonable causes of mucilage formation. *Marine Chemistry, 95*(3), 255–269. doi:10.1016/j.marchem.2004.09.005

Mehra, R., & Rojas, M. H. (2008). *A Significant Shift: Women, Food Security and Agriculture in a Global Marketplace. International Centre for Research on Women, citing FAO Focus on Women and Food Security.* FAO.

Meng, X. H., Zeng, S. X., Shi, J. J., Qi, G. Y., & Zhang, Z. B. (2014). The relationship between corporate environmental performance and environmental disclosure: An empirical study in China. *Journal of Environmental Management, 145,* 357–367. doi:10.1016/j.jenvman.2014.07.009 PMID:25113230

MEPC 75/INF.23 (2020). Progress report of the GESAMP Working Group on Sea-based Sources of Marine Litter.

Merino, G., Barange, M., Blanchard, J. L., Harle, J., Holmes, R., Allen, I., Allison, E. H., Badjeck, M. C., Dulvy, N. K., Holt, J., Jennings, S., Mullon, C., & Rodwell, L. D. (2012). Can marine fisheries and aquaculture meet fish demand from a growing human population in a changing climate? *Global Environmental Change, 22*(4), 795–806. doi:10.1016/j.gloenvcha.2012.03.003

Merriam, S. B. (1988). *Case study research in education: A qualitative approach.* Jossey-Bass.

Merriam, S. B., & Tisdell, E. J. (2016). *Qualitative research: A guide to design and implementation* (4th ed.). Jossey Bass.

Mertz, O., Mbow, C., Reenberg, A., & Diouf, A. (2009). A Farmers' Perceptions of Climate Change and Agricultural Adaptation Strategies in Rural Sahel. *Environmental Management, 43*(5), 804–816. doi:10.1007/s00267-008-9197-0 PMID:18810526

Meta. (2022). *2021 Sustainability Report.* Meta Inc. https://sustainability.fb.com/wp-content/uploads/2022/06/Meta-2021-Sustainability-Report.pdf

Meyer, A. L. S., Bentley, J., Odoulami, R. C., Pigot, A. L., & Trisos, C. H. (2022). Risks to biodiversity from temperature overshoot pathways. *Philosophical Transactions of the Royal Society B: Biological Sciences, 377*(1857). doi:10.1098/rstb.2021.0394

Mhlanga, W., Mashavave, T., & Dzvairo, S. (2017). A Study of the roles, opportunities and challenges of environmental sub-committees (ESCs): A case of Masvingo district, Zimbabwe. In South Africa Journal of Education, Science and Technology, 4(1).

Michalopoulos, L. T. M., Baca-Atlas, S. N., Simona, S. J., Jiwatram-Negrón, T., Ncube, A., & Chery, M. B. (2017). "Life at the River is a Living Hell:" a qualitative study of trauma, mental health, substance use and HIV risk behavior among female fish traders from the Kafue Flatlands in Zambia. *BMC Women's Health*, *17*(1). Advance online publication. doi:10.1186/s12905-017-0369-z

Microsoft. (n.d.). *Innovative cooling solutions reduce datacenter water consumption*. Microsoft. https://local.microsoft.com/wp-content/uploads/2021/11/Datacenter-water-consumption-fact-sheet.pdf

Miller, T. (2019). Explanation in artificial intelligence: Insights from the Social Sciences. *Artificial Intelligence*, *267*, 1–38. doi:10.1016/j.artint.2018.07.007

Minear, L. (2002). *The Humanitarian Enterprise: Dilemmas and Discoveries*. Kumarian Press.

Min, H. (2019). Blockchain technology for enhancing supply chain resilience. *Business Horizons*, *62*(1), 35–45. doi:10.1016/j.bushor.2018.08.012

Ministry of Fisheries. Animal Husbandry & Dairying (2023). *Department of Fisheries implements various schemes to develop fisheries sector in the country*. https://pib.gov.in/PressReleaseIframePage.aspx?PRID=1897083#:~:text=The%20Schemes%20being%20implemented%20by,and%20responsible%20development%20of%20fisheries

Ministry of Rural Development. (2012). *MGNREGS Sameeksha: An Anthology of Research Studies on the Mahatma Gandhi National Rural Employment Guarantee Act, 2005*. Government of India.

Ministry of Rural Development. (2021-2022). *Annual Report*. Government of India.

Ministy of Women and Child Development, Government of India. (2020). Malnutrition among Women. Ministry of Women and Child Development.

Mirowski, P. (2002). *Machine Dreams: Economics Becomes a Cyborg Science*. Cambridge University Press.

MIT. (2023). *Greenhouse Gases* MIT. https://climate.mit.edu/explainers/greenhouse-gases

Mitra, A., Abdel-Gawad, F. K., Bassem, S., Barua, P., Assisi, L., Parisi, C., Temraz, T. A., Vangone, R., Kajbaf, K., Kumar, V., & Guerriero, G. (2023). Climate change and reproductive biocomplexity in fishes: Innovative management approaches towards sustainability of fisheries and aquaculture. *Water (Basel)*, *15*(4), 725. doi:10.3390/w15040725

MOEF. (2010). *India: Greenhouse Gas Emissions 2007*. Indian Network for Climate Change Assessment.

Moeti, I. (2023). Sustainability and Consumer Education on Food Sustainability [Doctoral dissertation, London South Bank University.

Mohamed, K. S., Vijayakumaran, K., Zacharia, P. U., Sathianandan, T. V., Maheswarudu, G., Kripa, V., Narayanakumar, R., Rohit, P., Joshi, K. K., Sankar, T. V., Edwin, L., Kumar, A. K., Bindu, J., Gopal, N., & Puthra, P. (2017). Indian marine fisheries code: Guidance on a marine fisheries management odel for India. *Central Marine Fisheries Research InstituteMarine Fisheries Policy Series 4* (pp. 120). http://eprints.cmfri.org.in/11873/1/IMFC%20Code_Revised_2019.pdf

Mohamed, K. S., Zacharia, P. U., Maheswarudu, G., Sathianandan, T. V., Abdussamad, E. M., Ganga, U., Pillai, S. L., Sobhana, K. S., Nair, R. J., Josileen, J., Chakraborty, R. D., Kizhakudan, S. J., & Najmudeen, T. M. (2014). Minimum Legal Size (MLS) of capture to avoid growth overfishing of commercially exploited fish and shellfish species of Kerala. *The Marine Fisheries Information Service, Technical & Extension Series, 220*, 3–7. http://eprints.cmfri.org.in/10282/1/MFIS_220_1.pdf

Mohammed, E. Y., & Uraguchi, Z. B. (2013). Impacts of Climate Change on Fisheries: Implications for Food Security in Sub-Saharan Africa. In M. A. Hanja (Ed.), *Global Food Security*. Nova Science Publishers, Inc.

Mohanty, A. K. (2018). Impact of climate change: A curse to the shrimp farming in India. *International Journal of Advance Research, Ideas and Innovations in Technology, 4*(3), 1151–1158. https://www.ijariit.com/manuscripts/v4i3/V4I3-1559.pdf

Mohanty, A., & Shreya, W. (2021). *Mapping India's Climate Vulnerability: A District-Level Assessment.* Council on Energy, Environment and Water.

Mohanty, B. P., Mohanty, S., Sahoo, J., & Sharma, A. (2010). Climate change: Impacts on fisheries and aquaculture. In S. Simard (Ed.), *Climate Change and Variability* (pp. 199–138). INTECH Open Access Publisher., doi:10.5772/9805

Mohanty, P. C., Mahendra, R. S., Bisoyi, H., Tummula, S. K., Grinson, G., Nayak, S., & Sahu, B. K. (2013). Assessment of the coral bleaching during 2005 to decipher the thermal stress in the coral environs of the Andaman Islands using Remote Sensing. *European Journal of Remote Sensing, 46*(1), 417–430. doi:10.5721/eujrs20134624

Molari, L., & Garcia, J. (2021). On the radial variation of the transverse mechanical properties of bamboo. *Build. Eng.*, 33-34.

Mollen, A. (2023). Sustain-AI and the Challenge of Sustainability, a report on The Sustainability Index for Artificial Intelligence. *Algorithm Watch.* https://algorithmwatch.org/en/wp-content/uploads/2023/03/SustAIn-magazine-issue-2.pdf

Moosvi, S. (2010). Environmental concerns in Mughal era. *Journal of History & Social Sciences, 1*(1), 1–4.

Morgan, J. (2023). Sustainable Contradictions? The Prospects for an Eco-Socialist School Geography. In Understanding Sustainability with Pedagogical Practice: A Contribution from Geography Education (pp. 67-77). Springer Nature. 10.1007/978-981-99-2687-9_6

Morgan, H. (2022). Conducting a qualitative document analysis. *The Qualitative Report, 27*(1), 64–77. doi:10.46743/2160-3715/2022.5044

Mosse, D. (1999). Colonial and contemporary ideologies of 'community management': The case of tank irrigation development in South India. *Modern Asian Studies, 33*(2), 303–338. doi:10.1017/S0026749X99003285 PMID:22026039

Most polluting industries in 2023 revealed. (n.d.). Heatable. https://heatable.co.uk/boiler-advice/most-polluting-industries

Mpolomoka, D. L., Ikabongo, M. K., & Mushibwe, C. (2018). *Women's Education a Tool of Social Transformation: Experiences of female students in ODeL institutions of higher learning in Zambia.* Knowledge Solutions for the Society, the Economy and the Environment hosted by the Faculty of Education Complex, Egerton University, Njoro, Kenya.

Mpolomoka, D. L. (2021). Using Participatory Approaches to Improve Environmental Sustainability in selected parts of Zambia. *Mulungushi University Multidisciplinary Journal, 2*(1), 1–11.

Msabila, D. T., & Nalaila, S. G. (2013). *Research Proposal and Dissertation Writing: Principles and Practice.* Dar-es-Salaam: Nyambari Nyangwine Publishers.

Msuya, F. E., & Hurtado, A. Q. (2017). The role of women in seaweed aquaculture in the Western Indian Ocean and South-East Asia. *European Journal of Phycology, 52*(4), 482–494. doi:10.1080/09670262.2017.1357084

Mukherjee, U. P. (2010). *Post-colonial environments: Nature, culture and the contemporary Indian novel in English.* Palgrave. doi:10.1057/9780230251328

Mukundan, M. K. (2007). Safety for hazards and quality defects in fish processing industry. In S. Girija (Ed.), *Spearheading quality fish processing, Integrated Fisheries Project Souvenir* (pp. 1–12). Integrated Fisheries Project.

Mulenga, B., & Wineman, A. (2014). Climate Trends and Farmers' Perceptions of Climate Change in Zambia. (Working Paper, Number 8). Lusaka: IAPRI.

Munoz, M., Reul, A., Guijarro, B., & Hidalgo, M. (2023). Carbon footprint, economic benefits and sustainable fishing: Lessons for the future from the Western Mediterranean. *The Science of the Total Environment*, *865*, 160783. doi:10.1016/j.scitotenv.2022.16078

Muralidhar, M., & Vijayan, K. K. 2015. Climate change and coastal aquaculture: Impacts, mitigation and adaptation measures. In: Souvenir: Aqua Aquaria India. The Marine products Export Development Authority, Kochi, India, 7-13.

Muralidhar, M., Kumaran, M., & Ashok Kumar, J. SyamaDayal, J., Jayanthi, M., Jayanthi, R., Lalitha, N., Panigrahi, A., Sreenivasa Rao, A., Muniyandi, B., &Murugan, P. (2013a). Climate change and coastal aquaculture in West Godavari district, Andhra Pradesh: Impacts, vulnerability, adaptations and mitigations for resilience. Journal of Agrometeorology, 15(Special Issue-II), 116–122.https://krishi.icar.gov.in/jspui/bitstream/123456789/1305/1/76-Muralidhar%20M%20-%20J%20Agrometeorology%20paper.pdf

Muralidhar, M., Vasanth, M., Saraswathy, R., Syamadayal, J., Lalitha, N., & Nagavel, A. (2013b). Measurement of greenhouse gas emission from aquaculture. In H. Pathak, R. C. Upadhyay, M. Muralidhar, P. Bhattacharyya, & B. Venkateswarlu (Eds.), Measurement of greenhouse gas emission from crop, livestock and aquaculture (pp. 45–54). Indian Agricultural Research Institute (IARI). ISBN: 978-81-88708-98-7

Muralidhar, M., & Jayanthi, M. (2014). Climate change and its implications for brackishwater aquaculture. In *Proceedings of the policy meet on the impact of environmental changes on the livelihoods of coastal women in north Tamil Nadu.* India (pp. 21-31).

Muralidhar, M., & Vijayan, K. K. (2016). Risks associated with extreme weather in aquaculture and their management. In V. U. M. Rao (Ed.), *Agrometeorological techniques for risk assessment and management of extreme events* (pp. 60–67). ICAR-Central Research Institute for Dryland Agriculture.

Muslemani, H., Liang, X., Kaesehage, K., Ascui, F., & Wilson, J. (2021). Opportunities and challenges for decarbonizing steel production by creating markets for 'green steel'products. *Journal of Cleaner Production*, *315*, 128127. doi:10.1016/j.jclepro.2021.128127

Muyambo, T. (2018). *Indigenous religion and environmental challenges: Phenomenological reflections on the role of the Environmental Management Agency in Zimbabwe*. Routledge. doi:10.4324/9781351066501-13

Mwanangombe, N. S., Mpolomoka, D. L., & Banda, S. (2021). *An exploration of Financial Literacy of Women Marketeers Participating in Micro Finance Programmes in selected Communities in Lusaka, Zambia.* Paper presented at the 12th Bi-Annual Pan-African Literacy for All (PALFA) conference held at Mulungushi International Conference Centre (MICC), Lusaka, Zambia.

Mweemba, L. (2014). Climate Change in the Zambian Mind: Communicating Risk Perception of Climate Change and Variability in Zambia. *Zambia Social Science Journal*, *5*(1), 71–93.

Mytton, D. (2021). Data centre water consumption. *npj Clean Water, 4*(1), 11.

Nadjahi, H. L., & Lemasson, S. (2018, September). A review of thermal management and innovative cooling strategies for data center. *Sustainable Computing : Informatics and Systems*, *19*, 14–28. doi:10.1016/j.suscom.2018.05.002

Nag, O. S. (2017, April 25). Top fish and seafood exporting countries. *Economics*. https://www.worldatlas.com/articles/top-fish-and-seafood-exporting-countries.html

Nag, A., Vyas, H., Shah, P., & Nag, P. K. (2012). Risk factors and musculoskeletal disorders among women workers performing fish processing. *American Journal of Industrial Medicine*, *55*(9), 833–843. doi:10.1002/ajim.22075

Nagaraj, K. (2008). *Farmers' suicides in India: Magnitudes, trends and spatial patterns.* Madras Institute of Development Studies. https://www.macroscan.org/anl/mar08/pdf/Farmers_Suicides.pdf

Nagel, J., & Lies, T. S. (2022). Re-gendering Climate Change: Men and Masculinity in Climate Research, Policy, and Practice. *Frontiers in Climate, 4*, 856869. doi:10.3389/fclim.2022.856869

Nag, P. K., & Nag, A. (2007). Hazards and health complaints associated with fish processing activities in India—Evaluation of a low-cost intervention. *International Journal of Industrial Ergonomics, 37*(2), 125–132. doi:10.1016/j.ergon.2006.10.012

Nahavandi, S. (2019). Industry 5.0—A human-centric solution. *Sustainability (Basel), 11*(16), 4371. doi:10.3390/su11164371

Nandwani, P., & Verma, R. (2021). A review on sentiment analysis and emotion detection from text. *Social Network Analysis and Mining, 11*(1), 81. doi:10.1007/s13278-021-00776-6 PMID:34484462

Nasrida, N., Saediman, H., Rianse, I. S., & Hidrawati, H. (2023). Contribution of Home Gardening to Household Vegetable Consumption During COVID-19 Pandemic. International Journal of Research in Engineering. *Science and Management, 6*(5), 5–9.

Nath, A. J., & Das, A. K. (2012). Carbon pool and sequestration potential of village bamboos in the agroforestry system of northeast India. *Tropical Ecology*, 287–293.

Nath, A. J., Lal, R., & Das, A. K. (2015). Managing woody bamboos for carbon farming and carbon trading. *Global Ecology and Conservation, III*, 654–663. doi:10.1016/j.gecco.2015.03.002

Nath, J. A., Gitasree, G. D., & Ashesh, D. K. (2009). *Aboveground standing biomass and carbon storage in village bamboos in North East India.* Elsevier.

Nathwani, J., & Kammen, D. M. (2019). Affordable energy for humanity: A global movement to support universal clean energy access. *Proceedings of the IEEE, 107*(9), 1780–1789. doi:10.1109/JPROC.2019.2918758

National Determined Contribution. (2022). *India's updated Nationally Determined Contribution under Paris Agreement 2021-2030. Submission to UNFCC.* Government of India.

National grid. (2023). *What is carbon capture and storage?* National Grid. www.nationalgrid.com/stories/energy-explained/what-is-ccs-how-does-it-work

Nayar, P. K. (2012). *Writing wrongs: The cultural construction of human rights in India.* Routledge.

Nayar, P. K. (2013). Indigenous cultures and the ecology of protest: Moral economy and 'knowing subalternity' in Dalit and Tribal writing from India. *Journal of Postcolonial Writing, 50*(3), 291–303. doi:10.1080/17449855.2013.815127

Naylor, A., Ford, J., Pearce, T., & Van Alstine, J. (2020). Conceptualizing climate vulnerability in complex adaptive systems. *One Earth, 2*(5), 444–454. doi:10.1016/j.oneear.2020.04.011

NCDC. (2023). *Towards Climate-smart Hospitals: Insights from a National Hospital Energy Consumption Survey (85/NCDC/NPCCHH/2022-23/NatEnergySurvey).* National Centre for Disease Control, Directorate General of Health Services, Ministry of Health & Family Welfare, Government of India.

NDRC. (2022). *What Are the Causes of Climate Change?* NDRC. https://www.nrdc.org/stories/what-are-causes-climate-change#natural

Neto, J. A. B., Gaylarde, C., & Beech, I. (2019). Microplastics and attached microorganisms in sediments of the Vitória bay estuarine system in SE Brazil. *Ocean and Coastal Management, 169*, 247–253. doi:10.1016/j.ocecoaman.2018.12.030

Neumayer, E., & Plümper, T. (2007). The gendered nature of natural disasters: The impact of catastrophic events on the gender gap in life expectancy, 1981–2002. *Annals of the Association of American Geographers*, *97*(3), 551–566. doi:10.1111/j.1467-8306.2007.00563.x

Ng, A. W. (2018). From sustainability accounting to a green financing system: Institutional legitimacy and market heterogeneity in a global financial centre. *Journal of Cleaner Production*, *195*, 585–592. doi:10.1016/j.jclepro.2018.05.250

Ng, A. W., Nathwani, J., Fu, J., & Zhou, H. (2021). Green financing for global energy sustainability: prospecting transformational adaptation beyond Industry 4.0. *Sustainability: Science. Sustainability*, *17*(1), 377–390. doi:10.1080/15487733.2021.1999079

Ngarava, S., Zhou, L., Ningi, T., & Chari, M. M. (2022). Differentiated Intra-Household Food Utilisation in Raymond Mhlaba Local Municipality, South Africa. In Sustainable Agriculture and Food Security (pp. 87–106). Springer International Publishing. 10.1007/978-3-030-98617-9_6

Ngaruiya, F. W., Ogendi, G. M., & Mokua, M. A. (2019). Occupational health risks and hazards among the fisherfolk in KampiSamaki, Lake Baringo, Kenya. *Environmental Health Insights*, *13*(1), 1–11. doi:10.1177/1178630219881463

Ngcamu, B. S. (2023). Climate change effects on vulnerable populations in the Global South: A systematic review. *Natural Hazards*, *118*(2), 977–991. doi:10.1007/s11069-023-06070-2

Nguyen, T.-H., Nguyen, L. V., Jung, J. J., Agbehadji, I. E., Frimpong, S. O., & Millham, R. C. (2020). Bio-inspired approaches for Smart Energy Management: State of the art and Challenges. *Sustainability (Basel)*, *12*(20), 8495. doi:10.3390/su12208495

Nicholls, N., Butler, C. D., & Hanigan, I. (2006). Inter-annual rainfall variations and suicide in New South Wales, Australia, 1964–2001. *International Journal of Biometeorology*, *50*(3), 139–143. doi:10.1007/s00484-005-0002-y PMID:16235091

Nieuwkoop, M. V., Van de Velde, P., Huyer, S., & Freeman, K. K. (2022, March 8). Gender-smart agriculture: The only way forward for women and climate. *World Bank Blogs*. https://blogs.worldbank.org/climatechange/gender-smart-agriculture-only-way-forward-women-and-climate

Nikarthil, D. S., & Kejriwal, S. (2021). *India's Healthcare Systems Persistently Exclude LGBTQ+ People. This Needs To Change*. The Wire.

Nikola, F., & Radoica, L. (2022). *Climate Change & Quality of Life*. Research Gate. https://www.researchgate.net/publication/360996574_CLIMATE_CHANGE_AND_QUALITY_OF_LIFE.

Niranjanamurthy, M., Nithya, B. N., & Jagannatha, S. (2018). Analysis of Blockchain technology: Pros, cons and SWOT. *Cluster Computing*, *5*(2), 1–15.

Nishant, R., Kennedy, M., & Corbett, J. (2020). Artificial Intelligence for Sustainability: Challenges, opportunities, and a research agenda. *International Journal of Information Management*, *53*, 102104. doi:10.1016/j.ijinfomgt.2020.102104

Nixon, R. (2005). Environmentalism and postcolonialism. In A. Loomba, S. Kaul, M. Bunzl, A. Burton, & J. Esty (Eds.), *Postcolonial studies and beyond* (pp. 233–251). Permanent Black. doi:10.1215/9780822386650-011

Nolen, S., & Galdieri, D. (2022, August 16). The forgotten virus: Zika families and researchers struggle for support. *The New York Times*. https://www.nytimes.com/2022/08/16/health/zika-children-research.html

Nordander, C., Ohlsson, K., Balogh, I., Rylander, L., Palsson, B., & Skerfving, S. (1999). Fish processing work: The impact of two sex dependent exposure profiles on musculoskeletal health. *Occupational and Environmental Medicine*, *56*(4), 256–264. doi:10.1136/oem.56.4.256

Nova, P., Pinto, E., Chaves, B., & Silva, M. (2020). Urban organic community gardening to promote environmental sustainability practices and increase fruit, vegetables and organic food consumption. [PubMed]. *Gaceta Sanitaria, 34*(1), 4–9. doi:10.1016/j.gaceta.2018.09.001 PMID:30471837

Nualphan. (2013). *Types and Sources of Marine Debris in BangSaen Beach. Chonburi Province*, [Master Thesis, Chulalongkorn University, Thailand].

Nuttall, M. (2012). Imagining and governing the Greenlandic resource frontier. *The Polar Journal, 2*(1), 113–124. doi:10.1080/2154896X.2012.679563

Nyamadzawo, G., Gwenzi, W., Kanda, A., Kundhlande, A., & Masona, C. (2013). Understanding the causes, socio-economic and environmental impacts, and management of veld fires in tropical Zimbabwe. *Fire Science Reviews, 2*(1), 2. http://www.firesciencereviews.com/content/2/1/2. doi:10.1186/2193-0414-2-2

O'Hara, A. C., Miller, A. C., Spinks, H., Seifert, A., Mills, T., & Tuininga, A. R. (2022). The Sustainable Prescription: Benefits of Green Roof Implementation for Urban Hospitals. *Frontiers in Sustainable Cities, 4*, 798012. doi:10.3389/frsc.2022.798012

O'Mahony, T. (2021). Cost-benefit analysis and the environment: The time horizon is of the essence. *Environmental Impact Assessment Review, 89*, 106587. doi:10.1016/j.eiar.2021.106587

O'Reilly, C. M., Alin, S. R., Plisnier, P., Cohen, A. S., & McKee, B. A. (2003). Climate change decreases aquatic ecosystem productivity of lake Tanganyika, Africa. *Nature, 424*(6950), 766–768. doi:10.1038/nature01833

Occupational Safety and Health Administration (OSHA). (2023). *Personal Protective Equipment*. https://www.osha.gov/sites/default/files/publications/osha3151.pdf

Oda, J., & Akimoto, K. (2022). Carbon intensity of the Japanese Iron and steel Industry: Analysis of factors from 2000 to 2019. *Journal of Cleaner Production, 345*, 130920. doi:10.1016/j.jclepro.2022.130920

OECD. (2023). *Climate Change and Plastics Pollution*. OECD. https://www.oecd.org/environment/plastics/Policy-Highlights-Climate-change-and-plastics-pollution-Synergies-between-two-crucial-environmental-challenges.pdf

OECD. (n.d.). *Environmental Outlook to 2050: The Consequences of Inaction - Key Facts and Figures*. OECD. https://www.oecd.org/env/indicators-modelling-outlooks/oecdenvironmentaloutlookto2050theconsequencesofinaction-keyfactsandfigures.htm

Ogunmakinde, O. E. (2019). A review of circular economy development models in China, Germany and Japan. *Recycling, 4*(3), 27. doi:10.3390/recycling4030027

Ohlsson, K., Hansson, G.-A., Balogh, I., Stromberg, U., Palsson, B., Nordander, C., Rylander, L., & Skerfving, S. (1994). Disorders of the neck and upper limbs in women in the fish processing industry. *Occupational and Environmental Medicine, 51*(12), 826–832. doi:10.1136/oem.51.12.826

Okoffo, E. D., O'Brien, S., Ribeiro, F., Burrows, S. D., Toapanta, T., Rauert, C., O'Brien, J. W., Tscharke, B. J., Wang, X., & Thomas, K. V. (2021). Plastic particles in soil: State of the knowledge on sources, occurrence and distribution, analytical methods and ecological impacts. *Environmental Science. Processes & Impacts, 23*(2), 240–274. doi:10.1039/D0EM00312C PMID:33514987

Olafsdottir, H., & Rafnsson, V. (2000). Musculoskeletal symptoms among women currently and formerly working in fish-filleting plants. *International Journal of Occupational and Environmental Health, 6*(1), 44–49. doi:10.1179/oeh.2000.6.1.44

Olaoye, O. J., & Ojebiyi, W. G. (2021). Occupational health hazards associated with Nigerian fisheries. In Palaniappan, K., & McCauley, P. (Eds.), *Occupational Wellbeing*. https://doi.org/ doi:10.5772/intechopen.94205

Olaoye, O. J., Odebiyi, O. C., & Abimbola, O. T. (2015). Occupational hazards and injuries associated with fish processing in nigeria. *Journal of Aquatic Sciences*, *3*(1), 1–5. doi:10.12691/jas-3-1-1

Olivier, J. G. J., Schure, K. M., & Peters, J. A. H. W. (2017). *Trends in global CO2 and total greenhouse gas emissions. 2017 Report*. PBL.

Omvedt, G. (1997, June 24). Why Dalits dislike environmentalists. *The Hindu*. http://ces.iisc.ernet.in/hpg/envis/doc97html/envenv627.html

Omvedt, G. (2008). *Seeking Begumpura: The social vision of anticaste intellectuals*. Navayana Pub.

Önal, M., Neşer, G., & Gürsel, K. T. (2021). Environmental impacts of steel ship hulls building and recycling by life cycle assessment (LCA). *Ships and Offshore Structures*, *16*(10), 1061–1066. doi:10.1080/17445302.2020.1816706

OPM. (2012). *The 2010-2011 Integrated Rainfall Variability Impacts, Needs Assessment and Drought Risk Management Strategy*. Department of Disaster Management.

Orderud, H., Harkonen, J., Harsaker, C. T., & Bogren, M. (2022). Floods and maternal healthcare utilization in Bangladesh. *Population and Environment*, *44*(3-4), 193–225. doi:10.1007/s11111-022-00410-3

Organisation for Economic Co-operation and Development (OECD). (2000). *Reducing the risk of policy failure: Challenges for regulatory compliance*.https://www.oecd.org/gov/regulatory-policy/46466287.pdf

Orhon, D., Sözen, S., Kirca, V. S. O., Duba, S., Mermutlu, R., & Sumer, B. M. (2021). Pollutant Dynamics between The Black Sea and The Marmara Sea: Basis for wastewater management strategy. *Marine Pollution Bulletin*, *168*, 112388. doi:10.1016/j.marpolbul.2021.112388 PMID:33940368

Orighoye, O., Apekey, T. A., & Maynard, M. J. (2023). Informing Diet and Physical Activity Interventions with Family Involvement in an Urban Setting: Views of Children and Adults in Lagos, Nigeria. *Sustainability (Basel)*, *15*(10), 7850. doi:10.3390/su15107850

Orman ve Su İşleri Bakanlığı (OSB), (2013). Su Yönetimi Daire Başkanlığı. *Marmara Denizi kirlilik raporu.*

Ortiz Cuadra, J. J. (2023). *Learning for Crisis: Improving food security in Uppsala County through participative localized food production*. Research Gate.

Osman-Elasha, B. (2010). *Women ... in the Shadow of Climate Change*. United Nations.

Oudraat, C. J., & Brown, M. E. (2022, January 25). *Gender, Climate Change, and Security: Making the Connections*. Wilson Center. Wilson Center. https://www.wilsoncenter.org/article/gender-climate-change-and-security-making-connections

Ounifi, H.-A., Gherbi, A., & Kara, N. (2022, August). Deep machine learning-based power usage effectiveness prediction for sustainable cloud infrastructures. *Sustainable Energy Technologies and Assessments*, *52*, 101967. doi:10.1016/j.seta.2022.101967

Outlook. (2022). *Report At COP27: India Records Highest Emission Increase Among Top Global Contributors*. Outlook. Www.Outlookindia.Com/International/Report-At-Cop27-India-Records-Highest-Emission-Increase-Among-Top-Global-Contributors-News-236452

Owen, G. (2020). What makes climate change adaptation effective? A systematic review of the literature. *Global Environmental Change*, *62*, 102071. doi:10.1016/j.gloenvcha.2020.102071

Oxfam (Great Britain). (2002). *Natural Resource Management and Gender: A global Source Book*. KIT Publishers.

Oyekale, J., Petrollese, M., Tola, V., & Cau, G. (2020). Impacts of renewable energy resources on effectiveness of grid-integrated systems: Succinct review of current challenges and potential solution strategies. *Energies*, *13*(18), 4856. doi:10.3390/en13184856

Palmer, T. N. (1998). Climate change from a nonlinear dynamical perspective. *Journal of Climate*, *12*, 575–591. doi:10.1175/1520-0442(1999)012<0575:ANDPOC>2.0.CO;2

Palmer, T. N. (2000). Uncertainty in Forecast of Weather and Climate. *Reports on Progress in Physics*, *63*(2), 71–116. doi:10.1088/0034-4885/63/2/201

Pal, P., Gopal, P. R. C., & Ramkumar, M. (2023). Impact of transportation on climate change: An ecological modernization theoretical perspective. *Transport Policy*, *130*, 167–183. doi:10.1016/j.tranpol.2022.11.008

Palsson, B., Stromberg, U., Ohlsson, K., & Skerfving, S. (1998). Absence attributed to incapacity and occupational disease/accidents among female and male workers in the fish-processing industry. *Occupational Medicine*, *48*(5), 289–295. doi:10.1093/occmed/48.5.289

Pan, C., Zhou, G., Shrestha, A. K., Chen, J., Kozak, R., Li, N., & Wang, G. (2023). *Bamboo as a Nuture-Based Solution (NbS) for Climate Change Mitigation: Biomass, Products, and Carbon Credits*. MDPI.

Panda, S., Panigrahi, G. K., & Padhi, S. N. (2016). *Earning animals*. Anchor Academic Publishing.

Panigrahy, S., Murthy, T. V. R., Singh, J. G., & Patel, T. S. (2012). Wetlands of India: Inventory and assessment at 1:50,000 scale using geospatial techniques. Current Science, 102(6), 852–856. https://currentscience.ac.in/Volumes/102/06/0852.pdf

Paskarini, I., Alwi, M. N. M., & Martiana, T. Mahmudah, Arini, S. Y., &Dwiyanti, E. (2022). The decrease in the quality life of fishermen due to Covid-19 widespread. *Malaysian Journal of Medicine and Health Sciences, 18*, 79-85. https://scholar.unair.ac.id/en/publications/the-decrease-in-the-quality-life-of-fishermen-due-to-covid-19-wid

Pasqualino, M., Kennedy, G., Longley, K., & Thilsted, S. H. (2014). *Food and Nutrition Security in the Barotse Floodplain System*. GGIAR Research, WorldFish, Research Program on Aquatic Agricultural Systems and Biodiversity International.

Patel, B., Gami, B., & Patel, P. (2017). Carbon sequestration by bamboo farming on marginal land and sustainable use of wood waste for bioenergy: case studies from Abellon clean energy. In *Wood is Good: current trends and future prospects in wood utilization* (pp. 451–467). Springer Singapore. doi:10.1007/978-981-10-3115-1_43

Pathak, H., Mishra, J. P., & Mohapatra, T. (2022). *Indian agriculture after independence*. Indian Council of Agricultural Research.

Patrick, A. E. S. (2016). Influence of rainfall and water level on inland fisheries production: A review. *Archives of Applied Science Research*, *18*(6), 44–51. https://www.researchgate.net/publication/313529635

Patterson. (1995). Foraminiferal Evidence of Younger Dryas Age Cooling on the British Colombia Shelf. *Geography et Quaternaries*, *49*(3), 409.

Paudyal, B. R., Chanana, N., Khatri-Chhetri, A., Sherpa, L., Kadariya, I., & Aggarwal, P. (2019). Gender Integration in Climate Change and Agricultural Policies: The Case of Nepal. *Frontiers in Sustainable Food Systems*, *3*, 3. doi:10.3389/fsufs.2019.00066

Paul, A. (2011). Carbon Credit and Carbon Trading: An Overview. *Business Studies*, 136-146.

Paul, T. T., Dennis, A., & Grinson George, G. (2016). A review of remote sensing techniques for the visualization of mangroves, reefs, fishing grounds, and molluscan settling areas in tropical waters. In C. W. Finkl & C. Makowski (Eds.), *Seafloor mapping along continental shelves* (pp. 105–123). Springer Climate., doi:10.1007/978-3-319-25121-9

Peel, J., & Lin, J. (2019). Transnational Climate Litigation: The contribution of the Global South. *The American Journal of International Law, 113*(4), 679–726. doi:10.1017/ajil.2019.48

Peel, J., & Osofsky, H. M. (2020). Climate Change Litigation. *Annual Review of Law and Social Science, 16*(1), 21–38. doi:10.1146/annurev-lawsocsci-022420-122936

Pelage, L., Domalain, G., Lira, A. S., Travassos, P., & Fredou, T. (2019). Coastal land use in northeast Brazil: Mangrove coverage evolution over three decades. *Tropical Conservation Science, 12*, 194008291882241. doi:10.1177/1940082918822411

Pemberton, D., Brothers, N. P., & Kirkwood, R. (1992). Entanglement of Australian fur seals in man-made debris in Tasmanian waters. *Wildlife Research, 19*(2), 151–159. doi:10.1071/WR9920151

Penang Institute. (2021). Strengthening Food Security in Light of Covid-19. Penang Institute. https://penanginstitute.org/publications/monographs/strengthening-food-security-in-light-of-covid-19

Penang2030. (2023). Unity in Diversity: Penang Experience and Approach. Penang 2030. https://penang2030.com/2023/03/unity-in-diversity-penang-experience-and-approach

Penna, N., Berluti, S., Penna, A., & Ridolfi, F. (2000). Study and monitoring of mucilage in the Adriatic Sea. *Water Science and Technology, 42*(1-2), 299–304. doi:10.2166/wst.2000.0329

Perdan, S., & Azapagic, A. (2011). Carbon trading: Current schemes and future developments. *Energy Policy, 39*(10), 6040–6054. doi:10.1016/j.enpol.2011.07.003

Phimmachanh, S., Ying, Z., & Beckline, M. (2015). Bamboo resources utilization: A potential source of income to support rural livelihoods. *Applied Ecology and Environmental Sciences, 3*(6), 176–183.

Phooi, C. L., Azman, E. A., Ismail, R., & Tongkaemkaew, U. (2022). *Call home gardening for enhancing food in the urban area*. Research Gate.

Picciariello, A., Colenbrander, S., Bazaz, A., & Roy, R. (2021, June). *The costs of climate change in India: A review of the climate-related risks facing India, and their economic and social costs*. Overseas Development Institute. https://cdn.odi.org/media/documents/ODI-JR-costclimatechangeindia-final.pdf

Pillai, S. M., & Muralidhar, M. (2006). Survey and demarcation of seawater inundated areas for eco-friendly aquaculture in Andaman and Nicobar Islands (pp. 69). Central Institute of Brackishwater Aquaculture (CIBA).

Pinto-Bazurco, J. F. (2022, June). *Still Only One earth: Lessons from 50 years of UN Sustainable Development Policy*. International Institute for Sustainable Development. https://www.iisd.org/projects/still-only-one-earth-lessons-50-years-un-sustainable-development-policy

Pinto, G., Wang, Z., Roy, A., Hong, T., & Capozzoli, A. (2022). Transfer learning for smart buildings: A critical review of algorithms, applications, and future perspectives. *Advances in Applied Energy, 5*, 100084. doi:10.1016/j.adapen.2022.100084

PlasticsEurope. (2019). *Plastics – the Facts 2019: An analysis of European plastics production, demand and waste data*. Plastics Europe. https:// www.plasticseurope.org/application/files/9715/7129/9584/FINAL_ web_version_Plastics_the_facts2019_14102019.pdf. Accessed 21 June 2021.

Plumwood, V. (2002). *Environmental culture: The ecological crisis of reason*. Routledge.

Ponniah, A. G., & Muralidhar, M. (2009). Research requirements to understand impact of climate change on brackish water aquaculture and develop adaptive measures. In *Proceedings of 96th Session of the Indian Science Congress, Part II: Session of Animal Veterinary and Fisheries Sciences* (pp. 46–47). The Indian Science Congress Association (ISCA).

Poortinga, W., Fisher, S., Bohm, G., Steg, L., Whitmarsh, L. and Ogunbode, C. (2018). *European Attitudes to Climate Change and Energy: Topline Results from 8 of the European Social Survey.* London: European Social Survey.

Power technology. (2018). *Turning waste into power: the plastic to fuel projects*. Power Technology. https://www.power-technology.com/comment/plastic-to-fuel/

Prakash, A., McGlade, K., Roxy, M. K., Roy, J., Some, S., & Rao, N. (2022). Climate Adaptation Interventions in Coastal Areas: A Rapid Review of Social and Gender Dimensions. *Frontiers in Climate*, *4*, 785212. doi:10.3389/fclim.2022.785212

Prasad, I. (2022). Towards Dalit ecologies. *Environment and Society*, *13*(1), 98–120. doi:10.3167/ares.2022.130107

Preston, N. P., Jackson, C., Thompson, P. A., Austin, M., Burford, M. A., & Rothlisberg, P. (2000). *Prawn farm effluent: composition, origin and treatment*. Fisheries research and development corporation report (FRDC). *CRC for Aquaculture*. https://www.frdc.com.au/sites/default/files/products/1995-162-DLD.PDF

Preston, B. J. (2018). The evolving role of environmental rights in Climate Change Litigation. *Chinese Journal of Environmental Law*, *2*(2), 131–164. doi:10.1163/24686042-12340030

Prusty, A. K., Meena, D. K., Mohapatra, S., Panikkar, P., Das, P., Gupta, S. K., & Behera, B. K. (2015). Synthetic pyrethroids (Type II) and freshwater fish culture: Perils and mitigations. *International Aquatic Research.*, *7*(3), 163–191. doi:10.1007/s40071-015-0106-x

Ptak-Wojciechowska, A., Januchta-Szostak, A., Gawlak, A., & Matuszewska, M. (2021). The Importance of Water and Climate-Related Aspects in the Quality of Urban Life Assessment. *Sustainability (Basel)*, *13*(12), 6573. doi:10.3390/su13126573

Puczko, K., & Jekatierynczuk-Rudczyk, E. (2020). Extreme hydro-meteorological events influence to water quality of small rivers in urban area: A case study in Northeast Poland. *Scientific Reports*, *10*(1), 10255. doi:10.1038/s41598-020-67190-4 PMID:32581301

Purohit, G. K., Mahanty, A., Suar, M., Sharma, A. P., Mohanty, B. P., & Mohanty, S. (2014). Investigating hsp gene expression in liver of Channastriatus under heat stress for understanding the upper thermal acclimation. *BioMed Research International*, *2014*(381719), 1–10. doi:10.1155/2014/381719

Purwanti, S. H. (2013). From Bali Bombing to Sukhoi Tragedy. Jakarta Timu, 17–25. Rayyana Komunikasindo.

Qo, S. K., Dhatterwal, J. S., Baliyan, A., & Rani, S. (2023). *Cyborg Human and Machine Communication Paradigm.* CRC press.

Race to Zero. (n.d.). *Global Climate action.* Race to Zero. https://racetozero.unfccc.int/system/race-to-zero/

Rahaman, Z. A., Kafy, A. A., Saha, M., Rahim, A. A., Almulhim, A. I., Rahaman, S. N., Fattah, M. A., & Rahman, M. T., S, K., Faisal, A.-A., & Al Rakib, A. (2022). Assessing the impacts of vegetation cover loss on surface temperature, urban heat island and carbon emission in Penang City, Malaysia. *Building and Environment*, *222*, 109335. doi:10.1016/j.buildenv.2022.109335

Rahman, A., & Karim, M. M. (2015). Green shipbuilding and recycling: Issues and challenges. *International Journal of Environmental Sciences and Development*, *6*(11), 838–842. doi:10.7763/IJESD.2015.V6.709

Raihan, A. (2023). A review of the global climate change impacts, adaptation strategies, and mitigation options in the socio-economic and environmental sectors. *Journal of Environmental Science and Economics*, *2*(3), 36–58. doi:10.56556/jescae.v2i3.587

Rajan, R., Rana, N. P., Parameswar, N., Dhir, S., & Dwivedi, Y. K. (2021). Developing a modified total interpretive structural model (M-TISM) for organizational strategic cybersecurity management. *Technological Forecasting and Social Change*, *170*, 120872. doi:10.1016/j.techfore.2021.120872

Raman, S. (2022, March 9). Climate change set to upend global fishery agreements, study warns. *Mongabay*. https://news.mongabay.com/2022/03/climate-change-set-to-upend-global-fishery-agreements-study-warns

Rao, N., Patil, S., Singh, C., Roy, P., Pryor, C., Poonacha, P., & Genes, M. (2022). Cultivating sustainable and healthy cities: A systematic literature review of the outcomes of urban and peri-urban agriculture. *Sustainable Cities and Society*, *85*, 104063. doi:10.1016/j.scs.2022.104063

Rao, R. G. (2009). Climate change mitigation through reforestation in Godavari mangroves in India. *International Journal of Climate Change Strategies and Management*, *1*(4), 340–355. doi:10.1108/17568690911002870

Rath, M., Darwish, A., Pati, B., Pattanayak, B. K., &; Panigrahi, C. R. (2020). Swarm intelligence as a solution for technological problems associated with internet of things. *Swarm Intelligence for Resource Management in Internet of Things*, 21–45. doi:10.1016/B978-0-12-818287-1.00005-X

Rathour, R., Kumar, H., Prasad, K., Anerao, P., Kumar, M., Kapley, A., Pandey, A., Kumar Awasthi, M., & Singh, L. (2022). Multifunctional applications of bamboo crop beyond environmental management: An Indian prospective. *Bioengineered*, *13*(4), 8893–8914. doi:10.1080/21655979.2022.2056689 PMID:35333141

Ravikumar, V. M. (2014). Green democracy: Relevance of Ambedkar's ideas for Indian environmentalism. *The Journal of Tribal Intellectual Collective India*, *2*(1), 24–34.

Ravindra, K., Rattan, P., Mor, S., & Aggarwal, A. N. (2019). Generalized additive models: Building evidence of air pollution, climate change and human health. *Environment International*, *132*, 104987. doi:10.1016/j.envint.2019.104987 PMID:31398655

Ravindrnath, N. H., & Indu, K. M. (2018). Estimation of Carbon Sequestration under MGNREGA: Achievement and Potential. India Indian Institute of Science, Bangalore..

Reddy, K. K., Babu, S. V., & Gavali, R. S. (2019). *Sustainable Livelihood and Adaptation to Climate Change. Training Manual*. National Institute of Rural Development and Panchayath Raj, Ministry of Rural Development, Government of India.

Ren, X., Tang, J., Liu, X., & Liu, Q. (2020). Effects of microplastics on greenhouse gas emissions and the microbial community in fertilized soil. *Environmental Pollution*, *256*, 113347. doi:10.1016/j.envpol.2019.113347 PMID:31672352

Report I. P. C. C. (2007). Synthesis Report. Contribution of Working Groups I, II and III to the Fourth Assessment Report of the Intergovernmental Panel on Climate Change. Switzerland.

Report I. P. C. C. (2013). The Physical Science Basis. Contribution of Working Group I to the Fifth Assessment Report of the Intergovernmental Panel on Climate. Cambridge University Press, Cambridge, United Kingdom and New York.

Report, I. P. C. C. (2001). *Intergovernmental Panel on Climate Change*. Swiss.

Research & Markets. (2019, February 07). Global Commercial Fishing Industry Market to 2026 - Exhaustion of Marine Resources are Inhibiting Market Growth. *PR Newswire*. https://www.prnewswire.com/news-releases/global-commercial-fishing-industry-market-to-2026---exhaustion-of-marine-resources-are-inhibiting-market-growth-300791788.html

REUTERS. (2019). *Accenture Microsoft Team up on Blockchain-Based Digital ID Network.* Reuters. https://www.reuters.com/article/us-microsoftaccenture- digitalid/accenture-microsoft-team-up-on-blockchain-baseddigital-id-network- idUSKBN19A22B.

Reva Dhingra, E. F., & Simeon, K. (2022, March 9). *The climate crisis, migration, and refugees.* Brookings. https://www.brookings.edu/articles/the-climate-crisis-migration-and-refugees/

Revi, A., Satterthwaite, D., Aragón-Durand, F., Corfee-Morlot, J., Kiunsi, R. B., Pelling, M., Roberts, D., Solecki, W., Gajjar, S. P., & Sverdlik, A. (2014). Towards transformative adaptation in cities: The IPCC's Fifth Assessment. *Environment and Urbanization, 26*(1), 11–28. doi:10.1177/0956247814523539

Reygondeau, G., Cheung, W. W. L., Wabnitz, C. C. C., Lam, V. W. Y., Frolicher, T., & Maury, O. (2020). Climate change-induced emergence of novel biogeochemical provinces. *Frontiers in Marine Science, 7*. https://doi.org/10.3389/fmars.2020.00657

Reynolds, J. L. (2019). Solar geoengineering to reduce climate change: A review of governance proposals. *Proceedings - Royal Society. Mathematical, Physical and Engineering Sciences, 475*(2229), 20190255. doi:10.1098/rspa.2019.0255 PMID:31611719

Richardson, K., Hardesty, B. D., & Wilcox, C. (2019a). Estimates of fishing gear loss rates at a global scale: A literature review and metaanalysis. *Fish and Fisheries, 20*(6), 1218–1231. doi:10.1111/faf.12407

Riche, B. (2007). *Climate Change Vulnerability Assessment in Zambia: IUCN, Climate Change and Development Project, Pilot Phase*. Geneva: IUCN. http://cmsdata.iucn.org/downloads/climate_change_vulnerability_assessment_zambia.pdf

Richmond, R. (2015). Coral reefs: Present problems and future concerns resulting from anthropogenic disturbance. *Integrative and Comparative Biology, 33*(6), 524–536. https://academic .oup.com/icb/article/33/6/524/2107143

Rillig, M. C., Lehmann, A., de Souza Machado, A. A., & Yang, G. (2019). Microplastic effects on plants. *The New Phytologist, 223*(3), 1066–1070. doi:10.1111/nph.15794 PMID:30883812

Rimmer, M. A., Larson, S., Lapong, I., Purnomo, A. H., Pong-Masak, P. R., Swanepoel, L., & Paul, N. A. (2021). Seaweed aquaculture in Indonesia contributes to social and economic aspects of livelihoods and community wellbeing. *Sustainability, 13*(19), 10946. doi:10.3390/su131910946

Riondato, E., Pilla, F., Basu, A. S., & Basu, B. (2020). Investigating the effect of trees on urban quality in Dublin by combining air monitoring with i-Tree Eco model. *Sustainable Cities and Society, 61*, 102356. doi:10.1016/j.scs.2020.102356

Rios, L. M., Moore, C., & Jones, P. R. (2007). Persistent organic pollutants carried by synthetic polymers in the ocean environment. *Marine Pollution Bulletin, 54*(8), 1230–1237. doi:10.1016/j.marpolbul.2007.03.022 PMID:17532349

Rising, J. A., Taylor, C., Ives, M. C., & Ward, R. E. T. (2022). Challenges and innovations in the economic evaluation of the risks of climate change. *Ecological Economics, 197*, 107437. doi:10.1016/j.ecolecon.2022.107437

Rissman, J., Bataille, C., Masanet, E., Aden, N., Morrow, W. R. III, Zhou, N., Elliott, N., Dell, R., Heeren, N., Huckestein, B., Cresko, J., Miller, S. A., Roy, J., Fennell, P., Cremmins, B., Koch Blank, T., Hone, D., Williams, E. D., de la Rue du Can, S., & Helseth, J. (2020). Technologies and policies to decarbonize global industry: Review and assessment of mitigation drivers through 2070. *Applied Energy, 266*, 114848. doi:10.1016/j.apenergy.2020.114848

Rizaoglu, E. (2011). *Classical Mechanics*. Seçkin Press.

Rizzello, A. (2022). Financing the Green Recovery: The New Directions of Finance After the COVID-19 Crisis. In *Green Investing: Changing Paradigms and Future Directions* (pp. 133–158). Springer International Publishing. doi:10.1007/978-3-031-08031-9_6

RMI. (2012). *Project Case Study.* Rocky Mountain Institute. https://rmi.org/wp-content/uploads/2017/04/Pathways-to-Zero_Cleveland-Clinic-Retrofit_Case-Study_2012.pdf

Robinson, J., & Yeh, E. (2011). Transactional sex as a response to risk in Western Kenya. *American Economic Journal. Applied Economics*, *3*(1), 35–64. doi:10.1257/app.3.1.35

Robinson, L. (2018). *Sustainable Agriculture and Farming*. Larsen and Keller Education.

Rochdane, S., Reichert, B., Messouli, M., Babqiqi, A., & Khebiza, M. Y. (2012). Climate Change Impacts on Water Supply and Demand in Rheraya Watershed (Morocco) with Potential Adaptation Strategies. *Water (Basel)*, *4*(1), 28–44. doi:10.3390/w4010028

Rode, J., Pinzon, A., Stabile, M. C., Pirker, J., Bauch, S., Iribarrem, A., Sammon, P., Llerena, C. A., Muniz Alves, L., Orihuela, C. E., & Wittmer, H. (2019). Why 'blended finance'could help transitions to sustainable landscapes: Lessons from the Unlocking Forest Finance project. *Ecosystem Services*, *37*, 100917. doi:10.1016/j.ecoser.2019.100917

Rodrigues, D. E., & Kiran, U. (2013). A Pilot Study on knowledge & practice regarding prevention of occupational hazards and attitude towards utilisation of safety measures among fishermen working at a selected HA. B.R. *Journal of Health and Allied Sciences NU, 03*(03), 068–071. doi:10.1055/s-0040-1703681

Rodrigues, J. P., Duarte, A. C., Santos-Echeandía, J., & Rocha-Santos, T. (2018). Signifcance of interactions between microplastics and POPs in the marine environment: A critical overview. *Trends in Analytical Chemistry*, *111*, 252–260. doi:10.1016/j.trac.2018.11.038

Rodríguez-Artalejo, F., Guallar-Castillón, P., Pascual, C. R., Otero, C. M., Montes, A. O., García, A. N., Conthe, P., Chiva, M. O., Banegas, J. R., & Herrera, M. C. (2005). Health-related quality of life as a predictor of hospital readmission and death among patients with heart failure. *Archives of Internal Medicine*, *165*(11), 1274–1279. doi:10.1001/archinte.165.11.1274 PMID:15956007

Roelfsema, M., van Soest, H. L., Harmsen, M., van Vuuren, D. P., Bertram, C., den Elzen, M., Höhne, N., Iacobuta, G., Krey, V., Kriegler, E., Luderer, G., Riahi, K., Ueckerdt, F., Després, J., Drouet, L., Emmerling, J., Frank, S., Fricko, O., Gidden, M., & Vishwanathan, S. S. (2020). Taking stock of national climate policies to evaluate implementation of the Paris Agreement. *Nature Communications*, *11*(1), 2096. doi:10.1038/s41467-020-15414-6 PMID:32350258

Rogelj, J., Shindell, D., Jiang, K., & Beşta, S. (2018). Mitigation Pathways Compatible with 1.5°C in the Context of Sustainable Development. *IPCC SR1 5,* ss. 93-174.

Rojith, G., & Zacharia, P. U. (2015). Enhancing Regional Climate Resilience of Indian Fisheries Through Wetland Restoration and Scientific Fish Farming. *International Journal of Tropical Agriculture*, *33*(4), 3439–3445. https://www.researchgate.net/publication/298788364

Roome, J. (2015). *Rapid, climate–informed Development is needed to keep climate change from pushing more than 100 million people into poverty by 2030.* World Bank.

Rosevelt, C., Los Huertos, M., Garza, C., & Nevins, H. M. (2013). Marine debris in central California: Quantifying type and abundance of beach litter in Monterey Bay, CA. *Marine Pollution Bulletin*, *71*(1-2), 299–306. doi:10.1016/j.marpolbul.2013.01.015 PMID:23499538

Rout, B., & Pradhan, J. P. (2023). Odisha's Extractive Industries: Present Status and Future Prospects. *Productivity*, *64*(1), 107–114. doi:10.32381/PROD.2023.64.01.10

Roy, S., Tandukar, S., & Bhattarai, U. (2022). Gender, Climate Change Adaptation, and Cultural Sustainability: Insights From Bangladesh. *Frontiers in Climate*, *4*, 2022. doi:10.3389/fclim.2022.841488

Rudolph, L., & Gould, S. (2015). Climate change and health inequities: A framework for action. *Annals of Global Health*, *81*(3), 432–444. doi:10.1016/j.aogh.2015.06.003 PMID:26615079

Rulia, A. (2023). The Importance Of Sustainable Agriculture. Bernama. https://www.bernama.com/en/thoughts/news.php?id=2199261

Rus, T., Cruciat, G., Nemeti, G., Mare, R., & Muresan, D. (2022). Thermal comfort in maternity wards: Summer vs. winter conditions. *Journal of Building Engineering*, *51*, 104356. doi:10.1016/j.jobe.2022.104356

Saevarsdottir, G., Kvande, H., & Welch, B. J. (2020). Aluminum production in the times of climate change: The global challenge to reduce the carbon footprint and prevent carbon leakage. *Journal of the Minerals Metals & Materials Society*, *72*(1), 296–308. doi:10.1007/s11837-019-03918-6

Safa, H. (1998). Female Headed Households in the Caribbean: Sign of Pathology of Alternative Form of Family Organization? *The Brown Journal of World Affairs*, *5*(2), 203–214.

Safari, A., Salehzadeh, R., Panahi, R., & Abolghasemian, S. (2018). Multiple pathways linking environmental knowledge and awareness to employees' green behavior. *Corporate Governance: The international journal of business in society*, *18*(1), 81-103.

Sahrawat, R., & Garg, A. P. (2023). Rice straw degradation with mixed culture of microfungi and their enzymes. *International Journal of Environment and Climate Change.*, *13*(8), 2097–2104. doi:10.9734/ijecc/2023/v13i82167

Saifi, S., Mogul, R., Coren, A., Sidhu, S., Iqbal, J., & Mandhro, S. (2022, September 26). *First came the floods. Now, Pakistan's children face a new disaster.* CNN. https://edition.cnn.com/2022/09/25/asia/pakistan-floods-children-water-borne-disease-intl-hnk-dst/index.html

Sajesh, V. K., Suresh, A., Mohanty, A. K., Singh, V., & Ravishamkar, C. N. (2021). Skill development in marine fisheries: Some reflections on the issues and way outs. *The Indian Journal of Animal Sciences*, *91*(7), 518–524. https://doi.org/ 10.56093/ijans.v91i7.115894

Salaria, N. (2012). Meaning of the Term-Descriptive Survey Research Method. *International Journal of Transformations in Business Management.*, *1*(6), 1–7.

Salim, S. S., & Anuja, A. R. (2022). *Marine fisheries policies in India: Opportunities and challenges.* In: Souvenir 1st Indian fisheries outlook 2022. ICAR-Central Inland Fisheries Research Institute, Kolkata (pp. 215-224). http://eprints.cmfri.org.in/15861/

Salimah, N., Siallagan, M. A., Banjarnahor, H. E., & Matondang, K. (2023). Comparison of Capitalist and Socialist Economic Models in International Trade. *Journal of Business Management and Economic Development*, *1*(01), 63–69. doi:10.59653/jbmed.v1i01.39

Salzer, C., Wallbaum, H., Lopez, L. F., & Kouyoumji, J. L. (2016). Sustainability of social housing in Asia: A holistic multi-perspective development process for bamboo-based construction in the Philippines. *Sustainability (Basel)*, *8*(2), 151. doi:10.3390/su8020151

Samantray, K., & Pathak, S. C. (2001). An overview of women in the Indian fisheries sectors and suggestions for improvement of their socioeconomic status. In M. J. Williams, M. C. Nandeesha, V. P. Corral, E. Tech, & P. S. Choo (Eds.), *International Symposium on Women in Asian Fisheries* (pp. 121–126). ICLARM- The World Fish Centre. ISBN:983-2346-02-9

Samarasiri, M., Chai, K. F., & Chen, W. N. (2023). Forward-looking risk assessment framework for novel foods. *Food and Humanity*, *1*, 500–513. doi:10.1016/j.foohum.2023.06.020

Sampath, B., Jensen, M., Lenoci-Edwards, J., Little, K., Singh, H., & Sherman, J. D. (2022). *Reducing Healthcare Carbon Emissions: A Primer on Measures and Actions for Healthcare Organizations to Mitigate Climate Change.* (Prepared by Institute for Healthcare Improvement under Contract No. 75Q80122P00007.) AHRQ Publication No. 22-M011. Rockville, MD: Agency for Healthcare Research and Quality.

Samuel, P. (2010). *Final report: Psychometric evaluation of the Family Quality of Life Survey for low-income, minority families*. Wayne State University, Detroit, MI.

Sandra, M., Devriese, L., De Raedemaecker, F., Lonneville, B., Lukic, I., & Altvater, S. (2020). *Knowledge wave on marine litter from aquaculture sources. D2.2 Aqua-Lit project*. Oostende.

Santos, E. (2022, June 28). *[Opinion] Why climate justice is queer justice*. Rappler. https://www.rappler.com/voices/ispeak/opinion-why-climate-justice-is-queer-justice/

Santos, M., Moreira, H., Cabral, J. A., Gabriel, R., Teixeira, A., Bastos, R., & Aires, A. (2022). Contribution of home gardens to sustainable development: Perspectives from a supported opinion essay. [PubMed]. *International Journal of Environmental Research and Public Health, 19*(20), 13715. doi:10.3390/ijerph192013715 PMID:36294295

Sanyé-Mengual, E., Gasperi, D., Michelon, N., Orsini, F., Ponchia, G., & Gianquinto, G. (2018). Eco-efficiency assessment and food security potential of home gardening: A case study in Padua, Italy. *Sustainability (Basel), 10*(7), 2124. doi:10.3390/su10072124

SAPEA (Science Advice for Policy by European Academies). (2019). *A Scientific Perspective on Microplastics in Nature and Society*. SAPEA. doi:10.26356/microplastics

Savun-Hekimoğlu, B., & Gazioğlu, C. (2021). Mucilage problem in the semi-enclosed seas: Recent outburst in the Sea of Marmara. [IJEGEO]. *International Journal of Environment and Geoinformatics, 8*(4), 402–413. doi:10.30897/ijegeo.955739

Sazima, I., Gadig, O. B. F., Namora, R. C., & Motta, F. S. (2002). Plastic debris collars on juvenile carcharhinid sharks (Rhizoprionodon lalandii) in southwest Atlantic. *Marine Pollution Bulletin, 44*(10), 1147–1149. doi:10.1016/S0025-326X(02)00141-8 PMID:12474977

Scales, K. L., Moore, T., Sloyan, B. M., Spillman, C. M., Paige Eveson, J., Patterson, T. A., Williams, A. B., Hobday, A. J., & Hartog, J. R. (2023). Forecast-ready models to support fisheries' adaptation to global variability and change. *Fisheries Oceanography, 32*(4). Advance online publication. doi:10.1111/fog.12636

Scarascia-Mugnozza, G., Russo, G., & Sica, C. (2012). Plastic materials in European agriculture: Actual use and perspectives. *Journal of Agricultural Engineering, 3*(3), 15–28. doi:10.4081/jae.2011.28

Schanbacher, W. D., & Cavendish, J. C. (2023). The effects of COVID-19 on Central Florida's community gardens: Lessons for promoting food security and overall community well-being. [PubMed]. *Frontiers in Public Health, 11*, 11. doi:10.3389/fpubh.2023.1147967 PMID:37483929

Schecter, A., Colacino, J., Haffner, D., Patel, K., Opel, M., Päpke, O., & Birnbaum, L. (2010). Perfluorinated compounds, polychlorinated biphenyls, and organochlorine pesticide contamination in composite food samples from Dallas, Texas, USA. *Environmental Health Perspectives, 118*(6), 796–802. doi:10.1289/ehp.0901347 PMID:20146964

Schiaparelli, S., Castellano, M., Povero, P., Sartoni, G., & Cattaneo-Vietti, R. (2007). A benthic mucilage event in North-Western Mediterranean Sea and its possible relationships with the summer 2003 European heatwave: Short term effects on littoral rocky assemblages. *Marine Ecology (Berlin), 28*(3), 341–353. doi:10.1111/j.1439-0485.2007.00155.x

Schneider, C. R., Zaval, L., & Markowitz, E. M. (2021). Positive emotions and climate change. *Current Opinion in Behavioral Sciences*, *42*, 114–120. doi:10.1016/j.cobeha.2021.04.009

Schneider, Y., Beier, M., & Rosenwinkel, K. H. (2012). Effect of substrate availability on nitrous oxide production by deammonification processes under anoxic conditions. *Microbial Biotechnology*, *5*(3), 415–424. doi:10.1111/j.1751-7915.2011.00328.x

Schnurr, R. E., & Walker, T. R. (2019). *Marine transportation and energy use. Reference Module in Earth Systems and Environmental Sciences*. Elsevier.

Scholes, M., & Andreae, M. O. (2000). Biogenic and pyrogenic emissions from Africa and their impact on Global atmosphere. *Ambio*, *29*(1), 23–26. doi:10.1579/0044-7447-29.1.23

Schrand, T. G. (2020). Utopianism and the equity path to sustainability. *Journal of Environmental Studies and Sciences*, *10*(4), 457–466. doi:10.1007/s13412-020-00619-4

Schranz, M., Umlauft, M., Sende, M., & Elmenreich, W. (2020). Swarm robotic behaviors and current applications. *Frontiers in Robotics and AI*, *7*, 36. doi:10.3389/frobt.2020.00036 PMID:33501204

Schwarz, A. E., Ligthart, T. N., Boukris, E., & Van Harmelen, T. (2019). Sources, transport, and accumulation of different types of plastic litter in aquatic environments: A review study. *Marine Pollution Bulletin*, *143*, 92–100. doi:10.1016/j.marpolbul.2019.04.029 PMID:31789171

Scott, A. (1999). Nonlinear Science. *Oxford University Pres, Oxford*, *01985017*, 2.

Seddon, N., Chausson, A., Berry, P., Girardin, C. A., Smith, A., & Turner, B. (2020). *Understanding the value and limits of nature-based solutions to climate change and other gloabl challenges*. The Royal Society Publisher.

See Khoo, L., Soon Lai, P., Hafizam Hasmi, A., & Shah Mahmood, M. (2016). Secondary identifier for positive identification in DVI. *Forensic Science and Criminology*, *1*(1), 1–3. doi:10.15761/FSC.1000102

Şeker, M., & Yalçın, H. (2021). An analysis on mucilage (sea snot) research. I. Ozturk & M. Şeker (Eds.). In Ecology of the Marmara Sea: Formation and Interactions of Marine Mucilage, and Recommendations for Solutions (pp.69-84). Tek Ses Ofset Matbaacılık. Ankara.

Selby, D. (2015). Climate Change: Reorienting the Development Agenda. In G. McCann & S. McCloskey (Eds.), From the Local to the Global (3rd edition): Key Issues in Development Studies (pp. 113–131). London: Pluto Press. doi:10.2307/j.ctt183p5fn.11

Semenza, J. C. (1996). Deaths in the Chicago heat wave. *The New England Journal of Medicine*, *335*(24), 1848–1849. doi:10.1056/NEJM199612123352418 PMID:8649494

Sen, S. (2022). Law and social policy in India: From growth-based welfare to welfare entitlement. 102-131.

Senn, R. D., & Weems, R. A. (2013). *Manual of Forensic Odontology* (5th ed.). CRC Press.

Shahbandeh, M. (2022). Leading exporting countries of fish and fishery products worldwide in 2020. *Main exporting countries of fish and fishery products worldwide 2020*. https://www.statista.com/statistics/268269/top-10-exporting-countries-of-fish-and-fishery-products/

Shahidullah, A. K. M., Islam, A., & Kendall, L. (2023, July). Women Agro-Entrepreneurship Promoting Vegetables at a Family Level: A Healthcare Approach towards Non-Communicable Disease Risk Reduction. []. MDPI.] [PubMed]. *Health Care*, *11*(15), 2165. PMID:37570405

Shahzad, A., Ullah, S., Dar, A. A., Sardar, M. F., Mehmood, T., Tufail, M. A., Shakoor, A., & Haris, M. (2021). Nexus on climate change: Agriculture and possible solution to cope future climate change stresses. *Environmental Science and Pollution Research International, 28*(12), 14211–14232. doi:10.1007/s11356-021-12649-8 PMID:33515149

Shanmugam, S. P., Nurni, V. N., Manjini, S., Chandra, S., & Holappa, L. E. (2021). Challenges and outlines of steelmaking toward the year 2030 and beyond—Indian perspective. *Metals, 11*(10), 1654. doi:10.3390/met11101654

Shao, X., Zhang, Z., Song, P., Feng, Y., & Wang, X. (2022). A review of energy efficiency evaluation metrics for data centers. *Energy and Building, 271*, 112308. doi:10.1016/j.enbuild.2022.112308

Sharma, A., Prusty, S., Rathod, R. K., Arthi, R., Watterson, A., & Cavalli, L. S. (2023b). Occupational hazards of Indian shrimp farm workers. *Aquaculture and Marine Biology, 16*(1), 2225762. doi:10.1080/26895293.2023.2225762

Sharma, M., Sahni, Y. P., Mandloi, A. K., & Pandey, G. (2012). Toxicity in fish polluted with heavy metals, chemicals or drugs. *Jigyasa, 6*(2), 67–71. https://www.researchgate.net/publication/270213954_Toxicity_in_fish_polluted_with_heavy_metals_chemicals_or_drugs

Sharma, A., Upadhayay, V. K., & Ghorai, T. (2023a). Possible impact of climate change on fisheries. In A. Sinha, S. Kumar, & K. Kumari (Eds.), *Outlook of climate change and fish nutrition* (pp. 41–47). Springer., doi:10.1007/978-981-19-5500-6

Sharma, M. (2012). Dalits and Indian environmental politics. *Economic and Political Weekly, 47*(23), 46–52.

Sharma, M. (2017). *Caste and nature: Dalits and Indian environmental politics*. Oxford University Press.

Sharma, R., Kannan, D., Darbari, J. D., & Jha, P. C. (2022). Analysis of Collaborative Sustainable Practices in multi-tier food supply chain using integrated TISM-Fuzzy MICMAC model: A supply chain practice view. *Journal of Cleaner Production, 354*, 131271. doi:10.1016/j.jclepro.2022.131271

Shatokha, V. (2016). Environmental sustainability of the iron and steel industry: Towards reaching the climate goals. *European Journal of Sustainable Development, 5*(4), 289–289. doi:10.14207/ejsd.2016.v5n4p289

Shaw, A., Burch, S., Kristensen, F., Robinson, J., & Dale, A. (2014). Accelerating the sustainability transition: Exploring synergies between adaptation and mitigation in British Columbian communities. *Global Environmental Change, 25*, 41–51. doi:10.1016/j.gloenvcha.2014.01.002

Shehabi, Smith, S., Sartor, D., Brown, R., Herrlin, M., Koomey, J., Masanet, E., Horner, N., Azevedo, I., & Lintner, W. (2016, June). United States Data Center Energy Usage Report. *LBNL, 1005775*, 1372902. doi:10.2172/1372902

Sherrington, C.S., Hann, G., Cole, G., & Corbin, M. (2016). Study to support the development of measures to combat a range of marine litter sources. *Report for European Commission DG Environment.*

Shetty, D. (2021). *Why air pollution conversations leave out women's health.* Devex. https://www.devex.com/news/why-air-pollution-conversations-leave-out-women-s-health-102011

Shiva, V., & Mies, M. (2014). *Ecofeminism.* Zed Books.

Siemon, C., Rueckel, D., & Krumay, B. (2020). *Blockchain technology for emergency response. Proceedings of the 53rd Hawaii International Conference on System Sciences*, Maui, Hawaii.

Silva, B. K. R., Figueiredo, F. W. dos S., Maciel, E. da S., Quaresma, F. R. P., & Adami, F. (2019). Factors associated with perceived quality of life in artisanal fishermen: A cross-sectional study. *BMC Research Notes, 12*(1), 479. doi:10.1186/s13104-019-4525-4

Silvestri, S., Bryan, E., Ringler, C., Herrero, M., & Okoba, B. (2012). Climate Change Perception and Adaptation of Agro-Pastoral Communities in Kenya. *Regional Environmental Change, 12*(4), 791–802. doi:10.1007/s10113-012-0293-6

Simpson, N. P., Mach, K. J., Constable, A., Hess, J., Hogarth, R., Howden, M., Lawrence, J., Lempert, R. J., Muccione, V., Mackey, B., New, M. G., O'Neill, B., Otto, F., Pörtner, H.-O., Reisinger, A., Roberts, D., Schmidt, D. N., Seneviratne, S., Strongin, S., & Trisos, C. H. (2021). A framework for complex climate change risk assessment. *One Earth*, *4*(4), 489–501. doi:10.1016/j.oneear.2021.03.005

Singhal, A., Dube, P., & Jain, V. K. (2022). Modelling barriers of artificial intelligence in banking sectors using total interpretive structural modelling. *International Journal of Applied Decision Sciences*, *15*(3), 311–335. doi:10.1504/IJADS.2022.122649

Singh, D. (2019). Implementation of technology innovation in MSMEs in India: Case study in select firms from Northern region. *Journal of Science and Technology Policy Management*, *10*(3), 769–792. doi:10.1108/JSTPM-06-2018-0065

Singh, N. S., Aryasinghe, S., Smith, J., Khosla, R., Say, L., & Blanchet, K. (2018). A long way to go: A systematic review to assess the utilization of sexual and reproductive health services during humanitarian crises. *BMJ Global Health*, *3*(2), e000682. doi:10.1136/bmjgh-2017-000682 PMID:29736272

Singh, P., Pradhan, V. S., & Patil, Y. B. (2024). Modeling drivers and barriers of climate change mitigation strategies in Indian iron and steel industry: A TISM-based approach. *Management of Environmental Quality*, *35*(1), 38–60. doi:10.1108/MEQ-04-2023-0097

Singleton, B. E., Rask, N., Magnusdottir, G. L., & Kronsell, A. (2022). Intersectionality and climate policy-making: The inclusion of social difference by three Swedish government agencies. *Environment and Planning C. Politics and Space*, *40*(1), 180–200.

Sinha, B., Basu, A., & Katiyar, A. S. (2011). *Adapting to Climate Change: Opportunities under MGNREGA*. Presented in the conference on Sustaining Commons: Sustaining Our Future, the Thirteenth Biennial Conference of the International Association for the Study of Common Property, Hyderabad. (https://dlc.dlib.indiana.edu/dlc/handle/10535/7144)

Slegers, M. F. W. (2008). If only it would rain: Farmers' perceptions of rainfall and drought in semi-arid central Tanzania. *Journal of Arid Environments*, *72*(11), 2106–2123. doi:10.1016/j.jaridenv.2008.06.011

Slovic, S., Rangarajan, S., & Sarveswaran, V. (2015). *Ecocriticism of the Global South*. Lexington Books.

Smith, L. (2022). Healthcare and Climate Change. The Nexus between Climate Change and Healthcare. *Health Policy Partnership*. https://www.healthpolicypartnership.com/the-nexus-between-climate-change-and-healthcare/.

Smith, C., & Clay, P. (2010). Measuring Subjective and Objective Well-being: Analyses from Five Marine Commercial Fisheries. *Human Organization*, *69*(2), 158–168. doi:10.17730/humo.69.2.b83x6t44878u4782

Smith, G. S., Anjum, E., Francis, C., Deanes, L., & Acey, C. (2022). Climate change, environmental disasters, and health inequities: The underlying role of structural inequalities. *Current Environmental Health Reports*, *9*(1), 80–89. doi:10.1007/s40572-022-00336-w PMID:35338470

Snow, C. C., & Hambrick, D. C. (1980). Measuring organizational strategies: Some theoretical and methodological problems. *Academy of Management Review*, *5*(4), 527–538. doi:10.2307/257458

Sobhana, K. S. (2009). *Diseases of seabass in cage culture and control measures*. National Fisheries Development Board., http://eprints.cmfri.org.in/6087/1/15._Sob.pdf

Sofo, A., & Sofo, A. (2020). Converting home spaces into food gardens at the time of Covid-19 quarantine: All the benefits of plants in this difficult and unprecedented period. [PubMed]. *Human Ecology*, *48*(2), 131–139. doi:10.1007/s10745-020-00147-3 PMID:32317823

Solomon, C. G., & LaRocque, R. C. (2019). Climate change—A health emergency. *The New England Journal of Medicine, 380*(3), 209–211. doi:10.1056/NEJMp1817067 PMID:30650319

Solomon, T. (2014). ENVIRONMENTAL ISSUES DUE TO IMPROPER PLASTIC DISPOSAL IN INDIAN CITIES. *International Journal of Research in Engineering and Technology, 03*(23), 286–289. doi:10.15623/ijret.2014.0323061

Song, S., Cheong, J. C., Lee, J. S., Tan, J. K., Chiam, Z., Arora, S., Png, K. J. Q., Seow, J. W. C., Leong, F. W. S., Palliwal, A., Biljecki, F., Tablada, A., & Tan, H. T. (2022). Home Gardening in Singapore: A feasibility study on the utilization of the vertical space of retrofitted high-rise public housing apartment buildings to increase urban vegetable self-sufficiency. *Urban Forestry & Urban Greening, 78*, 127755. doi:10.1016/j.ufug.2022.127755

Sonsale, A. N., Pohekar, S. D., & Purohit, J. K. (2023). Drivers to energy efficiency measures in recycled paper and pulp industry in India: An interpretive structural modelling-based framework. *Sustainable Energy Technologies and Assessments, 55*, 102961. doi:10.1016/j.seta.2022.102961

Soto, D., Aguilar-Manjarrez, J., & Hishamunda, N. (2008). Building an ecosystem approach to aquaculture. FAO/ Universitat de les IllesBalears Expert Workshop. In *Proceedings of FAO Fisheries and Aquaculture* (pp. 14). Food and Agriculture Organization of the United Nations (FAO). ISBN: 978-92-5-106075-9

Sou, G., Ooi, C. S., & Zhang, Y. (2023). Resilience and Resistance. In Islands and Resilience: Experiences from the Pandemic Era (pp. 53-64). Springer Nature Singapore. 10.1007/978-981-19-9964-2_4

Sridevi, B., Sabira, S., & Sarma, V. (2023). Impact of ocean warming on net primary production in the northern Indian Ocean: Role of aerosols and freshening of surface ocean. *Environmental Science and Pollution Research International, 30*(18), 53616–53634. doi:10.1007/s11356-023-26001-9

Srinivas, C. R., & Sethy, M. (2023). Occupational dermatoses. *Indian Dermatology Online Journal, 14*(1), 21. doi:10.4103/idoj.idoj_332_22

Srishti. (2021). *Green Hospitals – Key to a healthy future*. Srishti. https://www.architectandinteriorsindia.com/insights/green-hospitals-key-to-a-healthy-future.

Statista. (2023). *Global plastics industry*. Statista, https://www.statista.com/topics/5266/plastics-industry/#topicOverview

Statista. (2023). *Plastic industry in India*. Statista. https://www.statista.com/topics/6902/plastic-industry-in-india/#topicOverview

Statista. (2023). *Plastic waste worldwide*. Statista. https://www.statista.com/topics/5401/global-plastic-waste/#topicOverview

Steinmetz, Z., Wollmann, C., Schaefer, M., Buchmann, C., David, J., Troger, J., Muñoz, K., Frör, O., & Schaumann, G. E. (2016). Plastic mulching in agriculture. Trading shortterm agronomic benefits for long-term soil degradation? *The Science of the Total Environment, 550*, 690–705. doi:10.1016/j.scitotenv.2016.01.153 PMID:26849333

Stiepani, J., Gillis, L. G., Chee, S. Y., Pfeiffer, M., & Nordhaus, I. (2021). Impacts of urbanization on mangrove forests and brachyuran crabs in Penang, Malaysia. *Regional Environmental Change, 21*(3), 69. doi:10.1007/s10113-021-01800-3

Streck, C., Keenlyside, P., & Von Unger, M. (2016). The Paris Agreement: A new beginning. *Journal for European Environmental & Planning Law, 13*(1), 3–29. doi:10.1163/18760104-01301002

Strogatz, S. H. (1994). *Nonlinear Dynamics and Chaos: with applications to physics, biology, chemistry, and engineering*. Perseus Books.

Subrat Kumar Mishra & Firoz Khan. (2023). Guaranteeing employment in Rural India during Covid-19 Pandemic. Economic and Political Weekly.

Sudhakara, K., & Jijeesh, C. (2015). *Bamboos: Emerging carbon sink for global climate change mitigation. Thrissur.* Department of Forestry.

Sultana, F. (2021). Critical climate justice. *The Geographical Journal, 188*(1), 118–124. doi:10.1111/geoj.12417

Suman, D. O. (2019). Mangrove management: Challenges and guidelines. In Gerardo M. E. Perillo, Eric W., Donald R. C., Charles S. H. (Eds.), Coastal wetlands (Second Edition) (pp. 1055-1079). Elsevier.

Sumer, G. (2022). The Problem of Climate Change in Terms of Different Disciplines, Kahramanmaraş Sütçü İmam University. *Journal of Social Sciences.*

Sushil. (2018). How to check correctness of total interpretive structural models? *Annals of Operations Research, 270*(1-2), 473-487.

Sussarellu, R., Suquet, M., Thomas, Y., Lambert, C., Fabioux, C., Pernet, M. E. J., Le Goïc, N., Quillien, V., Mingant, C., Epelboin, Y., Corporeau, C., Guyomarch, J., Robbens, J., Paul-Pont, I., Soudant, P., & Huvet, A. (2016). Oyster reproduction is afected by exposure to polystyrene microplastics. *Proceedings of the National Academy of Sciences of the United States of America, 113*(9), 2430–2435. doi:10.1073/pnas.1519019113 PMID:26831072

Sustainable lifestyles . (2023). UNEP - UN Environment Programme. https://www.unep.org/explore-topics/resource-efficiency/what-we-do/sustainable-lifestyles.

Swaen, G. M. H., Langendam, M., Weyler, J., Burger, H., Siesling, S., Atsma, W. J., & Bouter, L. (2018). Responsible epidemiologic research practice: A guideline developed by a working group of the Netherlands Epidemiological Society. *Journal of Clinical Epidemiology, 100*, 111–119. doi:10.1016/j.jclinepi.2018.02.010 PMID:29432862

Swaminathan, M. S. (2012). Aquaculture and sustainable nutrition security in a warming planet. In *Proceeding of the Global Conference on Aquaculture 2010- Farming the Waters for People and Food*, Food and agriculture organization (pp. 3-895).ISBN: 978-92-5-107233-2

Swedish National Board of Occupational Safety and Health. (1991). *An excerpt from NewsletterBilaga 4*, A quarterly review, provided for ILO's Food and Drink Industries Committee.

Sydeman, W. J., Garcia-Reyes, M., Schoeman, D. S., Rykaczewski, R. R., Thompson, S. A., Black, B. A., & Bograd, S. J. (2014). Climate change and wind intensification in coastal upwelling ecosystems. *Science, 345*(6192), 77–80. https://doi.org/10.1126/science.1251635

Sze, J. (2006). From environmental justice literature to the literature of environmental justice. In J. Adamson, M. M. Evans, & R. Stein (Eds.), *The environmental justice reader: Politics, poetics and pedagogy*. University of Arizona Press.

Szemplinska Stupnicka, W., & Iooss, G. ve Moon F. C. (1988). Chaotic Motions in Nonlinear Dynamical Systems. Springer Verlag Wien.

Tabary, Z. (2018, March 8). *Women, LGBT people 'invisible' victims of disasters – experts*. Thomson Reuters Foundation. https://news.trust.org/item/20180308114259-otaio/

Takeuchi, I., Miyoshi, N., Mizukawa, K., Takada, H., Ikemoto, T., Omori, K., & Tsuchiya, K. (2009). Biomagnification profiles of polycyclic aromatic hydrocarbons, alkylphenols and polychlorinated biphenyls in Tokyo Bay elucidated by δ13C and δ15N isotope ratios as guides to trophic web structure. *Marine Pollution Bulletin, 58*(5), 663–671. doi:10.1016/j.marpolbul.2008.12.022

Tang, G., Liu, M., Zhou, Q., He, H., Chen, K., Zhang, H., Hu, J., Huang, Q., Luo, Y., Ke, H., Chen, B., Xu, X., & Cai, M. (2018). Microplastics and polycyclic aromatic hydrocarbons (PAHs) in Xiamen coastal areas: Implications for anthropogenic impacts. *The Science of the Total Environment, 634*, 811–820. doi:10.1016/j.scitotenv.2018.03.336 PMID:29653425

Tatira, L. (2000b). The role of *Zviera* in socialisation. In Chiwome,E. Gambahaya, Z. and Tatira, L. (2013). *Some Aspects of Ubuntu/Unhu as encapsulated in Zviera (Taboos).* Harare: Bhabhu books.

Tatira, L. (2000). *Zviera zvaVaShona*. Mambo Press.

Tel, T., & Gruiz, M. (2006). *Chaotic Dynamics*. Cmabridge University Press. doi:10.1017/CBO9780511803277

Temporosa, R. G. (1979). A study of the manpower skills of fishery workers in the commercial fisheries industry considered important by respondents in Western Visayas. *A thesis presented to the faculty of the school of graduate studies. Central Philippine University*. 6-10. https://repository.cpu.edu.ph/handle/20.500.12852/1701?locale-attribute=ru

Tennison, I., Roschnik, S., Ashby, B., Boyd, R., Hamilton, I., Oreszczyn, T., Owen, A., Romanello, M., Ruyssevelt, P., Sherman, J. D., Smith, A. Z. P., Steele, K., Watts, N., & Eckelman, M. J. (2021). Health care's response to climate change: A carbon footprint assessment of the NHS in England. *The Lancet. Planetary Health, 5*(2), e84–e92. doi:10.1016/S2542-5196(20)30271-0 PMID:33581070

Tewabe, D. (2014). Climate change challenges on fisheries and aquaculture. *International Journal of Aquaculture and Fishery Sciences, 1*(1), 006-011. doi:10.17352/2455-8400.000002

Thailand U. N. D. P. (2023). Policy Brief: Gender Responsive And Socially Inclusive Public Climate Finance. UN. https://www.undp.org/thailand/publications/policy-brief-gender-responsive-and-socially-inclusive-public-climate-finance

The Borneo Post. (2021). Covid-19 pandemic sees more Malaysians getting into home gardening. The Borneo Post. https://www.theborneopost.com/2021/12/22/covid-19-pandemic-sees-more-malaysians-gettinginto home-gardening/

The CityMaker. (2020). Kebun Kita(r): Penang's 1st Self-Sustaining Community Urban Farm Opens at Penang Digital Library. The City Maker. https://thecitymaker.com.my/kebun-kitar-penang-urbancommunity-digital/

The Daily Star. (2008, April 9). Staff Correspondent. Women more vulnerable to violence during floods. *The Daily Star.* https://www.thedailystar.net/news-detail-31477

The Economist Newspaper. (n.d.). Holding polluting sectors accountable for the Climate Crisis. *The Economist*. https://impact.economist.com/sustainability/circular-economies/how-to-hold-polluting-sectors-accountable-for-the-climate-crisis

The European Commission Director General for Research and Innovation, (2021). *Horizons: New-materials-make-ships-more-sustainable-and-less-noisy-marine-life.* OECD.

The Guardian. (2023, April 10). Sea-level rise poses existential threat to Miami and New Orleans, studies find. *The Guardian*. https://www.theguardian.com/environment/2023/apr/10/sea-level-rise-climate-crisis-miami-new-orleans

The Herald. (2021). Veld fire destroys environment. *The Herald.* https://www.herald.co.zw/veld-fires-destroy-environment/

The Hindu. (2023, June 18). Heatwave in U.P. and Bihar kill nearly 100. *The Hindu*. https://www.thehindu.com/news/national/other-states/heatwave-up-bihar-kill-100-india-heatwave/article66982736.ece

The Indian Express. (2023, July 14). Woman missing, 28 villages inundated after cloudburst in Uttarakhand's Pithoragarh. *The Indian Express*. https://indianexpress.com/article/india/woman-missing-28-villages-inundated-cloudburst-uttarakhands-pithoragarh-8142486/

The Organisation for Economic Co-operation and Development. (2015). *Green bonds Mobilising the debt capital markets for a low-carbon transition*. OECD. https://www.oecd.org/environment/cc/Green%20bonds%20PP%20%5Bf3%5D%20%5Blr%5D.pdf

The Organisation for Economic Co-operation and Development. (2022). *Shipbuilding-policy-and-market-developments-in-selected-economies-2022*. OECD. https://www.oecd.org/ocean/topics/ocean-shipping/

The Star. (2020). MCO boosts demand for seeds, gardening tools. The Star. https://www.thestar.com.my/news/nation/2020/04/09/mco-boosts-demand-for-seeds-gardening-tools

The Times of India. (2022). *Now, India is the world's largest producer of solar energy*. Ornate Solar.

The Wire . (2022). India's Per Capita Greenhouse Gas Emissions Far Below World Average: UNEP Report. UN. https://thewire.in/environment/india-per-capita-greenhouse-gas-emissions-unep

The World Bank. (2023). *Green Bonds*. The World Bank. https://www.worldbank.org/en/news/feature/2023/04/10/from-india-to-indonesia-green-bonds-help-countries-move-toward-sustainability

The World Economic Forum. (2022). *Net Zero Tracking*. WEF. https://www.weforum.org/reports/the-net-zero-industry-tracker/

The World Health Organization. (2021). *Energy Progress*. WHO. https://www.who.int/news/item/07-06-2021-global-launch-tracking-sdg7-the-energy-progress-report

Theofilou, P. (2013). Quality of life: Definition and measurement. *Europe's Journal of Psychology*, *9*(1), 150–162. doi:10.5964/ejop.v9i1.337

Thomas, A., & Suresh, M. (2022). Readiness for sustainable-resilience in healthcare organisations during Covid-19 era. *The International Journal of Organizational Analysis*, *31*(1), 91–123. doi:10.1108/IJOA-09-2021-2960

Thompson, P. B. (2017). *The spirit of the soil: Agriculture and environmental ethics*. Taylor & Francis., doi:10.4324/9781315559971

Thompson, R. C., Olsen, Y., Mitchell, R. P., Davis, A., Rowland, S. J., John, A. W. G., McGonigle, D., & Russell, A. E. (2004). Lost at sea: Where is all the plastic? *Science*, *304*(5672), 838–838. doi:10.1126/science.1094559 PMID:15131299

Thorvaldsen, T., Salomonsen, C., Ranum, S. A., Trædal, P., Misund, A., & Holmen, I. M. (2023). Prepared for the worst? Emergency preparedness in Norwegian fish farming – status and further improvements. *Aquaculture (Amsterdam, Netherlands)*, *577*, 739921. https://doi.org/10.1016/j.aquaculture.2023.739921

Thurston, A. M., Stockl, H., & Ranganathan, M. (2021). Natural hazards, disasters and violence against women and girls: A global mixed-methods systematic review. *BMJ Global Health*, *6*(4), e004377. doi:10.1136/bmjgh-2020-004377 PMID:33958379

Tiejun, D. (2011). The influence of iron flow on iron resource efficiency in the steel manufacturing process. *Resources, Conservation and Recycling*, *55*(8), 760–771. doi:10.1016/j.resconrec.2011.03.002

Tom, A. P., Jayakumar, J. S., Biju, M., Somarajan, J., & Ibrahim, M. A. (2021). Aquaculture wastewater treatment technologies and their sustainability: A review. *Energy Nexus*, *4*, 100022. doi:10.1016/j.nexus.2021.100022

Tomoda, S. (1997). *Safety and health of meat, poultry and fish processing workers*. International Labour Office. https://www.ilo.org/wcmsp5/groups/public/---ed_protect/---protrav/---safework/documents/publication/wcms_219576.pdf

Tong, Y., Cai, J., Zhang, Q., Gao, C., Wang, L., Li, P., Hu, S., Liu, C., He, Z., & Yang, J. (2019). Life cycle water use and wastewater discharge of steel production based on material-energy-water flows: A case study in China. *Journal of Cleaner Production*, *241*, 118410. doi:10.1016/j.jclepro.2019.118410

Tran, T. P. (2020). Ho Chi Minh thought on solidarity and the social sustainable development in Vietnam. In *E3S Web of Conferences* (Vol. 164, p. 12004). EDP Sciences., doi:10.1051/e3sconf/202016412004

Troell, M., Henriksson, P. J. G., Buschmann, A. H., Chopin, T., & Quahe, S. (2022). Farming the ocean – Seaweeds as a quick fix for the climate? *Reviews in Fisheries Science & Aquaculture*, *4*, 1–11. doi:10.1080/23308249.2022.2048792

Trudel, D., Scheringer, M., von Goetz, N., & Hungerbühler, K. (2011). Total consumer exposure to polybrominated diphenyl ethers in North America and Europe. *Environmental Science & Technology*, *45*(6), 2391–2397. doi:10.1021/es1035046 PMID:21348481

Turkish Academy of Sciences. (2021). *TÜBA Müsilaj-Deniz Salyası Değerlendirme Raporu (3 Haziran 2021).* Turkish Academy of Sciences. https://www.tuba.gov.tr/tr/yayinlar/suresiz-yayinlar/bilim-ve-dusunce/marmara-denizinin-ekolojisi-deniz-salyasi-olusumu-etkilesimleri-ve-cozum-onerileri-1.

Turner, B., Devisscher, T., Chabaneix, N., Woroniecki, S., Messier, C., & Seddon, N. (2022). The role of nature-based solutions in supporting social-ecological resilience for climate change adaptation. *Annual Review of Environment and Resources*, *47*(1), 123–148. doi:10.1146/annurev-environ-012220-010017

Turnšek, M., Gangenes Skar, S. L., Piirman, M., Thorarinsdottir, R. I., Bavec, M., & Junge, R. (2022). Home Gardening and Food Security Concerns during the COVID-19 Pandemic. *Horticulturae*, *8*(9), 778. doi:10.3390/horticulturae8090778

U.S. Food and Drug Administration (USFDA). 2022. *Fish and fishery products hazards and controls guidance.* https://www.fda.gov/media/80637/download

UN. (2015). *The Paris Agreement.* UN. https://www.un.org/en/climatechange/paris-agreement

UN. (2020). *Causes and Effects of Climate Change.* UN. https://www.un.org/en/climatechange/science/causes-effects-climate-change

UNCTAD. (2020). COVID-19 has changed online shopping forever, survey shows. UN. https://unctad.org/news/covid-19-has-changed-online-shopping-forever-survey-shows

UNDP. (2023, February 28). *What does gender equality have to do with climate change?* [Blog post]. UN. https://climatepromise.undp.org/news-and-stories/what-does-gender-equality-have-do-climate-change#:~:text=Giving%20women%20increased%20access%20to,benefit%20the%20needs%20of%20women

UNEP. (2021). *Addressing Single-Use Plastic Products Pollution using a Life Cycle Approach.* UNEP. https://www.unep.org/resources/publication/ addressing-single-use-plastic-products-pollution-using-life-cycleapproach. Accessed 6 July 2021.

UNEP. (2023). *Sustainable Roadmap.* UNEP. https://unpartnerships.un.org/news/2023/roadmap-sustainable-future-ending-plastic

UNESCAP. (n.d.). *Climate Action.* ESCAP. https://www.unescap.org/sites/default/d8files/event- documents/SDG%2013%20Goal%20Profile_Final%20Web_EDIT%20v3.pdf, as retrieved on 18th March 2023.

UNFCC. (2015). *The Paris Agreement.* UNFCC. https://unfccc.int/process-and-meetings/the-paris-agreement

UNFCC. (2022). *Making the pivot toward implementation.* UNFCC. https://unfccc.int/making-the-pivot-toward-implementation

UNFCCC. (2007). *Climate Change: Impacts, Vulnerability and Adaptations in Developing Countries*. UNFCCC.

UNFPA East and Southern Africa Regional Office. (2021). *The impact of the climate crisis on sexual and reproductive health and rights (SRHR)*. UN. https://esaro.unfpa.org/sites/default/files/pub-pdf/advocacy_-_impact_of_climate_crisis_on_srhr.pdf

UNFPA. (2009). *State of the World Population: Women, Population and Climate Change*. UNFPA.

UNFPA. (2015, December 3). *Essential health needs of women often neglected in assistance after natural disasters, conflicts* [Press release]. UN. https://www.unfpa.org/press/essential-health-needs-women-often-neglected-assistance-after-natural-disasters-conflicts

United Nations (UN). (n.d.). *What Is Climate Change?* United Nations. https://www.un.org/en/climatechange/what-is-climate-change.

United Nations climate change. (n.d.). *Gender & Climate Change: an important connection*. UN. https://unfccc.int/gender

United Nations Environment Programme, UN Women, UNDP, & UNDPPA/PBSO. (2020). *Gender, Climate and Security: Sustaining inclusive peace on the frontiers of climate change*. UN.

United Nations Environment Programme. (2021). *Drowning in Plastics – Marine Litter and Plastic Waste Vital Graphics*. UN.

United Nations Industrial Development Organization. (2023). *Who We Are*. UN. https://www.unido.org/about-us/who-we-are

United Nations. (2017). *Voluntary National Review Report on the implementation of the sustainable development goals*. UN.

United Nations. (2019). *United Nations Sustainable Development Goals: 17 Goals. SDG Progress Report*. Department of Economic and Social Affairs, Government of India.

United Nations. (2020). *High-Level Political Forum on Sustainable Development*. Sustainable Development Goals Knowledge Platform.

United Nations. (2020). *Sustainable Development Goals*. New York: United Nations. https://www.un.org/sustainabledevelopment/energy/

United Nations. (2020). *Sustainable Development Goals*. New York: United Nations. https://www.un.org/sustainabledevelopment/goal-of-the-month

United Nations. (2021). "*What Is the Kyoto Protocol?*" UN.https://unfccc.int/kyoto_protocol

United Nations. (2021). *COP26: Together for our Planet*. UN. https://www.un.org/en/climatechange/cop26

United Nations. (2021). *The Guidelines to Implement the Kyoto Protocol: the Marrakesh Accords and the 5,7&8 Implications*. UN. https://unfccc.int/process-and-meetings/transparency-and-reporting/reporting-and-review-under-the-kyoto-protocol/overview/background-and-resources/the-guidelines-to-implement-the-kyoto-protocol-the-marrakesh-accords-and-the-578-implications

United Nations. (2022). Sustainable Development Goal 2 Zero Hunger. UN https://malaysia.un.org/en/sdgs/2

United Nations. (2023). *United Nations SDG Progress Report*. Department of Economic and Social Affairs, Government of India.

University of Stavenger. (2020). *RISKSEC 2.0*. UIS. https://www.uis.no/en/research/climate-change-adaptation-strategies

Unnikrishnan, A. S., & Shankar, D. (2007). Are sea-level-rise trends along the coasts of the north Indian Ocean consistent with global estimates? *Global and Planetary Change*, *57*(3-4), 301–307. doi:10.1016/j.gloplacha.2006.11.029

URL-1. (2023). *Marmara denizi koruma eylem planı.* URL-1. https://webdosya.csb.gov.tr/db/marmarahepimizin/eylemplani/ Retrieved Date: 29.05.2023

URL-2. (2023). *Marmara belediyeler birliği sonuç bildirgesi.* URL2. https://www.marmara.gov.tr/ Retrieved Date: 29.05.2023

Ushakov, S., Stenersen, D., & Einang, P. M. (2019). Methane slip from gas fuelled ships: A comprehensive summary based on measurement data. *Journal of Marine Science and Technology*, *24*(4), 1308–1325. doi:10.1007/s00773-018-00622-z

Van Dijk, K. (2012). *Empowerment of Pastoralist Women.* PENHA narrative report. https://oxfamilibrary.openrepository.com/bitstream/handle/10546/222733/cs-empowerment-pastoralist-women-sudan-penha-2012-en.pdf;jsessionid=A74E516013AED6518687A3C1C29F8874?sequence=1

Vasanth, M., Muralidhar, M., Saraswathy, R., Nagavel, A., Dayal, J. S., Jayanthi, M., Lalitha, N., Kumararaja, P., & Vijayan, K. K. (2016). Methodological approach for the collection and simultaneous estimation of greenhouse gases emission from aquaculture ponds. *Environmental Monitoring and Assessment*, *188*(12), 671. doi:10.1007/s10661-016-5646-z

Veen, E. J. (2015). *Community gardens in urban areas: a critical reflection on the extent to which they strenghten social cohesion and provide alternative food.* Wageningen University and Research.

Veenhoven, R. (2000). The four qualities of life. *Journal of Happiness Studies*, *1*(1), 1–39. doi:10.1023/A:1010072010360

Velmurugan, A. V., Roy, S. D., Krishnan, P., Swarnam, T. P., Jaisankar, I., Singh, A. K., & Biswas, T. K. (2015). Climate change & Nicobar Islands: Impacts and adaptation strategies. *Andaman Science Association*, *20*(1), 7–18. ISSN: 0970-4183

Velzeboer, I., Kwadijk, C. J. A. F., & Koelmans, A. A. (2014). Strong sorption of PCBs to nanoplastics, microplastics, carbon nanotubes, and fullerenes. *Environmental Science & Technology*, *48*(9), 4869–4876. doi:10.1021/es405721v PMID:24689832

Verma, R., Nellemann, C., & Hislop, L. (2011). *Women at the Frontline of Climate Change: Gender Risks and Hopes.*

Verma, N. (2023). *India's first bamboo-based biorefinery to start operations in March.* Reuters.

Versey, H. S. (2021). Missing pieces in the discussion on climate change and risk: Intersectionality and compounded vulnerability. *Policy Insights from the Behavioral and Brain Sciences*, *8*(1), 67–75. doi:10.1177/2372732220982628

Vicedo-Cabrera, A. M., Scovronick, N., Sera, F., Royé, D., Schneider, R., Tobias, A., Astrom, C., Guo, Y., Honda, Y., Hondula, D. M., Abrutzky, R., Tong, S., Coelho, M. S. Z. S., Saldiva, P. H. N., Lavigne, E., Correa, P. M., Ortega, N. V., Kan, H., Osorio, S., & Gasparrini, A. (2021). The burden of heat-related mortality attributable to recent human-induced climate change. *Nature Climate Change*, *11*(6), 492–500. doi:10.1038/s41558-021-01058-x PMID:34221128

Vijayan, K. K. (2018). Brackishwater aquaculture in India: An overview. https://krishi.icar.gov.in/jspui/bitstream/123456789/20212/1/Training%20Manual%20-%20Final%20-%20Water%20and%20soil%20treatment%20applications%20in%20aquaculture.pdf

Vilchis, L. I., Tegner, M. J., Moore, J. D., Friedman, C. S., Riser, K. L., Robbins, T. T., & Dayton, P. K. (2005). Ocean warming effects on growth, reproduction, and survivorship of southern california abalone. *Ecological Applications*, *15*(2), 469–480. doi:10.1890/03-5326

Vivekanandan, E., Rajagopalan, M., & Pillai, N. G. K. (2009). Recent trends in sea surface temperature and its impact on oil sardine. In A. Surendran (Ed.), *Global Climate Change and Indian Agriculture* (pp. 89–92). Indian Council of Agricultural Research., http://eprints.cmfri.org.in/8435/

Voeten, H. A. C. M., Vissers, D. C. J., Gregson, S., Zaba, B., White, R. G., de Vlas, S. J., & Habbema, J. D. F. (2010). Strong association between in-migration and HIV prevalence in urban sub-Saharan Africa. *Sexually Transmitted Diseases*, *37*(4), 240–243. doi:10.1097/OLQ.0b013e3181c3f2d0 PMID:19959971

Volcovici, V. (2017, August 22), *Energy Transfer sues Greenpeace over Dakota pipeline.* Reuters. https://www.reuters.com/article/us-energy-transf-lawsuit-idUSKCN1B2245

Wadhawan, S. (2021). *Three steps that India should take to Climate –Proof Infrastructure. Council of Energy Environment and Water.* CEEW.

Wainwright, J. (2005). The geographies of political ecology: After Edward Saïd. *Environment & Planning A*, *37*(6), 1033–1043. doi:10.1068/a37166

Walljasper, C., & Polansek, T. (2020). Home gardening blooms around the world during coronavirus lockdowns. Reuters. https://www.reuters.com/article/us-health-coronavirus-gardensidUSKBN2220D3

Walsh, N. (2022). Sharing the latest improvements to efficiency in Microsoft's datacenters. *Azure Blog*. https://azure.microsoft.com/en-us/blog/sharing-the-latest-improvements-to-efficiency-in-microsoft-s-datacenters/

Walsh, J. E., Ballinger, T. J., Euskirchen, E. S., Hanna, E., Mård, J., Overland, J. E., Tangen, H., & Vihma, T. (2020). Extreme weather and climate events in northern areas: A review. *Earth-Science Reviews*, *209*, 103324. doi:10.1016/j.earscirev.2020.103324

Wang, R. (2022). *Energy saving technologies and optimisation of energy use for decarbonised iron and steel industry.*

Wang, F., Wong, C. S., Chen, D., Lu, X., Wang, F., & Zeng, E. Y. (2018). Interaction of toxic chemicals with microplastics: A critical review. *Water Research*, *139*, 208–219. doi:10.1016/j.watres.2018.04.003 PMID:29653356

Wang, W., Ge, J., Yu, X., & Li, H. (2020a). Environmental fate and impacts of microplastics in soil ecosystems: Progress and perspective. *The Science of the Total Environment*, *708*, 134841. doi:10.1016/j.scitotenv.2019.134841 PMID:31791759

Warfield, J. N. (1974). Developing interconnection matrices in structural modeling. *IEEE Transactions on Systems, Man, and Cybernetics*, *SMC-4*(1), 81–87. doi:10.1109/TSMC.1974.5408524

Warrier, M. V. S. (2001). Women at work: Migrant women in fish processing industry. *Economic and Political Weekly*, *36*(37), 3554–3562. https://www.epw.in/journal/2001/37/special-articles/women-work.html

Water Footprint. (n.d.). *Water Footprint Network.* Water Footprint. https://www.waterfootprint.org/water-footprint-2/

Waterman, P., Chapman, S., Sargent, S., Laves, G., English, K., & Morvell, G. (2009, October). Climate Proofing as a Regional Development Strategy. In *2009 Sustainable Economic Growth for Regional Australia Conference: Refereed Papers* (pp. 1-22). Management Solutions (QLD) Pty Ltd.

Watson, R. T., Zinyowera, M. C., Moss, R. H., & Dokken, D. J. (1998). *The Regional Impacts of Climate Change.* IPCC.

Watts, N., Amann, M., Arnell, N., Ayeb-Karlsson, S., Belesova, K., Boykoff, M., Byass, P., Cai, W., Campbell-Lendrum, D., Capstick, S., Chambers, J., Dalin, C., Daly, M., Dasandi, N., Davies, M., Drummond, P., Dubrow, R., Ebi, K. L., Eckelman, M., & Montgomery, H. (2019). The 2019 report of the Lancet Countdown on health and climate change: Ensuring that the health of a child born today is not defined by a changing climate. *Lancet*, *394*(10211), 1836–1878. doi:10.1016/S0140-6736(19)32596-6 PMID:31733928

Webb, H. K., Arnott, J., Crawford, R. J., & Ivanova, E. P. (2013). Plastic Degradation and Its Environmental Implications with Special Reference to Poly(ethylene terephthalate). *Polymers*, *5*(1), 1–18. doi:10.3390/polym5010001

WEF. (2022). *This is how cities can reduce emissions with waste reduction solution.* WEF. https://www.weforum.org/agenda/2022/11/waste-emissions-methane-cities/

Weigand, M., Wurm, M., Dech, S., & Taubenböck, H. (2019). Remote Sensing in environmental justice research—A Review. *ISPRS International Journal of Geo-Information*, *8*(1), 20. doi:10.3390/ijgi8010020

Wei, J., Zhang, Y., Hansen, A., Liu, Q., Sun, Y., & Bi, P. (2014). Perception, Attitude and Behaviour in Relation to Climate Change: A Survey among CDC Health Professionals in Shanxi Province, China. *Environmental Research*, *134*, 301–308. doi:10.1016/j.envres.2014.08.006 PMID:25199970

Wei, N., Liu, S., Jiao, Z., & Li, X. C. (2022). A possible contribution of carbon capture, geological utilization, and storage in the Chinese crude steel industry for carbon neutrality. *Journal of Cleaner Production*, *374*, 133793. doi:10.1016/j.jclepro.2022.133793

Weiskopf, S. R., Rubenstein, M. A., Crozier, L. G., Gaichas, S. G., Griffis, R., Halofsky, J. E., Hyde, K. J. W., Morelli, T. L., Morisette, J. T., Munoz, R. C., Pershing, A. J., Peterson, D. L., Poudel, R., Staudinger, M. D., Sutton-Grier, A. E., Thompson, L., Vose, J., Weltzin, J. F., & Whyte, K. P. (2020). Climate change effects on biodiversity, ecosystems, ecosystem services, and natural resource management in the United States. *The Science of the Total Environment*, *733*, 137782. doi:10.1016/j.scitotenv.2020.137782

Wendiro, D., Wacoo, A. P., & Wise, G. (2019). Identifying indigenous practices for cultivation of wild saprophytic mushrooms: Responding to the need for sustainable utilization of natural resources. *Journal of Ethnobiology and Ethnomedicine*, *15*(1), 1–15. doi:10.1186/s13002-019-0342-z PMID:31831075

Wentz, J. A., & Franta, B. (2022, December). Liability for Public Deception: Linking Fossil Fuel Disinformation Liability for Public Deception: Linking Fossil Fuel Disinformation to Climate Damages to Climate Damages. *Environmental Law Reporter*, *52*, 10995. https://scholarship.law.columbia.edu/cgi/viewcontent.cgi?article=1196&context=sabin_climate_change

Wesdrop, K. (2022). Health Care has a Carbon Problem. Health Tech can Help Fix It. *State News.* https://www.statnews.com/2022/06/22/health-tech-help-solve-health-care-carbon-emissions-problem/.

Westlund, L., & Poulain, F. (2007). *Disaster response and risk management in the fisheries sector. Food and Agriculture Organization of the United Nations.* FAO., https://www.researchgate.net/publication/242239330

WFP. (2016). *The Impacts of Climate Change on Food Security and Livelihoods in Karamoja, Uganda.* WFP.

WFP. (2017). *Vulnerability Assessment and Mapping (VAM). VAM Bulletin Number 8.* WFP.

White-Newsome, J. L., Meadows, P., & Kabel, C. (2018). Bridging climate, health, and equity: A growing imperative. *American Journal of Public Health*, *108*(S2), S72–S73. doi:10.2105/AJPH.2017.304133 PMID:29072940

Whitley, C. T., & Bowers, M. M. (2023). Queering Climate Change: Exploring the Influence of LGBTQ+ Identity on Climate Change Belief and Risk Perceptions. *Sociological Inquiry*, *93*(3), 1–27. doi:10.1111/soin.12534

Whitman, S., Good, G., Donoghue, E. R., Benbow, N., Shou, W., & Mou, S. (1997). Mortality in Chicago attributed to the July 1995 heat wave. *American Journal of Public Health*, *87*(9), 1515–1518. doi:10.2105/AJPH.87.9.1515 PMID:9314806

WHO. (2003). *Climate Change and Human Health: Risks and Responses.* WHO.

WHO. (2012). *WHOQOL User Manual* (Vol. 03). WHO. https://apps.who.int/iris/handle/10665/77932. Retrieved: 20 June 2023.

WHO. (2014). *Gender, Climate Change and Health.* WHO.

WHO. (2017). *Framework for action in building health systems resilience to climate change in South-East Asia Region 2017-2022.* WHO. https://apps.who.int/iris/handle/10665/258953.

WHO. (2021). *Climate change and health.* WHO. https://www.who.int/news-room/fact-sheets/detail/climate-change-and-health

WHO. (2021). *Climate Change and Health.* WHO. https://www.who.int/news-room/fact-sheets/detail/climate-change-and-health.

WHO. (2023). *Climate-change-and-health, June 2023.* WHO. https://www.who.int/news-room/fact-sheets/detail/climate-change-and-health..

WHO/UNICEF Joint Monitoring Programme (JMP). (2023). *Progress on household drinking water, sanitation and hygiene 2000-2022: Special focus on gender* WHO. [https://cdn.who.int/media/docs/default-source/wash-documents/jmp-2023_layout_v3launch_5july_lowreswhowebsite.pdf?sfvrsn=c52136f5_3&download=true]

Wilkins, J. L. (2020). Challenges and opportunities created by the COVID-19 pandemic. [PubMed]. *Journal of Nutrition Education and Behavior, 52*(7), 669–670. doi:10.1016/j.jneb.2020.05.005 PMID:32646596

Williams, M. J., & Syddall, V. (2022). Women, fisheries technology and development: Toward new research approaches. *Gender, Technology and Development, 26*(1), 1–28. doi:10.1080/09718524.2022.2125456

Windsor, F. M., Durance, I., Horton, A. A., Thompson, R. C., Tyler, C. R., & Ormerod, S. J. (2019). A catchment-scale perspective of plastic pollution. *Global Change Biology, 25*(4), 1207–1221. doi:10.1111/gcb.14572 PMID:30663840

Winter, N. (2022). Renewables 2022. *Global Status Report United States of America Factsheet.*

WIPO. (2020, January), *Climate Action and Sustainability: Indigenous Peoples Are Part of the Solution.* WPO. https://www.wipo.int/wipo_magazine/en/2020/01/article_0007.html

Wiseman, A. W. (2008). A Culture of (in)Equality?: A Cross-National Study of Gender Parity and Gender Segregation in National School Systems. *Research in Comparative and International Education, 3*(2), 179–201. doi:10.2304/rcie.2008.3.2.179

Wisner, B. (2004). *At risk: Natural hazards, people's vulnerability and disasters* (2nd ed.). Routledge.

Wisner, B. (2004). *At Risk: Natural Hazards, Peoples' Vulnerability and Disasters* (2nd ed.). Routledge.

WMO. (2018). *IPCC Special Report on Global Warming of 1.5 °C* (V. Masson-Delmotte, Eds.). WMO..

Women U. N. (2022a, November 18). *Stepping up for women and girls during the flood crisis in Pakistan.* UN Women. https://www.unwomen.org/en/news-stories/feature-story/2022/11/stepping-up-for-women-and-girls-during-the-flood-crisis-in-pakistan

Women U. N. (2022b). Rapid Gender Analysis of Flood Situation in North and North-Eastern Bangladesh: Gender in Humanitarian Action Working Group Bangladesh. *Asia Pacific.* https://asiapacific.unwomen.org/sites/default/files/2022-07/bd-Rapid-gender-analysis-north-northeaster-flood-2022.pdf

Women U. N. (n.d.). Cl*imate Change and Women in Agriculture. EmPower project.* UN. https://www.uncclearn.org/wp-content/uploads/library/02-issue-brief_climate-change-and-women-in-agriculture.pdf

Womenwatch, U. N. (n.d.). *Women, gender equality and climate change: Fact Sheet*. UN. https://www.un.org/womenwatch/feature/climate_change/

Wong, T. A., Bustami, M. R., & Salleh, S. M. (2022). Applying Social Capital Theory on Conservation Policies of Protected Areas: A Case Study of Penang, Malaysia. *International Journal of Business and Society*, *23*(1), 172–187. doi:10.33736/ijbs.4607.2022

World Fish Center (2007). *The Threat to Fish and Aquaculture from Climate Change. Policy Brief.* Penang: WorldFishCenter.

World Health Organization (WHO). (n.d.). *Occupational hazards in the health sector*. https://www.who.int/tools/occupational-hazards-in-health-sector

World Health Organization. (2011). *Gender and health*. World Health Organization. https://www.who.int/health-topics/gender#tab=tab_1.

World Health Organization. (2014). *Gender, climate change and health*. WHO. https://apps.who.int/iris/bitstream/handle/10665/144781/9789241508186_eng.pdf

World Health Organization. (2023, May 24). *Climate Change*. WHO. https://www.who.int/health-topics/climate-change#tab=tab_1

World Health Organization. (n.d.a). *Improving LGBTIQ+ health and well-being with consideration for SOGIESC*. WHO. https://www.who.int/activities/improving-lgbtiq-health-and-well--being-with-consideration-for-sogiesc

World Health Organization. (n.d.,$_b$). *Air quality and health*. WHO. https://www.who.int/teams/environment-climate-change-and-health/air-quality-and-health/health-impacts

Worldsteel. (2018). *Steel facts*. World Steel Association, Brussels, Belgium. htt ps://www.worldsteel.org/en/dam/jcr:ab8be93e-1d2f-4215-9143-4eba6808bf0 3/20190207_steelFacts.pdf.

Wu, R., & Lin, B. (2022). Environmental regulation and its influence on energy-environmental performance: Evidence on the Porter Hypothesis from China's iron and steel industry. *Resources, Conservation and Recycling*, *176*, 105954. doi:10.1016/j.resconrec.2021.105954

Wu, W., Liu, Q., Zhu, Z., & Shen, Y. (2014). *Managing Bamboo for Carbon Sequestration Bamboo*. Springer.

Xiang, X., Li, Q., Khan, S., & Khalaf, O. I. (2021). Urban water resource management for sustainable environment planning using artificial intelligence techniques. *Environmental Impact Assessment Review*, *86*, 106515. doi:10.1016/j.eiar.2020.106515

Xiao, Y., Chai, J., Wang, R., & Huang, H. (2022). Assessment and key factors of urban liveability in underdeveloped regions: A case study of the Loess Plateau, China. *Sustainable Cities and Society*, *79*, 103674. doi:10.1016/j.scs.2022.103674

Xu, P., Zhu, J., Li, H., Xiong, Z., & Xu, X. (2023). Coupling analysis between cost and carbon emission of bamboo building materials: A perspective of supply chain. *Energy and Building*, *280*, 112718. doi:10.1016/j.enbuild.2022.112718

Xu, R., Yu, P., Abramson, M. J., Johnston, F. H., Samet, J. M., Bell, M. L., Haines, A., Ebi, K. L., Li, S., & Guo, Y. (2020). Wildfires, global climate change, and human health. *The New England Journal of Medicine*, *383*(22), 2173–2181. doi:10.1056/NEJMsr2028985 PMID:33034960

Xu, X., Lu, Y., Vogel-Heuser, B., & Wang, L. (2021). Industry 4.0 and Industry 5.0—Inception, conception and perception. *Journal of Manufacturing Systems*, *61*, 530–535. doi:10.1016/j.jmsy.2021.10.006

Yadav, J., & Gupta, N. (2021, June). Urban quality of life: Domains, dimensions and indicators for Indian cities. [). IOP Publishing.]. *IOP Conference Series. Earth and Environmental Science*, *796*(1), 012032. doi:10.1088/1755-1315/796/1/012032

Yager, K. (2015). Satellite imagery and community perceptions of climate change impacts and landscape change. *Climate Cultures*, 146–168. doi:10.12987/yale/9780300198812.003.0007

Yakoub, W. A., Eleinen, O. M. A., Mahmoud, M. F., & Elrayies, G. M. (2021). Developing a holistic green urban meter: An analytical study of global assessment tools for urban sustainability. *International Journal of Sustainable Development and Planning*, *16*(2), 263–275. doi:10.18280/ijsdp.160206

Yanarateş, E., & Yılmaz, A. (2020). Metaphorical perceptions of prospective teachers towards the concept of environmental sensitivity. *Gazi University Journal of Gazi Educational Faculty*, *40*(3), 1019–1050. doi:10.17152/gefad.699406

Yang, J. (2021). How Does State Policy Shape East Asia's Steel Industry? A Selective Review. *Growth Mechanisms and Sustainability: Economic Analysis of the Steel Industry in East Asia*, 11-42.

Yang, X.-S., &; Karamanoglu, M. (2013). Swarm intelligence and bio-inspired computation. *Swarm Intelligence and Bio-Inspired Computation*, 3–23. doi:10.1016/B978-0-12-405163-8.00001-6

Yang, F., & Gu, S. (2021). Industry 4.0, a revolution that requires technology and national strategies. *Complex & Intelligent Systems*, *7*(3), 1311–1325. doi:10.1007/s40747-020-00267-9

Yavinsky, R. (2012). *Women more Vulnerable than Men to Climate Change*. Population Reference Bureau.

Yea, S., & Stringer, C. (2023). The informalisation of precarious work in fishing crew: Experiences of Fijian fishers on distant water vessels. *Marine Policy*, *155*, 105709–105709. doi:10.1016/j.marpol.2023.105709

Yee, S. H., Paulukonis, E., & Buck, K. D. (2020). Downscaling a human well-being index for environmental management and environmental justice applications in Puerto Rico. *Applied Geography (Sevenoaks, England)*, *123*, 102231. doi:10.1016/j.apgeog.2020.102231 PMID:34326563

Yellishetty, M., Mudd, G. M., Ranjith, P. G., & Tharumarajah, A. (2011). Environmental life-cycle comparisons of steel production and recycling: Sustainability issues, problems and prospects. *Environmental Science & Policy*, *14*(6), 650–663. doi:10.1016/j.envsci.2011.04.008

Yen, T.-M. (2016). Culm height development, biomass accumulation and carbon storage in an initial growth stage for a fast-growing moso bamboo (Phyllostachy pubescens). *Botanical Studies*, *57*(1), 10. doi:10.1186/s40529-016-0126-x PMID:28597419

Yılmaz, A., & Yanarateş, E. (2020). Determination of metaphorical perceptions of prospective teachers on the concept of "water pollution" through triangulation. *Kastamonu Education Journal*, *28*(3), 1500–1528. doi:10.24106/kefdergi.722554

Yiping, L., Yanxia, L., Henley, K., & Zhou, G. (2010). *Bamboo and Climate Change Mitigation:a comparative analysis of carbon sequestration*. International Network for Bamboo and Rattan.

YKA. (2019). *How 5 Initiatives are helping india reduce its carbon emission*. YKA. https://www.youthkiawaaz.com/2019/11/5-initiatives-helping-india-reduce-carbon-emissions/

Yoon Yi, C. (2022). Thermal comfort modelling of older people living in care homes: An evaluation of heat balance, adaptive comfort, and thermographic methods. *Building and Environment, 207*, 108550.

Younger, M., Morrow-Almeida, H. R., Vindigni, S. M., & Dannenberg, A. L. (2008). The built environment, climate change, and health: Opportunities for co-benefits. *American Journal of Preventive Medicine*, *35*(5), 517–526. doi:10.1016/j.amepre.2008.08.017 PMID:18929978

Yuan, F., Yao, R., Sadrizadeh, S., Li, B., Cao, G., Zhang, S., & Short, C. A. (2022). Thermal comfort in hospital buildings–A literature review. *Journal of Building Engineering*, *45*, 103463. https://www.sciencedirect.com/science/article/pii/S2352710221013218. doi:10.1016/j.jobe.2021.103463

Yuniarsih, P. D. S. E. T., Hanifa, P. D. A. P., Haryati, M. S. Y., Lestari, I. P., Saleh, M. S. Y., Nurmalinda, M. S. N., & Lesmayati, M. S. S. (2023). *Perspective Chapter: How Important Is Urban Farming in Indonesia to Support Food Sovereignty?* Research Gate.

Yun, K., Lurie, N., & Hyde, P. S. (2010). Moving Mental Health into the Disaster-Preparedness Spotlight. *The New England Journal of Medicine*, *363*(13), 1193–1195. doi:10.1056/NEJMp1008304 PMID:20860498

Yurkevicius, B. R., Alba, B. K., Seeley, A. D., & Castellani, J. W. (2021). Human cold habituation: Physiology, timeline, and modifiers. *Temperature*, *9*(2), 1–36. doi:10.1080/23328940.2021.1903145

Yusoff, M. M. (2020, August). Improving the quality of life for sustainable development. [). IOP Publishing.]. *IOP Conference Series. Earth and Environmental Science*, *561*(1), 012020. doi:10.1088/1755-1315/561/1/012020

Zacharia, P. U., Dineshbabu, A. P., Thomas, S., Shoba, J. K., Vivekanandan, E., Pillai, L. S., Sivadas, M., Ghosh, S., Ganga, U., Rajesh, K. M., Nair, R. J., Najmudeen, T. M., Koya, M., Chellappan, A., Dash, G., Divipala, I., Akhilesh, K. V., Muktha, M., & Sen, D. S. (2016). *Relative vulnerability assessment of Indian marine fishes to climate change using impact and adaptation attributes* (pp. 192). CMFRI Special Publication (125). http://eprints.cmfri.org.in/id/eprint/11297

Zain, R. M., Kamarudin, M. K. A., & Saad, M. H. M. (2018). Assessment of Quality of Life on fishermen community in Kuala Terengganu, Malaysia: A Review. *International Journal of Academic Research in Business & Social Sciences*, *8*(10), 640–650. doi:10.6007/ijarbss/v8-i10/4770

Zasada, I., Weltin, M., Zoll, F., & Benninger, S. L. (2020). Home gardening practice in Pune (India), the role of communities, urban environment and the contribution to urban sustainability. *Urban Ecosystems*, *23*(2), 403–417. doi:10.1007/s11252-019-00921-2

Zeng, X., Yu, Y., Yang, S., Lv, Y., & Sarker, M. N. I. (2022). Urban resilience for urban sustainability: Concepts, dimensions, and perspectives. *Sustainability (Basel)*, *14*(5), 2481. doi:10.3390/su14052481

Zhang, M. (2022). *Data Center Water Usage: Billions of Gallons Every Year*. Dgtl Infra. https://dgtlinfra.com/data-center-water-usage/

Zhang, Y., Tang, K. W., Yang, P., Yang, H., Tong, C., Song, C., Tan, L., Zhao, G., Zhou, X., & Sun, D. (2022). Assessing carbon greenhouse gas emissions from aquaculture in China based on aquaculture system types, species, environmental conditions and management practices. *Agriculture, Ecosystems & Environment*, *338*, 108110. doi:10.1016/j.agee.2022.108110

Zhao, M., Tyzack, M., Anderson, R., & Onoakpovike, E. (2013). Women as visible and invisible workers in fisheries: A case study of Northern England. *Marine Policy*, *37*, 69–76. doi:10.1016/j.marpol.2012.04.013

Zheng, Z., Xie, S., Dai, H., Chen, X., & Wang, H. (2017). *An overview of blockchain technology: Architecture, consensus, and future trends*. International Congress on Big Data (Big Data Congress), Honolulu, HI, USA.

Zhou, C., Shen, C., Ma, G., & Yan, X. (2022, August). Power usage effectiveness analysis of a liquid-pump-driven hybrid cooling system for data centers in subclimate zones. *Sustainable Energy Technologies and Assessments*, *52*, 102277. doi:10.1016/j.seta.2022.102277

Zhou, X., Fang, J., Shi, S., Yang, L., Xie, B., & Huang, L. (2022). Ecosystem services of ecosystem approach to mariculture: Providing an unprecedented opportunity for the reform of China's sustainable aquaculture. *Frontiers in Marine Science*, *9*, 909231. doi:10.3389/fmars.2022.909231

Ziska, L., Crimmins, A., Auclair, A., DeGrasse, S., Garofalo, J. F., Khan, A. S & Walls, I. (2016). Ch. 7: Food safety, nutrition, and distribution. *The impacts of climate change on human health in the United States: a scientific assessment,* 189-216. Research Gate.

Zollet, S. (2023). Hybrid food networks and sustainability transitions: Shared and contested values and practices in food relocalization and resocialization. *Sociologia Ruralis*, *63*(1), 117–139. doi:10.1111/soru.12391

Zulu, D. and Sichikwenkwe, P. (2011). Zambia: The Impact of Climate Change on Women. *Times of Zambia Newspaper.* Ndola: Times of Zambia.

About the Contributors

Kasturi Shukla, is presently working as Associate Professor at Symbiosis Institute of Health Sciences, Symbiosis International University, Pune, India. She is a trained Hospital Administrator with a Bachelor's and Master's degree in Hospital Administration from IMS, Devi Ahilya University, Indore, India and Ph.D. from King George's Medical University, Lucknow, India. She was awarded Senior Research Fellowship (SRF) by Indian Council of Medical Research (ICMR) for her research work during 2009-2011 and she was the recipient of the Vice Chancellor's Gold Medal for Best PhD thesis in 2012. Dr. Shukla has more than 18 years' experience in academia and industry and is a trainer, researcher and academician. Having published more than 30 original research articles in journals of national and international repute, she has published 2 books and has 3 registered patents to her credit. She has presented research papers in reputed conferences and is on the review board of some of the reputed health journals. She is a Life Member of Academy of Hospital Administration and CAHO, India and other reputed professional bodies. Besides, hospital management and healthcare quality, her core research interest includes Health-Related Quality of Life, Work-Related Quality of Life, Emotional Intelligence, Resilience, Climate change and Sustainability and Green business.

Yogesh Patil is a Professor and Head - Research Publications at the Symbiosis International University, Pune, India. Dr. Patil holds his Doctorate degree in Environmental Sciences from Pune University, India and has over twenty-four years of Post Graduate teaching and research experience in Environmental Science, Management & Technology. His research areas of interest include - waste management, bioremediation, sustainability, climate change and industrial ecology. He has over 60 research papers in national and international journals of repute and has also published two edited books namely, Applied Bioremediation – Active and Passive Approaches (InTech, Croatia) and Reconsidering the Impact of Climate Change on Global Water Supply, Use, and Management (IGI Global, USA). Dr. Patil has undertaken several research/consultancy projects funded by UGC, IFS (Sweden) & OPCW (The Netherlands), World Bank DST and HEMRL. He is recipient of several honours/fellowship awards like DST, CSIR, Best PhD Research Award, UGC-Post Doctoral Fellowship to name a few. He is reviewer of several journals indexed in Scopus and SCI.

Ronald C. Estoque is a Senior Researcher at the Center for Biodiversity and Climate Change, Forestry and Forest Products Research Institute (FFPRI), Japan. He is a forester and geoenvironmental scientist by training. His research interests are at the interface of nature and society, cutting across different fields and topics, including land change science (forest cover change, land use/cover change, urbanization), sustainability science, the forest transition theory, landscape/urban ecology, urban heat islands, ecosystem services, social-ecological systems, quality of life, climate change mitigation, vulnerability, risk, and adaptation, and nexus assessment. He applies various tools and techniques, including GIS, remote sensing, and social-ecological approaches. Dr. Estoque was a contributing author (CA) to the Intergovernmental Panel on Climate Change's (IPCC) sixth assessment report (AR6) (WG II, Chapter 10). Currently, he serves as an expert for the Scoping for the IPCC Special Report on Climate Change and Cities, a review editor (RE) for the Intergovernmental Science-Policy Platform for Biodiversity and Ecosystem Services' (IPBES) nexus assessment (Chapter 2), a section Editor-in-Chief for Sustainability (section: Sustainable Forestry), a subject editor for *Ecosystem Health and Sustainability*, an associate editor for *Frontiers in Forests and Global Change*, and a member of the editorial board for *Remote Sensing* and *Ecological Civilization*.

Pedro Antonio López de Haro is a Research Professor at the Autonomous Indigenous University of Mexico, he has a PhD in Economics and a Master's degree in Industrial Engineering. He has participated in collaborations with several higher learning institutions and has publications on the topics of sustainable development, technological innovation, economics, production, supply chain and research methodology. He is the author of two books, "The art of Modelling, introduction to Operations Research" and " Mexican Sayings, A Hermeneutical Analysis of Popular Wisdom".

***

Hesham Mohammed Ali Abdullah is Assistant Professor (Ph.D. in Computer Science)-Jouf University- KSA . He has obtained his BSc Degree in July 2006 from The National University, Yemen, completed his Master of Computer Applications (MCA) degree in 2013 from Bharathiar University, Coimbatore, India. He received Ph.D degree in Computer Science from Bharathiar University, Coimbatore, India in 2018. His research interests include Cognitive Radio Ad Hoc Network, Wireless Communication, network security, big data,IoT, Machine learning, and Data Visualization. He has publication several papers in national and international journals.

Selina Banda is a nutritionist and lecturer in Hospitality and Tourism Management. She is in the Department of Social Studies under the School of Humanities and Social Sciences at the Copperbelt University in Kitwe, Zambia. Likewise, she has worked three years, during which period Selina acted as head of department for two years. Before that, she also worked for more than 10 years as an educationist, nutritionist and researcher at Zambian Open University. During her working period, she also acted in different positions at supervisory levels. She has a background in primary and secondary education, which provided her an opportunity to serve as a teacher in various government schools in Lusaka. Selina worked as a teacher for 19 years. Her interest in primary school has resulted in conducting some research studies among children at different age levels. Because of her background in education and nutrition, Selina has been involved in related research activities. She has published some research papers in education, theatre for development and nutrition. Currently, she is involved in lecturing and supervising university students at under and post graduate levels, where she is engaged on full-time and part-time basis. Adult education, is another area of interest in which she has immerse, specifically, in alleviating illiteracy among adults in communities. Additionally, Selina is passionate with theatre for development grounded in community mobilization. This combination of nutrition and education has awakened in her a passion to be involved in community nutrition, which caters for a wide range of people. Selina is currently a doctoral student in food and nutrition at the University of Zambia - Zimbabwe Open University.

Bianca Chereches is a lecturer at the Department of English and German Philology of the University of Zaragoza (Spain). She graduated in English Philology from the University of Zaragoza and she holds a Master's Degree in English Studies in Secondary Education from the same university. She wrote her PhD thesis on the Dalit community and their representation in Indian literature written in English. She is currently the Director of New Voices in Postcolonial Studies international network for graduate and post-graduate students. She is also a member of the Spanish research project "Literature Of(f) Limits: Pluriversal Cosmologies and Relational Identities in Present-Day Writing in English", granted by the Spanish Ministry of Education and Science. Her current interests are Postcolonial Studies, Feminist Studies, contemporary Dalit Studies, Ecocriticism, and Trauma.

Ranjana Choudhari, MBBS,CENCD, is former Scientist E of Clinical Epidemiology Dept., Health Science Div., ICMR-National Institute of Occupational Health. She contributed to this book chapter along with the corresponding author during her tenure at ICMR-NIOH. A qualified medical professional, currently employed as a senior medical scientist (of Deputy director grade) in the Department of Clinical Epidemiology at ICMR- National AIDS Research Institute, Pune, under the aegis of Indian Council of Medical Research, Department of health research, Ministry of Health and Family Welfare, Government of India. She has more than 18 years' research experience in planning and executing research on various occupational and environmental health issues of public health significance at the national level. Additionally, she has rich experience in public health care delivery in rural areas of India and clinical practice in urban areas of India. She has also served for a period about three years as a Medical officer for Government of Maharashtra State (India), delivering health care services at the grass-root level, including curative medical services, Child immunization, Epidemics investigations, Planning & implementation of prevention and control strategies for epidemics, Antenatal and perinatal health care, Execution of national programs on prevention and control of various diseases of public health importance, etc. She is an alumna of Government Medical College, Aurangabad during her professional courses on medicine studies and an awardee of WHO fellowship on "Epidemiology of non-communicable diseases" at All India Institute of Hygiene and Public health, Kolkata. Her areas of interest include Human biomonitoring, Heavy metal toxicity, Air pollution, Environmental and health risk assessment, Dust exposure and Occupational diseases, etc. She has credit for publishing research papers in various national and international journals and is a member of several national and international organizations. As a lead researcher, she has successfully completed research projects on occupational and environmental issues, including Hon'ble Supreme court and High court of Gujarat directive research projects, also projects sponsored by the Government of Gujarat and State Pollution control boards.

Ravindra S. Gavali is an academician, researcher, and a trainer with over 19 years of experience in the area of Natural Resources Management, Water quality -watershed management, Climate change, Disaster mitigation, Rural Development, and Geoinformatics. Currently, he is working at National Institute of Rural Development & Panchayati Raj in Hyderabad. He is involved in capacity building & policy advocacy for major flagship program in Rural Development in India. His insight on climate proofing of the villages and the determination to explore novel approaches to address the Climate change issues from rural perspective is commendable. He is the member of National Advisory Committee on Climate Resilient Agriculture, DST-TIFAC.

Goundla Pradeep Goud is a Young Researcher with more than 5 years of experience in climate change and ecology. Holding an integrated master's degree in Environmental science with a specialization in ecology, Mr. Goundla's academic journey paved the way for a remarkable career. His professional journey began at the Institute of Forest Biology (IFB-ICFRE, MOEFCC), where he played an important role in an all-India coordinated research project focusing on the climate-driven effects on Indian forests. From establishing permanent ecological monitoring plots to conducting biodiversity estimations, Mr. Goundla's contributions were invaluable. Subsequently, he joined the National Institute of Rural Development and Panchayati Raj as a Research Associate, where his focus shifted towards estimating carbon sequestration through MGNREGA plantations. His professional journey showcases his dedication to sustainable development and environmental conservation. Prior to this, he also worked in the MS Swaminathan Research Foundation, adding another layer of diversity to his impactful career.

Shashikala Gurpur, with an illustrious career over 32 years in teaching, research and industry, has been with Symbiosis Law School since 2007. She is a distinguished academician and orator, national merit scholar, University topper in LLM, holds PhD in international law and taught law in NLSIU, Manipal University and National University Ireland. Dr Gurpur has more than 85 publications (20 Scopus/WoS) and presented more than 330 invited lectures, workshops on Gender literacy, Human rights, Media literacy and seminars across India, Thailand, US, UK, Ireland, Germany, Australia, Canada and UAE. She is a pioneer of EU Legal Studies in India . She is awarded with the most coveted professional achievement "Jean Monnet Chair Professor for EU Climate Justice Law, Governance, Management and Policy (EUC-LAMP)" Co-funded by the Erasmus+ program of the European Union. She is team member of global evaluation and accreditation IALS, USA.

Soledad Ibarra is a professor at the Universidad Autónoma Indígena de México – UAIM (Autonomous Indigenous University of Mexico) since 2013, attached to the Academy of Ingeniería en Sistemas de Calidad - ISC (Quality Systems Engineering Academy). For the past six years, he has developed programs aimed at fostering 'Community Collaborative Projects' that promote the preservation and sustainable use of the local biocultural heritage resources; he is also a promoter of the SDGs and bioenergy use. He was the initiator of the institutional program of scientific dissemination for children and youth, called 'Laboratorio para todos (Laboratories for all)'; he teaches subjects such as: Chemistry, Community Diagnosis, Sampling Techniques, Design and Analysis of Experiments, among others.

Nissy John has a strong background in Environmental Education and Research with over 10 years of experience. She completed her Ph.D from Kerala University focusing on hydrochemistry and microbiology of freshwater sources. She has worked closely with multicultural and heterogeneous groups in a diverse array of organizational settings, including academia, research-based organisations, government bodies, non-government organisations, community-based organisations and industry. Her areas of interest include Water quality Assessment and pollution control, Biodiversity Conservation, Rural Development, Environmental Impact Assessment. She is currently working with National Institute of Rural Development & Panchayati Raj in Hyderabad, as Fellow in Natural Resources Management, Climate Change and Disaster Mitigation.

A.V.Senthil Kumar is working as a Director & Professor in the Department of Research and PG in Computer Applications, Hindusthan College of Arts and Science, Coimbatore since 05/03/2010. He has to his credit 11 Book Chapters, 265 papers in International and National Journals, 25 papers in International Conferences, 5 papers in National Conferences, and edited Nine books (IGI Global, USA). He is an Editor-in-Chief for various journals. Key Member for India, Machine Intelligence Research Lab (MIR Labs). He is an Editorial Board Member and Reviewer for various International Journals. He is also a Committee member for various International Conferences. He is a Life member of International Association of Engineers (IAENG), Systems Society of India (SSI), member of The Indian Science Congress Association, member of Internet Society (ISOC), International Association of Computer Science and Information Technology (IACSIT), Indian Association for Research in Computing Science (IARCS), and committee member for various International Conferences

Sneha Kumari is Assistant professor at Symbiosis School of Economics. Prior to that, she worked as Assistant professor at Vaikunth Mehta National Institute of Cooperative Management (VAMNICOM), Ministry of Cooperation, Government of India, Pune. She was a Visiting Faculty at the National Institute of Food Technology and Entrepreneurship Management, Ministry of Food Processing Industries.

Pavitira Manogaran is a senior lecturer at Universiti Malaysia Kelantan. She completed her PhD at the Graduate School of Business, Universiti Sains Malaysia. Her research interests encompass the fascinating realms of business information technology, smart cities, and digital identities, reflecting her passionate dedication to advancing understanding and practical applications in these fields. Building on her Master's in Business Administration from the same esteemed university, her contributions are poised to impact the ever-evolving landscape of business, technology integration, and sustainability practices.

Nur Aliah Mansor is a dedicated researcher whose expertise spans the dynamic intersection of social media, tourism, hospitality, and wellbeing, with a keen interest in integrating resilience-focused approaches. With a profound understanding of the digital landscape, Nur Aliah explores the evolving role of social media in fostering positive change at the grassroots level.

Sandra C. Maputa holds a Diploma in Social work (University of Zambia), Bachelor of Arts Degree in Social Work (Rusangu University) and a Master of Arts Degree in Transformative Community Development (Zambia Open University). Sandra has worked as a social worker in various communities across Zambia. She worked for an organisation called Bwafwano Central Board in Mporokoso District, Northern Province, Zambia as a community facilitator. In 2007 she joined the Young Women Christian Association (YWCA) as a field officer and later become the regional coordinator (Provincial Coordinator) Western Province. Since then, Sandra has conducted various capacity building training programmes and courses such as Sexual Gender Based violence, HIV/AIDS, Proposal Writing, Strategic Planning, Financial Management, Communication and Reporting, Humanitarian Practices, Resource Mobilization, Data Analysis, Behavioral Centered Programming, Good Governance, Social Accountability, Child Safeguarding.

Daniel L. Mpolomoka is the Dean of the School of Education, Humanities and Social Sciences at Unicaf University Zambia. He has 14 years of professional experience as a lecturer and researcher. Daniel taught at various government secondary schools before taking up lectureship at Zambian Open University

(ZAOU), were he saved in many key administrative positions. He once served as a Masters Coordinator for a Project dubbed the Transformative Engagement Network (TEN) Project, which involved Mynooth University (Ireland), Muzuzu University (Malawi), Mulungushi University (Zambia) and Zambian Open University (Zambia). He has won scholarships and awards too, one of them being the Harmonization, Accreditation and Quality Assurance in African Higher Education (HAQAA2), October, 2022–January, 2023, which culminated into a Diploma certification. His research interests include: Literacy Studies, Technological Education, Open and Distance Education in Higher Institutions of Learning, Adult Education, Special Education and Early Education Studies. Daniel has made paper presentations at local and international conferences. He has also published various articles in local and international refereed journals. He is an editorial board / reviewer for many reputable scientific journals.

Menard Musendekwa (PhD) is an Old Testament Scholar and has also become an interdisciplinary scholar who has published papers in Old Testament, Church History, Gender Studies, Education and Political Science. He is Vice Principal of Murray Theological college and Chair of Theology at Reformed Church University.

Christine Phiri Mushibwe is the current Vice Chancellor at Unicaf University Zambia. She has a Ph.D. from University of Huddersfield, UK; where she worked as a Lead Tutor and taught in Universities in China. She has also worked as an associate Lecturer at Leeds Trinity University as a Research Methodology Tutor. She is currently a Visiting Research Fellow at University of Huddersfield. Christine has served in various Higher Education Institutions as a Lecturer and in leadership positions in the United Kingdom and Zambia. She earned the prestigious award of 'Africa's Most Influential Woman in Business and Government' by CEO Global, (in the education category), as regional and Country winner in 2017, 2018/19 respectively. Christine is a founder member of Training Teachers for Community Schools in Zambia (TTCSZ) [] and Lusaka Open Business College. She is the Chair of the Advisory Body at Campaign for Female Education, Board Member of COLD and GARA respectively. Christine is the author of the highly acclaimed books titled "The Effect of Cultural Traditions on the Education of Women" (2010), "Followers as a Key Variable in Organization Success" (2020) and "The Basics of Research: Making Research Easy for Student" (2022). She has published a number of journals and contributed chapters to Books. She is a supervisor and examiner of PhD students in various universities, a reviewer of books and journals. She is an established Corporate Trainer, outstanding public speaker as well as Keynote Speaker at various local and International Conferences.

Nibedita Naha, PhD, is Scientist 'E' and Head of the Biochemistry Department under Biological Sciences Division, ICMR-National Institute of Occupational Health (Ministry of Health and Family Welfare, Department of Health Research, Govt. of India). She received PhD in Science from Jadavpur University, India; and Postdoctorate in prostate cancer cell signalling, tumour xenografting, and neurobiochemistry of substance abuse from Gyeongsang National University, South Korea. She also served as a faculty of Govt. Degree Colleges, Medical Colleges, and Universities in India and abroad. She has vast experience on mechanism-based approach of different techniques using pre-clinical animal models, cell lines and human subjects (patient's samples) with respect to occupational and environmental exposure since 2000. Dr. Naha is a recognized PhD Guide of Gujarat University, India; and a member of several PhD Research Committees (RDC) and Board of Studies (BoS) of different universities, as well as the CPCSEA nominee of Animal Ethics Committee (Ministry of Fisheries, Animal Husbandry and Dairying, Govt. of India) for monitoring animal experiments in several research organizations and pharmaceutical companies, and Chairperson of Human Ethics Committee (IHEC, under Dept. of Health Research, Govt. of India) of different universities; and Department of Biotechnology nominee of the Institute Biosafety Committee (Ministry of Science and Technology, Govt. of India). Apart from that, Dr. Naha is also a reviewer and editorial member of several international and national peer-reviewed journals; and invited speakers of different national and international conferences. She is also a trained NABL internal auditor of ISO/IEC17025:2017 and ISO/IEC15189:2012. She has authored more than 45 research articles, invited reviews and book chapters along with patents 2 patents on her research work. Dr. Naha is an elected member of the Physiological Society of India (PSI) since 2013, and the advisory board member of some national conferences sponsored by Govt. of India. She has also honoured as a life member of several scientific societies in India and US such as, SRBCE, ISSRF, NESA, ISNA, ISCA, SfN and AAAS etc.; and awardees of many national and international scientific reputes including SfN (USA), Brain Korea 21 (South Korea), PB Sen Gold Medal (Physiological Society of India), Young Scientist Award (Indian Science Congress), National Scholarship (Govt. of India) etc. Dr. Naha considers herself extremely fortunate to contribute in improving the health and wellbeing through her research findings in the respective fields. Dr. Nibedita Naha is available for her PGV and PhD students, and colleagues for any scientific help in the related fields at naha.n@gov.in or nibedita.72@yahoo.com.

Rudrajit Paul is a consultant Physician at Ruby General Hospital, Kolkata. A former medical professor and HIV master trainer of NACO, he is keenly interested in clinical research and has published numerous articles in indexed international journals. His research interests include climate change and human health, infectious diseases etc.

Deepsi Rathore is currently a CSIR-UGC-NET Fellow and PhD student in the Biochemistry Department, under Biological Sciences Division of ICMR-National Institute of Occupational Health, Ministry of Health and Family Welfare, Department of Health Research, Govt. of India. She has a Masters in Biotechnology from Banasthali Vidyapith, India. Her recent systematic review work entitled "Phenolic compounds and nanotechnology: Application during biotic stress management in agricultural sector and occupational health impacts" recently published as a book chapter in the book namely, 'Phenolic compounds and nanotechnology: application during biotic stress management' from Springer, USA.

Somi N. Sharma is currently a PhD student in the Biochemistry Department, under Biological Sciences Division of ICMR-National Institute of Occupational Health, Ministry of Health and Family Welfare, Department of Health Research, Govt. of India. She has a Masters in Biochemistry from Gujarat University, India. Recently she published her systematic review work in the form of a book chapter entitled "Impact of heavy metal-based nanomaterials on environment and health: Small things having big impacts!" in a book namely, 'Implications of Nanoecotoxicology on Environmental Sustainability' from IGI-Global, USA.

Puja Singh is a full time PhD scholar at Symbiosis International University, Pune India. She has qualified UGC-NET and published two research papers based on her research work in the field of climate change mitigation strategies adopted by the industries. With a combined background in Zoology and Forestry, she is committed to finding sustainable solutions to safeguard the environment.

Meera Suresh is an Associate Professor in the Dept of Homeopathy at Lincoln University College, Malaysia. She was also the Program Manager of KDPP-WDF project at SCTIMST, Trivandrum which was aimed to reduce Diabetes incidence in Kerala. She is a Homeopathic physician with interest in public health research.

Liveson Tatira teaches both undergraduate and post graduate courses as well as supervising PhD students in the Faculty of Education, Department of Languages and Arts Education at the University of Zimbabwe. He has served as Chairperson of the same Department. Currently he is the Deputy Dean of the Faculty of Education. He has published in Indigenous knowledge, Literature, Onomastics, Language Education, Education, Folklore and Linguistics. He is also a published Shona poet.

V.G. Venkatesh is a corporate practitioner/ an academic specializing in Supply Chain Management and Logistics domain with versatile experience and background. He is a Certified Supply Chain Professional (CSCP) from APICS-USA and has International industry experience from Honduras, Sri Lanka, the USA, Hong Kong, and Bangladesh in international supply chain practices. He has over 22 years of experience from Industry and academics.

Adem Yilmaz is an academic, analyst, educational researcher, and a measurement and evaluation specialist. He holds a PhD in Measurement and Evaluation. His fields of expertise are as follows: AI, teaching and learning of science, science education, scale development, technologies in K-12 education, climatology, digital literacy, environmental science, measurement and evaluation, SPSS, LISREL, AMOS, EFA/DFA analyses. In addition, he concentrates upon topics like teacher training, quality in higher education, and accreditation. Dr. Yilmaz is a Measurement, Learning & Evaluation (MLE) specialist and is the editor of several international books on Measurement and Evaluation. He is currently a faculty member and is working as an associate professor of education in Kastamonu University, Türkiye. He is also a member of several committees in the same university.

Index

D

E

F

G

H

I

L

M

N

O

P

Q

R

S

T

V

W

www.ingramcontent.com/pod-product-compliance
Lightning Source LLC
LaVergne TN
LVHW060350300926
849162LV00006B/1239
* 9 7 8 1 6 6 8 4 9 8 6 3 7 *